Lecture Notes in Computer Science 12999

More information about this subseries at http://www.springer.com/series/7409

Chunxiao Xing · Xiaoming Fu · Yong Zhang ·
Guigang Zhang · Chaolemen Borjigin (Eds.)

Web Information Systems and Applications

18th International Conference, WISA 2021
Kaifeng, China, September 24–26, 2021
Proceedings

 Springer

Editors
Chunxiao Xing (iD)
Tsinghua University
Beijing, China

Yong Zhang (iD)
Tsinghua University
Beijing, China

Chaolemen Borjigin
Renmin University of China
Beijing, China

Xiaoming Fu (iD)
Institute of Computer Science
University of Göttingen
Goettingen, Germany

Guigang Zhang
Chinese Academy of Sciences
Beijing, China

ISSN 0302-9743   ISSN 1611-3349 (electronic)
Lecture Notes in Computer Science
ISBN 978-3-030-87570-1   ISBN 978-3-030-87571-8 (eBook)
https://doi.org/10.1007/978-3-030-87571-8

LNCS Sublibrary: SL3 – Information Systems and Applications, incl. Internet/Web, and HCI

This Springer imprint is published by the registered company Springer Nature Switzerland AG
The registered company address is: Gewerbestrasse 11, 6330 Cham, Switzerland

Preface

It is our great pleasure to present the proceedings of the 18th Web Information Systems and Applications Conference (WISA 2021). WISA 2021 was organized by the China Computer Federation Technical Committee on Information Systems (CCF TCIS) and Henan University. WISA 2021 provided a premium forum for researchers, professionals, practitioners, and officers closely related to information systems and applications to discuss the theme of artificial intelligence and information systems, focusing on difficult and critical issues, and the promotion of innovative technology for new application areas of information systems.

WISA 2021 was held in Henan, Kaifeng, China, during September 24–26, 2021. WISA 2021 focused on intelligent cities, government information systems, intelligent medical care, fintech, and network security, emphasizing the technology used to solve the difficult and critical problems in data sharing, data governance, knowledge graphs, and block chains.

This year we received 206 submissions, each of which was assigned to at least three Program Committee (PC) members to review. The peer review process was double-blind. The thoughtful discussions on each paper by the PC resulted in the selection of 49 full research papers (an acceptance rate of 23.79%) and 18 short papers. The program of WISA 2021 included keynote speeches and topic-specific invited talks by famous experts in various areas of artificial intelligence and information systems to share their cutting-edge technologies and views about the state of the art in academia and industry. The other events included industrial forums, the CCF TCIS salon, and a PhD forum.

We are grateful to the general chairs, Baowen Xu (Nanjing University), Ge Yu (Northeastern University) and Xin He (Henan University), as well as all the PC members and external reviewers who contributed their time and expertise to the paper reviewing process. We would like to thank all the members of the Organizing Committee, and the many volunteers, for their great support in the conference organization. Especially, we would also like to thank publication chairs, Yong Zhang (Tsinghua University), Guigang Zhang (Chinese Academy of Sciences), and Chaolemen Borjigin (Renmin University of China), for their efforts on the publication of the conference proceedings. Last but not least, many thanks to all the authors who submitted their papers to the conference.

August 2021

Chunxiao Xing
Xiaoming Fu

Organization

Steering Committee

Baowen Xu	Nanjing University, China
Ge Yu	Northeastern University, China
Xiaofeng Meng	Renmin University of China, China
Yong Qi	Xi'an Jiaotong University, China
Chunxiao Xing	Tsinghua University, China
Ruixuan Li	Huazhong University of Science and Technology, China
Lizhen Xu	Southeast University, China
Xin Wang	Tianjin University, China

General Chairs

Baowen Xu	Nanjing University, China
Ge Yu	Northeastern University, China
Xin He	Henan University, China

Program Committee Co-chairs

Chunxiao Xing	Tsinghua University, China
Xiaoming Fu	University of Göttingen, Germany

Local Chairs

Fang Zuo	Henan University, China
Dong Wang	Henan University, China

Forum Chairs

Ruixuan Li	Huazhong University of Science and Technology, China
Haofen Wang	Tongji University, China

Publicity Chairs

Xin Wang	Tianjin University, China
Derong Shen	Northeastern University, China

Publication Chairs

Yong Zhang	Tsinghua University, China
Guigang Zhang	Institute of Automation, Chinese Academy of Sciences, China
Chaolemen Borjigin	Renmin University of China, China

Website Chairs

Weifeng Zhang	Nanjing University of Posts and Telecommunications, China
Bin Xu	Northeastern University, China

Program Committee

Tiecheng Bai	Tarim University, China
Bin Cao	Beijing Small & Medium Enterprises Information Service Co., Ltd., China
Kong Chao	Anhui University of Engineering, China
Chaolemen Borjigin	Renmin University of China, China
Ling Chen	Yangzhou University, China
Yanping Chen	Guizhou University, China
Lin Chen	Nanjing University, China
Yuefeng Du	Liaoning University, China
Zaobin Gan	Huazhong University of Science and Technology, China
Tianlong Gu	Guilin University of Electronic Science and Technology, China
Wenzhong Guo	Fuzhou University, China
Qinming He	Zhejiang University, China
Tieke He	Nanjing University, China
Xin He	Henan University, China
Mengxing Huang	Hainan University, China
Shujuan Jiang	China University of Mining and Technology, China
Weijin Jiang	Hunan Industrial and Commercial University, China
Cheqing Jin	East China Normal University, China
Shenggen Ju	Sichuan University, China
Xiangjie Kong	Zhejiang University of Technology, China
Yue Kou	Northeastern University, China
Bin Li	Yangzhou University, China
Chunying Li	Guangdong Technical Normal University, China
Lin Li	Wuhan University of Technology, China
Qingzhong Li	Shandong University, China
Weimin Li	Shanghai University, China
Yuhua Li	Huazhong University of Science and Technology, China
Zhenxing Li	Beijing AgileCentury Information Technology Co., Ltd., China
Ye Liang	Beijing Foreign Studies University, China
Bingxiang Liu	Jingdezhen Ceramic Institute, China
Chen Liu	Northern Industrial University, China

Genggeng Liu	Fuzhou University, China
Qing Liu	Renmin University of China, China
Xu Liu	SAP (China) Co., Ltd., China
Shan Lu	Nanjing FiberHome Starrysky Co., Ltd., China
Youzhong Ma	Luoyang Normal College, China
Xiangke Mao	Tsinghua University, China
Weiwei Ni	Southeast University, China
Baoning Niu	Taiyuan University of Technology, China
Jun Pang	Wuhan University of Science and Technology, China
Zhiyong Peng	Wuhan University, China
Weiguang Qu	Nanjing Normal University, China
Jiadong Ren	Yanshan University, China
Yonggong Ren	Dalian Liaoning Normal University, China
Derong Shen	Northeastern University, China
Ming Sheng	Tsinghua University, China
Xiaohua Shi	Shanghai Jiao Tong University, China
Baoyan Song	Liaoning University, China
Jie Song	Northeastern University, China
Wei Song	Wuhan University, China
Chenchen Sun	Tianjin University of Technology, China
Haojun Sun	Shantou University, China
Zhongbin Sun	Xi'an Jiaotong University, China
Yong Tang	South China Normal University, China
Buyu Wang	Inner Mongolia Agricultural University, China
Dong Wang	Henan University, China
Guojun Wang	Guangzhou University, China
Haofen Wang	Tongji University, China
Xin Wang	Tianjin University, China
Xingce Wang	Beijing Normal University, China
Yanlong Wen	Nankai University, China
Feng Xia	Dalian University of Technology, China
Chunxiao Xing	Tsinghua University, China
Bin Xu	Northeastern University, China
Lei Xu	Nanjing University, China
Lizhen Xu	Southeast University, China
Zhuoming Xu	Hohai University, China
Zhongmin Yan	Shandong University, China
Nan Yang	Renmin University of China, China
Shiyu Yang	Guangzhou University, China
Ding Yanhui	Shandong Normal University, China
Hua Yin	Guangdong University of Finance and Economics, China
Jinguo You	Kunming University of Science and Technology, China
Ge Yu	Northeastern University, China
Hong Yu	Dalian Ocean University, China
Mei Yu	Tianjin University, China
Fang Yuan	Hebei University, China

Guan Yuan	China University of Mining and Technology, China
Xiaojie Yuan	Nankai University, China
Chun Zeng	Tsinghua University, China
Guigang Zhang	Institute of Automation, Chinese Academy of Sciences, China
Mingxin Zhang	Changshu Institute of Technology, China
Rui-Ling Zhang	Luoyang Normal College, China
Weifeng Zhang	Nanjing University of Posts and Telecommunications, China
Ying Zhang	Nankai University, China
Yong Zhang	Tsinghua University, China
Zhiqiang Zhang	Harbin Engineering University, China
Erping Zhao	Tibet University for Nationalities, China
Feng Zhao	Huazhong University of Science and Technology, China
Xiang Zhao	National Defense University of Science and Technology, China
Jiantao Zhou	Inner Mongolia University, China
Junwu Zhu	Yangzhou University, China
Mingdong Zhu	Henan Institute of Technology, China
Qingsheng Zhu	Chongqing University, China
Fang Zuo	Henan University, China

Contents

Data Mining

Data Privacy and Security

Knowledge Graph

Machine Learning

Natural Language Processing

Recommendation

Architecture and Systems

World Wide Web

An Active Learning Approach for Identifying Adverse Drug Reaction-Related Text from Social Media Using Various Document Representations

Jing Liu[1,2(✉)], Lihua Huang[1], and Chenghong Zhang[1]

[1] School of Management, Fudan University, Shanghai 200433, People's Republic of China
[2] School of Management Science and Engineering, Tianjin University of Finance and Economics, Tianjin 300222, People's Republic of China

Abstract. Adverse drug reaction (ADR) is a major health concern. Identifying text that mentions ADRs from a large volume of social media data discussing other topics is a key preliminary but nontrivial task for drug-ADR pair detection. This task suffers from severe imbalance issue. Moreover, prior studies have overlooked the simultaneous use of high-level abstract information contained in data and the domain-specific information embedded in knowledge bases. Therefore, we propose a novel multi-view active learning approach, in which a selection strategy is tailored to the imbalanced dataset and various document representations are regarded as multi views. We capture data-driven and domain-specific information by resorting to deep learning methods and handcrafted feature engineering, respectively. Experimental results demonstrate the effectiveness of our proposed approach.

Keywords: Adverse drug reaction-related text identification · Multi-view active learning · Document representation

1 Introduction

Adverse drug reaction (ADR) is a major health problem, causing irreversible health damage, millions of deaths and hospitalizations, and considerable financial loss [1, 2]. Timely and accurate detection of ADRs is of vital importance for patients, regulatory authorities, and pharmaceutical companies. However, existing drug safety monitoring channels have limitations. For example, pre-marketing clinical trials suffer from homogeneous participants and short durations. Spontaneous reporting systems (SRS), one of important post-marketing surveillance channels, suffer from heavy underestimation (up to 90% of cases are unreported) [3].

With the rapid development of Web 2.0 technology, social media has been a fertile and supplemental data source for post-marketing drug safety monitoring. In the United States, 72% of the population actively uses social media,[1] and 43.55% of adults seek and obtain health information in medical-related online platforms [4]. However, mining

---

[1] https://www.pewresearch.org/internet/fact-sheet/social-media/.

© Springer Nature Switzerland AG 2021
C. Xing et al. (Eds.): WISA 2021, LNCS 12999, pp. 3–15, 2021.
https://doi.org/10.1007/978-3-030-87571-8_1

ADRs from social media is a non-trivial task. First, social media is generally character-ized as "big data but less information" because the majority of user-generated content is irrelevant to the target task. Second, annotation is a time-consuming and domain knowledge-intensive process. Moreover, the heavily skewed data distribution resulted from information overload may impose detrimental effects on the predictive ability of text classification models. Third, users on social media tend to convey the same semantic meaning with diverse expressions, and share their experiences using creative phrases and colloquial terms. Moreover, misspellings are inevitable in social media data.

To tackle the first challenge, this study focuses on identifying ADR-related text from a large volume of ADR-irrelevant text on social media. To address the cost annotation challenge and the imbalance issue, we exert effort to build a corpus that is as balanced as possible with the aid of active learning, which exploits unlabeled data with the original labeled data. Instead of adopting the same section criterion for all instances indiscrimi-nately, we tailor respective criteria to separately query instances that potentially belong to different classes for annotation (i.e., to select the most *confident* potentially *ADR-related* text and most *informative* potentially *ADR-irrelevant* text). To deal with the text irregularity and diversity of social media data and ADR expressions, we implement deep learning methods to derive respective document representation capturing high-level and hierarchical abstract features [5–7]. Moreover, handcrafted feature engineering is also conducted to fully take advantage of external knowledge bases and human ingenuity in this field. A research question naturally arises, that is, how to simultaneously use these different strands of information (i.e., data-driven and domain knowledge-incorporated). To answer this question, inspired by the work in [8], we propose a novel multi-view active learning approach, in which various document representations serve as different views. The main contributions of our work are three-fold.

1. We provide a new view-generation mechanism for multi-view active learning (i.e., regarding various document representations as different views), and thus offering an alternative method to fuse different types of features.
2. We propose a novel selection strategy, which is suitable for implementing active learning on imbalanced data. The selection strategy can help alleviate imbalance issue of the annotated dataset.
3. We conduct extensive experiments. The experimental results demonstrate the effec-tiveness of our approach in terms of improving the predictive performance and reducing the demand for manual annotation effort.

2   Related Work

2.1   Identifying Adverse Drug Reactions from Social Media

Using social media to automatically detect ADR mentions has recently received signif-icant attention. According to different research objectives, related studies can be cate-gorized into four groups: ADR-related text classification [2, 5–7, 9, 10], ADR mention extraction [6], concept relation extraction [11], and concept normalization [12]. We limit our literature review to research on automatic identification of text mentioning an ADR.

Concerning techniques, machine learning methods have dominated existing studies, ranging from conventional machine learning algorithms, such as the support vector machine (SVM) [2, 9, 10] to deep learning methods, such as the bi-directional long short-term memory (Bi-LSTM) with an attention mechanism [5, 13], convolutional neural network (CNN) [5], and bi-directional encoder representations from transformers (BERT) [6]. In terms of features, prior studies have focused on two streams of feature generation methods (i.e., handcrafted feature engineering and deep neural network-based methods). The effectiveness of domain-specific knowledge-derived features has been verified in [10]. Yang et al. [9] constructed a feature space model using Latent Dirichlet Allocation (LDA). Fan et al. [6] exploited sentences embedding with BERT model. Some studies have considered both distributed embedding features and domain-specific features. For example, Wu et al. [5] used a concatenation of word embedding, character-based representation, part-of-speech tag embedding, and additional handcrafted features comprising lexicon appearance and sentiment scores. Likewise, Dai and Wang [7] derived sentence representation based on word embedding, term frequency and inverse document frequency (tf-idf), domain knowledge features, and negation features. Zhang et al. [2] generated holistic deep linguistic representations based on extracted predicate-ADR pairs and then combined these deep linguistic features with shallow features.

To cope with the difficulty of annotating data, to the best of our knowledge, only the work in [9] implemented partially supervised learning. The authors augmented the training dataset with reliable positive and negative instances that were extracted automatically by adopting different strategies.

2.2 Document Representation

Document representation learning aims to transform a document into a fixed-length vector. The traditional technique is the vector space model [14], which can capture word co-occurrence information. When conducting feature engineering, experts can design features and exploit domain-specific information. Given textual data, dimensionality reduction techniques, such as principal component analysis, are usually adopted to obtain low-dimensional vectors [15]. Additionally, LDA, a topic modeling method, has been widely used for feature space generation [8, 9].

With the development of deep learning techniques and the availability of large-scale corpora, two neural network-based methods are developed to obtain distributed representations of words (i.e., Word2vec [16]). As an extension of Word2vec, Doc2vec [17] has been proposed to learn distributed representation of sentences, paragraphs, and documents. These methods can capture semantic information contained in data. At present, other deep learning-based methods have achieved outstanding performance in many NLP tasks [5, 6, 13]. For example, Cao et al. [18] applied a deep neural network model to learn fine-grained features, combined with LDA-based coarse-grained features, for deceptive review detection. Guo et al. [19] proposed a text feature representation model based on CNN and variational auto encoder (VAE) for text classification.

Several prior studies have attempted to simultaneously exploit data-driven semantic information captured by applying deep learning and other aspects of information, such as domain-specific features. In addition to simply concatenating all features [5, 15], which is the most intuitive method, some research has explored more effective fusion methods. Kim et al. [8] fused three document representations obtained using tf-idf, LDA, and Doc2vec in a co-training-style semi-supervised learning manner. Li et al. [20] transformed each sentence into a set of concepts identified by an external taxonomy knowledge base (i.e., Probase) and afterward used a deep neural network to learn numeric vectors of conceptualized sentences.

3 Our Proposed Approach

3.1 Framework

The framework of our proposed approach is described in Fig. 1. In the first step, we transform collected social media data into various document representations and regard each representation as a view. When performing deep learning-based representation learning, we forgo the use of supervised techniques because their premise is to prepare a large volume of labeled data for training, which is contrary to the original intention of active learning. Because collecting large amounts of unlabeled data is more feasible and cost-effective than generating a large-scale labeled corpus, we explore several unsupervised deep neural network-based methods. Then, given the original labeled data, a balanced dataset is derived by performing resampling for each view. On this basis, different classifiers are trained. Afterward, active learning is conducted iteratively. In each iteration, the most valuable instances are automatically selected by classifiers and then annotated by experts. In the end, for each view, based on the balanced data in the current iteration, an augmented dataset is generated for the next iteration by adding a newly balanced dataset tagged with ground-truth labels. The newly balanced dataset is the result of performing resampling on the union of instances selected by classifiers in other views. Unlike prior methods (e.g., co-testing) in which multiple views share all newly labeled data, we add selected instances with ground-truth labels in a co-training style [21]. Specifically, for each view v_i, the newly added instance set is the output of performing resampling on the union of instances selected by classifiers in all other views (i.e., $v_j (j = 1, 2..N \& j \neq i)$, where N denotes the number of views). For example, to augment the training data for the view v_1 (i.e., the doc2vec view), we add balanced data derived from the newly annotated instances selected by other views (i.e., the average Word2vec view, the stacked autoencoder view, and the feature engineering view). In this way, the newly augmented data for the four views are different at each iteration. The augmentation strategy is designed to increase the degree of diversity, which is an important factor for a good ensemble model, and ensure close cooperation between multiple views.

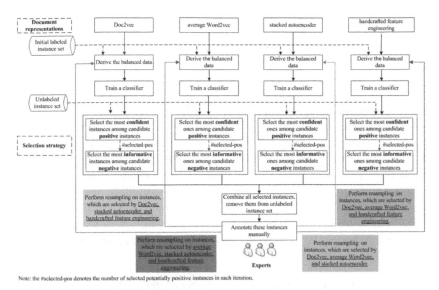

Note: the #selected-pos denotes the number of selected potentially positive instances in each iteration.

Fig. 1. Framework of our proposed approach.

3.2 View-Generation Mechanism Using Various Document Representations

Deep learning-based methods can learn task-independent and high-level abstract information from large-scale unlabeled data, whereas handcrafted feature engineering explores task-dependent and context-aware features. Therefore, we argue that document representations obtained using the abovementioned two groups of methods can complement each other, which is partially supported in the prior literature [8].

Stacked Autoencoder. The autoencoder consists of two parts (i.e., encoder and decoder). The *encoder* transforms an input vector I into a hidden representation R. The *decoder* is responsible for mapping the hidden representation R back to I in the input space. The objective of training an autoencoder model is to minimize the reconstruction error, retaining much of the information presented in the input data. Compared with the traditional autoencoder, stacked autoencoder (SAE) has multiple layers, improving the ability to represent nonlinearities and capture more abstract information. During the pretraining phase, the greedy layer-wise training strategy is adopted [22]. The input at each layer is the resulting hidden representation at its former layer. For each instance, the hidden representation obtained at the last layer is used as its final document representation, which can serve as input features for the target task.

Average Word2vec. In terms of word-vector generation methods, the traditional strategy is one-hot encoding. However, this method suffers from two main weaknesses. It generally exhibits high-dimensionality and fails to consider the semantic relationships between words. To overcome these drawbacks, Word2vec has been proposed [16], drawing on the distributional hypothesis that words tend to have close numeric representations if they have similar linguistic contexts [23]. The numeric vector dv of each document D is computed based on vectors $\mathbf{wv}_w$ corresponding to the words in D (with stop words removed), where tf-idf is adopted to weight each word:

$$dv = \frac{\sum_{w \in D} \mathbf{wv}_w tf - idf(w)}{n}. \tag{1}$$

Doc2vec. Inspired by the idea of learning word vectors [16], Doc2vec has been proposed for learning distributed and continuous vectors of variable-length instances, such as paragraphs, sentences, and documents [17]. The distributed memory model of paragraph vectors (PV-DM) structure is employed in our work because the authors have shown that PV-DM generally works well for most tasks. In PV-DM, the paragraph representations can be derived by predicting the next word based on the concatenation of the paragraph vectors and the vectors of the context words. During the training phase, word vectors and paragraph vectors of seen paragraphs are trained simultaneously. Each paragraph is transformed into a unique vector, whereas word vectors are shared among paragraphs. In the inference stage, the vectors of new paragraphs are trained until convergence while holding word vectors fixed.

Handcrafted Feature Engineering. The document representation obtained by feature engineering is task-dependent and relies heavily on domain expertise. Following [10], we provide a brief introduction of explored features.

Shallow Linguistic Features. We use n-grams with tf-idf as the feature value. The n-grams refer to contiguous n tokens in the text ($n = 1, 2, 3$).

Domain-specific Knowledge-Based Features. The feature set consists of medical semantic features, the ADR lexicon match-based features, and the negation features. Medical semantic features include the Unified Medical Language System (UMLS)[2] semantic types and concept IDs, which respectively represent broad and fine categories of medical concepts. The UMLS is a compendium that encompasses many medical vocabularies. These features can be identified using MetaMap.[3] The ADR lexicon match-based feature group consists of two features. One feature is an indication of whether the instance contains ADR mentions in a predefined lexicon. The other feature is the relative number of ADR mentions. The lexicon is derived from four domain-specific knowledge bases: COSTART,[4] SIDER,[5] MedEffect,[6] and Consumer Health Vocabulary (CHV).[7] We explore CHV because users on social media tend to describe ADRs using colloquial language rather than technical terms. The negation features refer to negated concepts. These are recognized and tagged using NegEx,[8] which is incorporated in MetaMap.

---

[2] https://www.nlm.nih.gov/research/umls/index.html.

[3] https://mmtx.nlm.nih.gov/.

[4] https://www.nlm.nih.gov/research/umls/sourcereleasedocs/current/CST/index.html.

[5] http://sideeffects.embl.de/.

[6] https://www.canada.ca/en/health-canada/services/drugs-health-products/medeffect-canada.html.

[7] https://www.nlm.nih.gov/research/umls/sourcereleasedocs/current/CHV/index.html.

[8] https://code.google.com/p/negex/.

Other Discriminative Features. These features include synonym expansion features, change phrase-related features, Sentiword score feature, and topic-based features. Synonym expansion features refer to synonyms for each adjective, noun, and verb in the text. These are identified via WordNet,[9] which is a universal knowledge base. In addition, the change phrase-related feature set comprises four features: less-good, less-bad, more-good, and more-bad. For example, if a bad thing (e.g., dizziness) was reduced, the value of the feature less-bad is 1, indicating a positive outcome. The Sentiword score feature is computed as the total sentiment score of a text based on a sentiment lexicon divided by the text length. The topic-based features include topic terms generated using Mallet,[10] and the total relevance score of all topic terms.

3.3 Selection Strategy in Active Learning

We first provide a brief introduction of the selection strategy in co-testing [24], which is a popular multi-view active learning method. In co-testing-style algorithms, the model first identifies the contention instances on which a certain degree of disagreement exists concerning predicted labels in different views. Subsequently, different query strategies, such as naive, aggressive, and conservative, can be adopted to determine the final selected instances for manual annotation.

In this paper, for the generation of the candidate positive dataset, the selection strategy adopted in the disagreement-based semi-supervised learning is a good reference. This is because confident instances are required, and multiple views are available. In disagreement-based semi-supervised learning, multiple diverse classifiers are first trained and then the "majority teaches minority" strategy is performed to measure the confidence of an unlabeled instance [11]. Specifically, for each view i, the confidence of an unlabeled instance x is computed as follows:

$$\varphi(x, i) = \frac{\max(m, N - 1 - m)}{N - 1}, \tag{2}$$

where m represents the number of other classifiers that label x as positive, and N is the number of classifiers (i.e., the number of views in this study).

Instances are selected as candidate positive instances if they satisfy two conditions. First, $\varphi(x, i) > \varphi$, where φ is a predefined threshold. Second, $m \geq \frac{N-1}{2}$, indicating the other classifiers label the instance x as positive using a majority voting scheme.

Among candidate positive instances, we further strengthen the guarantee of confidence by selecting the most confident instances based on an additional confidence measure [8]. For each view i, the confidence of an unlabeled instance x is the minimum value of its $N - 1$ confidence values measured by the other classifiers. In this way, information conveyed by different views can complement each other during the active learning process:

$$C(x, i) = argmin_{j=1,2,...,N \ \& \ j \neq i}\left(-E_{hj} \times H(x,j)\right), \tag{3}$$

---

[9] https://wordnet.princeton.edu/.
[10] http://mallet.cs.umass.edu/.

where E_{h_j} represents the error of classifier h_j for view j, and $H(x,j)$ is the following entropy function:

$$H(x,j) = -\sum_{c=1}^{2} P_{jc}(x)logP_{jc}(x).$$ (4)

Afterward, we sort the candidate positive instances into descending order according to the confidence computed using Eq. (3). Then, we select the top-ranked instances as potentially positive instances, waiting for subsequent manual annotation.

To generate the candidate negative dataset for view i, we select instances whose pseudo labels predicted by h_i are negative. Subsequently, given the candidate negative dataset, the confidence of an instance x is measured as the maximum value of its $N-1$ confidence values measured by the other classifiers:

$$C'(x,i) = argmax_{j=1,2,...,N \& j \neq i}\left(-E_{h_j} \times H(x,j)\right).$$ (5)

Finally, we sort the candidate negative instances into ascending order according to the confidence computed using Eq. (5) and select the top-ranked instances as potentially negative instances, preparing for manual annotation.

4 Experimental Dataset and Settings

We used an open-source dataset,[11] released in prior work [10]. The dataset consists of 10,822 instances that are collected from Twitter. However, actual tweets are not directly contained in the released dataset due to privacy. Instead, the dataset contains three fields (i.e., tweet id, user id, and the label indicating ADR related or ADR irrelevant). Because some tweets have been removed and are no longer accessible, we obtained 7,060 instances, including 756 positive instances and 6,304 negative instances. To learn document representations using unsupervised deep learning methods (i.e., stacked autoencoder, average word2vec, and doc2vec), we collected approximately 2 million sentences from a health-related forum. Furthermore, we performed data preprocessing, such as converting text to lowercase, conducting tokenization, performing lemmatization, removing sentences made up of less than 5 tokens.

To evaluate the performance of our proposed approach, we used F-Score as the evaluation metric. A series of experiments were conducted to investigate the effectiveness of our approach. For all experiments, we performed 10-fold cross-validation to minimize the influence of the variability of data. We adopted the SVM as the classification learning algorithm following [7, 10]. Word2Vec and Doc2vec were implemented using the genism[12] package. We used an open source code[13] to pre-train our SAE model. The handcrafted feature engineering was conducted using a released source code.[14] Our proposed method was implemented in-house using the Waikato Environment for

---

[11] http://diego.asu.edu/Publications/ADRClassify.html.
[12] https://radimrehurek.com/gensim/.
[13] https://github.com/rajarsheem/libsdae-autoencoder-tensorflow.
[14] http://diego.asu.edu/Publications/ADRClassify.html.

Knowledge Analysis (WEKA)[15] package. In addition, the SVM was employed using LibSVM in WEKA. Parameter tuning was conducted to determine the values of several parameters. Some related abbreviations are as follows:

D2V_AW2V_SAE_FE: Using all representations explored in Sect. 3.2.

D2V_AW2V_SAE: Using representations derived by deep learning methods.

Same: Separately selecting potentially positive instances and potentially negative instances with the same selection criterion.

WithoutPosNeg: Selecting instances using the same selection criterion, without concerning their potential labels using the same selection criterion.

MV_SL: A majority voting ensemble fusing classifiers obtained by performing supervised learning using all representations. D2V_SL, AW2V_SL, SAE_SL, FE_SL represent performing supervised learning using representation obtained by Doc2vec, average Word2vec, stacked autoencoder, and feature engineering, respectively.

5 Results and Discussion

5.1 Effectiveness of Different View Configurations

We first analyze the results from the perspective of views in multi-view active learning to examine the complementary nature of various document representations. Specifically, we treat each document representation as a view and test two view configurations. In addition to reporting results achieved by our proposed method and co-testing, with the same view configuration, we also list the results obtained by performing supervised learning with equal numbers of the total labeled instances (referred to as MV_SL). As shown in Table 1, exploring all document representations with our approach improves the F-Score by 18.67% compared to MV_SL. For both our method and co-testing, D2V_AW2V_SAE, in which only domain-independent representations are used, obtains lower F-Score and F-Score improvement. The results lend strong support to the complementary nature of domain-independent and domain-specific document representations.

Table 1. Performance of our approach and co-testing with different view configurations.

Method	View configurations	Active learning	MV_SL	Improvement
Our approach	D2V_AW2V_SAE_FE	**49.39%**	41.62%	**18.67%**
	D2V_AW2V_SAE	47.47%	41.47%	14.47%
Co-testing	D2V_AW2V_SAE_FE	40.53%	39.14%	3.54%
	D2V_AW2V_SAE	39.19%	37.89%	3.42%

---

[15] http://www.cs.waikato.ac.nz/ml/weka/.

5.2 Effectiveness of Different Selection Strategies

The custom selection strategy described in Sect. 3.3 (i.e., selecting the most confident instances among candidate positive instances and selecting the most informative instances among candidate negative instances) is a core component of our approach. The results in Fig. 2 imply that the proposed selection strategy can surpass other selection strategies.

5.3 Comparison between the Proposed Approach and Other Methods

As illustrated in Fig. 3, we gain comparable F-Score values using individual dense document representations obtained by performing deep learning methods. The performance achieved by the feature engineering-based method is limited, especially when only small amounts of labeled instances are available. Conceivably, an ensemble combining classifiers using different representations can enhance the predictive power, which is validated by the result that MV_SL gains improved F-Score compared to D2V_SL, AW2V_SL, SAE_SL, and FE_SL. With the same human-labeling effort, our approach performs best. All three our approach-family methods achieve improved F-Scores compared to MV_SL. As shown in Fig. 3, to gain the same level of predictive power, the required annotation effort in our method is substantially reduced. Moreover, the F-Score of our approach has significant improvement potential by adding additional unlabeled data.

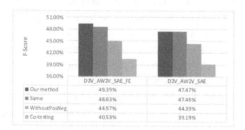

Fig. 2. F-Score of different selection strategies.

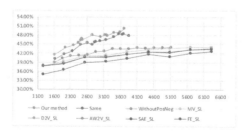

Fig. 3. F-Score with varying numbers of labeled instances

5.4 Performance Comparison with Baselines

We compared our proposed approach with two state-of-the-art baselines (i.e., feature engineering-based [10] and BERT). The BERT-based model was fine-tuned using the available labeled data. Moreover, two fusing strategies (i.e., adopting majority voting and using concatenation of all features) were conducted as baselines. For all baselines, we addressed the imbalance issue using SMOTE. As shown in Table 2, our proposed approach achieved the highest F-Score with significantly reduced annotation effort compared to baselines.

Table 2. Performance comparison with baselines.

	Method	# Labeled data	F-Score
State-of-the-art methods	Feature engineering-based	7,060	44.90%
	BERT (Fine-tuning)	7,060	44.70%
Fusing strategy	Majority voting	7,060	44.96%
	All features	7,060	43.64%
Our approach		3,978	49.39%

6 Conclusions

In this study, we proposed a scalable and generalized multi-view active learning approach to identify ADR-related text from social media. We developed a view-generation mechanism that regards various document representations as multiple views. Moreover, we proposed a novel selection strategy that separately selects potentially positive instances and potentially negative instances using confidence-oriented and informativeness-oriented measures. The experimental results demonstrate that our method can obtain improved predictive capability with reduced annotation effort.

The implications of this work are two-fold. First, from the perspective of methodological implications, our method can be generalized to many other text classification tasks. Moreover, the proposed view-generation mechanism can be applied to almost all text classification tasks. In addition to the generalization ability across applications, the proposed view-generation mechanism may benefit other multi-view learning, such as multi-view semi-supervised learning. Second, from the perspective of practical implications, our experimental results have shown that our proposed method can achieve enhanced predictive capability over existing approaches, even with less amounts of labeled instances. This advantage improves the feasibility of conducting social media-based drug safety monitoring. The effective and efficient ADR-related text detector offers a solid foundation for the subsequent task of extracting ADR mentions. This work provides valuable decision-making support for key stakeholders.

Our future work can focus on three points. First, we consider verifying the effectiveness of our proposed method with a larger volume of unlabeled data to further exploit its potential. Second, we could apply the proposed method to other tasks to verify the generalizability of the method, such as fake news identification from social media and comparative sentence identification from reviews. Third, we would like to further assess the scalability of our method, such as incorporating features obtained by other feature extractors, and adopting other active learning strategies.

Acknowledgments. This work is partially supported by the National Natural Science Foundation of China (Nos. 71701142 and 71971067), and China Postdoctoral Science Foundation (No. 2018M640346).

References

1. Sarker, A., et al.: Utilizing social media data for pharmacovigilance: a review. J. Biomed. Inform. **54**, 202–212 (2015)
2. Zhang, Y., Cui, S., Gao, H.: Adverse drug reaction detection on social media with deep linguistic features. J. Biomed. Inf. **106**, 103437 (2020)
3. Hazell, L., Shakir, S.A.: Under-reporting of adverse drug reactions. Drug Saf. **29**(5), 385–396 (2006)
4. Amante, D.J., et al.: Access to care and use of the Internet to search for health information: results from the US national health interview survey. J. Med. Internet Res. **17**(4), e106 (2015)
5. Wu, C., et al.: Detecting tweets mentioning drug name and adverse drug reaction with hierarchical tweet representation and multi-head self-attention. In: Empirical Methods in Natural Language Processing (2018)
6. Fan, B., et al.: Adverse drug event detection and extraction from open data: A deep learning approach. Inf. Process. Manage. **57**(1), 102131 (2020)
7. Dai, H., Wang, C.: Classifying adverse drug reactions from imbalanced twitter data. Int. J. Med. Inf. **129**, 122–132 (2019)
8. Kim, D., et al.: Multi-co-training for document classification using various document representations: TF–IDF, LDA, and Doc2Vec. Inf. Sci. **477**, 15–29 (2019)
9. Yang, M., Kiang, M., Shang, W.: Filtering big data from social media–building an early warning system for adverse drug reactions. J. Biomed. Inf. **54**, 230–240 (2015)
10. Sarker, A., Gonzalez, G.: Portable automatic text classification for adverse drug reaction detection via multi-corpus training. J. Biomed. Inf. **53**, 196–207 (2015)
11. Liu, J., Wang, G., Chen, G.: Identifying adverse drug events from social media using an improved semi-supervised method. IEEE Intell. Syst. **34**, 1 (2019)
12. Emadzadeh, E., et al.: Hybrid semantic analysis for mapping adverse drug reaction mentions in tweets to medical terminology. In: American Medical Informatics Association, Washington, D.C (2017)
13. Chowdhury, S., Zhang, C., Yu, P.S.: Multi-task pharmacovigilance mining from social media posts. In: Proceedings of the 2018 World Wide Web Conference (2018)
14. Salton, G., Wong, A., Yang, C.S.: A vector space model for automatic indexing. Commun. ACM **18**(11), 613–620 (1974)
15. Li, Y., et al.: Imbalanced text sentiment classification using universal and domain-specific knowledge. Knowl. Based Syst. **160**, 1–15 (2018)
16. Mikolov, T., et al.: Distributed representations of words and phrases and their compositionality. In: Advances in Neural Information Processing Systems, Harrahs and Harveys, Lake Tahoe (2013)

17. Le, Q., Mikolov, T.: Distributed representations of sentences and documents. In: International Conference on Machine Learning, Beijing, China (2014)
18. Cao, N., et al.: A deceptive review detection framework: combination of coarse and fine-grained features. Expert Syst. Appl. **156**, 113465 (2020)
19. Guo, C., Xie, L., Liu, G., Wang, X.: A text representation model based on convolutional neural network and variational auto encoder. In: Wang, G., Lin, X., Hendler, J., Song, W., Xu, Z., Liu, G. (eds.) WISA 2020. LNCS, vol. 12432, pp. 225–235. Springer, Cham (2020). https://doi.org/10.1007/978-3-030-60029-7_21
20. Li, Y., et al.: Incorporating knowledge into neural network for text representation. Expert Syst. Appl. **96**, 103–114 (2018)
21. Blum, A., Mitchell, T.: Combining labeled and unlabeled data with co-training. In: Proceedings of the Eleventh Annual Conference on Computational Learning Theory. ACM Madison (1998)
22. Bengio, Y., et al.: Greedy layer-wise training of deep networks. In: Advances in Neural Information Processing Systems. Vancouver, B.C., Canada (2007)
23. Bengio, Y., et al.: A neural probabilistic language model. J. Mach. Learn. Res. **3**, 1137–1155 (2003)
24. Muslea, I., Minton, S., Knoblock, C.A.: Active learning with multiple views. J. Artif. Intell. Res. **27**, 203–233 (2006)

DualLink: Dual Domain Adaptation for User Identity Linkage Across Social Networks

Bei Xu[1], Yue Kou[1(✉)], Guangqi Wang[2], Derong Shen[1], and Tiezheng Nie[1]

[1] Northeastern University, Shenyang 110004, China
{kouyue,shenderong,nietiezheng}@cse.neu.edu.cn
[2] Liaoning Provincial Higher and Secondary Education Enrollment Examination
Committee Office, Shenyang 110031, China

Abstract. User Identity Linkage (UIL) across social networks can be used to identify the accounts belonging to the same individual in multiple social networks, which is of great significance for user behavior analysis and network security supervision. By integrating deep learning into the process of UIL, the original intrinsic features of users can be retained more completely. However, the data distribution among different networks is different, and the existing UIL methods often ignore these differences, or only adjust a single step by domain adaptation mechanism, so it is difficult to ensure the accuracy of UIL. In this paper, we propose a new UIL model (called DualLink) which uses a dual domain adaptation mechanism to solve the problem of inconsistent data distribution. On one hand, a node embedding method based on adversarial domain adaptation is proposed to learn the node representation by considering the attributes, the topological structure and the difference between domains. On the other hand, a node matching method based on back-propagation domain adaptation is proposed to learn the suitable matching function by using back propagation neural network (BPNN). The feasibility and effectiveness of the key technology proposed in this paper are verified by experiments.

Keywords: User identity linkage · Adversarial domain adaptation · Back-propagation domain adaptation · Node embedding · Node matching

1 Introduction

Different social networks provide different types of services, and people join multiple networks according to their own needs, but multiple accounts belonging to the same person are separate from each other [1]. The typical aim of User Identity Linkage (UIL) is to detect that users from different social platforms are

This research is supported by National Natural Science Foundation of China (62072084, 62072086), Fundamental Research Funds for the Central Universities (N2116008, N180716010).

actually the same natural person [2]. It is an important prerequisite for many interesting cross-network applications, such as friend recommendation, etc. [3].

The data distribution among different networks is different, and this difference will affect the accuracy of cross-network user identity linkage. Therefore, the domain adaptation technology emerges as The Times require, which aims to use abundant data from the source domain to help the target domain learning [4]. A popular domain adaptation algorithm is feature-based [5,6], the purpose of it is to learn the feature representations with network-invariance to reduce the difference between networks.

Most of the existing network embedding algorithms are designed for a single network. When considering the cross-network case, the changing data distributions across networks can have an impact on the node representations when the model learned from the source domain is applied to the target domain [7]. Therefore, the single network embedding algorithm that does not solve domain difference will not be able to learn the network-invariant node representations across networks [6]. Although some studies have taken into account the difference of data distribution between different networks (e.g. [5,7,8]), as Fig. 1 shows, these studies only adjust a single step of UIL by domain adaptation mechanism (e.g. node embedding or node matching), and it is difficult to ensure the accuracy of UIL.

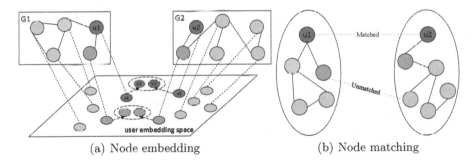

(a) Node embedding   (b) Node matching

Fig. 1. An example of user identity linkage across social networks.

In this work, this paper firstly proposes an adversarial deep network embedding module across networks, which innovatively combines deep network embedding with adversarial domain adaptation. The proposed deep network embedding module consists of two feature extractors, which are respectively based on the attributes of each node and those of its neighbors. In order to solve the distribution difference across networks, a domain discriminator is introduced to compete against the deep network embedding module. Finally, the network-invariant node representations can be learned to effectively solve the problem of the inconsistent data distribution. Then, based on these node representations, the Back Propagation Neural Network (BPNN) is applied to learn a suitable cross-network user matching function. The contributions of this paper can be summarized as follows:

- A cross-network user identity linkage model (DualLink) based on a dual domain adaptation is proposed, which includes two stages: node embedding and node matching. Different from the traditional UIL model, this model takes into account the difference of node embedding and matching function between different networks, and adopts the dual domain adaptation mechanism to solve the problem of inconsistent data distribution, which effectively ensures the accuracy of cross-network user identity linkage.
- In the stage of node embedding, a node embedding method based on adversarial domain adaptation is proposed to learn the node representation with network-invariance by considering the attributes, the topological structure and the difference between domains. Firstly, node embedding is generated by deep network embedding module. Then, the domain discriminator and deep network embedding module are used to play an adversarial game to learn node representation, so as to reduce the difference of node embedding among different networks.
- In the stage of node matching, a node matching method based on back-propagation domain adaptation is proposed. In order to reduce the difference of the matching function among different networks, the suitable matching function is learned by the back propagation neural network (BPNN).
- A large number of experimental results on real-world datasets show that the effectiveness of the proposed DualLink model for UIL problem.

2   Related Work

The existing UIL method is mainly based on the information of the matched user(anchor link), including the user's attribute information, behavior information and topology information [2]. For UIL methods based on topology structure, they can be specifically divided into user identity linkage based on traditional machine learning (such as classification) and user identity linkage based on deep learning.

User identity linkage based on traditional machine learning need to define and extract features in advance. For example, Jiangtao Ma et al. [9] firstly divide the large network into several small networks by the community partition algorithm, and then extract parameters to train the random forest model for UIL. Pedarsani et al. [10] identify users based on structural features by Bayesian algorithm. Olga Peled et al. [11] extract distance-based profile features and neighborhood-based network features and many popular classifiers are performed in the experiments such as Adaboost, Random Forest, etc. The effectiveness of such methods depends on whether the user's features are accurate and whether they can be extracted comprehensively. Because topology is difficult to extract and model, therefore these methods is not satisfactory [2].

There are many UIL methods based on deep learning. For example, GraRep [12] factorizes the positive point mutual information (PPMI) matrix through singular value decomposition(SVD). Yu [13] converts the UIL into the weight matching problem of graph. Shun Fu et al. [14] leveraged random walk (RW) to capture the higher order structural similarity, and heuristic edge weighting mechanism is used to capture structural attributes. However, most of such methods do not consider network heterogeneity. Therefore, their performances will be limited when learning cross-network node representations [1]. While some researches take the domain differences into account, it is limited to a single step (e.g. node embedding). For example, the statistics-based approaches [5,7] widely incorporate the maximum mean difference (MMD) [15] into deep neural networks to match the mean values of cross-domain distributions. Adversarial autoencoders [8], which can learn node representations without any supervision. Besides, the influence of domain differences on the accuracy of node matching should also be considered.

3   DualLink: The Proposed Model

The overall framework of the DualLink model is shown in Fig. 2. It mainly consists of two components: node embedding based on adversarial domain adaptation and node matching based on back-propagation domain adaptation.

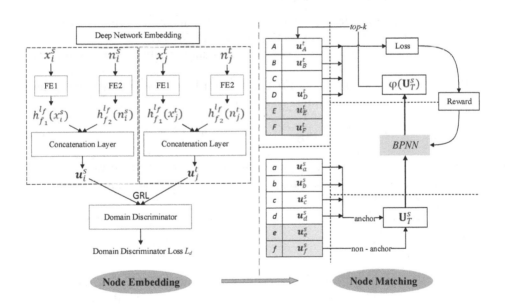

Fig. 2. The basic idea of DualLink model.

The first stage refers to node embedding based on adversarial domain adaptation: learning node representations of capturing the attribute similarity and topological proximity between nodes by deep network embedding module, then domain discriminator and deep network embedding play an adversarial game to learn network-invariant node representations across networks.

The second stage refers to node matching based on back-propagation domain adaptation: the back propagation neural network is used to adjust and learn the matching function, in order to reduce difference of mapping users in different domains. Through Dual domain adaptation, network-invariant node representations can be learned to reduce the influence of heterogeneous network, and a suitable matching function can be finally learned to improve the performance of UIL.

4   DualLink-Based User Identity Linkage Algorithm

In this section, a DualLink-based User Identity Linkage algorithm is proposed. Firstly, we introduce the first stage: Node Embedding Based On Adversarial Domain Adaptation. Then, we introduce the second stage: Node Matching Based On Back-propagation Domain Adaptation.

4.1   Node Embedding Based on Adversarial Domain Adaptation

This process consists of two components: deep network embedding module and adversarial domain adaptation learning.

Deep Network Embedding. The deep network embedding module consists of two feature extractors and a connection layer. The feature extractor includes a node attribute feature extractor (FE1) and a neighbor attribute feature extractor (FE2).

Firstly, the node attribute feature extractor (FE1) is constructed, which has l_f hidden layer and can be defined as Eq. (1):

$$h_{f_1}^{(k)}(\boldsymbol{x}_i) = ReLU(h_{f_1}^{(k-1)}(\boldsymbol{x}_i)W_{f_1}^{(k)} + b_{f_1}^{(k)}), \quad 1 \leq k \leq l_f \tag{1}$$

where $h_{f_1}^{(0)}(x_i) = \boldsymbol{x}_i$ represents the input attribute vector of node v_i, $h_{f_1}^{(k)}(\boldsymbol{x}_i) \in R^{1 \times f(k)}, 1 \leq k \leq l_f$ represents the latent attribute representation of v_i learned by the k-th hidden layer of FE1.

Then, the neighbor attribute feature extractor (FE2) can be constructed as Eq. (2):

$$h_{f_2}^{(k)}(\boldsymbol{n}_i) = ReLU(h_{f_2}^{(k-1)}(\boldsymbol{n}_i)W_{f_2}^{(k)} + b_{f_2}^{(k)}), \quad 1 \leq k \leq l_f \tag{2}$$

$h_{f_2}^{(0)}(n_i) = \boldsymbol{n}_i$ represents the input neighbor attribute vector of v_i, neighbors' attribute is integrated by assigning a higher weight to closer neighbor to calculate $\boldsymbol{n}_i$, as shown in Eq. (3):

$$n_{ik} = \sum_{j=1, j \neq i}^{n} \frac{a_{ij}}{\sum_{g=1, g \neq i}^{n} a_{ig}} x_{jk} \tag{3}$$

a_{ij} represents the topological proximity between two nodes v_i and v_j. $h_{f_2}^{(k)}(\boldsymbol{n}_i) \in R^{1 \times f(k)}$ represents the latent neighbor attribute representation of v_i learned by the k-th hidden layer of FE2. And $W_{f_2}^{(k)}, b_{f_2}^{(k)}$ are the trainable weight and bias parameters associated with the k-th hidden layer of FE2.

Finally, the deepest node attribute feature vectors $h_{f_1}^{(l_f)}(\boldsymbol{x}_i)$ and the deepest neighbor attribute feature vectors $h_{f_2}^{(l_f)}(\boldsymbol{n}_i)$ are integrated by a connection layer, which can be defined as Eq. (4):

$$\boldsymbol{u}_i = ReLU([h_{f_1}^{(l_f)}(\boldsymbol{x}_i), h_{f_2}^{(l_f)}(\boldsymbol{n}_i)]W_c + b_c) \tag{4}$$

$\boldsymbol{u}_i \in R^{1 \times d}$ represents the node representation of v_i finally learned by deep network embedding module, which preserves both attribute similarity and topology proximity.

Finally, the network-invariant node representation can be learned by the deep network embedding module, $\theta_e = \{\{W_{f_1}^{(k)}, b_{f_1}^{(k)}, W_{f_2}^{(k)}, b_{f_2}^{(k)}\}_{k=1}^{l_f}, W_c, b_c\}$ is all the trainable parameters associated with the deep network embedding module.

Adversarial Domain Adaptation Learning. In this paper, an adversarial domain adaptation approach is employed to make the node representations learned by the deep network embedding module network-invariant. Firstly, domain discriminator is fed with the node representation learned by the deep network embedding module to predict which network the node comes from, as defined in Eq. (5):

$$\begin{aligned} h_d^{(k)}(\boldsymbol{u}_i) &= ReLU(h_d^{(k-1)}(\boldsymbol{u}_i)W_d^{(k)} + b_d^{(k)}), \quad 1 \le k \le l_d \\ \hat{d}_i &= Softmax(h_d^{(l_d)}(\boldsymbol{u}_i)W_d^{(l_d+1)} + b_d^{(l_d+1)}) \end{aligned} \tag{5}$$

$h_d^{(0)}(\boldsymbol{u}_i) = \boldsymbol{u}_i, h_d^{(k)}(\boldsymbol{u}_i) \in R^{1 \times d(k)}$ represents the domain representation of node v_i learned by the k-th hidden layer of the domain discriminator. $\hat{d}_i$ represents the prediction probability that v_i comes from the target network. And $\theta_d = \{W_d^{(k)}, b_d^{(k)}\}_{k=1}^{l_d+1}$ is all the trainable parameters of the domain discriminator.

Then, the nodes from the source network and the target network are utilized for training, and the loss of domain discrimination is defined as Eq. (6):

$$L_d = -\frac{1}{n^s + n^t} \sum_{v_i \in \{V^s \cup V^t\}} (1 - d_i)log(1 - \hat{d}_i) + d_i log(\hat{d}_i) \tag{6}$$

d_i is the ground-truth domain label of the node v_i, $d_i = 1$ indicates that v_i comes from the target network and $d_i = 0$ indicates that v_i comes from the source network.

In order to make node representations network-invariant, the domain discriminator and the deep network embedding module are against each other in an adversarial manner. On the one hand, the domain discriminator can accurately distinguish the node representations of the source network from those of the target network by $min_{\theta_d}\{L_d\}$. On the other hand, $min_{\theta_e}\{-L_d\}$ makes the deep network embedding module is trained to fool the domain discriminator by generating indistinguishable node representations across networks.

Algorithm Description. Our node embedding algorithm is described as Algorithm 1. Firstly, in lines 2–9, the same deep network embedding module is utilized

Algorithm 1: Node Embedding

Input: networks $G^s = (V^s, E^s)$ and $G^t = (V^t, E^t)$, adaptation weight λ;

Output: network-invariant node representations u_i;

1   While not max iteration do;

2   For $v_i^s \in V^s$;

3   Learn node attribute representation $h_{f_1}^{(l_f)}(x_i^s)$ by FE1 and

             neighbor attribute representation $h_{f_2}^{(l_f)}(n_i^s)$ by FE2;

4   Learn node embedding u_i^s by concatenation layer;

5   End for

6   For $v_j^t \in V^t$;

7   Learn node attribute representation $h_{f_1}^{(l_f)}(x_j^t)$ by FE1 and

             neighbor attribute representation $h_{f_2}^{(l_f)}(n_j^t)$ by FE2;

8   Learn node embedding u_j^t by concatenation layer;

9   End for

10   Domain discriminator predicts which network the node comes from by

         $h_d^{(l_f)}(u_i)$ and $\hat{d}_i$;

11   Compute domain discrimination loss L_d by u_i and d_i;

12   Update parameters θ_e and θ_d by SGD;

13   End while

---

to learn the node representations of the two networks respectively; Then, in line 10, Domain discriminator distinguishes nodes. Then, in line 11, discrimination loss are computed; Next, in line 12, update the trainable parameters θ_e, θ_d by the stochastic gradient descent algorithm (SGD). After the adversarial iteration training converges finally, the network-invariant node representations u_i can be generated based on the optimal network embedding parameters θ_e.

4.2   Node Matching Based on Back-Propagation Domain Adaptation

After learning the network-invariant node representations from source and target networks by adversarial domain adaptation, DualLink applies back propagation neural network (BPNN) to learn a suitable matching function.

Back-Propagation Domain Adaptation Learning. Given any pair of anchor nodes (v_i^s, v_j^t) and their vector representations (u_i^s, u_j^t). In this paper, cosine distance is selected to define the loss function as Eq. (7):

$$l(u_i^s, u_j^t) = 1 - cos(\varphi(u_i^s), u_j^t) \tag{7}$$

The known anchor links set is T, and the sub-vector spaces composed of anchor nodes are $\mathbf{U}_T^s$ and $\mathbf{U}_T^t$ respectively. The objective function of node matching learning can be defined as Eq. (8):

$$l(\mathbf{U}_T^s, \mathbf{U}_T^t) = arg\ min_{W,b}(1 - cos(\varphi(\mathbf{U}_T^s), \mathbf{U}_T^t); W, b) \tag{8}$$

W and b are respectively the weight and bias parameters. Using the known anchor links as the supervised information, the SGD can learn the optimal W, b to minimize the loss function.

Algorithm 2: Node Matching

Input: the sub-vector spaces of anchor nodes in two networks, $\mathbf{U}_T^s, \mathbf{U}_T^t$;
Output: optimized weight parameter and bias parameter, W, b;
1   Initialize weight parameters W and threshold θ of objective function l;
2   While loss is higher than threshold θ;
3   Input and propagate $\mathbf{U}_T^s, \mathbf{U}_T^t$ forward;
4   Output layer calculate loss and reward by the back-propagation;
5   Update parameters W and b by SGD;
6   End while
7   Output optimized weight parameters W and bias parameters b;

Algorithm Description. Our node matching algorithm is described as Algorithm 2. Firstly, input and forward propagate anchor node embedding vectors $\mathbf{U}_T^s, \mathbf{U}_T^t$. Then, the output layer computes the matching loss and feeds back. Next, update and adjust the parameters of objective function by SGD. When the matching loss is not higher than the threshold θ, the iterative process is terminated. At this time, the optimal parameters of the objective function is obtained, and a suitable matching function for UIL is finally learned.

5   Experiments

5.1   Datasets

This experiment utilizes the real-world dataset provided by AMiner, including four social networks, Twitter-Foursquare (T-F) and Douban-Weibo (D-W). The information of this dataset is summarized in Table 1.

Table 1. Statistics of T-F and D-W datasets.

Dataset	#Users	#Relations	#Attributes	#Anchor users
Twitter	6210	175987	248647	2154
Foursquare	6523	84549	209757	
Douban	9734	260467	219754	4598
Weibo	9514	196978	198424	

5.2   Baselines and Evaluation Metrics

In this paper, we compare DualLink with four baseline methods:

- **ULink** [8]: Using basic profile features and a projection algorithm is proposed.
- **PALE** [16]: Using potential network features, a projection method is adopted.
- **Peled'13** [17]: Extracting profile features and network features, many common classifiers are used (e.g. random forest).
- **DeepLink** [18]: Samples networks by random walks and learns to encode network nodes into vector representations to capture the local and global network structures. Finally, a deep neural network model is trained through the dual learning to realize user identity linkage.

In this paper, *Precision*, *Recall* and F1 are used as the evaluation metrics of UIL. The most common method is to use F1-measure, and when F1 is high, it indicates that the experimental method is more effective.

$$Precision = \frac{|TP|}{|TP|+|FP|}$$
$$Recall = \frac{|TP|}{|TP|+|FN|} \tag{9}$$
$$F1 = \frac{2 \times Precision \times Recall}{Precision + Recall}$$

5.3   Parameter Setup

In the stage of node embedding, we set the K-step as 3 when computing the topology proximity, both FE1 and FE2 are constructed with two hidden layers whose dimensionalities are $F(1) = 512$ and $F(2) = 128$. The dimensionality of node representations is set to $d = 128$. In addition, the domain discriminator consists of two hidden layers whose dimensionalities are $d(1) = d(2) = 128$, and the domain adaptation weight is $\lambda = \frac{2}{1+exp(-10p)} - 1, p \in (0,1)$ is the training progress linearly changing. In the stage of node matching, the number of iterative training is set as $i = 1000000$, and the percentage of anchor nodes used for training is set as $r = 0.8$.

5.4   Performance Evaluation

Firstly, we evaluate the influence of the parameters on the performance of node matching, as shown in Fig. 3. As can be seen from the Figure on the left, there is no overfitting problem in DualLink. In contrast, DualLink can not only get better results, but also have faster convergence speed. The Figure in the middle shows the effect of different training ratio on the performance of all methods. As r increases from 0.1 to 0.9, DualLink consistently outperforms the other comparison methods. When r is only 0.1 or 0.2, DualLink still performs significantly better than the other methods. The influence of vector dimensionality d is shown in the figure on the right. PALE, Peled'13, DeepLink and DualLink all perform well in low-dimensional vector space. DeepLink performs best when the dimensionality is lower than 80, but DualLink is significantly better than other methods when the dimensionality reaches 100.

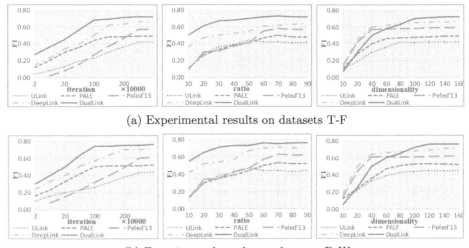

(a) Experimental results on datasets T-F

(b) Experimental results on datasets D-W

Fig. 3. Parameter setting.

Next, we selected the most appropriate parameters to conduct experiments on all methods: anchor nodes training ratio is set as $r = 0.8$, vector dimensionality is set as $d = 128$, and number of training iterations is set as $i = 3000000$. We use a bar chart to intuitively compare and analyze the experimental results of all methods, as shown in Fig. 4.

We can see that ULink's F1 is the lowest, this is because ULink only learns node attribute feature representations. PALE, Peled'13, and DeepLink all consider topology between nodes. Among them, PALE has poor performance, probably because PALE considers only a single network structure. Then Peled'13 considers both attribute feature and structural feature, so its performance is better than ULink and PALE. DeepLink has best performance among them, because it considers both local and global network structures when learning node representations, and trains deep neural network module through dual learning. It can be seen intuitively from Fig. 4 that DualLink has a higher performance than DeepLink. This is because DualLink takes into account the structural difference among networks and learn the network-invariant node representations by adversarial domain adaptation technology. In addition, the comparison of the experimental results between DualLink and Peled'13 also shows that leveraging back-propagation domain adaptation to learn matching function can improve accuracy for UIL.

Finally, we conducted ablation experiments. The experimental results are shown in Fig. 5. Compared with DualLink, in the absence of FE1 or FE2, the F1 on both datasets will decrease significantly, indicating the effectiveness of leveraging two feature extractors in the deep network embedding module. In addition, compared to DualLink, performance without domain discriminator(no-D) is also poor, which indicates that reducing domain difference between networks

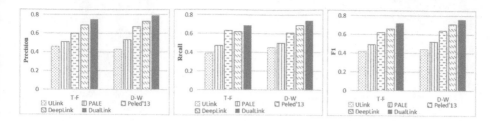

Fig. 4. Comparisons of different methods on precision, recall and F1.

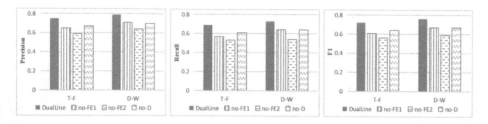

Fig. 5. Comparisons of variants on precision, recall and F1.

is indeed necessary for cross-network node representations. The results of the ablation experiment once again indicate that it is necessary to generate network-invariant node representations by adversarial domain adaptation technology, and also indicate that the node representations based on multiple features has better performance in UIL problem.

Based on the above experimental results in real-world datasets, it can be seen that DualLink method proposed in this paper demonstrated a distinctive performance over the state-of-the-art baseline methods.

6   Conclusions

In this paper, we propose a Dual Domain Adaptation (DualLink) method for UIL. This paper emphasizes the importance of network embedding in UIL and the influence of heterogeneous network on cross-network node embedding. This paper firstly leveraged deep network embedding module to generate node representations with both attribute similarity and topological proximity for networks, then leveraged adversarial domain adaptation approach to make the deep embedding network module against with the domain discriminator to generate network-invariant node representations. Secondly, leveraging back-propagation domain adaptation algorithm to optimize the parameters of objective function based on learned node representations, in order to learn a suitable matching function for UIL. Experimental results in real-world datasets show that the proposed DualLink method is more effective than the non-domain adaptation method for UIL, and DualLink can be extended to large datasets. Next, we can continue to study the effect of domain adaptation approach for UIL in multi-network based on this paper.

References

1. Li, X., et al.: RLINK: deep reinforcement learning for user identity linkage. World Wide Web **24**(1), 85–103 (2021). https://doi.org/10.1007/s11280-020-00833-8
2. Zhou, T., Lim, E.-P., Lee, R.K.-W., Zhu, F., Cao, J.: Retrofitting embeddings for unsupervised user identity linkage. In: Lauw, H.W., Wong, R.C.-W., Ntoulas, A., Lim, E.-P., Ng, S.-K., Pan, S.J. (eds.) PAKDD 2020, Part I. LNCS (LNAI), vol. 12084, pp. 385–397. Springer, Cham (2020). https://doi.org/10.1007/978-3-030-47426-3_30
3. Kou, Y., Li, X., Feng, S., Shen, D., Nie, T.: A cross-network user identification model based on two-phase expansion. In: Ni, W., Wang, X., Song, W., Li, Y. (eds.) WISA 2019. LNCS, vol. 11817, pp. 397–403. Springer, Cham (2019). https://doi.org/10.1007/978-3-030-30952-7_39
4. Li, J., Li, Z., Lü, S.: Feature concatenation for adversarial domain adaptation. Expert Syst. Appl. **169**, 114490 (2021)
5. Tzeng, E., Hoffman, J., Saenko, K., Darrell, T.: Adversarial discriminative domain adaptation. In: Proceedings of the IEEE Conference on Computer Vision and Pattern Recognition, pp. 7167–7176 (2017)
6. Shen, X., Chung, F.L.: Deep network embedding for graph representation learning in signed networks. IEEE Trans. Cybern. **50**(4), 1556–1568 (2018)
7. Ganin, Y., et al.: Domain-adversarial training of neural networks. J. Mach. Learn. Res. **17**(1), 2030–2096 (2016)
8. Makhzani, A., Shlens, J., Jaitly, N., Goodfellow, I., Frey, B.: Adversarial autoencoders. arXiv preprint arXiv:1511.05644 (2015)
9. Ma, J., et al.: De-anonymizing social networks with random forest classifier. IEEE Access **6**, 10139–10150 (2017)
10. Pedarsani, P., Figueiredo, D.R., Grossglauser, M.: A Bayesian method for matching two similar graphs without seeds. In: 2013 51st Annual Allerton Conference on Communication, Control, and Computing (Allerton), pp. 1598–1607. IEEE (2013)
11. Peled, O., Fire, M., Rokach, L., Elovici, Y.: Entity matching in online social networks. In: 2013 International Conference on Social Computing, pp. 339–344. IEEE (2013)
12. Levy, O., Goldberg, Y.: Neural word embedding as implicit matrix factorization. Adv. Neural Inf. Process. Syst. **27**, 2177–2185 (2014)
13. Yu, M.: Entity linking on graph data. In: Proceedings of the 23rd International Conference on World Wide Web, pp. 21–26 (2014)
14. Fu, S., Wang, G., Xia, S., Liu, L.: Deep multi-granularity graph embedding for user identity linkage across social networks. Knowl.-Based Syst. **193**, 105301 (2020)
15. Gretton, A., Borgwardt, K., Rasch, M.J., Scholkopf, B., Smola, A.J.: A kernel method for the two-sample problem. arXiv preprint arXiv:0805.2368 (2008)
16. Mu, X., Zhu, F., Lim, E.P., Xiao, J., Wang, J., Zhou, Z.H.: User identity linkage by latent user space modelling. In: Proceedings of the 22nd ACM SIGKDD International Conference on Knowledge Discovery and Data Mining, pp. 1775–1784 (2016)
17. Man, T., Shen, H., Liu, S., Jin, X., Cheng, X.: Predict anchor links across social networks via an embedding approach. IJCAI. **16**, 1823–1829 (2016)
18. Zhou, F., Liu, L., Zhang, K., Trajcevski, G., Wu, J., Zhong, T.: DeepLink: a deep learning approach for user identity linkage. In: IEEE INFOCOM 2018-IEEE Conference on Computer Communications, pp. 1313–1321. IEEE (2018)

Multimodal Topic Detection in Social Networks with Graph Fusion

Yuhao Zhang[1,3], Kehui Song[2,3], Xiangrui Cai[1,3(✉)], Yierxiati Tuergong[4],
Ling Yuan[4], and Ying Zhang[2,3]

[1] College of Cyber Science, Nankai University, Tianjin 300350, China
zhangyuhao@dbis.nankai.edu.cn, caixr@nankai.edu.cn
[2] College of Computer Science, Nankai University, Tianjin 300350, China
songkehui@dbis.nankai.edu.cn, yingzhang@nankai.edu.cn
[3] Tianjin Key Laboratory of Network and Data Security Technology,
Tianjin 300350, China
[4] School of Computer Science and Technology, Kashi University, Kashi, China

Abstract. Social networks have become a popular way for Internet users
to express their thoughts and exchange real-time information. The increasing number of topic-oriented resources in social networks has drawn more
and more attention, leading to the development of topic detection. Topic
detection of pure texts originates from text mining and document clustering, aiming to automatically identify topics from massive data in an
unsupervised manner. With the development of mobile Internet, user-
generated content in social networks usually contains multimodal data,
such as images, videos, etc. Multimodal topic detection poses a new challenge of fusing and aligning heterogeneous features from different modalities, which has received limited attention in existing research studies. To
address this problem, we adopt a Graph Fusion Network (GFN) based
encoder and a multilayer perceptron (MLP) decoder to hierarchically fuse
information from images and texts. The proposed method regards multimodal features as vertices and models the interactions between modalities with edges layer by layer. Therefore, the fused representations contain rich semantic information and explicit multimodal dynamics, which
are beneficial to improve the performance of multimodal topic detection.
Experimental results on the real-world multimodal topic detection dataset
demonstrate that our model performs favorably against all the baseline
methods.

Keywords: Multimodal · Topic detection · Graph fusion · Social
networks

1 Introduction

Nowadays, social networks have become the main way for Internet users to
exchange information and share opinions [17]. With the increasing number of
user-generated resources in social networks, researches on topic detection have

© Springer Nature Switzerland AG 2021
C. Xing et al. (Eds.): WISA 2021, LNCS 12999, pp. 28–38, 2021.
https://doi.org/10.1007/978-3-030-87571-8_3

drawn more and more attention from governments, enterprises, and academic institutions [13]. For example, government departments need to monitor widely-spread public topics and provide proper guidance in real-time. Enterprises want to understand users' preferences based on social topics and develop sales strategies accordingly. Therefore, it is valuable to detect topics in social networks.

Topic detection is derived from text mining and document clustering, aiming to automatically identify topics from large amounts of data in an unsupervised manner [2]. To determine whether a document is relevant to a specific topic, the crucial problem to be addressed is how to represent a document [9]. One mainstream of methods [10,14] is based on the vector space model (VSM). They use TF-IDF and its improved calculations to map texts into vector spaces for feature measuring. Another mainstream of approaches is typically probabilistic topic models, such as the latent Dirichlet allocation (LDA) model and its extended implementations [6,12,15]. They determine the relevance of documents to different topics through the probability distribution and mathematical statistics of the vocabulary. Besides, some other topic detection approaches have explored the optimization of clustering algorithms to obtain an optimal performance [26,29,30]. However, these topic detection methods commonly rely on bag-of-words (BoW) representations. They ignore the semantic and compositional structures of documents, which is crucial for summarizing meaningful topics [16]. Therefore, extensive works [4,8,22,23] construct graph-based representations to describe the topic structures. Different topics are identified by applying novel graph analysis techniques. Due to the graphical representations, graph-based approaches could yield higher performance on topic detection.

While topic detection of pure texts has achieved great success, the rapid growth of multimodal data in social networks poses new challenges for this task. Unlike the traditional methods, the fundamental challenge of multimodal topic detection is the fusion and alignment of heterogeneous data, which has received limited attention in existing studies. To address this problem, Li *et al.* presented a Multimodal And-Or Graph (MT-AOG) to model the visual and textual features of news stories [16]. The hierarchical compositions of news topics (e.g., time, place, and people) are efficiently described with the three different types of nodes in the MT-AOG. Then the Swendsen-Wang Cuts (SWC) algorithm is employed to traverse the solution space and obtain optimal topic clustering solutions. However, different from news stories, user-generated texts in social networks are commonly short, grammatically confusing, and contain a lot of noisy information. Hence, the graph construction algorithm of MT-AOG is not applicable under the social network scenario. Recently, researches based on multimodal fusion methods have made great progress in the social network data mining community. Xiong *et al.* proposed a multimodal deep fusion method based on semantic feature matrices to achieve fine-grained event detection in social networks [28]. Pang *et al.* introduced a coupled Poisson deconvolution to jointly handle topic detection and topic description for online videos on streaming platforms [20]. However, due to the goal of unsupervised learning and the alignment between modalities, the proposed methods in [20, 28] are not applicable to image-text based multimodal topic detection.

In this paper, we focus on multimodal topic detection of image-text social network data. Inspired by the Graph Fusion Network (GFN) introduced in [19], we adopt a GFN-based encoder and a multilayer perceptron (MLP) decoder to address the fusion problem in multimodal topic detection. Different from the original GFN, we design a reconstruction loss function to constrain the learning process of the encoder and decoder in an unsupervised manner. The encoder hierarchically fuses multimodal information by constructing vertices layer by layer, where features are regarded as vertices and multimodal dynamics are considered as edges. Therefore, the fused representations contain informative structures, providing explicit multimodal dynamics and rich semantic information to improve multimodal topic detection performance. Experimental results on the *Yelp* dataset demonstrate that our proposed method performs favorably against baseline approaches.

The remainder of this paper is organized as follows. In Sect. 2, we give a brief review of topic detection and multimodal fusion methods. We introduce our model in Sect. 3 and present experiment settings, baseline methods, experimental results, and discussions in Sect. 4. Section 5 is the conclusion and future work of this paper.

2   Related Work

2.1   Topic Detection

Topic detection is a study of organization and utilization of massive topic-oriented resources, aiming to automatically identify topics in an unsupervised manner. Research in topic detection began in 1996 when the original technology was proposed by the U.S. Defense Advanced Research Projects Agency (DARPA) for discovering topics from news data streams [1]. Each year since 1998, with the support of DARPA, an international conference has been held by the National Institute of Standards and Technology (NIST) to report and evaluate the advances in topic detection studies. In recent years, with the rapid development of the Internet and the gradual deepening of research, topic detection in social networks has drawn more and more attention from government departments, enterprises, and academic institutions [13]. Advanced topic detection techniques have been applied to opinion monitoring and preference analysis.

The crucial problem of text topic detection is how to represent a document with rich semantic information [9]. Extensive methods [10, 14] are based on the vector space model (VSM). They used TF-IDF and its improved calculations to map texts into vector spaces and split vectors according to time, places, people, or other key information for feature measuring. Due to the independence assumption of VSM, all the connections among features are lost in the transformation from texts to vectors [9]. This limitation lowers the performance of VSM-based approaches.

Some other methods are typically probabilistic topic models [6,12,15]. They model the probability distribution and mathematical statistics of the vocabulary to determine the relevance of documents to different topics. Blei *et al.* proposed a hierarchical Bayesian method, namely the latent Dirichlet allocation (LDA), to represent the topic information with the vocabulary probability of documents [6]. Lau *et al.* introduced the Online-LDA model to extend the effectiveness of the original LDA model to Twitter data streams [15]. Hofmann proposed a generative latent class model, which conducts a probabilistic mixture decomposition to perform semantic analysis [12].

Recently, A large number of works [4,8,22,23] have used graph-based representations to describe the topic structure. Different topics are identified by applying novel graph analysis techniques. Compared to traditional methods with bag-of-words (BoW) representations, graph-based approaches can model the semantic and compositional structures of documents, which is crucial for summarizing meaningful topics [16]. Therefore, graph models yield better performance on topic detection.

In addition to the research on document representation, plenty of works have been devoted to the optimization of clustering algorithms to improve topic detection performance [26,29,30].

2.2 Multimodal Fusion

With the rapid development of the mobile Internet, topic-oriented resources in social networks usually contain multimodal data. Different from topic detection of pure texts, detecting topics from multimodal data poses a new challenge of fusing and aligning heterogeneous features from different modalities, which has received limited attention in the topic detection community. Li *et al.* presented a Multimodal And-Or Graph (MT-AOG) to hierarchically model the visual and textual features of news stories (e.g., time, place, and people) with the three different types of nodes in the MT-AOG [16]. However, because texts in social networks are commonly short, grammatically confusing, and contain a lot of noisy information, the graph construction algorithm of MT-AOG is not applicable under our social network scenario. Hence, the graph construction algorithm of MT-AOG is not applicable under our social network scenario.

Based on deep learning models, multimodal fusion has made great progress in recent years. Studies on multimodal fusion focused on early fusion and late fusion strategies. Early fusion methods directly fuse the extracted multimodal features at the input level [27], while late fusion approaches fuse the decisions made by each modality by weighted averaging [31]. Truong *et al.* introduced a Visual Aspect Attention Network (VistaNet) that considers images as a source of alignment of texts through a weighting matrix based on visual information [27]. Zadeh *et al.* proposed a Tensor Fusion Network (TFN), to model the intra- and inter-modality interactions dynamically [31]. Mai *et al.* developed a Graph Fusion Network (GFN) that hierarchically fuses multimodal features by constructing vertices and edges layer by layer [19]. The fused representations contain informative structures and explicit multimodal dynamics, which are beneficial for topic detection. Therefore, we adopt a GFN-based encoder and an MLP decoder to learn rich semantic information of the input data to improve the performance of multimodal topic detection.

3   Methodology

3.1   Overview

In this section, we give an overview of our multimodal topic detection method. As shown in Fig. 1, the proposed model consists of a feature embedding module F, a graph fusion module G, and a clustering module C. F takes raw images and documents as input and extracts visual and textual feature vectors. Then G hierarchically fuses multimodal information and outputs representations with informative structures through graph fusion. Based on the fused representations, C employs clustering algorithms to conduct topic detection.

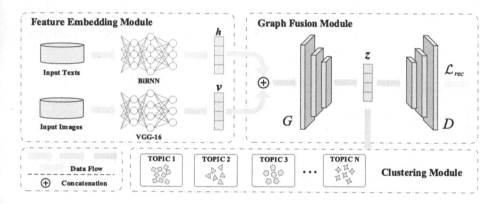

Fig. 1. The framework of the proposed model.

3.2   Feature Embedding Module

F is composed of a visual encoder and a textual encoder, extracting visual and textual feature vectors from the input images and documents.

Visual Encoder. We employ a pre-trained VGG-16 [25] to obtain the visual feature vectors v from the input images I, where $v \in \mathbb{R}^f$ are the outputs of the last fully-connected layer (FC7).

Textual Encoder. We employ NLTK [5] for text tokenization. Then the tokenized sequences $seq = \{e_1, e_2, ..., e_L\}$ are initialized with pre-trained word embeddings from GloVe [21] to obtain vector representations, where $e_i \in \mathbb{R}^{D_e}$, and L is the length of the sequences. Finally, we use a bidirectional recurrent neural network (BiRNN) [24] with LSTM cells [11] to obtain hidden state representations of the embedded sequences, where $h \in \mathbb{R}^f$.

3.3   Graph Fusion Module

G consists of a GFN-based encoder E and an MLP decoder D to hierarchically fuse multimodal representations with informative structures in an unsupervised manner.

Inspired by [19], E regards multimodal features as vertices and models interactions between modalities with edges layer by layer. In the first layer, a Modality Attention Network $MAN \in \mathbb{R}^f \to \mathbb{R}^1$ is employed to reweight the importance of multimodal features. Then we average the reweighted feature vectors to obtain the final outputs of the first layer, which are the unimodal dynamics U:

$$
\begin{aligned}
\alpha_v &= MAN(\boldsymbol{v}; \theta_{MAN}), \\
\alpha_h &= MAN(\boldsymbol{h}; \theta_{MAN}), \\
\boldsymbol{U} &= \frac{1}{2}(\alpha_v \cdot \boldsymbol{v} + \alpha_h \cdot \boldsymbol{h}),
\end{aligned}
\tag{1}
$$

where α_v and α_h are weighting matrices, and θ_{MAN} are the parameters of MAN.

In the second layer, a multilayer network $MLP_{bi} \in \mathbb{R}^{2f} \to \mathbb{R}^f$, which is composed of two dense layers and activated by *LeakyReLU* and *tanh*, is used to construct bimodal vertices $\boldsymbol{V}_{v,h}$:

$$
\boldsymbol{V}_{v,h} = MLP_{bi}(\boldsymbol{v} \oplus \boldsymbol{h}; \theta_{MLP_{bi}}),
\tag{2}
$$

where $\oplus$ denotes concatenation, $\theta_{MLP_{bi}}$ are the parameters of MLP_{bi}. As introduced in [19], the closer two vectors are to each other, the less complementary information exists between them. Therefore, a matrix $S_{v,h}$ is employed to describe the similarity of the multimodal feature vectors:

$$
S_{v,h} = \widetilde{\boldsymbol{v}}^T \, \widetilde{\boldsymbol{h}},
\tag{3}
$$

where $\widetilde{\boldsymbol{v}}$ and $\widetilde{\boldsymbol{h}}$ denote the soft-normalized vectors of $\boldsymbol{v}$ and $\boldsymbol{h}$ respectively, T represents vector transport operation. Hence, the weights of edges that link vertices in the first and second layers are defined as:

$$
\alpha_{v,h} = \frac{\alpha_v + \alpha_h}{S_{v,h} + 0.5}.
\tag{4}
$$

The final outputs of the second layer, i.e., the combined bimodal dynamics $\boldsymbol{B}$, are calculated through reweighting:

$$
\boldsymbol{B} = \alpha_{v,h} \cdot \boldsymbol{V}_{v,h}.
\tag{5}
$$

Based on $\boldsymbol{U}$ and $\boldsymbol{B}$, the fused representations $\boldsymbol{z}$ that encoded by E are:

$$
\boldsymbol{z} = MLP_z(\boldsymbol{U} \oplus \boldsymbol{B}; \theta_{MLP_z}),
\tag{6}
$$

where $MLP_z \in \mathbb{R}^{2f} \to \mathbb{R}^t$ is a multilayer network parameterized with θ_{MLP_z}.

The MLP decoder $D : \mathbb{R}^t \rightarrow \mathbb{R}^{2f}$ transforms the fused representations z into $\hat{x}$ and updates parameters together with E under a reconstruction loss function $\mathcal{L}_{rec}$:

$$x = v \oplus h,$$
$$\hat{x} = D(z; \theta_z), \qquad (7)$$
$$\mathcal{L}_{rec} = MSE(\hat{x}, x; \theta),$$

where $MSE(\cdot)$ is a mean squared error function, $\theta = \{\theta_{MAN}, \theta_{MLP_{bi}}, \theta_{MLP_z}, \theta_z\}$ denotes the parameters of our model. In practice, D consists of three dense layers activated by *LeakyReLU*.

3.4 Clustering Module

Based on z, the clustering module C conducts clustering algorithms to identify latent topics of input data automatically:

$$Tp = C(z), \qquad (8)$$

where $Tp \in \{tp_1, tp_2, ..., tp_N\}$, and N is the total number of topics. In practice, we use the Birch algorithm [32] to implement the clustering module.

4 Experiments

In this section, we describe our experiment settings, introduce the baseline methods, and present experimental results and further discussions.

4.1 Experiment Settings

Dataset. We use an open dataset of online reviews released in *Yelp.com*[1]. Reviews in the *Yelp* dataset cover 5 different topics, namely: *inside, food, outside, drink*, and *menu*. We selected samples from the original dataset that contain both images and texts for our experiments. Specifically, 60% of the selected data is used for training, 20% for validation, and 20% for testing. Detailed statistics of our dataset are shown in Table 1.

Table 1. Statistics of multimodal samples with 5 different topics in training, validation, and testing sets.

Topics	Training	Validation	Test	Total
Inside	13, 306	4, 409	4, 448	22, 163
Food	35, 132	11, 596	11, 669	58, 397
Outside	3, 336	1, 055	1, 104	5, 495
Drink	3, 469	1, 159	1, 084	5, 712
Menu	507	157	149	813

---

[1] https://www.yelp.com/dataset.

Implementation Details. For text preprocessing, we use NLTK [5] for word stemming and tokenization. We employ the pre-trained GloVe [21] to initialize the embedding matrix with dimensionality $D_e = 300$, and the dimension of BiRNN outputs is $D_f = 4096$. For images, we use a pre-trained VGG-16 model [25] for feature extraction and fix its parameters during training. The visual feature vectors are drawn from the FC7 layer with the same dimensionality $D_f = 4096$ as the BiRNN outputs. The proposed method is implemented using PyTorch and Scikit-learn, and the deep learning models are trained on an NVIDIA GeForce RTX 2080Ti. We apply the gradient descent optimization to update the parameters of our model on the training set, and conduct model selection on the validation set. All the results are collected from the experiments conducted on the testing set, which has empty intersections with the training and the validation sets.

Evaluation Metrics. In this paper, we follow the evaluation protocol introduced in [7,16], i.e., accuracy and normalized mutual information, to evaluate the performance of baselines and our proposed method on topic detection.

4.2 Baselines

We compare our method with two categories of approaches: document clustering algorithms and topic models. For the first category, we employ K-Means [18], Agglomerative Clustering [3], Birch [32] with Scikit-learn; for the second category, we employ the LDA model [6]. Besides, we implement several deep learning models as simplified versions of our method to testify the effectiveness of the proposed modules.

- **BiRNN.** A simplified implementation with only textual feature vectors drawn from the outputs of BiRNN. C conducts the clustering algorithm based on h.
- **VGG+BiRNN.** A simplified implementation with concatenated visual and textual feature vectors, which are drawn from the outputs of the VGG-16 and BiRNN respectively. C conducts the clustering algorithm based on $v \oplus h$.

4.3 Results

We implemented all the baselines and compare them to our proposed model. The experimental results are demonstrated in Table 2. As shown, our method outperforms baselines in both clustering accuracy (ACC) and normalized mutual information (NMI). Specifically, it can be seen that traditional document clustering algorithms have the worst performance, especially in terms of the NMI metric. Due to the short length of texts in social networks, the embedded textual feature vectors are fairly sparse and not able to convey efficient semantic information. Hence, clustering algorithms become significantly inferior because of the sparse vectors.

Table 2. Experimental results

Methods	Clustering accuracy (%)	Normalized mutual information (%)
K-Means	26.3	1.1
Agglomerative	26.8	1.2
Birch	31.9	1.2
LDA	29.6	4.6
BiRNN	52.9	27.5
VGG+BiRNN	54.0	37.8
Ours	**54.9**	**55.0**

Compared to the clustering algorithms, the LDA model has a slight improvement in topic detection performance. By constructing a probabilistic model for the vocabulary, the LDA model can better extract the semantic features from texts, which are crucial for topic detection. However, a large number of texts in social networks are grammatically confusing and contain noisy information, leading to a less accurate probabilistic model and limiting the performance of the LDA model.

Since BiRNN can model deep semantic features as well as temporal information, topic detection based on BiRNN hidden features achieves relatively high performance. Moreover, images in social networks also contain topic information. Therefore, based on the concatenated multimodal feature vectors $v \oplus h$, VGG+BiRNN obtains relatively competitive results.

We can also observe that deep neural networks generally work better than other approaches. In general, among all the deep learning models, the proposed method outperforms other models. This is because our method can hierarchically fuse multimodal representations with graph fusion. The fused representations contain explicit multimodal dynamics and rich semantic information, thus significantly improving the performance of multimodal topic detection.

5   Conclusion

In this paper, we deal with the image-text-based multimodal topic detection task. Different from traditional topic detection of pure texts, the fundamental challenge of multimodal topic detection is the fusion and alignment of heterogeneous multimodal data. To address this problem, we adopt a GFN-based encoder and an MLP decoder to hierarchically fuse information from different modalities. The proposed modules encourage the fused representation to carry rich semantic information and explicit multimodal dynamics, which are crucial for multimodal topic detection. A reconstruction loss is employed to constrain the learning process of our model in an unsupervised manner. Experimental results conducted on the real-world multimodal topic detection dataset demonstrate the effectiveness of our proposed method.

In the future, we will consider extending our existing work to topic detection datasets with more than two modalities, and we will continue to explore deep learning models to solve multimodal topic detection more effectively.

Acknowledgements. This research is supported by the NSFC-Xinjiang Joint Fund (No. U1903128), NSFC General Technology Joint Fund for Basic Research (No. U1836109, No. U1936206), Natural Science Foundation of Tianjin, China (No. 18ZXZNGX00110, No. 18ZXZNGX00200), and the Fundamental Research Funds for the Central Universities, Nankai University (No. 63211128).

References

1. Allan, J.: Introduction to topic detection and tracking. In: Allan, J. (ed.) Topic Detection and Tracking. The Information Retrieval Series, vol. 12, pp. 1–16. Springer, Boston (2002). https://doi.org/10.1007/978-1-4615-0933-2_1
2. Allan, J., Carbonell, J., Doddington, G., Yamron, J., Yang, Y.: Topic detection and tracking pilot study final report (1998)
3. Beeferman, D., Berger, A.: Agglomerative clustering of a search engine query log. In: ACM Knowledge Discovery and Data Mining, pp. 407–416 (2000)
4. Berrocal, J., Figuerola, C.G., Rodríguez, Z.: Reina at replab2013 topic detection task: Community detection. reina.usal.es (2013)
5. Bird, S., Loper, E.: NLTK: the natural language toolkit. In: ACL, pp. 214–217 (2004)
6. Blei, D.M., Ng, A., Jordan, M.I.: Latent Dirichlet allocation. J. Mach. Learn. Res. **3**, 993–1022 (2003)
7. Cai, D., He, X., Han, J.: Locally consistent concept factorization for document clustering. IEEE Trans. Knowl. Data Eng. **23**(6), 902–913 (2010)
8. Cataldi, M., Di Caro, L., Schifanella, C.: Emerging topic detection on twitter based on temporal and social terms evaluation. In: Proceedings of the Tenth International Workshop on Multimedia Data Mining, pp. 1–10 (2010)
9. Chen, Y., Liu, L.: Development and research of topic detection and tracking. In: IEEE International Conference on Software Engineering and Service Science, pp. 170–173 (2017)
10. Connell, M., Feng, A., Kumaran, G., Raghavan, H., Shah, C., Allan, J.: UMass at TDT 2004. In: Topic Detection and Tracking Workshop Report, vol. 19 (2004)
11. Hochreiter, S., Schmidhuber, J.: Long short-term memory. Neural Comput. **9**(8), 1735–1780 (1997)
12. Hofmann, T.: Unsupervised learning by probabilistic latent semantic analysis. Mach. Learn. **42**(1), 177–196 (2001). https://doi.org/10.1023/A:1007617005950
13. Huang, F., Zhang, S., Zhang, J., Yu, G.: Multimodal learning for topic sentiment analysis in microblogging. Neurocomputing **253**, 144–153 (2017)
14. Kuo, Z., Juan-zi, L., Gang, W., Ke-hong, W.: A new event detection model based on term reweighting (2008)
15. Lau, J.H., Collier, N., Baldwin, T.: On-line trend analysis with topic models:# Twitter trends detection topic model online. Proc. COLING **2012**, 1519–1534 (2012)
16. Li, W., Joo, J., Qi, H., Zhu, S.C.: Joint image-text news topic detection and tracking by multimodal topic and-or graph. IEEE Trans. Multimedia **19**(2), 367–381 (2016)

17. Liu, W., Zhang, M.: Semi-supervised sentiment classification method based on Weibo social relationship. In: Ni, W., Wang, X., Song, W., Li, Y. (eds.) WISA 2019. LNCS, vol. 11817, pp. 480–491. Springer, Cham (2019). https://doi.org/10.1007/978-3-030-30952-7_47

18. MacQueen, J., et al.: Some methods for classification and analysis of multivariate observations. In: Proceedings of the Fifth Berkeley Symposium on Mathematical Statistics and Probability, vol. 1, pp. 281–297 (1967)

19. Mai, S., Hu, H., Xing, S.: Modality to modality translation: an adversarial representation learning and graph fusion network for multimodal fusion. In: AAAI Conference on Artificial Intelligence, pp. 164–172 (2020)

20. Pang, J., Tao, F., Huang, Q., Tian, Q., Yin, B.: Two birds with one stone: a coupled Poisson deconvolution for detecting and describing topics from multimodal web data. IEEE Trans. Neural Netw. Learn. Syst. **30**(8), 2397–2409 (2018)

21. Pennington, J., Socher, R., Manning, C.: GloVe: gobal vectors for word representation. In: EMNLP, pp. 1532–1543 (2014)

22. Petkos, G., Papadopoulos, S., Aiello, L., Skraba, R., Kompatsiaris, Y.: A soft frequent pattern mining approach for textual topic detection. In: International Conference on Web Intelligence, Mining and Semantics, pp. 1–10 (2014)

23. Sayyadi, H., Raschid, L.: A graph analytical approach for topic detection. ACM Trans. Internet Technol. **13**(2), 1–23 (2013)

24. Schuster, M., Paliwal, K.K.: Bidirectional recurrent neural networks. IEEE Trans. Signal Process. **45**(11), 2673–2681 (1997)

25. Simonyan, K., Zisserman, A.: Very deep convolutional networks for large-scale image recognition. In: Bengio, Y., LeCun, Y. (eds.) 3rd International Conference on Learning Representations, ICLR 2015, San Diego, CA, USA, 7–9 May, 2015, Conference Track Proceedings (2015)

26. Trieschnigg, D., Kraaij, W.: TNO hierarchical topic detection report at TDT 2004. In: Topic Detection and Tracking Workshop Report (2004)

27. Truong, Q.T., Lauw, H.W.: VistaNet: visual aspect attention network for multimodal sentiment analysis. In: AAAI, vol. 33, pp. 305–312 (2019)

28. Xiong, Y., Zhang, Y.F., Feng, S., Wang, D.L.: Event detection and tracking in microblog stream based on multimodal feature deep fusion. Control and Decision (2019)

29. Yang, Y., Carbonell, J.G., Brown, R.D., Pierce, T., Archibald, B.T., Liu, X.: Learning approaches for detecting and tracking news events. IEEE Int. Syst. Appl. **14**(4), 32–43 (1999)

30. Yu, H., Zhang, Y., Ting, L., Sheng, L.: Topic detection and tracking review. J. Chin. Inf. Process. **6**(21), 77–79 (2007)

31. Zadeh, A., Chen, M., Poria, S., Cambria, E., Morency, L.P.: Tensor fusion network for multimodal sentiment analysis. In: EMNLP, pp. 1103–1114 (2017)

32. Zhang, T., Ramakrishnan, R., Livny, M.: BIRCH: an efficient data clustering method for very large databases. In: Jagadish, H.V., Mumick, I.S. (eds.) ACM Conference on Management of Data, pp. 103–114 (1996)

A New Novel Label Propagation Algorithm

Jie Zhang and Zhengyou Xia[✉]

College of Computer Science and Technology,
Nanjing University of Aeronautics and Astronautics, Nanjing 210016, China
zhengyou_xia@nuaa.edu.cn

Abstract. Community detection is an enduring research hotspot in the field of complex networks. The label propagation algorithm is a semi-supervised learning method, which has the advantages of close to linear time complexity, simplicity and ease of implementation. However, LPA has two significant shortcomings in dividing communities: poor accuracy and strong randomness, which seriously affect the performance of the algorithm. This paper proposes a new label propagation algorithm to solve these two problems. In the initialization stage, a new node importance metric is proposed, which simultaneously considers the importance both of the node itself and its neighbor nodes to rank the importance of the nodes. In the label propagation stage, We also propose a new node similarity metric and the label is updated according to the similarity between the current node and neighbor nodes. Our experiments on real networks and artificial synthetic networks show that this algorithm can effectively find community structure and has better stability and accuracy than some existing LPA improved algorithms, and this advantage is more obvious on large networks.

Keywords: Label propagation · Community detection · Node importance · Node similarity

1 Introduction

There are many applications in the field of complex networks [1], community detection is one of the typical representatives. Recent years, many algorithms have emerged to detect communities. Among them, label propagation algorithm has gained the favor of scholars because of its close to linear time complexity and simplicity. Easy to realize and other features. However, as we said in the abstract, LPA has the disadvantages of high randomness and poor accuracy. Therefore, many improvements have been proposed to improve the accuracy and robustness of tag propagation algorithm. However, most algorithms generally have two shortcomings: (1) dependence on input parameters, (2) the need for global network structure, which is very unsuitable for large-scale networks [2].

In this paper, a new algorithm based on LPA is proposed to improve its accuracy and stability. In the initialization stage, We define a new measure of node

© Springer Nature Switzerland AG 2021
C. Xing et al. (Eds.): WISA 2021, LNCS 12999, pp. 39–46, 2021.
https://doi.org/10.1007/978-3-030-87571-8_4

importance. In the label update stage, We also propose a new node similarity metric and a new label propagation rule. The specific improvement measures can be seen in Sect. 3.

2   Related Work

In 2007, Raghavan proposed LPA algorithm [3], which is almost linear in time and space complexity. However, its biggest disadvantage is that the randomness is too strong, which leads to the instability and accuracy of the algorithm. Therefore, in recent years, researchers have mainly improved the LPA algorithm for these two points. The literature [4–6] has improved LPA from different perspectives, but these algorithms only judge the influence of labels based on the number of labels, and do not consider the influence of node information and closeness between nodes on label selection. [7] Comprehensively consider the influence of node degree and edge weight on label selection, but does not consider the influence of neighboring information and the degree of aggregation of neighboring nodes on label selection.

Based on the above considerations, this paper proposes a new novel label propagation algorithm KC_LPA mainly based on node degree (k_i) and clustering coefficient (c_i), which not only the information of the node itself, but also the influence of the attributes of its neighbors on the algorithm is considered. The specific process and effect of the algorithm will be described in detail in the next few sections.

3   KC_LPA

3.1   Movitation of KC_LPA

"Degree" is a simple and important concept in the attributes of a single routine, and it is a basic parameter describing the local characteristics of the network. The bigger, the more important the node. The expression formula of degree is as follows:

$$K_i = \sum_{j \in N, j \neq i} A_{ij} \qquad (1)$$

Among them, A_{ij} represents the connection between nodes, when there is an edge between v_i and v_j , $A_{ij} = 1$; otherwise, $A_{ij} = 0$.

The "clustering coefficient" is a local feature quantity, which reflects the closeness of the connection between the node and the neighboring nodes. The larger the clustering coefficient c, the more likely it is to belong to a community. It is defined as follows:

$$C_i = \frac{2\,l_i}{k_i\,(k_i - 1)} \qquad (2)$$

Where l_i represents the number of edges between k_i neighbors of node i

In the literature [8], the author models people as nodes and relationships as edges, and believes that the importance of a person is constituted by the importance of himself and the importance of friends. Based on this inspriation, we can make the analogy that the importance of a node is made up of its own importance and the importance of its neighbors. Therefore, we propose the following formula as the importance measure of nodes

$$N_{vi} = C_{vi} \times K_{vi} + \alpha \sum_{vj \in N(vi)} C_{vj} \times K_{vj} \tag{3}$$

The first term in the formula is the importance of node v_i itself, and the second term is the importance of its neighbors. The impact factor α is an adjustable parameter, The experiment shows that the effect is best when α is 0.5.

We calculate the importance of each node according to formula (3), In order to solve the instability problem, then rank the nodes in ascending order of importance, because the nodes with low influence are easy to be affected by other nodes.

The update rule of the LPA algorithm is to randomly select a label when the number of times the label appears is the same. By default, this update rule regards the similarity between each neighbor node and the current node as the same. In fact, they are not the same. If treated equally, it will lead to the problem of low accuracy and poor stability of the algorithm. So we need to calculate the similarity between the node and its neighbors. When multiple labels have the maximum number at the same time, we update the label of the current node by selecting the label with the highest similarity among neighbor nodes. We choose the HPI node similarity index to measure the similarity between the current node and their neighbor nodes, HPI is defined as follows:

$$sim(i, j) = \frac{|\Gamma(i) \cap \Gamma(j)|}{\min(k_i, k_j)} \tag{4}$$

It can be seen from the formula that this is a favorable similarity index for nodes with large degree. We have made some improvements to this formula, We consider the degree and clustering coefficient of the common neighbor nodes of the node pair (i, j), noted as C_z, K_z , It's worth noting that, we also consider The effect of the degree of the second-order common neighbor node of the node pair(i,j) on the similarity measure, denoted as k_t, because research has shown that in the single node degree phase at the same time, the second node degree of nodes with high similarity. In fact, in the experiment, also proved the effectiveness of our innovation. The improved similarity calculation formula between nodes is as follows:

$$sim(i, j) = \frac{\sum_{\substack{z \in \Gamma_i \cap \Gamma_j \\ t \in \{\Gamma[\Gamma(i) \cap \Gamma(j)] \cup \Gamma[\Gamma(j)] \cap \Gamma(i)\}}} C_z \times K_z + \frac{k_t}{k_z}}{\min(k_i, k_j)} \tag{5}$$

We define the label update rule as follows: For each node in the network, we update the label of the current node with the label that appears most frequently among its neighbor nodes. If there are multiple labels that appear most frequently, then we choose the label with the most similarity among the current neighbor nodes as the optimal label of the current node. If there are multiple optimal labels, we randomly calculate from the multiple optimal labels. One label is updated. The formula for obtaining the maximum similarity node is as follows, Where N is the current node and v is the set of neighbor nodes:

$$NodeLabel(N) = \{u \in v | \arg\max\{sim(i, j)\}\} \qquad (6)$$

3.2 KC_LPA Model

Aiming at the defects of LPA algorithm, we propose a new heuristic improved algorithm for label propagation, which mainly consists of two parts: (1) Initialize the nodes and sort them in the order of importance of the nodes. (2) Update the sequence of the step 1 according to certain label propagation rules.

The pseudocode of the KC_LPA algorithm is shown as follows:

Algorithm 1. Algorithm KC_LPA

Input: $G = (V,E)$
Output: $C=C_1,C_2,C_3...,C_n$
 1: loadgraphfromefile()
 2: **for** v in V **do**
 3:     v.label = v.id
 4: **end for**
 5: calculate importance() //according to the formula in 3.1
 6: sortNodeByNodeImportanceInAscending(V)//Sort nodes in ascending order
 7: calculate similarity() //according to the formula (5)
 8: **while** StopCondiction()not true and numIter< maxIter **do**
 9:     numIter← numIter+1
10:     **for** v in V **do**
11:         fremaxlabel = mostFrequentLabs(v) // get neighbors of v which have the most frequent labels
12:         **if** len(fremaxlabel>1) **then:**
13:             neigwithmaxsim = findneigwithmaxsim(fremaxlabel)
14:             **if** len(neigwithmaxsim>1) **then:**
15:                 v.label = random(neigwithmaxsim)
16:             **else:**
17:                 v.label = neigwithmaxsim.label
18:             **end if**
19:         **else:**
20:             v.label = fremaxlabel.label
21:         **end if**
22:     **end for**
23:     **if** the label of all nodes are in stable state **then:**
24:         break
25:     **end if**
26:     iter+=1
27: **end while**
28: Communities=getCommunities(V)
29: return Communities

---

4   Experimental Results and Analysis:

4.1   Real Network Experiment

We choose to take The real data sets—Zachary's karate club; dophlins; football and Political blogs as the open standard test dataset. Use modularity —Q to measure the performance of community division. Newman proposed modularity Q metric [9] based on the definition of community to evaluate community discovery. Theoretically, the larger the modularity is, the more reasonable the community is; the smaller the modularity is, the poorer the effect of community division is. Modularity Q is defined as follows:

$$Q = \frac{1}{2E} \sum_{ij} \left(A_{ij} - \frac{K_i\,K_j}{2|E|} \right) \bullet \emptyset\,(C_i, C_j) \tag{7}$$

In the above formula, The specific meaning of each section will not be described here due to the length of the paper.readers can know it in reference.

We compare our algorithm with the basic improved LPA algorithm, namely LPA, NIBLPA, KBLPA and AHLPA [10]. During the experiment, each algorithm was run 100 times to ensure the accuracy of the experimental results. Table 1 shows these five types.

Table 1. ModularityQ comparison results of five algorithms

Network	Q(LPA)	Q(KBLPA)	Q(NIBLPA)	Q(AHLPA)	Q(KC_LPA)
Karate	0.21 ± 0.21	0.20 ± 0.20	0.12	0.41	0.42
Dophlins	0.39 ± 0.12	0.43 ± 0.09	0.50	0.51	0.52
Polbooks	0.56 ± 0.04	0.54 ± 0.06	0.60	0.60	0.62
Football	0.48 ± 0.05	0.45 ± 0.02	0.51	0.52	0.55

It can be seen from Table 1 that LPA performs very poorly in karate, with Q even worse than 0.3. This is because in a small network, the label propagation is very fast. Because the network structure is relatively simple, the algorithm is very unstable and the division effect is very poor, but this shortcoming will be improved as the network scale expands. From Table 1, it is not difficult to see that the performance of our algorithm is slightly better than other algorithms on any data set. This may be because our algorithm considers more comprehensive considerations no matter in the node importance ranking stage or the label update stage.

4.2   Synthetic Network Experiment

In this section, we use the LFR benchmark network [11] as our experimental network to test the performance of the network community division. The parameters commonly used when generating the LFR benchmark network are introduced in

Table 2, Among them, mu represents the probability that the node is connected to the outside of the community, and its value is between 0 and 1, and the minimum value of mu indicates that the less obvious the structure of the community, the more difficult it is to find the community.

Table 2. LFR baseline network parameters

Parameters	Meaning
N	The numbers of nodes
k	The average degree
maxk	The maximum degree
mu	The mixing parameter of the topology
minc	The minimum for the community size
maxc	The maximum for the community size

NMI is usually used to measure the community partition of the algorithm result, which value is usually in the range of [0,1]. The closer to 1, the more similar the two communities are. If it's close to zero, that means the two communities are very different.

The calculation formula of NMI is as follows:

$$NMI(A,B) = \frac{-2 \sum_{i=1}^{C_A} \sum_{j=1}^{C_B} N_{ij} \log \left(\frac{N_{ij} N}{N_i N_j} \right)}{\sum_{j=1}^{C_B} N_j \log \left(\frac{N_j}{N} \right) + \sum_{i=1}^{C_B} N_i \log \left(\frac{N_i}{N} \right)} \tag{8}$$

The meaning of each section is not described here due to space limitations, reader can get it in the reference [12].

We set up three groups of experiment, with network size of 500,1000 and 5000. When the network scale is 500 and 1000, the network parameters are set as k = 15, maxk = 50, minc = 20 and maxc = 50, while the network size is 5000, we adjust the parameters appropriately to k = 15, maxk = 50, minc = 20 and maxc = 100. Each algorithm is executed for 100 times individually.

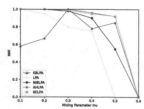

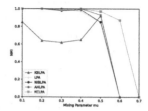

Fig. 1. Result of NMI at network scale N = 500

Fig. 2. Result of NMI at network scale N = 1000

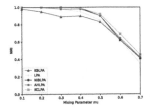

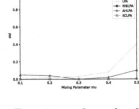

Fig. 3. Result of NMI at network scale
N = 5000

Fig. 4. Experimental result of std at
network scale N = 500

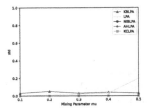

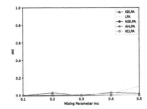

Fig. 5. Experimental result of std at
network scale N = 1000

Fig. 6. Experimental result of std at
network scale N = 5000

Figures 1, 2, and 3 are graphs showing the change of NMI value with mu
under the network scale of 500, 1000, and 5000. Overall, the performance of the
KC_LPA algorithm is better than the other four algorithms. Especially when the
network scale is large, such as when mu is equal to 0.6 in Fig. 5, the other four
algorithms cannot work, but KC_LPA can still divide the community better. The
decline of KC_LPA in Fig. 6 is the most stable, which means that despite the
increasing complexity of the network structure, KCLPA can still maintain a relatively high accuracy rate for discovering communities. Reflects the superiority
of KC_LPA

Stability is another evaluation indicator based on the label propagation algorithm. The smaller the std value, the more stable the community recognition
result of the algorithm. We run each of these five algorithms 100 times on the
artificial synthesis network, and take the standard deviation std of the 100 times.
The specific information is shown in the Figs. 4, 5 and 6. It can be seen that,
our algorithm can basically obtain the smallest standard deviation in different
networks, indicating that KC_LPA is stable and robust.

5   Summary

Aiming at the shortcomings of LPA algorithm with strong randomness and poor
accuracy, this paper proposes a new heuristic algorithm KC_LPA. In the initialization phase, not only the importance of the node itself is considered, but
the importance of its neighbors, integration and clustering are also considered.
The two indexes of coefficient calculate the importance of nodes and sort them

in order from high to low, avoiding less important nodes from influencing more important nodes, and reducing the occurrence of backflow phenomenon. In the label update stage, the similarity between the node and its neighbor nodes is calculated, and the degree, clustering coefficient, and the degree of the second-order neighbor node are comprehensively considered to affect the label selection. This will increase the complexity of the algorithm to a certain extent, but it is obvious Improve the accuracy of label selection, thereby improving the stability and accuracy of the algorithm. We have set up comparative experiments on 4 real data sets and on artificial synthetic networks of three scales. The experimental results have proved the effectiveness of our algorithm, especially on large-scale networks. Our algorithm is more prominent.

References

1. Liu, Y., Shen, D., Kou, Y., Nie, T.: Link prediction based on node embedding and personalized time interval in temporal multi-relational network. In: Ni, W., Wang, X., Song, W., Li, Y. (eds.) WISA 2019. LNCS, vol. 11817, pp. 404–417. Springer, Cham (2019). https://doi.org/10.1007/978-3-030-30952-7_40
2. Berahmand, K., Bouyer, A.: A link-based similarity for improving community detection based on label propagation algorithm. J. Syst. Sci. Complexity **32**(3), 737–758 (2019). E69, 066133 (2004)
3. Raghavan, U.N., Albert, R., Kumara, S.: Near linear time algorithm to detect community structures in large-scale networks. Phys. Rev. E **76**(3), 036106 (2007)
4. Gregory, S.: Finding overlapping communities in networks by label propagation. New J. Phys. **12**(10), 103018 (2010)
5. Barber, M.J., Clark, J.W.: Detecting network communities by propagating labels under constraints. Phys. Rev. E **80**(2), 026129 (2011)
6. Leung, I.X., Hui, P., Lio, P., Crowcroft, J.: Towards real-time community detection in large networks. Phys. Rev. E **79**(6), 066107 (2009)
7. Zhuoxiang, Z., Yitong, W., Jiatang, T., Zexu, Z.: A novel algorithm for community discovery in social networks based on label propagation. J. Comput. Res. Dev. **3**, 8–15 (2011)
8. Zhang, Y., Liu, Y., Zhu, J., Yang, C., Yang, W., Zhai, S.: NALPA: a node ability based label propagation algorithm for community detection. IEEE Access **8**, 46642–46664 (2020)
9. Newman, M.E.J., Girvan, M.: Finding and evaluating community structure in networks. Phys. Rev. E **69**(2), 026113 (2004)
10. Zhu, X., Xia, Z.: Label Propagation Algorithm Based on Adaptive H Index. Springer, Cham (2018)
11. Lancichinetti, A., Fortunato, S., Radicchi, F.: Benchmark graphs for testing community detection algorithms. Phys. Rev. E **78**(4), 046110 (2008)
12. Danon, L., Diaz-Guilera, A., Duch, J., et al.: Comparing community structure identification. J. Stat. Mech. Theory Exp. **2005**(09), P09008 (2005)

Query Processing and Algorithm

A Coreset Based Approach for Continuous k-regret Minimization Set Queries over Sliding Windows

Wei Ma[1], Jiping Zheng[1,2]($\boxtimes$) (iD), and Zhiyang Hao[1]

[1] College of Computer Science and Technology,
Nanjing University of Aeronautics and Astronautics, Nanjing, China
{mawei,jzh,haozhiyang}@nuaa.edu.cn
[2] Collaborative Innovation Center of Novel Software Technology
and Industrialization, Nanjing, China

Abstract. Extracting a few tuples to represent the whole database is an important problem in real-world applications. In the literature, there are three representative tools: top-k, skyline and k-regret queries. Among these, the k-regret query has received much attention in recent decades for it does not require any preferences from users and the output size is controllable. However, almost all existing algorithms aim at the static databases while data streams are becoming more and more popular in many applications. In this paper, we propose continuous k-regret minimization set queries on data streams where tuples are valid in a sliding window. Further, we develop an algorithm to maintain a tiny coreset over sliding windows such that traditional static algorithms for k-regret queries can be applied on the coreset by sacrificing a little accuracy but improving the efficiency. We conduct experiments to show the effectiveness and efficiency of our proposed algorithm compared with existing ones.

Keywords: k-regret queries · Coreset · Data streams · Sliding windows

1 Introduction

To select a few representative tuples from a large database is an important functionality in many applications, e.g., multi-criteria decision making, recommendation systems and web searching. Consider an example where there are two attributes, HP (Horse Power) and MPG (Miles Per Gallon) in a car database. When Alice wants to buy a car, it is not a practical way to return all cars in the database to Alice since she will be overwhelmed when facing so many options. Hence, how to select and return a few representative cars in the database to Alice has become an urgent problem needed to be solved.

One approach to solve this problem is the top-k query [10]. Based on the utility function provided by the user which is a weighted vector expressing the user's preferences on the attributes, the scores of all tuples in the database are computed. According to these scores all tuples are ranked and the k tuples with the largest scores are returned to the user. For example, Alice gives weights 20% to HP and

© Springer Nature Switzerland AG 2021
C. Xing et al. (Eds.): WISA 2021, LNCS 12999, pp. 49–61, 2021.
https://doi.org/10.1007/978-3-030-87571-8_5

80% to MPG. We assume that there are two cars A, B with 110, 130 (in HP) and 40, 50 (in MPG) respectively. Thus the score of car A is 54 ($0.2 \times 110 + 0.8 \times 40$) which is less than that of car B ($0.2 \times 130 + 0.8 \times 50 = 66$). Thus Alice prefers B to A. For the top-k query, the utility function should be provided for further processing.

Another approach is the skyline query [5], which does not need the user to provide a utility function. The skyline query is based on the concept of *domination*. A tuple p is dominated by another tuple q if q is better than p in at least one attribute and not worse than p in all attributes. The skyline query returns all tuples which are not dominated by any tuple in the database. Recall the car example, since the HP and MPG values of A are worse than those of B, A will not be returned by the skyline query since it is dominated by B. For the skyline query, the size of the results is uncontrollable which increases exponentially with the dimensionality and it is as large as that of the whole database in the worst case.

Recently, an approach named the k-regret query [14] is proposed which neither asks the users to provide utility functions nor overwhelms the users. It introduces the concept of the maximum regret ratio to measure how well a subset to represent the whole database. The smaller the maximum regret ratio, the more representative the subset is. Chester et al. [7] extend the k-regret query to the k-regret minimization set query which is based on the concept of the maximum k-regret ratio. To minimize the maximum k-regret ratio, the top-k tuples in the database for any possible utility function are selected into the result set. Due to the NP-hardness of the k-regret minimization set query, more and more methods are proposed to obtain a subset that minimizes the maximum k-regret ratio.

However, most of existing algorithms for the regret minimization queries aim at static databases. In many real applications, the tuples in the database are not static but arrive in a streaming manner, such as web tracking, medical monitoring and financial monitoring. These data have several characteristics as follows: a) Large volume of data. Thus, it is impractical to save and process all tuples. Existing algorithms for static databases return the result set based on all tuples in the database, which cannot be directly applied to data streams since it will never get all data. b) High frequency of tuples arriving. It's a naive and time-consuming approach to rerun static algorithms based on the up-to-date database after each tuple arriving. Moreover, an inefficient algorithm may cause congestion even breakdown in the system since the next tuple arrives while the last tuple has not been processed. c) Insertion-only. Existing dynamic algorithms such as FD-RMS [19] with tuple insertions and deletions are unnecessary since deletions never happen in data streams. Many other problems with similar data model are investigated recently [8]. However, there is still no specific algorithm for continuous k-regret minimization set queries on data streams. In brief, it is necessary to develop effective and efficient algorithms for continuous k-regret minimization set queries on data streams.

In addition, data is time-sensitive in many applications, e.g., sensor database, network traffic engineering and various anomaly monitoring systems. Users are only interested in the data recently collected, and it is also unnecessary to save all the data arrived. For example, in medical monitoring, the system should respond to exceptions in time, and the outdated, normal data is not concerned.

For another example, a network administrator wants to gather statistics about packets processed by a set of routers over the last day instead of the days before yesterday. Therefore, to only consider the valid data, a sliding window is applied. The data stream comes as the sliding window moves and only the recent data in the sliding window are considered.

Nevertheless, it is inefficient to answer the k-regret minimization set query over a sliding window when the size of the window becomes large. We cannot focus on all tuples in the sliding window due to the large size of the data even in the window. In this paper, we first define the continuous k-regret minimization set query on data streams. Then we introduce our proposed algorithm based on sliding windows and coresets to tackle the problem. The coreset is a "small subset" of the input "large dataset", and every possible query has approximately the same answer on the subset as to the whole dataset. Instead of processing on the whole dataset in the window, we perform the k-regret minimization set query on the coreset with little loss of accuracy. After constructing the initial coreset over a sliding window, we maintain the coreset along with tuples arriving and recompute the result set whenever the coreset changes. The result set is obtained by running a heuristic greedy algorithm on the coreset. At length, we evaluate and compare our algorithm with existing algorithms for k-regret minimization set queries to verify the advantage of our proposed algorithm on data streams.

Our contributions are summarized as follows.

- We provide the formal definition of the continuous k-regret minimization set query on the data streams.
- Based on the constructed coreset, we develop an efficient algorithm to tackle the k-regret minimization set query over sliding windows.
- We introduce a heuristic strategy to reduce the space and time complexity of our proposed algorithm.
- We conduct experiments on both synthetic and real-world datasets to verify the efficiency and effectiveness of our proposed algorithm.

Road Map. The rest of the paper is organized as follows. We introduce the related work in Sect. 2. Preliminaries and the definition of our problem are provided in Sect. 3. In Sect. 4, we propose our algorithm based on coresets and sliding windows. We conduct experiments on several datasets to evaluate our algorithm in Sect. 5. Lastly, we conclude our work in Sect. 6.

2   Related Work

Motivated by the drawbacks of the top-k and skyline queries, Nanongkai et al. [14] propose the k-regret query where users want a set of k tuples such that the maximum regret ratio of the set is minimized. Chester et al. [7] extend the problem to kRMS where k means users want the tuple with the k-th largest utility instead of the tuple with the largest utility. Due to the NP-hardness of kRMS (the k-regret query is 1RMS problem), a number of algorithms [3,4,6,11, 15,20] are proposed to obtain a subset from a static dataset with the maximum

k-regret ratio as small as possible. The most relevant work is [3,6]. They show the relationship between ϵ-*kernel* (a.k.a. coreset) and the k-regret query and they construct and return a fixed-size coreset as the result of kRMS directly. Also, a number of variants of the k-regret query are studied in the literature such as k-regret queries with other kinds of utility functions [9,16,17], the interactive k-regret query [13], the average regret minimizing query [21].

However, existing algorithms for k-regret minimization set queries focus on static datasets, they must recompute the result for the current dataset when the dataset changes, i.e., insertions and deletions of some tuples. The continuous recomputing process costs heavily both in time and space, and it's also unnecessary since an insertion of one tuple usually causes part or even only a tuple rather than all tuples of result set to change. Most recently, [19] proposed FD-RMS to solve the problem in a fully dynamic setting where tuples are inserted or deleted. They transform the problem to a dynamic set cover problem and maintain the result set under each tuple insertion or deletion. However, there still lack the efficient and effective algorithms for continuous k-regret minimization set queries on data streams which only consider the insertions of tuples.

3   Preliminaries

In this section, we introduce concepts related to the k-regret minimization set query, the formal definition of the studied problem, as well as the concepts of sliding window and coreset.

3.1 k-Regret Minimization Set Queries

Let D be a set of d-dimensional tuples[1] of size n, and the i-th dimensional attribute value of a tuple $p \in D$ is denoted by $p[i]$ where $i \in [1, d]$.

Definition 1. Utility Function and Utility. *A utility function f is a mapping $f \colon \mathbb{R}_+^d \to \mathbb{R}_+$. The utility of a tuple p for a user with utility function f is $f(p)$.*

Along with the setting in many related works [7,13–15], we study the k-regret minimization set query with linear utility functions, namely, $f(p) = u \cdot p = \sum_{i=1}^d u[i] \cdot p[i]$, where u is a d-dimensional non-negative vector and $u[i]$ measures the user's preference of the i-th dimensional attribute value. Moreover, we assume the L1-norm of the utility vector equals to 1, i.e., $\sum_{i=1}^d u[i] = 1$. Recall the car example in Sect. 1, the utility function of Alice is $< 0.2, 0.8 >$ and the utility $f(p)$ is $0.2 \times p[1] + 0.8 \times p[2]$.

Let $\max_{p \in D}^{(k)} f(p)$ denote the k-th largest utility among $p \in D$ for utility function f. k is omitted when $k = 1$ for brevity.

Definition 2. k-regret ratio [7]. *The k-regret ratio of a subset $S \subseteq D$ for f is denoted as $krr_D(S, f)$ and defined by*

---

[1] We use tuples and points interchangeably in the paper.

$$krr_D(S, f) = \frac{\max\{0, \max_{p\in D}^{(k)} f(p) - \max_{p\in S} f(p)\}}{\max_{p\in D}^{(k)} f(p)} = \max\left\{0, 1 - \frac{\max_{p\in S} f(p)}{\max_{p\in D}^{(k)} f(p)}\right\}$$

The k-regret ratio of S measures how regretful a user is with utility function f when facing the subset S instead of the whole dataset D, which ranges from 0 to 1. The closer k-regret ratio gets to 1, the more regretful the user is since the maximum utility of S is much different from the k-th maximum utility of D. However, the utility function is unknown for a specific user, thus we define the maximum k-regret ratio based on the assumption that the user's utility function lies in a utility function class $\mathcal{F}$.

Definition 3. *Maximum k-regret ratio [7]*. *Given a utility function class $\mathcal{F}$, the maximum k-regret ratio of a subset $S \subseteq D$ with respect to $\mathcal{F}$ is defined as*

$$mkrr(S, \mathcal{F}) = krr_D(S, \mathcal{F}) = \max_{f\in\mathcal{F}} krr_D(S, f)$$

It can be regarded as the k-regret ratio in the worst case when $f \in \mathcal{F}$. We replace the notion with $mkrr(S)$ for brevity.

Definition 4. *The k-regret minimization set query* *(also known as k-regret minimizing set problem, kRMS for short [7]). A k-regret minimizing set of size r is a subset of D such that $mkrr(R)$ is minimized where r is a positive integer provided by users, i.e.,*

$$R^* = \arg\min_{R\subseteq D\wedge|R|=r} mkrr(R)$$

The goal of the k-regret minimization set query is to find the subset R^* minimizing $mkrr(R)$. Moreover, we define the continuous k-regret minimization set query on data streams.

Definition 5. *The continuous k-regret minimization set query*. *Given a sequence of (possibly infinite) tuples $D_t = \langle p_1, p_2, p_3, \cdots \rangle$ arriving as a data stream where p_t is the tuples arriving at time t i.e., the t-th tuple. The goal of the continuous k-regret minimization set query is to find a k-regret minimizing set of size r at time t where*

$$R_t^* = \arg\min_{R_t\subseteq D_t\wedge|R_t|=r} mkrr(R_t)$$

If the number of tuples that have already arrived is less than r, we return all the tuples as the result set.

3.2 Sliding Window Model

We focus on maintaining a result set for the continuous k-regret minimization set query on data streams by applying the sliding window model, namely the most recent arrived tuples in the sliding window are considered. There are two versions of sliding windows which are widely used in existing works: time-based and count-based sliding windows [12,18]. A time-based sliding window contains all tuples that arrive in a fixed time period recently while a count-based sliding window contains the fixed number of tuples that arrive most recently. Since the two versions of sliding windows are interchangeable, we only focus on the count-based sliding window model in this paper.

3.3   Coreset

We first introduce a generalized concept of the *k-th directional width* [1] used in the definition of coresets. Given a set of points D of dimensionality d and a direction v which is a d-dimensional vector, the k-th directional width $w_k(v, D)$ is shown as follows

$$w_k(v, D) = \max_{p \in D}^{(k)} \langle v, p \rangle - \min_{p \in D}^{(k)} \langle v, p \rangle$$

$w_k(v, D)$ measures the extent of the set of points in a specific direction, based on which the coreset is defined. A subset $Q \subseteq D$ is called a (k, ϵ)-coreset if $w(v, Q) \geq (1 - \epsilon)w_k(v, D)$ holds for each direction v in d-dimensional space and it can be regarded as a weaker definition of (k, ϵ)-coreset introduced in [1] which is defined as $w_i(v, Q) \geq (1 - \epsilon)w_i(v, D)$ for every $i \leq k$. We use the notion ϵ-coreset [2] when $k = 1$ for brevity. A (k, ϵ)-coreset is a small subset of points extracted from the original dataset such that the extent of original dataset in each direction is approximately preserved.

4   The Coreset Based Algorithm

In this section, we introduce how to obtain a sequence of result sets for the continuous k-regret minimization set query over sliding windows. We first construct a coreset and then maintain the coreset for each arriving tuple. Based on the maintained coreset, we obtain the result set of fixed size r for the k-regret minimization set query.

4.1   Coreset Construction

As stated above, we only consider $|W|$ tuples that recently arrived in a sliding window W. Since the attribute values of a tuple are non-negative in our problem, we only focus on the non-negative quadrant instead of the whole coreset in its original definition as shown in Fig. 1. Moreover, we assume that all attribute values of points of D are in $[0, 1]$. Otherwise, each attribute value should be normalized by dividing the maximum value in each attribute among the point set D. Let S_+ be the non-negative quadrant of the sphere with radius $1 + \sqrt{d}$ and centered at the origin. We construct a set $\mathcal{G}$ of points on S_+, so that for any point $x \in S_+$ there exists a point $s \in \mathcal{G}$ such that $||x - s||$ is not larger than a small constant. Next, for each point $s \in \mathcal{G}$, we obtain kNN(s), the k nearest neighbors of s in the sliding window W and store the distance to its k-th nearest neighbor (denoted as NN$_k(s)$) which will be used for the coreset maintaining phrase. Finally, we obtain the coreset $C = \bigcup_{s \in \mathcal{G}} k$NN($s$). Figure 1 shows a concrete example where $k = 1$ and the kNNs search turns to the NN search. Specifically, the blue points as well as the black points are the normalized points such that all attribute values of these points are in $[0, 1]$, i.e., all points are in the box $[0, 1]^2$. Then the set $\mathcal{G}$ is shown in yellow points on S_+. For each point $s \in \mathcal{G}$, the yellow line connects s and the nearest neighbor of s which is marked by black point with the dotted line circle. Hence, the coreset is composed of all the black points.

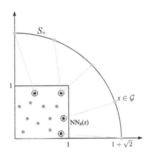

Fig. 1. Coreset construction by NN search (Color figure online)

4.2   Coreset Maintaining

Since the coreset is constructed based on kNNs of points in $\mathcal{G}$, thus our objective is to maintain the kNNs along with the data stream, and continuously report the new coreset and the up-to-date result set once the coreset changes. The whole process of maintaining the coreset is shown in Algorithm 1.

Specifically, for an arriving point p, we check and find a set I among $\mathcal{G}$ whose kNNs include p in Lines 2–4. I contains the points in $\mathcal{G}$ whose kNNs will be changed due to the arrival of p. We do nothing with the coreset if I is empty since p is not one of the kNNs for any point $s \in \mathcal{G}$ and p will not be selected into the coreset. Otherwise, the kNNs of these points in I along with the coreset are updated. Specifically, for each $s \in \mathcal{G}$, if the recorded distance to its k-th nearest neighbor $||s - NN_k(s)||$ is larger than $||s - p||$, i.e., the arriving point p will be one of the kNNs of s, then the previous k-th nearest neighbor of s should be replaced by p. In addition, point p should be selected into the coreset if I is not empty and a point should be removed from the coreset if it is no longer one of the kNNs for any point $s \in \mathcal{G}$.

In terms of a point q leaving W, we will do nothing if it is not in the coreset. Otherwise, we also find the set I among $\mathcal{G}$ that q is one of their kNNs. Then their k-th nearest neighbors should be searched and selected into the coreset if it is not in the coreset yet as shown in Lines 19–21 because one of their kNNs, q is no longer in W, let alone in the coreset.

A Heuristic Strategy. We propose a heuristic strategy that utilizes the properties of skyline and k-regret minimization set queries in order to improve space and time efficiency by removing points as early as possible (maybe they are still in the sliding window). The detailed specifications of the strategy are listed as follows.

- If p is dominated by q, then p will never be selected into result set since users always obtain a smaller k-regret ratio on q than p for any utility function $f \in \mathcal{F}$.
- All tuples in the coreset dominated by an arriving tuple p can be safely discarded since they will not be selected into the result set of the k-regret minimization set query.
- Even if an arriving tuple p is dominated by some tuple p in the coreset, it cannot be directly discarded since q leaves W before p thus p may be selected into the coreset and further into the result set.

Algorithm 1: CoresetMaintaining(W, C, p)

Input: Current sliding window W, Current coreset C, Arriving point p.
Output: Updated coreset C.

1 $I \leftarrow \emptyset$;
2 **foreach** $s \in \mathcal{G}$ **do**
3   **if** $\|s - p\| \leq \|s - NN_k(s)\|$ **then**
4          $I \leftarrow I \cup s$;
5 **if** $I \neq \emptyset$ **then**
6   **foreach** $s \in I$ **do**
7          $q \leftarrow NN_k(s)$;
8   **if** $\bigwedge_{s \in \mathcal{G}} q \notin kNN(s)$ **then**
9              $C \leftarrow C \backslash q$;
10     $C \leftarrow C \cup p$;
11 $q \leftarrow$ the point leaving W;
12 **if** $q \in C$ **then**
13     $I \leftarrow \emptyset$;
14   **foreach** $s \in \mathcal{G}$ **do**
15   **if** $q \in kNN(s)$ **then**
16             $I \leftarrow I \cup s$;
17     $C \leftarrow C \backslash q$;
18   **foreach** $u \in I$ **do**
19         $p \leftarrow NN_k(s)$;
20   **if** $p \notin C$ **then**
21             $C \leftarrow C \cup p$;
22 **return** C;

Therefore, for an arriving point p, we first find the points in W which is dominated by p and remove them from the sliding window. Since these points which will never be selected into the coreset are not stored, the space consumption is significantly reduced. Moreover, the efficiency for maintaining the kNNs is also improved since we only search kNNs in the subset instead of the whole sliding window.

4.3   Result Set Computation

As mentioned before, existing static algorithms can be directly applied on the coreset which is constructed and maintained as described above. In general, the greedy algorithm [14] performs well for the k-regret minimization set query. The greedy algorithm obtains a high quality solution but takes too much time with a large dataset. Instead, we run the greedy algorithm on our constructed coreset rather than the whole dataset thus the efficiency is improved.

Our greedy algorithm is much similar to the one proposed in [14]. The only difference is that we select points from the coreset rather than whole dataset until r points are selected. Once the coreset changes when a new tuple arrives, we rerun the greedy algorithm on the new coreset to obtain the updated result set.

5   Experiments

In this section, we implement our coreset based algorithm as well as some baseline algorithms and evaluate the performance of these algorithms on both synthetic and real-world datasets.

Dataset. The experiments are conducted on 1 synthetic and 2 real-world datasets which are widely adopted by the researches in the literature [3,7,11,14, 15,20], and the information of the datasets is shown as follows.

- Anti-correlated. The anti-correlated dataset is a 6-dimensional synthetic dataset of size 10,000 created by the dataset generator in [5].
- NBA[2]. The NBA dataset is an 8-dimensional real-world dataset consisting of 21,961 tuples, each of which shows NBA players' game statistics such as points, steals and rebounds, etc.
- Color[3]. The color dataset is a 9-dimensional real-world dataset of size 68,040 which contains the mean, standard deviation and skewness of each H, S and V in the HSV color spaces of a color image.

Algorithm. We compare several static algorithms as well as a fully dynamic algorithm listed as follows.

- GREEDY. The greedy algorithm for 1RMS and kRMS proposed in [14] and [7] respectively.

---

[2] https://www.basketball-reference.com.

[3] https://archive.ics.uci.edu/ml/datasets/corel+image+features.

- CORESET. Our proposed algorithm for the continuous k-regret minimization set query based on coresets over sliding windows. The ANN library[4] is used to help us for the k nearest neighbors search to construct the coreset.
- FD-RMS. The fully dynamic algorithm for the kRMS with tuple insertions and deletions proposed in [19].
- KERNEL. A static algorithm which returns an ϵ-kernel directly for the kRMS proposed in [3].

Implementation and Workload. All algorithms are implemented in C++ and all experiments are conducted on a PC running Ubuntu 18.04 LTS with a 3.00 GHz processor and 32 GB memory. We use the datasets to simulate data streams. For a specific dataset, the tuples arrive one after another until all tuples have arrived. We record the total CPU time and the maximum k-regret ratio of result set after 10% tuples have arrived. We report the average CPU time for each tuple and the average value of the maximum k-regret ratios of those 10 records.

Experimental Results

Varying r. The results for varying r, i.e., the sizes of outputs are shown in Fig. 2. We set $k = 1$ and vary r from 10 to 50. The maximum k-regret ratios decrease as r increases for all algorithms, because a larger result set contains more tuples and the maximum k-regret ratio becomes smaller. The CORESET and GREEDY obtain smaller maximum k-regret ratios. Moreover, the maximum k-regret ratios of CORESET are similar to those of GREEDY, both of which are smaller than those of other two algorithms. In terms of the CPU time, all algorithms take more time to get a larger result set. CORESET and FD-RMS run much faster than the static algorithms, i.e., GREEDY and KERNEL. Although CORESET

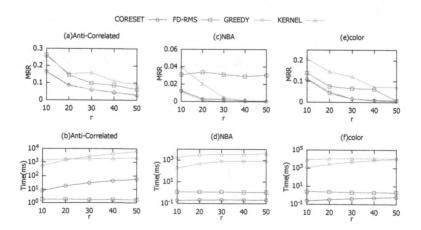

Fig. 2. The maximum k-regret ratio and the CPU time of each algorithm on the datasets by varying r

---

[4] https://www.cs.umd.edu/~mount/ANN.

runs the greedy algorithm on the coreset and reruns it whenever the coreset is changed, CORESET takes much less time than GREEDY because of the small size of the coreset compared to that of the whole dataset. But the maximum k-regret ratio of CORESET is not much worse than that of GREEDY (and even better than GREEDY in most cases). In addition, our heuristic strategy helps us avoid many unnecessary maintainings and CORESET runs much faster than FD-RMS since FD-RMS updates its set system when each tuple arrives. However, CORESET runs slower than FD-RMS on the Anti-correlated dataset. This is because there are more skyline points in the Anti-correlated dataset, thus the size of the coreset is large and the coreset needs to be updated frequently. Thus it takes more time to obtain the result set with the greedy algorithm once the coreset changes.

Varying k. As shown in Fig. 3, we present the results when varying k from 1 to 5 where r is set to be 20, 20 and 10 for the Anti-correlated, Color and NBA datasets respectively. In general, the maximum k-regret ratios decrease with k, but there are special cases such as FD-RMS and KERNEL which obtain smaller maximum k-regret ratios with a larger k because FD-RMS and KERNEL do not return the same result set for different ks. Moreover, it's similar to the case of varing r that the maximum k-regret ratios of CORESET and GREEDY are also better than those of the other two algorithms. The CPU time increases with k both on the Anti-correlated and real-world datasets for all algorithms. CORESET needs more time to update kNNs since the size of the coreset becomes larger as k increases. FD-RMS needs to update a larger set system when k is large. The static algorithms take nearly 10 s for each arriving tuple which is intolerable. CORESET takes more time than FD-RMS on the Anti-correlated dataset due to the same reason mentioned above.

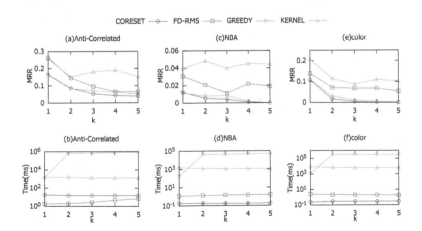

Fig. 3. The maximum k-regret ratio and the CPU time of each algorithm on the datasets by varying k

Summary. CORESET and FD-RMS perform better than the static algorithms in the CPU time and obtain a comparable result quality. CORESET takes less time than FD-RMS in real-world datasets since it just maintains a coreset which is much smaller than the whole dataset. In addition, CORESET uses the greedy algorithm to obtain a result set, hence CORESET takes much less time than GREEDY while the result quality of CORESET is comparable and even better than that of GREEDY.

6   Conclusion

In this paper, we study the continuous k-regret minimization set query on data streams. Instead of rerunning the static algorithms when a new tuple arrives, we develop an algorithm by maintaining a coreset to obtain a sequence of result sets over sliding windows. We introduce a heuristic strategy to further improve the time and space efficiency. Experiments on the synthetic and real-world datasets are conducted to show the efficiency and effectiveness of our proposed CORESET algorithm. In our future work, it would be more convincing to investigate whether the proposed algorithm performs well comparing with the existing works on the error bound and the time complexity.

Acknowledgment. This work is partially supported by the National Natural Science Foundation of China under grant U1733112 and the Fundamental Research Funds for the Central Universities under grant NS2020068.

References

1. Agarwal, P.K., Har-Peled, S., Har-Peled, S., Yu, H., Yu, H.: Robust shape fitting via peeling and grating coresets. Discret. Comput. Geom. **39**(1–3), 38–58 (2008)
2. Agarwal, P.K., Har-Peled, S., Varadarajan, K.R.: Approximating extent measures of points. JACM **51**(4), 606–635 (2004)
3. Agarwal, P.K., Kumar, N., Sintos, S., Suri, S.: Efficient algorithms for k-regret minimizing sets. In: SEA, pp. 7:1–7:23 (2017)
4. Asudeh, A., Nazi, A., Zhang, N., Das, G.: Efficient computation of regret-ratio minimizing set: a compact maxima representative. In: SIGMOD, pp. 821–834 (2017)
5. Börzsöny, S., Kossmann, D., Stocker, K.: The skyline operator. In: ICDE, pp. 421–430 (2001)
6. Cao, W., et al.: K-regret minimizing set: efficient algorithms and hardness. In: ICDT, pp. 11:1–19 (2017)
7. Chester, S., Thomo, A., Venkatesh, S., Whitesides, S.: Computing k-regret minimizing sets. VLDB **7**(5), 389–400 (2014)
8. Fan, Y., Shi, Y., Kang, K., Xing, Q.: An inflection point based clustering method for sequence data. In: Ni, W., Wang, X., Song, W., Li, Y. (eds.) WISA 2019. LNCS, vol. 11817, pp. 201–212. Springer, Cham (2019). https://doi.org/10.1007/978-3-030-30952-7_22
9. Faulkner, T.K., Brackenbury, W., Lall, A.: K-regret queries with nonlinear utilities. VLDB **8**(13), 2098–2109 (2015)

10. Ilyas, I.F., Beskales, G., Soliman, M.A.: A survey of top-k query processing techniques in relational database systems. ACM Comput. Surv. (CSUR) **40**(4), 11:1–58 (2008)
11. Kumar, N., Sintos, S.: Faster approximation algorithm for the k-regret minimizing set and related problems. In: ALENEX, pp. 62–74 (2018)
12. Mouratidis, K., Bakiras, S., Papadias, D.: Continuous monitoring of top-k queries over sliding windows. In: SIGMOD, pp. 635–646 (2006)
13. Nanongkai, D., Lall, A., Das Sarma, A., Makino, K.: Interactive regret minimization. In: SIGMOD, pp. 109–120 (2012)
14. Nanongkai, D., Sarma, A.D., Lall, A., Lipton, R.J., Xu, J.: Regret-minimizing representative databases. VLDB **3**(1), 1114–1124 (2010)
15. Peng, P., Wong, R.C.: Geometry approach for k-regret query. In: ICDE, pp. 772–783 (2014)
16. Qi, J., Zuo, F., Samet, H., Yao, J.: K-regret queries using multiplicative utility functions. TODS **43**(2), 1–41 (2018)
17. Soma, T., Yoshida, Y.: Regret ratio minimization in multi-objective submodular function maximization. AAAI, 905–911 (2017)
18. Tao, Y., Papadias, D.: Maintaining sliding window skylines on data streams. TKDE **18**(2), 377–391 (2006)
19. Wang, Y., Li, Y., Wong, R.C., Tan, K.L.: A fully dynamic algorithm for k-regret minimizing sets. arXiv (2020)
20. Xie, M., Wong, R.C.W., Li, J., Long, C., Lall, A.: Efficient k-regret query algorithm with restriction-free bound for any dimensionality. In: SIGMOD, pp. 959–974 (2018)
21. Zeighami, S., Wong, R.C.W.: Minimizing average regret ratio in database. In: SIGMOD, pp. 2265–2266 (2016)

Configurable In-Database Similarity Search of Electronic Medical Records

Yuewen Wu[1,2], Yong Zhang[2(✉)], and Jiacheng Wu[2]

[1] Princeton University, Princeton, NJ 08544, USA
[2] Tsinghua University, Beijing 100084, China
zhangyong05@tsinghua.edu.cn

Abstract. With the development of technology, Electronic Medical Records (EMRs) are widely used for medical analysis through methods such as similarity search. Typical EMRs contain attributes of different data types including string, enumeration and numeric data, and are commonly stored in a database. However, many EMR similarity search algorithms neither separate different data types nor conduct search directly in the database. In addition, for researchers and doctors who need similarity search but do not have strong programming background, a user-friendly interface is missing. Therefore, we design a tool "SIR" to solve the aforementioned problems. SIR can conduct configurable similarity search in high dimensions (within 0.0931 and 0.7824 s respectively using its basic and advanced version), can be embedded directly in the database, and has an intuitive interface.

Keywords: EMR · UDF · GUI · Similarity search · Database

1 Introduction

Real-World Evidence (RWE) is widely used by those who develop medical products or who study, deliver, or pay for health care [11]. With the development of technology, Electronic Medical Records (EMRs), one category of RWE, are frequently archived and subsequently extracted for analysis [2]. EMR is proven to be useful in many aspects. Besides making better judgements by comparing similar patients, researchers even manage to leverage similar EMRs to predict possible future medical concepts (e.g., disorders) [7]. The global electronic health records market size was valued at USD 26.8 billion in 2020 and is expected to witness a compound annual growth rate (CAGR) of 3.7% from 2021 to 2028 [10]. This expanding market shows both the potential and necessity of EMR research. One application of EMR is patient similarity search, which uses collected medical data to find a cohort of patients with similar attributes to any given patient [3]. Sometimes, patients share similarity in terms of a combination of different attributes rather than only one. With similarity search, multiple dimensions of an EMR can be taken into account and reflected as a straightforward number, allowing doctors and lab researchers to find the hidden connections between patients. Therefore, the goal of this paper is

© Springer Nature Switzerland AG 2021
C. Xing et al. (Eds.): WISA 2021, LNCS 12999, pp. 62–73, 2021.
https://doi.org/10.1007/978-3-030-87571-8_6

to present a tool "SIR", standing for "Similarity In Records", which helps users to conduct configurable in-database similarity search of EMRs.

Despite the great progress made by EMR similarity search, existing researches still have their own defects. A typical EMR consists of diverse attributes including patient ID, gender, diagnosis and so on. These attributes are categorized into three main data types: string, enumeration and numeric data. However, early academic researches barely consider this multi-dimensional characteristic. For example, Gravano et al. [5] investigate the distance calculation algorithm for EMR's string data, but they leave the other data types undiscussed. A decade later, He et al. [6] design a system that accepts EMRs collected from hospitals as input, goes through a series of process, and eventually calculates the similarity of any two EMRs. Nevertheless, most hospitals and labs store EMRs in databases, while He's work conducts similarity search outside the database. Tashkandi and Wiese [13] notice this synthetic condition and thus propose to processes the majority of the workload within the database. Nonetheless, they calculate the distance between every EMR pair, while in reality, only the distances between the query patient and other EMRs are needed. Furthermore, McGinn et al. [9] point out that real-world users are concerned about the ease to use similarity search. Hence, a user-friendly interface that hides complicated calculation details for real-world users is essential, but current academic work rarely recognizes this need.

Our work consists of three major contributions: 1) an EMR similarity search algorithm that separates different data types, 2) an in-database search algorithm through MySQL's UDF and 3) an user-friendly interface[1] that eliminates technical barriers.

The rest of the paper is organized as follows: Sect. 2 defines the problem and Sect. 3 introduces the architecture of SIR. Section 4 presents the basic design of SIR, which utilizes nmslib to conduct external similarity search, while Sect. 5 presents the advanced design of SIR, which utilizes UDFs to conduct in-database similarity search. The implementation details are explained in Sect. 6. Section 7 explains the evaluation of both versions of SIR. Section 8 discusses some related work on patient similarity search. Section 9 concludes the paper with directions for further improvements.

2 Problem Definition

An EMR is a list that consists of n attributes, with n_s string attributes, n_e enumeration attributes and n_n numeric attributes. In order to calculate the distance between each data type, an EMR is partitioned into three small lists, each containing attributes of one data type. For string data, the list contains n_p long strings and n_q short strings. When calculating the distance between an EMR pair, each attribute has a weight value w_i. In addition, each data type has a weight value

---

[1] SIR is user-friendly in the sense that no heavy coding is required, but users are supposed to have knowledge in patient similarity search and know how to make their own customized search settings, such as weight configurations.

w_s, w_e, w_n, so users can calculate distance based on one data type. In this way, the distance D between two EMRs are defined as:

$$D = w_s \cdot d_s + w_e \cdot d_e + w_n \cdot d_n \tag{1}$$

where d_s, d_e and d_n are distances between string, enumeration and numeric data of two EMRs.

Each string within the string list will be partitioned into tokens (word separated by space, comma or period) and put into a string set. For the convenience of illustration, A_{pi} and A_{qi} represents the query EMR's strings sets containing long strings and short strings; B_{pi} and B_{qi} represents other EMR's strings sets. We use J and E to represent Jaccard distance and Edit distance. Jaccard distance is the disjunctive union of A, B divided by the size of the union of the sample sets[2],

$$J(A, B) = 1 - \frac{|A \cap B|}{|A \cup B|} = 1 - \frac{|A \cap B|}{|A| + |B| - |A \cap B|} \tag{2}$$

and edit distance is the minimum number of operations required to transform one string into the other. Hence, d_s is defined as:

$$d_s = \sum_{i=0}^{n_p} w_i \cdot J(A_{pi}, B_{pi}) + \sum_{i=0}^{n_q} w_i \cdot E(A_{qi}, B_{qi}) \tag{3}$$

d_e measures the distance between enumeration data, where distance is 0 if two labels are the same and 1 if not.

d_n measures the distance between numeric data, where x_i is the ith attribute of the numeric data:

$$d_n = \sum_{i=0}^{n_n} w_i \cdot \frac{x_i}{max(x_i) - min(x_i)} \tag{4}$$

In order to aggregate the distances of each data type into a weighted sum D, d_s, d_e and d_n need to be normalized to the same range $[0, \omega]$. In our work, we set $\omega = 10$. If a range is too small, the difference between distances will not be obvious, but if a range is too large, computing spaces will be wasted.

3   System Architecture

The architecture of SIR is demonstrated in Fig. 1. The grey area represents the system of SIR, which consists of a EMR database, an attribute level, an EMR level and a service level. Users such as doctors and lab researchers only interact with the service level. They specify the query EMR ID, select a Top-K Value (number of similar patients they want) and adjust each attribute's weight (details explained in Sect. 6.3). Afterwards, SIR passes these information to the attribute

---

[2] Reference: https://deepai.org/machine-learning-glossary-and-terms/jaccard-index.

level and then the EMR level to find the distance between the query item and other EMRs in the database. Based on the distances, SIR sorts the records in the EMR database in ascending order, and the closest neighbors are returned to users as search results.

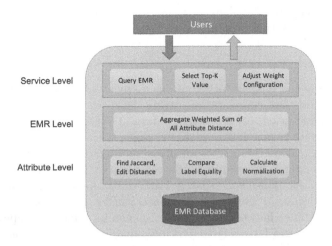

Fig. 1. System architecture of SIR

SIR has two versions that both accomplish the expected tasks. The first version (Fig. 2.a) is a basic design, which uses the Non-metric Space Library (nmslib) in an external program to perform similarity search outside the database. EMR data are first downloaded to local file for the external program, and the search results are then passed back to the interface for demonstration (details explained in Sect. 4). The second version (Fig. 2.b)) is an advanced one, which embeds the search algorithm inside the database system by User-Defined Functions (UDFs) in MySQL. SIR passes search specifications to predefined UDFs, and finds similar patients within the database (details explained in Sect. 5).

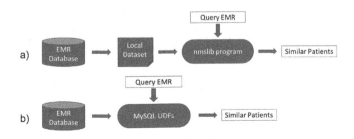

Fig. 2. Comparison of SIR's two versions

4   Basic Design

The essential idea behind the basic design is combining an external program with the user interface. The external program is in charge of similarity search while the user interface receives search specifications from users and demonstrate search results. Section 4.1 explains how the similarity search is done, and Sect. 4.2 explains how the external program cooperates with the interface.

4.1   Non-metric Space Library (nmslib) for Similarity Search

As aforementioned, a typical EMR contains many attributes about a patient. For instance, diagnosis and complaint are string data; sex and gender are enumeration data; age and heartbeat are numeric data. These attributes are categorized into different data types and need to be dealt with separately. This need of multi-dimension calculation makes nmslib a suitable choice.

nmslib is an open source similarity search library [1,8] for evaluation of k-NN methods for generic non-metric spaces. Each data type of EMR can be treated as a new dimension, or even each attribute can be treated as a new one. In this way, the distance between each dimension can be calculated separately with different methods and later aggregated into an overall value. Upon completion of the calculation, all the EMRs will be sorted in ascending order.

4.2   External Program Embedding

The decision to do similarity search as an external program is made after experiments with alternative plans. It turns out that nmslib needs to call many other local files and libraries, and several failed attempts to embed it into MySQL suggest that running nmslib as an external program is more realistic.

The external program is embedded in SIR like a specialized toolkit. It takes in the query EMR ID, the Top K value (number of similar patients needed), an EMR data file (in .csv format) and a configuration file. Then it returns the EMR ID of similar patients. Since the external program has no access to EMRs stored in the database, all the data need to be downloaded to local in .csv format in advance. When a similarity search is triggered, SIR passes the query EMR ID and the Top K value to the external program. In terms of the configuration file, there exists an initial default file, and SIR updates the file based on customized settings by users. Now, all four parameters needed for the external program become available and similarity search can be done within seconds. Upon completion of the search, the external program sends the results back to SIR for demonstration.

5   Advanced Design

5.1   User-Defined Function (UDF) for Similarity Search

Although the basic design of SIR could perform EMR similarity search, it cannot conduct synchronous search. In order to solve this problem, MySQL's UDF is employed in the advanced design of SIR. UDF stands for User-Defined Function, which is a customized function that can be called directly by MySQL like a built-in function. A UDF can return string, integer, or real values and accept arguments of those same types. Simple functions operate on a single row while aggregate functions operate on a group of rows of a database table.

Three UDF "strd", "equa" and "norm" are created to calculate the distance between string, enumeration and numeric attributes respectively. Users can specify each attribute's weight by passing parameters to the corresponding UDF. Each UDF then takes in the customized weight and returns a value from 0 to 10 that reflect the distance. Afterwards, SIR adds up the three distance values based on the overall weight proportion to get an aggregate final distance. Based on this value, SIR can sort all the records in the database to determine the most similar EMRs. Note that we assume the structure of the EMR, such as the number of string attributes, is known and fixed. The following subsections introduce the detailed algorithm of each UDF.

UDF "strd" for String Data. The UDF "strd" returns the Jaccard distance and Edit distance between string data. Edit distance calculates the distance between strings by characters, so it tolerates typos or different writing styles. However, this algorithm is relatively slow, so it only applies to strings shorter than 10 phrases. Jaccard distance is used for longer strings.

The function works as follows. Initially, each string attribute of the EMR is partitioned into string tokens (word separated by space, comma or period). Tokens from the same string attribute are then put into a string set. Next, for each string set, "strd" finds the distance between the query's set and other EMRs'. Finally, "strd" combines the distance between each set into one single value based on customized weight. Note that empty strings in the query EMR are ignored in the calculation by setting its distance with any other string set to 0, because empty strings contain no information.

UDF "equa" for Enumeration Data. The UDF "equa" returns the distance between enumeration data. Enumeration data are labels such as "male" and "female", so the distance between these data is done directly through equality comparison. Basically, the distance between two labels is 0 if they are equal, 1 if not. Likewise, "equa" finds the overall enumeration distance by adding up all the distances according to their weight.

UDF "norm" for Numeric Data. The UDF "norm" returns distance between numeric data. For each numeric attribute, "norm" first finds the largest possible range. Then, it uses the range to divide the difference between the query and the compared EMR. The ratio is considered as the distance for this numeric attribute. The overall distance is found similar to the previous two UDFs.

5.2   UDF Embedding and Combination

After the three UDFs are created, they need to be embedded into MySQL. As instructed by MySQL's manual[3], UDF is initially written in C++ as a dynamically loadable file. Next, the programming file needs to be complied into a sharable library file. Then, the shared object needs to be copied into the server's plugin directory so that MySQL can locate it and install the corresponding UDF.

After the installation of the UDFs, SIR can call these functions through its interface directly. In this way, SIR receives the distance for string data, enumeration data and numeric data separately, and then combines them according to their weight proportion. Based on this overall distance, SIR then instructs MySQL to sort all the EMRs in the database and thus finds the most similar EMRs. A sample MySQL instruction is shown below:

```
sql = "select ID, Age, Sex, Diagnosis from translate_data order by" + "
    (" + enum_prop + "*equa('" + query_sex + ",,,,," + query_race +
    ",,,,," + query_allergy + ",,,,," + query_blood + ",,,,," +
    query_condition + ",,,,," + query_status + "', " + " concat(Sex,
    ',,,,,', Race, ',,,,,', Allergy, ',,,,,', Blood_type," + " ',,,,,',
    Health_condition, ',,,,,', Heartbeat_status), '" +
    enum_weight_string + "') + ......"
```

*Instructions in purple are normal sql command. Parameters in black are weight proportions specified by users (or by default). Attributes of the same data type form one string separated by five commas (',,,,,'). Due to space limitation, only enumeration data's distance calculation is shown.

6   Implementation

6.1   Environment

System. All of the programs are written, complied and run on Ubuntu 20.04.

MySQL. Initially, MySQL 8.0 is used, but it turns out that UDF works more stably in older versions, so MySQL is degraded to 5.7 instead. Theoretically, both versions could work, but some coding details of UDF are different in the two versions.

Programming Language. The interface is programmed using Tkinter, a Python binding to the Tk GUI toolkit. Python version is 3.8.5. nmslib is an open source library [1,8]. Most of its programs are written in C++, so SIR's similarity search algorithm is written in C++ as well. As required by MySQL's manual, all UDF codes are written in C++. Complier g++-9 is used, but theoretically, any version higher than g++-7 could work.

---

[3] https://dev.mysql.com/doc/extending-mysql/5.7/en/udf-features.html.

6.2   Dataset

The dataset used contains 1908 EMRs obtained from Beijing Tsinghua Changgeng Hospital. Each EMR includes the following attributes (missing attributes treated as NULL):

- 5 string attributes: diagnosis, complaint, medical history, smoking history, drinking history
- 6 enumeration attributes: sex, race, allergy, blood type, health condition, heartbeat status
- 3 numeric attributes: age, number of visits to the hospital, heartbeat

6.3   Interface

The interface consists of three different windows as demonstrated in Fig. 3.

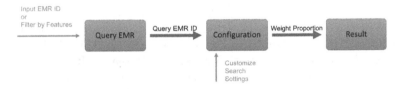

Fig. 3. The process of interface usage

EMR ID Search Window. This window is the first level that gets the query EMR ID. If users already know the query EMR ID, they can enter the ID and go to next level. Otherwise, they can select up to three features to locate the query ID. All potential EMRs filtered by the database will then pop up on the screen and users can click on one candidate to enter the next level. As soon as users click on "next", the query EMR ID will be stored backstage for further usage.

Configuration Window. This window allows the users to make customized configuration settings (shown in Fig. 4). As aforementioned, each EMR has three different type of data (string, enumeration and numeric), and each data has several attributes. When calculating distance for similarity search, each attribute contributes in a different proportion. As demonstrated, users can make adjustments by dragging the slider bars on the basis of a default setting. As soon as they click on "search", the current configuration file will be updated backstage for similarity search.

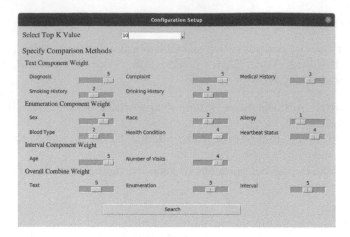

Fig. 4. Configuration Window

Result Window. This window conducts similarity search backstage and then displays the search results (shown in Fig. 5). Upon entering this window, SIR has collected all the necessary information for similarity search. It passes the query EMR ID and weight configuration to either the external executable (basic design) or MySQL (advanced design), and finds the most similar EMRs. Afterwards, SIR returns the result with a few key information, and users can click on the record of their interest to see more details.

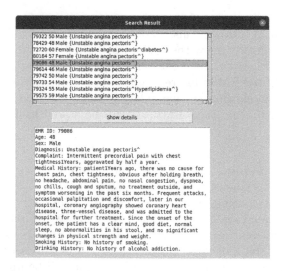

Fig. 5. Result window

7   Evaluation

7.1   Running Time Evaluation

In order to test the running time of SIR comprehensively, both versions conduct similarity search under different situations. Overall, the running time for both versions takes less than 1 s, but basic version of SIR runs much faster than the advanced version, because nmslib is a mature library for high-dimensional computation, while the UDFs for this work use relatively naive algorithms. The rest of this section discusses detailed analysis of the running time data.

On one hand, the number of attributes taken into consideration does not have distinct effect on the basic version of SIR, and it takes around 0.0931 s to find 5 most similar patients to a query patient. On the other hand, the running time of the advanced version varies a lot, and the data is shown in Table 1. The dataset consists of 5 string attributes, 6 enumeration attributes and 2 numeric attributes, and 10 different situations are considered. The distance calculation for pure enumeration data and pure numeric data almost takes time (both smaller than or around 0.01 s), while string attributes require much longer time to compare their distance. For the purpose of demonstration, String Attribute, Enumeration Attribute and Numeric Attribute are abbreviated as SA, EA and NA.

Table 1. Running time of the advanced version of SIR under different situations.

Combination of attributes	Running time (second)
1 SA	0.4018
1 EA	0.0093
1 NA	0.0077
2 SA	0.4151
2 EA	0.0101
1 SA + 1 EA + 1 NA	0.4121
2 SA + 2 EA + 1 NA	0.4258
2 SA + 3 EA + 1 NA	0.4258
3 SA + 3 EA + 1 NA	0.7624
All Attributes	0.7824

7.2   Interface Evaluation

Three doctors from Beijing Tsinghua Changgeng Hospital are invited to try out SIR and give feedback accordingly. They use both traditional methods and SIR to locate 5 similar patients to a query patient. For evaluation purpose, a small dataset consisting of 100 EMRs is used.

It turns out that SIR is at least 20 times faster than manual filtering methods. The doctors only took 3 min to set up the similarity search, and received the search results within seconds. However, if they manually looked for similar patients, it could took them 1 h or more.

Moreover, the doctors are especially satisfied with the interactive interface. They said that without strong programming background, using MySQL or other similar tools to filter similar patients used to be a huge challenge. However, SIR was very intuitive and required no technical experience. They only needed to input their requirements and could leave the complicated calculation to the computer. Also, SIR allowed them to assign weights to each attribute, which greatly enhanced the accuracy of the results and made the search more specific.

8 Related Work

Besides the methods introduced in this paper, EMR similarity search can be done in many other ways. Graph representation is a popular approach, as it is fast and requires very few storage spaces [4]. When the reference database is very large or the distance computation is expensive due to hardware limitation, it is more practical to find the approximate nearest neighbor [12]. nmslib is an open source similarity search library [1,8] that finds approximate nearest neighbors, though it does not have direct application in EMR similarity search yet.

Although many studies have made great progress in EMR similarity search, a few shortcomings still exist. To begin with, Gravano et al. [5] discuss the distance calculation for string attributes within the database. They develop a technique for building approximate string join capabilities on top of commercial database. However, they only consider the string attributes and give up using UDF for in-database calculation due to algorithm inefficiency. A few years later, He et al. [6] design a more comprehensive system that calculates the similarity of any two EMRs. However, the calculation is done in synthetic conditions: data are extracted from the EMR database and processed in external tools. Therefore, Tashkandi et al. [13] propose to processes the majority of the workload within the database. However, their method finds the distance between all EMR pairs inside the database. In reality, the query patient is usually known, so only its distance with other patients is needed. Furthermore, McGinn et al. [9] conducts a survey among real-world users, and finds that they are hesitant to adopt EMR similarity search due to design and technical concerns, ease of use and interoperability. Hence, a user-friendly interface is necessary if researchers want to make real-world impact, but not much work has covered this topic yet.

9 Conclusion

In this paper, both basic and advanced designs of SIR are introduced. SIR is invented as a tool for EMR similarity search that comes with a user-friendly interface. Its basic version conducts similarity search using the external nmslib, while the advanced version employs UDFs for in-database similarity search. The interface requires almost no technical background to use. It helps users to locate their query patient, allows them to make customized search configuration and shows them search results with detailed information.

However, SIR can still be improved in a few aspects. First, an adaptive interface that can automatically fit with different EMR structures is needed. In addition, the efficiency of UDFs in basic version still needs enhancement. Lastly, SIR's compatibility with other applications requires further study.

Acknowledgement. This work was supported by National Key R&D Program of China (2020AAA0109603), State Key Laboratory of Computer Architecture (ICT, CAS) under Grant No. CARCHA202008.

References

1. Boytsov, L., Naidan, B.: Engineering efficient and effective non-metric space library. In: Brisaboa, N., Pedreira, O., Zezula, P. (eds.) SISAP 2013. LNCS, vol. 8199, pp. 280–293. Springer, Heidelberg (2013). https://doi.org/10.1007/978-3-642-41062-8_28
2. Celi, L.A., Charlton, P., Ghassemi, M.M., et al.: Secondary Analysis of Electronic Health Records. Springer, New York (2016)
3. Eteffa, K.F., Ansong, S., Li, C., Sheng, M., Zhang, Y., Xing, C.: Application of patient similarity in smart health: a case study in medical education. In: Ni, W., Wang, X., Song, W., Li, Y. (eds.) WISA 2019. LNCS, vol. 11817, pp. 714–719. Springer, Cham (2019). https://doi.org/10.1007/978-3-030-30952-7_72
4. Eteffa, K.F., Ansong, S., Li, C., Sheng, M., Zhang, Y., Xing, C.: An experimental study of time series based patient similarity with graphs. In: Wang, G., Lin, X., Hendler, J., Song, W., Xu, Z., Liu, G. (eds.) WISA 2020. LNCS, vol. 12432, pp. 467–474. Springer, Cham (2020). https://doi.org/10.1007/978-3-030-60029-7_42
5. Gravano, L., Ipeirotis, P.G., Jagadish, H.V., et al.: Approximate string joins in a database (almost) for free. VLDB 1, 491–500 (2001)
6. He, Z., Yang, J., Wang, Q., Li, J.: A method of electronic medical record similarity computation. In: Xing, C., Zhang, Y., Liang, Y. (eds.) ICSH 2016. LNCS, vol. 10219, pp. 182–191. Springer, Cham (2017). https://doi.org/10.1007/978-3-319-59858-1_18
7. Le, N., Wiley, M., Loza, A., et al.: Prediction of medical concepts in electronic health records: similar patient analysis. JMIR Med. Inform. 8(7), e16008 (2020)
8. Malkov, Y.A., Yashunin, D.A.: Efficient and robust approximate nearest neighbor search using hierarchical navigable small world graphs. CoRR abs/1603.09320 (2016). http://arxiv.org/abs/1603.09320
9. McGinn, C.A., Grenier, S., Duplantie, J., et al.: Comparison of user groups' perspectives of barriers and facilitators to implementing electronic health records: a systematic review. BMC Med. 9(1), 1–10 (2011)
10. Grand Viewer Research: Electronic health records market size, share and trends analysis report, 2021–2028, April 2021. https://www.grandviewresearch.com/industry-analysis/electronic-health-records-ehr-market
11. Sherman, R.E., Anderson, S.A., Dal Pan, G.J., et al.: Real-world evidence–what is it and what can it tell us. N. Engl. J. Med. 375(23), 2293–2297 (2016)
12. Sundari, P.S., Subaji, M., Karthikeyan, J.: A survey on effective similarity search models and techniques for big data processing in healthcare system. Res. J. Pharm. Technol. 10(8), 2677–2684 (2017)
13. Tashkandi, A., Wiese, I., Wiese, L.: Efficient in-database patient similarity analysis for personalized medical decision support systems. Big Data Res. 13, 52–64 (2018)

Database Native Approximate Query Processing Based on Machine-Learning

Yang Duan[1,2], Yong Zhang[3(✉)], and Jiacheng Wu[3]

[1] University of Illinois Urbana-Champaign, Illinois, USA
yangd4@illinois.edu
[2] BNRist, Tsinghua University, Beijing, China
[3] BNRist, Department of Computer Science and Technology, RIIT,
Institute of Internet Industry, Tsinghua University, Beijing, China
zhangyong05@tsinghua.edu.cn

Abstract. With the worldwide digital transformation, many databases with large volumes appear and provide interesting insights analyzed by data scientists through all kinds of tools. The large volumes of them inevitably increase the workload of calculation, lengthen the response time of applications and negatively impact the user experience. Approximate Query Processing (AQP) is proposed to alleviate this issue. Although many researchers continuously improve the performance of AQP with the help of Machine Learning, there are few studies on embedding Machine Learning based AQP inside the relational database through User Defined Functions (UDF). In this paper, we focus on one specific kind of aggregate queries and present two different implementations to embed one Machine Learning based AQP inside Relational Database Management System (RDBMS) by taking advantage of UDF. Both implementations are able to calculate estimates with acceptable errors, and the implementation with external training and internal query processing has even better performance in term of response times.

Keywords: AQP · Machine learning · UDF · RDBMS · Database

1 Introduction

With continuously soaring data in databases, decision support applications are frequently used to do analyses [7]. However, these applications commonly involve aggregation queries, and the cost of executions of these queries is often expensive and requires a large amount of resources [7,13]. Unlike the traditional method of computations, Approximate Query Processing (AQP) can be used to efficiently answer aggregation queries with limited resources in a short period of time by computing approximate answers within acceptable ranges of errors to lower the workload of computations [12]. Due to its high-performance property, AQP is considered as a promising technique and already has plenty of applications in some areas, such as data exploration and visualization [19].

© Springer Nature Switzerland AG 2021
C. Xing et al. (Eds.): WISA 2021, LNCS 12999, pp. 74–86, 2021.
https://doi.org/10.1007/978-3-030-87571-8_7

Machine Learning is well developed after decades. It has been already used in many areas [10]. One novel technique for AQP is to take advantage of Machine Learning to further reduce the execution time, improve the accuracy, and support most aggregate functions [14]. With this technique, data scientists are able to train models with existing data and obtain approximate answers as predictions with acceptable errors from trained models. In this paper, training models and answering queries are separated, and the main idea of this paper is to embed Machine Learning based AQP inside database servers to do aggregation queries.

Many researchers have already proposed new methods and new algorithms of AQP with huge improvements. Agarwal et al. [4] proposed BlinkDB, a parallel, sampling-based approximate query engine to trade-off query accuracy for response time with bounded errors. Olma et al. [16] proposed Taster, a self-tuning, elastic, online AQP engine that has both advantages of online and offline AQP. Including these two mentioned pieces of research, most early researches did not employ UDF and Machine Learning. Google and Microsoft [8,15] integrated AQP into their SQL servers to do approximate computations. The usages of these functions are very similar to the usages of UDF, but they cannot be used to calculate the percentage that our AQP module estimates. Zhang et al. [20] applied AQP techniques, extended SQL grammar by UDF and proposed a system Parrot that uses Stanford NLP and Jieba. Their goal is to do analysis of the term frequency on large text data, which is also different from ours. In the later researches, Thirumuruganathan et al. [19] applied Deep Learning to learn the data distribution so that answering arbitrary AQP queries can be done without contacting the database server. Ma and Triantafillou [14] adopted regression and density-estimator models to create an SML-model-based AQP engine, which has incredible speedups with high accuracy. However, although these AQP engines they proposed took advantage of Machine Learning, they were not embedded inside database servers through User Defined Functions (UDF). Embedding the Machine Learning based AQP engine inside MySQL or other RDBMS is crucial in some cases that the volume of databases is increasing but the pattern of data changes slowly in a long period of time, and a portable AQP engine is needed.

In this paper, we mainly employ MySQL's UDF [1] to embed the Machine Learning based AQP engine inside MySQL. In summary, we make three contributions as the following. 1) We design a neural network that learns the distributions of data. 2) We use two different designs to implement an embedded Machine Learning based AQP engine inside the relational database through UDF. 3) We did experiments on a synthetic dataset and the TPC-H dataset [2], and the results show a dramatic speedup on executing the AQP UDF External Training and Internal Query compared to the AQP UDF External Training and External Query.

The organization of this paper is the following. We define the problem in Sect. 2. We show the overview and architecture of this system in Sect. 3. We present the details of how Machine Learning models work inside these AQP UDFs in Sect. 4. We demonstrate the details of embedding two different AQP UDFs in Sect. 5. We discuss the details of the implementation and evaluation in

Sect. 6. We review the related work in Sect. 7. We discuss the drawbacks of our work and make the conclusion in Sect. 8.

2   Problem Definition

Given a database relation R with N records and a column set (an attribute set) A with n elements where any element in A is a numerical column containing real numbers, our embedded Machine Learning based AQP engine can be used to answer the following aggregation query.

SELECT COUNT(*) / N FROM R
WHERE $(b_{1L} < A_1$ AND $A_1 < b_{1U})$ AND $(b_{2L} < A_2$ AND $A_2 < b_{2U})$
$\cdots$ AND $\cdots$ $(b_{nL} < A_n$ AND $A_n < b_{nU})$;

where b_{iL} is used to denote the lower bound of the attribute A_i and b_{iU} is used to denote the upper bound of the attribute A_i. This query can be easily rewritten in other forms as the following to use our AQP engine in MySQL, and the order of these parameters matters.

a. External Training and External Query:
   SELECT sys_eval("/mnt/c/Users/d/AQP b_{1L} b_{1U} b_{2L} b_{2U} $\cdots$ b_{nL} b_{nU}");

b. External Training and Internal Query:
   SELECT myAQP$(b_{1L}, b_{1U}, b_{2L}, b_{2U}, \cdots, b_{nL}, b_{nU})$;

3   System Overview

Figure 1 shows the architecture of the AQP module. There are two layers in the system, User and Database Server. In the layer of Users, people send the COUNT aggregation query with selection conditions mentioned above to database servers and receive the approximate responses from these servers. In the layer of Database Server, there are several components. After receiving queries from users, the query engine starts to set up the environment and call AQP UDFs. When these UDFs are called, they begin to do their work inside the execution environment of database servers. To enable these AQP UDFs utilize Machine Learning, the external training program is an important component in Database Server Layer. It takes advantage of the neural network to learn the distribution of data provided by the training dataset generator so that the trained Machine Learning model can be used to answer the specified queries of users for the true distribution. Due to the separation of tasks, AQP UDFs consist of two main tasks which are loading the trained Machine Learning model and answering queries.

In addition to the overview and architecture we present above, the algorithms we adapt in our work to train the Machine Learning model and answer queries are discussed in Sect. 4, and the details of how we embed AQP inside database servers by UDFs are discussed in Sect. 5.

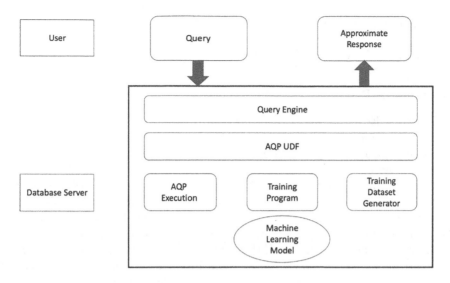

Fig. 1. Architecture of AQP system

4 Machine Learning Based AQP

Machine Learning plays an important role in our work. It helps us lower the workload of computations and reduce response times so that we are able to install this AQP inside MySQL. In Sect. 4.1, we present the mathematical theories and algorithms used in the training process. In Sect. 4.2, we discuss the steps of answering queries.

4.1 Neural Network Training

We use a Convolutional Neural Network (CNN) [11] to learn the distribution of data based on Inclusion–exclusion principle.

$$
\begin{aligned}
P[&(b_{1L} < A_1 < b_{1U})\ AND\ (b_{2L} < A_2 < b_{2U})] \\
&= (P[A_1 < b_{1U}] - P[A_1 < b_{1L}]) * (P[A_2 < b_{2U}] - P[A_2 < b_{2L}]) \\
&= P[A_1 < b_{1U}\ AND\ A_2 < b_{2U}] - P[A_1 < b_{1L}\ AND\ A_2 < b_{2U}] \\
&\quad - P[A_1 < b_{1U}\ AND\ A_2 < b_{2L}] + P[A_1 < b_{1L}\ AND\ A_2 < b_{2L}]
\end{aligned}
\tag{1}
$$

Equation 1 demonstrates the main idea of our work on an example relation which has two attributes. First, we create a list composed of two upper bounds for corresponding attributes, such as b_{1U} and b_{2U}. In addition to these two upper bounds, we also append one probability that we obtain the data point in the range, shown as $P[A_1 < b_{1U}\ AND\ A_2 < b_{2U}]$ in Eq. 1. With many inputs composed of different bounds and probabilities in the training set, the Machine Learning model is able to return the predicted probability that the data points satisfying other selection conditions such as $P[A_1 < b_{1L} \ AND \ A_2 < b_{2L}]$ in this example. In our work, we extend this model to support any relations containing finite numerical columns with real numbers.

4.2   Queries Process

As shown by Eq. 1, calculating the probability that the data points are satisfying all selection conditions is equivalent to do arithmetical operations with probabilities that data points are lower than some upper bounds. Therefore, we decompose selection conditions of queries into all possible combinations as inputs of the trained model after shuffling. For relations with 3 attributes, it generates 2^3 combinations since every attribute has two bounds, 2^4 combinations for 4 attributes and so on. More calculations are needed for more attributes. The purpose of shuffling these combinations is to prevent overfitting by making the connections more arbitrary. With outputs of the trained model, we calculate every corresponding sign and compose them into one real number as the output of this AQP module. This procedure is illustrated by Algorithm 1.

Algorithm 1. Decomposition and composition

Input: a list with bounds for corresponding columns in the order mentioned before
Output: Composed prediction r
 1: Initialization: load trained model M
 2: d ← Decompose Input
 3: s ← Shuffle d
 4: r ← Obtain the prediction from M with s

---

5   Embedding in Database by UDF

We choose MySQL as the platform to embed our AQP engine through MySQL's UDF. There are two different implementations demonstrated in Fig. 2 that return the same results with different performance. External Training and External Query (ETEQ) shown in Fig. 2. a) utilizes an open source MySQL UDF library lib_mysqludf_sys[1] to execute the external AQP module with a pre-trained Machine Learning model. The details of ETEQ are discussed in Sect. 5.1. External Training and Internal Query (ETIQ) shown in Fig. 2. b) embeds the whole AQP module inside MySQL as one UDF with the same pre-trained Machine Learning model. The details of ETIQ are discussed in Sect. 5.2.

---

[1] Github repository of lib_mysqludf_sys, https://github.com/mysqludf/lib_mysqludf_sys.

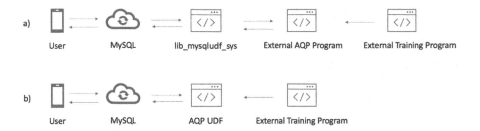

Fig. 2. Two different implementations of AQP UDFs

5.1   ETEQ

In this implementation, the training dataset generator and Machine Learning training program are designed externally. The training dataset generator is used to find the distribution of data. With the information it generates, Machine Learning training program starts to calculate the optimal parameters of the model that the AQP UDFs will use later. Another part of this implementation is mainly to execute the external program by MySQL through the open source MySQL UDF library lib_mysqludf_sys. In our work, we mainly use the UDF sys_eval in the library. The basic idea of sys_eval is to fork a child process in the execution environment of MySQL to execute the command and read its output. This UDF takes in one argument that is a string consisting of the path of our AQP module and selection conditions.

5.2   ETIQ

ETEQ and ETIQ share the same training dataset generator and Machine Learning training program, but they differ in the execution of AQP UDFs. The AQP UDF ETEQ needs to utilize the open source MySQL UDF library lib_mysqludf_sys to execute the external AQP module, which produces unnecessary resource waste. Besides, its usage is inconvenient for users since the return value of the AQP UDF with ETEQ is a string that requires more work to do the conversion if users want to do other operations. Moreover, it only accepts one formatted string as the argument. In order to alleviate these issues, we implements the AQP module as one whole UDF so that MySQL can directly execute the AQP module without the UDF library lib_mysqludf_sys. This implementation helps users easily combine functionalities of the AQP UDF with their programs. Since the AQP module is directly executed by MySQL, it can take advantage of MySQL's optimization, which provides a huge speedup compared to other implementations.

6   Implementation and Evaluation

6.1   Settings

Hardware. All of our work are done on the computer with one CPU Intel i7-6820HK, one GPU NVIDIA GeForce GTX1070 and 2400 MHz 32GB memory.

System. All of our work are done on Windows Subsystem for Linux 2 (WSL2) with Ubuntu 18.04.

Programming Language. The external Machine Learning training program is written with Python 3.8.5. The external AQP module and the AQP UDF are programmed as specified by MySQL official documents in C++14.

Package. The build system is CMake 3.19.2. The compiler is gcc 7.5.0. Pytorch Stable 1.8.1, Cudatoolkit 11.0 on WSL2, and Cudnn 8.2.0 are used. In our work, Boost 1.69.0 and Numcpp 2.4.2 [17] are also used.

The code has been published on Github[2].

6.2   Datasets

In the experiment, we use the function numpy.random.rand to randomly sample ten million data points from a uniform distribution over [0, 1), and these data points form a synthetic dataset containing 1 million unique records with 10 numerical columns as the distribution of data. In this way, the synthetic dataset is already normalized so that every element of this dataset is in the range between 0 and 1 for convenience. Besides, we also use the TPC-H dataset. For this dataset, we convert two columns containing dates to integers representing amounts of seconds starting from 1970 1-1 00:00:00, and normalize every column so that every element in each column is in the range between 0 and 1. In addition to these two columns, we choose another eight numerical columns together to form the population. To deal with the diversity of queries, we create the training set by randomly generating many queries with their exact results.

6.3   Training Process

In our work, we adopt Least Absolute Deviations as the loss function:

$$Loss(x,y) = \frac{1}{n} \sum_{i=1}^{n} |y_i - f(x_i)| \qquad (2)$$

where n is the batch size, x is the input and y is the target. We select 16 as the batch size. The initial learning rate is set to 0.1. We use the Adadelta optimizer with epochs from 1 to 1000 for training. We set the dropout rate as 0.5 in order to prevent overfitting.

---

[2] Github repository of AQP UDFs, https://github.com/thu-west/Learned-AQP.

6.4   Experiment Results

We randomly generate 1000 queries as specified in Sect. 2, and use them on both the synthetic dataset and TPC-H dataset to test performances of AQP UDFs with ETEQ and ETIQ. In addition to these 1000 queries, we also randomly generate five example queries to demonstrate the process of evaluation.

Metrics. In the evaluation, we use two metrics: 1) the error of the estimate in every query; 2) the response times of exact processing, approximate processing with ETEQ, and approximate processing with ETIQ. The error ϵ is the absolute value of the absolute error calculated by $\epsilon = |\hat{\theta} - \theta|$ where $\hat{\theta}$ is the estimate of AQP and θ is the exact result. The response times in seconds are measured by calculating when we receive return values.

Example Queries. There are 5 example queries that are randomly generated by numpy. The following is the original form of Query 1. In Table 1, we show lower bounds and upper bounds of corresponding attributes by open intervals. Table 2 shows results of them for the AQP UDF ETEQ and ETIQ on the synthetic dataset in terms of return values and execution times. Table 3 shows results of them for the AQP UDF ETEQ and ETIQ on the TPC-H dataset in terms of return values and execution times.

Query 1: select count(*) * 100/1000000 from Data where (0.145 < attr1 and attr1 < 0.74) and (0.165 < attr2 and attr2 < 0.51)and (0.145 < attr3 and attr3 < 0.31) and (0.08 < attr4 and attr4 < 0.44)and (0.09 < attr5 and attr5 < 0.57)and (0.43 < attr6 and attr6 < 0.99)and (0.22 < attr7 and attr7 < 0.62)and (0.12 < attr8 and attr8 < 0.48)and (0.02 < attr9 and attr9 < 0.97)and (0.01 < attr10 and attr10 < 0.82);

Table 1. Five example queries

Item	Query 1	Query 2	Query 3	Query 4	Query 5
Attribute 1	(0.145,0.74)	(0.09,0.44)	(0.19,0.84)	(0.025,0.9)	(0.255,0.96)
Attribute 2	(0.165,0.51)	(0.34,0.87)	(0.01,0.42)	(0.32,0.64)	(0.15,0.58)
Attribute 3	(0.145,0.31)	(0.28,0.79)	(0.28, 0.63)	(0.15,0.87)	(0.075,0.38)
Attribute 4	(0.08,0.44)	(0.04, 0.61)	(0.0,0.77)	(0.035,0.46)	(0.18,0.81)
Attribute 5	(0.09,0.57)	(0.18,0.55)	(0.395,0.88)	(0.23,0.65)	(0.255,0.78)
Attribute 6	(0.43,0.99)	(0.065,0.16)	(0.315,0.97)	(0.19,0.87)	(0.14,0.58)
Attribute 7	(0.22,0.62)	(0.15,0.64)	(0.04,0.45)	(0.4,0.96)	(0.405,0.84)
Attribute 8	(0.12,0.48)	(0.165,0.72)	(0.095,0.3)	(0.32,0.97)	(0.03,0.32)
Attribute 9	(0.02,0.97)	(0.235,0.62)	(0.155,0.45)	(0.065,0.98)	(0.3,0.75)
Attribute 10	(0.01.0.82)	(0.1,0.4)	(0.04,0.91)	(0.05,0.33)	(0.185,0.41)

ETEQ and ETIQ on the Synthetic Dataset. The statistics for the AQP UDF ETEQ on the synthetic dataset with the mentioned 1000 queries are the following. The average error between the estimates our AQP UDF ETEQ calculates and the exact results is 0.084. The minimum error is 0, and the maximum error is 2.082738.

The statistics for the AQP UDF ETIQ on the synthetic dataset with the mentioned 1000 queries are the following. Since both of these two AQP UDFs share the same algorithm, they return the same estimate when they receive the same query. Therefore, the average, minimum and maximum of errors are almost the same. The tiny differences are due to the data type conversion.

The response time of ETEQ is approximately 26 times longer than that of ETIQ on average even though the first execution of ETIQ will take about 4 s to write data to MySQL's buffer, which shows a huge speedup.

Table 2. Comparison of ETEQ and ETIQ on the synthetic dataset

Item	Query 1	Query 2	Query 3	Query 4	Query 5
Exact result	0.0406	0.0059	0.0496	0.2246	0.0178
Estimate	0.05264	0.00767	0.0205	0.0236	0.01216
Error	0.01204	0.00177	0.0291	0.201	0.00564
Duration of exact result	1.51 s	0.60 s	0.72 s	0.67 s	0.59 s
Duration of estimate (ETEQ)	3.64 s	3.51 s	3.99 s	3.30 s	3.38 s
Duration of estimate (ETIQ)	0.42 s	0.22 s	0.24 s	0.23 s	0.24 s

ETEQ and ETIQ on the TPC-H Dataset. The statistics for the AQP UDF ETEQ on the TPC-H dataset with the mentioned 1000 queries are the following. The average error between the estimates our AQP UDF ETEQ calculates and the exact results is 0.090. The minimum error is 0, and the maximum error is 2.3835.

The statistics for the AQP UDF ETIQ on the TPC-H dataset with the mentioned 1000 queries are the following. Similar to the AQP UDFs on the synthetic dataset, the average, minimum and maximum of errors are almost the same as these of the AQP UDF ETEQ, and the tiny differences are due to the data type conversion.

The response time of ETEQ is approximately 22 times longer than that of ETIQ on average even though the first execution of ETIQ will take about 5 s to write data to MySQL's buffer, which is consistent with the performances of ETEQ and ETIQ on the synthetic dataset.

Table 3. Comparison of ETEQ and ETIQ on the TPC-H dataset

Item	Query 1	Query 2	Query 3	Query 4	Query 5
Exact result	0.0039	0.0100	0.0628	0.1950	0.0181
Estimate	0.0421	0.0032	0.02066	0.0346	0.0144
Error	0.0382	0.0068	0.04214	0.1604	0.0037
Duration of Exact result	1.80 s	0.88 s	0.92 s	0.87 s	0.87 s
Duration of estimate (ETEQ)	10.49 s	5.43 s	5.50 s	5.72 s	5.52 s
Duration of estimate (ETIQ)	0.37 s	0.28 s	0.30 s	0.28 s	0.28 s

Figure 3 shows the performance of ETIQ is dramatically better than that of ETEQ on the TPC-H dataset in the long run. In other words, ETEQ needs much more time to execute with incoming queries. There are several possible reasons to explain this trend. First, ETIQ does not need to utilize MySQL UDF library lib_mysqludf_sys so that it has a lower workload. Second, MySQL identifies the repeated computations so that it optimizes the work of ETIQ after its first execution. Lastly, MySQL stores the loaded Machine Learning model or other necessary files inside buffers, which avoids a part of executions of IO operations.

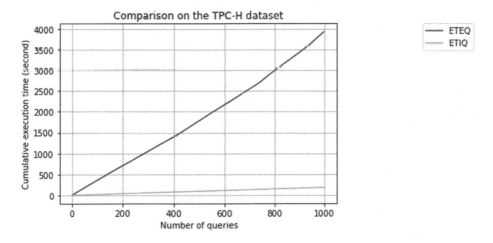

Fig. 3. Cumulative execution times of the AQP UDF ETEQ and ETIQ on the TPC-H database

7   Related Work

Early researches on AQP focus on sampling and synopses of data on the algorithms of AQP. Chakrabarti et al. [6] identified the limitations of these early researches, and proposed a new approach based on building wavelet-coefficient synopses of the data to solve them. Rather than proposing new approaches, Babcock et al. [5] revised the technique of sampling to provide approximate answers with higher quality by dynamically constructing an appropriately biased sample. Agarwal et al. [3] discovered a problem that error bars generated by the existing work in the sampling-based AQP are inaccurate for real query workloads, and proposed a query approximation pipeline to deal with it. With Machine Learning being popular, more researchers study on Machine Learning based AQP. Savva et al. [18] developed a lightweight ML-led system that learns previously answered queries to answer new queries so that it has low response times, low workloads of computation and hence low usage of energy. Thirumuruganathan et al. [19] utilized trained deep generative models with a rejection sampling based approach to improve accuracy. In our work, we collect information of the distribution of data to train the Machine Learning model, and use the pre-trained model to answer queries. In this way, we can make the size of our model very small so that we are able to embed the Machine Learning based AQP inside RDBMS.

UDF provides a good interface for users to extend the functionalities of databases. The SQL servers of Google and Microsoft [8,15] provide AQP functions. These functions have some limitations although their usages are similar to these of UDF. The SQL servers of Microsoft can only approximately count distinct values. In addition to counting distinct values, these of Google are able to approximately calculate boundaries and top elements. However, these functions cannot compute the percentage that we want. MADlib Analytics Library [9] takes advantage of UDF to embed a plenty of Machine Learning algorithms. These functions are written in C++ and Python to do in-database analysis. In our work, we implement the external training program and internal AQP module to reduce the workload of server. In the later research, Zhang et al. [20] proposed a system Parrot that uses Stanford NLP and Jieba and utilized AQP techniques and UDF. Their work is similar to ours, but we have different goals. They focused on analysis of the term frequency on large text data, and our goal is to approximately calculate the percentage of data points that satisfy all selection conditions. Another difference is that they did not embed the whole module inside database, but we aim to do so.

8   Conclusion

This paper proposes an idea of embedding AQP inside database servers. In our work, we present two different implementations to embed AQP inside MySQL through UDFs to help users execute AQP inside database servers and pre-train the Machine Learning model externally. ETEQ utilizes a MySQL UDF library lib_mysqludf_sys to execute the external AQP module inside MySQL. In ETIQ, the AQP module is designed as one UDF function fully installed on MySQL and works in the same way as built-in functions. Besides, the AQP UDF with ETIQ takes advantage of MySQL's optimization with a huge speedup compared to ETEQ. Both two implementations enable users to avoid unnecessary work by approximately answering the specified kind of aggregation queries inside databases, so users can easily merge the functionalities of them in other tools of databases such as stored procedures, triggers, and so on to achieve further goals.

Although these proposed designs have their advantages, future work is needed to improve them. The first direction is the synchronization of the training of the Machine Learning model with the continuously updated databases. Currently, we separate the training of Machine Learning models and the AQP module. It will be more convenient for users if the AQP module can call the external training program to update the existing model when necessary. Secondly, the algorithms of the Machine Learning training program could be improved, so models can give closer estimates for extreme boundaries.

Acknowledgments. This work was supported by National Key R&D Program of China(2020AAA0109603), State Key Laboratory of Computer Architecture (ICT, CAS) under Grant No. CARCHA202008.

References

1. MySQL::extending MySQL 5.7::adding functions to MySQL::adding a loadable function. https://dev.mysql.com/doc/extending-mysql/5.7/en/adding-loadable-function.html
2. TPC-H benchmark. http://www.tpc.org/tpch/
3. Agarwal, S., Milner, H., Kleiner, A., et al.: Knowing when you're wrong: building fast and reliable approximate query processing systems. In: Proceedings of 2014 ACM SIGMOD International Conference on Management of Data, pp. 481–492 (2014)
4. Agarwal, S., Mozafari, B., Panda, A., et al.: BlinkDB: queries with bounded errors and bounded response times on very large data. In: 8th Eurosys Conference 2013, pp. 29–42 (2013)
5. Babcock, B., Chaudhuri, S., Das, G.: Dynamic sample selection for approximate query processing. In: Proceedings of 2003 ACM SIGMOD International Conference on Management of Data, pp. 539–550 (2003)
6. Chakrabarti, K., Garofalakis, M., Rastogi, R., et al.: Approximate query processing using wavelets. VLDB J. **10**(2), 199–223 (2001)
7. Chaudhuri, S., Das, G., Narasayya, V.: Optimized stratified sampling for approximate query processing. ACM Trans. Database Syst. **32**(2), 9-es (2007)
8. Google: Approximate aggregate functions in standard SQL. https://cloud.google.com/bigquery/docs/reference/standard-sql/approximate_aggregate_functions
9. Hellerstein, J., Re, C., Schoppmann, F., et al.: The MADlib analytics library or mad skills, the SQL (2012). http://arxiv.org/abs/1208.4165
10. Huang, S., Huang, M., Zhang, Y., et al.: Under water object detection based on convolution neural network. In: WISA, pp. 47–58 (2019)
11. Krizhevsky, A., Sutskever, I., Hinton, G.E.: Imagenet classification with deep convolutional neural networks. Commun. ACM **60**(6), 84–90 (2017)
12. Li, K., Li, G.: Approximate query processing: what is new and where to go? - A survey on approximate query processing. Data Sci. Eng. **3**(4), 379–397 (2018)
13. Li, K., Zhang, Y., Li, G., et al.: Bounded approximate query processing. IEEE Trans. Knowl. Data Eng. **31**(12), 2262–2276 (2019)

14. Ma, Q., Triantafillou, P.: DBEst: revisiting approximate query processing engines with machine learning models. In: Proceedings of 2019 International Conference on Management Data, SIGMOD Conference 2019, pp. 1553–1570 (2019)
15. Microsoft: intelligent query processing in SQL databases. https://docs.microsoft.com/en-us/sql/relational-databases/performance/intelligent-query-processing?view=sql-server-ver15
16. Olma, M., Papapetrou, O., Appuswamy, R., et al.: Taster: self-tuning, elastic and online approximate query processing. In: 2019 IEEE 35th International Conference on Data Engineering, ICDE, pp. 482–493 (2019)
17. Pilger, D.: NumCpp: a templatized header only C++ implementation of the python NumPy library. https://dpilger26.github.io/NumCpp
18. Savva, F., Anagnostopoulos, C., Triantafillou, P.: ML-AQP: query-driven approximate query processing based on machine learning (2020)
19. Thirumuruganathan, S., Hasan, S., Koudas, N., et al.: Approximate query processing for data exploration using deep generative models. In: 2020 IEEE 36th International Conference on Data Engineering, ICDE, pp. 1309–1320 (2020)
20. Zhang, Y., Zhang, H., He, Z., Jing, Y., Zhang, K., Wang, X.S.: Progressive term frequency analysis on large text collections. In: Nah, Y., Cui, B., Lee, S.-W., Yu, J.X., Moon, Y.-S., Whang, S.E. (eds.) DASFAA 2020. LNCS, vol. 12113, pp. 158–174. Springer, Cham (2020). https://doi.org/10.1007/978-3-030-59416-9_10

Improving Medical Record Search Performance by Particle Swarm Optimization Based Data Fusion Techniques

Qiuyu Xu and Shengli Wu[✉]

School of Computer Science, Jiangsu University, Zhenjiang 212013, China
swu@ujs.edu.cn

Abstract. In this paper, we aim to improve the performance of electronic medical record search by fusing results from multiple search engines. We propose a particle swarm optimization-based data fusion method. To evaluate it effectiveness, experiments are carried out with two data sets from the TREC medical track in 2011 and 2012. We find that on average the proposed method outperforms the two traditional data fusion methods CombSum and CombMNZ and three different linear combination methods: multiple linear regression, learning to rank, and the genetic algorithm. An analysis is given to explain why the particle swarm optimization algorithm outperforms the others.

Keywords: Information search · Medical record search · Data fusion · Particle swarm optimization

1 Introduction

In the last two decades, with the prevalence of the paperless office in the medical domain, Electronic Medical Records (EMRs) also known as Electronic Health Records (EHRs) or Electronic Patient Records (EPRs) are very popular. These refer to digital documents produced by doctors and practitioners, which often include information relating to many aspects of the patients in their care: conditions, various kinds of examination results, medical treatments, therapeutic effects, etc. Such information can be used for clinical decision support, various types of secondary uses, and so on. Although many parts of EMRs can be structured, it is still impossible to entirely do away with free text. How to access and retrieve those free text fields is an important function. With the rapid growth of EMRs, it is desirable to find the required information in them. Information search engine is a very useful tool that enables medical staff and other users to retrieve free texts in massive EMRs effectively and efficiently.

Some early medical retrieval systems are implemented with underlying database management systems [15, 26]. The limitation of this kind of retrieval systems is that it only works with structured data in databases. Therefore, some EMR retrieval systems are enhanced with a full-text search engine. One such example is EMERSE [16, 17]. Afterwards, different kinds of methods such as classification and weighting of different types of contents [18], medical ontology-based query expansion [5] and meta-data based query expansion [1] amongst others have been investigated so as to improve retrieval performance. As we know, query reformulation and expansion has been widely used in many information search tasks [36].

Data fusion, also known as meta-search, is a good approach for text retrieval [12, 13, 24, 32, 33]. For a given query, data fusion combines several resultant lists from different component retrieval systems into a new one. In this work, we investigate its use in the medical search task. Especially, we focus on linear combination and using four optimization methods, multiple linear regression, learning to rank, generic algorithm and particle swarm optimization, to find suitable weights for linear combination. According to our knowledge, this is the first time that a large number of results are fused for medical retrieval; and the particle swarm algorithm is the first time to be used for such a task. Extensive experimentation shows that data fusion can be used to for achieving improved performance in EMR retrieval. The Particle Swarm optimization is a very good option for weight assignment of the linear combination method.

The rest of this paper is organized as follows. Section 2 reviews related work. Section 3 describes the data fusion methods used in this study. Section 4 discusses experimental setups and results. Section 5 is the conclusion.

2 Related Work

We review previous work in two separate parts: EMR retrieval and data fusion in information retrieval.

2.1 EMR Retrieval

Research on EMR retrieval began in the 1990s. Some early work concerns the frequent ambiguity of and relationships amongst medical terms. Evans et al. [11]. aimed to establish the foundation for the canonical representation of medical concepts. Campbell et al. [6] proposed a general framework for representing clinical data, thus enabling developers to define a declarative semantics of terms to express the relationships among terms or combinations of terms. Some other systems were developed in different organizations [15, 16, 26].

Different measures have been investigated to improve retrieval performance. Díaz-Galiano et al. medical ontology-based approach was taken in both [9] and [4]. Wang et al. proposed a set of features to capture the importance of the terms and assigned different weights for different query words [31]. Wang et al. tried both pseudo-relevance feedback and structured knowledge bases to expand user queries [30]. Amini et al. investigated a meta-data based approach for EMR search [1] utilising the International Classification of Diseases (**ICD**), which is a medical classification list defined by the World Health Organization [1]. Soldaini et al. investigated an approach of adding more appropriate expert expressions to queries for medical information [27]. Durão et al. utilized tags to enhance information retrieval capabilities and expanded users' original queries with qualified tag neighbors [10].

Apart from retrieval effectiveness, some other aspects have also been investigated. Natarajan et al. [22] carried out user study by analysing user log files from an EMR-based, free-text search engine. Quantin et al. [26] proposed a distributed approach that is able to provide confidentiality of patients rather than creating a national centralised medical record system.

Medical record retrieval has been the subject of some information retrieval evaluation conferences. Both TREC and CLEF have held medical records-related evaluation workshops for several years. TREC had medical tracks in 2011 and 2012 [29] and had clinical decision support tracks in 2014 and 2015. CLEF has an e-health track for many years.

54 research groups altogether participated in the 2011 and 2012 TREC medical track over the two years, trying a myriad of different tactics to enhance performance. In this work, we use two groups of results (which are referred to as runs in TREC) submitted to the medical track in 2011 and 2012.

2.2 Data Fusion

Data fusion, which combines multiple ranked lists of documents into one, is used to improve retrieval performance [34]. Let us suppose there is a collection of documents C and n information search systems $ir_1, ir_2, \ldots, ir_n$. For a given query q, each of n information search systems ir_i provides a ranked list of documents L_i ($1 \leq i \leq n$). These multiple ranked lists of documents henceforth referred to as component results. Correspondingly, the retrieval systems that generated the component results are referred to as component systems. According to their treatment of component results, data fusion methods can be classified into equally treated and biased methods. As their names suggest, the former treats all component results equally, while the latter may treat every component system in a different manner. Borda [2], Condorcet [21], CombSUM [13, 20] and CombMNZ [13, 20] are equally treated methods, while linear combination [28, 33], Weighted-Borda [2], and Weighted Condorcet [35] are biased methods. In an ideal situation that the all component results perform equally well, and the difference between each pair of the ranked lists is the same, then equally-treated methods should perform well; otherwise, biased methods are better choices [34]. Linear combination is a typical biased method and affords a high flexibility for dealing with more complex situations. However, a key issue for linear combination is weight assignment for all the component systems involved.

In this paper, we investigate how to apply the data fusion technique to medical record retrieval. Especially we focus on linear combination and compare four different weight assignment methods: multiple linear regression, learning to rank, genetic algorithm, and particle swarm optimization algorithm.

3 Data Fusion for Information Search

In this section we introduce all the data fusion methods that we use, including CombSUM, CombMNZ and linear combination. For Linear Combination, four weight assignment methods are discussed.

3.1 Several Fusion Methods

Suppose there is a collection of documents C and a group of n retrieval systems IR $=$ $\{ir_i\}$ $(1 \leq i \leq n)$. For a given query q, every retrieval system ir_i search C and provide a ranked list of documents $L_i = \; < d_{i1}, d_{i2}, ..., d_{il} >$. Each document is associated with a relevance related score, which is used to rank the documents in the resultant list.

- CombSUM is defined as $g(d) = \sum_{i=1}^{n} s_i(d)$, where n represents the number of component results that need to be fused and $s_i(d)$ is the score that retrieval system ir_i assigns to document d. $g(d)$ is the sum of scores that document d obtains. Every document d will be ranked in the fused list according to its $g(d)$.
- CombMNZ is defined as $g(d) = v * \sum_{i=1}^{n} s_i(d)$, where v refers to the number of component retrieval systems for which document d is retrieved.
- The linear combination method is defined as $g(d) = \sum_{i=1}^{n} w_i * s_i(d)$, where w_i is the weight assigned to retrieval system ir_i. Linear combination is a generalized format of CombSUM. In Eq. 3, if all weights w_i $(1 \leq i \leq n)$ are set to 1, then linear combination becomes CombSUM. All the weights need to be set before data fusion takes place.

3.2 Weight Assignment for the Linear Combination Method

How to set weights for those retrieval systems is a key question for the success of linear combination. In the following we discuss four different ways of weights assignment.

Multiple Linear Regression. It can be used for weights assignment [33]. Suppose there are m queries and n information retrieval systems. For every query q_j each information retrieval system ir_i retrieves l documents and each retrieved document d_k $(1 \leq k \leq l)$ has a relevance-related score s_{ik}^{j}. y_k^{j} is the human-judged relevance score of document d_k for query q_j. If binary relevance judgment is used, then 1 is assigned to relevant documents and 0 to irrelevant documents. Note that all the values of s_{ik}^{j} $(1 \leq i \leq n, 1 \leq j \leq m, 1 \leq k \leq l)$ and y_k^{j} $(1 \leq j \leq m, 1 \leq k \leq l)$ are known within a training data set. By minimizing u in the following equation

$$u = \sum_{j=1}^{m} \sum_{k=1}^{l} \left[y_k^{j} - \left(\hat{\beta}_0 + \hat{\beta}_1 s_{1k}^{j} + \hat{\beta}_2 s_{2k}^{j} + \cdots + \hat{\beta}_i s_{ik}^{j} + \cdots + \hat{\beta}_n s_{nk}^{j} \right) \right]^2 \quad (1)$$

We are able to calculate weights $\hat{\beta}_i (1 \leq i \leq n)$ for all component resultant lists. These weights are optimum in the sense of least squares that make the most accurate estimation of the relevance scores of all the documents in all the queries as a whole. After this process the weights obtained for all component retrieval systems can be used for fusion. Linear combination with such a weight assigment method is referred to as MR-Fusion later in this paper.

Learning to Rank Approach. It has been used in many applications. They can be divided into three categoris: pointwise, pairwise, and listwise. In this study, we also apply a listwise method, ListNet, [7] to the problem.

For a given query q, let us assume that there are α relevant documents in the whole collection C. Cross Entropy-based loss funtion is defined

$$L\left(y^j, z^j\right) = -\sum_{k=1}^{l} y_k^j log\left(z_k^j\right) \tag{2}$$

in which y_k^j is the jueged score and z_k^j is the score obtained by a linear combination of scores from all component retrieval systems, or $z_k^j = \hat{\beta}_0 + \hat{\beta}_1 s_{1k}^j + \hat{\beta}_2 s_{2k}^j + \cdots + \hat{\beta}_i s_{ik}^j + \cdots + \hat{\beta}_n s_{nk}^j$. y_k^j equals to $1/\alpha$ if document d_k is relevant; otherwise, it is zero. Any component results are normalized by the sum-to-one method. Set the objective of learing as minimizing $L(y^j, z^j)$, we may use a linear neural network model to find the parameters required: $\hat{\beta}_0, \hat{\beta}_1, \hat{\beta}_2, \ldots, \hat{\beta}_i, \ldots, \hat{\beta}_n$.

The Genetic Algorithm. It can be used to train the weights for all component retrieval systems [3, 14]. Genetic algorithms belong to the larger class of evolutionary algorithms, which generate solutions to optimization problems using techniques inspired by natural evolution. In the genetic algorithm, each solution is referred to as an individual and is represented by a binary string. We need to define a fitness function. Thus each individual has a fitness score, which indicates how good the solution is. First a given number of individuals are initialized, then we generate new generations from existing individuals by a group of operations including selection, crossover, mutation and elitism. See [14] for more details.

Linear combination with such a weight assignment method is referred to as GA-Fusion later in this paper.

3.3 Particle Swarm Optimization Based Linear Combination

The Particle Swarm Optimization Algorithm [23] Suppose there is a n-dimensional space which is the solution space of the problem to be optimized. We have a group of candidate solutions, or particles in the solution space. These particles can move around in the search space based on simple mathematical formulae over the position and velocity of these particles. Each particle's movement is influenced by its local best-known position and the best-known positions of other particles in the search space. It is expected that the swarm of particles move toward the best solutions.

At the beginning of the algorithm, a set of t particles is initialized randomly, which are, for the sake of brevity, denoted by their starting positional vectors $X_i = (x_1^i, x_2^i, \ldots, x_n^i)$ for $(1 \leq i \leq t)$. These particles move in the search space at velocity $V_i = \left(v_1^i, v_2^i, \ldots, v_n^i\right)$ for $(1 \leq i \leq t)$. Each particle's performance is evaluated by a fitness function, which is related to the problem to be solved. The best previous position of particle X_i is represented as $P_i = (p_1^i, p_2^i, \ldots, p_n^i)$. The best particle among all particles is represented as $P_g = (p_1^g, p_2^g, \ldots, p_n^g)$. The velocity and position of particle X_i is updated by the equations below

$$v_j^i(u + 1) = \omega * v_j^i(u) + r_1 * c_1\left(p_j^i - x_j^i(u)\right) + r_2 * c_2\left(p_j^g - x_j^i(u)\right) \tag{3}$$

$$x_j^i(u + 1) = x_j^i(u) + v_j^i(u + 1) \tag{4}$$

where $i = 1, 2, \cdots, t, j = 1, 2, \cdots, n$, and u is the number of generations that have been produced. $v_j^i(u)$ represents the j-th velocity component of particle X_i after u generations. $x_j^i(u)$ represents the position component of particle X_i on dimension j after u generations. $random_1$ and $random_2$ are two random numbers in the range of $[0, 1]$. c_1 and c_2 are learning factors and ω is the inertia weight. c_1, c_2 and ω control the behaviour and effectiveness of PSO. At each generation, all t particles need to be evaluated by the fitness function to decide their goodness.

Careful consideration is needed for the setting of ω. It is better to start with a large value of ω and reduce it gradually over generations. At an earlier stage, a large value of ω leads to more exploration in the solution space for a possibly globally optimized solution. And at later stages, a small value of ω leads to the focus on the neighbouring area for a locally optimized solution. We use the following equation to set a dynamic value for ω:

$$\omega(u) = \omega_{max} - (\omega_{max} - \omega_{min}) * \frac{u}{u_{max}} \tag{5}$$

where ω_{max} and ω_{min} represent the maximum and minimum values of ω, respectively. u_{max} represents the maximum number of iteration.

To guarantee the particle's searching in the valid search space, the maximum value of x_j^i should be set to 1. The maximum velocity, V_{max}, affects the convergence of the algorithm. If V_{max} is too small, then it may lead to slow convergence and stop at a local optimum; If V_{max} is too large, then it may lead to the particle overshooting the global optimum solution and fail to converge. The value of V_{max} is normally set to a range between 10% and 20% of x_j^i. The flowchart of the particle swarm optimization algorithm is shown in Fig. 1.

Just as for GA-fusion, we also use MAP as the measure for the fitness function of PSO. For any particle X, we may map its position to a set of weights $w = (w_1, w_2, \ldots, w_n)$. Using the same training data set as GA-fusion, we may fuse them by linear combination to obtain result R over m queries and $MAP(R)$ is defined as the fitness score of X. In the remaining part of this paper this method is referred to as PSO-Fusion.

4 Experiments

4.1 Experimental Setup

The Text REtrieval Conference (TREC, whose website is located at http://trec.nist.gov/) is an on-going series of workshops focusing on a variety of different information retrieval tasks. The electronic medical record retrieval task was held in the two consecutive years of 2011 and 2012. A corpus of medical documents, a group of queries and a collection of relevance judgments are provided. In both years, the corpus is a set of de-identified clinical reports that are obtained from the University of Pittsburgh NLP Repository. According to Voorhees and Hersh [29], this corpus contains one month of reports from multiple hospitals; and there are nine types of reports: Radiology Reports, History and Physicals, Consultation Reports, Emergency Department Reports, Progress Notes, Discharge Summaries, Operative Reports, Surgical Pathology Reports, and Cardiology Reports. Some example queries are:

136: Children with dental caries

137: Patients with inflammatory disorders receiving TNF-inhibitor treatment

152: Patients with Diabetes exhibiting good Hemoglobin A1c Control (<8.0%)

Several dozens of research groups participated in the tasks and submitted their retrieval results, which are called runs in TREC. In our study, we chose 24 top-ranked runs (measured by their MAP values) from all the runs submitted to the 2011 and 2012 electronic medical record retrieval tasks, respectively. In TREC, each research group is allowed to submit multiple runs to the same task. Those multiple runs submitted by the same research group are much similar than those submitted by different research groups. In order to reduce the unfavourable impact of similar results on fusion performance, we only take one run from each research group. In 2011, there are 35 queries from number 101 to number 135. In 2012, there are 50 queries from number 136 to number 185. Number130, 138, 159 and 166 are deleted because no relevant documents are found for these queries. Retrieved documents by any of the runs submitted are examined by medical experts to judge their relevance.

For all the documents in any given run, their raw scores are normalized by using the reciprocal scheme: $1/(rank + 60)$, where $rank$ is the ranking position of the document in the resultant list. According to Cormack et al., [8] this is a very good score normalization scheme. Note that the learning to rank approach is an exception. It uses the sum-to-one method.

For the linear combination method with any of the three different weight assignment methods, we divide all the queries into two groups A and B. The queries in each group are randomly selected and we make A and B include the same number of queries insofar as possible. First we use group A for weight training and B for fusion test. Then we swap the positions of A and B. This is referred to as the two-fold cross validation in machine learning and statistical analysis. All the fusion results are evaluated using four metrics: MAP, P@10, P@30 and RP. During the training process, we use the top 100 documents in the resultant list of each query.

GA-Fusion needs to set a few parameters. In our experiment the probability for the crossover operation p_c is set to a typical value of 0.7. Another parameter is the mutation operation probability p_m. Usually, p_m is set in the range of 0.01 and 0.1. In our experiment we start with $p_m = 0.2$ and reduce it by a factor of 0.8 every 10 generations. The size of the population is set to 30.

PSO-Fusion also needs to set some parameters. In Eq. 3, both c_1 and c_2 are both set to 2. In Eq. 5, ω_{max} is set to 0.9 and ω_{min} is set to 0.4. V_{max} is set to 0.2, while the number of particles t is set to 30.

4.2 Experimental Results

First of all, we combine all 24 runs selected in 2011 and 2012 with six different fusion strategies. Tables 1 and 2 show the results of the 2011 and 2012 groups, respectively. For both GA-Fusion and PSO-Fusion, we set 200 as the number of generations. For all the linear combination methods involved, we run 20 times and the figures in Tables 1 and 2 are the average of these 20 runs.

We have a few observations from Tables 1 and 2. First of all, both traditional data fusion methods CombSum and CombMNZ do not perform well. In TREC 2012, they are worse than the best component result for all four evaluation metrics. In TREC 2011, they are better than the best component result in three metrics including MAP (Mean average precision over all relevant documents), P@30 (Precision at 30 document level), and RP (Recall-level precision), but worse than the best result in P@10. Linear combination methods are more successful than both CombSum and CombMNZ. For MR-fusion, GA-fusion, and PSO-fusion, they are better than the best component result in both years. The only exception is in 2012, when none of them perform as well as the best component results in terms of P@10. ListNet is the least successful in all four linear combination methods. If we compare all four metrics, then data fusion, especially linear combination, is successful with MAP, RP, and P@30, but P@10 proves to be especially challenging.

We combine different numbers of component results to assess the impact of this on fusion performance. To enable fair comparisons, we fuse various number of component results but keep other conditions as constant as possible. For any given number t between 2 and 20, we randomly select a set of t runs 100 times with replacement for fusion test. The results are averaged and presented in Fig. 1 (MAP, 2011). We can see a common tendency of all the curves: they increase with the number of runs. This phenomenon shows that the number of component results has a positive effect on fusion performance. Due to space limitation, some other figures are not presented. It is noticeable some differences exist among different metrics.

If we look at the performance of each individual data fusion method, then PSO-fusion consistently achieves very good performance. In all the cases, it is always one of the

Table 1. Performance of fusing 24 runs submitted to the TREC 2011 medical record retrieval task (the best performance for each metric is shown in bold)

Method	MAP	P@10	P@30	RP
Best	0.5071	0.7265	0.5059	0.4999
CombSum	0.5484 (+8.14%)	0.6941 (−4.46%)	0.5706 (+12.79%)	0.5136 (+2.74%)
CombMNZ	0.5336 (+5.23%)	0.6853 (−5.67%)	0.5578 (+10.26%)	0.5008 (+0.18%)
ListNet	0.5537 (+9.19%)	0.7000 (−3.65%)	0.5765 (+13.96%)	0.5248 (+4.98%)
MR-fusion	0.5525 (+8.95%)	0.7400 (+1.86%)	0.5784 (+14.33%)	0.5307 (+6.16%)
GA-fusion	0.5465 (+7.77%)	**0.7412** (**+2.02%**)	0.5618 (+11.05%)	0.5237 (+4.76%)
PSO-fusion	**0.5841** (**+15.18%**)	**0.7412** (**+2.02%**)	**0.5961** (**+17.83%**)	**0.5574** (**+11.50%**)

best. When fewer than ten runs are fused, GA-fusion performs as well as PSO-fusion. However, when more than ten runs are fused, GA-fusion falls behind PSO-fusion and the difference between them increases with the number of runs. MR-fusion is the best in 2011 but not as good as PSO-fusion in 2012. ListNet is quite successful in 2011 but is not good in 2012. CombSum and CombMNZ are close, although more often CombSum outperforms CombMNZ.

Table 2. Performance of fusing 24 runs submitted to the TREC 2012 medical record retrieval task (the best performance for each metric is shown in bold)

Method	MAP	P@10	P@30	RP
Best	0.4610	**0.7489**	0.5376	0.4770
CombSum	0.4684 (+1.61%)	0.6447 (−13.91%)	0.5071 (−5.67%)	0.4489 (−5.89%)
CombMNZ	0.4614 (+0.09%)	0.6489 (−13.35%)	0.5128 (−4.61%)	0.4512 (−5.41%)
ListNet	0.4720 (+2.39%)	0.6383 (−14.77%)	0.5121 (-4.61%)	0.4562 (−5.41%)
MR-fusion	0.4990 (+8.24%)	0.7266 (−2.98%)	0.5801 (+7.91%)	0.4980 (+4.40%)
GA-fusion	0.5178 (+12.32%)	0.7298 (−2.55%)	**0.5950 (+10.68%)**	0.4985 (+4.51%)
PSO-fusion	**0.5297 (+14.90%)**	0.7298 (−2.55%)	0.5809 (+8.05%)	**0.5204 (+9.10%)**

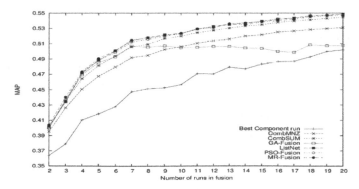

Fig. 1. Performance of fusing different number of runs (2011, MAP)

4.3 Analysis of the Four Learning Algorithms

For regression-based weight assignment, its optimization goal is to make the most accurate estimation of the relevance scores for all the documents involved in the fused results.

The goal of ListNet is similar to that of regression-based weight assignment. While for GA-based and PSO-based weight assignment, their optimization goal is to achieve the highest MAP value for the fused results. Because the goal of regression-based weight assignment and ListNet is not directly optimizing the MAP value, their effectiveness in MAP may not be as good as those that directly optimize the MAP value. However, one good property of regression-based weight assignment method is its efficiency over the others. On the other hand, ListNet does not have this good property.

Both GA-based and PSO-based weight assignment methods optimize MAP values directly. But we observe that PSO-fusion is more effective than GA-fusion in our experiment. It is interesting to explain why this happens. After careful consideration, we find there is one hidden problem in GA-based weight assignment. Recall in that process, we use binary strings of length $u*n$ for the search. They need to be mapped to weights of n component retrieval system: mapping every binary substring of u digits into a binary number [14]. There are two spaces: the space for search and the space for weighting. The relationship between these two spaces is not straightforward. Let us look at a simple example. Consider two strings 10000,00000 and 00000,10000 in the search space. They are quite similar because 8 out of 10 digits are the same. Note that all the digits in these two strings are equally important. However, if we map them to the weights of 2 component search systems, then the weights obtained for the former string are 32(10000) and 0(00000), and the weights for the latter string are 0(00000) and 32(10000). These two weightings are very different. Such an inconsistency between these two spaces may affect the effectiveness of crossover and mutation. In the end, the effectiveness and efficiency of the genetic algorithm are affected.

In PSO-based weight assignment method, each particle's position in the n dimensional search space can be transformed to a group of n weights naturally. This can explain why PSO-fusion is better than GA-fusion in both effectiveness and efficiency in our experiment.

5   Conclusion

In this paper, we have investigated the application of data fusion methods to medical record search. Experiments have been carried out with the data sets available from TREC. In our experiment, linear combination is able to beat the best component result in most cases, while traditional data fusion methods CombSum and CombMNZ do not generally perform as well as linear combination. Among the four weight assignment methods, particle swarm optimization is likely the best option. From the experiments, we conclude that data fusion is a useful technique to improve medical record search if used properly.

References

1. Amini, I., Martínez, D., Li, X., Sanderson, M.: Improving patient record search: a meta-data based approach. Inf. Process. Manage. **52**(2), 258–272 (2016). https://doi.org/10.1016/j.ipm.2015.07.005

2. Aslam, J.A., Montague, M.H.: Models for Metasearch. In: Proceedings of SIGIR, 2001, pp. 275-284 (2001). https://doi.org/10.1145/383952.384007

3. Bartell, B.T., Cottrell, G.W., Belew, R.K.: Automatic combination of multiple ranked retrieval systems. In: Proceedings of SIGIR, 1994, pp. 173-181 (1994). https://doi.org/10.1007/978-1-4471-2099-5_18

4. Bhatt, M., Rahayu, J.W., Soni, S.P., Wouters, C.: Ontology driven semantic profiling and retrieval in medical information systems. J. Web Sem. **7**(4), 317–331 (2009). https://doi.org/10.1016/j.websem.2009.05.004

5. Bhogal, J., MacFarlane, A., Smith, P.: A review of ontology based query expansion. Inf. Process. Manage. **43**(4), 866–886 (2007). https://doi.org/10.1016/j.ipm.2006.09.003

6. Campbell, K.E., Das, A.K., Musen, M.A.: Research paper: a logical foundation for representation of clinical data. JAMIA **1**(3), 218–232 (1994). https://doi.org/10.1136/jamia.1994.95236154

7. Cao, Z., Qin, T., Liu, T., Tsai, M., Li, H.: Learning to rank: from pairwise approach to list-wise approach. In: Proceedings of ICML, 2007, pp. 129-136 (2007). https://doi.org/10.1145/1273496.1273513

8. Cormack, G.V., Clarke, C.L.A., Büttcher, S.: Reciprocal rank fusion outperforms Condorcet and individual rank learning methods. In: Proceedings of SIGIR, 2009, pp. 758-759 (2009). https://doi.org/10.1145/1571941.1572114

9. Díaz-Galiano, M.C., Martín-Valdivia, M., López, L.A.U.: Query expansion with a medical ontology to improve a multimodal information retrieval system. Comp. Bio. Med. **39**(4), 396–403 (2009). https://doi.org/10.1016/j.compbiomed.2009.01.012

10. Durao, F., Bayyapu, K., Xu, G., Dolog, P., Lage, R.: Expanding user's query with tag-neighbors for effective medical information retrieval. Multimedia Tools Appl. **71**(2), 905–929 (2012). https://doi.org/10.1007/s11042-012-1316-5

11. Evans, D.A., Cimino, J.J., Hersh, W.R., Huff, S.M., Bell, D.S.: Position paper: toward a medical-concept representation language. JAMIA **1**(3), 207–217 (1994). https://doi.org/10.1136/jamia.1994.95236153

12. Farah, M., Vanderpooten, D.: An outranking approach for information retrieval. Inf. Retr. **11**(4), 315–334 (2008). https://doi.org/10.1007/s10791-008-9046-z

13. Fox, E.A., Shaw, J.A.: Combination of multiple searches. In: Proceedings of TREC, 1993, pp. 243-252 (1993)

14. Ghosh, K., Parui, S.K., Majumder, P.: Learning combination weights in data fusion using genetic algorithms. Inf. Process. Manage. **51**(3), 306–328 (2015). https://doi.org/10.1016/j.ipm.2014.12.002

15. William, M.G., Jim, J., Nancy, M.L., Dario, A.G.: StarTracker: an integrated, web-based clinical search engine. In: AMIA (2003)

16. David, A.H.: EMERSE: the electronic medical record search engine. In: AMIA (2006)

17. David, A.H., Qiaozhu, M., James, L., Ritu, K., Kai, Z.: Supporting information retrieval from electronic health records: a report of University of Michigan's nine-year experience in developing and using the electronic medical record search engine (EMERSE). J. Biomed. Inf. **55**, 290–300 (2015). https://doi.org/10.1016/j.jbi.2015.05.003

18. He, Y., Hu, Q., Song, Y., He, L.: Estimating probability density of content types for promoting medical records search. In: Ferro, N., et al. (eds.) ECIR 2016. LNCS, vol. 9626, pp. 252–263. Springer, Cham (2016). https://doi.org/10.1007/978-3-319-30671-1_19

19. King, B., Wang, L., Provalov, I., Zhou, J.: Cengage learning at TREC 2011 medical track. In: Proceeding of TREC (2011)

20. Lee, J.H.: Combining multiple evidence from different properties of weighting schemes. In: Proceeding of SIGIR, 1995, pp. 180-188 (1995). https://doi.org/10.1145/215206.215358

21. Montague, M.H., Aslam, J.A.: Condorcet fusion for improved retrieval. In: Proceeding of CIKM, 2002, pp. 538-548 (2002). https://doi.org/10.1145/584792.584881

22. Natarajan, K., Stein, D.M., Jain, S., Elhadad, N.: An analysis of clinical queries in an electronic health record search utility. I. J. Medical Informatics **79**(7), 515–522 (2010). https://doi.org/10.1016/j.ijmedinf.2010.03.004
23. Pedersen, M.E.H., Chipperfield, A.J.: Simplifying particle swarm optimization. Appl. Soft Comput. **10**(2), 618–628 (2010). https://doi.org/10.1016/j.asoc.2009.08.029
24. Prasath, R., Duane, A., O'Reilly, P.: Topic assisted fusion to re-rank texts for multi-faceted information retrieval. In: Banchs, R.E., Silvestri, F., Liu, T.-Y., Zhang, M., Gao, S., Lang, J. (eds.) AIRS 2013. LNCS, vol. 8281, pp. 97–108. Springer, Heidelberg (2013). https://doi.org/10.1007/978-3-642-45068-6_9
25. Quantin, C., Jaquet-Chiffelle, D., Coatrieux, G., Benzenine, E., Allaert, F.: Medical record search engines, using pseudonymised patient identity: an alternative to centralised medical records. Int. J. Med. Inf. **80**(2), e6–e11 (2011). https://doi.org/10.1016/j.ijmedinf.2010.10.003
26. Scully, K.W., et al.: Development of an enterprise-wide clinical data repository: merging multiple legacy databases. In: AMIA (1997)
27. Soldaini, L., Yates, A., Yom-Tov, E., Frieder, O., Goharian, N.: Enhancing web search in the medical domain via query clarification. Inf. Retr. J. **19**(1–2), 149–173 (2015). https://doi.org/10.1007/s10791-015-9258-y
28. Vogt, C.C., Cottrell, G.W.: Fusion via a linear combination of scores. Inf. Retr. **1**(3), 151–173 (1999). https://doi.org/10.1023/A:1009980820262
29. Voorhees, E.M., Hersh, W.R.: Overview of the TREC 2012 medical records track. In: Proceeding of TREC (2012)
30. Wang, H., Zhang, Q., Yuan, J.: Semantically enhanced medical information retrieval system: a tensor factorization based approach. IEEE Access **5**, 7584–7593 (2007). https://doi.org/10.1109/ACCESS.2017.2698142
31. Wang, Y., Lu, K., Fang, H.: Learning2extract for medical domain retrieval. In: Proceeding of AIRS, 2017, pp. 45-57 (2017). https://doi.org/10.1007/978-3-319-70145-5_4
32. Wei, F., Li, W., Liu, S.: iRANK: a rank-learn-combine framework for unsupervised en-semble ranking. JASIST **61**(6), 1232–1243 (2011). https://doi.org/10.1002/asi.21296
33. Wu, S.: Linear combination of component results in information retrieval. Data Knowl. Eng. **71**(1), 114–126 (2012). https://doi.org/10.1016/j.datak.2011.08.003
34. Shengli, W.: Data Fusion in Information Retrieval. Springer, Heidelberg (2012)
35. Wu, S.: The weighted Condorcet fusion in information retrieval. Inf. Process. Manage. **49**(1), 108–122 (2012). https://doi.org/10.1016/j.ipm.2012.02.007
36. Yan, W., Wang, Y., Huang, C., Wu, S.: Word embedding-based reformulation for long queries in information search. In: Wang, G., Lin, X., Hendler, J., Song, W., Xu, Z., Liu, G. (eds.) WISA 2020. LNCS, vol. 12432, pp. 202–214. Springer, Cham (2020). https://doi.org/10.1007/978-3-030-60029-7_19

Real-Time Aggregation Approach for Power Quality Data

Jun Fang[1,2(⊠)], Wentao Bai[1,2], and Xiaodong Xue[1,2]

[1] School of Information, North China University of Technology, Beijing 1000144, China
fangjun@ncut.edu.cn
[2] Beijing Key Laboratory on Integration and Analysis of Large-Scale
Stream Data, Beijing 100144, China

Abstract. Compliance verification and performance analysis of grid power quality are the main targets of state-wide power quality monitoring and analysis system in China. Real-time aggregation of power quality data is a prerequisite to achieve the targets. Since power quality data generated by over 10,000 monitors are extremely massive, data aggregations of different indicators meet great challenges for time-consuming. An aggregation framework with the incremental computing and approximate computing engine is proposed. The incremental computing methods of maximum, minimum, mean and variance functions are presented, as well as two different approximate computing methods for 95% probability value function. Performance analyses are carried out with real data.

Keywords: Power quality · Data aggregation · Incremental computing · Approximate computing

1 Introduction

Aiming to assess the power quality performance of power systems at different voltage levels and broader area [1], State Grid Corporation of China (SGCC) has constructed a state-wide Power Quality Monitoring and Analysis System (PQMAS) based on big data technologies. The objectives of PQMAS [2, 3] include compliance verification and performance analysis. Compliance verification compares aggregation values of special PQ indicators with limits given by national standards and regulatory specification.

The PQ indicators [4] include harmonic current, harmonic voltage, frequency deviation, voltage deviation, negative sequence current, long-time flicker, etc., with a total number of 2555. The aggregation dimension also includes time and space. The aggregate functions include maximum, minimum, mean and 95% probability value. As the amount of PQ data collected from over 10,000 monitors are increasing, the real-time processing has encountered great challenges. Taking 95% probability value aggregation as an example, values are sorted from large to small, the 5% maximum value is removed, and the remaining maximum value is the 95% probability value [5]. During the process, the sorting cost of massive data is great, especially when many indicators are involved.

© Springer Nature Switzerland AG 2021
C. Xing et al. (Eds.): WISA 2021, LNCS 12999, pp. 99–106, 2021.
https://doi.org/10.1007/978-3-030-87571-8_9

The emergence of real-time distributed computing platforms, such as Spark and Flink, provides scalable processing capacity. However, the performance is still poor when sorting big data. Incremental and approximate computations [6, 7] are increasingly being adopted for data analytics to achieve low-latency execution and efficient utilization of computing resources. Both paradigms rely on computing over a subset of data items instead of computing over the entire dataset.

A real-time aggregation approach for power quality data is proposed based on incremental and approximate computations. In Sect. 2, we propose a data aggregation computing framework which aims to facilitate data process. In Sect. 3, we present the details how to realize power quality data aggregation with special incremental and approximate computation methods. Some experiments results are discussed in Sect. 4.

2   Data Aggregation Computing Framework

Incremental computing relies on the reusing of intermediate results of sub-computations, and these memorized results across jobs. Approximate computing relies on representative sampling of the entire dataset to compute over a subset of data items. Such intermediates results and sampling data should be memorized in our framework.

According to the system requirements, the window type of real-time processing is time rolling window. The rolling time cycle includes one day, one month and one year. The aggregation indicator includes various current and voltage indexes. Based on the Spark real-time computing framework, the following data aggregation computing framework is designed, shown in Fig. 1.

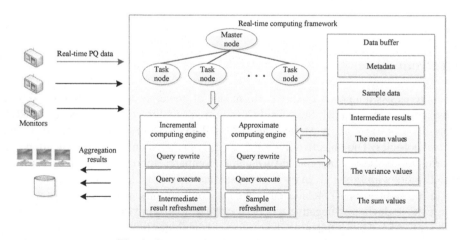

Fig. 1. Data aggregation computing framework

The master node receives real-time PQ data from multiple monitors, and then distributes the data evenly to multiple task nodes for parallel processing according to the hash algorithm. Each task node has a built-in computing module, which uses the incremental computing and approximate computing engine to read the cached data of aggregate results from data buffer and realize the aggregate functions of different indicators with different time range. The module compares the aggregation results with limits given by national standards and saves the results to the database or displays them in the foreground interface in real time. Here incremental computing engine and approximate computing engine make real-time data queries. Incremental computing engine updates the output incrementally instead of re-computing everything from scratch for successive runs of a job with input changes. Approximate computing engine returns an approximate output for a job instead of the exact output. The two engines are also responsible for updating intermediate result and refreshing samples.

3 Data Aggregation Approach

Definition 1 (aggregate query) given a relational dataset R, $\mathbf{D} = |\mathbf{R}|$ is the number of relational entries, $\mathbf{C} = \{c_i | i \in [1, m]\}$ is its attribute set. Given the aggregate column $C' \subseteq C$, its grouping is expressed as $\{g_i | i \in [1, n]\}$. Define th aggregate query Q as $\mathbf{R} \rightarrow R'$, where $R' = \{(g_i, Q(g_i))\}$. The input relationship is divided into groups according to the value of one or more attributes, and the aggregate function is applied in each group. For simplification, this paper does not consider the query predicate condition. As shown in the following example, given the relationship table powerQuality (location, parameter, value), the daily average value of 7th harmonic current of each location is formulized as.

Select location, avg(value) from powerQuality where parameter = 'hc7' group by location

In power quality data, the query aggregation Q mainly includes mean value, maximum value, minimum value, variance and 95% probability value. Next, we detailed the method of using incremental computing and aggregate computing to realize the above aggregate query.

3.1 Aggregate Query by Incremental Computing

Aggregation Functions such as maximum, minimum, and average values can be achieved by incremental computing. Suppose there are two groups of data values, one is historical value, $h_1, h_2, \cdots, h_M$, the other is incremental value, $h_{M+1}, h_{M+2}, \cdots, h_N$. Table 1 shows the formulas of indicators concerned of the two groups of sample values.

Then the maximum of the total values, $h_1, h_2, \cdots, h_M, h_{M+1}, h_{M+2}, \cdots, h_N$, is $\max(h_i, h_m)$. The minimum is $\min(h_j, h_n)$. The mean of all values is

$$\mu = \overline{A} + \frac{M(\overline{H} - \overline{A})}{N} \tag{1}$$

and the variance of the total values is

$$\sigma^2 = \frac{M\left[\sigma_H^2 + (\mu - \overline{H})^2\right] + (N - M)\left[\sigma_A^2 + (\mu - \overline{A})^2\right]}{N} \tag{2}$$

As mentioned, the window type of real-time processing is time rolling window, which includes one day, one month and one year. After the incremental computing engine

Table 1. The results of normal distribution method

Indicator	Historical sample	Incremental sample
Maximum	$h_i, i \in [1, M]$, for $\forall k \in [1, M]$, $k \neq i, h_k \leq h_i$	$h_m, m \in [M + 1, N]$, for $\forall k \in [M + 1, N]$, $k \neq m, h_k \leq h_m$
Minimum	$h_j, j \in [1, M]$, for $\forall l \in [1, M]$, $l \neq j, h_j \leq h_l$	$h_n, n \in [1, M]$, for $\forall l \in [1, M], l \neq n, h_n \leq h_l$
Mean	$\overline{H} = \frac{1}{M} \sum\limits_{i=1}^{M} h_i$	$\overline{A} = \frac{1}{N-M} \sum\limits_{j=M+1}^{N} h_j$
Variance	$\sigma_H^2 = \frac{1}{M} \sum\limits_{i=1}^{M} (h_i - \overline{H})^2$	$\sigma_A^2 = \frac{1}{N-M} \sum\limits_{j=M+1}^{N} (h_j - \overline{A})^2$

starts, the intermediate results are initialized. The engine dynamically maintains the intermediate results, and the incremental computing results are obtained quickly based on the above incremental calculation methods. When the window slides, the engine reinitializes the intermediate results.

3.2 95% Probability Value Aggregation by Approximate Computing

In PQMAS, each indicator can have up to 1440 pieces of data (one piece per minute) in a day. When computing 95% probability value aggregation for a month, there are 1440 * 31 pieces of data at most for each indicator. In this case, the sorting process is too slow to make a real-time response. Because the calculation of 95% probability value involves data sorting, incremental computing has little effect on improving performance. Approximate computing relies on representative sampling of the entire dataset to compute over a subset of data items. By analyzing the distribution characteristics of indicators, two approximate computing methods are given as follows:

1) based on probability distribution.

It can be observed that some power quality indicators (such as voltage deviation and frequency deviation) have normal distribution characteristics. In this case, 95% probability value can be calculated by formula 1 with the mean and variance of normal distribution [5].

$$value = 1.645 * \sigma + \mu \tag{3}$$

In addition, PQMAS uses the incremental method presented in Sect. 3 to calculate mean value and variance value, which further improves the real-time computing efficiency.

2) based on data sampling.

Data sampling is an effective way to deal with large amount of data. Because the total amount of data cannot be determined in advance in PQMAS, and the real-time computing needs to be based on the sliding window, the "reservoir" sampling algorithm is applied, which can sample data in an equal probability way when the total number is unknown. Assuming that there are n pieces of data, the sliding window capacity is m, $n \gg m$, and the data are sent in sequence, the steps of "reservoir" sampling algorithm are as follows:

(i)   If the amount of data received is less than m, it will be put into the reservoir in turn.

(ii) When the ith data is received, $i \geq m$, take a random number d in the range of [0, i], if d falls in the range of [0, m−1], replace the dth data in the reservoir with the ith data received.

(iii) Repeat step (ii) until the data is sent out. It can be proved that when all data are processed, every data in the reservoir is obtained with the probability of m/n.

4 Experiments

In this section, the feasibility of our approach, as well as time performance and accuracy, is verified by experiments.

A server cluster composed of 10 machines is used, each of which is configured with 4-core inter i5 CPU, main frequency of 3.10 ghz, 32 GB memory, 1.2TB SAS hard disk, and big data software platform deployed is cdh2.6 + spark1.6. The dataset used in the experiment is the power quality monitoring dataset measured by the PQMAS. The dataset includes the power quality measurement data (such as fundamental voltage, harmonic current, etc.) collected from nearly 10000 monitoring points from January to March 2018. The data scale is about 4T, and its main data structure is (monitor_id, index, timestamp, value).

4.1   Comparation of Incremental and Full Computing

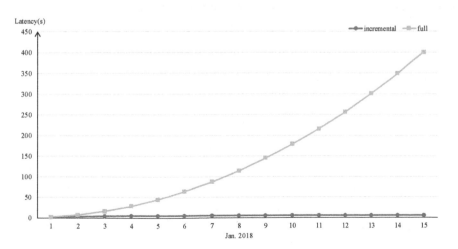

Fig. 2. Comparation of incremental and full computing

The first experiment is the performance comparison between incremental computation and full computation. Taking PQ data of January 2018 in PQMAS as the test data, the mean value of 2555 indicators within the month is calculated and the time performance is evaluated each day. The mean value is calculated by formula (1) during the incremental computing process while full computing calculates the mean value with full data involved. The experiment result is shown as Fig. 2.

It can be seen clearly that the latency of full computing is larger than the incremental computing. And as the amount of data increases, the time of full computing increases gradually, while the time of incremental computing is stable in a certain range and close to zero.

4.2 Comparison of Approximate Computing and Full Computation

The second experiment is the performance comparison between approximate estimation based on normal distribution and full computation. The accuracy of approximate computing results is an important evaluation index. This section not only evaluates the time performance, but also uses the relative error index to evaluate the accuracy. The formula of relative error is as follows:

$$\epsilon = \frac{\left|Q - Q'\right|}{Q} * 100 \tag{4}$$

Where Q is the exact value and Q' is the approximate value.

Taking monthly average data of frequency deviation as the test data, the results of 95% probability value aggregation of the monthly data are calculated. The accuracy and time delay are recorded, shown in Table 2. The first column of the table is the dataset from January to March, the second column is the 95 probability value calculated by using normal distribution approximation, its query time and query error, and the third column is the 95 probability value and query time by full computing.

Table 2. The results of normal distribution method

Month	Normal distribution			Full computation	
	Query result	Query time(ms)	ϵ	Query result	Query time(ms)
Jan	0.0154	0.37	1.315	0.0152	38.22
Feb	0.0231	0.35	1.316	0.0228	33.13
Mar	0.0134	0.38	1.471	0.0136	36.05

Table 2 shows that the relative error is about 1.3% compared with the real value, and the time performance is improved nearly 100 times. It can be concluded that if the data conform to the normal distribution, the accuracy of approximate results may be acceptable while time delay gains much improvement.

To evaluate the performance of the approximate computing with sampling data, the third experiment for the "reservoir" sampling is presented. Sampling rate is 10%. Also time delay and relative error are our concerns. Firstly, just as the second experiment, the results of 95% probability value of frequency deviation are calculated. The results are shown in Table 3. Secondly, the results and time performance of the monthly data of A-phase 3th harmonic voltage are recorded in Table 4.

Table 3. 95% Probability value of frequency deviation.

Month	Data sampling			Full computation	
	Query result	Query time(ms)	ϵ	Query result	Query time(ms)
Jan	0.0171	6.31	12.50	0.0152	38.22
Feb	0.0179	6.26	21.49	0.0228	33.13
Mar	0.0169	6.42	24.26	0.0136	36.05

Table 4. 95% Probability value of a-phase 3th harmonic voltage.

Month	Data sampling			Full computation	
	Query result	Query time(ms)	ϵ	Query result	Query time(ms)
Jan	2.2536	6.55	16.61	1.9325	37.15
Feb	1.9673	6.48	0.6	1.9804	34.23
Mar	1.9837	6.67	1.60	1.9524	36.17

Table 3 shows that the estimation error is over 10% compared with the real value, and the time performance is improved over 5 times. Table 4 shows that the estimation error varies from 0.6% to 16.61%, and the time performance is also improved over 5 times.

It can be seen that in terms of time delay, the improvement of approximate computing based on sample is relatively obvious, but in terms of query error, the average error is either large, or the average error is small but the variance is large. No matter what the above situation, it will have a greater impact on business decisions.

5 Conclusions and Future Work

In response to the challenges encountered by PQMAS in real-time aggregation of PQ data, incremental computing methods for maximum function, minimum function, average function and variance function are introduced, as well as approximate computing methods for 95% probability value function. An aggregation framework based on these methods is also proposed. Performance analysis and application results are presented with real data. Future work includes data preprocessing and performance improving of data sampling methods.

References

1. Kilter, J., Meyer, J., Elphick, S., et al.: Guidelines for power quality monitoring - results from CIGRE/CIRED JWG C4.112, In: International Conference on Harmonics and Quality of Power, pp. 703–707 (2014)
2. Wang, T., Li, Y., Deng, Z., et al.: Implementation of state-wide power quality monitoring and analysis system in China. In: 2018 IEEE Power & Energy Society General Meeting, pp.57–61 (2018)
3. Liu, S., Fang, J.: Fast dynamic density outlier detection algorithm for power quality disturbance data. In: Wang, G., Lin, X., Hendler, J., Song, W., Zhuoming, X., Liu, G. (eds.) Web Information Systems and Applications: 17th International Conference, WISA 2020, Guangzhou, China, September 23–25, 2020, Proceedings, pp. 194–201. Springer International Publishing, Cham (2020). https://doi.org/10.1007/978-3-030-60029-7_18
4. SGCC.: requirements for vertical exchange interface of power quality monitoring and analysis system (2016)
5. Ning, L., Xin, W., Haorui, Y.: Research on 95 probability value of power quality index based on normal distribution theory. Guizhou Electr. Power Technol. **17**(4), 28–30 (2014)
6. Wen, H., Kou, M., et al.: A spark-based incremental algorithm for frequent itemset mining. In: Proceedings of the 2nd International Conference on Big Data and Internet of Things, pp. 53–58 (2018)
7. Mittal, S.: A survey of techniques for approximate computing, ACM Comput. Surv. **48**(4), 62:1–62:33 (2016)
8. Krishnan, D.R., Quoc, D.L., Bhatotia, P., et al.: IncApprox: a data analytics system for incremental approximate computing. In: Proceedings of the 25th International Conference on World Wide Web (WWW 2016). International World Wide Web Conferences Steering Committee (2016)
9. Ma, S., Huai, J.: Approximate computation for big data analytics. ACM SIGWEB Newsl. 1–8 (2021)
10. Sheng, J., Fang, J., et al.: Implementation of multidimensional aggregate query service for time series data. J. Chongqing Univ. **43**(7), 121–128 (2020)
11. Cormode, G., Garofalakis, M., Haas, P.J., et al.: Synopses for massive data: samples, histograms, wavelets, sketches. Found. Trends Databases **4**(1–3), 1–294 (2012)

Data Mining

Decision Behavior Based Private Vehicle Trajectory Generation Towards Smart Cities

Qiao Chen[1], Kai Ma[1], Mingliang Hou[1], Xiangjie Kong[2(✉)], and Feng Xia[3]

[1] School of Software, Dalian University of Technology, Dalian 116620, China
[2] College of Computer Science and Technology, Zhejiang University of Technology, Hangzhou 310023, China
xjkong@ieee.org
[3] School of Engineering, IT, and Physical Sciences, Federation University Australia, Ballarat, VIC 3353, Australia

Abstract. In contrast with the condition that the trajectory dataset of floating cars (taxis) can be easily obtained from the Internet, it is hard to get the trajectory data of social vehicles (private vehicles) because of personal privacy and government policies. This paper absorbs the idea of game theory, considers the influence of individuals in the group, and proposes a decision behavior based dataset generation (DBDG) model of vehicles to predict future inter-regional traffic. In addition, we adopt simulation tools and generative adversarial networks to train the trajectory prediction model so that the private vehicle trajectory dataset conforming to social rules (e.g., collisionless) is generated. Finally, we construct from macroscopic and microscopic perspectives to verify dataset generation methods proposed in this paper. The results show that the generated data not only has high accuracy and is valuable but can provide strong data support for the Internet of Vehicles and transportation research work.

Keywords: Smart cities · Spatial-temporal interaction · Dataset generation · Generative adversarial networks

1 Introduction

Recent years, the related technologies of smart cities continue to proliferate, including Internet of Things (IoT) [1], cyber-physical systems (CPS) [2], Intelligent Transportation Systems (ITS) [7], etc. One important application scenario in a smart city is the ITS which aims at improving the public services in an urban environment and dealing with problems in urbanization such as traffic congestion, energy consumption, and environment pollution [3,4]. Trajectory data represents a significant part of the ITS [11]. It is the basis for researchers to research to study the mobility of urban. The trajectory data refers to the data information obtained by collecting the movement process of one or more moving objects in a spatial-temporal environment.

© Springer Nature Switzerland AG 2021
C. Xing et al. (Eds.): WISA 2021, LNCS 12999, pp. 109–120, 2021.
https://doi.org/10.1007/978-3-030-87571-8_10

However, in the era of big data, the security of personal information becomes even more important. Trajectory data is important private information [12]. Due to personal privacy and social security limitations, the real GPS trajectory data of social vehicles (private vehicles) mentioned are not available in large quantities to ordinary researchers. Therefore, it is key points in the current research to combine the existing information, propose a feasible private car track data generation method, and more accurately generate the track information in line with the movement law. In summary, our main contributions are:

- We construct a decision behavior based dataset generation (DBDG) model of vehicles from the perspective of interaction behavior and travel decision game.
- We combine deep learning and generative adversarial networks, using existing taxi trajectory data to get vehicle trip characteristics and patterns.
- We propose a validation model from both macro and micro perspectives to validate our proposed method.

2   Related Work

The theory of spatial-temporal interaction, which in turn is important for smart city construction, economic and demographic policy formulation, transportation planning, and urban traffic flow forecasting. The intervening opportunity model is a typical decision model for travel decision behavior [5]. Another commonly used behavioral decision model is called the "radiation model" [6]. However, current models of spatial-temporal interaction are all about maximizing individual utility as the ultimate goal, making the choice of a multi-option optimal combination of solutions, or predicting the probability of each option being chosen only takes into account the destination selection behavior of individuals and lacks consideration of the actual interactions between individuals.

For trajectory generation, there are many related techniques in neural network-generated data research, Altché et al. [8] proposed an LSTM neural network-based data generation model. Because of its ability to retain the memory of previous inputs, LSTM is particularly effective for time series prediction, and in the past few years has been widely used to predict pedestrian trajectories [9,10] or to predict vehicle destinations at intersections [13]. However, all of the above studies using deep learning for data generation have certain limitations. For example, when conducting training, one must rely on a large amount of existing data, which requires the researcher to obtain vehicle trajectories in advance, which is difficult to achieve for policy reasons.

3   Method

3.1   Framework

Figure 1 is the framework of our method. The first stage of the DBDG method is to construct a travel decision model oriented to a crowd game called Regional

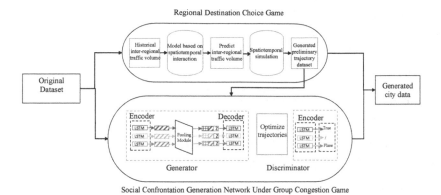

Fig. 1. Framework for decision behavior based vehicle trajectory dataset generation.

Destination Choice Game (RDCG) model. Then, simulation tools are used to generate private vehicle trajectory simulation data based on the predicted inter-regional private vehicle traffic volumes obtained. The second stage of the DBDG approach is to construct a socially generated adversarial network based on the congestion game. The third stage of the DBDG method is the validation of the generated dataset.

3.2 Regional Destination Choice Game Model

In RDCG model, we construct a utility function is constructed to quantify the benefits of the decision behavior of traveling individuals. We use T_{ij} to represent the inter-regional traffic from region i to region j. In conjunction with the literature [14], define the utility function U_{ij} from region i to region j, calculated as follows:

$$U_{ij} = h\left(A_j\right) - f\left(D_j\right) - c(d_{ij}) - g\left(T_{ij}\right) \tag{1}$$

As shown in Eq. (1), the benefit U_{ij} to any individual's choice of travel destination is influenced by four factors. (1) The payoff $h\left(A_j\right)$ received for choosing to go to different destinations. (2) Destination region j due to the agglomeration effect of the influx of people $f\left(D_j\right)$. (3) Trip cost $c(d_{ij})$, which is a monotonically increasing function of the distance d_{ij} of the trip. (4) The congestion effect $g\left(T_{ij}\right)$ along the route of the trip.

We draw on the "Weber-Fechner law" of psychology. The general formula for its logarithmic form is $S = k \log I$ where S is the sensory quantity, k is the proportionality constant, and I is the physical quantity. According to Weber-Fechner's law, in Eq. (1), $h\left(A_j\right) = \alpha \ln A_j$, $c\left(d_{ij}\right) = \beta \ln d_{ij}, f\left(D_j\right) = \gamma \ln D_j, g\left(T_{ij}\right) = \delta \ln T_{ij}$. Consider further, since for regional influence A_j, the population, regional area, and regional GDP are considered in the study, i.e., $A_j = \left\{A_j^1, A_j^2, A_j^3\right\}$, which also corresponds to three different values of α. And in order to simplify the calculation without affecting the final result, it can be assumed that $g(T_{ij}) = \ln T_{ij}$. Ultimately, a usable gain utility function is obtained:

$$U_{ij} = \alpha_1 \ln A_j^1 + \alpha_2 \ln A_j^2 + \alpha_3 \ln A_j^3 - \beta \ln d_{ij} - \gamma \ln D_j - \ln T_{ij} \qquad (2)$$

3.3   Predict the Private Vehicle Traffic

In combination with the earnings utility function, we use the method of successive averages (MSA) for iterative prediction of T_{ij}. The MSA based inter-regional traffic prediction process is iterated through the following steps:

(1) Initialize and set the iteration index $n = 1$. Calculate the initial solution for the traffic from region i to j using the following equation:

$$T_{ij}^{(n)} = O_i \frac{\left(A_j^1\right)^{\alpha_1} \left(A_j^2\right)^{\alpha_2} \left(A_j^3\right)^{\alpha_3} d_{ij}^{-\beta} \left[D_j^{(n)}\right]}{\Sigma_j \left(A_j^1\right)^{\alpha_1} \left(A_j^2\right)^{\alpha_2} \left(A_j^3\right)^{\alpha_3} d_{ij}^{-\beta}} \qquad (3)$$

where O_i is the independent variable representing the number of trips from area i. A_j^1, A_j^2, A_j^3 respectively represent the attractiveness generated by destination j in terms of population, area, and gross product. d_{ij} represents the attractiveness from region i to j distance. $O_i, A_j^1, A_j^2, A_j^3, d_{ij}$ these five data are quantities that can be obtained directly as input data.

(2) Calculate the solution of the traffic additions from area i to j:

$$F_{ij}^{(n)} = O_i \frac{\left(A_j^1\right)^{\alpha_1} \left(A_j^2\right)^{\alpha_2} \left(A_j^3\right)^{\alpha_3} d_{ij}^{-\beta} \left[D_j^{(n)}\right]^{-\gamma}}{\Sigma_j \left(A_j^1\right)^{\alpha_1} \left(A_j^2\right)^{\alpha_2} \left(A_j^3\right)^{\alpha_3} \alpha_{ij}^{-\beta} \left[D_j^{(n)}\right]^{-\gamma}} \qquad (4)$$

where $D_j^{(n)} = \Sigma_i T_{ij}^{(n)}$ represents the sum of trips taking region j as the destination.

(3) Calculate the average solution of the traffic volume from region i to j:

$$T_{ij}^{(n+1)} = \frac{T_{ij}^{(n)} + F_{ij}^{(n)}}{2} \qquad (5)$$

when $\left|T_{ij}^{(n+1)} - T_{ij}^{(n)}\right| < \varepsilon$ (ε is an extremely small value, taking 0.1 in this experiment), the current solution is considered the final approximation of the solution, the algorithm stops; otherwise, the iteration index $n = n + 1$ and tunes to step (2) to continue. After the completion of the iteration, the matrix (i.e., the Origin/Destination matrix) of the inter-regional traffic T_{ij} estimated from the RDCG model computation can be obtained.

3.4   Trajectory Simulation and Data Generation

We predict the traffic volumes between all the functional areas within Beijing's fifth ring road, arranging these data into a matrix named is the traffic Origin/Destination (O/D) matrix. Then, we utilize the OD2TRIPS plugin in the SUMO tool to import the O/D matrix and break it down into individual vehicle trips. With the help of OD2TRIPS, we generate the origin and destination information for each vehicle trip. Besides, the DUAROUTER plug-in, which uses the shortest path to calculate the possible vehicle paths used by SUMO, is used to generate information about the vehicle's path trajectory. We consider dynamic route planning with the shortest distance to plan and use the Dijkstra algorithm as the routing algorithm.

3.5   Optimization of Private Vehicle Trajectory Data

Due to the dataset generated by SUMO cannot consider the impact of vehicle interactions, diversified travel, etc., we propose an optimized method based on the Social GAN [15]. However, all the experimental scenarios mentioned in the literature [15] involve less than a thousand trips. In contrast, in the actual scenario of this study, the number of private cars exceeded 2 million times (a day's traffic volume in Beijing's Fifth Ring Road). If you put the data directly into the SGAN, the pooling module will not be able to execute massive trajectories at the same time due to the limited functions of the experimental equipment. On the other hand, The group influence of vehicle driving in a straight line is much smaller than that of human walking, and most of the locations affected by vehicle driving occur at intersections.

4   Experiments

4.1   Dataset Preprocessing and Region Division

In the experiment, we utilize the floating car, the taxi trajectory data from Beijing, China in November 2012, as the original data, which contained more than 10 billion GPS records of about 27,000 taxis. The format of the dataset is shown in Table 1. We deleted the record with the exercise status of 0 and the data outside the Beijing Fifth Ring Road. Then, we extracted the same vehicle ID into a file and sorted it by driving time to get a single trajectory of a taxi. In addition, we combine the administrative boundaries and provincial highway network within the region to divide the city, merge too small areas to form 50 independent areas. As shown in Fig. 2, areas are numbered from 0 to 49.

4.2   Traffic Forecast

According to the MSA method mentioned in the Sect. 3.3, the T_{ij} matrix estimated by the RDCG model can be obtained after iteration(namely O/D matrix). To verify the accuracy of the generated traffic volume, we calculate

Table 1. The format of the original taxi GPS trajectory dataset.

Attribute name	Notes	Sample
ID	The ID of the vehicle	1351023253
Operation State	0 = get off,1 = pick up	1
GPS time	Travel time record (GMT)	20121130001658
GPS longitude	Vehicle longitude	116.4243011
GPS latitude	Vehicle latitude	40.0727348
GPS speed	Vehicle speed (km/h)	53
GPS state	0 = invalid, 1 = valid	1

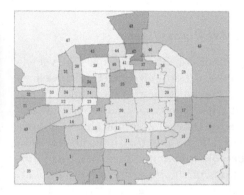

Fig. 2. The research regions after division.

Fig. 3. TOP 20 SSI of RDCG model.

the Sørensen Similarity Index (Sørensen Similarity Index, SSI) between the real traffic volume T'_{ij} and T_{ij}. The calculation method is as follows:

$$SSI = SSI = \frac{1}{N(N-1)} \sum_{i}^{N} \sum_{j \neq i}^{N} \frac{2 \min\left(T_{ij}, T'_{ij}\right)}{T_{ij} + T'_{ij}} \tag{6}$$

where N is the total number of regions divided. Through this formula, it is easy to obtain that when $T_{ij} = T'_{ij}$, that is, when the predicted traffic volume is the same as the real traffic volume, $SSI = 1$. The SSI tends to be 0, indicating a big difference.

Due to $\alpha^1, \alpha^2, \alpha^3, \beta$, and γ are unknown, we use a grid search to estimate them. As a comparison experiment, the DCG model proposed in [14] is copied and compared with the RDCG model. By comparing Figs. 3 and 4, it can be seen that RDCG is more dominant in SSI indicators, and its model is more effective and stable in the test set than the DCG model.

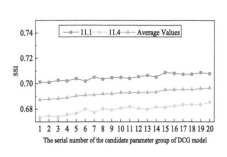

Fig. 4. TOP 20 SSI of DCG model.

Fig. 5. Traffic flow of major roads in Beijing.

4.3   Trajectory Simulation

We use SUMO to simulate the trajectory of private vehicles based on the O/D matrix. The input is the road network file and O/D matrix file, and the output is the driving track file of the vehicle. In the format shown in Table 2, we generated the dataset of vehicle trajectory information, including vehicle ID, travel time, and road condition.

4.4   Trajectory Optimization

Model Training. In this subsection, we introduced how to use existing taxi trajectory data and SUMO tools to simulate trajectory data and how to train the SGAN model to predict future trajectories.

We took the taxi GPS trajectory data on November 1, 2012, as the training set, and the dataset on November 4, 2012, as the verification set. According to literature [15], we construct and train the social GAN model. After the model training is completed, we process the trajectory data obtained by SUMO simulation according to the method proposed in Sect. 3.4 and input it into the trained SGAN model for trajectory optimization.

Table 2. The format of the generated trajectory dataset.

Attribute name	Notes	Sample
ID	Vehicle ID	3901
depart	Departure time	59900.64
from taz	Starting area number	103
to taz	Terminal area number	95
Route edges	A list of passing roads	201526561♯2,...247814291♯0

Table 3. Quantitative results of all prediction across methods.

Index	Linear	LSTM	S-SLTM	SGAN 1V1	SGAN 1V20	SGAN 20V20
ADE(10-6)	4.86	4.05	3.87	4.41	3.82	3.51
FDE(10-6)	8.82	8.19	8.22	8.71	8.37	7.02

Model Evaluating. We also use the following two indicators as the evaluation indicators of the results: ADE (Average Displacement Error) and FDE (Final Displacement Error). As a comparison experiment, we use Linear, LSTM, S-LSTM [10] as baselines.

SGAN is named as SGAN-kV-N in the experiment, k represents the diversification loss function of $k = N$ used for training ($k = 1$ means unused). N means take N samples in the test. The encoder's hidden state is 16, and the decoder's hidden state is 32. Embed the input coordinates as a 16 dimensional vector. The generator and discriminator with a batch size of 64 were trained iteratively, with an initial learning rate of 0.001 and iterative training of 200 times. Besides, some other parameters are set and adjusted according to the actual situation. The result is in Table 3. Due to the diversification loss function is conducive to the generation of multi-modal samples, SGAN-20V-20 is significantly better than all other models.

5   Results Verification

5.1   Evaluation of Macroscopic Experimental Results

Traffic Flow Comparison. We analyze the actual traffic flow data and generated traffic flow data of main roads in Beijing. Figure 5 is a comparison diagram of traffic flow on the main road of the fifth ring road in Beijing. From the overall comparison results, the DBDG model is the most consistent with the actual situation, while the gravity model is very inaccurate in describing the south fifth ring, east fourth ring, east third ring, and east second ring.

Range of Human Travel. Travel time distribution and distance distribution are two important parameters in human travel motion analysis. Figure 6 shows the distribution ratio of residents' travel time under different models and real data. The results show that the gravity model's description of the 0:00–7:00 travel distribution is very weak, while the PWO model's description of the morning peak and evening peak is incorrect. Figure 7 shows the distribution of human travel distance. The DBDG model is closest to the true distribution.

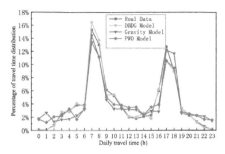

Fig. 6. Travel time distribution in Beijing.

Fig. 7. Travel distance distribution in Beijing.

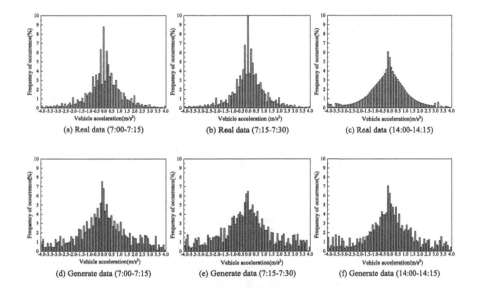

(a) Real data (7:00-7:15)

(b) Real data (7:15-7:30)

(c) Real data (14:00-14:15)

(d) Generate data (7:00-7:15)

(e) Generate data (7:15-7:30)

(f) Generate data (14:00-14:15)

Fig. 8. Acceleration frequency of real dataset and generated dataset.

5.2 Evaluation of Microscopic Experimental Results

In the microscopic verification process, we analyzed and evaluated the authenticity of trajectory data generated by the DBDG model from acceleration and relative spacing perspectives. In this part, the morning peak time (especially 7:00–7:30) and off-peak time (14:00–14:15) are selected as the analysis period.

Acceleration Analysis. Figure 8 shows the acceleration frequency of the real dataset and the generated dataset, from which it can be seen that the frequency distribution of the vehicle acceleration is normally distributed both in the real data and the generated data. The statistical results obtained by the acceleration gradient error analysis are shown in Table 4. If the trajectory data error is less than 10%, the dataset is relatively accurate. In addition, the irrationality of the maximum and minimum acceleration gradients is also within the understandable range.

Table 4. Jerk error statistics of real and generated trajectories.

Index	Original real trajectory data			DBDG generated data		
	7:00–7:15	7:15–7:30	14:00–14:15	7:00–7:15	7:15–7:30	14:00–14:15
$|j| > 3\,\mathrm{m/s^3}$	3.276272	2.994038	5.472061	5.977154	3.934523	4.784231
Max(j)	7.275132	25.000003	8.173139	10.252591	9.758725	10.991288
Min(j)	−10.648148	−27.500009	−20.624181	−11.907857	−12.563471	−13.127592

Consistency Analysis. Vehicles in the process of driving must keep a reasonable distance from other vehicles, otherwise prone to traffic accidents. So, we verify the authenticity of the generated dataset by vehicle spacing. The statistical results of analysis consistency are shown in Table 5. We can see that the trajectory data generated based on the DBDG model is similar to the real data in peak and off-peak hours. Figure 9 shows the frequency of the distance between vehicles. Obviously, the distributions of the six observation data sets are very similar, which means that the generated data sets have similar vehicle motion patterns to the real data.

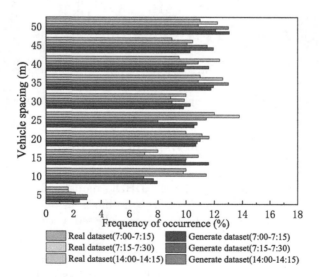

Fig. 9. Inter-vehicle spacing occurrence frequency.

Table 5. Platoon consistency (spacing indicators) error statistics of real and generated trajectories.

Index	Original real trajectory data			DBDG generated data		
	7:00–7:15	7:15–7:30	14:00–14:15	7:00–7:15	7:15–7:30	14:00–14:15
Total number of vehicles	4248	4416	3823	3677	3813	3071
Total number of vehicles with spacing less than 5 m	24	18	16	78	43	33
The proportion of vehicles with spacing less than 5 m	0.56	0.41	0.55	1.89	1.31	0.91
Avg vehicle spacing (m)	19.62	24	20.33	22.83	23.49	24.81
Maximum vehicle spacing (m)	49.87	49.85	49.77	50.02	49.88	50.02

6   Conclusions

We build the DBDG model from the perspective of interaction behavior and travel decision game, which uses taxi historical trajectory data to predict future cross-regional traffic volume. And taking into account the impact of traveler interaction and congestion effects, we use SUMO simulation tools to simulate trajectory data and combine with Social GAN to generate urban private car spatio-temporal trajectory data that conforms to social laws. Besides, we also use different methods from the macro and micro levels to verify the proposed method of generating a private vehicle trajectory dataset. The results show that the trajectory data generated by the DBDG model has high precision and application value, which can provide strong data support for Smart Cities and ITS research.

Acknowledgments. This work is partially supported by the National Natural Science Foundation of China (62072409), Zhejiang Provincial Natural Science Foundation (LR21F020003), and Fundamental Research Funds for the Provincial Universities of Zhejiang (RF-B2020001).

References

1. Kong, X., et al.: Mobile edge cooperation optimization for wearable internet of things: a network representation-based framework. IEEE Trans. Ind. Inform. **17**(7), 5050–5058 (2021). https://doi.org/10.1109/TII.2020.3016037
2. Silva, E.M., Jardim-Goncalves, R.: Cyber-physical systems: a multi-criteria assessment for Internet-of-Things (IoT) systems. Enterprise Inf. Syst. **15**(3), 332–351 (2021). https://doi.org/10.1080/17517575.2019.1698060
3. Zhu, H., Yang, X., Wang, B., Wang, L., Lee, W.-C.: Private trajectory data publication for trajectory classification. In: Ni, W., Wang, X., Song, W., Li, Y. (eds.) WISA 2019. LNCS, vol. 11817, pp. 347–360. Springer, Cham (2019). https://doi.org/10.1007/978-3-030-30952-7_35
4. Kong, X., et al.: A federated learning-based license plate recognition scheme for 5G-enabled internet of vehicles. IEEE Trans. Ind. Inform. **17**(12), 8523–8530 (2021). https://doi.org/10.1109/TII.2021.3067324

5. Barbosa, H., et al.: Human mobility: models and applications. Phys. Rep. **734**, 1–74 (2018). https://doi.org/10.1016/j.physrep.2018.01.001
6. Piovani, D., Arcaute, E., Uchoa, G., Wilson, A., Batty, M.: Measuring accessibility using gravity and radiation models. Roy. Soc. Open Sci. **5**(9), 171668 (2018)
7. Xia, F., Wang, J., Kong, X., Zhang, D., Wang, Z.: Ranking station importance with human mobility patterns using subway network datasets. IEEE Trans. Intell. Transp. Syst. **21**(7), 2840–2852 (2020). https://doi.org/10.1109/TITS.2019.2920962
8. Altché, F., de La Fortelle, A.: An LSTM network for highway trajectory prediction. In: 2017 IEEE 20th International Conference on Intelligent Transportation Systems (ITSC), pp. 353–359 (2017). https://doi.org/10.1109/ITSC.2017.8317913
9. Xin, L., Wang, P., Chan, C.Y., Chen, J., Li, S.E., Cheng, B.: Intention-aware long horizon trajectory prediction of surrounding vehicles using dual LSTM networks. In: 2018 21st International Conference on Intelligent Transportation Systems (ITSC), pp. 1441–1446 (2018). https://doi.org/10.1109/ITSC.2018.8569595
10. Alahi, A., Goel, K., Ramanathan, V., Robicquet, A., Fei-Fei, L., Savarese, S.: Social LSTM: human trajectory prediction in crowded spaces. In: 2016 IEEE Conference on Computer Vision and Pattern Recognition (CVPR), pp. 961–971 (2016). https://doi.org/10.1109/CVPR.2016.110
11. Han, X., Shen, G., Yang, X., Kong, X.: Congestion recognition for hybrid urban road systems via digraph convolutional network. Transp. Res. Part C: Emerg. Technol. **121**, 102877 (2020). https://doi.org/10.1016/j.trc.2020.102877
12. Zou, Y., Bao, X., Xu, C., Ni, W.: Top-k frequent itemsets publication of uncertain data based on differential privacy. In: Wang, G., Lin, X., Hendler, J., Song, W., Xu, Z., Liu, G. (eds.) WISA 2020. LNCS, vol. 12432, pp. 547–558. Springer, Cham (2020). https://doi.org/10.1007/978-3-030-60029-7_49
13. Phillips, D.J., Wheeler, T.A., Kochenderfer, M.J.: Generalizable intention prediction of human drivers at intersections. In: 2017 IEEE Intelligent Vehicles Symposium (IV), pp. 1665–1670 (2017). https://doi.org/10.1109/IVS.2017.7995948
14. Yan, X.Y., Zhou, T.: Destination choice game: a spatial interaction theory on human mobility. Sci. Rep. **9**(1), 1–9 (2019)
15. Gupta, A., Johnson, J., Fei-Fei, L., Savarese, S., Alahi, A.: Social GAN: socially acceptable trajectories with generative adversarial networks. In: 2018 IEEE/CVF Conference on Computer Vision and Pattern Recognition, pp. 2255–2264 (2018). https://doi.org/10.1109/CVPR.2018.00240

Self-learning Tags and Hybrid Responses for Deep Knowledge Tracing

Shuang Li, Lei Xu, Yuchen Wang, and Lizhen Xu$^{(\boxtimes)}$

School of Computer Science and Engineering, Southeast University, Nanjing 211189, China
{220171621,lzxu}@seu.edu.cn

Abstract. Knowledge Tracing, as a classic task of evaluating student knowledge mastery and predicting performance by modeling student response sequences, has become one of the key motivations for stimulating the vigorous development of personalized online education. With the support of Recurrent Neural Network, Deep Knowledge Tracing (DKT) and its variants demonstrate remarkable KT performance on account of their excellent learning ability and knowledge state representation. However, the drawbacks of these models have gradually emerged with the surge in student interaction data scale and the variety of interaction forms, which include employing the pre-defined knowledge point tags and the single response feature input. In this paper, we advocate a brand-new LTDKT-HR model, which constructs exercise embedding mapped from exercise space to tag space by self-training and optimizes input by adding an effective feature of first response time to the original feature of whether answer is correct or not. Sufficient experiments on two open datasets prove that LTDKT-HR outperforms the general DKT in student performance prediction, that is, self-learning tags are superior to existing manual tags. In addition, the proposed model can dig out the influence between exercises, which provides the mathematical basis for further setting exercise relationship constraints.

Keywords: Knowledge tracing · Deep learning · Student response sequences

1 Introduction

In the development of education informatization and professionalization, the online education model supported by information technology and artificial intelligence has become the main trend. More and more students prefer to complete their learning by online learning platforms [1]. With the deepening of learning process, the platforms tend to collect massive online data. It is of great significance to build a dynamic student model. Knowledge tracing (KT) was proposed in 1995 as a pivotal technology to effectively evaluate students' learning status and predict their performance [2]. By modeling student response sequences, KT fully diagnoses the mastery of each knowledge point and predict future interactions. Recently, some typical theories and models have been proposed to measure students' performance, such as Bayesian Knowledge Tracing (BKT) and its variants, Deep Knowledge Tracing (DKT) and its variants, etc.

© Springer Nature Switzerland AG 2021
C. Xing et al. (Eds.): WISA 2021, LNCS 12999, pp. 121–132, 2021.
https://doi.org/10.1007/978-3-030-87571-8_11

However, most of the existing models are lack of detailed processing of exercise space because they usually use knowledge point tags of exercises directly while these tags are predefined by domain experts which requires great effort. As a result, it is necessary to find an end-to-end knowledge point discovery strategy [3].

Another classic problem of the current models is that they neglect the full use of student response sequences which include but are not limited to results, time spent, first response time, and attempts and etc. It would be incomplete and inaccurate to use exercise results simply when assessing students' knowledge states [4].

In order to solve the above challenges, we propose an innovative model called Self-Learning Tags Deep Knowledge Tracing with Hybrid Responses (LTDKT-HR) to predict student performance by taking full advantage of their response sequences. Specifically, an exercise embedding matrix would be constructed by conducting self-training on the exercises answered to perform the mapping from exercise space to tag space smoothly and effectively. Then, we shall analyze the response features during answer process in detail and add a second key feature – first response time to the basis of primary feature, that is whether the answer is correct or not. The model input is optimized by using dimension reduction operation on hybrid responses. Experiments on two open datasets show that LTDKT-HR model shows better prediction performance than the original DKT model with a small amount of added computational cost.

The main contributions of this work are summarized as follows:

1. A complete LTDKT-HR model that abandons manual labeling is proposed, which expands the application scope of KT from student response sequences containing predefined tags to broader learning behavior data.
2. A glossy mapping from exercise space to tag space is explored by defining the self-training process of exercise embedding.
3. A detailed analysis of multiple response features is performed to improve accuracy and interpretability from model input perspective.

2   Related Work

Once KT is put forward, many mathematical and computational frameworks are proposed and researched in depth, which can be divided into models based on Bayesian method and models based on deep learning [5].

Bayesian Knowledge Tracing (BKT) model was proposed in 1995 and applied to Intelligent Tutoring System (ITS) [2]. The model expresses student knowledge state as a set of binary variables, that is, knowledge points mastered or not, and update its probability distribution through whether student answer exercise correctly or not by first-order Markov [6]. Later, a transitional state – student may master knowledge point is added on the basis to improve the evaluation accuracy [7]. However, the oversimplification and idealization of BKT doomed that it could not be pushed for a long time [8].

With the rise of artificial intelligence and the continuous development of machine learning, Piech et al. claimed a Deep Knowledge Tracing (DKT) model using Recurrent Neural Network (RNN) in 2015 [9], which is compared with BKT to discover its exhaustive processing of student response sequences and knowledge points [10]. Some scholars tried to explain the model with specific theorems [11], or announced the role of deep learning and RNN in the model [12]. A classic discussion was brought up by Chun et al. which became another major benchmark for further re-search [13]. Later, DKT was expanded by introducing more factors and information from exercise level and student level [4, 14]. Chen et al. took prerequisite relation-ship between knowledge points as model constraints and raised PDKT-C model to solve the sparse problem of student data [15]. The E2E-DKT model given in takes the corresponding relationship between exercise and tags as a binary vector matrix and predicts knowledge mastery with response sequences. DKT-DSC model utilizes unsupervised clustering of students with similar learning abilities to strengthen the personalized KT ability [16].

There are many other models using deep learning. [17] applied improved Memory-Augmented Neural Network (MANN) and proposed DKVMN which com-presses and encodes the underlying concepts of exercises. [18] suggested a novel Exercise-Enhanced Recurrent Neural Network (EERNN) framework to fully excavate student response sequences and exercise. KQN, which models student learning interaction as a dot prod-uct of knowledge state vector and knowledge point vector, was carried out to improve model interpretability and intuitiveness [19]. GKT de-scribes KT as node classification problem of Graph Neural Network (GNN) and models complex relationship between knowledge points [20].

Based on the analysis of the current models, LTDKT-HR model proposed in this paper fully mines student response sequences to improve model input, and introduces a novel learning process to enhance the extensive automation of KT process.

3 Proposed Model

3.1 Model Student Response Sequences

Exercise Embedding
In an integral online student learning behavior dataset, we denote student space as S, exercise space as E, and knowledge point tag space (hereinafter called tag space) as K. For a student $s \in S$, his chronological answer-exercise response sequence $x^s \in X$ can be normalized to $\{(e_1^s, q_1^s), (e_2^s, q_2^s), \ldots, (e_t^s, q_t^s)\}$, where $t \in T$ indicates time step and $q \in Q$ indicates student responses to each exercise. In order to express and calculate conveniently, we analyze $x_t = (e_t, q_t)$ as a unit in this work.

M is defined as embedding matrix from exercise space to tag space, by which the d_e dimensional vector $e_i \in E$ could be mapped to d_k dimensional vector $k_i \in K$. In traditional DKT, M is annotated by experts or predefined rules, whereas this paper will learn it from student response sequences.

The embedding matrix M is a binary matrix, $m_{e,k} = 1$ when an exercise contains a certain knowledge point, otherwise, $m_{e,k} = 0$. The knowledge point vector can be obtained by multiplying exercise vector by M (Fig. 1).

$$k_t = e_t \cdot M \tag{1}$$

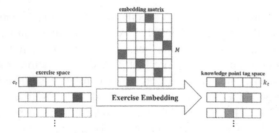

Fig. 1. Overview of exercise embedding method

Hybrid Responses

Student responses q_t should contain several features. Most previous studies merely conduct data extraction of answer results as the main basis for evaluating student mastery of knowledge point corresponding to exercise. Such approaches are one-sided and do not utilize student responses fully. In this work, the typical features of student responses are analyzed as follows firstly (Table 1):

Table 1. Typical features of student responses

Feature name	Description
c_t	Whether the answer is correct or not, the value is 1 if the answer is correct, or 0 otherwise
a_t	Attempts to answer the exercise
p_t	The first action made on exercise, such as attempt, hint, empty
g_t	Time between start time and first student action
l_t	Time to complete the exercise

After investigating the influence of the above main features on the assessment of student knowledge mastery, and in order to achieve the best performance of the current model, this paper selects the time between start time and first student action g_t as the second key feature during answering exercise, except for c_t, and extends it to:

$$\widetilde{g_t} = \begin{cases} g_t \oplus O, & if\ c_t = 1 \\ O \oplus g_t, & if\ c_t = 0 \end{cases} \tag{2}$$

Similarly, the knowledge point vector after exercise embedding is extended as:

$$\widetilde{k_t} = \begin{cases} k_t \oplus O, & if\ c_t = 1 \\ O \oplus k_t, & if\ c_t = 0 \end{cases} \tag{3}$$

Then, the model input can be expressed as:

$$\widetilde{x_t} = pca(\widetilde{k_t} \oplus \widetilde{g_t}) \tag{4}$$

Where $\oplus$ is the operation that concatenates two vector, O is a zero vector with the same dimension as the concatenated vector. pca is a dimension reduction function of Principal Component Analysis (PCA). The primary vector k_t and g_t are d_k and d_x dimensions respectively, the expanded vector $\widetilde{k_t}$ and $\widetilde{g_t}$ are $2d_k$ and $2d_x$ dimensions respectively, and the concatenated vector x_t is $2d_k + 2d_x$ dimension which is an oversized vector and may lead to excessive computational cost. Function pca can solve the problem and acquire a vector with the same $2d_x$ dimension as the traditional DKT.

By modeling student response sequences, we present the architecture of LTDKT-HR model. A hidden layer u_t is added between the student response sequences x_t and the input layer $\widetilde{x_t}$ to complete the dimension reduction of exercise embedding and hybrid responses. Another hidden layer v_t is added between the student hidden state h_t and the output layer y_t to complete the mapping from tag space to exercise space (Fig. 2).

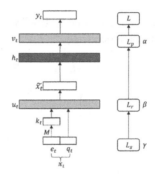

Fig. 2. LTDKT-HR architecture drawn at time t

3.2 Knowledge State

The traditional RNN model fails to transmit effective information from an earlier time step to a current or subsequent time step due to the affection by short-term memory. In order to avoid the above situation, this work adopts two improved RNN frameworks to trace student knowledge state and give prediction.

Long Short-Term Memory (LSTM)
The cell state and three gates are defined in LSTM to characterize temporary knowledge state to improve model accuracy [21]. Interaction information is added or removed to the cell state through different gates, thereby determining which memories are discarded or retained. In the following calculation, W, U are taken as the weight matrixes of each parameter and b as the bias term of each calculation formula.

Input gate: At the current time step, i_t determines how much information from the input vector v_t and the previous hidden state h_{t-1} will be saved to the cell state. σ is *sigmoid* function which compresses data to the value between 0 and 1.

$$i_t = \sigma(W_i\widetilde{x_t} + U_i h_{t-1} + b_i) \tag{5}$$

Forget gate: This gate determines which information should be discarded.

$$f_t = \sigma\left(W_f \tilde{x}_t + U_f h_{t-1} + b_f\right) \tag{6}$$

New memory: This indicates the temporary cell state of the current time step. *tanh* is activation function which compresses data to the value between -1 and 1.

$$c_t' = tanh(W_c \tilde{x}_t + U_c h_{t-1} + b_c) \tag{7}$$

Final memory: This indicates the final cell state of the current time step. The abandoned information is given by multiply the last cell state and f_t, and the reserved one is given by multiply new memory and i_t. Where $\otimes$ is an element-wise multiplication.

$$c_t = f_t \otimes c_{t-1} + i_t \otimes c_t' \tag{8}$$

Output gate: Unlike GRU, this gate is unique to LSTM and resolves which part of v_t and h_{t-1} should be transmitted to Hidden state.

$$o_t = \sigma(W_o \tilde{x}_t + U_o h_{t-1} + b_o) \tag{9}$$

Hidden state: Based on Output gate and Final memory, student hidden state at the current time step is as follows:

$$h_t = o_t \otimes tanh(c_t) \tag{10}$$

Gate Recurrent Unit (GRU)

GRU is a variant of LSTM, which reduces the number of parameters and avoids overfitting by combining Input gate and Forget gate into Update gate and mixes cell state and hidden state [22].

Update gate: It determines the weight of previous hidden state and newly hidden state.

$$z_t = \sigma(W_z \tilde{x}_t + U_z h_{t-1} + b_z) \tag{11}$$

Reset gate: The discarded and retained parts of past information and new information are calculated at the same time.

$$r_t = \sigma(W_r \tilde{x}_t + U_r h_{t-1} + b_r) \tag{12}$$

New memory: This gate gives the new hidden state at the current time step.

$$\tilde{h}_t = tanh(W_h \tilde{x}_t + U_h[r_t \otimes h_{t-1}] + b_h) \tag{13}$$

Hidden state: Based on *Update gate* and two hidden states, the final result is obtained.

$$h_t = (1 - z_t) \otimes h_{t-1} + z_t \otimes \tilde{h}_t \tag{14}$$

3.3 Prediction Output

The student knowledge state h_t is given through the calculation of the above two neural network model, and the prediction output is as follows:

$$y_t = \sigma(W_y h_t + b_y) \tag{15}$$

3.4 Model Training

This paper will train the proposed model from the following three aspects:

Prediction Performance: The main training goal of the model is to minimize the negative log likelihood of the observed student response sequences.

$$L_p = \sum_t \ell\left(y_t^T \delta(e_{t+1}), c_{t+1}\right) \tag{16}$$

Reconstruction Regularization Term of Model Input: This model advocates conducting training on exercise vector itself to obtain knowledge point tag and adding first response time as another main input feature. Consequently, if student mastery of underlying knowledge point and first response time have a positive impact to the situation of answering exercise, the prediction accuracy will get improved once we train these two factors well. In order to verify such a hypothesis, the following loss function is defined:

$$x_t' = \sigma(W_{x'} k_t + U_{x'} g_t + b_{x'}) \tag{17}$$

$$L_r = \sum_t \ell\left(x_t' \delta(e_t), c_t\right) \tag{18}$$

Binary Regularization Term of Exercise Embedding Matrix: We will get an exercise embedding matrix after model training, which is not binary in theory. Therefore, a loss function is defined during performing matrix binarization operation:

$$L_b = \sum_t (0.5 - |k_t - 0.5|) \tag{19}$$

Total Objective Function: The final loss function obtained by integrating the above aspects is as follows:

$$L = \alpha L_p + \beta L_r + \gamma L_b \tag{20}$$

Where α, β, γ are the weight parameters to be trained. The training goal of the model is to minimize L.

4 Experiments

4.1 Datasets

Considering that exercises should be labeled manually, the following datasets are chosen and their detailed data statistics are shown in Table 2. Assistments09–10 skill builder(b) removes repeated logs and scaffolding exercises and duplicates response sequences involving multiple knowledge points to ensure one-to-one correspondence between exercise and knowledge point. KDD Cup2010 has exercises consisting of several steps corresponding to different knowledge point, but students only need to answer steps with unmastered knowledge points, which is friendly to our experiment.

Table 2. Dataset statistics

Dataset	Assistments09–10 skill builder(b)	KDDCup2010
#of records	328,292	607,026
#of students	4,217	574
#of knowledge point tags	124	436

Moreover, for the first response time g_t added to the model in this paper, the two datasets have their relevant feature descriptions, as shown in Table 3.

Table 3. Feature description of g_t

Dataset	Feature name of g_t
Assistments09–10 skill builder(b)	first_response_time
KDD Cup2010	First Transaction Time-Step Start Time

4.2 Implementation Details

Experiment Settings
In each experiment, 80% of the data is randomly selected as a training set to gain exercise embedding, 10% is a validation set, and the remaining 10% is used to test. Furthermore, 5-fold cross-validation is applied and Area Under ROC Curve (AUC) is measured for model evaluation. An AUC value of 0.5 indicates that the prediction performance is equivalent to random guess, and the larger the value, the better the model performance. Since the main purpose of the experiment is the acquisition of exercise embedding matrix, the dimension of knowledge point vector, that is, the number of the learned tags is directly set to the number of existing tags in the datasets, which are 124 for Assistments09–10 skill builder(b) and 436 for KDD Cup2010.

Since the model needs to be trained on mapping from exercise space to tag space, a comparison with the models employing predefined knowledge point tags, including BKT, DKT, DKT+, is carried out to measure the correctness of the learned ones. EM is used to train BKT and the limit of iteration is 200. A single-layer LSTM with 200 states is used by DKT and DKT+, while in LTDKT-HR, the state size of LSTM is 200, the state size of GRU is 400, and the added hidden layer u_t and v_t is 100.

During the training process, the initial weights of models randomly follow a Gaussian distribution with zero mean and small variance. PCA is utilized to reduce the dimension of student hybrid responses. The Adam optimization method with the learning rate is 0.01 and other parameters consistent with the default is used to minimize the total loss. In addition, we apply Dropout to prevent overfitting and set the drop rate of h_t in three models is 0.5 and the drop rate of u_t and v_t in LTDKT-HR is 0.2.

Hyperparameter Search
The exercise embedding matrix M is extracted from the model with the highest AUC value, which is learned from training data and validated by validation data. The elements in M are continuous values between 0 and 1. In order to achieve the mapping from exercise space and tag space, matrix binarization operation is defined as follows:

$$\widetilde{M_{i,j}} = \begin{cases} 1, & if\ M_{i,j} = \max(M_i)\ or\ M_{i,j} \geq \varepsilon \\ 0, & else \end{cases} \tag{21}$$

Where i and j are the indexes of rows and columns of M, and the ranges are exercise space dimension d_e and tag space dimension d_k respectively. The parameter ε is a threshold that indicates the relevance degree between exercise and knowledge point.

We perform hyperparameter search on the weights of final loss function α, β, γ and binarization threshold ε. Specifically, train the model with training data and then adjust the hyperparameters with validation data. The initial value of α, β, γ are set as 1.0, keep two of them unchanged and the other one is increased to 2.0 or decreased to 0.0 with a interval of 0.5. The threshold ε wavies from 0.5 to 1.0 with an increment of 0.1.

4.3   Results and Discussion

Student Performance Prediction
Figure 3 represents the results of our comparison of four models on the two datasets. we find that the prediction performance of the model with predefined tags has not surpassed that of LTDKT-HR with self-learning tags at almost the same computational cost. This indicates that using self-learning tags is desirable in evaluating student knowledge state. The model also improves prediction accuracy and interpretability to a certain extent by employing the hybrid responses when processing model input.

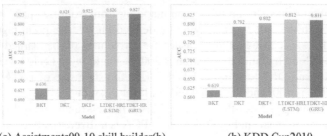

(a) Assistments09-10 skill builder(b)   (b) KDD Cup2010

Fig. 3. Overall results of AUC of four models on two datasets

Another noteworthy finding is the application of two RNNs. As shown in Table 4, the performance of LSTM on Assistments09–10 skill builder(b) is slightly inferior to that of GRU, which is just the opposite on KDD Cup2010. The phenomenon is in line with the characteristics of both RNNs. That is to say, LSTM is more suitable for large datasets while GRU performs better on small ones. The way to distinguish the size of datasets is the number of knowledge point tags and student response sequences.

Table 4. Prediction performance of LTDKT-HR with different RNNs

Model	AUC	
	Assistments09–10 skill builder(b)	KDD Cup2010
LTDKT-HR (LSTM)	0.826 ($\alpha = 1.0, \beta = 1.0, \gamma = 0.5, \varepsilon = 0.7$)	**0.812** ($\alpha = 2.0, \beta = 1.0, \gamma = 1.5, \varepsilon = 0.7$)
LTDKT-HR (GRU)	**0.827** ($\alpha = 1.0, \beta = 1.0, \gamma = 0.5, \varepsilon = 0.7$)	0.811 ($\alpha = 2.0, \beta = 1.0, \gamma = 1.5, \varepsilon = 0.7$)

Exercise Association Discovery

The proposed model should further be applied to discover the underlying relationship and structure in the data. As defined in Piech's lecture, we calculate the influence $\varphi_{i,j}$ between every directed pair of exercises e_i and e_j.

$$\varphi_{i,j} = \frac{y(j|i)}{\sum_k y(j|k)} \tag{22}$$

Where $y(j|i)$ refers to the predicted probability that a student answers exercise e_j correctly immediately after answering exercise e_i correctly. Considering the mapping from exercise to knowledge point, the influence between knowledge points is obtained by this formula. Taking the response sequences of one student in two datasets as an example, Fig. 4 shows the influence graphs of the existing tags and the learned tags mapped from exercises answered. The nodes represent tags and the larger the node, the greater the influence of corresponding tag on other tags. The edges with arrows represent the direction of influence between tags and their thickness is on behalf of influence degree. Obviously, our model can generate more evidence-based and more organized exercise association.

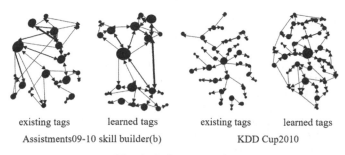

existing tags	learned tags	existing tags	learned tags
Assistments09-10 skill builder(b)		KDD Cup2010	

Fig. 4. Influence graphs

5   Conclusion

In this paper, we take the task of KT as the research objective, analyze its tracing process and propose a novel model named LTDKT-HR based on deep learning, which completes the self-learning process of knowledge point tags by exercise embedding. The model gets rid of the limitation of manual labeling and demonstrates the superior performance both in student performance prediction and exercise association discovery. The addition of student hybrid responses also results in a more interpretable model.

During the implementation of self-learning work of exercise tags, we set the dimension of tag space as the number of existing knowledge points. On further study, some algorithms could be applied to find an appropriate dimension to achieve thoroughness of self-learning process and dynamism of self-learning tags under the allowance of computational cost. For describing the influence between exercises, more convincing calculation methods such as cosine similarity, Euclidean distance, etc. need to be discussed and adopted to analyze the relevance of exercise from several distinct perspectives, which provides a basis for adding prerequisite and inclusion and other specific relations to knowledge point mapped from exercise or exercise itself.

References

1. Means, B., Toyama, Y., Murphy, R., Bakia, M., Jones, K.: Evaluation of evidence-based practices in online learning: a meta-analysis and review of online learning studies. Centre for Learning Technology (2009)
2. Corbett, A.T., Anderson, J.R.: Knowledge tracing: modeling the acquisition of procedural knowledge. User Model. User Adap. Inter. (1995). https://doi.org/10.1007/BF01099821
3. Chen, H., Dong, Y., Gu, Q., Liu, Y.: An end-to-end deep neural network for truth discovery. In: Wang, G., Lin, X., Hendler, J., Song, W., Xu, Z., Liu, G. (eds.) WISA 2020. LNCS, vol. 12432, pp. 377–387. Springer, Cham (2020). https://doi.org/10.1007/978-3-030-60029-7_35
4. Zhang, L., Xiong, X., Zhao, S., Botelho, A., Heffernan, N.T.: Incorporating rich features into deep knowledge tracing. In: Proceedings of the Fourth (2017) ACM Conference on Learning Scale, New York (2017). https://doi.org/10.1145/3051457.3053976
5. LeCun, Y., Bengio, Y., Hinton, G.: Deep learning. Nature (2015). https://doi.org/10.1038/nature14539
6. van de Sande, B.: Properties of the bayesian knowledge tracing model. J. Educ. Data Min. (2013). https://doi.org/10.5281/zenodo.3554629
7. Zhang, K., Yao, Y.: A three learning states bayesian knowledge tracing model. Knowl. Based Syst. (2018). https://doi.org/10.1016/j.knosys.2018.03.001

8. Mao, Y., Lin, C., Chi, M.: Deep learning vs. bayesian knowledge tracing: student models for interventions. J. Educ. Data Min. (2018). https://doi.org/10.5281/zenodo.3554691

9. Piech, C., et al.: Deep knowledge tracing (2015). http://arxiv.org/pdf/1506.05908v1

10. Montero, S., Arora, A., Kelly, S., Milne, B., Mozer, M.: Does deep knowledge tracing model interactions among skills?. In: Proceedings of the 11th International Conference on Educational Data Mining (2018)

11. Lu, Y., Wang, D., Meng, Q., Chen, P.: Towards interpretable deep learning models for knowledge tracing. In: Bittencourt, I.I., Cukurova, M., Muldner, K., Luckin, R., Millán, E. (eds.) AIED 2020. LNCS (LNAI), vol. 12164, pp. 185–190. Springer, Cham (2020). https://doi.org/10.1007/978-3-030-52240-7_34

12. Khajah, M., Lindsey, R.V., Mozer, M.C.: How deep is knowledge tracing? http://arxiv.org/pdf/1604.02416v2 (2016)

13. Yeung, C.-K., Yeung, D.-Y.: Addressing two problems in deep knowledge tracing via prediction-consistent regularization. http://arxiv.org/pdf/1806.02180v1 (2018)

14. Yeung, C.-K., Lin, Z., Yang, K., Yeung, D.-Y.: Incorporating features learned by an enhanced deep knowledge tracing model for STEM/Non-STEM job prediction. http://arxiv.org/pdf/1806.03256v1 (2018)

15. Chen, P., Lu, Y., Zheng, V.W., Pian, Y.: Prerequisite-driven deep knowledge tracing. In: 2018 IEEE International Conference on Data Mining (ICDM), Singapore, 17 November 2018–20 November 2018, pp. 39–48. IEEE (2018). https://doi.org/10.1109/ICDM.2018.00019

16. Minn, S., Yu, Y., Desmarais, M.C., Zhu, F., Vie, J.-J.: Deep knowledge tracing and dynamic student classification for knowledge tracing. In: 2018 IEEE International Conference on Data Mining (ICDM), Singapore, 17 November 2018–20 November 2018, pp. 1182–1187. IEEE (2018). https://doi.org/10.1109/ICDM.2018.00156

17. Zhang, J., Shi, X., King, I., Yeung, D.-Y.: Dynamic key-value memory networks for knowledge tracing. In: Proceedings of the 26th International Conference on World Wide Web, Perth Australia, 03 April 2017–07 April 2017, pp. 765–774. International World Wide Web Conferences Steering Committee, Republic and Canton of Geneva, 3 April 2017. https://doi.org/10.1145/3038912.3052580

18. Liu, Q., et al.: EKT: exercise-aware knowledge tracing for student performance prediction. IEEE Trans. Knowl. Data Eng. (2021). https://doi.org/10.1109/TKDE.2019.2924374

19. Lee, J., Yeung, D.-Y.: Knowledge query network for knowledge tracing. In: Proceedings of the 9th International Conference on Learning Analytics & Knowledge, Tempe AZ USA, 04 March 2019–08 March 2019, pp. 491–500. ACM, New York, 4 March 2019. https://doi.org/10.1145/3303772.3303786

20. Nakagawa, H., Iwasawa, Y., Matsuo, Y.: Graph-based knowledge tracing: modeling student proficiency using graph neural network. In: IEEE/WIC/ACM International Conference on Web Intelligence, Thessaloniki Greece, 14 October 2019–17 October 2019, pp. 156–163. ACM, New York, 14 October 2019. https://doi.org/10.1145/3350546.3352513

21. Hochreiter, S., Schmidhuber, J.: Long short-term memory. Neural Comput. (1997). https://doi.org/10.1162/neco.1997.9.8.1735

22. Chung, J., Gulcehre, C., Cho, K., Bengio, Y.: Empirical evaluation of gated recurrent neural networks on sequence modeling (2014)

Shortest Path Distance Prediction Based on CatBoost

Liying Jiang[1], Yongxuan Lai[1(✉)], Quan Chen[2], Wenhua Zeng[1], Fan Yang[3], and Fan Yi[4]

[1] School of Informatics, Xiamen University, Shenzhen Research Institute Xiamen University, Shenzhen, China
jiangliying@stu.xmu.edu.cn, {laiyx,whzeng}@xmu.edu.cn
[2] Department of Computer Science, Xiamen University Malaysia, Sepang, Malaysia
[3] Department of Automation, Xiamen University, Xiamen, China
yang@xmu.edu.cn
[4] School of Mathematics and Statistics, Key Laboratory of Complex Systems and Intelligent Computing, Qiannan Normal University for Nationalities, Dunyun, China

Abstract. Shortest path distances between node pairs on road networks are essential for many applications. Traditional methods, such as breadth first search (BFS) and Dijkstra algorithm, focus on precise result. However, they are difficult to apply to the large-scale road network because of the high time cost. And some methods precomputed the shortest path of all node pairs and stored them, then answer distance queries by simple lookups. But these methods need high space cost. For some applications, such as finding nearest point of interest (POI) for travel recommendations, which only need the approximate distances. Therefore, it is important to find a method to estimate the shortest path distance timely with low time and space cost. In this paper, we use CatBoost, a machine learning method based on gradient boosting decision trees, to estimate the shortest distance. We first obtain the node features based on landmarks and combine them with the longitude and latitude of the node and the real Euclidean distance between node pairs as the node features. Then we fed them into CatBoost model to train the model. The space complexity is $O(k|V|)$ and the query time complexity is $O(1)$. We conduct experiments on the real road network in Xiamen and New York City. Experiments demonstrate that our model can estimate the shortest distance with low error and small time cost and space cost.

Keywords: Shortest path prediction · Machine learning · CatBoost

This work was supported in part by the Natural Science Foundation of Guangdong under Grant 2021A1515011578; in part by the Natural Science Foundation of China under Grant 61672441 and Grant 61673324; in part by the Natural Science Foundation of Fujian under Grant 2018J01097; in part by the Shenzhen Basic Research Program under Grant JCYJ20170818141325209 and Grant JCYJ20190809161603551.

C. Xing et al. (Eds.): WISA 2021, LNCS 12999, pp. 133–143, 2021.
https://doi.org/10.1007/978-3-030-87571-8_12

1   Introduction

The shortest path distance estimation in the road network aims to quickly estimate the shortest distance between a pair of specified nodes. It is a classic problem in many applications such as traffic engineering and traffic planning. Related solutions have also been proposed by many researchers.

For a road network with n nodes and m edges, Dijkstra algorithm can calculate the shortest path distance between a pair of nodes with $O(n^2)$ time cost. The A* [5] algorithm is a simple extension of Dijkstra algorithm, which uses heuristic techniques to calculate the shortest distance. In fact, A* algorithm is at least as fast as Dijkstra algorithm, but the time cost is still $O(n^2)$. These precise methods are effective in small graphs, but the time cost is unbearable in big graphs. For example, with the development of economy and urban construction, the scale of urban road networks continues to expand. In a graph with millions of nodes, the time cost is unacceptable. Considering the high space cost of storing pre-computed shortest path distances [9], researchers have limited options. In most practical applications, sub-graph can be sampled or approximate results can be sought. In recent years, researchers have proposed a variety of preprocessing techniques, including heuristic strategies [1], hierarchical strategies, landmark strategies [11], etc., which are generally used to obtain important information on the network. These strategies are hope to reduce the time cost and space cost.

In this paper, we propose a new method that uses CatBoost [2] to combine the landmark strategy and static information of the graph to estimate the shortest path distance between two nodes. We first use the landmark strategy to calculate the global features of the nodes. We follow the "landmark-based" method proposed in [11] to select a small number of nodes as landmarks, then we calculate the shortest distance from all other nodes to the landmarks as the global space feature of each node. Besides, we calculate the similarity and the actual Euclidean distance between node pairs as the local features of each node pair. Then, we combine these features, and add some actual road network information, such as latitude and longitude as the finally feature of a node. Finally, we use CatBoost for node pairs' shortest path distance estimation.

The main contributions of this paper are summarized as follows:

- We propose an effective landmark-based method to obtain the global spatial features of nodes. First, we select k nodes as landmarks by using K-Means. Then, for each node, we calculate the shortest distance to all landmarks as the features, which can be describe as $DistEmb(v) \in R^k$.
- We apply the CatBoost based model to estimate the shortest distance. We select part of node pairs as training data to train the model. In the offline (training) stage, the space complexity is $O(k|V|)$. In the online (predicting) stage, the space complexity is only relating to the parameters of the model and has nothing to do with the size of the road network and the predicting time complexity is $O(1)$.
- We conduct experiments on two real road network datasets. Compared with traditional methods, our prediction time cost on the big graph is much lower.

Compared with landmarks based methods and machine learning based methods, our model has higher accuracy.

The remainder of this paper is organized as follows. Section 2 describes the related works. Section 3 introduces our problem and shows the details of our method. Section 4 introduces the datasets and evaluate the predictive performance of our model by two real-world road network datasets XM and NY. We conclude the paper and outline future works in Sect. 5.

2   Related Work

2.1   Shortest Path Distance Prediction

On a graph, the Dijkstra algorithm can calculate the shortest path distance between nodes precisely and A* [5] algorithm is a heuristic search algorithm, it time cost is generally outperformes precise algorithms such as Dijkstra and BFS. However, these are still exist following problems: 1) On large graph, the precise methods take lot of time for every query execution. 2) In some applications, only need approximate distances first while the precise paths may be computed later, precise methods cannot respond quickly to requests. 3) A real road network graph with practical meaning may have more meaningful features, which can effectively help estimate the shortest path distance. The traditional methods didn't make full use of this information.

In recent years, the researchers have proposed methods such as heuristic strategies, jumping strategies, hierarchical strategies, landmark strategies, etc. [13]. to solve the problems above. These strategies are generally used to obtain important information from the network through certain preprocessing and to improve the computing efficiency. For example, Chow et al. [1] proposed a heuristic algorithm to search the shortest path on undirected networks. But this algorithm relies on the heuristic function whose quality affects the efficiency and accuracy greatly. Rattgan et al. [9] designed a network structure index (NSI) algorithm to estimate the shortest path in the network by storing data. However, the initial construction leads to high time cost and space cost. Tang et al. [12] proposed a CDZ (Center Distance to Zone) algorithm based on local centrality and the existing path through the central node (10% of all nodes). It used the Dijkstra algorithm to calculate the shortest path between central nodes to approximate the distance. Although the CDZ algorithm has achieved a high accuracy rate on some social networks within a reasonable time, its performance poor on large-scale networks due to the large number of central nodes. Besides, landmark based methods [4,7] use a subset of k ($k \ll |V|$) vertices as the landmarks. Every vertex v_i stores its distances to these landmarks as its distance label, i.e., <d(v_i, l_1), d(v_i, l_2),..., d(v_i, l_k)> where d(v_i, l_k) means the shortest distance of vertex v_i and landmark l_k. For one query, scan the distance labels of the two query vertices v_i and v_j, then sum up the distances to the same landmark. And select the smallest distance sum as the shortest path distance. Though landmark based method reduces the space cost to O(k|V|), it may not return the exact distance. How the landmarks are chosen plays a critical role in the distance accuracy.

With the development of deep neural networks in recent years, there have also been some attempts to apply machine learning and neural network based methods to shortest path prediction problems. The neural network-based methods are inspired by Word2Vec [6] to learn about graph representation. For example, Rizi et al. [10] used node2vec [3], a graph embedding technology, to learn vertex embeddings and trains an MLP to predict vertex distances given the learned embeddings. But node2vec based method only focuses on the neighborhood of the vertices can not model the global spatial dependencies. Qi et al. [8] combine one-hot and landmarks to get the representations of each vertex. However, the real road network is usually sparse so the one-hot vectors requires a lot of space. And these methods ignore the information about the vertex itself, such as latitude and longitude.

To preserve more information of vertexes and obtain higher accuracy, we propose to learn an embedding for every vertex as its features. Our idea is motivated by [8,10]. This motivates us to map the vertices into a latent space to compute their shortest distance. Different from methods mentioned above, we combine landmark based distance labeling and the information about the vertex itself as total features.

2.2 CatBoost

CatBoost [2] is a gradient boosting algorithm that can handle categorical features well. It automatically handles categorical features in a special way. It first does some statistics on category features, then calculates the frequency of occurrence of a certain category feature, and then adds hyperparameters to generate new numerical features. It also uses combined category features, which can utilize the relationship between features, which greatly enriches the feature dimension. And it has also made optimizations in other areas, and these improvements can prevent the model from overfitting.

In this paper, we use CatBoost based model to handle vertexes features well and to better predict the shortest path well.

3   Shortest Distance Prediction Based on CatBoost

In this paper, we used CatBoost to predict the shortest distance between node pairs by using the coordinate information of nodes and the distance of a node to local landmarks.

In the rest of this section, first we will formulate the research problem, then we will describe the proposed model in detail.

3.1   Problem Formulation

We consider a road network graph $G = (V, E, W)$ be an weighted directed graph with $|V|$ nodes and $|E|$ edges, where V is the set of vertices (road intersections),

E is the set of edges (roads), W is the distance of node pairs. A vertex $v_i \in V$ has a pair of geographical coordinates (x_i, y_i), where x_i represents the latitude of the node, and y_i represents the longitude of the node. An edge $e_{i,j} \in E$ connects two vertices v_i and v_j, and $w_{i,j}$ represents the distance of vertices v_i and v_j.

For any pair of vertex v_i and v_j, we represent $d(v_i, v_j)$ as shortest distance (calculated by Dijkstra Algorithm). We calculate the shortest distance of each node to landmarks, and called it DistEmb $distemb(v) \in R^k$, then calculate the similarity $s_{i,j}$ and Euclidean distance $d_{i,j}$ of node v_i and v_j, then concatenate $distemb(v_i)$, $distemb(v_j)$, $s_{i,j}$, $d_{i,j}$, x_i, y_i, x_j, y_j as the final feature to predict the shortest distance of node v_i and v_j. In general, our goal is to approximate the distance as $\hat{d}$ using a CatBoost model. Formally, we define $\hat{d}$ as function:

$$\hat{d} : f(distemb(v_i), distemb(v_j), s_{i,j}, d_{i,j}, x_i, y_i, x_j, y_j) -> R^{2*k+4} \tag{1}$$

3.2 CatBoost Based Shortest Distance Prediction Model

DistEmb. As shown in Fig. 1, the proposed solution is formulated as follows: We first use K-Means by using the coordinate information to choose k landmarks, where $k \ll |V|$. Then, we compute the actual shortest distances from each node to all landmarks using Dijkstra Algorithm. For every node, we get a vector called $DistEmb \in R^k$, where k is the number as landmarks.

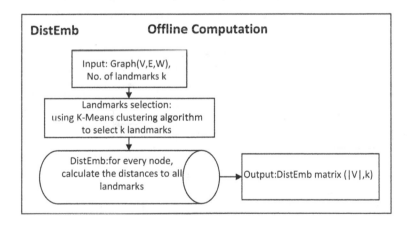

Fig. 1. Landmarks selection and DistEmb generation of proposed methodology.

Proposed Model. The framework of the proposed method is shown in Fig. 2. Our CatBoost based model consists of two CatBoost models. In the offline stage, we calculate the DistEmb vector as the global spatial information, then we concatenate the local spatial information, i.e. geographical coordinates as the total input node feature. As Fig. 2 shows, we takes the total feature of node pairs

(v_i, v_j) as the input of first CatBoost model. The output of first CatBoost model is $\hat{cat}_1 \in R^1$, to adjust the error adaptively, we concatenate the output of first CatBoost model and the original input feature as the new input $f'(f(v_i, v_j), \hat{cat}_1) \in R^{2*k+4+1}$. The output of second CatBoost model is the final prediction $\hat{d}$.

For CatBoost1 and CatBoost2, we both used grid search to find the best model. After training, we get two CatBoost models, then we used them to prediction the shortest path model online. Such costs depend mainly on the model size rather than the Graph size.

To better understand the model, we illustrate it with an example. For a weighted graph with 100 nodes and 120 edges. Wen select 3 landmarks $[v_6, v_8, v_{12}]$. Every node v_i stores its shortest distances to these landmarks as its global features, i.e., a k-dimensional vector $<d(v_i, v_6), d(v_i, v_8), d(v_i, v_{12})>$, where d($\cdot$) represents the distance. Then we concatenate latitude x_i, longitude y_i and the k-dimensional vector as the total feature of one node, it can be described as $V_{feat} = <d(v_i, v_6), d(v_i, v_8), d(v_i, v_{12}), x_i, y_i>$. At offline stage, we random select 500 pair as training data, for nodes v_i and v_j, we scan the store vector V_{feat} and concatenate V^i_{feat} and V^j_{feat}, we also calculate the similarity and Euclidean distance of v_i and v_i, then concatenate them to V^i_{feat} and V^j_{feat}, we describe the final feature as V_{final}. Finally, we fed V_{fianl} to CatBoost1. We get the output $\hat{cat}_1$ of CatBoost1, we concatenate $\hat{cat}_1$ to V_{final}, and fed it to Cat-Boost2 to get the final output $\hat{d}$. After training, we store the model CatBoost1 and CatBoost2. Therefore, in Online stage, the query time is $O(1)$.

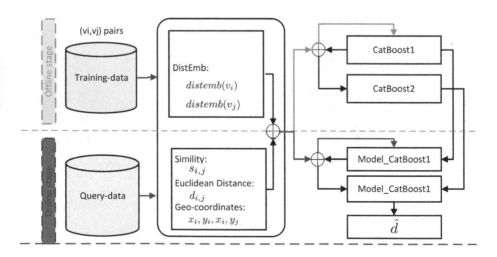

Fig. 2. CatBoost based shortest distance prediction model.

Cost Analysis. We consider an CatBoost based model to have an O(1) space cost for its parameters and O(1) time cost for query. And in DistEmb stage, we take O(k|V|) ($k \ll |V|$) space cost to stored DistEmb vector. Compared to store the distances of all node pairs, which take O($|V|^2$) space cost, our model greatly reduce space complexity. And in online stage, it only cost O(1) while the traditional methods such as BFS and DFS cost O($|V|^2$).

4 Experimental Evaluation

4.1 Datasets

To evaluate the effectiveness of our model, we carried out comparative experiments on two real-world road networks as shown in Table 1, where $\overline{dgr}$ denotes the average degree. In XM dataset, we use 4% $|V|$ node pairs for total training and testing, we randomly select 650,000 pairs of nodes and calculate their shortest path as training data and testing data. And in NY dataset, we use 2% $|V|$ node pairs for training and testing, we select 3000,000 pairs of nodes as training data and testing data. And in both datasets, we set training size as 0.7 and testing size as 0.3.

Table 1. Datasets

Dataset	$|V|$	$|E|$	$\overline{dgr}$	Training pairs	Testing pairs
Xiamen City, China (XM)	19K	22K	2.22	455,000	195,000
New York City, USA (NY)	264K	366K	2.76	2,100,000	900,000

4.2 Hyperparameters

For our model, the two CatBoost models both use grid search to find the best parameters. The training data is randomly shuffled, and the training size is 0.7. The number of landmarks is set as 3, and in Sect. 4.5, we will explore the impact of the number of landmarks on the experiment.

4.3 Evaluation Metrics

We predict the shortest distance between every tow vertices in each dataset and evaluate the performance by three widely used metrics, i.e., Root Mean Square Error (**RMSE**), Mean Absolute Error (**MAE**) and Mean Absolute Percentage Error (**MAPE**). They are defined as follows:

$$RMSE = \sqrt{\frac{1}{n} * \sum_{i=1}^{n}(d_i - \hat{d}_i)} \qquad (2)$$

$$MAE = \frac{1}{n} * |\sum_{i=1}^{n}(d_i - \hat{d}_i)| \tag{3}$$

$$MAPE = \frac{100\%}{n} * |\sum_{i=1}^{n} \frac{(d_i - \hat{d}_i)}{d_i}| \tag{4}$$

where d is the actual distance measured by Dijkstra algorithm and $\hat{d}$ the prediction distance.

Besides, we also use average distance prediction (query) time (**QT**) as the important evaluation metric. The ground truth distance are precomputed using the Dijkstra Algorithm.

4.4   Baselines

We compare with four baselines, among which the distance computed by Dijkstra algorithm is as ground truth distance. For Dijkstra algorithm, we mainly focus on the query time.

Dijkstra. Dijkstra algorithm uses breadth-first search method to solve the single-source shortest path problem in weighted graphs. Its time cost is $O(|V|^2)$ which is intolerable in large graphs.

landmark-bt [11]. A landmark based distance labeling method, it uses the top-k vertices passed by the largest numbers of shortest paths between the vertex pairs as the landmarks.

landmark-km. A landmark based distance labeling method, it uses the k vertices that are the closest to the vertex K-Means centroids as the landmarks.

node2vec [10]. it uses node2vec [3] to learn vertex embeddings and trains an MLP to predict vertex distances given the learned embeddings (we use its recommended settings).

4.5   Results

Overall Results. Table 2 shows the prediction errors. We used the shortest path distance calculated by Dijkstra as the ground truth. Our models outperform the three baseline models across all two datasets and greatly reduce the RMSE, MAE and MAPE. On XM dataset, landmark-km has a slightly lower RMSE than ours, while our MAE and MAPE are still lower (by more than 45%).

The advantage of models comes from the ability to capture the local spatial dependency and spatial dependency and the strong ability of CatBoost to handle the features. Landmark-bt and landmark-km rely on the landmarks which may not preserve the distance for all vertices and landmarks only use the global information of nodes and ignore the local information on nodes. Node2vec focuses on the neighborhood of the nodes i.e. the local information of nodes and ignore the global information of nodes. It also suffers on large road networks such as NY, it cannot train on one NVIDIA TITANX GPU which is marked as "−".

In terms of query time (QT), the landmark approaches and node2vec are much faster than our model, but the MAE and MAPE error are much larger than our model. In XM dataset, our model has no advantage on the query time compare to Dijkstra algorithm. But in larger road network dataset NY, our model reduces query time (QT) by 93% with only 0.01% MAPE error compare with Dijkstra.

Table 2. Performance comparison of different approaches for shortest distance prediction on XM and NY datasets

Method	XM				NY			
	RMSE	MAE	MAPE (%)	QT	RMSE	MAE	MAPE (%)	QT
Dijkstra	−	−	−	0.03	−	−	−	1.84
landmark-bt [11]	726.48	1808.37	0.33	7.01e−5	−	−	−	−
landmark-km	**77.57**	259.86	0.05	7.04e−5	4505.50	8926.03	0.03	**6.63e−5**
node2vec [10]	984.81	733.97	0.16	**3.04e−10**	−	−	−	−
Catboost Based (ours)	89.57	**141.25**	**0.02**	0.18	**4398.73**	**3300.98**	**0.01**	0.12

Effect of Number of Landmarks. In our model, we use landmarks to obtain the global features of nodes. Hence, if the number of landmarks is too small, for the model the information is limited so that it cannot predict the shortest path distance accurately. However, if the number of landmarks is too large, the training time of the model will increase and it will also bring some noise. In the experiment, we fount that the model can achieve good results while the number of landmarks greater than or equal three. Therefore, we choose some different landmark numbers from 3 to 50 in XM dataset and 3 to 300 in NY dataset to explore the influence of the number of landmarks on the experimental results.

Figure 3 visualized the MAE of different number of landmarks of our model on the two datasets. On the XM dataset, when k is approximately equal to 20 it obtained the best result, and on the NY dataset, when k is approximately equal to 150 it obtained the best result. It can be seen that at the beginning, as k increases, the MAE error decreases, but as k sustainable increases, the MAE error slightly decreases. Therefore, it is not that the larger the better. If the number of landmarks is too large, it will bring some noise. In the experiment, we should choose the appropriate value of k according to the dataset.

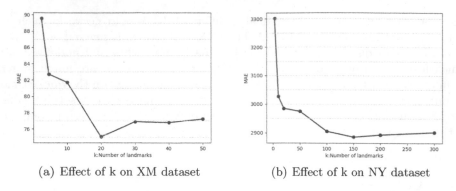

(a) Effect of k on XM dataset   (b) Effect of k on NY dataset

Fig. 3. The effect of number of landmarks k on different datasets (a) XM (b) NY.

5   Conclusion

We propose a landmark-based method to extract the features of the nodes on the road network and combine them with the real information of the road network as the total feature of nodes. Then we use CatBoost to predict the shortest path distance between node pairs. In the offline stage, the space complexity is $O(k|V|)$. In the online prediction stage, the space complexity is only relating to the parameters of the model and has nothing to do with the size of the road network and the time complexity is $O(1)$. The experimental results show the effectiveness of our method. However, this method still has many limitations. For example, the extraction of node features is separated from model training. Is there a better way to combine feature extraction and model training so that the prediction results of the model forward propagation to affect the feature extraction to improve model accuracy. In future work, we will focus on how to better extract the node features of the nodes.

References

1. Chow, E.: A graph search heuristic for shortest distance paths. Citeseer (2005)
2. Dorogush, A.V., Ershov, V., Gulin, A.: CatBoost: gradient boosting with categorical features support. arXiv preprint arXiv:1810.11363 (2018)
3. Grover, A., Leskovec, J.: node2vec: scalable feature learning for networks. In: Proceedings of the 22nd ACM SIGKDD International Conference on Knowledge Discovery and Data Mining, pp. 855–864 (2016)
4. Gubichev, A., Bedathur, S., Seufert, S., Weikum, G.: Fast and accurate estimation of shortest paths in large graphs. In: Proceedings of the 19th ACM International Conference on Information and Knowledge Management, pp. 499–508 (2010)
5. Maleki, S., Nguyen, D., Lenharth, A., Garzarán, M., Padua, D., Pingali, K.: DSMR: a parallel algorithm for single-source shortest path problem. In: Proceedings of the 2016 International Conference on Supercomputing, pp. 1–14 (2016)
6. Mikolov, T., Chen, K., Corrado, G., Dean, J.: Efficient estimation of word representations in vector space. arXiv preprint arXiv:1301.3781 (2013)

7. Potamias, M., Bonchi, F., Castillo, C., Gionis, A.: Fast shortest path distance estimation in large networks. In: Proceedings of the 18th ACM Conference on Information and Knowledge Management, pp. 867–876 (2009)
8. Qi, J., Wang, W., Zhang, R., Zhao, Z.: A learning based approach to predict shortest-path distances (2020)
9. Rattigan, M.J., Maier, M., Jensen, D.: Using structure indices for efficient approximation of network properties. In: Proceedings of the 12th ACM SIGKDD International Conference on Knowledge Discovery and Data Mining, pp. 357–366 (2006)
10. Rizi, F.S., Schloetterer, J., Granitzer, M.: Shortest path distance approximation using deep learning techniques. In: 2018 IEEE/ACM International Conference on Advances in Social Networks Analysis and Mining (ASONAM), pp. 1007–1014. IEEE (2018)
11. Takes, F.W., Kosters, W.A.: Adaptive landmark selection strategies for fast shortest path computation in large real-world graphs. In: 2014 IEEE/WIC/ACM International Joint Conferences on Web Intelligence (WI) and Intelligent Agent Technologies (IAT), vol. 1, pp. 27–34. IEEE (2014)
12. Tang, J.T., Wang, T., Wang, J.: Shortest path approximate algorithm for complex network analysis. Ruanjian Xuebao/J. Softw. **22**(10), 2279–2290 (2011)
13. Zhang, H., Xie, X., Wen, Y., Zhang, Y.: A twig-based algorithm for top-k subgraph matching in large-scale graph data. In: Wang, G., Lin, X., Hendler, J., Song, W., Xu, Z., Liu, G. (eds.) WISA 2020. LNCS, vol. 12432, pp. 475–487. Springer, Cham (2020). https://doi.org/10.1007/978-3-030-60029-7_43

Traffic Prediction Based on Multi-graph Spatio-Temporal Convolutional Network

Xiaomin Yao, Zhenguo Zhang$^{(\boxtimes)}$, Rongyi Cui, and Yahui Zhao

Department of Computer Science and Technology, Yanbian University,
Yanji 133002, China
{2020050074,zgzhang,cuirongyi,yhzhao}@ybu.edu.cn

Abstract. In the Intelligent Traffic System (ITS), accurate prediction of the state of the traffic at the next moment is of great significance to transportation planning. Existing researches mainly focus on the research of the topological structure of the road network in space, and consider the temporal dependence at the same time. However, we have noticed that it is not only important to consider the dependence of time and space at the same time, but also other organizational relationships between the road network will also affect the forecast results. In this paper, we reconsider the correlation between roads and capture their correlation in both space and time. More specifically, we first encode the road network into two graphs (connect graph and similar graph) based on the connectivity and historical pattern similarity, and merge the graphs to make the connected edges carry more information. Then graph convolution is used for the fused graph. In order to capture the temporal dependence, we use one-dimensional convolution to first convolve the information before the graph convolution, and then perform the one-dimensional convolution on the information after the fusion graph, which can achieve fast spatial-state propagation from graph convolution through fast spatial-state propagation. We evaluate the predictive performance of our model by real-world traffic dataset and experiments prove that the addition of multi-graph is effective. Compared with the relatively new baseline, *RMSE* has dropped by about 10%.

Keywords: Traffic prediction · Multi-graph · Spatial dependence · Temporal dependence

1 Introduction

With the development of society and the increase in traffic demand, traffic forecasting has become a widespread concern. The key to traffic flow forecasting is to capture information related to the forecast. However, the traffic flow changes dynamically. In terms of time, the state of the previous moment and the next moment are different, and there are similarities at the same moment in different period. For space, the relationship between road networks has a different organizational structure. There are strong constraints between adjacent sections, and

© Springer Nature Switzerland AG 2021
C. Xing et al. (Eds.): WISA 2021, LNCS 12999, pp. 144–155, 2021.
https://doi.org/10.1007/978-3-030-87571-8_13

sections with long distances but similar historical patterns actually have a certain correlation. We must take into account as many factors as possible.

By considering temporal dependence, both statistical methods and machine learning methods have achieved good results. Among them, the time series model is the most typical statistical method. Such as Autoregressive Integrated Moving Average (ARIMA), Seasonal Autoregressive Integrated Moving Average (SARIMA), and Kalman filtering methods. However, they ignore the spatial dependence in road networks.

Machine learning-based models have strong ability to rely on feature extraction, including Linear Regression (LR), K-nearest Neighbor Regression (KNN), Support Vector Regression (SVR), Long Short-Term Memory (LSTM), etc. The above methods all rely on historical data and only consider the temporal dependence. However, the traffic flow between different roads will affect each other (see Fig. 1). When the road 1 is congested, the traffic flow on the road 2 will also be affected. Therefore, it is necessary to consider spatial dependence in the scope of research. In recent years, the Graph Neural Network (GCN) [7] has achieved good results on the traffic prediction problem. This kind of method generally uses the adjacency matrix of the node as input, and the error has also been reduced to a certain extent. However, this method does not take into account the complicated spatial relationship between long-distance roads.

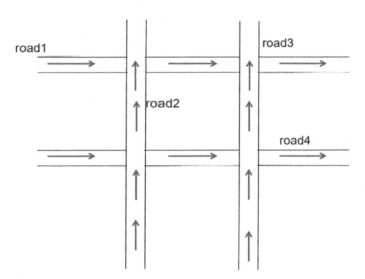

Fig. 1. The figure of road state influence. When road 1 is congested, road 2 will also be affected, which indicates that the topological relationship between the road networks makes the road states affect each other.

Based on this background, this paper proposes a spatio-temporal prediction model based on multiple graphs. In terms of space, we consider the various relationships between roads, and use graph convolution to extract features in spatial dimensions. In the temporal dependence, we use one-dimensional convolution to model the temporal dependence. Our model achieves stable results in both short-term and long-term predictions, which proves that this model has certain advantages in short-term, medium-term and long-term task predictions.

2 Related Work

2.1 Statistical Models

As for traditional statistical models, ARIMA [5] is the most common time series prediction model in statistical models. It regards the data sequence formed by the traffic state over time as a random sequence, and uses a certain mathematical model to approximate this sequence. Once the model is identified, it can predict future values of the time series. *Hamed* and others use ARIMA to predict urban traffic flow. In the past few decades, many time series models based on ARIMA have been used for traffic forecasting.

Billy et al. [15]proposed an ARIMA model with periodic components, namely the SARIMA model, they assuming that the weekly seasonal check of traffic data can produce a week of smooth conversion, the univariate traffic data flow can be simulated as a SARIMA process.

Okutani and *Stephanedes* [9] applied the Kalman filter method to traffic prediction for the first time, and then methods based on Kalman filter have been proposed one after another, and its applications in the field of transportation are increasing. Although a certain accuracy rate can be achieved, most traditional statistical methods can only capture linear relationships, but not nonlinear relationships. In addition, the system model is static and cannot reflect the uncertainty of traffic data.

2.2 Machine Learning Methods

In recent years, machine learning methods have been widely used in the field of traffic prediction. Long Short-Term Memory network (LSTM) [20] has achieved good results in time series data modeling. It can solve the problem of vanishing or exploding gradients when the standard recurrent neural network (RNN) simulates long-term dependence, that is, the gradient may increase or decay exponentially on a long sequence. *Wang* et al. [14] suggested to use the tree-based LSTM model to simulate long-term dependence based on the characteristics of natural language that combines words into phrases through grammar, and achieved good results, but it does not consider the period of traffic flow. And ignore the spatial dependence of the road network.

However, this attempt only considered the temporal correlation of traffic changes, and did not consider its spatial correlation from the perspective of the network. In order to make up for this shortcoming. *Sainath* et al. [11] proposed CNN, it is an image-based method to represent network traffic as an image, and use the deep learning structure of convolutional neural network (CNN) to extract the spatio-temporal traffic features contained in the image. It converts the traffic information into a two-dimensional spatio-temporal matrix, and then processes it through CNN. The accuracy is improved, and the result is obtained in a reasonable time with fewer parameters, which is suitable for large-scale transportation networks. Although the method of converting traffic information into images has been improved to a certain extent, the local receptive field of CNN cannot reflect the actual correlation between road networks and does not well represent the spatial features of traffic information.

With the development of GCN [7,18], its application field has also been extended to traffic prediction tasks. Since the relationship between road networks is very consistent with the definition of graphs, the introduction of graph convolution provides new ideas for traffic prediction problems. To solve the shortcomings of traditional convolution that can only deal with european space, GCN models the graph structured data of the road network to better obtain the spatial dependence. *Zhao* et al. [19] used graph neural network to model spatial dependence and Gate Recurrent Unit (GRU) to model temporal dependence, and the effect was significant.

3   Methodology

We use traffic speed as traffic information, and predict short, medium and long-term traffic information based on the historical traffic information of the road. First, we define the traffic network as a graph structure, $G\left(V, E\right).V$ represents roads, each road represents a vertex, and E represents the edges between roads and then formulate the traffic prediction problem.

$$[X_{t+h}] = f\left(G_f; \left(X_{t-k}, \cdots, X_{t-1}, X_t\right)\right) \tag{1}$$

where G_f represents the graph after fusion, X_{t+h} represent the traffic information at the $t + h$ time, where k is the length of historical time series.

3.1   Architecture

In this section, we describe how to use a multi-graph convolution model to complete the task of traffic prediction. Specifically, the model includes a graph convolutional layer and two gated convolutional layers, as shown in Fig. 2.

We first use the historical time series as the input of the one-dimensional convolutional layer, and then capture the spatial features through graph convolution, then input the sequence with both time and space features into the one-dimensional convolutional layer, and finally output the prediction results.

3.2   Multi-graph Generation

The generation of the graph is the key to the graph convolution model. If the constructed graph cannot play a role in the traffic prediction problem, the application

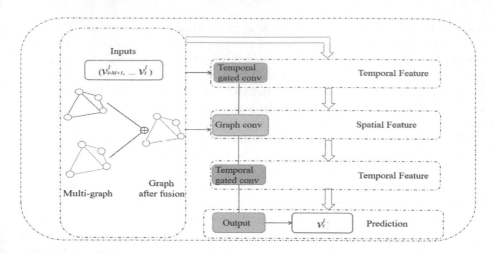

Fig. 2. The whole structure include the temporal dependence module, and the spatial dependence module. Among them, two layers of one-dimensional convolution and a graph convolution in the middle. It can achieve fast spatial-state propagation from graph convolution through fast spatial-state propagation.

of the graph convolution model will not help to improve the prediction effect, and it will even reduce the prediction performance during training.

Connect Graph. Generally, we believe that two roads that are adjacent in space must have a spatial correlation of traffic flow. Based on this, we define the adjacent graph $G\left(V_n, E\right)$, and weight on edges is defined as follows:

$$Weight = \begin{cases} 1; & i \text{ and } j \text{ is connected} \\ 0; & i \text{ and } j \text{ is not connected} \end{cases} \tag{2}$$

Similar Graph. According to Tobler's First Law of Geography [13], everything is related to other things, but similar things are more closely related. Although some roads are far apart in space, if the historical traffic patterns are similar, then the traffic state at the next moment is likely to be the same. For example, the roads close to the school will have varying degrees of congestion at the end of school. Based on this idea, we using Pearson's correlation coefficient [10], the following formula calculates the similarity of historical patterns between roads, which can better measure the correlation of historical traffic flow between two road sections. We define the similarity graph as $G\left(V_s, E\right)$, the weight on edges calculation is as follows:

$$\rho(X, Y) = \frac{E(XY) - E(X)E(Y)}{\sqrt{E\left(X^2\right) - E^2(X)}\sqrt{(Y^2) - E^2(Y)}} \tag{3}$$

Where X and Y represents the historical pattern of any two roads. Through this step of calculation, we get the similarity between the roads, which lays the foundation for the next step of generating similarity graphs.

Graph Fusion. After the multi-graph construction is completed, we use the weighted fusion method to fuse the multi-graphs. In order to maintain the standardization of the matrix after fusion, we have added the softmax operation to the calculation.

$$W_1{}', W_2{}' = softmax(W_1, W_2) \tag{4}$$

$$F = M_1 \circ W_1{}' + M_2 \circ W_2{}' \tag{5}$$

where F represents the graph after fusion; M_1 and M_2 represents two feature matrices; where $\circ$ is the element-wise product; W_1 and W_2 represents learnable parameters.

3.3 Spatial Dependence Modeling

In traffic prediction tasks, capturing complex spatial dependence has always been a difficult problem worth challenging. Convolution operations can accurately extract local spatial relationships, but standard CNN can only be used in Euclidean space [3].

In traffic prediction problems, the relationship between roads is non-European, more precisely in the form of a graph rather than a two-dimensional grid, so CNN cannot accurately model the spatial dependence of traffic prediction. Graph convolution(GCN) [7] has advantages in capturing spatial features. We use Chebyshev graph convolution [4] to model spatial dependence, and the process of Chebyshev graph convolution can be expressed as follows.

$$g \otimes x = \sum_{k=0}^{k} \beta_k T_k \left(\hat{L} \right) x \tag{6}$$

Where $\hat{L} = 2/\lambda_{max} - I_N$; The introduction of $\hat{L}$ is because the independent variable of the Chebyshev polynomial needs to be between -1 and $+1$, λ_{max} represents the largest eigenvalue in the Laplacian matrix L; β_k represents Chebyshev coefficient that is updated and learned in the graph network. $T_k (\cdot)$ is a Chebyshev polynomial of order k. Chebyshev polynomials are defined recursively.

$$T_k (x) = 2x T_{k-1} - T_{k-2} (x) \tag{7}$$

where $T_0 (x) = 1$; $T_1 (x) = x$.

Ordinary GCN includes Fourier transform. In order to reduce the complexity of calculation, Chebyshev polynomial plays a very good role. It not only has numerical stability, but also calculates more efficiently.

3.4 Temporal Dependence Modeling

Causal Convolution. In the traffic prediction problem, obtaining temporal dependence is another important issue. In recent years, Recurrent Neural Network (RNN) networks have been widely used to model sequence data [17]. However, the gradient vanishing problem and gradient exploding problem of RNN are very unfavorable for long-term prediction.

In this paper, we use an improved CNN one-dimensional convolution, that is, causal convolution, is used to model the time correlation. We denote the time series as $x_1, x_2, ..., x_k$, and the prediction process is shown in Eq. 8.

$$p(y_k) = \prod_{t=1}^{T} p\left(x_k \mid x_{k-1}, \ldots, x_1\right) \tag{8}$$

We calculate the output value at time k as y_k. Knowing all the eigenvalues before time k, we can calculate the eigenvalues output at time k faster through causal convolution. The causal convolution process is shown in Fig. 3

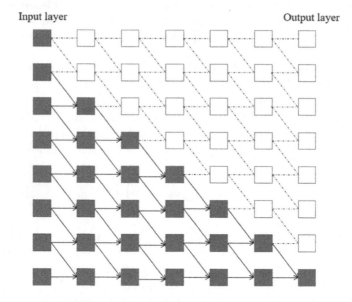

Fig. 3. The process of causal convolution. 11111111111 The output at a certain moment only depends on the input at the previous moment, and does not depend on the input at a certain moment and later.

Gating Mechanism. However, there are still some problems. When considering the variable x a long time ago, the convolutional layer needs to be increased accordingly, which will cause problems such as gradient vanishing problem, so a gating mechanism is introduced to solve this problem. If the general deep convolution is expressed as Eq. 9.

$$h_l(\mathbf{X}) = \mathbf{X} * \mathbf{W} + \mathbf{b} \tag{9}$$

Where W represents the convolution kernel,b represents bias. And the Gated Convolutional Networks(GLU) [2] can describe as follows:

$$h_l(\mathbf{X}) = (\mathbf{X} * \mathbf{W} + \mathbf{b}) \otimes (\mathbf{X} * \mathbf{V} + \mathbf{c}) \tag{10}$$

It performs two sets of convolution operations on the same layer input of the sequence. One output does not go through a nonlinear function, and the other is convolution through a sigmod activation function. Gated convolution can effectively reduce the problem of gradient vanishing problem and accelerate convergence. In LSTM [6], the input gate and the forget gate can achieve long-term storage, so that the information is not lost in the conversion process. The convolutional network does not have this problem, so only the output gate is used in the GLU, which can control which information can be passed to the next layer, which has achieved good results in extracting temporal dependence.

4   Experiments

4.1   Dataset

In this section, we evaluate the prediction performance of the our model on real-world datasets: SZ-taxi set. Road information is represented by speed. SZ-taxi set collects taxi trajectories in Luohu District, Shenzhen from Jan. 1 to Jan. 31, 2015. Including 156 roads. The data set includes three parts. One is a 156 * 156 adjacency matrix, which expresses the connection relationship between roads. One is a 156 * 156 similarity matrix, which represents the historical pattern similarity between roads. There is also a characteristic matrix, which records the speed change of each road at different times, it collect the speed value every 15 min.

We divide the data set into three parts: the data from January 1 to January 25 is used as the training set (about 80%), the data from January 25 to January 27 is the validation set (about 10%), the data from 28th to January 31st is the test set (about 10%). In addition, the data is normalized using the standard score method, and the processing method is shown in the Eq. 11.

$$Z = \frac{x - \mu}{\delta} \tag{11}$$

Then we measured the similarity of 156 roads according to the feature matrix. We used the calculation method of Pearson coefficient to generate a 156*156 matrix, which is also an important part of multi-graph construction.

4.2   Evaluation Metrics and Baselines

We compare the multi-graph convolution model with other methods. Root Mean Square Error (RMSE) represents the sample standard deviation of the difference between the predicted value and the observed value. Mean Absolute Error (MAE) represents the average value of the absolute error between the predicted value and the observed value. R-squared predicts the degree of agreement between the results and the actual situation. In the experimental results, the closer to 1 the better. Variance Score (Var) explain the variance score of the model. The closer it is to 1, the more the independent variable can explain the variance of the dependent variable. We have selected the following baselines as the result of comparison with this model. Among them, ST-GCN is a newer model that also considers temporal and spatial dependence.

(1) History Average Model (HA) [8], use the average of historical traffic information as the prediction condition.
(2) Autoregressive Integrated Moving Average model (ARIMA) [1], the time series is regarded as random and then converted into a stationary series, and then described by a mathematical model.
(3) Support Vector Regression (SVR) [12], take traffic information as input, select the kernel function, train the support vector regression machine, and then predict the traffic information at the next moment.
(4) Graph convolutional networks (GCN), it use the topology between roads Nature, modeling spatial dependence, predicting traffic information.
(5) ST-GCN[16], a method to predict traffic information considering both temporal and spatial dependence.

4.3   Experimental Results

As shown in Table 1, our method has achieved better results than baselines in traffic predictions. We think this is because the overall model has a better ability to extract spatial dependence and temporal dependence.

In order to verify whether the multi-graph convolution part of the model actually reduces the error, we set up an ablation experiment, as shown in Fig. 4.

Table 1. Performance comparison of traffic prediction methods (the data in bold font is the best experimental result, * means negligible).

Time	Metric/Method	HA	ARIMA	SVR	GCN	ST-GCN	**Our model**
15 min	RMSE	7.92	8.22	7.53	9.27	5.12	**4.56**
	MAE	5.51	6.22	5.12	7.26	3.69	**3.19**
	R-squared	0.81	0.08	0.81	0.61	0.90	**0.92**
	Var	0.81	*	0.81	0.61	0.90	**0.92**
30 min	RMSE	7.92	8.21	7.50	9.35	5.18	**4.59**
	MAE	5.51	6.22	4.91	7.32	3.71	**3.23**
	R-squared	0.82	0.08	0.82	0.61	0.90	**0.91**
	Var	0.82	*	0.81	0.61	0.90	**0.91**
45 min	RMSE	7.92	8.21	7.50	9.41	5.20	**4.62**
	MAE	5.51	6.22	5.03	7.37	3.78	**3.25**
	R-squared	0.79	0.08	0.81	0.60	0.90	**0.91**
	Var	0.79	*	0.81	0.60	0.90	**0.92**

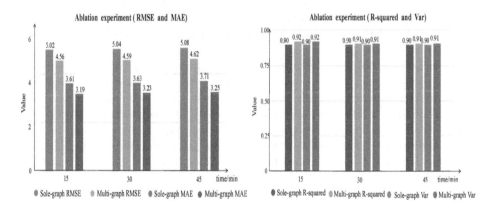

Fig. 4. It can be seen that the addition of the multi-graph significantly reduces the RMSE and MAE of the model, and Var and R-squared are slightly better than the single graph.

It can be seen from the experimental results that the design of multiple graphs significantly reduces the prediction error, the RMSE is reduced by 8.9% to 9.1 %.

In Fig. 5, we calculated the magnitude of changes in our model and GCN's different indicators over time, expressed as a percentage. The results show that our method is relatively stable in short-term, mid-term, and long-term forecasts.

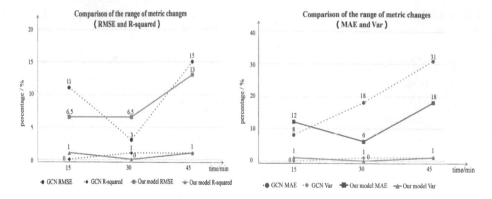

Fig. 5. This figure compares the trend of the various indicators of this model and the GCN model over time. Obviously, as time increases, the trend of this model is more relaxed, which is more suitable for traffic forecasting problems.

5   Conclusion and Future Work

In this paper, we propose a novel method which combines multi-graph convolution and one dimensional convolution. We perform multi-graph convolution in terms of spatial dependence, and applies one-dimensional convolution in terms of temporal dependence.

Our model has obtained better results than other methods on real-world data sets. The performance of the model is analyzed from the short-term, medium-term, and long-term respectively. The results show that the model can not only play a good role in short-term prediction tasks but also performed very well in the medium and long-term forecasts. Since this paper only considers single-step time prediction, the next step will focus on multi-step time prediction, and the method in this paper can also be further applied to other spatio-temporal prediction tasks.

Acknowledgements. This work is supported by the school-enterprise cooperation project of Yanbian University [2020-15], State Language Commission of China under Grant No. YB135-76 and Doctor Starting Grants of Yanbian University [2020-16].

References

1. Ahmed, M., Cook, A.: Analysis of freeway traffic time series data by using Box-Jenkins techniques. Transp. Res. Rec. **773**, 1–9 (1979)
2. Dauphin, Y.N., Fan, A., Auli, M., Grangier, D.: Language modeling with gated convolutional networks (2016)
3. Defferrard, M., Bresson, X., Vandergheynst, P.: Convolutional neural networks on graphs with fast localized spectral filtering (2016)
4. Defferrard, M., Bresson, X., Vandergheynst, P.: Convolutional neural networks on graphs with fast localized spectral filtering (2016)

5. Hamed, M., Al-Masaeid, H., Said, Z.: Short-term prediction of traffic volume in urban arterials. J. Transp. Eng. ASCE - J. Transp. Eng. ASCE **121** (05 1995). https://doi.org/10.1061/(ASCE)0733-947X(1995)121:3(249)

6. Hochreiter, S., Schmidhuber, J.: Long short-term memory. Neural Computation **9**(8), 1735–1780 (1997)

7. KipF, T.N., Welling, M.: Semi-supervised classification with graph convolutional networks (2016)

8. Liu, J., Wei, G.: A summary of traffic flow forecasting methods. J. Highw. Transp. Res. Dev. **3**, 82–85 (2004)

9. Okutani, I., Stephanedes, Y.J.: Dynamic prediction of traffic volume through Kalman filtering theory. Transp. Res. Part B Methodol. **18**(1), 1–11 (1984). https://doi.org/10.1016/0191-2615(84)90002-X. https://www.sciencedirect.com/science/article/pii/019126158490002X

10. Pearson, K.: VII. Note on regression and inheritance in the case of two parents. Proc. Roy. Soc. Lond. **58**, 240-242 (1895)

11. Sainath, T.N., et al.: Deep convolutional neural networks for large-scale speech tasks. Neural Netw. **64**, 39–48 (2015)

12. Smola†, A., Lkopf‡, B.: A tutorial on support vector regression. Stat. Comput. **14**(3), 199–222 (2004)

13. Tobler, W.R.: A computer movie simulating urban growth in the Detroit region. Econ. Geogr. **46**(2) (1970)

14. Wang, Y., Huang, M., Zhu, X., Zhao, L.: Attention-based LSTM for aspect-level sentiment classification. In: Proceedings of the 2016 Conference on Empirical Methods in Natural Language Processing, pp. 606–615. Association for Computational Linguistics, Austin, Texas, November 2016. https://doi.org/10.18653/v1/D16-1058. https://www.aclweb.org/anthology/D16-1058

15. Williams, B.M., Hoel, L.A.: Modeling and forecasting vehicular traffic flow as a seasonal ARIMA process: theoretical basis and empirical results. J. Transp. Eng. **129**(6), 664–672 (2003)

16. Yan, S., Xiong, Y., Lin, D.: Spatial temporal graph convolutional networks for skeleton-based action recognition (2018)

17. Zhang, J., Man, K.F.: Time series prediction using RNN in multi-dimension embedding phase space. In: SMC98 Conference IEEE International Conference on Systems (2002)

18. Zhao, B., Xu, Z., Tang, Y., Li, J., Liu, B., Tian, H.: Effective knowledge-aware recommendation via graph convolutional networks. In: Wang, G., Lin, X., Hendler, J., Song, W., Xu, Z., Liu, G. (eds.) WISA 2020. LNCS, vol. 12432, pp. 96–107. Springer, Cham (2020). https://doi.org/10.1007/978-3-030-60029-7_9

19. Zhao, L., Song, Y., Zhang, C., Liu, Y., Li, H.: T-GCN: a temporal graph convolutional network for traffic prediction. IEEE Trans. Intell. Transp. Syst. **PP**(99), 1–11 (2019)

20. Zhuo, Q., Li, Q., Han, Y., Yong, Q.: Long short-term memory neural network for network traffic prediction. In: 2017 12th International Conference on Intelligent Systems and Knowledge Engineering (ISKE) (2017)

Online Runtime Prediction Method for Distributed Iterative Jobs

Xiaofei Yue[1], Lan Shi[1], Yuhai Zhao[1(✉)], Hangxu Ji[1], and Guoren Wang[2]

[1] School of Computer Science and Engineering, Northeastern University,
Shenyang 110819, China
zhaoyuhai@mail.neu.edu.cn
[2] School of Computer Science and Technology, Beijing Institute of Technology,
Beijing 100081, China

Abstract. Predicting the runtime of distributed iterative jobs can help reduce the deployment cost of clusters and optimize their resource allocation and scheduling strategies, but the runtime depends on various factors which are difficult to be acquired before execution. In this paper, we propose a generalized online prediction method for the runtime of distributed iterative jobs, which is centered on a series of online machine learning models. The method consists of three phases: 1) estimating the number of iterations for the current iterative job. 2) predicting the runtime metrics of each iteration by an online polynomial regression model. 3) Runtime metrics sequence is analyzed using an LSTM trained with online learning to predict the runtime of each iteration. We conducted experiments on typical Flink iterative jobs, and the experimental results show that our method improves the accuracy by 4.79% compared to the state-of-the-art methods, while for the improvement in accuracy for delta iterative jobs is even more than 15%.

Keywords: Iterative job · Online runtime prediction · LSTM ·
Polynomial regression · Flink

1 Introduction

In the past few years, we have seen a rapid growth of large-scale analytic jobs, especially in the field of artificial intelligence, such as in the direction of distributed image recognition [1], etc. Moreover, the core algorithms of these operations are often iterative, i.e., one or more steps are executed repeatedly until convergence criteria are met. Current big data computing platforms like Apache Flink [2] allow users to perform distributed iterative computation on clusters to analyze large-scale dataset.

Most of the distributed iterative jobs need to comply with service level objectives (SLO) in real production [3], because external applications usually need timely feedback of analysis results, too high execution latency can violate service level agreements (SLA) with users [4]. However, in poor cluster environments, even using all the remaining computational resources can cause the job to take

© Springer Nature Switzerland AG 2021
C. Xing et al. (Eds.): WISA 2021, LNCS 12999, pp. 156–168, 2021.
https://doi.org/10.1007/978-3-030-87571-8_14

much longer to complete than the user expects, so runtime prediction is a very useful mechanism for solving the job feasibility analysis problem. In addition, based on the predicted runtime, the allocation of cluster resources and deployment costs can be rationalized and optimized [5–7]. Unfortunately, the runtime of iterative jobs depends on factors such as dataset characteristics, system configuration and algorithm parameters, there are also involve inter-iteration dependencies for delta iterative jobs, these information is difficult to obtain before job execution. Previous works tend to be a single model designed for a specific distributed framework and requires specialized isolated training that does not meet the user's needs for response time and prediction accuracy [8–10].

In this paper, we proposes a method for predicting the runtime of iterative jobs in distributed environment, which centers on fine-grained prediction using a series of online machine learning models. For distributed frameworks that support resource monitoring and pluggable components, our online model is generic, and once the model is built, it only needs to be fine-tuned to adapt to changes in the dataset and cluster.

The main contributions of this work can be summarized as follows:

- A generic approach to online predict the runtime of iterative jobs is proposed, which performs fast and fine-grained prediction through multiple phases.
- The runtime prediction of iterative jobs is enhanced by monitoring fine-grained resource consumption data and constructing runtime metrics sequences based on the order of iteration steps.
- The accuracy of the proposed method is validated by performing experiments on four typical Flink iterative jobs.

The rest of the paper is organized as follows. Section 2 presents the background on which our work is based. Section 3 elaborates our runtime prediction method. Section 4 validates the effectiveness of our method and analyzes the experimental results. Section 5 presents related work, and finally, Sect. 6 describes the conclusions and future research directions.

2 Preliminaries

2.1 Problem Definition

Definition 1 (Container). In resource management systems such as YARN [11], containers are requested by different frameworks as the basic unit of resource allocation and applied to the distributed jobs, and basic computing resources such as CPU cores, memory are encapsulated in containers, and the monitoring of system resources in this paper is carried out through containers.

Definition 2 (Runtime Prediction Model). In a distributed iterative job, assuming rsc is the current input and cluster resource state, $x^{(i)}(rsc)$ denotes the execution characteristics of the i-th iteration of the iterative job in that state, and $y^{(i)}$ is the time consumed by that iteration, then building a runtime prediction model is to find the functional relationship as shown below.

$$runtime = \sum_i f : rsc \rightarrow x^{(i)}(rsc) \rightarrow y^{(i)} \tag{1}$$

An iterative job consists of a series of steps $S = (S_1, \cdots, S_t, S_{t+1}, \cdots S_n)$ with certain dependencies between adjacent iterative steps S_t and S_{t+1}, and the execution of S_{t+1} will start only after the execution of S_t is completed. Our runtime prediction model does its work by analyzing multiple sequences consisting of execution features from iterative steps with absolute sequential order that from the resource management platform's monitoring of the consumption of resource utilization in the container assigned to the iterative job.

2.2   Background

Distributed Iterative Computation. There are many distributed systems that support users to build iterative programs using the provided iterative operators. The algorithmic logic of an iterative program is mainly composed of Multiple operators combined in different degrees of parallelism.

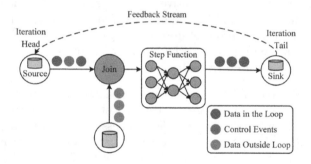

Fig. 1. The structure of iterative computing model in Apache Flink.

The Flink supports iterative computation by defining step function and embedding it in a special operator. As shown in Fig. 1, to maintain the DAG-based runtime and scheduler, Flink constructs iterative channels by creating iteration "head" and "tail" tasks and interconnected feedback edges [2].

Online Machine Learning. Online machine learning has gained wide attention from academia and industry [12]. Because it can stream training samples and incrementally update models, under the problem explored in this paper, the use of online learning methods can quickly incorporate data generated by periodically operating iterative jobs into the learning of models, and then give a near real-time [13]. Therefore, the online learning model is useful for creating a portable and general-purpose runtime prediction system.

3   Approach

In this section, we present an experimental method of online runtime prediction for distributed iterative jobs, as shown in Fig. 2, which consists of the following three key steps: 1) collecting iterative jobs and data required for training the model, 2) estimating the number of iterations for the current job, and 3) identifying the key features and building an online prediction model.

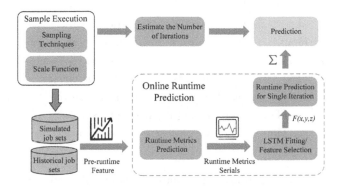

Fig. 2. The process of online runtime prediction for iterative jobs.

3.1   Data Collection

The iterative job sets collected for subsequent training of the runtime prediction model consist of two main types.

1) historical jobset. Since iterative jobs are typically executed periodically to analyse batches of data, we mainly use the relevant statistics and execution times of historical iterative jobs as training data.
2) Simulated jobset. For the case where historical data is scarce, we use sample execution to quickly generate simulated data, i.e., different iterative jobs are run on sample dataset generated with different sampling ratios.

Sampling Technique. Specifically, the convergence of the graph iteration algorithm is essentially determined by the vertices with higher out-degree, so we use the biased random jump (BRJ) sampling method proposed by Daniel et al. [9] which is more biased towards the vertices with higher out-degree when performing random wandering to preserve the key properties of the original graph.

3.2   Key Input Features

As shown in Tables 1 and 2, these key features are classified into two categories, namely, pre-runtime features and runtime features.

Pre-runtime features can be obtained statically before the execution of iterative jobs, and include job type, dataset size, and system configuration, which reflect the basic properties of the job and cluster environment. Runtime features, as the name implies, can only be determined when the iterative jobs are actually running, and they can be used to directly reflect the performance differences of the cluster environment during the iterative process and the dependencies between iterations, including the fine-grained resource usage of each iteration.

Table 1. Description of the pre-runtime features of an iterative job.

Type	Name	Description
Iterative job	id	Execution ID of the iterative job
	input	Relative size of the input dataset for the iterative job
Cluster status	memory	Available memory capacity of the cluster
	disk	Available disk capacity for the cluster
	vcpu	Number of CPU cores in the cluster
Time point	startTime	Start time of iterative job execution
	endTime	End time of iterative job execution

Table 2. Description of the runtime features of an iteration.

Resource	Name	Description
CPU	procs	Number of processes during this iteration
	uTime	Time spent in user mode of the CPU during this iteration
	threads	Number of threads during this iteration
	kTime	Time spent in kernel mode of the CPU during this iteration
Memory	rsSize	The size of the resident memory occupied during the iteration
	vmSize	The size of the virtual memory occupied during the iteration
IO	ioWait	Time spent waiting on IO during this iteration
	rChar	Number of bytes read using read-like syscall during this iteration
	rBytes	Number of bytes read from disk during this iteration
	wChar	Number of bytes written using write-like syscall during this iteration
	wBytes	Number of bytes written to disk during this iteration

3.3 Estimate the Number of Iterations

Our method for estimating the number of iterations relies on the sample execution described in Sect. 3.1.

Scale Function. For iterative algorithms where the convergence threshold changes according to the size of the dataset, let sr is the sampling ratio, the scale function T has the following form:

$$T = (Conf_S = Conf_G, Conv_S = Conv_G/sr) \tag{2}$$

where $Conf_S => Conf_G$ and $Conv_S => Conv_G$ denote the mapping of configuration parameters and convergence parameters.

By combining the sampling technique and scale function, it is possible to make the sample execution and the real execution invariant in terms of the number of iterations, thus perform the estimation of the number of iterations.

Example. PageRank is a classical iterative algorithm and converges when the change in the result of two adjacent iterations is less than a threshold τ. For example, the graph G shown in Fig. 3 and its three arbitrary samples S1-S3 (with a sampling ratio of 50%), samples S1 and S2 are sampled using BRJ, preserving the diameter $D = 2$ of the original graph G. However, the opposite is true for sample S3, but none of the above samples maintains the invariance of the number of iterations. Because after the first iteration, the average delata of their PageRank values change as: $\bar{\Delta}(s1,1) = d/8$, $\bar{\Delta}(s2,1) = d/8$, $\bar{\Delta}(s3,1) = 3d/16$, $\bar{\Delta}(G,1) = d/16$, where $d = 0.85$ is the damping factor. Assuming the threshold $\tau = d/16$, then the original graph G will converge after one execution, while the rest of the samples will continue to execute, but by applying the scale function $T = (Conf_S = Conf_G, Conv_S = Conv_G * 2)$ on samples S1 and S2 the corresponding convergence threshold can be adjusted to $d/16$, thus maintaining the same number of convergence as the graph G.

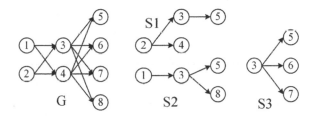

Fig. 3. Maintaining invariants for the number of iterations on sample graphs.

3.4   Online Runtime Prediction Model

The values of these runtime features cannot be determined before the execution, so a two-stage online machine learning model is built in this section, and the specific process is shown in Algorithm 1.

In the first stage, a model is generated for each runtime feature by iteration round, and the purpose of these models is to learn the impact of different inputs and cluster environments on the iterative process, and in the second stage, the final prediction of the runtime of each iteration step is made based on the runtime metric model generated in the first stage.

Algorithm 1: Online Runtime Prediction Algorithm

Input: J (Iterative job), S (Available resources), P (Input data scale)
Output: α (Runtime prediction value for Job J with resource S and input P)
1 $\tilde{N} \leftarrow IterNumEstimate\,(J)$;   // Estimate number of iterations
2 $\Phi \leftarrow \emptyset$;   // Initialize the set of pre-runtime features
3 **for** i=1:n **do**
4 $\quad\big|\quad \Phi \leftarrow \Phi \cup \Gamma_i\,(J,S,P)$;   // Collect the pre-runtime features
5 **end**
6 **for** i=1:$\tilde{N}$ **do**
7 $\quad\big|\quad$ **if** $isEmpty(r_i(J,S,P))$ **then**
8 $\quad\big|\quad\big|\quad R \leftarrow R \cup runtimeFeaturePredict\,(\Phi,i)$;
9 $\quad\big|\quad$ **else**
10 $\quad\big|\quad\big|\quad R \leftarrow R \cup \mathrm{r_i}\,(J,S,P)$; ;   // Collect the runtime features
11 $\quad\big|\quad$ **end**
12 **end**
13 $Rh \leftarrow serialize\,(R) = (r_1, \cdots, r_{\tilde{N}})$;   // Build the runtime metrics
   sequence
14 $\alpha \leftarrow 0$
15 **for** i=1:$\tilde{N}$ **do**
16 $\quad\big|\quad$ **if** $i < now$ **then**
17 $\quad\big|\quad\big|\quad \hat{t}_i = realTime(J,i)$; ;   // This iteration has been executed
18 $\quad\big|\quad$ **else**
19 $\quad\big|\quad\big|\quad \hat{t}_i = LSTM(R,i)$; ;   // Predict the runtime of every iteration
20 $\quad\big|\quad$ **end**
21 $\quad\big|\quad \alpha += t_i$;
22 **end**
23 **return** α

First Stage. Let $P=\{P_1, ..., P_n\}$ be the set of pre-runtime features, and for each pre-runtime feature P_i, $1 \leq i \leq n$, let Φ_i denote the set of possible values of that feature in the set of training jobs, then Φ is uniquely determined for each iterative job that has completed or is about to run. Therefore, we can assume that there exists a function Γ_i to represent the fetch values of the pre-runtime feature P_i for some iterative job in some cluster environment.

$$\Gamma_i : Job \times Input \times Resource \rightarrow \Phi_i, \quad \forall 1 \leq i \leq n \qquad (3)$$

Then an iterative job can be represented by a vector $P \in \mathbb{R}^n$, where n denotes the number of pre-runtime features. For each of the values of the runtime features shown in Table 2, we make predictions by building an online linear regression model with L2 regularization.

Specifically, considering the dependencies among the runtime features, we use a modified quadratic polynomial regression model (QPR) with a regression function of the form:

$$f\,(w, P) = w^T K\,(P), \quad w \in \mathbb{R}^{1+2n+\binom{n}{2}} \qquad (4)$$

where w is the weight vector to be learned and $K(\bullet)$ is the basis function of the polynomial regression model:

$$K(P) = \left(1, p_1, \cdots, p_n, p_1 p_2, \cdots, p_{n-1} p_n, p_1^2, \cdots, p_n^2\right)^T \tag{5}$$

Let $r_i^{(j)}$ denote the true value of the runtime feature x_j of the iterative job i, then the final form of the online regression model on the runtime feature x_j is

$$\underset{w}{argmin} \sum_{i=1}^{N} \eta\left(w^T K(P_i), r_i^{(j)}\right) + \lambda \|w\|_2^2 \tag{6}$$

where $\eta\left(w^T K(P_i), r_i^{(j)}\right)$ is the loss function and λ is the L2 regularization parameter. We choose the mean square error (MSE) as the loss function, so once online regression model finds the w that minimizes the current cumulative loss, the runtime features for each iteration can be predicted by Equation (4).

Second Stage. After the first stage, we can obtain a set of sequences of runtime indicators in the order of iterations. Based on the above considerations, the long and short term memory network (LSTM) becomes a suitable method to predict the runtime of each iteration based on runtime features, and the LSTM implemented in Keras supports online learning, and when the value of the $batchSize$ parameter is 1, it means that an online learning method is used.

The runtime feature records $R_t = \left\{r_t^{(1)}, \cdots, r_t^{(n)} | t \in \left[1, \tilde{N}\right]\right\}$ for each iteration can be obtained, where $\tilde{N}$ denotes the predicted value of the number of iterations of the current iterative job. These records are combined in the order of iterations to obtain a runtime metrics sequence $Rh_{current} = (R_1, ..., R_t, R_{t+1}, ..., R_{\tilde{N}})$. In particular, we assume that the iteration counts of training job set and current iterative job are same or similar. If the obtained sequence length is greater than $\tilde{N}$, the sequence is intercepted from the end of the sequence to the same length, and if the sequence length is insufficient, the end of the sequence is filled with zeros. Then we can obtain the matrix Rh consisting of the runtime metrics sequences of all iterative jobs in the training job set as shown in Fig. 4.

$$Rh = \begin{bmatrix} R_{1,1} & R_{1,2} & \cdots & R_{1,\tilde{N}} \\ R_{2,1} & R_{2,2} & \cdots & R_{2,\tilde{N}} \\ \vdots & \vdots & & \vdots \\ R_{M,1} & R_{M,1} & \cdots & R_{M,\tilde{N}} \end{bmatrix} \begin{matrix} \rightarrow Rh^{(1)} \\ \rightarrow Rh^{(2)} \\ \vdots \\ \rightarrow Rh^{(M)} \end{matrix}$$

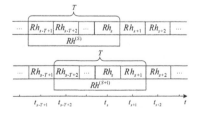

Fig. 4. Overview of the matrix composed of runtime metrics sequence.

Fig. 5. Schematic diagram of sliding window input.

The model inputs are updated by means of a sliding window, as shown in Fig. 5. The LSTM is used to learn a sequence of runtime features for each iteration step within a window of length T to predict the runtime of the relative $(T+1)$-st iteration, and then the prediction is cycled to the end of the iteration.

4   Evaluation

This section describes the experimental setup, which evaluates the accuracy of our method and compares it with the current state-of-the-art solutions.

4.1   Experimental Environment

Cluster Setup. We run experiments on a 7-nodes OMNISKY cluster (1 JobManager & 6 TaskManagers), and all nodes are connected with 10-Gigabit Ethernet. Each node has two Intel Xeon Silver 4210 CPUs @ 2.20 GHz (10 cores × 2 threads, 20 TaskNumberSlots), 128 GB memory, and 1 TB SSD. Hadoop version 2.7.0 (for storing data on HDFS) and Flink version 1.8.0 are chosen.

Evaluation Metrics. To validate the performance of our method, we use the relative absolute error (RAE) as an evaluation metric. Let e_{ij} denotes the predicted value of our method for r_{ij}, RAE is calculated as shown below.

$$RAE = \frac{\sum_{i=1}^{n} \sum_{j=1}^{m} |r_{ij} - e_{ij}|}{\sum_{i=1}^{n} \sum_{j=1}^{m} |r_{ij} - \frac{1}{n} \sum_{i=1}^{n} r_{ij}|} \tag{7}$$

where n is the number of predictions and m denotes the number of iterations.

Baselines. We compare our model with the most advanced baseline as follows.

- SMiPE [10]: It continuously estimates the progress of the current iterative dataflow based on the matched historical execution.
- Two-stage [14]: Pham et al.'s two-stage machine learning-based task offline runtime prediction model, which combines input data features with the cluster environment to predict task runtime.

4.2   Dataset

In the experiment, multiple datasets are used for the typical Flink iterative jobs, as detailed in Table 3.

The LiveJournal dataset is friendship relationship data between users from the LiveJournal online community, the Wiki dataset is an encyclopaedic page graph from the English Wikipedia, and the RoadNet-CA dataset is from a road network data in California, including intersection and road start information. The Point dataset is a two-dimensional dataset generated by sampling from five Gaussian mixture models with random clustering centres and equal variances.

Table 3. Benchmark jobs and datasets.

Benchmark Job	Dataset	Size	Parameter
PageRank	LiveJournal	1 GB	d = 0.85, 500 iterations
ConnectedComponent (CC)	Wiki	5.74 GB	500 iterations
SSSP	RoadNet-CA	6.9 GB	500 iterations
K-Means	Points	10 GB	5 clusters, 500 iterations

4.3 Parameter Selection

For the LSTM, we used sigmoid as the activation function, 10 hidden layers for the initial training of the model, and added dropout layers to prevent overfitting. Let the number of training *epochs* = 100 and the dropout ratio was 0.2, we use the RAE of the samples before and after prediction and the time required for the LSTM to complete one round of training as indicators to compare sliding windows of different lengths of time to select the optimal window size.

As shown in Fig. 6, the sliding window size fluctuates due to the large variability of different iterations. Considering the prediction limitation and accuracy, the final sliding window size is 5% of the number of iterations.

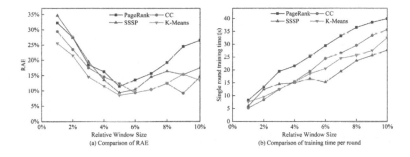

Fig. 6. Comparison of sliding window size under different iterative jobs.

4.4 Prediction Accuracy

We calculate the average RAE of the currently predicted iteration steps when the job runs to a certain time point, we use the relative iteration progress (RIP) to denote this time point. We implemented four iterative jobs with the bulk (Flink uses by default) and delta iterative operator to conduct experiments.

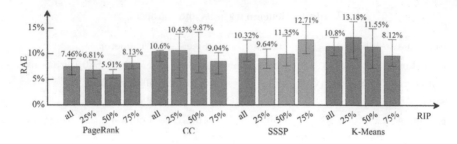

Fig. 7. Summary of prediction errors for runtime of bulk iterative job.

In Fig. 7 and 8, it can be seen that our method maintains between 5.91-12.71% and 10.16–15.03% throughout the prediction process for the bulk and delta iterative jobs, but the error at a RIP of 25% for the iterative jobs is generally larger than the other moments, which is due to the fact that little resource consumption information can be captured at the beginning of the iteration.

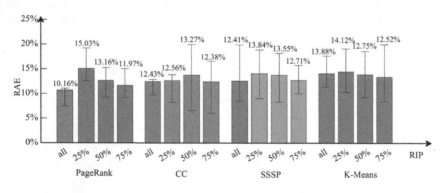

Fig. 8. Summary of prediction errors for runtime of delta iterative job.

4.5   Performance Comparison

In Table 4, the k indicates the size of the training data used and it is clear that our method outperforms the two-stage scheme when using 80% of the data, and when using the full amount of data, the error of our online prediction method is reduced by 4.79% compared to the superior two-stage scheme. In particular, the error of our method on the delta iterative job is reduced by about 15% compared to the two comparison schemes, which prefer to predict for the entire job.

Table 4. Comparison results of RAE for different prediction methods (%).

Type	Job	SMiPE	Two-stage	Our online prediction				
				k = 20%	k = 40%	k = 60%	k = 80%	k = 100%
Bulk	PageRank	23.57	18.26	31.51	28.63	19.54	13.57	10.72
	CC	27.13	17.63	35.33	32.58	21.66	18.52	12.87
	SSSP	19.54	12.87	29.88	24.62	18.51	14.35	9.83
	K-Means	17.88	14.54	27.44	22.92	14.23	13.47	10.85
	Average	22.03	15.86	31.04	27.19	18.49	14.98	**11.07**
Delta	PageRank	43.57	40.67	50.43	42.95	31.43	25.63	19.27
	CC	32.66	25.54	41.27	33.42	27.52	19.13	14.33
	SSSP	30.81	31.86	33.54	29.33	23.45	20.92	17.45
	K-Means	26.77	21.32	30.82	21.63	17.87	14.65	12.34
	Average	33.45	29.85	39.02	31.83	25.07	20.08	**15.85**

5   Related Work

There has been a lot of research such as [14–16] on how to predict the runtime of distributed jobs, as well as work for iterative jobs like [8–10].

Ellis [8] proposes the idea of modeling and predicting the runtime of Spark iterative jobs in stages. PREDIcT [9] captures the input features of each iteration and then build a cost model to predict the runtime of each iteration. SMiPE [10] is a system that captures runtime features to estimate the progress of iterative data flow by matching running jobs with previous executions.

In contrast to the above work based on statistics, Pham et al. [14] proposed a two-stage machine learning model-based prediction approach to predict the runtime of a continuous workflow task, and our approach is similar to it, which build a two-stage model to balance the response time and prediction accuracy.

6   Conclusion and Future Work

In this paper, we propose a online runtime prediction method for distributed iterative jobs, which core is a series of machine learning models that combine online learning and fine-grained resource consumption sequence data of each iteration to build a runtime prediction model by deeply analyzing the execution features of iterative jobs. In order to verify the effectiveness of the proposed method, we conduct many experiments to prove the superiority of our work. It is also worth considering how to improve job scheduling and resource allocation strategies based on runtime prediction in the future.

Acknowledgments. This research was supported by the National Key R&D Program of China under Grant No. 2018YFB1004402; and the National Natural Science Foundation of China under Grant No. 61772124.

References

1. Dean, J., et al.: Large scale distributed deep networks. In: Advances in Neural Information Processing Systems, pp. 1232–1240 (2012)
2. Carbone, P., et al.: Apache flink^{TM}: stream and batch processing in a single engine. IEEE Data Eng. Bull. **38**(4), 28–38 (2015)
3. Tumanov, A., et al.: TetriSched: global rescheduling with adaptive plan-ahead in dynamic heterogeneous clusters. In: Proceedings of the Eleventh European Conference on Computer Systems, pp. 35:1–35:16 (2016)
4. Wolf, J.L., et al.: FLEX: a slot allocation scheduling optimizer for mapreduce workloads. In: 11th International Middleware Conference, vol. 6452, pp. 1–20 (2010)
5. Thamsen, L., et al.: Selecting resources for distributed dataflow systems according to runtime targets. In: 35th IEEE International Performance Computing and Communications Conference, pp. 1–8 (2016)
6. Lama, P., Zhou, X.: AROMA: automated resource allocation and configuration of mapreduce environment in the cloud. In: 9th International Conference on Autonomic Computing, pp. 63–72 (2012)
7. Renner, T., et al.: Adaptive resource management for distributed data analytics based on container-level cluster monitoring. In: Proceedings of the 6th International Conference on Data Science, Technology and Applications, pp. 38–47 (2017)
8. Thamsen, L., et al.: Ellis: dynamically scaling distributed dataflows to meet runtime targets. In: IEEE International Conference on Cloud Computing Technology and Science, pp. 146–153 (2017). https://doi.org/10.1109/CloudCom.2017.37
9. Popescu, A.D., et al.: Predict: towards predicting the runtime of large scale iterative analytics. Proc. VLDB Endow. **6**(14), 1678–1689 (2013)
10. Koch, J., et al.: SMiPE: estimating the progress of recurring iterative distributed dataflows. In: 18th International Conference on Parallel and Distributed Computing, Applications and Technologies, pp. 156–163 (2017)
11. Kumar, V., et al.: Apache Hadoop YARN: yet another resource negotiator. In: ACM Symposium on Cloud Computing, pp. 5:1–5:16 (2013)
12. Hilman, M.H., et al.: Task runtime prediction in scientific workflows using an online incremental learning approach. In: 11th IEEE/ACM International Conference on Utility and Cloud Computing, pp. 93–102 (2018)
13. Gao, M., et al.: Online anomaly detection via incremental tensor decomposition. In: Ni, W., Wang, X., Song, W., Li, Y. (eds.) WISA 2019. LNCS, vol. 11817, pp. 3–14. Springer, Cham (2019). https://doi.org/10.1007/978-3-030-30952-7_1
14. Pham, T., et al.: Predicting workflow task execution time in the cloud using A two-stage machine learning approach. IEEE Trans. Cloud Comput. **8**(1), 256–268 (2020). https://doi.org/10.1109/TCC.2017.2732344
15. da Silva, R.F., et al.: Online task resource consumption prediction for scientific workflows. Parallel Process. Lett. **25**(3), 1541003:1–1541003:25 (2015)
16. Pumma, S., et al.: A runtime estimation framework for ALICE. Future Gener. Comput. Syst. **72**, 65–77 (2017). https://doi.org/10.1016/j.future.2017.02.040

Workload Prediction of Cloud Workflow Based on Graph Neural Network

Ming Gao[1,3]([✉]), Yuchan Li[1], and Jixiang Yu[2]

[1] School of Management Science and Engineering, Dongbei University of Finance and
Economics, Dalian, China
gm@dufe.edu.cn
[2] School of Data Science and Artificial Intelligence, Dongbei University of Finance and
Economics, Dalian, China
[3] Center for Post-Doctoral Studies of Computer Science, Northeastern University, Shenyang,
China

Abstract. With the continuous expansion of cloud computing market, the problem of low utilization rate of cloud computing resource has become increasingly prominent, because cloud computing vendors can not schedule a large number of server cluster effectively as before. Improving the utilization rate of cloud resources can not only improve the net profit of cloud computing manufacturers, but also reduce the time cost and economic cost of cloud computing users. In addition to resource scheduling, the current research on cloud workflow load is still focused on single task or single instance prediction, and even the data sets used are simulation data. This paper aims to predict workload of cloud workflow resources to make the cloud computing resources get better scheduling, and ultimately facilitate all relevant personnel in the cloud computing market. Firstly, compared with task and single instance, cloud workflow can get more context information. Secondly, in order to make this research more practical, this paper selects Alibaba cluster data V2018 released by Alibaba in 2018 as our research object. Thirdly, based on the graph structure characteristics of cloud computing workflow, this paper selects the Graph Neural Network (GNN) architecture which closely fits the graph structure to predict the load of cloud computing workflow, and specifically selects the homogeneous Graph Convolution Neural Network and Graph Attention Neural Network and heterogeneous GCN as our prediction algorithm. And it describes how cloud workflow is modeled as homogeneous graph and heterogeneous graph in detail. Finally, the algorithm in GNN is used to classify and predict Ali data with workflow length ranges from 4 to 12 separately and combined, and predicts the last and penultimate tasks of each length workflow. Besides, all the data from 4 to 12 are combined into one data to predict the last and penultimate tasks.

Keywords: Workflow in cloud computing · Performance prediction · DAG
structure · Deep learning · Workload prediction

© Springer Nature Switzerland AG 2021
C. Xing et al. (Eds.): WISA 2021, LNCS 12999, pp. 169–189, 2021.
https://doi.org/10.1007/978-3-030-87571-8_15

1   Introduction

Cloud computing [1, 2] is Internet-based computing,whereby shared resources, software and information are provided to computers and other devices on-demand, like a public utility. Strongly promoted by the leading industrial companies (e.g., Microsoft, Google, IBM, Amazon, etc.), cloud computing is quickly becoming popular in re-cent years. Cloud applications, which involve a number of distributed cloud components, are usually large-scale and very complex [3]. Cloud workflows, as typical software applications in the cloud, are composed of a set of partially ordered cloud software services to achieve specific goals [4]. Workload burstiness and spikes are among the main reasons for service disruptions and decrease in the Quality-of-Service (QoS) of online services [5]. They are hurdles that complicate autonomic resource management of data enters. With the increasing demand for cloud computing and the expansion of cloud computing market, cloud computing manufacturers have to reserve a large number of resources to meet the Quality of Service (QoS) requirements, so as to prevent the server overload caused by the sudden increase of resource usage. However, if there are too many reserved resources, there will be a waste problem caused by the low utilization rate of resources. In this way, the problem of low resource utilization has become increasingly prominent.

At present, the academia has provided a variety of research on load balancing [e.g., [6–8]], and the research on cloud resource load prediction is not mature. In order to achieve load balancing, the most important step is to predict the resources needed by the upcoming tasks in advance and allocate the resources reasonably. Due to the difficulty of obtaining the real environment data of cloud computing, many research data are based on simulation data [e.g., [9–11]]. However, the real cloud computing environment is complex and changeable, and experiments on simulation data can not reflect the practicability of the model. In addition, the research on cloud resource load predicting mainly focuses on single task predicting [e.g., [12–14]]. Because cloud workflow contains inter dependencies between tasks, it can obtain more context information than single task predicting. Therefore, this paper focuses on the resource load prediction of cloud workflow. Generally speaking, the function of a cloud workflow system and its role in a cloud computing environment, is to facilitate the automation of user submitted workflow applications where the tasks have precedence relationships defined by graph-based modelling tools such as DAG (directed acyclic graph) and Petri Nets [15, 16] or language-based modelling tools such as XPDL (XML Process Definition Language) [17]. The precedence relationships between tasks in Alibaba cluster-trace-v2018 is defined by graph-based modelling tool DAG. In order to make cloud workflow scheduling play a greater role, this paper predicts the resource load of cloud workflow to better serve the cloud workflow scheduling.

In the prediction method, traditional machine learning algorithms have become nearly mature in recent years. For example, algorithms such as SVM [18], Logistic Regression [19], and Random Forest [20] have good results in processing tabular data for classification, but they cannot effectively process data with multiple feature dimensions and has strict requirements for feature construction. Therefore, this paper investigates a new branch of machine learning: Deep learning [21] has been gradually applied to speech recognition, image recognition, natural language processing, and other fields with theoretical innovation, computing power growth, and data accumulation, and has achieved great success. However, the extension and integration of related technologies in cloud workflow scheduling are not deep enough. Cloud workflow can be abstracted as directed graph, while the emerging Graph Neural Network (GNN) [22] is a kind of original deep neural network supporting processing graph data structure, which is fit well with cloud workflow. The performance of cloud workflow resource load predicting on GNN is worth exploring.

In the data extraction phase, this paper first extracts the DAG information, task-based life cycle and resource occupancy data of each workflow whose length ranges from 4 to 12 in Alibaba data. The extracted data is used as benchmark data set to construct homogeneous graph and heterogeneous graph in Graph Neural Network. Then the last node and the penultimate node in the graph are classified and predicted respectively. Before prediction, we need to cover the features of subsequent nodes to be predicted.

The prediction of cloud workflow load not only enriches the academic research, but also benefits cloud computing stakeholders, because improving resource utilization benefits not only cloud computing vendors, but also cloud computing users. For cloud computing vendors, if they successfully predict the load of cloud resources, it means that cloud vendors can provide rental services for more people and get more revenue with the same number of servers. In addition, it will greatly reduce the problem of task failure and provide better services for customers. For cloud users, better resource allocation means more efficient work processing and less unnecessary time waste.

Nowadays, cloud computing how to allocate server related resources more effectively to improve resource utilization has become a new topic, the current cloud computing still has a lot of room for development. Therefore, this paper attempts to predict the resource load of cloud workflow based on graph neural network, so as to predict the resource level of tasks in advance and better allocate resources for them.

At present, the research faces the following challenges:

1. The real data of cloud computing environment is difficult to obtain, and the cloud computing data with workflow structure is more difficult to obtain.
2. After obtaining the data set, how to extract the features needed in this paper from a large number of features.
3. Due to the current research on cloud workflow prediction is limited, how to extract and embed graph structure data into the model is also a difficulty.

Our article is structured accordingly: First, we discuss the selection and extraction of cloud computing related data sets in Sect. 2. Subsequently, we go into more details about GCN and GAT algorithms in GNN in Sect. 3. In Sect. 4, we describe our study design, followed by the presentation of the results in Sect. 5. We summarize our findings and highlight potential for future work in Sect. 6.

2 Cloud Workflow Data Analysis

2.1 Dataset Selection

In order to make the research more practical, this paper compares and analyzes the real cluster data released by cloud computing vendors. At present, there are some large-scale enterprises that have disclosed some relevant data of cloud computing, such as

Google, azure, Alibaba, etc. The concise information of the data set is shown in the following Table1. According to the requirements of this paper, only Alibaba cluster-trace-v2018 can provide cloud workflow information. Therefore, Alibaba cluster-trace-v2018 is selected as the prediction data in this experiment.

Table 1. Cloud computing cluster data set

	Name	Year	Period	Content
Google [23]	ClusterData2011	2011	1 month	A single 12.5 k-machine Borg cell
	ClusterData2019	2019	1 month	Eight Borg cells
Azure [24, 25]	AzurePublicDatasetV1	2017		~2 M VMs and 1.2 B utilization readings
	AzurePublicDatasetV2	2019		~2.6 M VMs and 1.9 B utilization readings
	AzureFunctionsDataset2019	2019	2 weeks	A subset of applications running on Azure Functions
Alibaba Co-located cluster [26]	Cluster-trace-v2017	2017	12 h	About 1300 machines
	Cluster-trace-v2018	2018	8 days	About 4000 machines, the DAG information of our production batch workloads

2.2 Dataset Introduction

The cluster data released by Alibaba in 2018 contains the DAG information of workflow, which can obtain the interdependence between tasks, which is very consistent with the purpose of this study. Therefore, this experiment selects Alibaba cluster-trace-v2018 as the data set of this study.

In the workflow data of Alibaba cluster data V2018, there are two files that record offline batch processing. One file is the batch task table, which mainly describes the tasks in the batch workload. The other file is the batch instance table, which takes the instance in the batch workload as the main description object. The specific description information is shown in the following Tables 2 and 3.

A complete batch computing work in Ali cluster can be described by "work task instance" model. A job is usually composed of several tasks. The sequence of these tasks is represented by directed acyclic graph (DAG). There are several instances in each task, which are executed in parallel without sequence. Only when all the instances of a task are completed can the task be considered completed. That is, if task2 depends on Task1, only when all instances of Task1 are completed can the instance of task2 be started, as shown in the Fig. 1.

Table 2. Field description of batch-task file

Feature Name	Interpretation
Task_name	Id of task
Inst_nume	Number of instances included in the task
Task_type	Type of task
Job_name	The name of the job to which the task belongs
Status	Status of task
Start_time	Start time of task
End_time	End time of task
Plan_cpu	CPU requested for each instance of the task
Plan_mem	Memory requested for each instance of the task

Table 3. Field description of batch-instance file

Feature Name	Interpretation
Ins_nam	Name of instance
tid	The name of the task to which the instance belongs
jid	The name of the job to which the instance belongs
stime	Start time of instance
etime	End time of instance
mid	The name of the machine to which the instance belongs
Cpu_avg	The average CPU usage of the instance
Cpu_max	The max CPU usage of the instance
Mem_avg	The average memory usage of the instance
Mem_max	The max memory usage of the instance

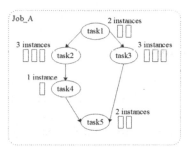

Fig. 1. Job-task-instanse DAG

Through the preliminary statistical analysis of the two tables, it is found that 4201013 batch computing workflows are tracked and recorded in this dataset information, 417576 different batch task types. A workflow can have as few as one task and as many as 1002 tasks. The number of tasks less than 10 (3980466) accounted for 94.75% of the total number of workflows, and the number of tasks less than 20 (4171498) accounted for 99.30% of the total number of workflows. It can be seen that there are too many tasks, and the work flows through complex computing work. Although there are, it is not common. Considering the number of samples that can be trained and the diversity of samples, this paper intends to study the batch computing workflow and its tasks with the number of tasks from 4 to 12.

2.3 Dataset Extraction

Firstly, the DAG information is deduced according to the task name field in the task table, and the work is grouped according to the derived DAG information. Then, all instances of the selected workflow are filtered from the instance table, and the instance records are counted to form task-based feature fields. Finally, according to the prediction target determined in the previous paper, the features that may contain the prediction target information are deleted.

There are six fields in the instance table that describe the life cycle and resource usage: instance start time, instance end time, average and maximum CPU usage, and average and maximum memory usage. The average value, variance, maximum value, minimum value, median value, skewness and kurtosis of the instance resource occupation data in the task are calculated as the characteristics to describe the resource occupation of the task; The average value, variance, maximum value and minimum value of the running time data of the instance are calculated, and the difference between the earliest time and the latest time of the whole task is calculated as the characteristics of describing the life cycle of the task.

The information contained in each row of data set after preprocessing is shown in the following way: each job is a row of data, and each row of data is composed of several tasks. Each task contains the characteristics of CPU usage, memory usage and task-based life cycle, followed by the in degree and out degree of each node. The next operations will be performed on this dataset.

3 Algorithm Introduction

Generally speaking, the function of a cloud workflow system and its role in a cloud computing environment, is to facilitate the automation of user submitted workflow applications where the tasks have precedence relationships defined by graph-based modelling tools such as DAG (directed acyclic graph) and Petri Nets [15, 16] or language-based modelling tools such as XPDL (XML Process Definition Language) [17]. The precedence relationships between tasks in Alibaba cluster-trace-v2018 is defined by graph-based modelling tool DAG.

The goal of this prediction is to classify and predict one of the tasks in the workflow. Although deep learning has achieved great success in the field of image and text, there are still some limitations. First, deep learning has been very good at processing Euclidean spatial data, but the feature extraction of non Euclidean spatial data is not satisfactory; Second, the core assumption of deep learning algorithm is that the data samples are independent of each other, but this is not the case for graphs [27]. There will be interdependence between data. If we treat each data as an isolated point, we will lose part of the feature information.

In recent years, in order to extend the deep learning method on graph, scholars have proposed many models on graph structure[e.g., [28–31]]. One of the most widely used is GNN, which is defined and designed the neural network structure for processing graph data by referring to the ideas of convolutional network, cyclic network and depth automatic encoder, and GNN came into being. GNN can be divided into five categories:

graph revolution networks, graph attention networks, graph autoencoders, graph general networks and graph spatial temporal networks.

3.1 Graph Convolution Networks (GCN)

GCN [32] is actually a feature extractor. Different from other classical models (such as CNN and RNN) in deep learning, GCN is a feature extraction method for data with graph structure, and its most obvious feature is the message passing and message aggregation for neighbor nodes.

The feature extraction of GCN is divided into two steps:

- Message passing: transfer several neighbor node characteristics of the target node to the target node's mailbox.
- Message reduce: some operations are performed on some features sent to the mailbox, such as additive aggregation in this paper, and the aggregated features are taken as the new features of the target node.

The propagation rules of GCN convolution layer are as follows:

$$H^{(l+1)} = \sigma\left(\tilde{D}^{-\frac{1}{2}} \tilde{A} \tilde{D}^{-\frac{1}{2}} H^{(l)} W^{(l)} \right)$$

Among them:

H^l is the network layer characteristic, $H^{(0)} = X$ (X is the initial feature), A is the adjacency matrix, W^l is the learning weight, $\sigma(\cdot)$ is the activation function.

$\tilde{A} = A + I_N$ is the adjacency matrix plus self join of an undirected graph G, I_N is the identity matrix, simultaneous interpreting the node's own characteristics and its neighbor node characteristics into the node.

$\tilde{D}^{-\frac{1}{2}} \tilde{A} \tilde{D}^{\frac{1}{2}}$ is the normalization of $\tilde{A}$.

3.2 Graph Attention Networks

GAT [33] is divided into global graph attention and mask graph attention. Global graph attention means that the current node and each node in the graph need to perform the attention operation, while mask graph attention only performs the attention operation on the neighbor nodes of the current node. This paper adopts the latter, and the following introduction is also about mask graph attention.

GAT feature extraction can also be divided into two steps:

- Calculate the attention coefficient: For vertex j ($j \in \mathcal{N}_i$), the similarity coefficient between its neighborsand itself is calculated one by one, which is expressed as $e_{ij} = a([Wh_i \| Wh_j]), j \in \mathcal{N}_i$, where h_i represents the feature vector of node i, $\mathcal{N}_i$ represents the neighbor of node i, and W is the shared parameter of node i and node j, and $a(\cdot)$ maps the stitched high-dimensional feature to a real number∘ Then e_{ij} is normalized by softmax, and the formula is expressed as $\alpha_{ij} = \frac{exp(LeakyReLU(e_{ij}))}{\sum_{k \in \mathcal{N}_i} exp(LeakyReLU(e_{ik}))}$.

- Message reduce: According to the calculated attention coefficient, the weighted sum of features is expressed as $h_i' = \sigma\left(\sum_{j \in \mathcal{N}_i} \alpha_{ij} Wh_j\right)$, and then multi head is added, $h_i'(K) = \|_{k=1}^{K}\sigma\left(\sum_{j \in \mathcal{N}_i} \alpha_{ij} Wh_j\right)$, where k is the number of spliced heads.

4 Study Design

The experimental purpose is that classify and predict the last and penultimate tasks of each length workflow. The experimental process was as follows:

1. Preprocess the data
2. Using K-Means clustering method to mark the target nodes
3. Divide training set and test set
4. Design the neural network architecture and predict the target nodes

4.1 Data Preparation

This paper selects nine data sets with workflow length ranges from 4 to 12 for research, and sets the length of workflow as variable L. Different lengths of workflow are in different files, and the amount of data in different files is shown in the Table 4 below. In each file, each line represents a job, and each job contains L tasks. Tasks are composed of resource data (node feature) and dependency relationship (edge feature), as shown in the Fig. 2 below.

Table 4. Cloud workflow datasets size

File name	Job number	Task number
sum_instance_to_task4_to_job.csv	215738	862952
sum_instance_to_task5_to_job.csv	101863	509315
sum_instance_to_task6_to_job.csv	61535	369210
sum_instance_to_task7_to_job.csv	67352	471464
sum_instance_to_task8_to_job.csv	54929	439432
sum_instance_to_task9_to_job.csv	45085	405765
sum_instance_to_task10_to_job.csv	28046	280460
sum_instance_to_task11_to_job.csv	29080	319880
sum_instance_to_task12_to_job.csv	25716	308592

The node feature size is a vector with the size of 1×35, including the number of instances in the task, and the average value, variance, maximum value, minimum value,

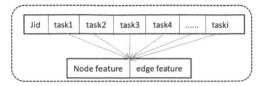

Fig. 2. Data set structure

median, skewness and kurtosis of the average CPU, the max CPU, the average memory and the max memory, and the average, variance, maximum and minimum running time of all instances in each task, and the task number.

The edge feature size is a vector with the size of $1 \times 2L$, including the relationship between each task and all tasks in the job. Using ao_b to express only when a ends, can b starts, and ai_b to express task a starts after task a ends (a,b = 1, ..., L). If the value of ao_b or ai_b is greater than 0, then the condition is true, otherwise it is not true.

4.2 Data Preprocessing

In order to construct a graph structure suitable for GNN, this paper performs the following operations on the data. Each job is constructed into a graph as shown in the Fig. 3 below.

In each file, do the following for each line of data. (The following operations are implemented in DGL).

- Step1: Reshape each row of data to make each task a row of data, we call this new data the reshaped data.
- Step2: The 36th to the last columns in the reshaped data are selected as edge feature vectors(which is an $L \times 2L$ matrix), the first L column data in the edge feature constitute the adjacency matrix. We use this adjacency matrix to construct the graph.
- Step3: The first 35 columns in the reshaped data are selected as node feature vectors(which is an $L \times 35$ matrix), then the node features of each row are embedded into the graph in order.
- Step4: Add the generated graph to the list in turn.

After the above steps, we get a graph set, which we call the original graph set. In the original graph set, we find that the node number is unreasonable. As shown in the Fig. 4 below, the nodes are not numbered according to their sequence. Because our prediction node is to predict the last node and the last node in relative time, the existing number can not reflect the relative time relationship between nodes. To solve this problem, this paper rearranges the node number according to the topological sequence.

The specific methods are as follows:

- Step 1: traverse each original graph to find the topological sequence of the original graph. For example, the topological sequence in the above figure is [1,6], [0,2,3,4], [5]
- Step 2: convert the above topological sequence into a list, such as list [1,6,0,2,3,4,5], and call this list LIST1

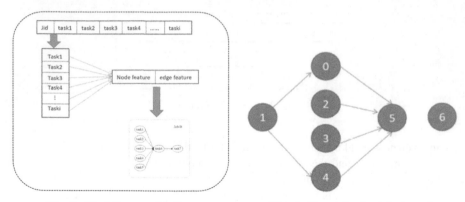

Fig. 3. Modeling graphic flow   **Fig. 4.** Example of original graph

- Step 3: arrange LIST1 from small to large, such as list [0,1,2,3,4,5,6], and call this list List2
- Step 4: map LIST1 to List2 to generate a new node number. The graph with the new node number is called topograph
- Step 5: reorder the rows of the matrix in the order of list and embed the node feature of the original graph into topograph

After the above steps, we get an ordered graph structure. On this basis, we further process topograph according to our own needs.

4.3 Label Value Processing

In this paper, K-Means [34] clustering is used to get data labels, and the clustering results are evaluated by the contour coefficient. K-means is a popular clustering analysis method based on partition and a typical unsupervised algorithm. K-means algorithm groups the data according to the different similarity between the data in the group and the data between the groups. This paper will use this algorithm to classify the task resource occupation. The basic idea of K-means algorithm is: clustering with k points in space as the center of mass, finding the Euclidean distance or Manhattan distance from any sample to K centers of mass, and dividing the sample into the cluster with the smallest center of mass. Through the iterative method, the centroid values of each cluster are updated step by step until satisfactory clustering results are obtained.

This paper attempts to cluster various pairwise combinations of CPU, and finally selects two attributes of average CPU: mean value and variance to cluster. The following table shows the results of three classification and five classification respectively in different workflows ranging in length from 4 to 12. Table 5 is the result of clustering 3 classification label. Table 6 is the result of clustering 5 classification label. SC means contour coefficient.

Table 5. 3 Classification label

	3 Classification label of the last task				3 Classification label of the penultimate task			
	0	1	2	SC	0	1	2	SC
task4	123633	48780	43325	0.666	108727	95365	11646	0.589
task5	48992	22124	30747	0.631	35778	26665	39420	0.575
task6	36053	10771	14711	0.661	30087	14778	16670	0.599
task7	29935	9700	27717	0.681	25094	14737	27521	0.585
task8	33357	9732	11840	0.661	30492	10159	14278	0.653
task9	26914	6858	11313	0.689	22560	8889	13636	0.655
task10	18448	3853	5745	0.711	15683	5150	7213	0.663
task11	17539	3593	7948	0.735	15725	4891	8464	0.67
task12	15023	4455	6238	0.719	12925	4849	7942	0.696
total	349894	119866	159584		297071	185483	146790	

4.4 Feature Value Processing

The prediction in this paper can be divided into the prediction of the last node and the prediction of the penultimate node. No matter which node is predicted, the information of the node to be predicted and the information of the node behind the time series of the node to be predicted need to be masked.

In this paper, we start with the nodes to be predicted and record the node numbers, which are called subsequence nodes. Then we locate the node features of subsequence nodes and zero them.

4.5 The Construction of Homogeneous Graph and Heterogeneous Graph

- Homogeneous graph [35]: it is a graph with only one node type and one edge type. We need to add reverse and self connected edges to each node in topograph to get more information.
- Heterogeneous graph [36]: it refers to the graph where the sum of node type and edge type is greater than two. In this paper, we divide the nodes into two types. The structure is shown in the Fig. 5. The node type of task to be predicted is named pred, and the node type of other nodes is named task. The specific edge connections in the figure are shown in the following table, which indicates that node1 type points to an edge of node2 type, and the edge type is determined by edge type. The edges between nodes and nodes are shown in the following Table 7.

4.6 Data Set Split

In this paper, we use StratifiedShuffleSplit to sample the graph data of different lengths according to the ratio of 9:1 (Training Set: Test Set). StratifiedShuffleSplit [37] can

Table 6. 5 Classification label

	5 Classification label of the last task						5 Classification label of the penultimate task					
	0	1	2	3	4	SC	0	1	2	3	4	SC
task4	103378	40495	31342	36837	3686	0.628	58077	53288	48815	47528	8030	0.55
task5	41951	20566	16579	22243	524	0.6	24536	20792	18877	33751	3907	0.574
task6	29708	11213	7306	12030	1278	0.62	21608	13377	9945	14340	2265	0.583
task7	26939	7293	8574	22679	1867	0.633	21895	12263	21492	10894	808	0.568
task8	26432	11283	6298	10369	547	0.626	23423	10706	6984	12825	991	0.612
task9	23373	7648	4811	8694	559	0.646	18980	6836	6416	11714	1139	0.628
task10	16207	4038	2819	4845	137	0.658	14910	5152	232	7519	233	0.678
task11	16030	3508	2937	6520	85	0.686	12796	4869	4003	7185	227	0.638
task12	14324	3045	505	3205	4637	0.687	11426	3457	4242	6583	8	0.67
total	298342	109089	81171	127422	13320		207651	130740	121006	152339	17608	

Fig. 5. The structure of heterogeneous graph

Table 7. Interpretation for edges

Node1 type	Edge type	Node2 type	Interpretation
'task'	'flow'	'task'	Original edge from one task to another
'task'	'flowed'	'task'	The inverse edge of the original edge from one task to another
'task'	'flow'	'pred'	Original edge from task to pred
'pred'	'flowed'	'task'	The inverse edge of the original edge from task to pred
'pred'	'in'	'task'	Original edge from pred to task
'task'	'from'	'pred'	The inverse edge of the original edge from pred to task
'task'	'all_in'	'pred'	A artificially created virtual edge where all tasks points to pred
'pred'	'all_from'	'task'	The inverse edge of the artificially created virtual edge
'pred'	'link'	'pred'	an edge from pred to pred

divide data into training set and test set, but it only provides the location index of training set/test set data in the original data set. The cross validation object generated by this class combines the functions of StratifiedKFold and ShuffleSplit. The object returns hierarchical random folds, and the object generates folds by retaining a certain proportion of samples for each class.

5 Result

The configuration of the computer used in the experiment is as follows: the processor is Intel Core CPU i7-10750h, the CPU frequency is 2.60 GHz, and the memory is 32.0 GB; the operating system is 64 bit win10 system; GPU is rtx2070, CUDA is v10.1, cudnn is v7.6.5; The integrated development environment is anaconda, python version 3.6.5.

In this paper, there are two evaluation indexes, micro precision score and macro precision score [38]. Micro precision score gives each class the same weight, The formula is expressed as $\frac{\sum_{i=0}^{n-1} TP_i}{\sum_{i=0}^{n-1}(TP_i+FP_i)}$ (n is the number of categories). Micro precision give each sample the same weight, the formula is expressed as $\sum_{i=0}^{n-1} \frac{TP_i}{TP_i+FP_i}$ (n is the number of categories). In addition, in order to make readers have a more intuitive understanding of the prediction results, this paper uses confusion matrix to show the final results.

After adjusting the parameters, the following results are obtained. The prediction results of validation sets of different length workflow in different models are shown in the following Table 8. The model of GCN based homogeneous graph is represented by homo-GCN. The model of GCN based heterogeneous graph is represented by hete-GCN. The model of GAT based homogeneous graph is represented by homo-GAT. Mi1 means micro precision of the last task. Ma1 means macro precision of the last task. Mi2 means micro precision of the penultima. Ma2 means macro precision of the penultima.

The amount of information that can be obtained by different length workflow is also different. For example, in the dataset with workflow length of 4, if the last node is predicted, the node can obtain information of up to 3 nodes. If the penultimate node is predicted, the node can obtain information of up to 2 nodes. In order to understand the impact of different length of workflow on the prediction results, this paper makes a broken line chart for the prediction results of the last node and the penultimate node of three categories, and the prediction results of the last node and the penultimate node of five categories respectively, and analyzes the change trend of the prediction results with the increase of workflow length.On the whole, the prediction results show an upward trend with the increase of workflow length. When the workflow length exceeds 10, the prediction results show an obvious downward trend. In the prediction of the penultimate node of three categories, the prediction results begin to decline when the length of the workflow is 9 (Figs. 6, 7, 8, 9, 10, 11, 12, and 13).

By exploring a more reasonable number of classification, it is necessary for the prediction model to get a more precise classification under the condition of ensuring the accuracy. Therefore, in order to analyze the differences of the prediction results under different categories, this paper constructs an area map (Figs. 14, 15, 16, and 17) for the prediction results of different evaluation indexes and different prediction targets of homo-GAT, and observes the differences of three categories and five categories under different conditions. On the whole, the prediction results of three categories are better than that of five categories. In Fig. 15, the results of task 7 and task 10 in 3 and 5 categories are similar or even coincident. In Fig. 14, the difference between the prediction results of three categories and five categories fluctuates about 5% in different length workflow, and the trend is consistent. In Fig. 16, when the workflow range is between 4 and 10, the difference between the prediction results of 3 categories and 5 categories fluctuates about 5% in different length workflow, and the trend is consistent with that of Fig. 16. When the workflow length is greater than 10, the prediction results of 5 categories decline sharply. The trend of Figs. 15 and 17 is basically the same, and there is a big difference in task 4.

In this paper, based on three models, the box plot (Figs. 18 and 19) is made for the prediction results of different models as a whole. The overall box diagram is based on all the evaluation results of each model, from which we can see homo-GCN was relatively stable, with a median value close to 85%, and its overall performance was better than that of concat. In homo-GAT, the dispersion degree is large, the median value is more than 85%, and the overall effect is better than the other two models. hete_ The dispersion of GCN is the largest, and the total is about 85%.

Table 8. Validation set evaluation.

Dataset	Model	3 Class				5 Class			
		mi1	ma1	mi2	ma2	mi1	ma1	mi2	ma2
task4	homo-GCN	84.12 ± 0.02	79.80 ± 0.03	79.51 ± 0.08	79.48 ± 0.23	77.49 ± 0.05	74.29 ± 0.16	63.39 ± 0.09	65.39 ± 0.21
	hete-GCN	83.61 ± 0.19	79.47 ± 0.20	79.31 ± 0.13	79.38 ± 0.20	77.23 ± 0.22	72.24 ± 0.51	64.27 ± 0.25	67.13 ± 0.20
	homo-GAT	85.55 ± 0.05	81.44 ± 0.14	82.22 ± 0.05	81.08 ± 0.47	80.88 ± 0.13	74.63 ± 0.52	67.37 ± 0.04	68.59 ± 0.19
task5	homo-GCN	82.53 ± 0.05	78.88 ± 0.05	80.07 ± 0.06	77.93 ± 0.09	75.47 ± 0.07	69.82 ± 0.10	72.30 ± 0.06	70.51 ± 0.11
	hete-GCN	82.16 ± 0.08	78.22 ± 0.08	80.80 ± 0.18	79.25 ± 0.21	75.06 ± 0.13	69.56 ± 1.33	73.33 ± 0.41	71.19 ± 0.40
	homo-GAT	83.96 ± 0.14	80.63 ± 0.20	82.42 ± 0.16	81.02 ± 0.17	77.25 ± 0.08	71.34 ± 0.77	75.85 ± 0.15	74.38 ± 0.09
task6	homo-GCN	87.03 ± 0.07	81.44 ± 0.08	83.31 ± 0.06	80.46 ± 0.04	79.00 ± 0.22	73.63 ± 0.21	75.29 ± 0.09	73.70 ± 0.20
	hete-GCN	87.26 ± 0.16	82.37 ± 0.40	84.20 ± 0.21	81.79 ± 0.14	81.35 ± 0.21	77.24 ± 0.46	77.03 ± 0.37	75.67 ± 0.45
	homo-GAT	88.33 ± 0.09	83.77 ± 0.27	85.77 ± 0.18	83.55 ± 0.24	82.23 ± 0.33	77.69 ± 0.49	79.47 ± 0.14	78.65 ± 0.15
task7	homo-GCN	86.38 ± 0.06	81.55 ± 0.13	84.81 ± 0.04	82.31 ± 0.06	82.48 ± 0.09	75.98 ± 0.15	84.49 ± 0.06	84.08 ± 0.19
	hete-GCN	89.32 ± 0.18	84.00 ± 0.25	86.19 ± 0.14	84.09 ± 0.15	83.55 ± 0.20	75.88 ± 0.44	84.81 ± 0.10	84.23 ± 0.55
	homo-GAT	88.80 ± 0.05	84.25 ± 0.12	85.90 ± 0.08	83.66 ± 0.14	84.10 ± 0.09	77.73 ± 0.24	85.57 ± 0.09	84.31 ± 0.43
task8	homo-GCN	88.01 ± 0.07	82.83 ± 0.07	86.73 ± 0.13	83.07 ± 0.17	84.27 ± 0.38	80.78 ± 0.38	80.82 ± 0.21	77.36 ± 0.50
	hete-GCN	88.05 ± 0.21	83.48 ± 0.32	87.69 ± 0.43	84.38 ± 0.61	84.20 ± 0.43	80.43 ± 0.78	80.75 ± 0.76	78.92 ± 0.76
	homo-GAT	89.13 ± 0.09	84.48 ± 0.09	89.28 ± 0.09	86.12 ± 0.20	85.31 ± 0.13	81.66 ± 0.56	83.94 ± 0.17	81.18 ± 0.69
task9	homo-GCN	90.20 ± 0.06	85.30 ± 0.16	89.21 ± 0.09	86.95 ± 0.13	84.78 ± 0.08	78.75 ± 0.28	83.88 ± 0.13	80.62 ± 0.31
	hete-GCN	89.94 ± 0.11	84.72 ± 0.22	91.04 ± 0.23	88.77 ± 0.25	84.56 ± 0.15	77.18 ± 0.33	85.89 ± 0.14	82.55 ± 0.56

(continued)

Table 8. (*continued*)

Dataset	Model	3 Class				5 Class			
		mi1	ma1	mi2	ma2	mi1	ma1	mi2	ma2
task10	homo-GAT	90.81 ± 0.11	86.71 ± 0.32	90.99 ± 0.09	88.97 ± 0.13	85.16 ± 0.09	78.60 ± 0.33	86.41 ± 0.11	84.21 ± 0.33
	homo-GCN	92.18 ± 0.06	85.71 ± 0.14	89.32 ± 0.14	86.30 ± 0.22	86.92 ± 0.09	81.92 ± 0.64	89.56 ± 0.16	86.28 ± 0.47
	hete-GCN	91.77 ± 0.31	85.61 ± 0.46	90.72 ± 0.17	87.80 ± 0.30	86.93 ± 0.15	80.98 ± 0.27	91.15 ± 0.17	88.62 ± 0.68
	homo-GAT	92.57 ± 0.07	86.86 ± 0.29	90.63 ± 0.19	88.17 ± 0.28	87.32 ± 0.16	81.47 ± 0.67	90.98 ± 0.15	89.75 ± 0.71
task11	homo-GCN	90.93 ± 0.10	84.33 ± 0.17	89.33 ± 0.09	85.67 ± 0.17	85.31 ± 0.19	75.43 ± 0.32	83.47 ± 0.24	80.15 ± 0.55
	hete-GCN	91.61 ± 0.09	85.22 ± 0.18	90.29 ± 0.23	86.43 ± 0.26	85.76 ± 0.17	74.65 ± 0.81	86.02 ± 0.18	82.09 ± 0.40
	homo-GAT	91.90 ± 0.27	86.27 ± 0.52	90.05 ± 0.11	86.53 ± 0.39	85.88 ± 0.23	77.74 ± 1.89	85.81 ± 0.21	81.74 ± 0.82
task12	homo-GCN	88.84 ± 0.10	83.41 ± 0.19	87.67 ± 0.16	83.90 ± 0.21	83.46 ± 0.20	69.36 ± 0.79	83.55 ± 0.14	78.24 ± 2.74
	hete-GCN	89.30 ± 0.25	84.20 ± 0.30	88.48 ± 0.35	84.66 ± 0.44	83.74 ± 0.15	71.93 ± 2.62	84.95 ± 0.38	81.56 ± 0.48
	homo-GAT	89.63 ± 0.27	84.70 ± 0.35	88.83 ± 0.07	85.65 ± 0.28	84.35 ± 0.26	69.27 ± 0.72	85.07 ± 0.33	81.83 ± 0.22

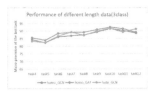

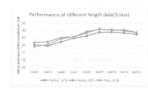

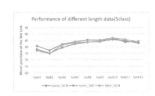

Fig. 6. The line chart of micro precition of the last task (3 class)

Fig. 7. The line chart of micro precition of the penultimate task (3 class)

Fig. 8. The line chart of micro precition of the last task (5 class)

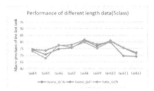

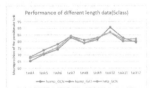

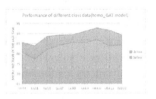

Fig. 9. The line chart of micro precition of the penultimate task (5 class)

Fig. 10. The line chart of macro precition of the last task (3 class)

Fig. 11. The line chart of macro precition of the penultimate task (3 class)

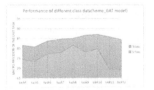

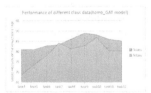

Fig. 12. The line chart of macro precition of the last task (5 class)

Fig. 13. The line chart of macro precition of the last task (5 class)

Fig. 14. The area map of micro precition of the last task

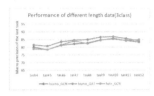

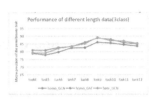

Fig. 15. The area map of micro precition of the penultimate task

Fig. 16. The area map of macro precition of the last task

Fig. 17. The area map of macro precition of the penultimate task

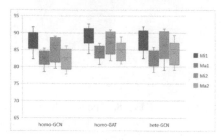

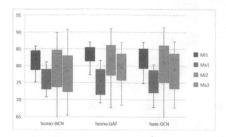

Fig. 18. Summary boxplots of validation results (3 class)

Fig. 19. Summary boxplots of validation results (5 class)

6 Discussion and Concluding Remarks

This paper studies from four aspects: first, three models are designed; second, the impact of different number of categories on the predict results; third, the impact of different length of workflow on the predict results; fourth, the impact of predicting in different positions of workflow on the predict results.

In experimenting we find out:

- In cloud workflow modeling, homo-GAT is slightly inferior to hete-GCN, and hete-GCN performs obviously better than homo-GCN.
- In different length workflow sequences, the longer the lead sequence, the better the workflow result when the length of leading sequence is less than 9. It conclude that the contrary of the above conclusion when the length of leading sequence is more than 9.
- In different number of categories, the result of three classification is about 5% higher than that of five classification.

The main contributions of this paper are as follows:

- Through K-means clustering, t-sen visualization and contour coefficient evaluation, it is found that in the CPU prediction of cloud workflow, the clustering result is better and the categories are more balanced by clustering the mean and variance of the average CPU.
- In the part of modeling graph, it provides experience for cloud workflow data modeling on GNN, and provides a method of cloud workflow modeling into homogeneous graph and heterogeneous graph. Besides, the tasks in the workflow are arranged and numbered in logical order through topological sequence to find the relative position information of each task.
- In the part of prediction, at present, the resource load prediction of cloud computing is mostly single task prediction, and the research of task prediction based on cloud workflow is almost blank. On this basis, this paper provides a method for the common prediction of different length workflow. This paper enriches the related research of cloud workflow resource load prediction.

Furthermore, the application of the research model in this paper is not only limited to cloud computing scenarios, but also can be applied to all industrial big data with similar graph structure information. Our research provides a new method for the data with graph structure in industrial big data. For example, in the future task prediction of multi location sensor and business log information, or in the prediction of missing sensors or specific indicators, there are similar complex data patterns.

Of course, this experiment is limited, and some new problems have been found in the experiment. On the one hand, this paper studies the data with the length of cloud workflow ranging from 4 to 12. Recently, we have extracted all the data with the length of workflow less than 50 from Alibaba cluster data. The author plans to predict and analyze the data with the length of workflow less than 50 in the next experiment, hoping to find more representative results. On the other hand, this study only explored three representative models in GNN, and found hete in the experiment_ GCN is better than Homo_ GCN, but not hete_GAT, the author infers hete_ The performance of GAT should be better. We plan to explore more GNN models, whether homogeneous or heterogeneous, in the next experiment.

Acknowledgement. This work is supported by the National Natural Science Foundation of China (71772033, 71831003), Natural Science Foundation of Liaoning Province, China (Joint Funds for Key Scientific Innovation Bases, 2020-KF-11-11), Scientific Research Project of the Education Department of Liaoning Province, China (LN2019Q14).

Authors' Contributions. All authors have been contributed equally to this work.

References

1. Armbrust, M., Fox, A., et al.: Above the clouds: a Berkeley view of cloud computing. Science (2009)
2. Creeger, M.: Cloud computing: an overview. Queue **7**(5), 2 (2009)
3. Zheng, Z., Zhang, Y., Lyu, M.R.: CloudRank: a QoS-driven component ranking framework for cloud computing. In: Reliable Distributed Systems, 2010 29th IEEE Symposium on IEEE (2010)
4. Liu, X., Yuan, D., Zhang, G., et al.: The Design of Cloud Workflow Systems. Springer, New York (2012)
5. Ali-Eldin, A., Seleznjev, O., Sara Sjöstedt-de, L., et al.: Measuring cloud workload Burstiness. In: IEEE/ACM International Conference on Utility & Cloud Computing. IEEE (2014)
6. Randles, M., Lamb, D.A., Taleb-Bendiab, A.: A comparative study into distributed load balancing algorithms for cloud computing. In: 24th IEEE International Conference on Advanced Information Networking and Applications Workshops, WAINA 2010, Perth, Australia, 20–13 April 2010. IEEE (2010)
7. Dhinesh, B., Krishna, P.V., Venkata Krishna, P.: Honey bee behavior inspired load balancing of tasks in cloud computing environments. Appl. Soft Comput. **13**(5), 2292–2303
8. Singh, A., Juneja, D., Malhotra, M.: Autonomous agent based load balancing algorithm in cloud computing. Procedia Comput. Sci. **45**, 832–841 (2015)
9. Buyya, R., Ranjan, R., Calheiros, R.N.: Modeling and Simulation of Scalable Cloud Computing Environments and the CloudSim Toolkit: Challenges and Opportunities. IEEE (2009)

10. Liu, D., Khoukhi, L., Hafid, A.: Prediction-based mobile data offloading in mobile cloud computing. IEEE Trans. Wireless Commun. **17**(7), 4660–4673 (2018)
11. Daetwyler, H.D., Calus, M.P.L., et al.: Genomic prediction in animals and plants: simulation of data, validation, reporting, and benchmarking. Genetics **193**(2), 347–365 (2013)
12. Da Bbagh, M., Hamdaoui, B., Guizani, M., et al.: Toward energy-efficient cloud computing: prediction, consolidation, and over commitment. Netw. IEEE **29**(2), 56–61 (2015)
13. Zhang, Y., Zheng, Z., Lyu, M.R.: Exploring latent features for memory-based QoS prediction in cloud computing. In: IEEE International Symposium on Reliable Distributed Systems. IEEE Computer Society (2011)
14. Jokhio, F.A., Ashraf, A., Lafond, S., et al. Prediction-based dynamic resource allocation for video transcoding in cloud computing. In: Euromicro International Conference on Parallel. IEEE (2013)
15. Kwok, Y.-K., Ahmad, I.: Static scheduling algorithms for allocating directed task graphs to multiprocessors. ACM Comput. Surv. **31**(4), 406–471 (1999)
16. Quaglini, S.: Workflow management—models, methods and systems. Artif. Intell. Med. **27**(3), 393–396 (2003)
17. Yu, J., Buyya, R.: A taxonomy of workflow management systems for grid computing. J. Grid Comput. **3**(3), 171–200 (2005)
18. Cortes, C., Cortes, C., Vapnik, V., et al. Support-vector networks (1995)
19. Kleinbaum, D.G., Klein, M. Logistic Regression (A Self-Learning Text). Springer, New York (2002). https://doi.org/10.1007/978-1-4419-1742-3
20. Mitchell, T.M.: Machine Learning. McGraw-Hill, New York (2003)
21. Schmidhuber, J.: Deep learning in neural networks: an overview. Neural Netw. **61**, 85–117 (2015)
22. Scarselli, F., Gori, M., Tsoi, A.C., et al.: The Graph Neural Network Model. IEEE Trans. Neural Netw. **20**(1), 61–80 (2009)
23. Hellerstein, J.L.: Google Cluster Data (2010)
24. Cortez, E., Bonde, A., Muzio, A., et al.: Resource Central: Understanding and Predicting Workloads for Improved Resource Management in Large Cloud Platforms Symposium. ACM, New York (2017)
25. Shahrad, M., Fonseca, R., Goiri, I., et al.: Serverless in the wild: characterizing and optimizing the serverless workload at a large cloud provider. In: 2020 USENIX Annual Technical Conference (USENIX ATC 20) (2020)
26. Lu, C., Ye, K., Xu, G., et al.: Imbalance in the cloud: an analysis on Alibaba cluster trace. In: 2017 IEEE International Conference on Big Data (Big Data). IEEE (2018)
27. Hao, X., Zhang, G., Ma, S.: Deep learning. Int. J. Semant. Comput. **10**(03), 417–439 (2016)
28. Zhao, B., Xu, Z., Tang, Y., Li, J., Liu, B., Tian, H.: Effective knowledge-aware recommendation via graph convolutional networks. In: Wang, G., Lin, X., Hendler, J., Song, W., Xu, Z., Liu, G. (eds.) WISA 2020. LNCS, vol. 12432, pp. 96–107. Springer, Cham (2020). https://doi.org/10.1007/978-3-030-60029-7_9
29. Cheng, B., Yang, J., Yan, S., et al.: Learning with l1-graph for image analysis. IEEE Trans. Image Process. Publ. IEEE Signal Process. Soc. **19**(4), 858–866 (2010)
30. Berberidis, D., Nikolakopoulos, A.N., Giannakis, G.B.: Adaptive diffusions for scalable learning over graphs. IEEE Trans. Signal Process. (2019)
31. Monti, F., Boscaini, D., Masci, J., et al.: Geometric deep learning on graphs and manifolds using mixture model CNNs. In: 2017 IEEE Conference on Computer Vision and Pattern Recognition (CVPR). IEEE (2017)
32. Kip, F.T.N., Welling, M.: Semi-Supervised Classification with Graph Convolutional Networks (2016)
33. Velikovi, P., Cucurull, G., Casanova, A., et al.: Graph Attention Networks (2017)

34. Hartigan, J.A., Wong, M.A.: Algorithm AS 136: A K-means clustering algorithm. Appl. Statis. **28**(1), 100 (1979). https://doi.org/10.2307/2346830
35. Taylor, P.D., Day, T., Wild, G.: Evolution of cooperation in a finite homogeneous graph. Nature **447**(7143), 469–472 (2007)
36. Bounova, G., Weck, O.D.: Overview of metrics and their correlation patterns for multiple-metric topology analysis on heterogeneous graph ensembles. Phys. Rev. E Stat. Nonlinear Soft Matter Phys. **85**(1), 016117 (2011)
37. Dean, J.A., Wong, K.H., Jones, A.B., Harrington, K.J., Nutting, C.M., Gulliford, S.L.: OC-0257: NTCP models for acute dysphagia resulting from (chemo)radiotherapy for head and neck cancer. Radiotherapy and Oncology **115**, S131 (2015)
38. Asch, V.V.: Macro- and micro-averaged evaluation measures

Prognosis Prediction of Breast Cancer Based on CGAN

Xi Liu, Runan Zhao, Yingqi Zhang, and Fan Zhang$^{(\boxtimes)}$

School of Computer and Information Engineering, Henan University,
Kaifeng 475004, China
zhangfan@henu.edu.cn

Abstract. Breast cancer is the malignancy with the highest morbidity and mortality rate in women worldwide, and prognosis prediction of breast cancer is of great practical importance for both patients and clinical practitioners. In this paper, we use a modified Conditional Generative Adversarial Networks (CGAN) to train the generators in a GAN into a predictive model that can perform prognosis of breast cancer on clinical data from patients and compare it with a multi-factor Cox proportional hazards model. In this paper, the accuracy of the prognostic model using CGAN was 0.950 with an AUC of 0.915; the AUC value of prognosis obtained using the multi-factor Cox proportional hazards model was 0.837. The experimental results demonstrated that the breast prognostic model based on CGAN can more accurately quantify and assess the prognosis of patients.

Keywords: Breast cancer · Prognosis · CGAN · Cox proportional hazards model

1 Introduction

Breast cancer is the most common malignancy with the highest incidence among women worldwide, accounting for 30% of all malignancies. According to the latest data released by the World Health Organization's International Agency for Research on Cancer (IARC), 2.26 million new cases of breast cancer will be diagnosed world-wide in 2020, exceeding the second highest number of lung cancer by 60,000 cases and accounting for 24.5% of all new cancers in women worldwide. Although breast cancer is curable in its early stages with a high probability, one third of women still die from the disease [1]. In clinical practice, biomarkers are necessary to predict the prognosis of breast cancer, but the prognosis of breast cancer patients is affected by several factors [2], and it remains a major challenge to accurately predict the prognosis of breast cancer and select the best therapy for patients [3].

This research was supported by the Natural Science Foundation of Henan Province (No. 202300410093).

C. Xing et al. (Eds.): WISA 2021, LNCS 12999, pp. 190–197, 2021.
https://doi.org/10.1007/978-3-030-87571-8_16

Wang Li et al. [4] analyzed the genetic related data of two types of such cancers, ER+ and ER−, respectively, and used a one-way Cox proportional hazards model for initial screening of genes, and then screened the genes again with the help of LASSO approach, using survival tree method, for patient prognosis. Wang Zhe et al. [5] compared the sensitivity of different variables on breast cancer prognosis by logistic regression model, decision tree and random forest algorithm analysis. Jing Du et al. [6] constructed a dynamic Cox proportional hazards model based on a Bayesian approach to prognosis analysis of breast cancer patients, and this model predicted the highest overall survival rate compared to other models. The research results of Song Xiaoqing et al. [7] showed that the LASSO regression method was used to screen the variables valuable for predicting patient prognosis and construct a multifactorial Cox proportional hazards model with good model discrimination and accuracy for higher prediction accuracy.

The estimation of survival period is one of the core research topics in prognosis. For the estimation and calculation of survival period, relevant algorithms or statistical methods are needed to analyze the characteristics of survival period in detail and build a survival prognosis model to guide the work [8]. In recent years, with the development of deep learning, the work on prognosis is no longer limited to traditional data analysis, and deep learning methods have been tried for cancer disease examination and prognosis with higher accuracy. Deep neural network-based cancer prognostic models, convolutional neural network-based cancer prognostic models, deep confidence network-based cancer prognostic models, etc., have all achieved good results in improving prognostic accuracy. However, new approaches should be sought continuously to break through. Generative adversarial networks have received a huge response since their introduction, and numerous researchers have devoted themselves to them. The research on generative adversarial networks has made great progress in the past few years, and many excellent variants and applications have been proposed, and the effectiveness of the generated data has been continuously improved. In this paper, we propose to use a deep learning model based on an improved CGAN to train the generators in generative adversarial networks into a predictive model that can perform prognosis, complete breast cancer prognosis based on patient clinical data, and compare it with the Cox proportional hazards model to predict the prognostic outcome of breast cancer patients.

2   Data and Methods

2.1   Dataset Processing

The data in this study were obtained from The Cancer Genome Atlas (TCGA) [9], the largest database of cancer genetic information available, which was established by the National Cancer Institute (NCI) and the National Human Genome Research Institute (HGRI). In addition to large sample data of various cancers, the TCGA database also has well-established clinical data. In this paper, clinical data of cancers are extracted from the TCGA database for prognostic analysis of breast cancer. The clinical information includes age, gender, whether smoker,

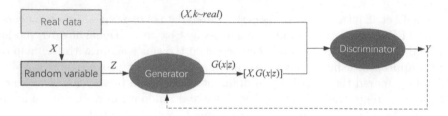

Fig. 1. CGAN model architecture diagram.

what drugs have been used, whether they have been treated before, cause of development, tumor period, and cancer staging for each cancer patient.

2.2   Conditional Generation Adversarial Network

Generative Adversarial Networks (GAN), proposed by Goodfellow et al. [10,11] in 2014, is a generative model. The basic idea of GAN comes from the two-person zero-sum game of game theory and consists of two parts: a generator and a discriminator. The generator inputs random noise and outputs fake sample data; the discriminator inputs the true data as well as the fake sample data generated by the generator to determine whether the input samples are true or false.

Conditional Generative Adversarial Networks (CGAN) simply changes the input of the GAN, where the original GAN only requires input data, while CGAN also requires input labels. This improvement, by attaching some information to the model guides the model to generate the sample data we want the objective function of CGAN is shown in the following equation.

$$\min_G \max_D V(D,G) = E_{x \sim Pdata(x)}[\log D(x|y)] + E_{z \sim Pz(z)}[1 - \log D(G(z|y))], \quad (1)$$

where G is the generator, D is the discriminator, $x \sim Pdata$ denotes sampling from real data, $z \sim Pz(z)$ denotes sampling from noise, and y denotes the corresponding sample label.

In this paper, we have made some improvements to the generator using the ideas of CGAN. In our improved CGAN, the generator not only serves to generate the samples, but also learns the features of the true samples and gives their predicted values based on the sample features. The input to the generator is the noise plus the true sample value and the output is the predicted value; the input to the discriminator is the true sample plus the predicted value and the true sample plus the true value and the output is true and error. The generator in our improved CGAN does not only serve to generate samples, but also learns the features of the true samples and gives their predicted values based on the sample features. In a traditional GAN network, the generator only inputs noise, which is mapped by the neural network to a false sample of the same size as the true sample, whereas in this GAN network the generator can be seen as a predictor, giving its prediction to the discriminator to determine whether it is correct or not; the discriminator also differs from the traditional GAN network in that the discriminator does not only receive the output of the generator and the true value, but also the combination of the two and the true sample The discriminator also differs from the traditional GAN network in that it does not only receive the output of the generator and the true value, but the combination of the two and the true sample, judging the true sample plus the predicted value as false and the true sample plus the true value as true. This improvement allows the generative adversarial network to perform regression analysis. The architecture of the improved CGAN model is shown in Fig. 1, and the objective function is shown as follows:

$$\min_{G} \max_{D} V(D,G) = E_{x \sim Pdata(x)}[\log D(k|x)] + E_{z \sim Pz(z)}[1 - \log D(G(x|z)|x)],$$
(2)

where G is the generator, D is the discriminator, $x \sim Pdata$ denotes sampling from real data, $z \sim Pz(z)$ indicates sampling from noise, k is the survival outcome of breast cancer patients corresponding to real data, and $G(x|z)$ is the survival outcome of breast cancer patients predicted by the generator.

2.3 Cox Proportional Hazards Model

Cox's proportional hazards model, a semi-parametric regression model proposed by the British statistician D.R. Cox [12], is a method of multifactorial survival analysis. The model is usually used in medical research to analyze the effect of one or more predetermined variables on the survival time of patients. For the Cox model it is defined as.

$$h(t, X) = h_0(t) \exp(\beta_1 X_1 + \beta_2 X_2 + \cdots + \beta_m X_m),$$
(3)

where $X = (X_1, \cdots, X_n)'$, are relevant factors that may affect survival time, that is, covariates that do not vary with time. $\beta_1, \beta_2, \cdots, \beta_m$ are regression coefficients, and $h_0(t)$ is a baseline risk function that does not require a specific distribution and is a nonparametric model part; the latter part is a parametric model, which is equivalent to a sub squares transformation of the output of the multiple linear regression, ensuring positivity and monotonicity.

3 Results and Discussion

3.1 Prognosis of Breast Cancer Based on Multivariate Cox Model

In this paper, 1098 tumor samples downloaded from TCGA database were used for prognostic analysis, and these 1098 samples were randomly grouped according to the survival status of the samples with a grouping ratio of 1:1, and finally the training set and test set data were obtained, containing 549 patients in each of the two data. The training set data were first screened from the clinical characteristics data using the LASSO [13] method for the construction of the multifactorial Cox model. glmnet package was used to implement the LASSO regression in this paper, and a 10-fold cross-validation method was used to estimate the adjustment parameters. In the regression equation, the regression coefficient indicates the parameter of the magnitude of the effect of the independent variable on the dependent variable, which can also be interpreted as the contribution of that variable. The cross-validation error rate was minimized when $\lambda = 0.013$ at $\ln \lambda = -4.34$, at which point 16 pathological features were screened.

For the 16 prognosis related pathological features screened by LASSO regression, the most significant prognostic pathological features were further analyzed using multifactor Cox model, from which the most significant prognostic pathological features were screened, and risk score models were constructed using these pathological features. In this paper, we used the SURVIVAL package to implement a multifactor Cox proportional hazards model, and the results of the analysis are shown in Table 1.

Table 1. Results of multi-factor Cox analysis.

	Coef.	Z-value	p-value
ajcc_staging_system_edition	3.583e−01	5.760	8.43e−09
ethnicity	1.766e−01	0.897	0.3695
race	3.522e−03	0.039	0.9693
year_of_diagnosis	−1.870e−04	−0.316	0.7517
days_to_last_follow_up	−6. 096e04	−5.287	1.24e−07
days_to_birth	−3.420e−05	−1.988	0.0468
tissue_or_organ_of_origin	5.098e−01	1.149	0.2507
age_at_diagnosis	N/A	N/A	N/A
ajcc_pathologic_n	2.349e−03	0.107	0.9145
prior_treatment	6.829e−01	1.091	0.2751
ajcc_pathologic_t	3.803e−02	0.985	0.3244
ajcc_pathologic_stage	−7.470e−03	−0.227	0.8202
icd_10_code	−4.217e−01	−1.304	0.1921
treatment_or_therapy	−2.366e−01	−1.891	0.0586
site_of_resection_or_biopsy	N/A	N/A	N/A
ajcc_pathologic_m	3.681e−03	0.022	0.9821

In the table, Coef. is the regression coefficient of covariates (clinicopathological characteristics), HR is the risk proportion coefficient, when $HR > 1$, it means that the influence factor is a risk factor; when $HR < 1$, it means that the influence factor is a protective factor. When $HR = 1$, it means that the influence factor is not related to the risk of disease. From the table, we can know that the p-values of some clinical characteristics are significantly less than 0.05, indicating that these six clinical characteristics can also be used as independent prognostic factors for the prognostic analysis of breast cancer.

There are various factors affecting the prognosis of breast cancer, and using a multifactorial Cox model to analyze prognosis is a better approach. In this study, a multifactorial Cox proportional risk regression model for breast cancer prognosis using these six clinicopathological characteristics yielded an AUC value of 0.837.

3.2   Prognosis of Breast Cancer Based on the Improved CGAN

The dataset used in the experiment is the dataset of clinical characteristics of breast cancer in TCGA. We removed many useless items and predictors that were too highly correlated with the prediction results in the original dataset, and set them all to 0 because some clinical features were only available in positive samples and some in negative samples. the values of each indicator were converted from strings to numerical values, and a total of 24 clinical indicators were used, including age, gender, disease age, race, cancer site, tumor stage, and so on. The values of each indicator were converted from strings to values and replaced by different numbers. 1098 samples were taken from this dataset, 896 samples were taken as the training set and 202 samples were taken as the test set.

The structure of the generator consists of three fully connected layers with input data of 26 Gaussian distributed noises and 24 predictors, which are processed by the neural network to output the predicted values. The discriminator consists of four fully connected layers with inputs of true samples consisting of 24 clinical indicators and the corresponding true values of the samples and false samples consisting of 24 clinical indicators and the predicted values output by the generator.

After repeated experiments, we finally determined that the learning rate is 0.0002, the batch_size is 128, the discriminator is trained 3 times per iteration, and the generator is trained 1 time when the best results are obtained on the test set. The model was first trained 10,000 times, and then the model was saved once for each iteration of training 1000 times, and then the model of each training was tested on the test set to find the model with the best results, and the model worked best around 15,000 times of training, and the process of loss change is shown in Fig. 2.

Out of 124 test samples, the accuracy was 0.950 with an AUC of 0.915. the experimental results are shown in Table 2.

Table 2. Experimental results.

	Precision	Recall	F1-score	Support
live	0.98	0.96	0.97	179
Dead	0.74	0.87	0.80	23
Macro avg.	0.86	0.92	0.89	202
Weighted avg.	0.96	0.95	0.95	202

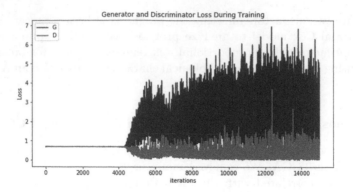

Fig. 2. The change process of loss.

3.3   Comparison of Improved CGAN-Based and Multi-factor Cox Prognostic Hazards Models

In this paper, a prognostic model for breast cancer and a prognostic model for breast cancer based on multifactorial Cox regression were developed using a CGAN-based network model. The datasets used for both models are the same dataset, both of which are clinicopathological datasets in TCGA. The discriminative ability of the models was evaluated by using ROC curves in the training set as well as the validation set, respectively. Comparison of the parameters of the CGAN-based and multi-factorial Cox proportional hazards models was obtained in Table 3.

Table 3. Comparison of parameters of two models.

Model	Precision	AUC
Cox proportional hazards model	N/A	0.837
CGAN	0.950	0.912

Therefore, it is possible to obtain a better breast prognosis model based on CGAN, which can more accurately quantify and assess the prognosis of patients and can provide intuitive information for physicians, patients, and healthcare policy makers.

4   Conclusions

In this paper, the generator is considered as a predictor that receives noise plus true sample values and the output is predicted based on CGAN; the discriminator receives true samples plus predicted values and true samples plus true values and the output is true and false. This improvement makes the generative adversarial network capable of regression analysis. The generators in the generative adversarial network are trained into a predictive model that can perform prognosis, allowing it to complete breast cancer prognosis based on patient clinical data. The metrics of each clinical feature are mapped into different numbers, and the data set is divided into a training set and a test set, and the experimental results are obtained after repeated experimental training. The AUC of the breast cancer prognostic model built by the modified CGAN was 0.912. The results of the LASSO method and the multifactor Cox proportional hazards were analyzed to obtain six clinicopathological characteristics to construct a multifactor Cox risk regression model as a breast cancer prognostic model to obtain an AUC value of 0.837. Both models are reliable for prognosis of breast cancer clinical data, and the modified CGNA breast cancer prognostic model was more accurate.

References

1. Keles, A., Keles, A., Yavuz, U.: Expert system based on neuro-fuzzy rules for diagnosis breast cancer. Expert Syst. Appl. **38**, 5719–5726 (2011)
2. Xu, W.H., Liu, Z.B., Yang, C., et al.: Expression of Dickkopf-1 and beta-catenin related to the prognosis of breast cancer patients with triple negative phenotype. PLoS ONE **7**, e37624 (2012)
3. Xu, Z., He, T., Lian, H., Wan, J., Wang, H.: Case facts analysis method based on deep learning. In: Ni, W., Wang, X., Song, W., Li, Y. (eds.) WISA 2019. LNCS, vol. 11817, pp. 92–97. Springer, Cham (2019). https://doi.org/10.1007/978-3-030-30952-7_11
4. Wang, L.: Prognosis of breast cancer based on cox proportional hazards regression model, lasso and survival tree. Stat. Appl. **07**, 99–110 (2018)
5. Zhe, W.: The study on prognostic factors of breast cancer based on machine learning method. China Digital Med. **14**, 18–20 (2019)
6. Jing, D.: Breast cancer patients prognostic analysis using Bayesian method. Comput. Eng. Appl. **56**, 146–151 (2020)
7. Xiaoqing, S.: Construction of prognostic evaluation model for patients with breast cancer. J. Dalian Med. Univ. **43**, 29–37 (2021)
8. Gripp, S., Moeller, S., Bälke, E., et al.: Survival prediction in terminally ill cancer patients by clinical estimates, laboratory tests, and self-rated anxiety and depression. J. Clin. Oncol. Official J. Am. Soc. Clin. Oncol. **25**, 3313–3320 (2007)
9. Koboldt, D., Fulton, R., Mclellan, M., et al.: Comprehensive molecular portraits of human breast tumours. Nature **490**, 61–70 (2012)
10. Rafael, S., Romero, A., Soua, S., et al.: Detection of gearbox failures by combined acoustic emission and vibration sensing in rotating machinery. Insight Non-Destr. Test. Condition Monit. **56**, 422–425 (2014)
11. Goodfellow, I., Pouget-Abadie, J., Mirza, M., et al.: Generative adversarial networks. Adv. Neural Inf. Process. Syst. **3**, 2672–2680 (2014)
12. Hsu, C.H., Yu, M.: Cox regression analysis with missing covariates via multiple imputation. Stat. Methods Med. Res. **28**, 1676–1688 (2017)
13. Tibshirani, R.: Regression shrinkage selection via the lasso. J. Roy. Stat. Soc. Ser. B **73**, 273–282 (2011)

Few-Shot Learning for Time Series Data Generation Based on Distribution Calibration

Yang Zheng, Zhenguo Zhang$^{(\boxtimes)}$, and Rongyi Cui

Department of Computer Science and Technology, Yanbian University,
977 Gongyuan Road, Yanji 133002, People's Republic of China
{2020050049,zgzhang,cuirongyi}@ybu.edu.cn

Abstract. Insufficient training data often makes the learning model prone to overfitting and bias in the selection of the sample leads to obtaining the wrong distribution. For this reason, few-shot learning has gained widespread attention as a challenging endeavor. Current work in few-shot learning is focused on developing stronger models, but these models does not have good generalization capabilities. In this paper, Our approach is find a similar base class with sufficient data for class with few-shot samples, then use statistical information to calibrate the distribution of class with few-shot samples. Time series are characterized by variability within the variance at each point in time and by overall statistical regularity and periodicity. So time series are extremely suitable for our approach. This approach do not require complex models and additional parameters. Our approach generate data that better match the actual distribution of the data. Validated with 9 time series data sets, the data generation for five samples led to some improvement in the classification accuracy. Moreover, it is found that this approach is not only applicable to the case of small data size, but also the classification effect is improved if the method of this paper is applied on the basis of sufficient data size.

Keywords: Data generation · Time series · Distribution calibration · Few-shot learning

1 Introduction

Time series is an important problem in the field of data mining. Time series can respond well to the changing characteristics of things and classify the time series according to the overall or local nature in the change. In classification of time series, deep learning has good results and is widely used [7]. But solving problems with deep learning methods relies on sufficient data, with more balanced classification of each category. If we don't get enough data in the process of deep learning, overfitting may often occur.

Nowadays, there is a growing interest in how to train in a limited number of labeled samples and get better results. Few-shot learning is mainly focused on the process of developing stronger models, but insufficient sample size still leads to

© Springer Nature Switzerland AG 2021
C. Xing et al. (Eds.): WISA 2021, LNCS 12999, pp. 198–206, 2021.
https://doi.org/10.1007/978-3-030-87571-8_17

overfitting, so that the model does not generalize well. Contrast learning, which projects samples into a space where like samples are closer together and non-like samples are further apart. Meta-learning [1] tries to have the machine learn a model on different tasks, and when a new task comes up, it only needs a limited number of iterations on the existing network to get good results. Our approach is to focus on the data itself, we expand the class of few-shot samples by prior knowledge and statistical information [4]. And we obtain results with broader adaptability.

We know from human's prior knowledge that the more similar time series should be more similar at all overall time points, but the covariance of each time point is somewhat different according to its own characteristics [5]. We put the time series into convolutional neural network for training to obtain the characteristics of time series [2]. Classes that exist in other related data sets with a certain amount of data and similarity are called base classes, and use few samples with labeled data in the current for data generation. The approach in this paper is to map the base class to a higher dimension by Encoder, so as to obtain more information, and to find out the mean and covariance of each dimension in the higher dimension [3,6]. We use the mean and covariance of the base class to calibrate the distribution of the class of few-shot samples, generate the calibrated data in the high dimension, then map back to the low dimension by Decoder. In this way to generate the data to match the distribution, in order to improve the accuracy of classification on the test set.

2 Main Approach

First of all, our approach is based on the existing base class with a certain amount of data to calibrate the class of few-shot samples distribution. The prerequisite of this experiment is that we need two data set can migrate and base class with a certain amount of data is needed, and few-shot samples and base classes are similar classification tasks. In this paper we follow the process shown in Fig. 1 to conduct the experiments.

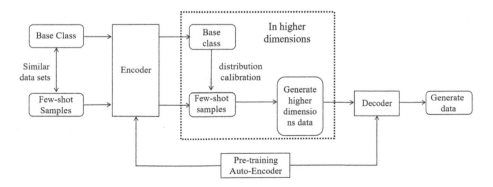

Fig. 1. The pipeline of our proposed. This figure can represent our whole experiment process, we will pre-train an Auto-encoder and fix the parameters of the network.

2.1 Pre-training of the Auto-Encoder

After obtaining two data sets that can be migrated, In order to ensure that the output and input are as similar as possible. The encoder structure consists of three *Conv1d* and a fully-connected layer that maps the x_{Input} data to the high dimension to become $x_{Highdim}$.

$$x_{Highdim} = Encoder(x_{Input}) \tag{1}$$

The high dimension is then reduced to the original dimension by the fully connected layer and three *ConvTranspose1d* to get x_{Output}.

$$x_{Output} = Decoder(x_{Highdim}) \tag{2}$$

We define the loss function of the whole network as (Fig. 2):

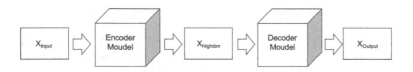

Fig. 2. The network structure of our pre-trained Auto-encoder.

$$MSEloss = \frac{1}{N} \sum_{t=1}^{N} (x_{Input} - x_{Output})^2 \tag{3}$$

In the above loss function, Xt is the input of the t dimension of Auto-Encoder. During the training process, the gradient descent of the loss function aims to have the same input and output as much as possible while having more information obtained in higher dimensions. We pre-train the network and should also choose the data set containing the base class that we will select afterwards.

2.2 Similar Class Searching for Few-Shot Samples

With a pre-trained Auto-Encoder, We map both the base class and the few-shot samples to high dimension to obtain statistical information of mean and covariance. The mean value can reflect the similarity between the few-shot sample and the base class.

In the base class task, the means and covariances of all classes have been obtained, and at this point we have also obtained the means and covariances from the few-shot samples. We want to use these two statistics to calibrate the class of few-shot samples distribution. In this paper, the Euclidean distance is used to measure the similarity of two time series.

$$dist\,(\boldsymbol{\mu}\mathbf{x}, \boldsymbol{\mu}\mathbf{y}) = \sqrt{\sum_{i=1}^{n} (\mu x_i - \mu y_i)^2} \tag{4}$$

Here μx and μy stands for the mean of the base class and the mean of the few-shot samples. If the difference between the mean of each dimension of the few-shot samples and the base class is smaller, we have reason to believe that the two time series are more similar. The class for which we obtain the smallest Euclidean distance is the base class used to calibrate the class of few-shot samples, then we calibrate the statistics according to the following equation:

$$\mu y_i{'} = \mu y_i \cdot \alpha + \mu x_i \cdot (1 - \alpha) \tag{5}$$

$$\sum y_i{'} = \sum y_i \cdot \alpha + \sum x_i \cdot (1 - \alpha) \tag{6}$$

After obtaining the mean $\mu y_i{'}$ and covariance $\sum y_i{'}$ with in higher dimensions by base class calibration of few-shot samples, the α defined in the above equation is an artificially set hyperparameter, and the α needs to be modified according to different data sets in order to make the data generated by the data set fit the true distribution better. After going through the above process, we can generate a large amount of data in high dimension based on the mean and covariance. The data generated in higher dimensions are reduced to the original data dimensions by the decoder part of the pre-trained autoencoder for the purpose of data generation. With few sample or biased data compared to the real part tend to improve better by calibration (Fig. 3).

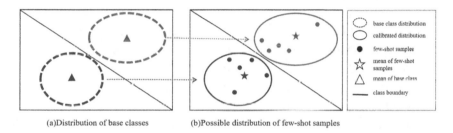

(a)Distribution of base classes   (b)Possible distribution of few-shot samples

Fig. 3. The figure of the distribution calibration process. (b) shows the distribution of few-shot samples, and the mean information is obtained according to the distribution, and the distribution is calibrated by selecting the base class from (a).

3   Experiments and Results

In this section, we need to verify that the accuracy of our classification task is indeed improved after we calibrate and generate the data according to the distribution. We validate this in three main ways:

1. The effect on classification accuracy improvement after calibration of the data in the class of few-shot samples.
2. Under the practice of generating data in high dimension and then mapping the data to low dimension, can we obtain better results than directly obtaining statistical information to generate data?
3. If there is a certain amount of data when the data generation, classification accuracy improvement effect.

The data generation methods used in this paper are not for a specific method. It is not only on a specific model, but all the traditional methods for better classification results. If the generated data can enhance the classification, then it is reasonable to assume that the data we generate will be similar to the actual data.

3.1 Experimental Setup

Data Sets. We use nine time-series data sets to validate the improvement of our idea in terms of classification effectiveness. Their commonality is the existence of similar data sets in terms of semantic or human prior knowledge that can be used to calibrate the current task distribution (Table 1).

Table 1. Summary of data sets used in the experiment.

Data sets	Train size	Test size	Length	Number of classes
DistalPhalanxOutlineAgeGroup	400	139	80	3
DistalPhalanxOutlineCorrect	600	276	80	2
GunPointMaleVersusFemale	135	316	150	2
GunPointMaleVersusFemale	135	316	150	2
MiddlePhalanxOutlineAgeGroup	400	154	80	3
MiddlePhalanxOutlineCorrect	600	291	80	2
ProximalPhalanxOutlineAgeGroup	400	205	80	3
ProximalPhalanxOutlineCorrect	600	291	80	2
PhalangesOutlinesCorrect	1800	858	80	2

Details of the Implementation. To confirm that the generation of few samples is generally effective, we selected five samples from each class from the larger data set, and the samples selected were completely random. A large number of samples were generated through these five samples, and then a large number of samples were used as the training set to see whether the Accuracy on the test set improved compared to that before the data generation. The selection of in the experiment varies for each data set, and the α values selected for the nine data sets in this paper range from 0.7 to 0.85.

3.2 Data Generation for Few-Shot Samples

According to several experiments, certain prerequisites need to be followed in order to get better results in data generation. This process, whether the extracted five data can respond to the most important is the distribution of few-shot samples is very important, if the extracted five data happen to respond to the overall data distribution, then generating a large number of data can solve the problem of over fitting; if the extracted data set happens to be all distribution edge, then it will lead to the selection of the wrong base class for the these samples and lead to poor results on the test set. The experimental results are shown in Table 2. In the experiments, the data generated by our method were improved in terms of both svm, and knn evaluation criteria, but the classification accuracy on logistic regression did not obtain a significant improvement and sometimes a decrease.

3.3   Introduction of Auto-Encoder

In this paper, we introduce Auto-Encoder, which adopts a self-supervised approach and without labeling. After the introduction of the anto-encoder and the experimental procedure described above by the manager, the higher dimension has more information and can be a more accurate distribution with a certain range of improvement; it can be imagined that samples that are difficult to classify in the lower dimension can be mapped to the higher dimension to achieve a clearer and more accurate classification; and more dimensions can allow few-shot samples to find the base class more accurately. It should be noted that the data set we use to pre-train the Auto-Encoder network should be the one containing the base class for better experimental results. The experimental results are shown in Table 3, We can see the impact on the accuracy rate with the introduction of Auto-Encoder and without Auto-Encoder. In the experiments, the data generated by introducing auto-encoder were improved in terms of both svm, and knn evaluation criteria, but the classification accuracy on logistic regression was sometimes improved and sometimes decreased depending on the dataset.

Table 2. Few samples for comparison experiments. Only five data each class are taken out for classification experiments, and then the data are generated for comparison experiments using the method in this paper.

Data sets	Original data			Generated data		
	KNN	SVM	LR	KNN	SVM	LR
MiddleP.C	61.5	48.9	70.4	**77.6**	**78.2**	**70.7**
ProximalP.C	49.1	46.7	**59.1**	**63.2**	**56.7**	54.9
DistalP.C	64.4	58.3	**58.6**	**66.6**	**59.4**	56.4
PhalangesO.C	55.4	45.2	58.7	**60.7**	**60.1**	**58.9**
MiddleP.A.G	61.4	61.2	**64.2**	**61.6**	61.2	49.3
DistalP.A.G	69.0	66.9	64.0	**69.9**	**71.5**	**66.2**
ProximalP.A.G	71.4	62.9	71.4	**78.2**	**76.4**	**72.2**
GunPointM.V.F	87.2	73.7	**89.2**	**88.0**	**92.7**	88.2
GunPointO.V.Y	98.9	99.3	**98.1**	**99.4**	**100**	78.5

Table 3. Comparison experiments with the introduction of auto-encoder. Comparison experiments with the introduction of auto-encoder. We want to investigate whether the introduction of Auto-encoder improves the classification accuracy, a comparison experiment was conducted to show that the average of the accuracy of multiple experiments, the introduction of Auto-encoder does improve the classification. In the comparison experiments, we bolded the results of the experiments with higher accuracy.

Data sets	Without auto-encoder			With auto-encoder		
	KNN	SVM	LR	KNN	SVM	LR
MiddleP.C	75.5	73.5	**77.9**	**77.6**	**78.2**	72.7
ProximalP.C	60.3	58.0	49.2	**62.5**	**58.7**	**50.8**
DistalP.C	61.2	59.7	**62.3**	**65.5**	**61.1**	60.5
PhalangesO.C	58.7	51.7	52.2	**60.2**	**54.3**	**54.1**
MiddleP.A.G	39.0	48.2	**49.2**	**45.1**	**52.2**	**49.7**
DistalP.A.G	65.4	58.1	49.6	**66.1**	**62.7**	**52.2**
ProximalP.A.G	63.9	62.9	68.7	**76.5**	**72.2**	**70.7**
GunPointM.V.F	79.4	86.7	83.5	**88.0**	**92.7**	**88.2**
GunPointO.V.Y	100	100	**61.3**	100	100	57.3

3.4   Data Generation for Sufficient Sample

Generating a large number of samples with validation for few-shot samples calibration does have some improvement on the accuracy. Two similar classification tasks with some calibration, and we use the base class to calibrate the classification of the current task, we will add a certain amount of information belonging to the base class to the current task. It should be able to have some positive effect on the current task. In this part of the experiment, ResNet was introduced as an accuracy evaluation metric due to the sufficient sample size. The experimental results are shown in Table 4.

Table 4. Sufficient samples for comparison experiments. Distribution calibration was performed on a data set with sufficient samples, and then a comparison experiment was conducted.

Data sets	Original data				Generated data			
	KNN	SVM	Resnet	LR	KNN	SVM	Resnet	LR
ProximalP	82.4	68.4	79.2	82.1	**83.1**	**80.0**	**81.4**	**82.4**
PhalangesO	76.1	62.9	67.8	**66.3**	**76.5**	**63.7**	**69.1**	66.0
DistalP.C	69.5	65.6	75.0	66.6	**71.3**	**68.4**	**78.2**	**72.1**
MiddleP.C	**78.8**	57.0	49.8	60.8	78.6	**57.3**	**53.2**	**62.1**
MiddleP.A.G	58.4	61.7	59.0	**59.0**	**59.6**	**63.0**	**59.7**	58.4
DistalP.A.G	64.0	66.9	74.2	66.1	**68.3**	**71.9**	**76.2**	**66.9**
ProximalP.A.G	84.3	83.3	81.9	85.8	**84.3**	**84.2**	**82.9**	**86.1**
GunPointM.V.F	95.8	89.2	95.5	**99.3**	**96.2**	**93.9**	**100**	96.8
GunPointO.V.Y	100	100	100	85.3	100	100	100	**86.3**

3.5 Generate Visualization of the Sample

We take out five data for each class in current task and generate a large amount of data according to our calibration method. Using the t-SNE method, we can see that the samples we selected and after distribution calibration, from two classes that are relatively already distinguishable. As the distribution is calibrated, the data we generate expresses a more comprehensive range and covers a larger scope, but relatively it also leads to some indistinguishable classes in the middle part of the intersection of the two classes (Fig. 4).

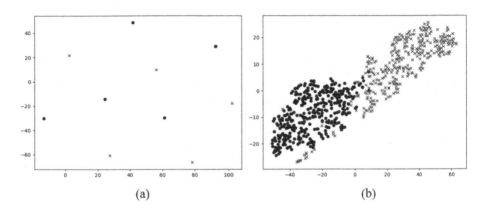

(a)   (b)

Fig. 4. Visualization of generated data. (a) shows the five samples we took for each category, and after calibration of the distribution, (b) shows that the resulting large amount of data forms clusters and is relatively easy to distinguish.

4 Conclusion

In this paper, we propose a method to calibrate the class of few-shot samples based on similar class. Our method generates a large amount of data, eliminates the occurrence of overfitting and trains a more stable model, improving the classification accuracy with a small training set. The method proposed in this paper not through a complex network structure, but by focusing on the data. On the basis of some improvement in data generation by few-shot samples, we also demonstrate that with a certain amount of data, our method has some improvement on the accuracy rate as well. The approach in this paper is to assume that both the few-shot samples and the base class fit a distribution and calibrate for it. However, in the real world, some data may obey multiple distributions, and our future work is to fit multiple distributions to reach the calibration of the generated data based on this paper, so that our method has a wider applicability. The generated data can be better fitted to the true distribution.

Acknowledgements. This work is supported by the school-enterprise cooperation project of Yanbian University [2020-15], State Language Commission of China under Grant No. YB135-76 and Doctor Starting Grants of Yanbian University [2020-16].

References

1. Finn, C., Abbeel, P., Levine, S.: Model-agnostic meta-learning for fast adaptation of deep networks (2017)
2. Guo, C., Xie, L., Liu, G., Wang, X.: A text representation model based on convolutional neural network and variational auto encoder. In: Wang, G., Lin, X., Hendler, J., Song, W., Xu, Z., Liu, G. (eds.) WISA 2020. LNCS, vol. 12432, pp. 225–235. Springer, Cham (2020). https://doi.org/10.1007/978-3-030-60029-7_21
3. Zhang, J., Shan, S., Kan, M., Chen, X.: Coarse-to-fine auto-encoder networks (CFAN) for real-time face alignment. In: Fleet, D., Pajdla, T., Schiele, B., Tuytelaars, T. (eds.) ECCV 2014. LNCS, vol. 8690, pp. 1–16. Springer, Cham (2014). https://doi.org/10.1007/978-3-319-10605-2_1
4. Martynenko, A.: Statistical analysis of medical time series (2020)
5. Tang, S., Chen, Z.Q.: Scale-space data augmentation for deep transfer learning of crack damage from small sized datasets. J. Nondestr. Eval. **39**(3), 1–18 (2020)
6. Wei, W., Yan, H., Wang, Y., Liang, W.: Generalized autoencoder: a neural network framework for dimensionality reduction. In: 2014 IEEE Conference on Computer Vision and Pattern Recognition Workshops (CVPRW) (2014)
7. Wen, Q., Sun, L., Song, X., Gao, J., Wang, X., Xu, H.: Time series data augmentation for deep learning: a survey (2020)

Sleep Analysis During Light Sleep Based on K-means Clustering and BiLSTM

Jiamin Xu and Haojun Sun[✉]

Shantou University, Shantou, China
{18jmxu,haojunsun}@stu.edu.cn

Abstract. Sleep staging are the basis of sleep analysis. Automatically and efficiently evaluating sleep stages is of great significance. A large number of researchers have come up with solutions to automate sleep staging. However, there are some deficiencies in these schemes. The sleep data is a kind of unbalanced data, and there is a high misjudgment rate of N1 stage in many previous works. This paper proposes a new method to solve this problem. It divides a sleep EEG sequence into continuous and shorter duration sequence segments (sequence segmentation), and obtains the feature symbol vector from the subdivided sleep sequence segments through clustering. Finally, the feature symbol vector is fed to the BiLSTM to mine the law of time series and realize sleep staging. Experiments have proved that our model can effectively improve the classification accuracy of N1 stage in the sleep staging task.

Keywords: Sleep staging · Sequence segmentation · Clustering · Deep learning · BiLSTM

1 Introduction

In the field of neuroscience, the research on sleep has always been a hot spot. Reasonable staging of sleep is the basis for studying sleep quality and diagnosing sleep diseases. According to the AASM standard, sleep is divided into five stages: Wake, REM, N1, N2, N3. Polysomnography (PSG) is divided into a stage every 30s. Professional sleep scoring can only be performed in the hospital, and then visual observation by experts, which is very tedious, time-consuming, and prone to subjective errors. Sleep staging automation can reach the same level as expert manual judgment, and at the same time can alleviate this task, so it helps families monitor sleep disorders.

Sleep EEG automatic staging is a cross-cutting frontier research field. Many experts and scholars have successively proposed a variety of automatic sleep EEG staging methods based on traditional machine learning. The core of the research lies in the extraction of features and the selection of classification algorithms. In recent years, some experts has adopted deep learning methods to automatically learn data features, and achieved better experimental results.

© Springer Nature Switzerland AG 2021
C. Xing et al. (Eds.): WISA 2021, LNCS 12999, pp. 207–214, 2021.
https://doi.org/10.1007/978-3-030-87571-8_18

Generally speaking, these existing methods can obtain higher classification accuracy as a whole, but the classification effect of N1 stage is poor. Generally, the classification accuracy of N1 is less than 50%. The light sleep stage, as a transitional period between the awake stage and the deep sleep stage, has important value for sleep research. Our goal is to improve the classification accuracy of light sleep stage. Therefore, we propose a sleep staging method based on clustering and BiLSTM to solve this problem. We obtained the characteristics of sleep EEG signals through the method of sequence segmentation and unsupervised learning, and at the same time reconsidered the classification labels of sleep stages. In other words, the conversion between stages (Conversions class) is also taken into consideration to further improve the feature expression of sleep EEG signals. Finally adopt the recognized deep learning recurrent neural network BiLSTM with strong learning ability as the classifier.

This article is divided into five sections. Section 1: Introduction, mainly introduces the standard and the meaning of sleep staging. The second chapter mainly introduces related work and achievements in the field of sleep staging. Section 3 will describe our proposed sleep staging model based on clustering and BiLSTM. Section 4 show experimental results. The fifth section is summary and prospect of our work.

2   Related Work

In the field of sleep staging, many researchers have proposed related data mining models, which have performed well on existing public data sets. Some authors use traditional machine learning algorithms to solve the problem. XIAO proposed to use the KMeans algorithm for sleep staging [3]. Zhu use SVM as a classifier [4]. Hassan et al. [5] used inherent modal functions to extract statistical time-domain features, and combined with the random downsampling promotion method. These sleep staging models have good interpretability, so they have more convincing predictions on the newly generated data sets, and are more acceptable to the public. However, it is a complicated process to build a machine learning model for sleep staging, which requires a complex manual feature extraction process.

In recent years, deep learning is a very popular method [6]. Some authors have used deep learning to carry out sleep staging research and achieved better classification results. Authors such as Sors used a one-dimensional sleep sequence as the input of the model [8], using a 14-layer convolutional neural network to extract features. Authors such as Supratak proposed to use the combination of CNN and LSTM [10] to process a single channel EEG signal. Perslev M got inspiration from the field of image segmentation and proposed a time full convolutional network U-time [13] based on the image segmentation U-net architecture. Deep learning algorithms have a strong ability to fit models and are widely used in the problem of automated sleep staging to achieve a high overall classification accuracy. However, these efforts still have shortcomings, and they failed to address the problem of high misscores for N1 stage.

N1 stage is the transitional stage of human beings from being awake to falling asleep. The ability to accurately distinguish the light sleep is important to the study of sleep disorders. Therefore, we propose a sleep staging model based on clustering and BiLSTM, which uses sequence segmentation, unsupervised learning and deep learning to mine features in sleep EEG signals, thereby improving the classification accuracy of light sleep stage.

3 Method

We propose a new sleep EEG sequence feature expression method, which uses sequence segmentation and unsupervised learning methods to construct a feature symbol vector of a sleep EEG signal. The learned vector is used as the feature expression of the sequence and is fed to BiLSTM for training. The principle framework of automatic sleep staging based on clustering and deep learning is shown in Fig. 1.

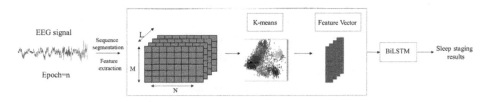

Fig. 1. The principle framework of automatic sleep staging based on clustering and deep learning

3.1 Sequence Segmentation

In our work on sleep staging, we found that the classification accuracy of N1 stage is often lower. On the one hand, it is because the data is a class of imbalanced data, in which the light sleep period accounts for a small proportion of the entire sleep cycle. On the other hand, we learned from the data that the N1 stage is characterized by a short duration, which is often a short-term transition between two different sleep stages with a longer duration. The N1 stage, on a scale of 30s as a marker, may confuse the characteristics of the sleep stage at the previous moment or the next moment.

Therefore, we screened out the data of the N1 stage from a large number of EEG signals for analysis. We perform frequency domain analysis on this segment of the signal under different segmentation scales si, where $si \in \{4, 5\}$. Firstly, we divide the N1 signal with a duration of 30 s into sequence fragments according to the scale si, and the duration of each fragment is S, $S = T/si$. Next, calculate the energy of δ wave, θ wave, α wave, β wave and their ratio. We can more intuitively understand the composition information of the frequency domain components of the data through the proportion. The result shows that the characteristics of each segment of the N1 are quite different. For the signals in the sleep period, the characteristics of the δ and θ waves are more important, so we can focus on them. In the later segment, the proportions of δ and θ waves have a greater increase compared to the previous moment, and this is one of the characteristics of the N2 stage. Different segmentation scales can observe different features. As the scale si increases, the detailed feature information of the time series becomes more and more obvious, which can be displayed in the case of segmentation.

The next stage of the N1 mentioned above is the signal labeled as N2 stage. We also analyze the data of this stage through frequency domain analysis. When the scale si $= 4$, the last sequence segment of the N1 stage and the first sequence segment the N2 stage have similar proportions of δ waves and θ waves, which are 49.0%, 17.8% and 47.1%, 15.5% respectively. In the case of scale si $= 5$, the last sequence segment of the N1 stage and the first sequence segment of the N2 stage have similar proportions of δ waves and θ waves, which are 45.5%, 18.8% and 48.3%, 15.8% respectively, which confirmed that there is a part of confusion between different sleep stages. This brings us inspiration. In sleep staging, the data is appropriately segmented and then further modeled, which can capture more detailed features in the sleep signal.

Observing data from different perspectives will result in different and even more valuable information. We try to segment the sleep EEG sequence to perform sleep staging and improve the feature expression of each sleep stage in the sequence. In the following we will show our work.

3.2   Unsupervised Learning Module

The purpose of the unsupervised learning module is to find which cluster the sequence segment is closer to based on their features. Then, the cluster to which segment belongs is recorded as a characteristic symbol. Finally, the characteristic symbols of consecutive sequence segments are combined into a characteristic symbol vector in chronological order as the characteristic expression of the sequence before segmentation.

Conversion Class

During a person's sleep period, the sleep stages alternately change. Davies HJ et al. proposed in the paper that the conversion of sleep stages is considered in the classification [7], so the original 5 classes (sleep stages) can theoretically develop to 25 classes, such as "W-N1", "N2-N1" and so on. We analyzed the conversion knowledge of types in the data set, and counted the combinations of adjacent labels. The following table will show some of the data analysis results. Figure 2 are the statistical results of the next time label of data set 1 and data set 8, and the statistical results of the frequency of the label at the next time. For the N1 stage, we can see that the frequency of "N1-N1" in Fig. 2(a) is 63.33%, and the frequency of "N1-N1" in Fig. 2(b) is 52.14%, which means that the duration of N1 period is relatively short.

In our work, we also consider these conversion classes into the classification of sleep stages. The emergence of the conversion category is not to break the original expert classification standards, but to help us dig out the inherent characteristics of data. This will derive the category from 5 to 15 categories (Subtract conversion classes with low occurrences, there are about 15 types of conversion). In this project, we use clustering to analyze the similarity of EEG data during sleep.

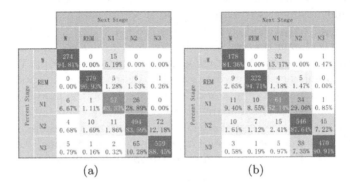

Fig. 2. Statistics of the conversion category: (a) data set 1, (b) data set 8

Clustering

Based on sequence segmentation and unsupervised learning, we obtain the characteristic expression of a piece of sleep EEG signal, and convert the high-dimensional original signal into a shorter, expressive characteristic symbol vector.

The process of obtaining feature expression is divided into four steps:

The second step is to segment the sequence. We need to choose a suitable segment length. Through the comparison experiment, we adopted the segmentation scale $si = 5$, that is to say, it is most appropriate to divide the original 30s sequence into 6s sequence. Secondly, we extract features from the 6s segment of the EEG sequence. We merge permutation entropy, fuzzy entropy, and SEF features to represent sequence fragments. Thirdly, we use the k-means algorithm to cluster them into K clusters according to the similarity of the features of the sequence fragments. We consider the conversion category as the sleep category. The actual category is 5–15 categories. So the selection range of K value is from 5 to 15, and the final K value is determined by the experimental results. We found that when k is 8, the best classification effect can be obtained. Finally, we treat the cluster of sequence fragments as a characteristic symbol, and then construct a symbol sequence. This symbol sequence is the characteristic expression of the 30 s EEG sequence.

The above process, through sequence segmentation and clustering, constitutes a new symbolic feature sequence of sleep EEG data.

3.3 BiLSTM Sleep Staging

We use BiLSTM to analyze the characteristic symbol sequence generated by the previous module to achieve sleep staging. The new feature symbol sequence is used as the input of the BiLSTM network to predict the sleep stage of the sequence. We constructed a two-layer BiLSTM network. Through sequence segmentation and unsupervised learning, feature expression vectors are obtained from the original signal, and these features are fed to the BiLSTM network to model the bidirectional time results. We not only use sleep information at the current moment to achieve sleep staging, but also use sleep information from previous stages to enhance the predictive ability of the model. In model training, we compare the impact of inputs of different lengths on the model effect, which are 30s (including sleep signal at time t), 60s (including sleep signal at time t-1, t), and 90s (including sleep signal at time t-2, t-1 and t). We add a dropout layer to the deep learning model to reduce overfitting during training. Finally, 5 categories of classification results are output through the fully connected softmax layer.

4 Experiments

In order to verify the judgment effect of our model on the problem of intelligent sleep staging, we constantly adjust the parameters to improve the predictive ability of the model, and compare our model with CNN, U-time, LSTM and other algorithms on the same data set to evaluate the accuracy of the proposed algorithm in dealing with sleep staging. The experimental data comes from the Sleep-EDF Expanded multi-channel sleep data set of MIT-BIH physiological information library[1], and each data has an expert label.

4.1 Experimental Results

We use 3 sets of data to carry out the experiment, and each set of data contains five randomly selected data sets. When training the model, the data set is divided into a training set and a test set, the proportion of the training set is 66.6%, and the proportion of the test set is 33.3%. Each group of experiments used a 10-fold cross-validation method. In the three groups of experiments, the average classification accuracy of Wake, N2, and N3 was the highest, which were 84.60%, 84.8%, and 86.26%, respectively. The average classification accuracy of REM stage was 74.8%, and the misclassification of REM stage was 25.2%. This article pays more attention to the accuracy of the shallow sleep stage 1 (N1 stage). For the N1 stage, the experimental test sets have obtained high precision which were 72.0%, 66.6%, 69.2% respectively. In the three sets of experiments, the recall rates of N1 period were 79.50%, 77.28%, 78.42% respectively. It can be seen that the model we proposed has a strong ability to identify the first period of light sleep. Comparing these three sets of experiments, the experimental results are

---

[1] https://www.physionet.org/.

relatively similar, and the overall average classification accuracy of the model is 80.64%. It can be seen that the fitting effect of our trained model is good.

Our model has an average classification accuracy of 69.27% for the light sleep period (N1). Although the classification accuracy of the N1 period is still the lowest among the five classes, compared to other models, our proposed clustering and BiLSTM model can improve the classification effect of light sleep stage. The innovation of our model lies in proposing a new way of expressing the characteristics of sleep EEG signals, using sequence segmentation and clustering methods to obtain the characteristic symbol vector of sleep EEG signals. This new method can effectively distinguish sleep EEG signals during different sleep stages. The comparison of the classification effect of our model with other models in the N1 stage is shown in the Table 1. We have consulted a large number of references, most sleep staging works have classification accuracy of N1 below 50%. In comparison, our model performs well, it can improve the classification accuracy rate of N1 period and reduce the misclassification rate.

Table 1. Classification accuracy of light sleep period of other models.

Authors	Classifier	Accuracy (%)	N1(%)
Shuyuan et al. [3]	K-means	76.00	27.67
Sors et al. [8]	CNN	87.00	34.92
Hsu et al. [14]	ELMAN RNN	76.80	36.70
R. Sharma et al. [15]	ITERATIVE FILTERING	77.04	30.40
S. Mishra et al. [11]	CNN RNN	86.00	46.02
Perslev M et al. [13]	U-TIME	85.00	47.00
Our work	K-means+BiLSTM	**80.64**	**69.27**

5   Conclusion

This paper proposes to use clustering and BiLSTM to achieve sleep staging. Compared with the methods proposed by other researchers, the model proposed in this paper can improve the N1 classification accuracy. This paper proposes a new method of constructing sleep EEG signal characteristics, which divides the original 30s EEG signal into equal lengths and appropriate lengths for further processing. The disadvantage is that although the classification accuracy of the N1 has been improved, the overall classification accuracy has been reduced. This may be due to the segmentation of the signals such as the awake period and the deep sleep period. In future work, we can study suitable adaptive sequence segmentation method. Adaptive sequence segmentation needs to pay attention to how to automatically identify the sequence that needs to be segmented. Only relatively unstable, fluctuating, and fast-changing sequences are segmented, which can improve the efficiency of the program and the overall classification accuracy.

References

1. Bai, S., Kolter, J.Z., Koltun, V.: An Empirical Evaluation of Generic Convolutional and Recurrent Networks for Sequence Modeling (2018)
2. Oord, A.V.D., Dieleman, S., Zen, H., et al.: WaveNet: A Generative Model for Raw Audio (2016)
3. Shuyuan, X., Bei, W., Jian, Z., Qunfeng, Z., Jun Zhong, Z., Nakamura, M.: An improved K-means clustering algorithm for sleep stages classification. In: 2015 54th Annual Conference of the Society of Instrument and Control Engineers of Japan, pp. 1222–1227 (2015)
4. Zhu, G., Li, Y., Wen, P.P.: Analysis and classification of sleep stages based on difference visibility graphs from a single-channel EEG signal. IEEE J. Biomed. Health Inform. **18**(6), 1813–1821 (2014)
5. Hassan, A.R., Bhuiyan, M.I.H.: Automated identification of sleep states from EEG signals by means of ensemble empirical mode decomposition and random under sampling boosting. Comput. Methods Programs Biomed. **140**, 201–210 (2017)
6. Chen, H., Dong, Y., Gu, Q., Liu, Y.: An end-to-end deep neural network for truth discovery. In: Wang, G., Lin, X., Hendler, J., Song, W., Xu, Z., Liu, G. (eds.) WISA 2020. LNCS, vol. 12432, pp. 377–387. Springer, Cham (2020). https://doi.org/10.1007/978-3-030-60029-7_35
7. Davies, H.J., Nakamura, T., Mandic, D.P.: A transition probability based classification model for enhanced N1 sleep stage identification during automatic sleep stage scoring. In: 2019 41st Annual International Conference of the IEEE Engineering in Medicine and Biology Society (EMBC). IEEE (2019)
8. Sors, A., Bonnet, S., Mirek, S., et al.: A convolutional neural network for sleep stage scoring from raw single-channel EEG. Biomed. Signal Process. Control **42**, 107–114 (2018)
9. Phan, H., Andreotti, F., Cooray, N., et al.: Joint classification and prediction CNN framework for automatic sleep stage classification. IEEE Trans. Biomed. Eng. **66**(5), 1285–1296 (2018)
10. Supratak, A., Dong, H., Wu, C., et al.: DeepSleepNet: a model for automatic sleep stage scoring based on raw single-channel EEG. IEEE Trans. Neural Syst. Rehabilitation Eng. **25**(11), 1998–2008 (2017)
11. Mishra, S., Birok, R.: Sleep classification using CNN and RNN on raw EEG single-channel. In: 2020 International Conference on Computational Performance Evaluation (ComPE), Shillong, India, pp. 232–237 (2020). https://doi.org/10.1109/ComPE49325.2020.9200002
12. Langkvist, M., Karlsson, L., Loutfi, A.: Sleep stage classification using unsupervised feature learning. In: Advances in Artificial Neural Systems 2012, pp. 1–9 (2012)
13. Perslev, M., Jensen, M.H., Darkner, S., et al. U-Time: A Fully Convolutional Network for Time Series Segmentation Applied to Sleep Staging (2019)
14. Hsu, Y.-L., Yang, Y.-T., Wang, J.-S., Hsu, C.-Y.: Automatic sleep stage recurrent neural classifier using energy features of EEG signals. Neurocomputing **104**, 105–114 (2013)
15. Sharma, R., Pachori, R.B., Upadhyay, A.: Automatic sleep stages classification based on iterative filtering of electroencephalogram signals. Neural Comput. Appl. **28**(10), 2959–2978 (2017)

Prognosis Analysis of Breast Cancer Based on DO-UniBIC Gene Screening Method

Xinhong Zhang[1], Tingting Hou[2], and Fan Zhang[2(✉)]

[1] School of Software, Henan University, Kaifeng 475004, China
[2] School of Computer and Information Engineering, Henan University,
Kaifeng 475004, China
zhangfan@henu.edu.cn

Abstract. A DO-UniBIC gene screening method was proposed. Firstly, Disease Ontology (DO) analysis was used to screen out breast cancer related genes from differentially expressed genes, and then UniBIC algorithm was used to find all gene clusters with the same changing trend based on the longest common subsequence. In addition, an eight-genes prognostic model was constructed to assess the prognostic risk of breast cancer patients. The prognostic analysis of the candidate gene set yielded eight genes that significantly related to the prognosis. The eight genes were ACSL1, CD24, EMP1, JPH3, CAMK4, JUN, S100B and TP53AIP1. Among them, ACSL1 was a new breast cancer potentially related gene screened by the DO-UniBIC method. More comprehensive cancer-related genes can be screened based on the DO-UniBIC method, which can be used as the candidate gene set for the prognostic analysis.

Keywords: Breast cancer · Differentially expressed gene · Disease ontology · Prognosis

1 Introduction

Breast cancer is the most common cancer among women in the world. Breast cancer is a disease caused by the malignant proliferation of cells in the breast epithelium. Most breast cancers begin in the lobules (mammary gland) or in the ducts that connect the lobules to the nipple. The induction of breast cancer is caused by many factors. At the molecular level, due to the regulatory changes in genes directly related to the prognosis of breast cancer, the accumulation of such regulatory changes can directly lead to abnormal cell growth and proliferation, thus leading to the occurrence of breast cancer [1]. It is of great significance to explore the genes closely related to breast cancer for the diagnosis and treatment of breast cancer. At the same time, finding genes that have a significant impact on the prognosis of breast cancer is helpful to improve the quality of prognosis prediction of breast cancer.

This research was supported by the Natural Science Foundation of Henan Province (No. 202300410093).

C. Xing et al. (Eds.): WISA 2021, LNCS 12999, pp. 215–222, 2021.
https://doi.org/10.1007/978-3-030-87571-8_19

Disease Ontology (DO) analysis method in bioinformatics can find genes associated with breast cancer from gene expression data [2]. However, Disease Ontology analysis is based on semantic annotation databases to obtain the association relationship between genes and diseases. These semantic annotation databases obtain all the annotation information through the integration of biological experiment verification and many medical genes data, that is, the breast cancer related genes obtained through Disease Ontology analysis are genes that have been mined and validation by researchers, and some potentially related genes in differentially expressed genes will be missed.

The bi-clustering method can identify genomes with similar expression patterns in specific condition subsets. By clustering gene dimension and condition dimension simultaneously, the relationship between genes and conditions can be dynamically used to obtain bi-clusters with similar expression patterns [3]. In view of the problem that disease ontology analysis cannot make full use of gene expression data to find out the potential related genes of breast cancer, this paper proposed the idea of combining the Disease Ontology analysis with the bi-clustering method to find a more comprehensive genes set for breast cancer prognosis analysis from the differentially expressed genes. Based on the study of the bi-clustering method, it was found that the bi-clustering method for finding change-related data is more biologically significance for the gene expression data. This paper combined the UniBIC (union bi-clustering) algorithm [4] with Disease Ontology analysis and proposed the DO-UniBIC method to find the related genes set of breast cancer. At the same time, multivariate Cox proportional-hazards regression analysis [5] was used to screen out genes that were significantly related to the prognosis of breast cancer.

2   Methods

2.1   Screening for Differentially Expressed Genes

In this paper, the transcriptome sequencing data of breast cancer in TCGA database (The Cancer Genome Atlas) were used as the research object. To eliminate the influence of different sampling time on the difference analysis, 110 pairs of paired breast cancer samples (normal tissue and paracancerous tissue of the same patient) were selected for the analysis and study [6].

The downloaded and organized gene expression data were analyzed by the R language's DESeq2 package, edgeR package and Limma package for the differential expression analysis, and the differentially expressed genes were screened through $Padj < 0.05$ and $|\log_2 FoldChange| > 1$. To further understand the molecular mechanism of the occurrence and development of breast cancer, the differential expression genes were analyzed by Gene Ontology (GO) analysis (including cellular component, biological process and molecular function) and KEGG (Kyoto Encyclopedia of Genes and Genomes) pathway enrichment analysis [7]. Among of them, $p < 0.01$ was used for GO enrichment analysis, and $p < 0.05$ was used for KEGG pathway analysis.

2.2   Selection of Breast Cancer Related Genes

Disease Ontology analysis is one of the enrichment analysis methods. In this paper, genes associated with breast cancer were screened from the gene expression data of breast cancer through the Disease Ontology analysis. This paper combined the Disease Ontology analysis with UniBIC algorithm and proposed the DO-UniBIC method to find a more comprehensive set of breast cancer related genes from the differential expressed genes data.

DO-UniBIC method firstly obtained the related gene sequence through the Disease Ontology analysis, and then looped through the bi-clustering results obtained by the UniBIC algorithm to determine whether it intersected the related gene sequence obtained from the DO analysis. The clusters that did not have intersection with the related gene sequence were deleted, and the clusters with intersection were retained. The genes in clusters were involved in regulation and they were the potentially related genes of breast cancer. Finally, the screened genes in the bi-clusters (including related genes and potential related genes) were selected as the candidate gene sets for the prognostic analysis.

2.3   Prognosis Analysis

The relevant genes set obtained was initially screened based on LASSO (Least absolute shrinkage and selection operator) regression analysis method [8], and then the multivariate Cox proportional hazards regression model (Proportional Hazards Model) was used to further screening genes and establishing a prognosis analysis model [9]. The Cox model is a semi-parametric regression model proposed by British statistician Cox [10]. This model takes survival outcome and survival time as dependent variables, can simultaneously analyze the influence of many factors on survival, can analyze data with censored survival time, and does not require estimation of the survival distribution type of the data. In this paper, 1,090 tumor samples downloaded from TCGA database were used for the prognosis analysis, and at the same time, the corresponding survival time and survival status of tumor samples were sorted out from the clinical data for analysis. The 1090 sample data were randomly grouped according to the survival status of the sample, and the grouping ratio was 1:1. Finally, the training set and the test set data were obtained. The patient samples contained 545 cases in each data set. LASSO method was used for the preliminary screening of related genes, and then multivariate Cox proportional hazard regression model was used to further screen genes and establish a multi-genes prognostic model.

3   Results

3.1   Differentially Expressed Genes

In this paper, three methods of DESeq2, edgeR and LIMMA were used for the differential expression analysis, and the intersection of these three methods was used as the final differential expression analysis result. DESeq2 screened out

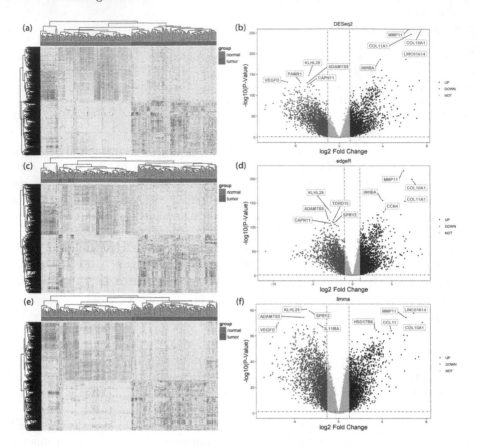

Fig. 1. Heat maps and volcano maps of the differential expression analysis results. (a) and (b): DESeq2 method screened out 4,235 differentially expressed genes; (c) and (d): edgeR method screened out 4,436 differentially expressed genes; (e) and (f): limma method screened out 4,291 differentially expressed genes.

4,235 differentially expressed genes, including 2,096 up-regulated genes and 2,139 down-regulated genes. EdgeR screened out 4,436 differentially expressed genes, including 2,422 up-regulated genes and 2,014 down-regulated genes. Limma screened out 4,291 differentially expressed genes, including 1,981 up-regulated genes and 2,310 down-regulated genes. Figure 1 visualizes the differential expression analysis results of the three methods in the forms of heat map and volcano map, respectively. Intersection of the differential expressed genes results screened by the three methods finally resulted in 3,249 differential genes, including 1,546 up-regulated genes and 1,703 down-regulated genes.

3.2   GO Enrichment Analysis and KEGG Pathway Analysis

The clusterProgiler package was used to perform Gene Ontology enrichment analysis and KEGG pathway analysis on the differential expressed genes. Most of these differential genes belong to extracellular matrix or extracellular matrix containing collagen. They have receptor regulation activity, transport protein activity, receptor protein tyrosine kinase activity, and are mainly involved in the regulation of ion transmembrane transport, membrane potential regulation, extracellular matrix organization, the regulation of synaptic transmission and other biological processes. Related studies have shown that the structural components of the extracellular matrix and the extracellular matrix were related to tumor metastasis and invasion [11].

3.3   DO-UniBIC Analysis

In this paper, DOSE package was used to implement DO analysis, by which the genes related to breast cancer were identified from the differentially expressed genes, and Benjamini-Hochberg multiple-testing correction method [12] was used to control the false positive rate under 0.05. The DO entries were sorted according to the corrected p-value, and the groups with statistical significance were selected.

UniBIC algorithm was used for bi-clustering analysis on the matrix of differential expressed genes, and 100 bi-clustering results can be obtained. The variation trend of the gene expression level in the two sub-matrices of bi-clustering is very poor in consistency, the bi-clustering results with poor consistency need to be further filtered and screened.

It can be seen from the reference [13] that Spearman correlation coefficient is superior to the quality evaluation of change-related bi-clustering. Therefore, this paper calculated the Spearman correlation coefficient for each sub-matrix of bi-clustering to measure the consistency level of the bi-clustering results, to further screened out better bi-clustering results. In this paper, the correlation coefficient greater than 0.95 was set as the screening condition of bi-clustering, and finally 56 bi-clusters were selected. The 56 bi-clustering results were sequentially compared with the list of related genes analyzed by DO analysis, the bi-clusters without intersection were deleted, and the bi-clusters with intersection were retained. After the circular comparison, it was found that at least one gene of 56 bi-clusters overlapped with the related gene listing by DO analysis. The intersection genes (related genes) and non-intersection genes (potentially related genes) in these 56 bi-clusters were extracted and integrated into a new gene list (99 genes in total). These new genes served as the candidate genes set for the prognostic analysis.

3.4   Prognosis Analysis

In this paper, 1,090 samples downloaded from the TCGA database were used for the prognostic analysis. At the same time, the corresponding survival time and survival status of the tumor samples from the clinical data were sorted out as the survival data for analysis. The data of these 1,090 samples were randomly divided into two groups according to the survival status of the samples. The grouping ratio was 1:1. Finally, the training data set and the testing data set were obtained. 545 samples were included in each data set.

For the training set data, the key genes were firstly screened out from the related genes by LASSO regression method, and then a multivariate Cox regression model was constructed. In this paper, the glmnet package was used to implement LASSO regression, and the 10-folds cross-validation method was used to estimate the adjustment parameter λ. According to the experiment in this paper, when $\ln \lambda = -4.34$ and $\lambda = 0.013$, the cross-validation error rate reached the minimum, and 19 genes were screened out at this time.

A multivariate Cox proportional regression analysis was performed on these 19 genes, and eight genes, ACSL1, C024, EMP1, JPH3, CAMK4, JUN, S100B and TP53AIP1, were screened. Among of them, ACSL1 was the new potential related gene of breast cancer screened by DO-UniBIC method. A multi-genes prognosis model was established according to the risk score formula.

After calculating the risk score, the patients in the training set and testing set were divided according to the median risk value. Patients with risk scores that is greater than the median risk value were classified as the high-risk groups, and patients with risk scores that is lower than the median risk value were classified as the low-risk groups.

Finally, an eight-genes prognostic model was established according to the risk scoring formula. According to the risk scoring formula, the median risk value of 545 patient samples was 0.985. Those with a risk value that was greater than 0.985 were included in the high-risk group, while those with a risk value that was lower than 0.985 were included in the low-risk group.

Figure 2 shows the results of multivariate Cox regression in the form of forest map, which is more intuitively. It can be seen from Fig. 2 that the p-values of most genes are significantly less than 0.01, which indicates that these eight genes can also be used as the independent prognostic factors for the analysis of breast cancer prognosis.

4   Discussion

The results of the Disease Ontology analysis and the UniBIC algorithm showed that there were genes associated with breast cancer in the overlapping gene clusters. A more comprehensive breast cancer-related gene can be screened from the differentially expressed genes of breast cancer as a candidate gene set for prognostic analysis.

In this paper, the breast cancer genes expressed data downloaded from the TCGA database were collected by the intersection of three differential analysis methods, DESeq2, edgeR and Limma, to obtain the differentially expressed genes. From the differentially expressed genes, the set of breast cancer-related genes was screened out using the DO-UniBIC method. The key genes were firstly preliminary screened out from the related genes to find out the prognostic-related genes by LASSO regression method, and then a multivariate Cox regression model was constructed to obtain eight significant prognostic related genes. The eight genes named ACSL1, C024, EMP1, JPH3, CAMK4, JUN, S100B and TP53AIP1. Among of them, ACSL1 is a new potential significance related gene

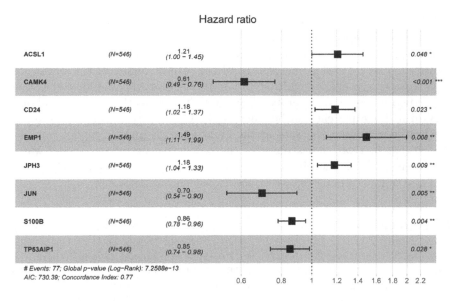

Fig. 2. Forest map of the multivariate Cox regression.

with prognostic of breast cancer that was screened by DO-UniBIC method, and this gene is worth further analysis.

In Fig. 2, taking HR = 1 as the dividing line, the four genes ACSL1, CD24, EMP1 and JPH3 are high-risk factors, and the high expression of these four genes will have an adverse effect on prognosis. On the contrary, the four genes CAMK4, JUN, S100B and TP53AIP1 are protective factors, and their high expression can play a good prognostic role. ACSL1 gene has a significant correlation with the prognosis of breast cancer. This gene is a poor prognostic gene, and its bio-molecular function is worthy of further research and exploration.

In this paper, Kaplan-Meier survival analysis [14] was used to compare the overall survival time of the high-risk group and the low-risk group in the training set and the testing set respectively, and logarithmic rank test was used to verify the classification results, in which p-value less than 0.01 was statistically significant. As shown in the Kaplan-Meier (KM) survival curve in Fig. 4a and Fig. 4b, in the sample data of training set, patients in the low-risk group have a higher median survival rate than patients in the high-risk group. Also, in the sample data of testing set, the median survival of the low-risk group is also higher than that of the high-risk group. The logarithmic rank test shows that the survival rate of the low-risk group and the high-risk group in the training set was significantly different (training set: $\chi^2 = 36.9$, $p < 0.001$), and the survival rate of the low-risk group and the high-risk group in the testing set is also significantly different (test set: $\chi^2 = 7.03$, $p < 0.008$). This indicates that the eight-genes prognostic model established by this paper can effectively evaluate the death risk of breast cancer patients.

In this paper, eight-gene prognostic model and pathological staging and TNM (Tumor, Node, Metastasis) staging of breast cancer patients were included as independent variables for Cox regression analysis. The results of univariate and multivariate Cox regression analysis. To further apply the model to clinical practice, this paper synthesized multiple predictive factors and simplified the statistical prediction model as the probability of result time, so that the survival probability of a single patient can be calculated. The experimental results showed that the predicted survival rate of the eight-gene prognostic model was close to the actual survival, indicating that the eight-gene prognostic model was an independent factor influencing the prognosis of breast cancer patients, and the model had a good prognostic ability for breast cancer patients.

References

1. Pondé, N., Zardavas, D., Piccart, M.: Progress in adjuvant systemic therapy for breast cancer. Nat. Rev. Clin. Oncol. **16**, 27–44 (2018)
2. Liu, X., Li, Y.: Is bigger data better for defect prediction: examining the impact of data size on supervised and unsupervised defect prediction. In: Ni, W., Wang, X., Song, W., Li, Y. (eds.) WISA 2019. LNCS, vol. 11817, pp. 138–150. Springer, Cham (2019). https://doi.org/10.1007/978-3-030-30952-7_16
3. Pontes, B., Giráldez, R., Aguilar-Ruiz, J.: Biclustering on expression data: a review. J. Biomed. Inform. **57**, 163–180 (2015)
4. Wang, Z., Li, G., Robinson, R., Huang, X.: Unibic: sequential row-based biclustering algorithm for analysis of gene expression data. Sci. Rep. **6**, 23466 (2016)
5. Su, J., Miao, L.F., Ye, X.H., et al.: Development of prognostic signature and nomogram for patients with breast cancer. Medicine **98**, e14617 (2019)
6. Zhu, Y., Qiu, P., Ji, Y.: TCGA-assembler: open-source software for retrieving and processing TCGA data. Nat. Methods **11**, 599–600 (2014)
7. Chen, J., Liu, C., Cen, J., et al.: KEGG-expressed genes and pathways in triple negative breast cancer: protocol for a systematic review and data mining. Medicine **99**, e19986 (2020)
8. Wang, H., Lengerich, B., Aragam, B., Xing, E.: Precision lasso: Accounting for correlations and linear dependencies in high-dimensional genomic data. Bioinformatics **35**, 1181–1187 (2018)
9. García, C., Camana, M., Koo, I.: Prediction of digital terrestrial television coverage using machine learning regression. IEEE Trans. Broadcast. **65**, 702–712 (2019)
10. Caputo, R., Cianniello, D., Giordano, A., et al.: Gene expression assay in the management of early breast cancer. Curr. Med. Chem. **26**, 2826–2839 (2019)
11. Suthers, G.: Comparing the performance of gene expression assays in breast cancer. Int. J. Cancer **145**, 1162–1169 (2019)
12. Chen, X., Sarkar, S.: On Benjamini-Hochberg procedure applied to mid p-values. J. Stat. Plan. Inference **205**, 34–45 (2020)
13. Flores, J., Inza, I., Larranaga, P., Calvo, B.: A new measure for gene expression biclustering based on non-parametric correlation. Comput. Methods Programs Biomed. **112**, 367–397 (2013)
14. Hess, A., Hess, J.: Kaplan-meier survival curves. Transfusion **60**, 670–672 (2020)

Data Privacy and Security

A Frequent Itemset Mining Method Based on Local Differential Privacy

Ning Wu, Yunfeng Zou[✉], and Chao Shan[✉]

State Grid Jiangsu Marketing Service Center Metrology Center,
Nanjing 210036, China

Abstract. As an important means of data analysis, frequent itemset mining is widely used in the field of big data. In recent years, local differential privacy has become a representative privacy protection technology in the field of frequent itemset mining due to its good mathematical theory, which has attracted the continuous attention of researchers. The existing frequent itemset mining methods based on local differential privacy have problems with insufficient data availability. Aiming at the existing binary coding-based perturbation method that causes large matching errors, an improved data perturbation method is proposed to enhance the availability of mining results while protecting data privacy. To solve the large privacy budget of existing methods, the hidden Markov model is introduced to avoid accessing a huge quantity of itemset. Thus, the candidate set can be quickly generated, which improves the efficiency of the algorithm. Experimental results show that the proposed method has a lower privacy budget and higher data accuracy.

Keywords: Local differential privacy · Frequent itemset · Hidden Markov model

1 Introduction

Frequent itemset mining plays an important role in the big data environment. By collecting and analyzing user behavior data, user behavior patterns can be provided for production and operation. However, user data contains sensitive information such as addresses and phone numbers. Mining user data directly will inevitably bring about user privacy leakage. In recent years, privacy-preserving frequent itemset mining has become a hot issue in the privacy protection area.

Differential privacy witnesses its thriving in privacy-preserving due to its strict mathematical definition. In centralized differential privacy, user data is collected and analyzed by a trusted third-party data center. However, it is difficult to find a trusted third-party data center in reality. Local differential privacy (LDP) is proposed. LDP is mainly designed to process distributed scenarios without any trusted data center. User data is disturbed locally and transmit into the data center. By reconstructing the statistical characteristics of the information on the data center, data analysis applications with privacy-preserving can be realized.

© Springer Nature Switzerland AG 2021
C. Xing et al. (Eds.): WISA 2021, LNCS 12999, pp. 225–236, 2021.
https://doi.org/10.1007/978-3-030-87571-8_20

In recent years, many methods are proposed in the area of frequent itemset mining. Erlingsson et al. [2] propose a Rappor algorithm with LDP. In this setting, data is binary encoded by hash function and then reconstructed by collectors to realize frequent items mining. When the candidate list is unknown, some researchers use a probability graph to restore the list and completes frequent itemset mining [3]. In addition, sampling and filling are used to find frequent itemset in set-valued data scenarios [5].

At present, LDP-based frequent itemset mining mainly aims at simple structured data. With the increasing complexity of applications, user data is more diversified. Potential combinations of the data are far exceeding the number of users. Existing methods may cause the following problems: 1) Traditional frequent itemset mining method samples one value at a time, which requires multiple interactions between, resulting in high communication cost; 2) The signal-to-noise ratio (SNR) of the perturbation protocol with LDP is larger, and the combined is greater than that of a single data; 3) The existing methods do not consider the issue of correlation between data. To solve the above problems, based on LDP, we propose a privacy protection frequent itemset mining method, HLDFIM, which uses a hidden Markov model and probability graph model to realize accurate mining of frequent itemset. Our contributions can be summarized as follows.

- We transform the frequency estimation problem into the parameter learning problem of the hidden Markov model, thus realizing the reduction of the privacy budget.
- A novel encoding protocol is designed to collect personal data. Besides, we propose to collect data based on bit characteristics to reduce errors.
- We conduct extensive experiments on multiple datasets to demonstrate the efficiency and effectiveness of our proposed method.

The rest of this paper is organized as follows. We describe and analyze existing methods in Sect. 2. We then go over the problem definition in Sect. 3. Section 4 presents our proposed methods. The experimental results are shown in Sect. 5, followed by a conclusion in Sect. 6.

2   Related Work

Frequent itemset mining based on privacy protection has received widespread attention. Recently, there are several works on DP-based frequent itemset mining [6,7,13]. Wang et al. [6] propose a superset-based method. This method avoids accessing a massive number of subsets by finding the largest frequent itemset directly. However, it requires a large privacy budget. Chen et al. [7] introduce a method based on sampling. However, sampling technology will inevitably result in the loss of data utility. Besides, centralized DP suffers from untrusted third-party which is a problem LDP can solve. In LDP, Erlingsson et al. [2] propose a RAPPOR algorithm for frequent items mining. Unfortunately, repeated disturbance data introduces a lot of noise, which reduces the accuracy of mining.

Aiming at the deficiencies of RAPPOR, Wang et al. [8] design a pure protocol to improve. It can improve the accuracy, but the degree is lower.

Sampling and grouping methods are often used to collect and mine frequent itemset in the set-valued scene, such as LDPMiner [9], SVIM, SVSM [5]. LDP-Miner has two stages. In the first stage, user data is sampled and filled to obtain frequent items and then send these items to users. Users can disturb frequent items less than the whole dataset by using more privacy budgets in the second stage. More privacy budget brings higher data utility. SVIM divides users into different groups, which avoids the segmentation of the privacy budget and enhances frequency estimation by estimating the length of frequent itemset in advance. SVSM obtains candidate sets by constructing smaller candidate sets and estimate the top-k candidate sets. The limitation of the above methods mainly lies in the scale of users, making the frequency estimation accurate enough. The increasingly complex user data bring about a large scale of candidate sets, making it difficult to achieve a reasonable number of users. Besides, multiple interactions with users lead to higher communication costs.

Reconstructing data based on probability distribution is an important method of frequent itemset mining. Fanti G et al. [3] propose the N-gram method to re-store the joint probability distribution under an unknown candidate value list. In a multidimensional data scenario, The large scale of the value list leads to an increase in the communication cost. Zhang et al. [10] use projection to reduce dimension. However, an excellent projection strategy cannot be determined in advance, and the correlation between attributes can easily be ignored. Ren et al. [4] propose LoPub, which uses regression analysis to reduce data dimension and publish the synthetic data set conforming to local differential privacy. The design of LoPub has two disadvantages. First, the adoption of general disturbance protocol leads to low accuracy. Second, applying this method directly to frequency itemset mining is equivalent to mining in noisy data with many candidate sets.

3   Problem Description and Related Definitions

3.1   Problem Description

We assume that N is the number of users. Given a collection of user data with d attributes. $D = X_1, X_2, X_3, ..., X_N$. Let $X_i = a_i^1, a_i^2, ..., a_i^k$ represents k attributes in a single record. a_i^j is the j-th attribute of i-th user. Ω_j is the range of A_j attribute where A_j is the attributes index. ω_j' represents all values in the dataset D. Our goal is to mine and publish the frequent itemset with LDP. The number of potential itemsets will be $\prod_{j=1}^{k} | \Omega_j |$. In summary, given the support threshold $\delta(0 < \delta < 1)$, we need to find all combinations satisfied the following condition.

$$P(\omega_i, \omega_j, ..., \omega_k) \geq \delta \qquad (1)$$

In Eq. (1), i, j, k are index of attributes $\omega_i \in \Omega_i, \omega_j \in \Omega_j, ..., \omega_k \in \Omega_k$, and $P(\omega_i, \omega_j, ..., \omega_k)$ is joint probability distribution.

3.2   Related Description

Definition 1. (*Adjacent Dataset*). *If only one data record in the datasets D_1 and D_2 is different, D_1 and D_2 are called adjacent datasets.*

In local differential privacy, dataset is in distributed settings and each user holds only one record. The datasets of any two users are adjacent to each other.

Definition 2. (*Local Differential Privacy*) *A randomized function M satisfies local differential privacy, if and only if for any two different data v_1, $v_2 \in$ Dom (A) and any possible output $o \in$ Range (A), we have*

$$Pr[A(v_1) = o] \le e^\epsilon \cdot Pr[A(v_2) = o],$$

where ϵ denotes the privacy budget, and the smaller the value, the higher the privacy-preserving degree of function M.

Theorem 1. (*Sequential Composition* [11]) *Suppose A_i satisfies ϵ_i-LDP, the algorithm A which simultaneously releases $A = \langle A_1, ..., A_t \rangle$ satisfies $\sum_{i=1}^{t} \epsilon_i$-LDP.*

Random Response (RR) [12] is a data disturbance technology that is often used in LDP, which is defined as follows.

Definition 3. (*Random Response*). *Given input X and output Y, we have*

$$P(Y = u \mid x = v) = p_u^v \tag{2}$$

Given the true value u, the probability that the output value vafter disturbance is p_u^v. Each random response mechanism can be determined by a matrix

$$p = \begin{pmatrix} p_{00} & p_{01} & \cdots \\ p_{10} & p_{11} & \cdots \\ \vdots & \vdots & \vdots \end{pmatrix}. \tag{3}$$

4   Frequent Itemset Mining Based on LDP

4.1   Basic

The main problem of frequent itemset mining under local differential privacy is the loss of data utility. A novel random response mechanism is proposed to satisfy local differential privacy. We encode data into bit string and flip each bit in a certain random response mechanism. Besides, a support degree can be seen as a joint probability distribution, which will spend much calculation cost with a huge candidate set. To solve the above problems, we propose HMM-based local differential privacy frequent itemset mining with the following stages.

- Local data protection. To improve the match error in the traditional encoding mechanism, we propose an adaptive strategy that is based on ones and zeros in the bit string. Thus, data utility can be higher than usual.
- Candidate sets generation. The Hidden Markov Model is applied to estimate the parameters of the disturbed dataset. Parameters can be used to construct the ini-tial candidate sets.
- Frequency itemset discovery. A probability graph is built by parameters obtained in step 2, which is used to find frequent itemsets by transcendental principle.

Figure 1 illustrates the frequent itemset mining process.

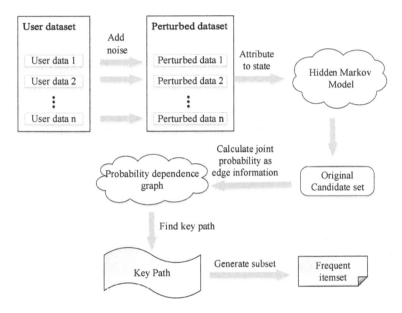

Fig. 1. Frequent itemset mining process

4.2   Data Disturbing

In local settings, sensitive data always be processed at the user side to ensure data privacy that is submitted to the third-party server. Therefore, it is essential to disturb data properly. Data disturbance is mainly divided into two steps. First, user data is encoded into predefined bit strings by some hash functions (e.g., ASCII or Bloom-Filter), which can fully represent the characteristics of user data and be easy to distinguish. Then, each bit in the string can be randomized by a certain probability. The length of the bit string is $\lceil log_2^n \rceil$. The string can represent a fixed number of candidate sets but some strings are not assigned to a certain record which may cause match error and reduction of data utility.

Definition 4. (*Match Error*). *Given a dataset X, disturb function $R(.)$, mathematic expectation function $E(.)$, the match error is defined as follows*

$$err(X, R(X)) = \frac{1}{k}(E(X) - E(R(X)))\qquad(4)$$

where k is the length of bit string. The match error is determined by difference between expectation in the origin dataset and disturbed dataset.

The steps of disturbing data locally include the following two parts. First, user encode their data by some hash function into bit strings. $s_i^j[t]$ represents the i-th users j-th attributes k-th bit (0 or 1). Second, each bit in the bit string will be flipped by the following rule

$$pr(\hat{s}_i^j[t] = 1) = \begin{cases} f[t]; & s_i^j[t] = 1 \\ 1 - f[t]; & s_i^j[t] = 0 \end{cases},\qquad(5)$$

where f is randomized probability array generated by Algorithm 1.

Algorithm 1: RPAG (Randomized Probability Array Generation)

Input: all candidate values $\Omega_1, \Omega_2, ..., \Omega_k$, set of hash function H, threshold μ.
Output: randomized probability array f
1   $S \leftarrow \emptyset$;
2   **for** *each* $\omega \in \Omega_1 \cup \Omega_2 \cup ... \cup \Omega_k$ **do**
3 | $S \leftarrow H(\omega)$;
4   **end**
5   $S \leftarrow S \cup \{s\}, f \leftarrow 0$;
6   **for** *each* $s \in S$ **do**
7 | **for** *each* $t \leftarrow 1, ..., |S|$ **do**
8 | | **if** $s[t] = 1$ **then**
9 | | | $f[t] = f[t] + 1$;
10 | | **end**
11 | **end**
12   **end**
13   compute $f[t] \leftarrow min(\frac{f[t]}{|S|}, \mu)$;
14   Return f;

In Algorithm 1, we use a threshold to control the privacy budget.

Theorem 2. *The algorithm RPAG satisfies ϵ_1-LDP.*

Proof. Bit strings are mainly modified in the disturbing stage, and the privacy budget is allocated to the randomized probability array. Since each bit is flipped independently in the bit string, according to Theorem 1, it can be deduced that the data disturbing stage satisfies the ϵ_1-LDP and $\epsilon_1 = ln\left(\frac{max(f)}{1-max(f)}\right)$. This completes the proof.

Theorem 3. *The algorithm RPAG has lower match error than Basic Rappor.*

Proof. Let R_1 and R_2 represent the mechanisms Basic Rappor and RPAG, respectively. Besides, p is the parameter of R_1 and f is the parameter of R_2. Then, we have

$$err(X, R_1) = \frac{1}{k}\left(E(X) - E(R_1(X))\right)$$

$$= \frac{1}{k}\left(\prod_{1 \leq i \leq k} E(X_i) - \prod_{1 \leq i \leq k} E(R_1(X_i))\right)$$

$$= \frac{1}{k}\left(\prod_{1 \leq i \leq k} f[i] - \prod_{1 \leq i \leq k} (1 - p + f[i] \cdot p)\right)$$

$$err(X, R_2) = \frac{1}{k}\left(E(X) - E(R_2(X))\right)$$

$$= \frac{1}{k}\left(\prod_{1 \leq i \leq k} E(X_i) - \prod_{1 \leq i \leq k} E(R_2(X_i))\right)$$

$$= \frac{1}{k}\left(\prod_{1 \leq i \leq k} f[i] - \prod_{1 \leq i \leq k} 1\right)$$

$$err(X, R_2) - err(X, R_1) = \frac{1}{k}\left(\prod_{1 \leq i \leq k} (1 - p + f[i] \cdot p) - \prod_{1 \leq i \leq k} 1\right) \leq 0;$$

As can be seen from the above, $err(X, R_2)$ is lower than $err(X, R_1)$. This completes the proof.

4.3   Candidate Set Generation

In the probability estimation stage, we estimate the dataset parameters and obtain the required joint probability distribution indirectly. Compared to the Laplace mechanism in centralized differential, the advantage of local differential privacy is that the input and output are composed of bit strings of length k. We apply the hidden Markov model to this problem because bit strings of length k represent 2^k states. By transforming the joint probability problem into parameter estimation problems of the hidden Markov model, accessing a massive number of candidates is avoided.

Theorem 4. *Disturbed dataset satisfy basic hypotheses of hidden Markov model.*

Proof. In the disturbing data stage, user data is processed by bit. Whether a bit is processed depends on the state of the previous bit regardless of time t and satisfies the homogeneous Markov hypothesis. When the collector receives the

i-th attribute of the user, the data is only determined by the current state and has nothing to do with other states. So it satisfies the observation independence hypothesis.

This completes the proof.

We use Baum-Welch algorithm to estimate the parameter of hidden Markov model and generate candidate set with these parameters as shown in Algorithm 2.

Algorithm 2: CGA (Candidate Generation Algorithm)

Input: disturbed user dataset D, set of attribute range Ω
Output: initial candidate set S

1   initialize $\lambda = (C, B, \pi)$, $S \leftarrow \varnothing$;
2   λ parameter iterates until it converges;
3   **for** *each* $\omega_i \in \Omega_1$ **do**
4     $\mid$ $P(\omega_i) = \pi_i$;
5   **end**
6   **for** *each* $i \in [2, k]$ **do**
7     $\mid$ **for** *each* $\omega_u \in \Omega_i$ **do**
8       $\mid$ $P(\omega_u) = \sum\limits_{\omega_v \in \Omega_{i-1}} P(\omega_v) C_{\omega_u \omega_v}$;
9       $\mid$ add $\{\omega_u, P(\omega_u)\}$ into S
10    $\mid$ **end**
11   **end**
12   Return S;

Parameters such as C, B, π can be obtained from Algorithm 2. C is the state transition matrix, B is the observation probability matrix, and π is the initial state vector. Initial candidate sets are being generated from these parameters. Because Algorithm 2 needs to initialize parameter that greatly influences the result, prior knowledge of dataset makes the parameters more meaningful. For example, the transition probability between different attributes in the same value range is zero $\{C_{ij} = 0 \mid \omega_i \in A_k \ \& \ \omega_j \in A_k\}$. Due to the order of attributes, only the attribute value of attribute A_1 tends to have a non-zero probability $(\pi_i = 0(\omega_i \notin A_1))$.

4.4   Candidate Set Generation

The generation of candidate frequent itemsets is mainly composed of the generation of 2-itemsets, the construction of probability maps, and the discovery of frequent itemsets. First part, state transition matrix of hidden Markov model de-scribes the probability of transition between different attributes. Multiplying the probability of the initial candidate set with the corresponding probability in the transition matrix can get 2-itemsets. Second part, a weighted directed acyclic graph is con-structed by taking different attribute values as nodes and

2-itemsets support degrees as weights. The last part, we take any point as start point to find all paths which meet the threshold. The set and subsets of nodes in these paths can be added into frequent itemset.

Theorem 5. *If the weight of an edge in the probability graph does not meet the threshold, all paths, including this edge, do not meet the threshold as well.*

Proof. Given an edge $e = (\omega_u, \omega_v)$, and the weight of this edge is $P(\omega_u, \omega_v) < \delta$. Paths including this edge is E. Then, the weight of this path is $P(\omega_u, \omega_v) \prod\limits_{E \backslash e}^{e'} P(e')$. The weight of each edge in E is $P \leq 1$. The weight of path excluding e is $\prod\limits_{E \backslash e}^{e'} P(e') \leq 1$. So the weight of whole path is $P(\omega_u, \omega_v) \prod\limits_{E \backslash e}^{e'} P(e') \leq P(\omega_u, \omega_v) < \delta$.

This completes the proof.

In this stage, we use initial candidate itemset combined with a state transition matrix to derive the follow-up itemset. Each derivation is equivalent to adding a new dimension to the candidate set. Theorem 5 avoids the calculation of unqualified itemsets and heuristically generates all possible candidate itemset.

The scale of the probability graph can be effectively reduced by Theorem 5. The condition of specific paths $\left(\prod\limits_e P(e) > \delta \right)$ can be transformed into $\sum\limits_e log(\frac{1}{P(e)}) \leq log(\frac{1}{\delta})$, which will improve calculation accuracy.

5   Experimental Results

In this section, we verify the utility of HLDPFIM. All the experiments were run on the machine with Intel i7-4770 3.8 GHZ and 8 GB RAM, using Windows7 and Python 3.7. We conducted the experiments in real-world datasets from the Frequent Itemset Mining Implementations Repository. The characters of datasets are shown in Table 1.

Table 1. Dataset characteristics.

Dataset	Type	Record(N)	Attr(k)
Pumsb	Integer	49047	49
Chess	Category	3196	32
Connect	Integer	Integer	42

This paper compares HLDPFIM with RAPPOR and SVSM. We use precision P and MAE (Mean Absolute Error) to evaluate each method. The P and MAE are defined as

$$P = \frac{FS \cap FS'}{FS'} \ and \ MAE(FS, FS') = \frac{1}{FS \cap FS'} \sum_{i=1}^{|FS \cap FS'|} |FS_i \cap FS_i'|,$$

where FS is actual frequent itemset, and FS' is protected frequent itemset.

5.1 Analysis of Mining Accuracy

Figure 2 shows the influence of support degree sup on the accuracy of frequent itemset mining. As sup changes from 0.02 to 0.06, we can see that the accuracy of the HLDPFIM method shows a non-decreasing trend and is better than other methods. With the same privacy budget, the accuracy of HLDPFIM is stable, while the other two algorithms are easily affected by the minimum support threshold. The reason is that candidate itemsets of RAPPOR and SVSM will change under different minimum support conditions, which will lead to fluctuations in the accuracy. RAPPOR performs better than SVSM in Pumsb and Connect dataset, but the op-posite is true in the Chess dataset because Pumsb and Connect have a large scale of candidate values. At the same time, SVSM cannot divide users into a reasonable number of groups, and each candidate value cannot get enough user support.

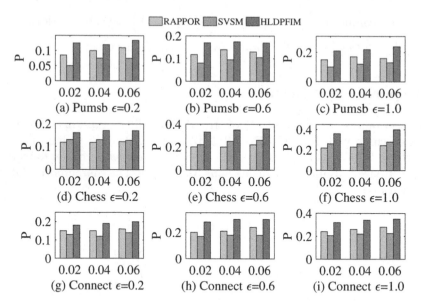

Fig. 2. Accuracy comparison of three algorithms for mining frequent itemset on different datasets

5.2 Candidate Set Generation

In this section, we fix the minimum support threshold sup to observe the influence of the privacy budget size on the average absolute error. In Fig. 3, with the increasing privacy budget, the errors of the three algorithms on different data sets all show a downward trend, and HLDPFIM is better than other algorithms. When the privacy budget is large ($\epsilon = 1.0$), the errors of the three algorithms are close. In other cases, SVSM has more significant mistakes than others. The reason is the same as mentioned in 5.1. Whats more, MAE is less affected by the minimum support threshold because the privacy budget is mainly allocated in the data disturbance stage in LDP. In contrast, the support threshold is only used to determine the candidate itemset.

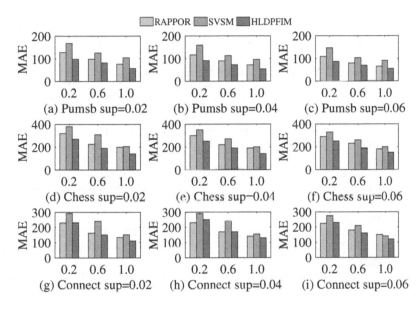

Fig. 3. Error comparison of three algorithms for mining frequent itemset on different datasets

6 Conclusion

In this paper, we propose a novel method, HLDPFIM. Aiming at data utility, we improve the disturbing mechanism and estimate the parameters by the hidden Markov model. A probability graph is constructed to discover frequent itemset in a distributed setting. Experimental results show that HLDPFIM is an efficient and effective method to mine frequent itemset. We will study local differential privacy technology in other data mining scenarios in future work.

Acknowledgement. This work is supported by the state Grid Jiangsu Electric Power Corporation Project (J2020113).

References

1. Dwork, C.: Differential privacy. In: Bugliesi, M., Preneel, B., Sassone, V., Wegener, I. (eds.) ICALP 2006. LNCS, vol. 4052, pp. 1–12. Springer, Heidelberg (2006). https://doi.org/10.1007/11787006_1
2. Erlingsson, U., Korolova, A., et al.: RAPPOR: randomized aggregatable privacy-preserving ordinal response. In: In Proceedings of the 2014 ACM Conference on Computer and Communications Security, pp. 1054–1067. ACM, New York (2014)
3. Fanti, G., Pihur, V., Erlingsson, U.: Building a RAPPOR with the unknown: privacy-preserving learning of associations and data dictionaries. In: Proceedings on Privacy Enhancing Technologies, no. 3, pp. 41–61 (2016)
4. Ren, X., Yu, C.-M., Yu, W., et al.: LoPub: high-dimensional crowdsourced data publication with local differential privacy. IEEE Trans. Inf. Forensics Secur. **13**(9), 2151–2166 (2018)
5. Wang, T., Li, N., Jha, S.: Locally differentially private frequent itemset mining. In: Proceedings of IEEE Symposium on Security and Privacy, San Francisco, pp. 127–143 IEEE (2018)
6. Wang, N., Xiao, X., Yang, Y., et al.: PrivSuper: a superset-first approach to itemset mining under differential privacy. In: Proceedings of the 33rd IEEE International Conference on Data Engineering, San Diego, pp. 809–820. IEEE (2017)
7. Chen, R., Xiao, Q., Zhang, Y., et al.: Differentially private high-dimensional data publication via sampling-based inference. In: In Proceedings of the 21th ACM International Conference on Knowledge Discovery and Data Mining, pp. 129–138. ACM, New York (2015)
8. Wang, T., Blocki, J., Li, N., et al.: Locally differentially private protocols for frequency estimation. In: Proceedings of USENIX Security Symposium, Vancouver, pp. 729–745. USENIX Association (2017)
9. Qin, Z., Yang, Y., Yu, T., et al.: Heavy hitter estimation over set-valued data with local differential privacy. In: Proceedings of the 2016 ACM SIGSAC Conference on Computer and Communications Security (CCS 2016), New York, NY, USA, pp. 192–203 (2016)
10. Zhang, Z., Wang, T., Li, N., et al.: CALM: consistent adaptive local marginal for marginal release under local differential privacy. In: Proceedings of the 2018 ACM SIGSAC Conference on Computer and Communications Security, pp. 212–229. ACM, New York (2018)
11. McSherry, F.: Privacy integrated queries: an extensible platform for privacy-preserving data analysis. Commun. ACM **53**(9), 89–97 (2010)
12. Ye, Q., Meng, X., Zhu, M., et al.: Survey on local differential privacy. J. Softw. **29**(7), 1981–2005 (2018). (in Chinese)
13. Zou, Y., Bao, X., Xu, C., Ni, W.: Top-k frequent itemsets publication of uncertain data based on differential privacy. In: Wang, G., Lin, X., Hendler, J., Song, W., Xu, Z., Liu, G. (eds.) WISA 2020. LNCS, vol. 12432, pp. 547–558. Springer, Cham (2020). https://doi.org/10.1007/978-3-030-60029-7_49
14. Jia, J., Gong, N.Z.: Calibrate: frequency estimation and heavy hitter identification with local differential privacy via incorporating prior knowledge. In: INFOCOM 2019, pp. 2008–2016 (2019)
15. Wang, T., Li, N., Jha, S.: Locally differentially private heavy hitter identification. IEEE Trans. Dependable Secur. Comput. **18**(2), 982–993 (2021)
16. Yan, J., Wang, Y., Li, W.: Behavior sequence mining model based on local differential privacy. IEEE Access **8**, 196086–196093 (2020)

Chain-AAFL: Chained Adversarial-Aware Federated Learning Framework

Lina Ge[1,3], Xin He[1,2(✉)], Guanghui Wang[1,2], and Junyang Yu[1,3]

[1] School of Software, Henan University, Kaifeng, China
[2] Henan International Joint Laboratory of Intelligent Network Theory and Key Technology, Henan University, Kaifeng, China
[3] Henan Provincial Engineering Research Center of Intelligent Data Processing, Kaifeng, China

Abstract. Federated learning (FL) distributes model training among multiple agents, who perform training locally but only exchange gradients due to privacy concerns. However, the threat of model poisoning attack and deep leakage from gradients still limit its practical use. We propose a novel privacy-preserving FL framework, termed as Chain-AAFL, based on a chained multi-party computing paradigm with the adversarial-aware gradient aggregation method to solve these problems. Specifically, the Chained-communication mechanism enables masked gradients to be transferred among participants. Adversarial-aware gradient aggregation could avoid model degradation caused by the model poisoning attack and gradually identify the adversary from all clients. Experimental results on CIFAR-100 demonstrate that our proposed Chain-AAFL can defense against the model poisoning attack and achieve privacy-preservation for federated learning without impairing the accuracy and training speed of the training model.

Keywords: Federated learning · Gradient leakage · Model poisoning attack · Security

1 Introduction

The initial stage of Federated Learning (FL) was proposed in 2016 by Google [11]. Federated Learning is a distributed machine learning technique that participants training their data locally under the central coordination server which play a role in aggregating the local model update of participants. The classic aggregation algorithm of the server is weighted averaging. Specifically, the owner of data exchange the parameter under the encryption mechanism in the federal system which establish a global sharing model. The global model only serve local target in their respective regions.

Federated Learning bulids a sharing and robust distributed machine learning setting that can address essential issues such as data privacy [14], data security and access to heterogeneous data. FL can support a large-scale training under a

C. Xing et al. (Eds.): WISA 2021, LNCS 12999, pp. 237–248, 2021.
https://doi.org/10.1007/978-3-030-87571-8_21

secure environment. Federated Learning System have been deployed in diverse fields (such as many major companies), including defense, telecommunications, IoT (Internet of Things), pharmaceutics and etc. Recent work by researcher in DLG (Deep Leakage from Gradients) [17] have shown that the FL faces challenges of privacy leakage. The adversary could reverse-deduce the user's original data by gradients since the gradient reveal some properties of the training data. Meanwhile, the model poisoning attack can also affect the training result of FL by pollutioning model.

To intensify the security of FL, investigators have proposed many methods and strategies, such as Differential-Privacy-based FL (DP-based FL), Secure Multi-party Computing (SMC), and Homomorphic Encryption (HE) algorithm. Nevertheless, these methods have limitations in practical applications. DP-based FL protects privacy by adding noise to data, hence it is necessary to strike a balance between accuracy and privacy. The computation and communication overhead of SMC and HE is relatively large since the encryption mechanism, which leads to higher requirements on the processing capacity of the devices.

In this paper, we present a novel Chained Adversarial-Aware Federated Learning framework termed as Chain-AAFL to address aforementioned problems. The main contributions of our work can be summarized as follows:

- *We introduce a Adversarial-aware Federated Learning framework based on Chain-PPFL to defense against gradient leakage and model poisoning attacks.*
- *We attach a confidence to each participant, Adversarial-aware gradient aggregation method identifies the adversary among participants based on the degree of confidence.*
- *Compare to DP-based FL and Chain-PPFL, our method could achieve higher security without lowering the model accuracy.*
- *Our experiment demonstrates that our method is superior to Chain-PPFL in terms of accuracy, and performance is close to Chain-PPFL.*

2   Related Work

In this section, we will explore exhaustively the challenges and recent work focused on Federated Learning from multiple perspectives, and summarize other related research.

2.1   Federated Learning

Federated Learning aims to solve the problem of isolated data islands by training distributed data stored in a large number of terminals and learning high-quality centralized machine learning models. The central server broadcasts the global model and participants perform local training by using the global model. Then the central server aggregates local model updates of participants to generate the global model of the next round with the Federated Averaging algorithm. Researchers have conducted in-depth research on the challenges facing federated learning and outline several research directions for future work [6,8].

Recently, kinds of literature have emerged that proposes current challenges of federated learning, including data and model poisoning attacks, privacy protection, communication efficiency, and heterogeneous data distribution. In this article, we mainly focus on gradient leakage and the model poisoning attack of federated learning. The process of distributed training needs to share gradient parameters, to achieve synchronization via exchanging gradients, and model updates, to optimize the global model. In federated learning, gradient exposes the attributes of participants which could cause gradient leakage. And there are usually some malicious agents who perform model poisoning attacks against the training of participants to undermine the global model. To address these issues, researchers attach importance to model updates defense and privacy protection, to establish a secure federal learning environment.

2.2   Attacks on Federated Learning

Federated learning is proposed to protect the private data of participants while training. In recent years, researchers have investigated a variety of approaches to improve security However, the attacker trying to learn and imitate the gradient to steal the training data from honest participants or pollute the model to mislead the training results.

2.2.1   Model Poisoning Attack

Model attacks on federated learning includes: model poisoning attacks [3] and backdoor attacks [1]. Model poisoning attacks reduce the performance of the global model. Backdoor attacks are to inject some characteristics expected by the attacker to make the wrong classification while maintaining the performance of the global model. Since both of them are perform the attack against the model, we simply simulate the model poisoning attack in this paper. In the federated learning system, the central server aggregates the local model uploaded by clients to perform global model updates. The purpose of model poisoning attacks is to transport local model updates trained by malicious clients to the central server to pollute or affect the correctness and convergence speed of the global model updates.

It has been observed that researchers have proposed many model poisoning attack strategies. Bhagoji et al. [2] proposed an alternate minimization strategy to alternately optimizes the training loss and the confrontation target. Fang et al. [5] regard this attack as an optimization problem and applies it to several recent Byzantine robust federated learning methods. In federated learning, participants have the authority to modify model parameters which makes them vulnerable to model poisoning attacks. On model poisoning attacks, researchers have conducted systematic research and proposed a new optimization-based model poisoning attack method to improve the attack success rate and sufficient concealment [17].

2.2.2   Gradients Leakage

Recently, some researches have shown that federated learning has security loopholes that can threaten data privacy [10]. In distributed training, participants train the model locally with different data. Participants exchange gradient to the server to optimize model parameters. Gradients leakage attacks aims to use back-propagation gradient information to infer the private features and label information of participants implicit in the gradient. So, the adversary can completely steal the private data of participants.

DLG [17] has shown that the adversary can obtain private data since the model structure and gradient parameters are shared in federated learning. In the training process, honest participants train private data locally, and the malicious attacker virtualizes an input and label. The attacker minimizes the gradient distance between the real gradient and the dummy gradient by continuously performing training updates. Then the attacker is able to infer the data of participants in limited rounds. To solve the problem of convergence and discovering basic fact labels consistently in DLG, Zhao et al. [16] proposed an improved DLG (iDLG) to accurately extract ground truth labels.

2.3   Defensive Strategies of Federated Learning

At the moment, the privacy preservation of federated learning mainly includes: Differential Privacy (DP) Homomorphic Encryption (HE), and Secret Sharing (SS). Differential privacy [4] is a privacy preservation method that generates noise with randomized mechanisms, mask the user's sensitive information by adding noise to data. The noise becomes larger, the level of privacy-preserving increases while the training accuracy decreases. Wei Et al. [15,18] propose a novel framework based on differential privacy (DP), which performs noise processing on parameters of participants before model aggregation. Homomorphic Encryption [12] realizes high-level privacy protection by exchanging parameters under the encryption mechanism. HE can still perform complex mathematical operations on encrypted data while ensuring confidentiality. For example, the use of homomorphic encryption in the medical field can alleviate the leakage issues of patients' sensitive data [7].

Secret sharing [13] is to split the secret value into multiple random shares, and each share is managed by different participants, namely, each participant can only have part of the secret. Therefore, only the cooperation of all or a certain number of participants can reconstruct the original secret. Recently, Chain-PPFL [9] proposes a scheme similar to secret sharing, a novel privacy-preserving FL framework based on a chained secure multi-party computing technique. In this framework, the Single-Masking and Chain-Communication mechanism are respectively used to protect information exchange between participants and enable masked information to be transferred between participants.

3   Our Method

In this section, we propose a novel Adversarial-Aware Chained Federated Learning framework based on Chain-PPFL against the Gradient Leakage and Model Poisoning Attack. The aim of this section is to describe our Chain-AAFL framework in details. Table 1 lists the symbols used in this paper and their meanings.

Table 1. List of symbols

Symbols	Description
K, k	The number of users and index
C	Confidence-score list
A	Adversary list
S	Suspect list
N	Number of training epoches
T	Threshold value
E_t	Index of training epoches
ACC	Test accuracy on verification dataset
AA	Adversary-aware
g_j	Training group j
w_G	Global model parameters
w_k^j	Local optimal parameters of user k in group j
U_0^j	Randomly initialized mask weight for group j
U_k^j	Mask weight computed by user k in group j
w_t	Temporal-optimal group training weight
η	Learning rate
$\nabla \ell(w; b)$	Gradient in each batch computed locally by user
PRG	Pseudo-ramdom generator on the aggregation node
ϵ	The privacy budget of DP
f	Clients Fraction rate
p	Model poisoning rate

3.1 Problem Preliminaries

The classic federated averaging algorithm (FedAvg) in FL is used to synchronize aggregation and model updates, and this paper uses it as the baseline algorithm. As we mentioned in Sect. 2, the user's private information and local model updates may be contaminated by adversaries due to gradient leakage and the model poisoning attack. To ensure the security of the training process, we proposed the Chain-AAFL algorithm.

Our Chain-AAFL algorithm mainly leverages two mechanisms: The Chained-Communication mechanism enables masked gradients to be transferred among participants; Adversarial-aware gradient aggregation could avoid model degradation caused by model poisoning attacks. And the Adversarial-aware Protocol distinguishes and identifies the adversary from all ordinary participants by the value of user confidence. In Adversarial-aware Protocol, we designed a user-divided grouping method to prevent model poisoning attacks from destroying the model and ensure the stability of the training process.

3.2   Proposed Architecture

Our Chain-AAFL uses a the Adversarial-aware gradient aggregation method and confidence as a symbol to identify the adversary among participants while improving the model training speed. The details of our scheme are shown in Fig. 1. The work of the server mainly includes: 1) broadcasting the global model parameter to participants; 2) create confidence and the random numbers as user's symbol and mask; 3) identify whether the participant is an ordinary user or adversary by confidence; 4) group participants according to their confidence; 5) aggregate the model parameters of all participants and update the global model.

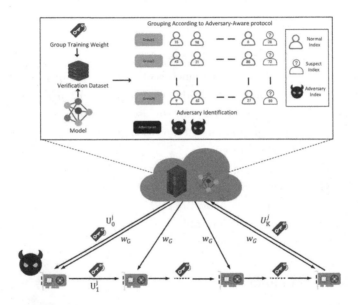

Fig. 1. The architecture of chain-AAFL with central server.

In the initial state of the Chain-AAFL algorithm, we set the total number of training rounds and initialize a global model w_G^t. In Fig. 1, the training method of participants is parallel within groups and serial between groups. The last node user at the end of each group of training transports the local aggregation model to the aggregation node. The aggregation node tests the accuracy of model weights on the private dataset. Compare the accuracy with the previous group, the aggregation node updates the global model and the confidence of the participants in the group. Participants in each group use the model update result of the previous group as the training model for the current group.

At the end of each round, the aggregation node first regards the participant whose confidence is lower than the threshold as an adversary, then eliminates the adversary and not allowed to participate in the subsequent training. The participant with lower confidence is selected as suspects. Then, participants are divided into N groups by their confidence value, where suspects are not divided into the same group. Under the influence of Adversarial-aware gradient aggregation and confidence identification, our scheme can realize secure federated learning training.

3.3 Chain-AAFL Algorithm

Preparation: Before the federted training get started, the aggregation node builds three lists for further usage. The *confidence score list C* denotes the confidence of each agent to behave well in the next training epoch. The *adversary list A* records those agents who have been determined as adversaries (the confidence score of these agents is less than a threshold T). And the *suspect list S* represents N agents with the lowest confidence scores but still above T.

Step 1:

Training Initialization: The aggregation node first broadcasts w_G to all agents for the current training epoch E_t, where w_G denotes the global weights propagated from the previous epoch (w_G is randomly initialized if $E_t = 1$). Then test w_G on a private dataset to obtain an initial classification accuracy ACC_0 Next, the aggregation node divides the agents into N training groups G $(g_1, ..., g_N)$ according to an adversary-aware protocol: (1) Agents that have already been determined as adversaries are not allowed to participate in following training epoches. (2) All suspects should be arranged to different groups.

Step 2:

Group Training: For federated training of the jth group, the aggregation node first uses the pseudo-random generator to generate a random weight U_0^j. The local update of user k in group g_j is termed as w_k^j. The training process of local parameters is the same as FedAVG. Next, the user k in group g_j computes

$$U_k^j = w_k^j + U_{k-1}^j \tag{1}$$

Group Aggregation: Suppose there are K agents in g_j, the Kth agent finally obtains a masked weight U_K^j. The aggregation node receives U_K^j and computes a temporal group training weight w_t

$$w_t = \frac{1}{K}(U_K^j - U_0^j) = \frac{1}{K}\sum_{i=1}^{K} w_i^j \tag{2}$$

Then the aggregation node tests w_t on the private dataset for a classification accuracy ACC_j. According to the performance of w_t and w_G, the aggregation node sets

$$w_G = \begin{cases} w_G & ACC_j < ACC_{j-1} \\ w_t & ACC_j > ACC_{j-1} \end{cases} \tag{3}$$

$$C_i = \begin{cases} C_i - 1 & ACC_j < ACC_{j-1} - 5\% \\ C_i + 1 & ACC_j > ACC_{j-1} \end{cases} (user_i \in g_j) \tag{4}$$

Step 3:

Aderversary Identification: The aggregation node fist puts agents whose confidence scores is less than threshold T to the adversary list A, then chooses N agents from the rest with the lowest confidence scores to the suspect list S.

Go to Step 1
The details of the proposed Chain-AAFL are given in Algorithm 1.

Algorithm 1. Chain-AAFL Algorithm

Preparation:
    Create confidence-score list C, adversary list A, suspect list S

Trainning Initialization:
    FOR *epoch* $i = 1, 2, ...$ DO
        Initialize and broadcast w_G //Global Model Patameter
        Test w_G on a verification dataset to get the task accuracy ACC_0
        divides agents into N groups $G(g_0, ..., g_N)$ according to AA protocol

Group Training:
    FOR *group* $j = 1, 2, ...$ DO
        $U_0^j \leftarrow PRG$ //Generate a pseudo-random mask weight U_0^j
        FOR *user* $k = 1, 2, ..., K$ DO
            Train and obtain local optimal weights w_k^j //$w \leftarrow w - \eta \nabla l(w; b)$
            $U_k^j \leftarrow w_k^j + U_{k-1}^j$

Group Aggregation:
    $w_t \leftarrow \frac{1}{K}(U_K^j - U_0^j)$ //Computes the average weights w_t
    Test w_G on a verification dataset to get the task accuracy ACC_j
    IF $ACC_j > ACC_{j-1}$ THEN
        $w_G \leftarrow w_t$
        $C_j \leftarrow C_j + 1$ //C_j denotes confidence scores of users in *group* j
    ELSE
        $w_G \leftarrow w_G$
    IF $ACC_j < ACC_{j-1} - 5\%$ THEN
        $C_j \leftarrow C_j - 1$

Adversary Identification:
    Put agents with confidence scores less than threshold T to list A
    Choose N agents from the rest with the lowest confidence scores to list S

---

4   Experiment

We simulate and compare the experimental results of three FL schemes, including our proposed algorithm, the Chain-PPFL algorithm, and the DP-based FedAVG algorithm with the Laplace mechanism. The baseline experiments are simple federated learning on MNIST and CIFAR100 datasets. To test the capability of leak-defense, we follow the idea of DLG method to verify the analysis of leak-defense experimentally. Then we conduct random model poisoning to demonstrate the effect of our proposed defense strategy.

4.1   Experiment Setup

The basic federated setting of our experiments is described as follows:

- Number of agents in each group $K = 10$
- Clients fraction rate $f = 0.1$
- Learning rate $\eta = 0.1$
- Model poisoning rate $p = 0.3$
- Hyperparameters for DP-based FL $\epsilon = 8$ and $\epsilon = 1$

In the simulation of gradient leakage and model poisoning, we use the public dataset CIFAR100 object classification. The training model is a CNN with two 5×5 convolution layers, a fully connected layer with ReLU activation, and a final softmax output layer. Besides, in the test of model poisoning, we set the momentum of SGD optimizer from 0.5 to 0.1, which alleviates the impact of abrupt gradient change. We choose PyTorch as the implementation platform. All experiments run in a server with the processor of Intel Core i7-9700 CPU @3.00GHz with 16GB RAM, and a single GPU RTX 3090.

4.2   Experimental Results

4.2.1   Baseline Training on CIFAR-100

The baseline results on the common CIFAR-100 benchmark of three training schemes are demonstrated in Fig. 2(a). The DP-based FL methods preserve privacy from DLG by adding noises into original gradients. Increasing the variance of noises could help improve the level of privacy protection. However, it also has a more notable negative impact on test accuracy. As is shown in Fig. 2(a), the models trained by our proposed Chain-AAFL are close to that of Chain-PPFL. Both outperform DP-based FL by 0.1% to 1% within 100 epochs. On the other hand, due to the test phase on 600 images for secure aggregation, our method consumes 2 more seconds than Chain-PPFL per round.

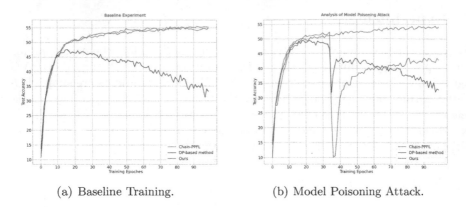

(a) Baseline Training.   (b) Model Poisoning Attack.

Fig. 2. Experiment on CIFAR100

4.2.2 Analysis of Deep Leakage from Gradients

The comparative results on the test of leak-defense are shown in Fig. 3. Following the idea of DLG, we use L-BFGS to match gradients from all trainable parameters with randomly initialized weights. The gradients' distance reflects the degree of discrimination for the raw training samples. The smaller the distance is, the more privacy leaked. When it is above a pre-defined threshold 0.15, no information is leaked. Figure 3 reports the effectiveness of different training schemes on the gradient leakage. Figure 3 suggests that both our Chain-AAFL method and Chain-PPFL can prevent the deep leakage from gradients, while the DP-based FL schemes can hardly achieve privacy-preservation.

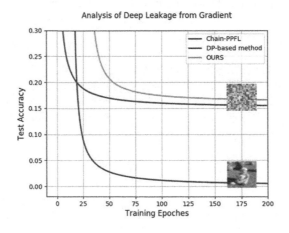

Fig. 3. The analysis of deep leakage from gradients.

4.2.3   Analysis of Model Poisoning

We explore this experiment to simulate model poisoning attacks in a multi-party training task and observe the performance of different FL schemes. We randomly set a client from all participants as the adversary. Whenever this node is selected to participate in a training epoch, it will output noisy weights to undermine the global model. As is shown in Fig. 2(b), model poisoning attacks can dramatically affect the test accuracy of Chain-PPFL and DP-based FL schemes. In addition, due to the constraints of learning rate and momentum, the global model cannot immediately recover from poisoning attacks. On the other hand, our Chain-AAFL framework carries out an effective defense strategy. With the designed secure-aggregation protocol, our scheme is able to notice the model poisoning attacks (e.g. at the 35 epoch) and gradually identify the adversary.

5   Conclusion

In this paper, we highlight the inadequacies of security in federated learning. To defense the gradient leakage and model poisoning attack, we propose the Chained Adversarial-Aware Federated Learning framework (Chain-AAFL). Then we present the proposed Chain-AAFL framework in details. The analysis of our experimental results indicates that our scheme is close to Chain-PPFL in terms of accuracy and privacy protection, and the defensiveness is much higher than other comparison methods. Therefore, our scheme can achieve a higher level of defense capability. In the future, we will concentrates on improving the security of Chain-AAFL and apply this scheme to real scenes to realize online timing detection.

Acknowledgment. This work was supported in part by Henan Provincial Major Public Welfare Project under Grant No. 201300210400, Elite Postgraduate Students Program of Henan University under Grant No. SYL20060174, China Postdoctoral Science Foundation under Grant No. 2020M672211, Scientific Research Projects of Henan Provincial Colleges and Universities under Grant No. 21A520003, and Key Technologies Research and Development Program of Henan under Grant No. 212102210078, 212102210090, 212102210094.

References

1. Bagdasaryan, E., Veit, A., Hua, Y., Estrin, D., Shmatikov, V.: How to backdoor federated learning. In: International Conference on Artificial Intelligence and Statistics, pp. 2938–2948. PMLR (2020)
2. Bhagoji, A.N., Chakraborty, S., Mittal, P., Calo, S.: Analyzing federated learning through an adversarial lens. In: International Conference on Machine Learning, pp. 634–643. PMLR (2019)
3. Blanchard, P., El Mhamdi, E.M., Guerraoui, R., Stainer, J.: Machine learning with adversaries: byzantine tolerant gradient descent. In: Proceedings of the 31st International Conference on Neural Information Processing Systems, pp. 118–128 (2017)

4. Dwork, C., Roth, A., et al.: The algorithmic foundations of differential privacy. Found. Trends Theor. Comput. Sci. **9**(3–4), 211–407 (2014)
5. Fang, M., Cao, X., Jia, J., Gong, N.: Local model poisoning attacks to byzantine-robust federated learning. In: 29th {USENIX} Security Symposium ({USENIX} Security 2020), pp. 1605–1622 (2020)
6. Kairouz, P., et al.: Advances and open problems in federated learning. arXiv preprint arXiv:1912.04977 (2019)
7. Kocabas, O., Soyata, T.: Towards privacy-preserving medical cloud computing using homomorphic encryption. In: Virtual and Mobile Healthcare: Breakthroughs in Research and Practice, pp. 93–125. IGI Global (2020)
8. Li, T., Sahu, A.K., Talwalkar, A., Smith, V.: Federated learning: challenges, methods, and future directions. IEEE Signal Process. Mag. **37**(3), 50–60 (2020)
9. Li, Y., Zhou, Y., Jolfaei, A., Yu, D., Xu, G., Zheng, X.: Privacy-preserving federated learning framework based on chained secure multi-party computing. IEEE Internet Things J. **8**(8), 6178–6186 (2020)
10. Lyu, L., Yu, H., Yang, Q.: Threats to federated learning: a survey. arXiv preprint arXiv:2003.02133 (2020)
11. McMahan, B., Moore, E., Ramage, D., Hampson, S., Arcas, B.A.: Communication-efficient learning of deep networks from decentralized data. In: Artificial Intelligence and Statistics, pp. 1273–1282. PMLR (2017)
12. Rivest, R.L., Adleman, L., Dertouzos, M.L., et al.: On data banks and privacy homomorphisms. Found. Secure Comput. **4**(11), 169–180 (1978)
13. Shamir, A.: How to share a secret. Commun. ACM **22**(11), 612–613 (1979)
14. Wang, G., He, J., Shi, X., Pan, J., Shen, S.: Analyzing and evaluating efficient privacy-preserving localization for pervasive computing. IEEE Internet Things J. **5**(4), 2993–3007 (2017)
15. Wei, K., et al.: Federated learning with differential privacy: algorithms and performance analysis. IEEE Trans. Inf. Forensics Secur. **15**, 3454–3469 (2020)
16. Zhao, B., Mopuri, K.R., Bilen, H.: IDLG: improved deep leakage from gradients. arXiv preprint arXiv:2001.02610 (2020)
17. Zhou, X., Xu, M., Wu, Y., Zheng, N.: Deep model poisoning attack on federated learning. Future Internet **13**(3), 73 (2021)
18. Zou, Y., Bao, X., Xu, C., Ni, W.: Top-k frequent itemsets publication of uncertain data based on differential privacy. In: Wang, G., Lin, X., Hendler, J., Song, W., Xu, Z., Liu, G. (eds.) WISA 2020. LNCS, vol. 12432, pp. 547–558. Springer, Cham (2020). https://doi.org/10.1007/978-3-030-60029-7_49

Differentially Private Linear Regression Analysis via Truncating Technique

Yifei Liu[1], Ning Wang[1(✉)], Zhigang Wang[1], Xiaodong Wang[1], Yun Gao[1], Xiaopeng Ji[1], Zhiqiang Wei[1], and Jun Qiao[2]

[1] Faculty of Information Science and Engineering, Ocean University of China, Qingdao, China
{wangning8687,wangzhigang,wangxiaodong,gaoyun,jxiaopeng, weizhiqiang}@ouc.edu.cn
[2] Shenyang Hangsheng Technology Co, Ltd., Shenyang 110000, Liaoning, China

Abstract. This paper discusses how to study the linear regression model accurately while guaranteeing ϵ-differential privacy. The parameters involved in linear regression are sensitive to one single record in database. As a result, a large scale of noise has to be added into the parameters to protect the records in database, which leads to inaccurate results. To improve the accuracy of published results, the existing works enforce ϵ-differential privacy by perturbing the coefficients in the objective function(loss function) of one optimization problem, which is constructed to derive parameters of linear regression, rather than adding noise to the parameters directly. And the scale of noise generated in the above technique is proportional to the square of dimensionality. Obviously, if the dimensionality is high, the scale of noise will be very large, i.e., curse of dimensionality. To settle this issue, this paper firstly studies a truncating length in a differential private way, where the length limits the maximal influence of one record on the coefficients of objective function. And then the noisy truncating coefficients are published with the truncating length limitation. Finally, the parameters involved in linear regression can be derived based on the objective function with noisy coefficients. The experiments on real datasets validate the effectiveness of our proposals.

Keywords: Differential privacy · Linear regression · Truncating length · Optimization · Noise

1 Introduction

As a classical optimization [15] problem, linear regression [13] has attracted a lot of attention in academia. However, since the training data may include individual sensitive information, it may incur privacy leakage to learn a linear regression model directly with sensitive training data. Besides, differential privacy [1–4, 6–11, 14, 18, 20, 24, 27] is the state-of-the-art model for releasing sensitive information while protecting privacy. Specifically, this model adds noise to the published results to make published results insensitive to any record in the database. And the noise

© Springer Nature Switzerland AG 2021
C. Xing et al. (Eds.): WISA 2021, LNCS 12999, pp. 249–260, 2021.
https://doi.org/10.1007/978-3-030-87571-8_22

scale is proportional to the maximum impact (i.e., sensitivity) of adding or deleting a record to the database on the published value. Considering this model can provide strict theoretical guarantee of privacy protection, we aim to learn a linear regression model while satisfying ϵ-differential privacy.

Linear regression [12,13] is a statistical analysis method to determine the quantitative relationship, between two or more variables. What's more, it is a classical problem which can be solved by the optimization technique. The parameters, i.e., optimal values, are very sensitive to one record in the database. Therefore, it is necessary to add noise to the parameters to protect privacy. And a lot of works have shown the scale of noise is very large, leading to useless published results [2,3,9,14,20]. Function mechanism [5] is a novel and effective method to derive parameters in a differentially private way. In order to improve the accuracy of derived parameters, this mechanism firstly perturbs the coefficients of objective function and then computes the optimal values, i.e., the parameters, rather than adding noise to the parameters directly. To guarantee differential privacy, the scale of adding noise in function mechanism is proportional to $2d + d^2$, where d is the dimensionality of database. As a result, for high-dimensional database, the noisy objective function deviates from the true one severely, due to too large noise involved in the coefficients. Further the derived parameters are meaningless.

To improve the accuracy of parameters involved in linear regression, this paper proposes FM_TC to publish the parameters with differential privacy. Based on the framework of function mechanism [24], we use the idea of truncation [6] used in frequent itemset mining. The core idea of FM_TC is to limit the contribution(truncating length) of each record to the coefficients involved in objective function. In this way, we can reduce the sensitivity effectively. In addition, the truncation operation imports information loss, due to ignoring some contribution of one record. A larger truncating length imports higher noise scale and lower information loss. So there is a tradeoff between these two parts. So we take them into consideration and design a technique to learn a better truncating length in a differential private way.

In the following, Sect. 2 reviews related work. Section 3 provides the necessary background on differential privacy and linear regression. Section 4 presents differentially private linear regression via truncating technique. Section 5 contains an extensive set of experiments. Finally, Sect. 6 concludes the paper.

2   Related Work

As a privacy protection model with strict theoretical guarantee, differential privacy has been widely used in the field of machine learning. Zhao et al. [25] empirically evaluate various implementations of differential privacy and measure its ability to protect privacy. Wu et al. [21] minimize the cost of machine learning models using stochastic gradient descent [22,23]. Gradient information satisfies differential privacy. Model pDP [19] is proposed to describe the actual loss of personal privacy in a linear regression task. It is like an analysis tool that helps people to materialize the loss of privacy. However, noise cannot be calibrated by setting a prescribed budget. Because the sensitivity of each individual is a data-dependent quantity. [16,17] study the problem of sparse linear regression, and present the solutions suitable for low-dimensional and high-dimensional situations, respectively. Zheng et al. [26] also consider privacy protection while studying machine learning models. However, these methods are designed for local differential privacy instead of differential privacy.

Zhang et al. [24] propose function mechanism, which adds noise to the coefficients of objective function. Based on objective function with noise, the optimal value can be solved. Function mechanism is better than traditional technique, because it can effectively reduce noise. [5] presents ϵ-differential privacy functional mechanisms of three regularized linear regression models, and proves the superiority of functional mechanism. However, they don't discuss the issue of high dimensionality.

For high dimensionality, noise proportional to $2d + d^2$ is added to coefficients of objective function. The large-scale noise makes objective function deviate from true one severely. Further, there is a large skew between true optimal value and noisy value. Wang et al. [20] study regression model in the scenario that some dimensions are sensitive, while others are non-sensitive. Based on functional mechanism, they proposed an optimization strategy for linear regression. Towards high dimensionality, lots of work is devoted to accurately publishing frequency under differential privacy. The noise scale is proportional to dimensionality. Day et al. [6] reduce the scale of noise by truncating high-dimensional records in the database. This gives us some inspirations to reduce noise scale by truncating coefficients of each individual.

3   Preliminary

3.1   Differential Privacy

The core idea of differential privacy is to ensure that the published data does not change too much due to the addition or deletion of a record in the database.

Definition 1 (Neighboring database [10]). There are two databases D_1 and D_2. Database D_2 can be obtained by adding or deleting a record t_m to D_1, that is, $D_1 = D_2 \cup \{t_m\}$ or $D_2 = D_1 \cup \{t_m\}$. In this way, D_1 and D_2 are called neighboring databases.

Definition 2 (ϵ-Differential Privacy [10]). Given two neighboring databases D_1, D_2, output O, random algorithm Θ will satisfy ϵ-differential privacy if and only if it satisfies the following constraints:

$$\frac{P[\Theta(D_1) = O]}{P[\Theta(D_2) = O]} \leq e^{\epsilon}. \tag{1}$$

In particular, $P[\Theta(D_1) = O]$ represents the probability that algorithm Θ outputs O based on the database D_1. ϵ is the privacy budget. A larger ϵ means lower privacy requirements. A smaller ϵ means that stricter privacy guarantee needs to be provided, and more random noise needs to be added to the published results.

In order to satisfy ϵ-differential privacy, Laplace Mechanism [7] and Exponential Mechanism [11] are usually adopted. Both of these mechanisms are closely related to sensitivity.

Definition 3 (Sensitivity [10]). Given a query Q, the sensitivity of algorithm is defined as $\Delta = \max_{D_1, D_2} ||Q(D_1) - Q(D_2)||_1$. $||Q(D_1) - Q(D_2)||_1$ refers to the L_1 distance between $Q(D_1)$ and $Q(D_2)$.

Specifically, if query Q outputs numeric data, Laplace Mechanism is usually used. In order to ensure ϵ-differential privacy, Laplace mechanism directly adds noise to the results. The noise is sampled from a Laplace distribution with a mean of 0 and a variance of Δ/ϵ. If query Q outputs categorical data, Exponential Mechanism is usually adopted. This mechanism requires user to specify a quality function $q : D \times \Omega \rightarrow R$. For any $\alpha \in \Omega$, $q(D, \alpha)$ represents the accuracy of output α based on D and Q. In order to ensure ϵ-differential privacy, Exponential Mechanism uses α as the output with the following probability from Ω, where Δ is the sensitivity of quality function q, $p[\alpha] \propto \exp(\epsilon q(\alpha, D)/\Delta)$.

Theorem 1. *(Sequential Composition [10]) Suppose that random algorithm $\Theta(D)$ is composed of multiple random algorithms $\{\Theta_1(D), \Theta_2(D), ..., \Theta_n(D)\}$. If $\Theta_i(D)$ satisfies ϵ_i-differential privacy, algorithm $\Theta(D)$ will satisfy $\sum_{i=1}^{n} \epsilon_i$-differential privacy.*

3.2   Problem Definition

Given the training data set D, there are n records $\{t_1, t_2, ..., t_n\}$ and $d + 1$ attributes $\{X_1, X_2, ..., X_d, Y\}$. Each record is expressed as $t_i = (x_i, y_i)$, where $x_i = (x_{i1}, x_{i2}, ..., x_{id})$. Without loss of generality, we defines $x_{ij}, y_i \in [0, 1]$. The goal of this paper is to obtain vector $w = (w_1, w_2, ..., w_d)$ composed of d parameters, so that $\sum_{j=1}^{d} w_j x_{ij}$ can predict y_i more accurately. That is to say, we need to get the linear regression parameters. Linear regression is defined as follows.

Definition 4 (Linear Regression). Based on training set D, linear regression outputs a prediction function $\rho(x_i) = x_i^T w^*$. w^* is a d dimensional vector that minimizes the error between predicted value $x_i^T w^*$ and true value y_i, such that

$$w^* = arg\,min_w \sum_{i=1}^{n} (y_i - x_i^T w)^2 \tag{2}$$

However, the training set contains individual private information. This paper proposes an optimization scheme based on the function mechanism framework and publishes w^* in a differential privacy way. The function mechanism is described below.

Function Mechanism (FM [24]): In order to protect privacy, function mechanism publishes the noisy coefficients of objective function in Eq. 2, and then

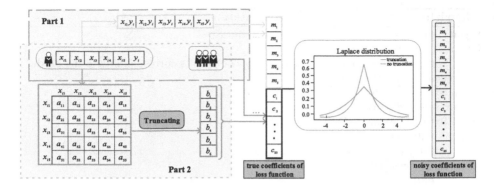

Fig. 1. Example for illustrating the truncating technique.

obtains the vector of d prediction parameters by solving the optimal problem with the given objective function. The objective function $\sum_{i=1}^{n}(y_i - x_i^T w)^2$ in Eq. 2 can be expressed as

$$\sum_{t_i \in D} y_i^2 - \sum_{j=1}^{d}(2 \sum_{t_i \in D} y_i x_{ij})w_j + \sum_{1 \le j,l \le d}(\sum_{t_i \in D} x_{ij}x_{il})w_j w_l \qquad (3)$$

Observing the above formula, we can see that $\sum_{t_i \in D} y_i^2$ is a constant, which has nothing to do with the optimal prediction parameter. Coefficients $2\sum_{t_i \in D} y_i x_{ij}$ and $\sum_{t_i \in D} x_{ij}x_{il}$ need to be published. In particular, the former corresponds to parameter w_j and the latter corresponds to the product $w_j w_l$. Since $x_{ij}, y_{ij} \in [0,1]$, the sensitivity is $2d + d^2$. In summary, in order to publish an objective function that satisfies ϵ-differential privacy, we need to add noise with a mean value of 0 and a variance of $(d^2 + 2d)/\epsilon$ to the coefficients in Eq. 3.

4   Differentially Private Linear Regression via Truncating Technique

Based on function mechanism, the sensitivity of publishing coefficients is $d^2 + 2d$. When d is large, too much noise needs to be added to the coefficients of objective function. This causes the published objective function to be unavailable. The sensitivity is mainly determined by the value of d^2. Every record produces a series of values $x_{ij}x_{il}(1 \le j,l \le d)$. Each value is added to the coefficients of product $w_j w_l$. In view of this situation, we only keep the top K $x_{ij}x_{il}$, and the rest of the values are set to 0. In this way, each record affects the coefficients corresponding to K products at most. The sensitivity of this part changes from d^2 to K. Here, K is called truncating length.

Figure 1 shows a simple example to illustrate the effectiveness of truncating technique. Suppose we have learned that the truncating length K is 6 in a differential private way. Dimensionality d of database is 5. User u_i possesses

---

Algorithm 1. Truncating length computation

Input: data of n users, dimensionality d of database.

Output: truncating length K.

    /*Matrix A represents the initial possible coefficients corresponding to the product $w_j w_l$ of parameter j and l. */

1: initial array $Q=0$;
2: **for** $k \in [1, d^2]$ **do**
3:    **for** each user $t_i \in D$ **do**
4:       initial each a to 0 for $a \in A$;
5:       **for** $j, l \in [1, d]$ **do**
6:          Compute $a_{jl} \leftarrow x_{ij} x_{il}$;
7:       **end for**
8:       Sort all possible coefficients in list a, and get top k coefficients, set other values to 0.

        // Q_k is the abbreviation of $Q(D, k, \epsilon_2)$.

9:       Compute $Q_k \leftarrow -\frac{1}{d^2} \sum a + \frac{2d+k}{\epsilon_2}$;
10:    **end for**
11: **end for**
12: Choose K from $[1, d^2]$ with probability $\text{Prob}[k]$ proportional to $\exp(-\epsilon_1 Q(D, k, \epsilon_2))$ for $k \in [1, d^2]$;
13: **return** K

---

data $t_i = \{x_{i1}, x_{i2}, x_{i3}, x_{i4}, x_{i5}, y_i\}$. As shown in Fig. 1, for part 1, each user generates 5 coefficients used for computing $\sum_{t_i \in D} y_i x_{ij} (1 \leq j \leq d)$ shown in Eq. 3. m_i stores the value of $\sum_{t_i \in D} y_i x_{ij}$. For part 2, u_i generates 25 coefficients used for computing $\sum_{t_i \in D} x_{ij} x_{il} (1 \leq j, l \leq d)$ shown in Eq. 3. c_{ij} stores the value of $\sum_{t_i \in D} x_{ij} x_{il}$. We store 25 coefficients in matrix A. Without truncating technique, each user needs to contribute 5 coefficients labeled by yellow line and 25 coefficients labeled by blue line to derive the objective function. With truncating technique, we keep top 6 coefficients with maximum values in A and set the remaining values to 0. In this way, each user needs to contribute 11 coefficients including the ones labeled by yellow line and the ones labeled by green line. Obviously, the latter method has lower sensitivity than the former one and needs a small scale of noise. Truncating technique can improve the accuracy of publishing results. Based on the idea of trunction, we design FM_TC framework. In the following, we discuss it in detail.

FM_TC Algorithm Framework: First, ϵ_1 is used to calculate truncating length. This way is based on the quality function proposed in Sect. 4.1. Then $\epsilon - \epsilon_2$ is used to publish truncated product coefficients. Finally, the regression parameters are released based on noisy objective function.

    In the following, Sect. 4.1 and Sect. 4.2 introduce the first two steps of FM_TC in detail.

4.1 Calculation of Truncating Length

Definition 5 (Truncating Error). For a record t_i, only the top K coefficients are retained, and the remaining values are set to 0. This will cause a skew between the calculated product coefficients and the coefficients without truncation. This part of the skew is called truncating error.

Definition 6 (Noise Error). In order to satisfy ϵ_2-differential privacy, FM_TC needs to add Laplace noise to the coefficients after truncation. The mean value of Laplace noise is 0 and the variance is $(2d + k)/\epsilon_2$. The error between noisy coefficients and true coefficients after truncation is called noise error.

We use $E(D, K, \epsilon)$ to represent the average error of the coefficients, such that

$$E(D, K, \epsilon_2) = \frac{1}{d^2} \sum_{j,l} (\sum_{t_i \in D} x_{ij}x_{il} - \sum_{\substack{t_i \in D \wedge \\ x_{ij}x_{il} \in top(t_i,K)}} x_{ij}x_{il} + \frac{2d + K}{\epsilon_2}). \tag{4}$$

Here, $top(t_i, K)$ represents the set of top K values of $x_{ij}x_{il}$. Our goal is to choose truncating length K with minimum $E(D, K, \epsilon_2)$ while ensuring ϵ_1-differential privacy. Observing the above formula, we can find that $(1/d^2)\sum_{j,l}\sum_{t_i \in D} x_{ij}x_{il}$ is a constant independent of K. Using $Q(D, K, \epsilon_2)$ to represent the remaining part of $E(D, K, \epsilon_2)$, we need to select a value from range $[1, d^2]$ that minimizes $Q(D, K, \epsilon_2)$ as truncating length K, such that $K = arg\,min_{K \in [1,d^2]}Q(D, K, \epsilon_2)$.

In order to ensure ϵ_1-differential privacy, the Exponential Mechanism is adopted. As shown in Algorithm 1, we specify the quality function corresponding to K as $-Q(D, K, \epsilon_2)$. Within the range $[1, d^2]$, each length k is returned as the truncating length with probability $P[k]$. Here, $P[k] \propto \exp(-\epsilon_1 Q(D, k, \epsilon_2))$.

Theorem 2. *The above-mentioned method for obtaining the truncating length satisfies ϵ_1-differential privacy.*

Proof. Given two neighbor databases D_1 and $D_2 = D_1 \cup \{t_m\}$, the sensitivity can be shown as follows:

$$|-Q(D_2, K, \epsilon_2) - [-Q(D_1, K, \epsilon_2)]|$$

$$= \left| \frac{1}{d^2} \{ \sum_{t_i \in D_2} (\sum_{x_{ij}x_{il} \in top(t_i,K)} x_{ij}x_{il}) - \sum_{t_i \in D_1} (\sum_{x_{ij}x_{il} \in top(t_i,K)} x_{ij}x_{il}) \} \right| \tag{5}$$

$$= \frac{1}{d^2} \{ \sum_{x_{mj}x_{ml} \in top(t_m,K)} x_{mj}x_{ml} \} \le \frac{K}{d^2} \le 1.$$

Θ represents the random algorithm for obtaining truncating length above. Furthermore, $P[\Theta(D_1) = K]$ and $P[\Theta(D_2) = K]$ represent the probability of outputting K based on database D_1 and D_2, respectively. Based on Eq. 5, we have

$$\frac{P[\Theta(D_2) = K]}{P[\Theta(D_1) = K]} = \frac{\exp[-\epsilon_1 Q(D_2, K, \epsilon_2)]}{\sum_{i=1}^{d^2} \exp[-\epsilon_1 Q(D_2, i, \epsilon_2)]} \cdot \frac{\sum_{i=1}^{d^2} \exp[-\epsilon_1 Q(D_1, i, \epsilon_2)]}{\exp[-\epsilon_1 Q(D_1, K, \epsilon_2)]} \le \exp(\epsilon_1).$$

Algorithm 2. Coefficients publication

Input: data of n users, dimensionality d of database, truncating length K.
Output: noisy coefficients.

/*Array m is used for storing coefficients of parameter w_j. Matrix A represents the initial possible coefficients corresponds to product $w_j w_l$ of the jth and lth parameter. Values in matrix A are sorted into array b. Array c is used for storing coefficients of product $w_j w_l$. $j, l \in [1, d]$. */

1: initial array m, c to 0;
2: **for** $j \in [1, d]$ **do**
3:    **for** each user $t_i \in D$ **do**
4:       Compute $m_j \leftarrow m_j + 2 y_i x_{ij}$
5:    **end for**
6: **end for**
7: **for** each user $t_i \in D$ **do**
8:    initial each a to 0 for $a \in A$; initial array b to 0;
9:    **for** $j, l \in [1, d]$ **do**
10:       Compute $a_{jl} \leftarrow x_{ij} x_{il}$
11:    **end for**
12:    Sort all possible coefficients in matrix a, and get list b.
13:    Keep top K coefficients in array b, set the remaining values to 0, and update array c.
14: **end for**
15: **for** $i \in [1, d]$ **do**
16:    Compute $\tilde{m}_i \leftarrow m_i + Laplace(0, \frac{2d+K}{\epsilon_2})$;
17: **end for**
18: **for** $i \in [1, d^2]$ **do**
19:    Compute $\tilde{c}_i \leftarrow c_i + Laplace(0, \frac{2d+K}{\epsilon_2})$;
20: **end for**
21: **return** list m and c

4.2 Publication of Objective Function Coefficients

Based on truncating length K obtained in the previous section, this section publishes coefficients of objective function (Eq. 3). Algorithm 2 shows approximate steps. The specific methods are as follows. For any record t_i, we first calculate the coefficient $2 y_i x_{ij}$ corresponding to parameter $w_j (1 \leq j \leq d)$. Next, we calculate coefficient $x_{ij} x_{il}$ corresponding to product $w_j w_l$. After sorting d^2 coefficients, we keep top K coefficients, and set the remaining coefficients to 0. Then, coefficients obtained from all records are added together to obtain final coefficients in Eq. 3. Here, the sensitivity of coefficients is $2d + K$. In order to ensure ϵ_2-differential privacy, it is necessary to add Laplace noise with a mean value of 0 and a variance of $(2d + k)/\epsilon_2$ to each coefficient. The corresponding probability density function (PDF) of Laplace distribution is $\frac{\epsilon_2}{2(2d+K)} \exp(-\frac{\epsilon_2 |x|}{2d+K})$. By adding noise to each coefficient in Eq. 3, we get noisy objective function and optimal parameter vector w of linear regression without revealing privacy.

Theorem 3. *The method for obtaining coefficients of objective function satisfies ϵ_2-differential privacy.*

Proof. Given two neighboring databases D_1 and D_2, dimensionality d, truncating length K, output W, and random algorithm Φ, let $||C(D_1) - C(D_2)||_1$ be L_1-distance between database D_1 and D_2, we have

$$\frac{P[\Phi(D_2) = W]}{P[\Phi(D_1) = W]} = \prod_{i=1}^{d+d^2} \frac{\exp(-\frac{\epsilon_2|C_i(D_2) - W_i|}{2d + K})}{\exp(-\frac{\epsilon_2|C_i(D_1) - W_i|}{2d + K})} \leq \prod_{i=1}^{d+d^2} \exp(\frac{\epsilon_2|C_i(D_1) - C_i(D_2)|}{2d + K})$$

$$= \exp(\frac{\epsilon_2 \cdot \sum_{i=1}^{d+d^2} |C_i(D_1) - C_i(D_2)|}{2d + K}) = \exp(\frac{\epsilon_2 \cdot ||C(D_1) - C(D_2)||_1}{2d + K}) \leq \exp(\epsilon_2).$$

$$(6)$$

Specifically, $||C(D_1) - C(D_2)||_1$ is not lager than $2d + K$ because one user affects $2d + K$ coefficients at most. Based on Theorem 1, FM_TC satisfies $(\epsilon_1 + \epsilon_2)$-differential privacy, i.e., ϵ-differential privacy.

5   Experimental Evaluation

In FM_TC, 0.1ϵ is used to calculate truncating length K, and 0.9ϵ is used to publish coefficients. We use two datasets: US and Brazil. US contains 370,000 records, while Brazil contains 190,000 records. Both datasets have dimensionality 13. We randomly select 80% data as training set, and select 20% data as testing set. FM_TC requires each user to generate d^2 coefficients and sort them. Considering that there are n users, time complexity of FM_TC is $O(nd^2 log d)$. FM has time complexity of $O(nd^2)$, since it does not sort coefficients. FM_TC is not as efficient as FM in time efficiency because of sorting coefficients. However, the accuracy of one algorithm is more important in the private scenario. So we just show the performance of our method with the accuracy metric. In this paper, mean absolute error (MAE) is used as the standard to measure efficiency of algorithm, which is defined as follows: $\sum_{t_i \in Test} \frac{|y_i - \tilde{y}_i|}{|Test|}$. We run each group of experiments 10 times and take the average value as result.

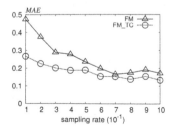

Fig. 2. Vary sampling rate (US).

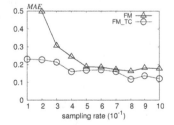

Fig. 3. Vary sampling rate (Brazil).

Figure 2 and Fig. 3 show the MAE corresponding to different sampling rates when the privacy budget is 0.0125. In order to show the relationship between MAE and the scale of data, we select a certain size subset from dataset at different sampling rates for testing. It can be seen from the figure that MAE of two algorithms decreases as the sampling rate increases. Because the scale of noise added to coefficients is independent of the size of dataset. A large dataset makes the true value of coefficients larger and further resists noise well. Therefore, noisy result is more accurate. In addition, compared with FM algorithm, the accuracy of FM_TC can be improved by 0.3. FM_TC reduces the sensitivity from $2d+d^2$ to $2d + K$ by truncating the product coefficients of the parameters, so that the objective function is closer to real objective function, and the predicted parameters are more accurate. Similarly, for this reason, FM_TC is better than FM under different privacy budgets. Specifically, Fig. 4 and Fig. 5 test the accuracy of these two algorithms under different privacy budgets {0.0125, 0.025, 0.05, 0.1, 0.2}.

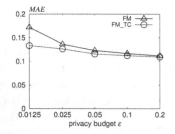 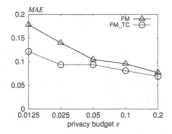

Fig. 4. Vary privacy budget (US).   **Fig. 5.** Vary privacy budget (Brazil).

Table 1 shows the number of coefficients retained under different privacy budgets. Respectively, the second row and the third row represent the number of coefficients based on FM_TC method. The forth row represents the number of coefficients based on traditional function mechanism [24]. As can be seen from the table, FM_TC retains fewer coefficients than FM. Since the scale of the noise added to the coefficients is proportional to the number of coefficients, FM_TC releases more accurate coefficients of objective function, which in turn leads to more accurate prediction parameters.

Table 1. The number of reserved coefficients (K)

eps	0.0125	0.025	0.05	0.1	0.2
K(US)	71	79	101	106	113
K(Brazil)	57	67	85	91	109
d^2	144	144	144	144	144

6   Conclusion

Based on the framework of function mechanism, this paper proposes algorithm FM_TC that satisfies differential privacy to release parameters in linear regression. Since the sensitivity of problem is proportional to the square of dimensionality in training set, this paper limits contribution of each record to the coefficients in objective function. In this manner, we reduces sensitivity and improves accuracy of published results.

Acknowledgement. This work was supported in part by the National Natural Science Foundation of China under Grant 61902365 and Grant 61902366, in part by the China Postdoctoral Science Foundation under Grant 2019M652473, Grant 2019M652474 and Grant 2020T130623.

References

1. Chamikara, M.A.P., Bertók, P., Khalil, I., Liu, D., Camtepe, S.: Privacy preserving face recognition utilizing differential privacy. CoRR abs/2005.10486 (2020)
2. Chaudhuri, K., Monteleoni, C.: Privacy-preserving logistic regression. In: Proceedings of NIPS, Vancouver, pp. 289–296. Curran Associates, Inc. (2008)
3. Chaudhuri, K., Monteleoni, C., Sarwate, A.D.: Differentially private empirical risk minimization. J. Mach. Learn. Res. **12**, 1069–1109 (2011)
4. Cheng, X., Tang, P., Su, S., Chen, R., Wu, Z., Zhu, B.: Multi-party high-dimensional data publishing under differential privacy. IEEE Trans. Knowl. Data Eng. **32**(8), 1557–1571 (2020). https://doi.org/10.1109/TKDE.2019.2906610
5. Dandekar, A., Basu, D., Bressan, S.: Differential privacy for regularised linear regression. In: Hartmann, S., Ma, H., Hameurlain, A., Pernul, G., Wagner, R.R. (eds.) DEXA 2018. LNCS, vol. 11030, pp. 483–491. Springer, Cham (2018). https://doi.org/10.1007/978-3-319-98812-2_44
6. Day, W., Li, N.: Differentially private publishing of high-dimensional data using sensitivity control. In: Proc of Asia'CCS, Singapore, pp. 451–462. ACM (2015). https://doi.org/10.1145/2714576.2714621
7. Dwork, C., McSherry, F., Nissim, K., Smith, A.: Calibrating noise to sensitivity in private data analysis. In: Halevi, S., Rabin, T. (eds.) TCC 2006. LNCS, vol. 3876, pp. 265–284. Springer, Heidelberg (2006). https://doi.org/10.1007/11681878_14
8. Ghane, S., Kulik, L., Ramamohanarao, K.: TGM: a generative mechanism for publishing trajectories with differential privacy. IEEE Internet Things J. **7**(4), 2611–2621 (2020). https://doi.org/10.1109/JIOT.2019.2943719
9. Lei, J.: Differentially private m-estimators. In: Proceedings of NIPS, Granada, pp. 361–369 (2011)
10. McSherry, F.: Privacy integrated queries: an extensible platform for privacy-preserving data analysis. In: Proceedings of SIGMOD, Providence, pp. 19–30. ACM (2009). https://doi.org/10.1145/1559845.1559850
11. McSherry, F., Talwar, K.: Mechanism design via differential privacy. In: Proceedings of FOCS, Providence, pp. 94–103. IEEE Computer Society (2007). https://doi.org/10.1109/FOCS.2007.41
12. Murphy, K.P.: Machine Learning - A Probabilistic Perspective. Adaptive Computation and Machine Learning Series. MIT Press, Cambridge (2012)

13. Rodney, A.: Using linear regression analysis and defense in depth to protect networks during the global corona pandemic. J. Inf. Secur. **11**, 261–291 (2020)
14. Smith, A.D.: Privacy-preserving statistical estimation with optimal convergence rates. In: Proceedings of STOC, San Jose, pp. 813–822. ACM (2011). https://doi.org/10.1145/1993636.1993743
15. Lafif Tej, M., Holban, S.: Determining optimal multi-layer perceptron structure using linear regression. In: Abramowicz, W., Corchuelo, R. (eds.) BIS 2019. LNBIP, vol. 353, pp. 232–246. Springer, Cham (2019). https://doi.org/10.1007/978-3-030-20485-3_18
16. Wang, D., Xu, J.: On sparse linear regression in the local differential privacy model. In: Proceedings of ICML, Long Beach, vol. 97, pp. 6628–6637. PMLR (2019)
17. Wang, D., Xu, J.: On sparse linear regression in the local differential privacy model. IEEE Trans. Inf. Theory **67**(2), 1182–1200 (2021). https://doi.org/10.1109/TIT.2020.3040406
18. Wang, Q., Li, Z., Zou, Q., Zhao, L., Wang, S.: Deep domain adaptation with differential privacy. IEEE Trans. Inf. Forensics Secur. **15**, 3093–3106 (2020). https://doi.org/10.1109/TIFS.2020.2983254
19. Wang, Y.: Per-instance differential privacy and the adaptivity of posterior sampling in linear and ridge regression. CoRR abs/1707.07708 (2017)
20. Wang, Y., Si, C., Wu, X.: Regression model fitting under differential privacy and model inversion attack. In: Proceedings of IJCAI, Buenos Aires, pp. 1003–1009. AAAI Press (2015)
21. Wu, N., Farokhi, F., Smith, D.B., Kâafar, M.A.: The value of collaboration in convex machine learning with differential privacy. In: Proceedings of SP, San Francisco, pp. 304–317. IEEE (2020). https://doi.org/10.1109/SP40000.2020.00025
22. Wu, X., Li, F., Kumar, A., Chaudhuri, K., Jha, S., Naughton, J.F.: Bolt-on differential privacy for scalable stochastic gradient descent-based analytics. In: Proceedings of SIGMOD, Chicago, pp. 1307–1322. ACM (2017). https://doi.org/10.1145/3035918.3064047
23. Xu, J., Zhang, W., Wang, F.: A(dp)^{2}sgd: asynchronous decentralized parallel stochastic gradient descent with differential privacy. CoRR abs/2008.09246 (2020)
24. Zhang, J., Zhang, Z., Xiao, X., Yang, Y., Winslett, M.: Functional mechanism: Regression analysis under differential privacy. Proc. VLDB Endow. **5**(11), 1364–1375 (2012). https://doi.org/10.14778/2350229.2350253
25. Zhao, B.Z.H., Kâafar, M.A., Kourtellis, N.: Not one but many tradeoffs: privacy vs. utility in differentially private machine learning. CoRR abs/2008.08807 (2020)
26. Zheng, H., Hu, H., Han, Z.: Preserving user privacy for machine learning: local differential privacy or federated machine learning. IEEE Intell. Syst. **35**(4), 5–14 (2020). https://doi.org/10.1109/MIS.2020.3010335
27. Zou, Y., Bao, X., Xu, C., Ni, W.: Top-k frequent itemsets publication of uncertain data based on differential privacy. In: Wang, G., Lin, X., Hendler, J., Song, W., Xu, Z., Liu, G. (eds.) WISA 2020. LNCS, vol. 12432, pp. 547–558. Springer, Cham (2020). https://doi.org/10.1007/978-3-030-60029-7_49

Efficient Privacy Preserving Single Anchor Localization Using Noise-Adding Mechanism for Internet of Things

Yajie Li[1,2], Guanghui Wang[1,2(✉)], and Fang Zuo[1,3(✉)]

[1] Henan International Joint Laboratory of Intelligent Network Theory and Key Technology, Henan University, Kaifeng 475000, China
`gwang@vip.henu.edu.cn, zuofang@henu.edu.cn`
[2] School of Software, Henan University, Kaifeng 475000, China
[3] Subject Innovation and Intelligence Introduction Base of Henan Higher Educational Institution-Software Engineering Intelligent Information Processing Innovation and Intelligence Introduction Base of Henan University, Kaifeng 475000, China

Abstract. The technological advantages of 5G networks facilitate the development of location technology for the Internet of Things. The single anchor localization algorithm has been designed using multiple antenna arrays, which enables the accurate and distributed localization of target nodes with a small number of anchor nodes. However, there are privacy breach issues in the localization process. To avoid the privacy breach, in this paper, an efficient privacy-preserving single anchor localization algorithm is proposed by adopting a novel noise-adding mechanism. A set of zero-sum noises has been carefully designed, which not only enables precise localization but also protects location privacy. The performance of the proposed algorithm is analyzed in terms of localization accuracy and privacy preservation. Simulations validate the correctness and theoretical results.

Keywords: Internet of Things · Localization algorithm · Privacy preservation · Noise-adding

1 Introduction

Smart homes, wearable devices, and smart cities (including smart surveillance and automated transportation) have become an important part of the Internet

Supported by Institute of Intelligent Network System of Henan University, in part by Key Scientific Research Project of Universities in Henan Province under Grant No. 19A520016, in part by Henan Higher Education Teaching Reform Research and Practice Project (Graduate Education) under Grant No. 2019SJGLX080Y, in part by China Postdoctoral Science Foundation under Grant No. 2020M672211, in part by Key Scientific Research Projects of Henan Provincial Colleges and Universities under Grant No. 21A520003, in part by Key Technologies Research and Development Program of Henan under Grant No. 212102210090.

C. Xing et al. (Eds.): WISA 2021, LNCS 12999, pp. 261–273, 2021.
https://doi.org/10.1007/978-3-030-87571-8_23

of Things (IoT). In order to provide smart services, these devices should provide accurate location information. For example, the authors in [1] improved indoor localization accuracy of smart buildings for IoT services. The authors in [2] focused on optimal localization of smart parking systems. The authors in [3] proposed a dummy-based trajectory privacy protection scheme. The authors in [4] proposed a method that addressed the challenges in IoT based smart device localization. Therefore, localization technologies play an important role in the IoT.

5G wireless communication technology starts to enter people's life. Existing deployments of large-scale MIMO technology have greatly ensured the accuracy of distance and angle measurement information [5–8]. Meanwhile, 5G networks allow large numbers of connected devices, including both stationary and mobile sensors, which supports the applications in the IoT [9]. Therefore, it has become an attractive research direction that uses MIMO and mmWare to achieve accurate localization in various scenarios, such as vehicle networks [10], indoor navigation [11], and industrial applications [12].

The risk of privacy breaches increases tremendously with billions of connected devices via the IoT [13]. In the location network scenario, the inherent nature of the IoT provides a favorable environment for privacy breaches. Some malicious nodes may gain the private information of other nodes, either for financial gains or to increase influence. Therefore, the privacy preservation issues in the localization process are crucial. In recent years, there has been a great deal of research on location privacy preservation in localization. For example, Shi et al. [14] proposed a privacy-preserving distributed localization algorithm (PP-DILOC) that employs noise-adding mechanism. The PP-DILOC can protect the privacy information while ensuring the localization accuracy. Sazdar et al. [15] proposed a privacy-preserving method uses Bloom filter. The method can create and preserve anonymity by using a Partial Radio Map (PRM) during the localization process. Zhang et al. [16] proposed a lightweight and privacy-preserving scheme, which protects both the private information of users and servers. Wang et al. [17] proposed an efficiently privacy-preserving algorithm that uses information hiding in adjacent subtraction-based localization. Therefore, it is an important topic to preserve privacy during the localization process in the IoT.

The multi-antenna technology enables the single anchor localization through both obtaining distance and angle information between the anchor and target nodes. There are existing researches on the localization of single anchor nodes. For example, the authors in [18] use a single Ultra-wideband anchor to achieve accurate localization both indoors and outdoors. The authors in [19] proposed a localization solution by using a single mobile anchor and mesh-based path planning model. The authors in [20] proposed a novel Single Anchor Localization (SAL) scheme employing target movement. Wang et al. [21] proposed an accurate and distributed localization algorithm, ADL, based on a multi-antenna single-anchor node localization. The ADL algorithm can achieve accurate localization, when only a small number of anchor nodes exist. However, the privacy protection still needs further study in the existing work. Therefore, it is necessary to conduct deep research on privacy-preserving single anchor localization.

To prevent others access to the node's private information, we propose a privacy-preserving single anchor localization scheme by adopting a zero-sum noise-adding mechanism. All anchor nodes cooperate to produce zero-sum noise. Target node estimates location according to the result after adding zero-sum noise, which ensures the location privacy of anchor nodes. The contributions of this paper are summarized as follows.

- The privacy-preserving accurate distributed localization algorithm (PP-ADL) is proposed based on single anchor localization. The proposed algorithm adds zero-sum noises into the existing ADL algorithm.
- The localization accuracy and privacy preservation of the PP-ADL algorithm are theoretical analyzed.
- Simulations demonstrate the superiority of the PP-ADL algorithm in terms of correctness, accuracy and efficiency.

The remainder of this paper is organized as follows. Section 2 gives the system model and problem formulation. Section 3 proposes the PP-ADL algorithm and analyzes the performance. Section 4 shows performance evaluation. Section 5 concludes the paper and presents the future work.

Table 1. Important notations

Symbol	Definition
m	The number of anchor nodes
X_i	The location of anchor node i
T_i	The estimated location of target node using anchor i
$\hat{X}_0$	The estimated location of target node
d_i	The measured distance between the anchor i and the anchor node
α_i	The measured angle of target node with horizontal orientation
δ_i	The noise term added by anchor node i
T_{i0}	The location information of the target node received from the anchor node i
w_{ik}	The set of zero-sum random numbers generated by the anchor node i
s_i	The actual distance between the anchor node i and the target node
β_i	The actual angle of target node with horizontal orientation
μ_i	The error during the measurement of the distance
η_i	The error during the measurement of the angle
ϵ_i	The localization error
ϵ_i^*	The upper bound on the localization error

2 System Model and Problem Formulation

This section first introduces the system model and the accurate distributed localization algorithm (ADL) based on single-anchor node localization (SAL). Then, the privacy-preserving problem in the ADL algorithm is considered.

2.1 System Model

In n-dimensional Euclidean space $\mathbb{R}^n$, consider a crowdsourcing location network that contains base stations or access points. Figure 1 shows the crowdsourcing location network. The publisher of crowdsourcing tasks is the user (the target node). The worker of crowdsourcing tasks is the access points (the anchor node). User posts a crowdsourcing location task. The anchor node receives the localization task and sends the estimated result to the user. The base stations and access points are equipped with massive antenna arrays and support high bandwidth. The access point can measure the distance and angle, and the user can calculate location based on the distance and angle information of a single anchor node.

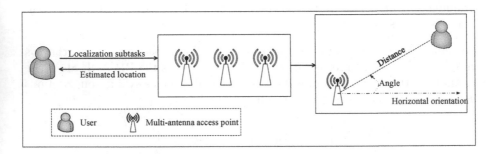

Fig. 1. The crowdsourcing localization process.

The number of anchor nodes is denoted by m. Let $X_i = (x_{i1}, x_{i2}, ..., x_{in})$ ($\forall i = 1, 2, ..., m$) denotes the location of the i-th anchor node. $X_0 = (x_{01}, x_{02}, ..., x_{0n})$ denote the location of the target node 0. d_i denotes the measured distance between the anchor node i and the target node 0. In two-dimensional space, the edge between the target node 0 and the anchor node i is denoted by $(i, 0)$. The angle between the edge $(i, 0)$ and the horizontal direction is denoted by α_i. The main notations are shown in Table 1. Therefore, the equation between the anchor node i and the target node 0 is satisfied as given follows:

$$\sqrt{(x_{01} - x_{i1})^2 + (x_{02} - x_{i2})^2} = d_i, \tag{1}$$

$$\arctan \frac{x_{02} - x_{i2}}{x_{01} - x_{i1}} = \alpha_i, \tag{2}$$

where x_{01} and x_{02} denote the location information of the target node 0. Denote the estimated location of the target node 0 by $T_i = (t_{i1}, t_{i2})$, that can be calculated by the following:

$$t_{i1} = x_{i1} + d_i \cos \alpha_i, \qquad t_{i2} = x_{i2} + d_i \sin \alpha_i, \tag{3}$$

where x_{i1} and x_{i2} denote the location information of the anchor node i.

The crowdsourcing localization process in the ADL algorithm includes two-step: In the first step, anchor node i sends the estimated location of the target node 0 calculated by (3) to the target node; in the second step, the target node 0 calculates its location by taking the mean value based on the received estimated location, as follows:

$$\hat{X}_0 = \frac{1}{m} \sum_{i=1}^{m} T_i, \tag{4}$$

where T_i is the estimated location of the target node by anchor node i, $\hat{X}_0$ is the estimated location of the target node.

2.2 Problem Formulation

In the crowdsourcing localization process of the ADL, each anchor node needs to send the estimated location to the target node 0. However, there are privacy breach issues in the network. For example, the target node can easily calculate the location of the anchor node. But the location of each node cannot be others know. Therefore, to prevent privacy breaches, a privacy preservation mechanism for the ADL is proposed to achieve the following goals.

- Accurate localization: All nodes can estimate own location through issues crowdsourcing localization task. The localization accuracy can achieve the same as the ADL.
- Privacy preservation: The private information of all nodes is preserved. All nodes cannot precisely get the private information of other nodes.

3 Privacy-Preserving ADL

This section firstly designs a privacy-preserving accurate distributed localization algorithm (PP-ADL). Then, the theoretical analysis is conducted regarding localization performance and privacy preservation.

3.1 PP-ADL

Privacy issues exist in the crowdsourcing localization network to breach the location privacy of the users. Therefore, to preserve location privacy, a noise term is added before each anchor node sends the estimated location information to the target node.

$$T_{i0} = T_i + \delta_i, \tag{5}$$

where T_{i0} is the noise-added estimated location received by the target node 0 from the anchor node i, δ_i is the noise term added by anchor node i. It is easy to affect the estimated location accuracy after adding noise. Therefore, the noise addition mechanism is carefully designed to ensure localization accuracy.

The PP-ADL algorithm can protect the location privacy of the anchor node while locating the target node accurately. Considering the localization performance, the sum of the added random noise terms is 0, i.e.

$$\sum_{i=1}^{m} \delta_i = 0, \tag{6}$$

where δ_i denotes the random noise term generated by each anchor node. A set of random numbers with zero-sum is a zero-sum random noise set, which needs to be generated cooperatively by each anchor node. If the number of anchor nodes selected in localization is m, these nodes cooperate generate the zero-sum random noise set. Therefore, to ensure the privacy of the added noise term, m needs to be larger than 2. When there are m anchor nodes, each anchor node generates m zero-sum random numbers. w_{ik} $(k = 1, 2, ..., m)$ denotes the set of zero-sum random numbers generated by the i-th anchor node, and satisfying $\sum_{k=1}^{m} w_{ik} = 0$. Then, each anchor node keeps one random number, e.g. w_{ii}, and sends the remaining m-1 random numbers to the other m-1 anchor nodes, respectively. Adding the random number kept by node i to the random numbers received from the other m-1 nodes, the result is the random noise term δ_i of node i, i.e. $\delta_i = \sum_{k=1}^{m} w_{ki}$ and $\sum_{i=1}^{m} \delta_i = 0$. Finally, the anchor node sends the result of (5) to the target node. So the procedure of this PP-ADL is shown in Algorithm 1.

Algorithm 1. PP-ADL.

Input: d_i; α_i; $X_i = (x_{i1}, x_{i2}, ..., x_{in})$ $(\forall i=1, 2, ..., m)$; $\delta_i(\forall i=1, 2, ..., m)$;
Output: $\hat{X}_0$;
1: **for** $i = 1$ to m **do**
2:     The anchor node i estimates the location of the target node 0 T_i according to (3);
3:     Update the estimated location of the target node 0 T_{i0} according to (5);
4:     Anchor node i sends T_{i0} to target node 0;
5: **end for**
6: Based on the result T_{i0} received from all anchor nodes, the target node 0 is calculated by (4) to obtain the estimated location $\hat{X}_0$.

Firstly, each anchor node i estimates the location T_i of the target node 0 according to (3). Then, each anchor node i add noise to update the estimated location T_{i0} of the target node 0 according to (5). They send T_{i0} to target node 0. After receiving T_{i0}, the target node 0 is calculated by (4) to obtain the estimated location $\hat{X}_0$. Therefore, the time complexity of the PP-ADL localization algorithm is $O(m)$. The execution time of the algorithm depends on the number of anchor nodes. The PP-ADL algorithm is efficient.

3.2 Performance Analysis

Correctness Analysis. To prove the correctness of the PP-ADL localization algorithm, it is necessary that deduce the difference between the estimated average value of the target node location and the actual location of the target node tends to 0 with a probability of 1. The following theorem gives the correctness of the PP-ADL localization algorithm:

Theorem 1. *The PP-ADL can achieve the equivalent location estimation results as the ADL.*

Proof. In [21], the correctness of the localization performance of the ADL algorithm has been proven. The error between the mean of the estimated location of the target node and the true location of the target node, tending to zero with probability 1. The equation is as follows:

$$Pr\left\{\lim_{K\to\infty}\left\|\left(\frac{1}{K}\sum_{k=1}^{K}\hat{X}_0^{(k)}\right) - X_0\right\| = 0\right\} = 1, \tag{7}$$

where X_0 is the actual location of the target node 0; $\hat{X}_0^{(k)}$ is the k-th location estimate of the target node 0 by the ADL localization algorithm; K is the total times the target node 0 location is estimated.

In the PP-ADL algorithm, anchor nodes send the estimated location to the target node. The sum of the added random noise is 0, i.e. $\sum_{i=1}^{m}\delta_i = 0$, we have:

$$\left(\frac{1}{K}\sum_{k=1}^{K}\hat{X}_0^{(k)}\right) - X_0 = \left(\frac{1}{K}\sum_{k=1}^{K}\left(\frac{1}{m}\sum_{i=1}^{m}T_i^{(k)}\right)\right) - X_0$$

$$= \left(\frac{1}{K}\sum_{k=1}^{K}\left(\frac{1}{m}\sum_{i=1}^{m}(T_i + \delta_i)^{(k)}\right)\right) - X_0 \tag{8}$$

$$= \left(\frac{1}{K}\sum_{k=1}^{K}\left(\frac{1}{m}\sum_{i=1}^{m}T_{i0}^{(k)}\right)\right) - X_0.$$

Therefore, the difference between the estimated average value of the target node location and the actual location of the target node tends to 0 with a probability of 1. The PP-ADL can achieve the equivalent location estimation results as the ADL.

Accuracy Analysis. To analyze the accuracy of the PP-ADL localization algorithm, it is necessary that analyze its localization error boundaries. Similar to [21], consider a bounded error model, i.e. $d_i = s_i + \mu_i$, $\alpha_i = \beta_i + \eta_i$. where s_i is the actual distance between the anchor node and the target node; μ_i is the error parameter that denotes the error in the process of measuring the distance; β_i is the angle between the actual location of the target node relative to the location of the anchor node and the horizontal line; η_i is the error parameter used to denote the error during the measurement of the angle information. They are satisfied with $|\mu_i| \leq \mu_b$, $|\eta_i| \leq \eta_b$. Where μ_b and η_b are non-negative boundaries for distance and angle measurements, respectively. We have the following theorem:

Theorem 2. *Considering the distance and angle measurements, d_i and α_i satisfied with $|\mu_i| \leq \mu_b$, $|\eta_i| \leq \eta_b$. The localization error ϵ of the PP-ADL algorithm satisfies:*

$$0 \leq \epsilon \leq \frac{1}{m}\sum_{i=1}^{m}\epsilon_i^*, \tag{9}$$

where ϵ_i^* is the upper bound on the localization error ϵ_i of the anchor node i.

Proof. The authors in [21] have proved the localization accuracy of the SAL. The localization error ϵ of the SAL using the anchor node i in a two-dimensional space satisfies:

$$0 \leq \epsilon_i \leq \epsilon_i^*. \tag{10}$$

The localization error bounds of the PP-ADL are analyzed based on the localization error bounds of the SAL. Because $\epsilon \geq 0$, and (4) (5) (10). We have:

$$
\begin{aligned}
\epsilon &= \left\| \left(\frac{1}{m} \sum\nolimits_{i=1}^{m} T_{i0} \right) - X_0 \right\| = \left\| \left(\frac{1}{m} \sum\nolimits_{i=1}^{m} (T_i + \delta_i) \right) - X_0 \right\| \\
&= \left\| \left(\frac{1}{m} \sum\nolimits_{i=1}^{m} T_i \right) - X_0 \right\| = \left\| \frac{1}{m} \sum\nolimits_{i=1}^{m} (T_i - X_0) \right\| \\
&\leq \frac{1}{m} \sum\nolimits_{i=1}^{m} \| T_i - X_0 \| = \frac{1}{m} \sum\nolimits_{i=1}^{m} \epsilon_i \leq \frac{1}{m} \sum\nolimits_{i=1}^{m} \epsilon_i^*.
\end{aligned}
\tag{11}
$$

Therefore, the Theorem 2 be validated. Additionally, the higher the number of anchor nodes, the more information they provide and the more accurate the positioning. The localization error bounds depend on the number of anchor nodes.

Privacy Preservation. To analyze the reliability of privacy protection of the PP-ADL, all nodes are honest but curious during the localization process. That means each node performs computation and communication according to the algorithm. However, each node is curious about whether it can obtain private information from the other nodes. We considered two scenarios, including independent and colluding nodes. For the independence node scenario, each node only accesses the private information of other nodes based on the legitimate information it obtains. For the colluding node scenario, the colluding nodes can interact with each other in order to obtain the private information of the other nodes. Therefore, we have the following theorem:

Theorem 3. *In the independence node scenario, when $m > 2$, the PP-ADL is privacy-protected. In other words, all anchor nodes cannot get the location of other anchor nodes, and the target node cannot obtain the location of the anchor node.*

Proof. During the PP-ADL localization, only the anchor node that performs the task can measure distance and angle information on the target node when the target node sends the crowdsourcing localization tasks. For independence nodes scenarios, anchor nodes cannot measure other anchor nodes. Therefore, the location information of other anchor nodes is privacy-protected. Additionally, as the anchor node sends T_{i0} to the target node after noise addition to the estimated location using Eq. (5). The target node cannot get information about its location from an anchor node. The target node cannot get the location of the anchor node, Even if the target node has distance and angle information.

Theorem 4. *In the colluding nodes scenario, when $m > 2$, the PP-ADL is privacy-protected. In other words, anchor nodes involved in collusion cannot obtain the location estimate of the target node.*

Proof. The anchor node in the PP-ADL algorithm takes the noise-added location T_{i0} as its private information interacts between colluding anchor nodes. However, with this information, the colluding node cannot obtain an estimate of the location of the target node.

4   Performance Evaluation

4.1   Simulation Setup

To evaluate and validate the PP-ADL localization algorithm, we conducted simulation experiments using MATLAB software running on a Dell desktop computer with Inter core i5-9500 CPU @ 3.00-GHz processor and 8.00 GB (7.81 GB available) RAM. We have set up a square two-dimensional space of $500 * 500$ m^2. All nodes are evenly distributed in this area, and the target node at the center. Anchor nodes are devices that support multi-antenna technology. These devices can obtain distance and angle information.

In the simulation experiments, the distance and angle measurements between the anchor node and the target node are got by simulating the actual distance and actual angle plus the measurement error, i.e. $d_i = s_i + \mu_i$, $\alpha_i = \beta_i + \eta_i$. When evaluating the performance of the PP-ADL localization algorithm, the measurement errors μ_i and η_i are assumed to follow a zero-mean normal distribution with standard deviations σ_1 and σ_2 respectively. The values of the above parameters do not affect the validation results of the simulation experiments. The results of all simulation experiments are an average of 10,000 independent experiments. The aim is to reduce the effect of the randomness of random variables on the results.

4.2   Evaluation Results

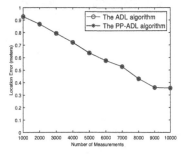
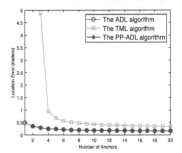

Fig. 2. Influence of the number of measurements on the localization error in three anchor nodes.

Fig. 3. Influence of the number of anchors on the localization error.

Location Error. To evaluate the effect of the number of distance and angle measurements on localization performance. The number of measurements is set to range from 1,000 to 10,000, each time increasing the number of measurements by 1,000. Figure 2 compares the change in location error for the ADL and the PP-ADL when there are three anchor nodes, and the number of measurements is increased. The location error of the PP-ADL is almost the same as the ADL. The more localization times, the error is closer to zero, which is due to both the distance measurement error and the angle measurement error are zero-mean errors.

To evaluate the impact of the number of anchor nodes on localization performance. The number of anchor nodes is set from 1 to 20, increasing 1 anchor node at a time. Figure 3 shows the location error of the ADL, the TML and the PP-ADL. Compare with the ADL and the traditional multilateration localization (TML) [22]. The location error of the PP-ADL is almost the same as the ADL and less than the TML. The more the number of anchor nodes, the more accurate localization. The more anchor nodes can provide more information to calculate the location of the target node.

These results demonstrate that the PP-ADL algorithm can achieve privacy-preserving while ensuring the correctness of the localization performance, validating the analysis of Theorem 1.

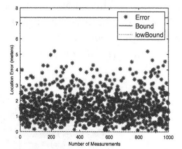

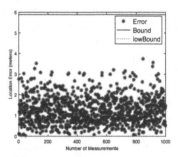

Fig. 4. The location error bound of the PP-ADL at 10 anchor nodes.

Fig. 5. The location error bound of the PP-ADL at 20 anchor nodes.

Error Bounds. Figure 4 and Fig. 5 respectively show the localization error bounds of the PP-ADL when the number of anchor nodes is 10 and 20. The results of the 1,000 independent runs demonstrate the accuracy of the PP-ADL algorithm in Theorem 2. The localization errors all lower than the upper bound, and higher than the lower bound. Additionally, Fig. 4 and Fig. 5 are shown that the latter's mean value and upper bound of the localization error are smaller than the former. The more anchor nodes contribute more information to the localization of the target node.

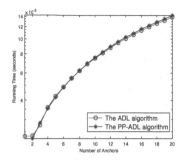

Fig. 6. Influence of the number of anchors on the running time.

Runtime. To evaluate the efficiency of the PP-ADL localization algorithm, Fig. 6 shows the running time of the algorithm when the number of anchor nodes varies from 1 to 20. The PP-ADL and the ADL have almost identical running times. As the number of anchor nodes gradually increases, the algorithm running time increases almost linearly. Therefore, our algorithm protects location privacy while ensuring no increase in the runtime.

5 Conclusion and Future Work

The development of multi-antenna technology has enabled single anchor node localization. In this paper, the privacy-preserving accurate and distributed localization algorithm (PP-ADL) is proposed to solve the location privacy problem of nodes during crowdsourcing localization. All of the anchor nodes involved in the localization cooperatively generate a zero-sum random noise set. The anchor node adds these noises respectively before sending the estimated location which protects the location privacy of the anchor node. The performance of the PP-ADL algorithm is analyzed in terms of localization accuracy and privacy preservation. Finally, simulations illustrate the accuracy and effectiveness of the proposed scheme.

In future work, consider the impact of various factors on localization results. For example, as time goes by, whether the nodes in the network are trusted, whether the node information changes.

References

1. Varma, P.S., Anand, V.: Random forest learning based indoor localization as an IoT service for smart buildings. Wirel. Pers. Commun. **117**(4), 3209–3227 (2021)
2. Ghorpade, S.N., Zennaro, M., Chaudhari, B.S.: GWO model for optimal localization of IoT-enabled sensor nodes in smart parking systems. IEEE Trans. Intell. Transp. Syst. **22**(2), 1217–1224 (2020)

3. Liu, X., Chen, J., Xia, X., Zong, C., Zhu, R., Li, J.: Dummy-based trajectory privacy protection against exposure location attacks. In: Ni, W., Wang, X., Song, W., Li, Y. (eds.) WISA 2019. LNCS, vol. 11817, pp. 368–381. Springer, Cham (2019). https://doi.org/10.1007/978-3-030-30952-7_37

4. Pandey, P., Tiwary, P., Kumar, S., Das, S.K.: Residual neural networks for heterogeneous smart device localization in IoT networks. In: 29th International Conference on Computer Communications and Networks (ICCCN), Honolulu, United States, Hawaii, pp. 1–9. IEEE (2020)

5. Shahmansoori, A., Garcia, G.E., Destino, G., Scco-Granados, G., Wymeersch, H.: Position and orientation estimation through millimeter wave MIMO in 5G systems. IEEE Trans. Wireless Commun. **17**(3), 1822–1835 (2017)

6. Abu-Shaban, Z., Wymeersch, H., Abhayapala, T., Seco-Granados, G.: Single-anchor two-way localization bounds for 5G mmwave systems. IEEE Trans. Veh. Technol. **69**(6), 6388–6400 (2020)

7. Wang, J., Xiong, J., Jiang, H., Chen, X., Fang, D.: D-watch: embracing "bad" multipaths for device-free localization with cots RFID devices. IEEE/ACM Trans. Netw. **25**(6), 3559–3572 (2017)

8. Vasisht, D., Kumar, S., Katabi, D.: Decimeter-level localization with a single WiFi access point. In: 13th {USENIX} Symposium on Networked Systems Design and Implementation ({NSDI} 2016), Santa Clara, CA, USA, pp. 165–178. USENIX Association (2016)

9. Wen, F., Wymeersch, H., Peng, B., Tay, W.P., So, H.C., Yang, D.: A survey on 5G massive MIMO localization. Digital Signal Process. **94**, 21–28 (2019)

10. Wymeersch, H., Seco-Granados, G., Destino, G., Dardari, D., Tufvesson, F.: 5G mmWave positioning for vehicular networks. IEEE Wireless Commun. **24**(6), 80–86 (2017)

11. Deng, Z., Zheng, X., Zhang, C., Wang, H., Yin, L., Liu, W.: A TDOA and PDR fusion method for 5G indoor localization based on virtual base stations in unknown areas. IEEE Access **8**, 225123–225133 (2020)

12. Witrisal, K., Antón-Haro, C., Peral-Rosado, J., Seco-Granados, G., Sackenreuter, B.: Whitepaper on New Localization Methods for 5G Wireless Systems and the Internet-of-Things (2018)

13. Strous, L., von Solms, S., Zúquete, A.: Security and privacy of the internet of things. Comput. Secur. **102**, 102148 (2020)

14. Shi, X., Tong, F., Zhang, W.A., Yu, L.: Resilient privacy-preserving distributed localization against dishonest nodes in internet of things. IEEE Internet Things J. **7**(9), 9214–9223 (2020)

15. Sazdar, A.M., Alikhani, N., Ghorashi, S.A., Khonsari, A.: Privacy preserving in indoor fingerprint localization and radio map expansion. Peer Peer Netw. Appl. **14**(1), 121–134 (2021)

16. Zhang, G., Zhang, A., Zhao, P., Sun, J.: Lightweight privacy-preserving scheme in Wi-Fi fingerprint-based indoor localization. IEEE Syst. J. **14**(3), 4638–4647 (2020)

17. Wang, G., He, J., Shi, X., Pan, J., Shen, S.: Analyzing and evaluating efficient privacy-preserving localization for pervasive computing. IEEE Internet Things J. **5**(4), 2993–3007 (2018)

18. Pandey, A., Tiwary, P., Kumar, S., Das, S.K.: Accurate position tracking with a single UWB anchor. In: 2020 IEEE International Conference on Robotics and Automation (ICRA), Paris, France, pp. 2344–2350. IEEE (2020)

19. Kaur, A., Kumar, P., Gupta, G.P.: A new localization using single mobile anchor and mesh-based path planning models. Wireless Netw. **25**(5), 2919–2929 (2019)

20. Tong, F., Wang, G., Shi, X.: A novel single anchor localization mechanism employing target movement. In: 2019 IEEE 20th International Conference on High Performance Switching and Routing (HPSR), Xi'an, China, pp. 1–5. IEEE (2019)
21. Wang, G., Xu, Y., Tong, F., Pan, J., Shen, S.: Modeling and analyzing single anchor localization for internet of things. In: ICC 2019–2019 IEEE International Conference on Communications (ICC), Shanghai, China, pp. 1–6. IEEE (2019)
22. Liu, Y., Yang, Z., Wang, X., Jian, L.: Location, localization, and localizability. J. Comput. Sci. Technol. **25**(2), 274–297 (2010)

Multi-user Fully Homomorphic Encryption Scheme Based on Policy for Cloud Computing

Taoshen Li[1,2(✉)], Qing Liu[2], and Ruwei Huang[2]

[1] School of Information Engineering, Nanning University, Nanning 530299, China
tshli@gxu.edu.cn

[2] School of Computer, Electronics and Information, Guangxi University,
Nanning 530004, China

Abstract. Cloud computing, a modern application that is very commonly utilized, has garnered considerable interest from researchers and developers. Data privacy security is the most urgent issue related to cloud computing. Studies have shown that fully homomorphic encryption (FHE) technology, which is related to traditional encryption and ciphertext evaluation, is an effective way of securing privacy in the cloud. This study proposes a policy-based, multi-user, FHE algorithm (PB-MUFHE) that accounts for user diversity in the cloud environment, in which, the encrypted data is set under an appropriate access policy, the key to which is a set attribute. Not only is homomorphic evaluation supported for ciphertexts among several users, but also fine-grained access control and sharing between users are supported. The security of the PB-MUFHE scheme was tested based on learning with errors (LWE) problems, and IND-CPA security analysis in the generic bilinear group random oracle model further verified the proposed algorithm's effectiveness. Performance assessment demonstrated that, PB-MUFHE can efficiently implement fully homomorphic evaluation of ciphertext, and effectively support access control and shared multi-user capability.

Keywords: Fully homomorphic encryption · Cloud computing · Attribute-based encryption · Access control · Multi-user

1 Introduction

Cloud computing is a model for enabling ubiquitous, convenient, on-demand network access to a shared pool of configurable computing resources that can be rapidly provisioned and released with minimal management effort or service provider interaction. Data privacy security is the most urgent issue related to cloud computing. For example, cipher text storage will lead to the inconvenience to data sharing and access [1], and the risk of data leakage is particularly concerning when data is being evaluated in the cloud.

Typically, user privacy can be protected effectively through encryption. Most encryption programs do not support ciphertext evaluations, however, making traditional encryption technology unable to meet the security needs of cloud environments. In theory, fully homomorphic encryption (FHE) technology is suited to addressing-privacy issues in the

© Springer Nature Switzerland AG 2021
C. Xing et al. (Eds.): WISA 2021, LNCS 12999, pp. 274–286, 2021.
https://doi.org/10.1007/978-3-030-87571-8_24

cloud environment by performing not only traditional encryption functions, but also supporting ciphertext evaluation. The features of FHE include evaluation results on plaintext equivalent to decrypted ciphertext results, support of ciphertexts evaluation, retrieval, and sorting, and secure transmission between memory and external memory.

The FHE scheme in the true sense was developed at first in [2]. Many subsequent, improved FHE schemes have been proposed, and the homomorphic encryption is now a rather hot topic in the cryptography field. Schemes previously proposed on the premise of the "Sparse Subset-Sum" problem (SSSP) assumption have yielded pure FHE based "squash" and bootstrapping technology [3, 4], which do support infinite full homomorphic operation, but are excessively making method and highly computational complex. Other schemes have obtained leveled FHE based on Learning With Error (LWE) problems or Ring-Learning With Error (R-LWE) problems [5–7], all of which show high computational efficiency and high safety, but oversize public key. Based on the LWE problem, [8] proposed a leveled certifycateless FHE (CLFHE) schemes in the random oracle model and a leveled CLFHE schemes in the standard model. Multi-key FHE schemes [9–11] have been proposed. The aforementioned schemes can serve only a single recipient. Actual cloud environment shave functions that may need to serve several users, such as access control, computing multi-user ciphertext, and multi-user sharing, so a FHE scheme for any single user receiving data in the cloud is still not quite sufficiently secure.

In [12], a bootstrappable, identity-based FHE scheme has been presented, which meets shared multi-user needs and supports ciphertexts evaluation with different identities or attributes, but the base relies heavily on in distinguish ability obfuscation, and is highly inefficient. [11] proposed a multi-identity and multi-key leveled FHE scheme, which does allow multiple-users and support ciphertext evaluation with different identities. Yet some efficient leveled identity-based FHE schemes have been proposed in [13–16], which is identity-based and attribute-based, supports shared multi-user functionality, and allows ciphertext computing according to similar identities or attributes.

In this paper, we propose a policy-based, multi-user, FHE scheme (PB-MUFHE) that performs homomorphic ciphertext evaluation, supports fine-grained access control, and allows for shared needs among many users. The scheme is based on the hardness of learning with errors problem, and was proven effective according to IND-CPA security in the generic bilinear group random oracle model. The following sections demonstrate the PB-MUFHE algorithm's capability to address privacy issues in cloud environments.

2 Preliminaries

2.1 Notations

Normally, vectors use lower-case letters in bold, (e.g., v.) The ith norm of v is denoted $v[i]$. Matrixes use upper-case letters in bold, (e.g., A.) Z_p is a set of integer modulus p, and Z_p^n is a set of integer modulus p with n dimensions. $\langle a,b \rangle$ denotes the dot product of a,b.

Let $\mathbb{G}_0$ and $\mathbb{G}_1$ be two multiplicative cyclic groups of prime order p, and let g be a generator of $\mathbb{G}_0$ and e be a bilinear map, $e\colon \mathbb{G}_0 \times \mathbb{G}_0 \to \mathbb{G}_1$. The bilinear map e has the following properties:

- Bilinearity: for all $u,v \in \mathbb{G}_0$ and $a,b \in Z_p$, there is $e(u^a, v^b) = e(u, v)^{ab}$.
- Non-degeneracy: $\exists g \in \mathbb{G}_0$, there is $e(g, g) \neq 1$.
- Computability: for all $g \in \mathbb{G}_0$, there is valid operation $e(g,g)$.
  Note: $e(*,*)$ is symmetrical, because $e(g^a, g^b) = e(g, g)^{ab} = e(g^b, g^a)$.

2.2 Learning with Errors (LWE) Problems

Definition 1. Decisional Learning with Errors (LWE) Problem [17]: For security parameter λ, let $n = n(\lambda)$ be an integer dimension, let $q = q(\lambda) \geq 2$ be an integer, and let $\chi = \chi(\lambda)$ be a distribution over Z. The $LWE_{n,q,\chi}$ problem is to distinguish two distributions. In the first distribution, one samples (a_i, b_i) uniformly from Z_q^{n+1}. In the second distribution, one first draws $s \leftarrow Z_q^n$ uniformly and then samples $(a_i, b_i) \in Z_q^{n+1}$ by sampling $a_i \leftarrow Z_q^n$ uniformly, $e_i \leftarrow \chi$, and setting $b_i = \langle a_i, s \rangle + e_i$. The $LWE_{n,q,\chi}$ assumption is that the $LWE_{n,q,\chi}$ problem is infeasible.

Definition 2. B-bounded Distributions Problem [17]: A distribution ensemble $\{\chi_n\}_{n \in N}$, supported over the integers, is called B-bounded if $\mathbf{Pr}_{e \leftarrow \chi_n}[|e| > B] = \mathbf{negl}(n)$.

Theorem 1 [18]. Let $q = q(n) \in N$ be either a prime power or a product of small, size $poly(n)$, distinct primes, and let $B \geq \omega(\log n) \cdot \sqrt{n}$. There then exist an efficient sampleable B-bounded distribution χ as such that if there is an efficient algorithm that solves the average-case $LWE_{n,q,\chi}$ problem, then, there is an efficient quantum algorithm that solves $GapSVP_{\tilde{O}(nq/B)}$ on any n-dimensional lattice, and if $q \geq (2^{n/2})$, then there is an efficient classical algorithm for $GapSVP_{\tilde{O}(nq/B)}$ any n-dimensional lattice.

Theorem 2. Solving n-dimensional LWE with $poly(n)$ modulus implies an equally efficient solution to a worst-case lattice problem (e.g., $GapSVP$,) in dimension $\sqrt{n}$.

2.3 Flatten Cphertext

As discussed in [17], the growth of the error vector after ciphertext computing is related to the correctness of the decryption. It is necessary, as such, to ensure bounds on error growth. Let a ciphertext C be B-strongly-bounded if its associated μ and the coefficients of C all have magnitude at most 1, while the coefficients of its e all have magnitude at most B. The following section describes an operation called ciphertext flattening that keeps ciphertexts strongly bounded.

Flattening uses some simple transformations that modify vectors without affecting dot products [17]. Let a, b be vectors of dimension k over Z_q, and let $l = \lfloor log_2 q \rfloor + 1$ and $N = k \cdot l$. Let $BitDecomp(a)$ be the N-dimensional vector $(a_{1,0}, ..., a_{1,l-1}, ..., a_{k,0}, ..., a_{k,l-1})$, where $a_{i,j}$ is the j-th bit in a_i's binary representation, and bits are ordered from least significant to most significant. For $a' = (a_{1,0}, ..., a_{1,l-1}, ..., a_{k,0}, ..., a_{k,l-1})$, let $BitDecomp^{-1}(a') = (\Sigma 2^j \cdot a_{1,j}, ..., \Sigma 2^j \cdot a_{k,j})$ be the inverse of $BitDecomp$, but well-defined even when the input is not a 0/1 vector. For N-dimensional a', let $Flatten(a') = BitDecomp(BitDecomp^{-1}(a'))$, an N-dimensional vector with 0/1 coefficients. When A is a matrix, let $BitDecomp(A)$, $BitDecomp^{-1}(A)$, and $Flatten(A)$ be the matrix formed by applying the operation to each row of A separately. Finally, let $Powerof2(\boldsymbol{b}) = (b_1, 2b_1, ..., 2^{l-1}b_1, ..., b_k, 2b_k, ..., 2^{l-1}b_k)$ be a N-dimensional vector. Keep in mind the following:

$$\langle BitDecomp(\boldsymbol{a}), Powerof2(\boldsymbol{b}) \rangle = \langle \boldsymbol{a}, \boldsymbol{b} \rangle.$$

For any N-dimensional $\boldsymbol{a'}$, $\langle \boldsymbol{a'}, Powerof2(\boldsymbol{b}) \rangle = \langle BitDecomp^{-1}(\boldsymbol{a'}), \boldsymbol{b} \rangle = \langle Flatten(\boldsymbol{a'}), Powerof2(\boldsymbol{b}) \rangle$.

3 Issues Description

3.1 Cloud Privacy Protection Support Model

Figure 1 models the privacy protection support scheme in a typical cloud computing environment, reflecting the interaction among data owners, users, certification center, and the clouds.

- The certification center uses the *setup* algorithm to generate the public parameters *pp* and the master key *msk*.
- When data encryption owners request certification center provide public key support, the certification center uses the *PubGen* algorithm to return the public key for the data owner. The data owners use the *Enc* algorithm encrypt sensitive data $m(i)$ to produce ciphertext $C(i)$, which is then stored in the cloud.
- Per user request, the certification center provides the private key or the computing key support, by using the *PrvGen* algorithm to return the private key *prvKey* to the user.
- Users upload operation f. According to the operation, the cloud uses the *Compute* algorithm to manage $C(i)$, get resulting inciphertext $C(result)$.
- Users use the private key *prvKey*, and the *Dec* algorithm to validatethe decryption authority, then decrypt the $C(result)$, get the plaintext$result$.

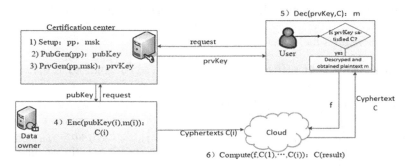

Fig. 1. Privacy protection support in cloud computing environment

During the aboveinteraction process, sensitive data is encrypted by the data owner, protecting user privacy while calculating ciphertext. There are a few problems inherent to this process, however.

- Cloud services can process any ciphertext without the authorization of the data owner, which leaves user data vulnerable to internal attack if the cloud provider has the decryption key.
- The system may not meet the multi-user sharing needs effectively, particularly if applied to a large organization.

The following section details our proposed solution to the above problems, the policy-based, multi-user, fully homomorphic encryption (PB-MUFHE) scheme.

3.2 Definitions of PB-MUFHE

Definition 3. Policy-based, Multi-user, FHE (PB-MUFHE) for Cloud Computing. Figure 2 models the cloud privacy protection support for PB-MUFHE.

The PB-MUFHE scheme is a tuple of probabilistic polynomial time (PPT) algorithms $\mathcal{E}_{pb\text{-}mufhe}\{Setup, PubGen, PrvGen, Enc, Dec, Compute\}$ with the following characteristics.

- The *Setup* algorithm allows all users to generate the public parameter *pp* and the master key *msk*, $(pp, msk) \leftarrow Setup(1^\lambda, 1^L, params)$, and *params* as a security parameter.
- The public key generation algorithm *PubGen* generates *pubKey* for the data owner, $pubKey \leftarrow PubGen(pp)$, where *pp* is the set of common parameters.
- The key generation algorithm *PrvGen* generates a decryption key *prvKey* and evaluation key *evalKey* for the user, $(prvKey, evalKey) \leftarrow prvGen(pp, msk, attrs)$, where *attrs* is a set of attribute strings.
- The encryption algorithm *Enc* is probabilistic, *D* is the domain of the plaintext data, and data $\mu \in D$, $C \leftarrow Enc(pubKey, policy, \mu)$, where *policy* is the access policy and μ is the plaintext data.
- The decryption algorithm *Dec* is deterministic, for the ciphertext *C*. $\mu \cup \{\bot\} \leftarrow Dec(prvKey, C)$, where $\bot$ represents no solution.

The ciphertext computation algorithm *Compute* may be probabilistic, for the ciphertextset $\{C_1, C_2, ..., C_l\}$, $C' \leftarrow Compute(op, C_1, C_2, ..., C_l, evalKey_1, evalKey_2, ..., evalKey_l)$, where *op* is the operator, and C' is the result.

Definition 4. The PB-MUFHE scheme's correctness can be validated as follows.

- $\forall \mu \in D, \exists Dec(prvKey, Enc(pubKey, policy, \mu)) = \mu$.
- $\forall \{\mu_1, \mu_2, ..., \mu_l\}(\mu_i \in D), \exists Compute'(op, \mu_1, \mu_2, ..., \mu_l) = Dec(prvKey, Compute(op, Enc(pubKey_1, policy_1, \mu_1), Enc(pubKey_2, policy_2, \mu_2), ..., Enc(pubKey_l, policy_l, \mu_l), evalKey_1, evalKey_2, ..., evalKey_l))$, where op is the operator and Compute' is the corresponding plaintext operation algorithms with Compute.

Definition 5. The PB-MUFHE scheme's IND-CPA security can be validated as follows.

Initialization: The challenger runs the *Setup* algorithm and gives the public parameters, *PP* to the adversary.

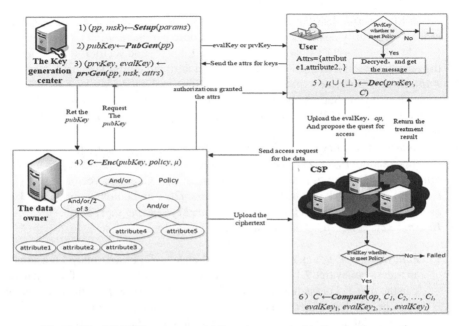

Fig. 2. The PB-FHE system model for privacy protect in the cloud computing

Phase 1: The adversary makes repeated attribute secret keys corresponding to sets of attributes $V_1,...,V_{q1}$.

Challenge: The adversary submits two equal length messages m_0 and m_1. In addition the adversary gives a challenge access structure T^* such that none of the sets $V_1,...,V_{q1}$ from Phase1 satisfy the access structure. The challenger flips a random coin b, and encrypts m_b under T^*. The access control ciphertext CT^* is given to the adversary.

Phase 2: Phase 1 is repeated with the restriction that none of the sets of attributes V_{q1+1}, $...,V_q$ satisfy the access structure corresponding to the challenge.

Guess: The adversary outputs a guess b' of b.

4   The Construction of PB-MUFHE

For $i \in Z_p$, the set is $S \leftarrow Z_p$ and the Lagrange coefficient defined as: $\Delta_{i,S(x)} = \prod_{j \in S, j \neq i}(x-j)/(i-j)$. Next, set the random field model. For hash function $H:\{0,1\}^* \to \mathbb{G}_0$, the function will map any attribute described as a binary string to a random group element.

The PB-MUFHE then runs in the following steps.

(1) **Setup**$(1^\lambda, 1^L, params)$: Choose a modulus q of $k = k(\lambda, L)$ bits, and a bilinear group $\mathbb{G}_0$ of prime order p with generator g, and two random exponents $\alpha, \beta \in Z_p$. Obtain $mpk = (\mathbb{G}_0, g, h = g^\beta, e(g,g)^\alpha)$ and $msk = (\beta, g^\alpha)$. Set lattice dimension parameter n

$= n(\lambda, L)$, and error distribution $\chi = \chi(\lambda, L)$ appropriately for an LWE that achieves at least 2^λ security against known attacks. Choose parameter $m = O(nlogq)$, $N = (n+1) \cdot (\lfloor logq \rfloor + 1)$. Generate matrix $E_r \in \{0, 1\}^{N \times N}$ and vector $t \leftarrow Z_q^n$, output the common parameters $pp = (mpk, t, E_r)$, and attribute master key msk.

(2) **PubGen**(pp): Generate a matrix $B \leftarrow Z_q^{m \times n}$ uniformly and a vector $e \leftarrow \chi^m$. Set vector $b = B \cdot t + e$. Set A to be the $(n+1)$-column matrix consisting of b followed by the n columns of B, set $A = (b|B)$. Output the public key $pubKey = (mpk, E_r, A)$.

(3) **PrvGen**$(pp, msk, attrs)$: First choose random $r \in Z_p$, for attribute $attr \in attrs$ and generate random $r_{attr} \in Z_p$. Then, set attribute secret key $abe\_sk = (D_{\alpha\beta} = {}^{(\alpha+\beta)/\beta}$, $\forall attr \in attrs: D_{attr} = g^r H(attr)^r_{attr}$, $D'_{attr} = g^r_{attr})$. Generate vector $s \leftarrow (1, -t_1, \dots, -t_n) \in Z_q^{n+1}$, where $t_i \in t$. Set $v = Powersof2(s)$, and output computation key $evalKey = (abe\_sk, E_r)$ and decryption key $prvKey = (abe\_sk, v, E_r)$.

(4) **Enc**$(pubKey, policy, \mu)$: That is implemented as follows.

   1) **PolicyEnc**$(mpk, policy, m_r)$: Encrypt random $m_r \in \mathbb{G}_1$ based policy. First, create an access policy tree T based policy, then choose a polynomial q_x for each node x in the tree T. For each node x in the tree, set the degree d_x of the polynomial q_x to be one less than the threshold value k_x of that node, that is $d_x = k_x - 1$. Starting with the root node r, choose a random $s \in Z_p$, sets $q_R(0, \dots, d_R - 1) = s$, for another node x, sets $q_x(0) = q_{parent(x)}(index(x))$, $q_x(1, \dots, d_x - 1) = s$. Let Y be the set of leaf nodes in T, then the ciphertext of random m_r is constructed by giving the tree access structure T and computing $C_t = (T, C_{mr} = m_r \cdot e(g, g)^{\alpha \cdot s}, C_s = h^s, \forall y \in Y: C_y = g^{q_y(0)}, C'_y = H(attr(y))^{q_y(0)})$.

   2) **PlaintextEnc**(m_r, E_r, A, μ): For encrypted message $\mu \in Z_q$, sample a uniform matrix $R \in \{0, 1\}^{N \times m}$, then the ciphertext of message μ computes $C_\mu = Flatten(\mu \cdot I_N + BitDecomp(R \cdot A) + m_r \cdot E_r) \in Z_q^{m \times n}$.

   3) Output the ciphertext $C = (C_t, C_\mu)$.

(5) **Dec**$(prvKey, C)$. That is implemented as follows.

   1) **policyDec**(abe_sk, C_t): Obtain m_r from decrypted ciphertext C_t. First, define a recursive algorithm $decryNode(C_t, abe\_sk, x)$. If the node x is a leaf node then let $i = attr(x)$, and if $i \in attrs$ define it as $decryptNode(C_t, abe\_sk, x) = e(D_i, C_x)/e(D'_i, C'_x) = e(g^r \cdot H(i)^r_i, h^q_x(0))/e(g^r_i, H(i)^{q_x(0)}) = e(g, g)^{r \cdot q_x(0)}$. If $i \notin attrs$, then define $decryptNode(C_t, abe\_sk, x) = \perp$. When x is a non-leaf node, call all nodes z that are children of x, $decryptNode(C_t, abe\_sk, z)$ and store the output as F_z. Let S_x be an arbitrary k_x-sized set of child nodes z such that $F_z \neq \perp$. If no such set exists, then the node is not satisfied and the function returns $\perp$. Otherwise, compute $F_x = \prod_{z \in Sx} F_z^{\Delta i, S'x(0)} = \prod_{z \in Sx}(e(g,g)^{r \cdot q_z(0)})^{\Delta i, S'x(0)} = \prod_{z \in Sx}(e(g,g)^{r \cdot q_{parent(z)}(index(z))})^{\Delta i, S'x(0)} = \prod_{z \in Sx} e(g,g)^{r \cdot q_x(i) \cdot \Delta i, S'x(0)} = e(g,g)^{r \cdot q_x(0)}$, where $i = index(z)$ and $S'_x = \{index(z): z \in S_x\}$. To decrypt C_t and obtain m_r, simply call the function on the root node r of the tree T. If the tree is satisfied by $attrs$, set $U = decryptNode(C_t, abe\_sk, r) = e(g,g)^{r \cdot q_r(0)} = e(g,g)^{r \cdot s}$. Then compute $C_{mr}/(e(C_s, D_{\alpha\beta})/U) = (m_r \cdot e(g,g)^{\alpha \cdot s})/(e(h^s, g^{(\alpha+r)/\beta})/e(g,g)^{r \cdot s})$ to reach the result m_r.

2) **decPlaintext**(C_μ, v, m_r, E_r). Obtain the message from decrypted ciphertext C_μ. First, set $C'_\mu = C_\mu - Flatten(m_r \cdot E_r)$ and $x = C'_\mu \cdot v = \mu \cdot v + R \cdot A \cdot s = \mu \cdot v + R \cdot e$. Following the treatment of vector x, $bit_i = (min(x_i, q - x_i) \geq |x_i - q/2|)$, to reach $result + = bit_i \cdot 2^{l-1-i}$, where $i = \{l - 2, ..., 0\}$.

3) **Compute**$(policy, op, C_1, C_2, evalKey_1, evalKey_2)$. That is implemented as follows.

    (a) First call the *policyDec* algorithm to confirm whether the computation keys *evalKey* meet the access control requirements of ciphertexts C_1 and C_2. If satisfied, then decrypt the access control ciphertext C_t in C_1 and C_2, and acquire r_1 and r_2; if not, output $\perp$.

    (b) If *policyDec* doesn't output $\perp$, use the *PolicyEncry* algorithm to encrypt random $r \in \mathbb{G}_1$ under *policy*, then ciphertext C'_t is the access control ciphertext of the resulting ciphertext.

    (c) If *policyDec* doesn't output $\perp$. Sets $C'\mu_1 = C\mu_1 - Flatten(r_1 \cdot E_r)$ and $C'\mu_2 = C\mu_2 - Flatten(r_2 \cdot E_r)$, then $C'\mu_1$ and $C'\mu_2$ all satisfy matrix operation, because $op = \{+, \cdot\}$, set $C_{add} = C'\mu_1 + C'\mu_2$, $C_{mul} = C'\mu_1 \cdot C'\mu_2$. Next, compute $C'_{add} = C_{add} + Flatten(r \cdot E_r)$, $C'_{mul} = C_{mul} + Flatten(r \cdot E_r)$.

Output the resultant ciphertext $C' = (C'_t, \{C'_{add}, C'_{mul}\})$ after computation.

5 The Proof of Correctness and Security

5.1 Correctness

Theorem 3. *PB-MUFHE* $= (Setup, PubGen, PrvGen, Enc, Dec, Compute)$ is correct, $\forall \mu \in D, \exists Dec(prvKey, Enc(pubKey, policy, \mu)) = \mu$.

Proof. Algorithm *Dec* includes two parts: Decrypting the access control ciphertext C_t and the message ciphertext C_μ. The decryption of C_t is the basis of the decryption of C_μ, as only when the correct m_r is obtained from C_t can C_μ be decrypted correctly.

First, using attribute key *abe_sk*, decrypting C_t is a recursive decryption process. The key to successful decryption is that D_{attr} in the attribute key must meet the access tree ciphertext C'_y $(C'_y = H(attr(y))^q y^{(0)}))$, providing the result set $U = e(g, g)^{r \cdot s}$. The attributes must meet the access policy requirements according to the symmetry of the billinear maps and polynomial interpolation, allowing random m_r from ciphertext C_t by $C_{mr}/(e(C_s, D_{\alpha\beta})/U)$.

Second, C_μ is a matrix, so compute $C'_\mu = C_\mu - Flatten(m_r \cdot E_r) = Flatten(\mu \cdot I_N + BitDecomp(R \cdot A))$. Because $C'_\mu \cdot v = Flatten(\mu \cdot v + R \cdot e)$, the elements of the matrix $\mu \cdot v + R \cdot e$ is small (not less than 1) according to the *Flatten* operations, so the secret key is v is an N-dimensional vector over Z_q with at least one "big" coefficient v_i. Because the secret key v is the approximate eigenvector of the ciphertext matrix $C'\mu$, and the message μ is the eigenvalue, extract the i-th row C_i from C'_μ and compute $x \leftarrow \langle C_i, v \rangle = \mu \cdot v_i + e_i$ to output $\mu = \lfloor x/v_i \rceil$.

So, $\forall \mu \in Z_q, \exists Dec(prvKey, Enc(pubKey, policy, \mu)) = \mu$.

Theorem 4. $PB\text{-}MUFHE = (Setup, PubGen, PrvGen, Enc, Dec, Compute)$ is correct, $\forall \{\mu_1, \mu_2,..., \mu_l\}(\mu_i \in D)$, $\exists Compute'(op, \mu_1, \mu_2, ..., \mu_l) = Dec(prvKey, Compute(op, Enc(pubKey_1, policy_1, \mu_1), Enc(pubKey_2, policy_2, \mu_2), ..., Enc(pubKey_l, policy_l, \mu_l), evalKey_1, evalKey_2, ..., evalKey_l))$.

Proof. Algorithm *Compute* includes three parts, ciphertext access control authentication, new access control establishment, and ciphertext addition and multiplication.

First, for ciphertext access control authentication, obtain the random $r_1, r_2,...,r_l$ from C_t as the access control ciphertext of ciphertext C_i. The process is similar as the above proof of the algorithm *Dec*.

Second, establish the new access control by using the new *policy* to encrypt the new random r that obtains access control ciphertext C'_t, then the proof of the process.

is the same as that of algorithm *Enc*.

Next, obtain the $r_1, r_2,..., r_l$ from above, then compute $C'_{\mu i} = C_{\mu i} - Flatten(r_i \cdot E_r) = Flatten(\mu_i \cdot I_N + BitDecomp(R_i \cdot A_i))$, and $C'_{\mu i} \cdot v = Flatten(\mu_i \cdot v + e_i)$, set $C^+ = C'_{\mu 1} + C'_{\mu 2}$,

$C^\times = C'_{\mu 1} \cdot C'_{\mu 2}$. For addition, $C^+ \cdot v = Flatten((\mu_1 + \mu_2) \cdot v + (e_1 + e_2))$; error vector e_i is small, and error vector $e_1 + e_2$ remains small after addition by *Flatten* operations. The sum of the ciphertext matrixes then encrypts the sum of the messages. For multiplication, $C^\times \cdot v = Flatten(C'_{\mu 1} \cdot (\mu_2 \cdot v + e_2)) = Flatten(\mu_2 \cdot (\mu_1 \cdot v + e_1) + C_1 \cdot e_2) = Flatten(\mu_1 \cdot \mu_2 \cdot v + \mu_2 \cdot e_1 + C'_{\mu 1} \cdot e_2)$, where, of course, the μ_2 and $C'_{\mu 1}$ are very small after *Flatten* operations, so $C^\times \cdot v = \mu_1 \cdot \mu_2 \cdot v + small$. The product of the ciphertext matrixes then encrypts the product of the messages.

So, $\forall \{\mu_1, \mu_2,..., \mu_l\}(\mu_i \in D)$, $\exists Compute'(op, \mu_1, \mu_2, ..., \mu_l) = Dec(prvKey, Compute(op, Enc(pubKey_1, policy_1, \mu_1), Enc(pubKey_2, policy_2, \mu_2), ..., Enc(pubKey_l, policy_l, \mu_l), evalKey_1, evalKey_2, ..., evalKey_l))$.

5.2 Security

Theorem 5. Any distinguishing algorithm with advantage ϵ against the IND-CPA security of the scheme with parameters n, m, q and χ can be converted to a distinguisher against $LWE_{n,m,q,\chi}$ with roughly the same running time and advantage of at least $p\epsilon/2m\zeta^2$, where ζ is a bound on the total number of group elements received from queries made to the oracle for the hash function, and p is the prime order from $\mathbb{G}_0$.

Proof. Let be a CPA-adversary that distinguishes between encryptions of messages of its choice with advantage ϵ. First construct a distinguisher with advantage of at least $\epsilon/2$ between the two distributions $\{(A, R \cdot A + m_r \cdot E_r): A \leftarrow (b|B), b \leftarrow B \cdot t + e, B \leftarrow Z_q^{m \times n}, e \leftarrow \chi^m, t \leftarrow Z_q^n, m_r \leftarrow \mathbb{G}_0 \times \mathbb{G}_0, E_r \leftarrow \{0, 1\}^{N \times N}\}$ and $\{Z_q^{m \times (n+1)}, Z_q^{N \times (n+1)}\}$. The distinguisher takes as input a pair of matrices $(A \in Z_q^{m \times (n+1)}, C \in Z_q^{N \times (n+1)})$, and runs the adversary $\mathscr{A}$ with A as the public key. Upon receiving message $I_N \cdot \mu_0, I_N \cdot \mu_1$ from the adversary, $\mathscr{D}$ chooses at random $i \in_R \{0, 1\}$, returns the challenge ciphertext $C + I_N \cdot \mu_i$ mod q, then outputs 1 if the adversary guesses the right i, and 0 otherwise.

On the one hand, if C is a uniformly random matrix, then the challenge ciphertext is also uniformly random regardless of the choice of i. In this case, outputs 1 with probability at most $1/2$. On the other hand, if $C = R \cdot A + m_r \cdot E_r \bmod q$, then the challenge ciphertext is $C + I_N \cdot \mu \bmod q$. This is identical to the output distribution of $PlaintextEnc(m_r, E_r, A, \mu)$, hence, by assumption, $\mathscr{A}$ will guess the right i with probability $(1 + \epsilon)/2$, which means that $\mathscr{D}$ outputs 1 with the same probability. To this effect, has advantage of at least $\epsilon/2$. In addition, $m_r \in \mathbb{G}_1 \leftarrow \mathbb{G}_0 \times \mathbb{G}_0$, ζ is a bound on the total number of group elements received from queries made to the oracle for the hash function, and p is the prime order from $\mathbb{G}_0$, so $\mathscr{D}$ has advantage of at least $p\epsilon/2\zeta^2$.

A standard hybrid argument can be used to convert the distinguisher $\mathscr{D}$ to a $LWE_{n,m,q,\chi}$ distinguisher with advantage $p\epsilon/2m\zeta^2$.

Theorem 6. Let two random encodings $\psi_{0,1}$ of the additive group $\mathbb{F}_p$ be injective maps $\psi_0, \psi_1 : \mathbb{F}_p \longrightarrow \{0, 1\}^k$, where $k > 3log(p)$. For $i = 0, 1$, write $\mathbb{G}_i = \{\psi(x) : x \in \mathbb{F}_p\}$. Let ζ be a bound on the total number of group elements received from queries made to the oracle for the hash function, groups $\mathbb{G}_0$ and $\mathbb{G}_1$ and bilinear map e, and from its interaction with the security game. The advantage of the adversary in the security game is then $O(\zeta^2/p)$.

Proof. In the security game, the access control challenge ciphertext has a component C_{mr} which is randomly either $m_0 e(g,g)^{\alpha \cdot s}$ or $m_1 e(g,g)^{\alpha \cdot s}$. Consider instead a modified game, in which C_{mr} is either $e(g,g)^{\alpha \cdot s}$ or $e(g,g)^{\theta}$, where θ is selected uniformly at random from $\mathbb{F}_p$, and the adversary must decide which is the case. Clearly, any adversary that has advantage ϵ in the original game can be transformed into an adversary that has advantage of at least $\epsilon/2$ in the modified game.

Initialization: The simulation chooses α, β at random from $\mathbb{F}_p$. The public parameters $h = g^{\beta}$, $e(g,g)^{\alpha}$ are sent to the adversary.

Phase 1: The adversary calls for the evaluation of H on any string i, a new random value w_i is chosen from $\mathbb{F}_p$, and the simulation provides g^{w_i} as the response to $H(i)$.

Phase 2: The adversary makes its j'th key generation query for the set S_j of attributes, a new random value $r^{(j)}$ is chosen from $\mathbb{F}_p$, and for every $i \in S_j$, new random values $r_i^{(j)}$ are chosen from $\mathbb{F}_p$. The simulator then computes $D_{\alpha\beta} = {}^{(\alpha + r(j))/\beta}$ and for each $i \in S_j$, has $D_i = g^{r^{(j)} + w_i r_i^{(j)}}$ and $D'_i = g^{r_i^{(j)}}$. These values are passed onto the adversary.

Challenge: Giving two messages $m_0, m_1 \in \mathbb{G}_1$, and the access tree T, the simulator does the following. First, it chooses a random s from $\mathbb{F}_p$, then it uses the linear secret sharing scheme associated with T to construct shares λ_i of s for all relevant attributes i. Finally, the simulation chooses a random $\theta \in \mathbb{F}_p$ and constructs the encryption as $C_{mr} = e(g,g)^{\theta}$ and $C_s = h^s$. For each relevant attribute i, there is $C_i = g^{\lambda_i}$, and $C'_i = g^{w_i \lambda_i}$. These values are sent to the adversary.

Guess: With probability $1 - O(\zeta^2/p)$, taken over the randomness of the choice of variable values in the simulation, the adversary's view of this simulation is identically distributed to what its view would have been if it had been given $C_{mr} = e(g,g)^{\alpha \cdot s}$. Therefore, the advantage of the adversary is at most $O(\zeta^2/p)$.

6   Performance Evaluation

The experiment was carried out on a campus cloud computing platform. The cloud platform takes kernel-based virtual machine (KVM) and Openstack as the infrastructure and is composed of 20 servers.The user terminal was a 32-bit Ubuntu microcomputer system (1.5GB memory, 2.27ghz CPU). We implemented a 160-bit elliptic curve group based on the supersingular curve $y^2 = x^3 + x$ over a 512-bit finite field. The purpose of the experiment is to test the PB-MUFHE's efficiency of encryption, decryption, ciphertext addition and multiplication operation.

In the access policy tree, the leaf node number is set to 2. In the attribute secret key, the number of attributes is set to 1. According to the values of the parameter n, Figs. 3, 4, 5, and 6 shows a comparison of the efficiency of encryption, decryption, ciphertext multiplication, and ciphertext addition between the proposed scheme and GSW.

Compared to GSW, PB-MUFHE showed similar encryption time, addition and multiplication operation time, gap in thedecryption time. The proposed scheme as opposed to GSW, not only to supports access control but also multi-user sharing and ciphertext computing among different users. Basically, the proposed scheme shows enhancedsecurity and function without sacrificing effectiveness.

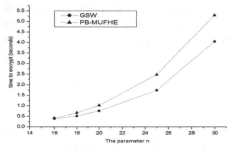

Fig. 3. Encryption time with parameter n for GSW and PB-MUFHE

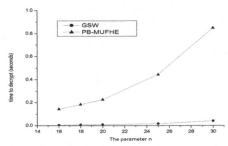

Fig. 4. Decryption time with parameter n for GSE and PB-MUFHE

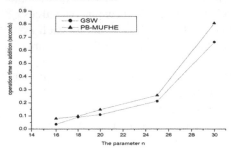

Fig. 5. Addition operation time with parameter n for GSW and PB-MUFHE

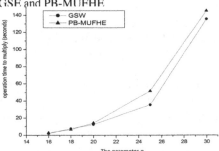

Fig. 6. Multiplication operation time with parameter n for GSW and PB-MUFH

7   Conclusions

The proposed PB-MUFHE scheme, which combines function encryption and FHE techniques, allows policy-based, multi-user homomorphic encryption and effectively solves problems related to ciphertext operations, access control, multi-user sharing, and other issues. The security of the proposed scheme is based on its performance solving LWE problems, and is further verified by testing its IND-CPA security in the generic bilinear group random oracle model. In short, the proposed scheme is shown to implement fully homomorphic evaluation of ciphertexts with multiple users securely and efficiently. Future potential improvements to the proposed scheme include adding key revocation mechanisms to facilitate the effective management of keys, as well as the use of multilinear mapping technology to replace the bilinear mapping technology to enhance access control efficiency. The use of bootstrapping techniques would also improve the degree of homomorphic evaluation.

References

1. Wu, Y., Lin, N., Song, W., et al.: Adaptive authorization access method for medical cloud data based on attribute encryption. In: Ni, W., Wang, X., Song, W., Li, Y. (eds.) WISA 2019. LNCS, vol. 11817, pp. 361–367. Springer, Cham (2019). https://doi.org/10.1007/978-3-030-30952-7_36
2. Craig, G.C.: Fully homomorphic encryption using ideal lattices. In: Proceedings of the 41st Annual ACM Syposium on Theory of Computing, pp. 169–178. ACM, New York (2009)
3. Brakerski, Z., Vaikuntanathan, V.: Fully homomorphic encryption from ring-LWE and security for key dependent messages. In: Rogaway, P. (ed.) CRYPTO 2011. LNCS, vol. 6841, pp. 505–524. Springer, Heidelberg (2011). https://doi.org/10.1007/978-3-642-22792-9_29
4. Coron, J.S., Lepoint, T., Tibouchi, M.: Batch fully homomorphic encryption over the integers. In: 32nd Annual International Conference on the Theory and Applications of Cryptographic Techniques, pp. 315–335, Springer Verlag, Athens (2013). https://doi.org/10.1007/978-3-642-38348-9_20
5. Brakerski, Z., Vaikuntanathan, V.: Efficient fully homomorphic encryption from (standard)LWE. In: Proceedings of the 52nd Annual Symposium on Foundations of Computer Science, pp. 97–106, IEEE, Washington (2011)
6. Bos, J.W., Lauter, K., Loftus, J., Naehrig, M.: Improved security for a ring-based fully homomorphic encryption scheme. In: Stam, Martijn (ed.) IMACC 2013. LNCS, vol. 8308, pp. 45–64. Springer, Heidelberg (2013). https://doi.org/10.1007/978-3-642-45239-0_4
7. Feng, Y.S., Ma, H., Chen, X.F.: Efficient and verifiable outsourcing scheme of sequence comparisons. Intellig. Automat. Soft Comput. 21(1), 51–63 (2015)
8. Lin, H.X.: Leveled certificateless fully homomorphic encryption schemes from learning with errors. IEEE Access 8, 26749–26763 (2020)
9. Peikert, C., Shiehian, S.: Multi-key FHE from LWE, revisited. In: Hirt, M., Smith, A. (eds.) TCC 2016. LNCS, vol. 9986, pp. 217–238. Springer, Heidelberg (2016). https://doi.org/10.1007/978-3-662-53644-5_9
10. Li, Z., Ma, C., Zhou, H.: Multi-key FHE for multi-bit messages. Sci. China Inf. Sci. 61(2), 029101:1-029101:3 (2018)
11. Clear, M., McGoldrick, C.: Multi-identity and multi-key leveled FHE from learning with errors. In: Gennaro, R., Robshaw, M. (eds.) CRYPTO 2015. LNCS, vol. 9216, pp. 630–656. Springer, Heidelberg (2015). https://doi.org/10.1007/978-3-662-48000-7_31
12. Clear, M., McGoldrick, C.: Bootstrappable identity-based fully homomorphic encryption. cryptology and network security. In: 13th International Conference on Cryptology and Network Security, pp. 1–14, Springer Verlag, Heraklion (2014)
13. Wang, F., Wang, K., Li, B.: An Efficient Leveled Identity-Based FHE. LNCS, pp. 303–315. Springer, Cham (2015). https://doi.org/10.1007/978-3-319-25645-0_20

14. Hu, M., Ye, Q., Tang, Y.: Efficient batch identity-based fully homomorphic encryption scheme in the standard model. IET Inf. Secur. **12**(6), 475–483 (2018)
15. Luo, F., Wang, K., Lin, C.: Leveled hierarchical identity-based fully homomorphic encryption from learning with rounding. In: 14th International Conference on Information Security Practice and Experience, pp. 101–115, Springer Verlag, Tokyo (2018)
16. Shen, T., Wang, F., Chen, K., et al.: Efficient leveled (multi) identity-based fully homomorphic encryption schemes. IEEE Access **7**, 79299–79310 (2019)
17. Gentry, C., Sahai, A., Waters, B.: Homomorphic encryption from learning with errors: conceptually-simpler, asymptotically-faster, attribute-based. In: Canetti, R., Garay, J.A. (eds.) CRYPTO 2013. LNCS, vol. 8042, pp. 75–92. Springer, Heidelberg (2013). https://doi.org/10.1007/978-3-642-40041-4_5
18. Brakerski, Z.: Fully homomorphic encryption without modulus switching from classical GapSVP. In: Safavi-Naini, R., Canetti, R. (eds.) Advances in Cryptology – CRYPTO 2012, pp. 868–886. Springer Berlin Heidelberg, Berlin, Heidelberg (2012). https://doi.org/10.1007/978-3-642-32009-5_50

Towards Efficient Learning Using Double-Layered Federation Based on Traffic Density for Internet of Vehicles

Xiaolin Hu[1,2], Guanghui Wang[1,2], Lei Jiang[1,2], Shuang Ding[1,2], and Xin He[1,2(✉)]

[1] Henan International Joint Laboratory of Intelligent Network Theory and Key Technology, Henan University, Kaifeng, China
[2] School of Software, Henan University, Kaifeng, China
hexin@vip.henu.edu.cn

Abstract. It is an important topic to research on the federal-learning based smart services to achieve data privacy preservation in the Internet of Vehicles field. However, model training in the vehicles is still confronting the challenge of low learning efficiency when applying the federal-learning concept into the scenario of dense road. To address the above issue, this paper presents a novel technique to enhance the learning efficiency based on traffic density for the Internet of vehicles. First, a double-layered federation architecture is built through coordinating multiple roadside units. The streams of traffic are divided into different regions, where the devices inside each region are federated for down-layer learning. The roadside units corresponding to each region layer are federated for up-layer learning. Second, based on the double-layered federation architecture, an efficient federal-learning algorithm is invented, where the computational overheads of dense traffic are decreased and the data privacy is still preserved during the model training process. Finally, the simulations are conducted using the real-world dataset from the Microscopic vehicular mobility trace of Europarc roundabout, Creteil, France. The simulation results show that the proposed efficient federal-learning algorithm can improve the learning performance and preserve data privacy in the scenario of intensive traffic.

Keywords: Double-layered federation · Internet of vehicles · Privacy preservation · Learning efficiency

Supported by Research on Henan Provincial Major Public Welfare Project under Grant (No. 201300210400), Key Technologies Research and Development Program of Henan under Grant (No. 212102210094, No.212102210090), China Postdoctoral Science Foundation under Grant (No. 2020M672211, No. 2020M672217) and Scientific Research Projects of Henan Provincial Colleges and Universities under Grant (No. 21A520003).

C. Xing et al. (Eds.): WISA 2021, LNCS 12999, pp. 287–298, 2021.
https://doi.org/10.1007/978-3-030-87571-8_25

1   Introduction

Nowadays with the continuous improvements of the Internet of Vehicles (IoV), there are more and more data interactions between vehicles. Vehicles can help drivers to have a better driving experience by using the road information obtained by onboard electronic equipment combined with their environmental perception abilities. Data sharing between vehicles is one of the effective means to improve data utilization and fully tap data value [1,2]. However, most drivers refuse to upload data to the data center because of personal information leakage, which reduces the effectiveness of the data sharing process and hinders the development of IoV.

Federal-learning technology has certain advantages in protecting data privacy. With the increasing awareness of public privacy protection, data privacy protection methods using federal-learning are gradually applied to various fields [3–8]. In federal-learning based IoV, vehicles carry out collaborative training without exposing local data. Vehicles only need to share the trained model parameters to Road Side Units (RSUs), because they do not need to share the entire original data-set, thereby reducing the privacy risk.

Federal-learning based IoV technology still faces challenges in vehicle training in dense areas. Intensive vehicle areas have heavier training tasks than other areas, putting RSU under greater communication and computing overheads. In dense areas such as crossroads, vehicles have the characteristics of high aggregations, high requirements for model training, short residence time, diverse models and rapidly changing road conditions. Traditional federal-learning methods, such as FedSGD and FedAvg algorithms [9], usually aggregate the update of the vehicle local model on a single central node. In dense traffic scenarios, a single central node faces more communication and computing overheads, which training time is large. Therefore, the traditional federal-learning methods are not suitable for dense traffic training because of the inefficiency of model training.

This paper proposes to apply a double-layered federation (DLF) architecture to improve the learning efficiency for the dense traffic. A double-layered federation architecture is presented through coordinating multiple RSUs. The dense traffic are divided into different regions, where each vehicle uploads model updates to the nearest RSU. The average aggregation is calculated among the RSUs in the up-layer federal-learning after the data is aggregated on the corresponding RSU in the down-layer federal-learning. Then, an efficient federal-learning algorithm is designed under the double-layered federation architecture. The learning efficiency of the vehicle is improved by reducing the average aggregation time while still preserving data privacy during the model training process. The contributions of this paper are summarized as follows:

- A novel double-layered federation (DLF) architecture is proposed to improve the efficiency of federal-learning in the dense traffic. Multiple RSUs are federated to train models in the up-layer learning and the vehicles are federated to train the model in the down-layer learning.
- An efficient federal-learning (EFL) algorithm is designed under the DLF architecture. The EFL algorithm reduces the average aggregation time during

the model training process to improve the efficiency of the learning efficiency while preserving the privacy of local data in vehicles.

- The Microscopic vehicular mobility trace of Europarc roundabout, Creteil, France, is utilized to conduct simulations to demonstrate the effectiveness of the proposed EFL algorithm.

The rest of this paper is structured as follows. Section 2 introduces the research of federal-learning in related fields. Section 3 presents the DLF architecture and problem setup. The EFL algorithm is designed in Sect. 4. The performance evaluation is conduct in Sect. 5. Section 6 summarizes the paper.

2   Related Work

As federal-learning technology has certain advantages in protecting data privacy, it is gradually accepted by the public and applied to various fields, such as the medical field [3,4], financial field [5,6], and intelligent transportation field [7,8]. And the application of privacy protection in other fields [10,11]. Brendan ct al. proposed that the client could reduce the number of communications through multiple rounds of local iterative training [9]. However, since it used a single central node for model aggregation, it could not effectively improve the learning efficiency in dense area scenarios. Q. Yang et al. proposed a classification of federal-learning based on different data features to internationalize it [12,13], focusing on the analysis of client data features. J. Konečna et al. proposed Google Federal Learning (GFL), selection model aggregation method, and dynamic sample selection optimization algorithm [14–17]. However, they did not propose solutions for efficient vehicle training in dense areas of the IoV.

H. Chai et al. applied federal-learning technology to the IoV and proposed distributed methods based on federal-learning, hierarchical federal-learning algorithms, efficient communication, and privacy-preserving federal-learning frameworks [18–25], respectively. However, they failed to effectively improve the efficiency of vehicle training in dense areas of the IoV [1,2]. To improve the efficiency of federal-learning, Lu et al. conducted the model aggregation process by grading the RSUs, but the RSUs were not placed in dense areas, which had limitations in improving the efficiency of vehicle training in dense areas. In the application scenario of the IoV, the learning efficiency of vehicle model training in dense areas is required to be high. For vehicles in dense regions, while federal-learning technology protects data privacy, a solution is still needed to solve the problem of how to effectively improve learning efficiency.

3   The DLF Architecture and Problem Setup

3.1   The DLF Architecture

The DLF is a distributed architecture for model aggregation of vehicles in dense intersection areas through multiple RSUs. Defines an $C_i(\forall i = 1, 2, ..., M)$ that represents the vehicle index collected by the RSUs to participate in the task. Where $\omega_j(\forall j = 1, 2, ..., K)$ represents the model state of the j-th RSUs, and K denotes the number of RSUs participate in task training. Define ω_i to represent model updates generated by local training of vehicles. When $T_t \leq T_{m1}$, the RSUs receives local updates sent by the vehicle.

The main symbols are shown in Table 1.

Table 1. Summary of main symbols.

Signs	Interpretation
i	Vehicle index
t	The number of local training sessions for vehicles
T	The aggregation times of the RSUs
M	The number of vehicles on mission collected by the RSUs at round T
D	Local vehicle data
η_i	The local learning efficiency of vehicles
K	The RSUs serial number
ω	Model state after the RSUs aggregation
ω_0	The initial Model State of the RSUs
ω_i	Vehicle model updating
ω_j	The RSUs Model State
ω_i^t	Local model state before vehicle training
ω_i^{t+1}	Model updating generated by local vehicle training
ω_T	The model state produced by the T round aggregation of the RSUs
ω'	The model state of the RSUs aggregation
T_{m1}	The specified time for the RSUs to collect local updates
T_t	Time spent collecting local updates in round T

Workflow of DLF. In the dense area of crossroads in the field of the IoV, task releases and local updates are first collected by the RSUs. Second, the vehicle uploads the resulting model updates to the nearest RSUs through local training. Each RSUs receives the local update content of the nearest vehicle within the set time, and finally aggregates the received content model average. After the aggregation of the RSUs model, the current state is transmitted to the mobile vehicle participating in the task of providing funds. The model state generated by the aggregation is shared periodically between the RSUs in the dense area. The model state is averaged again so that the model training can produce better results. The data interaction of the DLF is shown in Fig. 1.

Main Components of DLF. *Multilateral Cooperation.* Multiple RSUs establish communication to form a local area network. A temporary acting team leader was selected through a comprehensive assessment of the communication and computing capabilities of RSUs to determine training tasks and establish

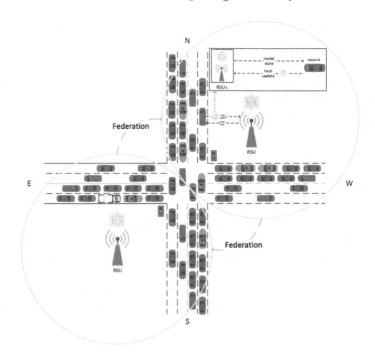

Fig. 1. Data interaction in DLF.

temporary vehicle index tables for vehicles participating in training. The temporary index table mainly records the contribution of vehicles participating in the task, which is convenient for rewarding vehicles with outstanding contributions.

Task Release. The RSUs with temporary acting group leader status select the only task of current joint training and releases the task to vehicles in dense areas through broadcasting, establishing the initial model state ω_0. The vehicle receives the task information released by the RSUs, and the owner decides whether to participate in the training task according to the vehicle status and reward. The vehicles involved in the task establish communication with the nearest RSUs, and the RSUs register the temporary index for it. The index is a dynamic set of C_i.

Local Training. Trained vehicles model through data features and current model state. Vehicles choose local single wheel iteration or local multi wheel iteration according to the calculation ability. The local single wheel iteration has lower requirements for vehicle computing power, the local multi wheel iteration has higher requirements for vehicle computing power, and the model updating trained by multi wheel iteration is relatively good. Finally, the vehicle transmits the trained model update to the nearest RSUs.

Model Aggregation. The RSUs receives the local update of the vehicle within the budget time and aggregates it on average. Multiple RSUs receive vehicle local updates at the same time. The RSUs feedback the aggregated model state to the vehicle participating in the task. The vehicle receives the model state and starts a new round of training. The model state is shared periodically between RSUs, and the model state is aggregated again.

Incentive Mechanism. Training generated models are provided free of charge to train vehicles to help owners predict road information. Vehicles that are not trained can provide some financial support to access the model service. These funds are used to reward participants who contribute greatly to the training process.

3.2   The Goals

This paper aims to design an EFL algorithm based on DLF architecture. The goals are:

- By proposing the DLF architecture, the efficiency of federal-learning on the dense traffic is improved.
- The DLF based EFL algorithm reduces the average aggregation time during model training.
- The vehicle uses the federal-learning method for model training and protects the data privacy of the vehicle by keeping the local data of the vehicle from leaving the local area.

4   EFL Algorithm

4.1   Basic Idea

The learning efficiency of vehicle model training in dense areas can be improved to a certain extent through the EFL algorithm. The key idea is that vehicles train models through local data and current model updates, uploading model updates to the nearest RSUs to participate in model aggregation. Each RSUs averages the update of the model uploaded by the collected vehicle within the specified time range and iterates repeatedly between the RSUs and the vehicle. Model state sharing is performed periodically between RSUs to average the model again to improve the learning efficiency of model training. Vehicles will select the nearest RSUs to upload local updates and spread communication and computing overheads to multiple RSUs. Repeated iterations of the training process until the completion of the entire model training, intensive vehicle learning efficiency can be greatly improved. Compared with the FedSGD algorithm and FedAVG algorithm, the EFL algorithm alleviates the communication and computing overheads of RSUs during task training in dense areas and greatly improves the efficiency of vehicle training. The flow chart of the EFL algorithm is shown in Fig. 2.

4.2   Local Calculation of Vehicles

The local calculation is carried out by combining local data and model state. The following objective function is used in model training, and the model parameter

ω is used to predict the loss of local data where D represents the local data of the vehicle.

$$\min_{\omega \in R^d} f(\omega) \tag{1}$$

$$f(\omega) = \ell(\omega, D) \tag{2}$$

Each vehicle i calculates the current gradient decline of the vehicle, and the specific calculation method is $g_i^t = \nabla f_i(\omega)$. ω represents the current model state of the vehicle receiving RSUs transmission. η_i represents the local learning efficiency of each vehicle. $\omega_i^{t+1} \leftarrow \omega_i^t - \eta_i g_i^t$ can be calculated by gradient descent, and the vehicle can get the latest local update. The local update is uploaded to the nearest RSUs to participate in model aggregation. The above formulas for local calculation of vehicles can be obtained.

$$\omega_i^{t+1} \leftarrow \omega_i^t - \eta_i \nabla(\omega, D) \tag{3}$$

Each client performs local gradient descent operations on the current model using local data.

4.3 Polymerization Model

Aggregate Local Updates. The RSUs receive and accumulate local updates uploaded by vehicles within the specified time, as shown in (3). Average the model state according to the total number of valid vehicles involved in the task, as shown in (4). Through this process, each RSUs iteratively trains a more efficient model.

$$T_s = \sum_{i=1}^{M} \omega_i \tag{4}$$

$$\omega_T = \frac{T_S}{M} \tag{5}$$

Aggregation Model State. The local model state is shared periodically between RSUs, and the shared model state is aggregated again to achieve better training results. As shown in (5).

$$\omega' = \sum_{j=1}^{K} \frac{\omega_j}{K} \tag{6}$$

Algorithm 1 gives complete pseudocode. Algorithm 1 shows the implementation process of multiple RSUs applied to DLF architecture in the dense regional environment to improve learning efficiency. Trained vehicle local data does not leave the local, multiple RSUs collaboratively participate in model aggregation to improve learning efficiency. It is not difficult to see that the time complexity of the EFL algorithm based on the DLF architecture is $O(n)$.

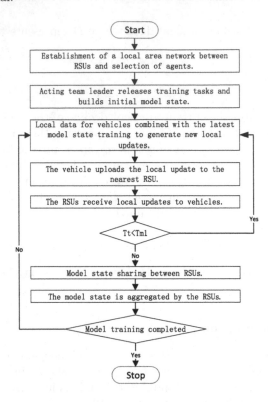

Fig. 2. The flowchart of the EFL algorithm.

Algorithm 1. EFL

Server Polymerization

    // Establishing initial model state

    Initialize ω_0

    For each round $T = 1, 2, ..., \infty$ do

        while$(T_t \leq T_{m1})$

            $\omega_i \leftarrow$ ClientUpdate(i, ω)

        $\omega_T = \sum\limits_{i=1}^{M} \frac{\omega_i}{M}$

        $\omega \leftarrow \omega_T$

End

Server Reintegration (ω_k)

    $\omega' = \sum\limits_{j=1}^{K} \frac{\omega_j}{K}$

    $\omega \leftarrow \omega'$

ClientUpdate (i, ω): // Run on client i

    $\omega_i^{t+1} \leftarrow \omega_i^t - \eta_i \nabla(\omega, D)$

    Return ω_i^{t+1} to server

---

5   Performance Evaluation

5.1   Simulation Setup

In this paper, a public and real proxy data-set is used to predict the vehicle speed at the intersection, to better verify the authenticity of the EFL algorithm. The proxy data-set is the vehicle driving data of two hours (7.00–9.00) in the morning and two hours (17.00–19.00) in the evening on Microscopic vehicular mobility trace of Europarc roundabout, Creteil, France. The sampling interval of the data-set is 1 s. The vehicle trajectory data ID, time, vehicle type, coordinate, vehicle speed, and other information [26]. In this paper, simulation verification is carried out on Intel Core i5 - 1035G1 CPU @ 1.00 GHz processor and Lenovo laptop computer with 16.0 GB (15.8 GB available) RAM configuration. In the simulation, the data collected at 7:00 am were used. After 3000 s, the RSUs collected the data generated by 130 cars and buses to complete the training task to test the authenticity of the scheme. Publishers allocate computing tasks to the IoV and select corresponding vehicles and RSUs to complete tasks. The simulation results are generated from 10000 independent runs. According to Algorithm 1, the vehicle selects the nearest RSUs to upload its training model update according to the distance between the RSUs and the vehicle. The vehicle uses a variety of local training methods to compare. Shared between multiple RSUs, averaging the model state again. This paper reports the effect comparison of the different numbers of RSUs participating in model aggregation in unit time. To verify this, the same vehicle learning ability is set for the experiment. When the vehicle conducts local training, the time t for each iteration of local learning is set to be 0.2 s, and the basic communication time consuming constant $k = 0.05$. According to previous work, FedAVG has the best effect when iterates locally five times [9], so $m = 5$. It is verified that the learning efficiency of multiple RSUs participating in model aggregation is significantly higher than that of single RSUs participating in aggregation.

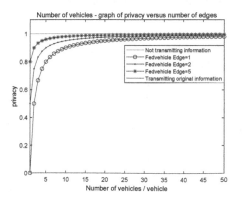
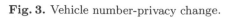
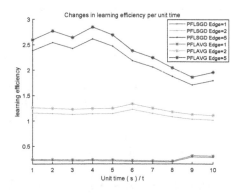

Fig. 3. Vehicle number-privacy change.   **Fig. 4.** Time-learning efficiency change.

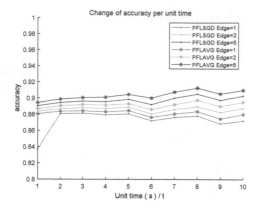

Fig. 5. Accuracy-communications.

5.2 Results

Vehicle data privacy security level is high in test performance. As shown in Fig. 3, the use of federal-learning technology can reduce the probability of privacy disclosure of vehicles in the process of data transmission, and protect the data privacy of vehicles to some extent. It can also be observed that with the increasing number of vehicles participating in training, the data privacy of vehicles will be better protected. With the increasing number of RSUs involved in aggregation tasks, the security of vehicle data privacy will be further guaranteed.

The learning efficiency of the EFL algorithm is evaluated in different local training methods and multiple RSUs involved in model aggregation. As shown in Fig. 4, the learning efficiency of the EFL algorithm is significantly higher than that of the single RSUs in the traditional federal-learning algorithm for aggregation. At the same time, it can be observed that with the increasing number of RSUs involved in the aggregation of the model, the learning efficiency will be further improved. The algorithm of FedAVG is better than that of FedSGD when the vehicle is trained locally. It can be concluded. EFL is more suitable for intensive area vehicle participation model training.

According to the local historical speed information of the vehicle, the vehicle speed and the possible road conditions in the next time are predicted. The relationship between the accuracy and time of the relevant prediction is shown in Fig. 5. The EFL algorithm is more accurate than the traditional federal-learning and training method in grasping the relevant prediction accuracy. Compared with the traditional FedSGD and FedAVG algorithms, the EFL algorithm using the DLF architecture can be better applied to the model prediction training in the dense regional environment.

6   Conclusions and Future Work

The application of federal-learning technology on the IoV is currently a research topic. Due to low efficiency, traditional federal-learning has challenges on the dense traffic scenarios. This paper proposes a double-layered federation architecture to improve the learning efficiency of the dense traffic. Multiple RSUs are federated to train models in the up-layer learning and the vehicles are federated to train the model in the down-layer learning. Secondly, an efficient federal-learning algorithm is proposed based on the double-layered federation architecture, which reduces the average aggregation time in the model training process, so as to maintain the local data privacy of vehicles and improve the learning efficiency. Finally, the simulations are conducted using the real-world dataset from the Microscopic vehicular mobility trace of Europarc roundabout, Creteil, France, are utilized to conduct simulations to demonstrate the effectiveness of the proposed efficient federal-learning algorithm.

The novel efficient federal-learning algorithm based on the double-layered federation architecture is capable of increasing the learning efficiency of traffic stream in dense areas for Internet of Vehicles research, nevertheless, the puzzles such as how to pick out the proper RSUs nodes or the way to combine them with the blockchain, are remain to be solved.

References

1. Lu, Y., Huang, X., Zhang, K., et al.: Blockchain empowered asynchronous federal learning for secure data sharing in internet of vehicles. IEEE Trans. Veh. Technol. **69**(4), 4298–4311 (2020)
2. Lu, Y.F.: Research on Data Privacy Protection and Sharing Method. Beijing Forestry University, Beijing (2020)
3. Brisimi, T.S., Chen, R., Mela, T., et al.: federal learning of predictive models from federated electronic health records. Int. J. Med. Informatics **112**, 59–67 (2018)
4. Lim, W.Y.B., Garg, S., Xiong, Z., et al.: Dynamic contract design for federal learning in smart healthcare applications. IEEE Internet Things J. (2020)
5. Su, J.M., Ye, H., Lv, B.L., et al.: The application of federal learning in the anti-fraud field of commercial banks. Financial Computer of China 02, pp. 39–42 (2021)
6. Zhang, Y.Y.: Federal learning and its application in finance. Rural Finance Res. **12**, 52–58 (2020)
7. Zhang, C., Liu, Y., Wang, L., et al.: Joint intelligence ranking by federated multiplicative update. IEEE Intell. Syst. **35**(4), 15–24 (2020)
8. Pokhrel, S.R., Choi, J.: federal learning with blockchain for autonomous vehicles: analysis and design challenges. IEEE Trans. Commun. **68**(8), 4734–4746 (2020)
9. McMahan, B., Moore, E., Ramage, D., et al.: Communication-efficient learning of deep networks from decentralized data. In: Artificial Intelligence and Statistics, pp. 1273–1282 (2016)
10. Wang, G.H., Pan, J.P.: Analyzing and evaluating efficient privacy-preserving localization for pervasive computing. IEEE Internet Things J. **5**(4), 2993–3007 (2018)
11. Wang, G.H., Pan, J.P., He, J.P., et al.: An efficient privacy-preserving localization algorithm for pervasive computing. In: Proceedings of the 26th International Conference on Computer Communication and Networks, IEEE ICCCN, pp. 1–9 (2017)
12. Yang, Q., Liu, Y., Cheng, Y.: Federal Learning, 1st edn. Electronic Industry Press, Beijing (2020)

13. Federal learning application guide. https://standards.ieee.org/standard/3652_1-2020.html. Accessed 27 Apr 2021
14. Ye, D., Yu, R., Pan, M., et al.: federal learning in vehicular edge computing: a selective model aggregation approach. IEEE Access **8**, 23920–23935 (2020)
15. Cai, L., Lin, D., Zhang, J., et al.: Dynamic sample selection for federated learning with heterogeneous data in fog computing. In: CONFERENCE 2020, ICC 2020-2020 IEEE International Conference on Communications (ICC), Dublin, Ireland, pp. 1–6 (2020). https://doi.org/10.1109/ICC40277.2020.9148586
16. Cao, J., Zhang, K., Wu, F., et al.: Learning cooperation schemes for mobile edge computing empowered internet of vehicles. In: 2020 IEEE Wireless Communications and Networking Conference (WCNC), Seoul, Korea (South), pp. 1–6. IEEE (2020).https://doi.org/10.1109/WCNC45663.2020.9120493
17. Konečný, J., McMahan, H.B., Ramage, D., et al.: Federated optimization: distributed machine learning for on-device intelligence, pp. 99–110 (2016)
18. Chai, H., Leng, S., Chen, Y., et al.: A hierarchical blockchain-enabled federal learning algorithm for knowledge sharing in internet of vehicles. IEEE Trans. Intell. Transp. Syst. **99**, 1–12 (2020)
19. Kong, X., Wang, K., Hou, M., et al.: A federal learning-based license plate recognition scheme for 5G-enabled internet of vehicles. IEEE Trans. Industr. Inf. (2021)
20. Samarakoon, S., Bennis, M., Saad, W., et al.: Distributed federal learning for ultra-reliable low-latency vehicular communications. IEEE Trans. Commun. **68**(2), 1146–1159 (2019)
21. Kang, J., Xiong, Z., Niyato, D., et al.: Training task allocation in federated edge learning: a matching-theoretic approach. In: 2020 IEEE 17th Annual Consumer Communications Networking Conference (CCNC) 2020, Las Vegas, NV, USA, pp. 1–6 (2020). https://doi.org/10.1109/CCNC46108.2020.9045112
22. Pokhrel, S.R., Choi, J.: Improving TCP performance over WiFi for internet of vehicles: a federal learning approach. IEEE Trans. Veh. Technol. **69**(6), 6798–6802 (2020)
23. Zhang, C., Liu, X., Zheng, X., et al.: Fenghuolun: a federal learning based edge computing platform for cyber-physical systems. In: 2020 IEEE International Conference on Pervasive Computing and Communications Workshops (PerCom Workshops) 2020, Austin, TX, USA, pp. 1–4 (2020). https://doi.org/10.1109/PerComWorkshops48775.2020.9156259
24. Tan, K., Bremner, D., Le Kernec, J., et al.: Federated machine learning in vehicular networks: a summary of recent applications. In: 2020 International Conference on UK-China Emerging Technologies (UCET) 2020, Glasgow, UK, pp. 1–4 (2020). https://doi.org/10.1109/UCET51115.2020.9205482
25. Jiang, W., Lv, S.: Inference acceleration model of branched neural network based on distributed deployment in fog computing. In: Wang, G., Lin, X., Hendler, J., Song, W., Xu, Z., Liu, G. (eds.) WISA 2020. LNCS, vol. 12432, pp. 503–512. Springer, Cham (2020). https://doi.org/10.1007/978-3-030-60029-7_45
26. Public data set. http://vehicular-mobility-trace.github.io/. Accessed 27 Apr 2021

Oblivious Data Structure for Secure Multiple-Set Membership Testing

Qin Jiang[1], Yanjun An[2], Yong Qi[1(✉)], and Hai Fang[3]

[1] Xi'an Jiaotong University, Xi'an, China
qjiang16@stu.xjtu.edu.cn,qiy@mail.xjtu.edu.cn
[2] Lanzhou Jiaotong University, Lanzhou, China
[3] Xi'an Institute of Space Radio Technology, Xi'an, China
hai_fang@yeah.net

Abstract. Searchable encryption (SE) assists untrusted cloud to provide various private query functionalities in an efficient way. Deploying SE schemes on the cloud directly can leak the locations of the data needed, namely, access pattern leakages. Statistical attacks have demonstrated the importance of sealing access-pattern leakages. To protect against such attacks, several works were made to design oblivious data structures by combining oblivious random access machine (ORAM) and SE. However, prior designs only work for private SQL or keywords queries such as range or boolean queries not for secure multiple-set membership testing which is also a fundamental service of the cloud. In this paper, we firstly design a new oblivious data structure, called OBTree, to overcome the problem of oblivious multiple-set membership testing over encrypted data. Specifically, OBTree is built from ORAM, bloom tree and the secure features of Intel SGX. Based on this novel oblivious data structure, we present a search algorithm that allows the users to do membership testing efficiently and securely.

Keywords: Searchable encryption · Membership testing · ORAM · Intel SGX

1 Introduction

With the development of the cloud service, both end-users and companies have been increasingly outsourcing their data and delegating computing services to the cloud such as Amazon EC2 and S3, Microsoft Azure for a lower cost and better performance. Outsourcing data to the untrusted cloud can cause many issues such as data access control [1,3,4], the privacy of the data [5,6]. More importantly, the utilization dilemma is a critical concern of the users when protecting the privacy of the data. Multiple-set membership testing (MMT) [7] is especially useful in traffic classification and packet processing and is an essential

Supported by National Key R&D Program of China (No. 2020YFB1808003) and the Strategic Research Program of Ministry of Education (China) (No. 2020KJ010801).

feature of the commonly used databases. For example, as shown in Fig. 1, the user sends a keyword such as IP address to the server, its associated value such as the group identifier *id* should be returned fast in deterministic time. MMT is especially useful when the dataset's size is large. Outsourcing dataset to the cloud causes the concern of data privacy. A naive approach is to employ standard encryption techniques on the dataset. Yet, it prevents the user from even simple operations (such as search), not to say performing MMT.

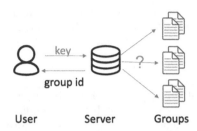

Fig. 1. An example of multiple-set membership testing

Recently, SE schemes [14,15,18–20] allow an untrusted cloud to privately search on the outsourced encrypted data for an authorized user. The state-of-the-art designs focus on solving secure SQL or keywords queries. It is achieved by the following steps: first, the user creates an encrypted index and encrypted datasets and then outsource them to the cloud. Later, the user can send encrypted queries named search token. Finally, the cloud searches on the encrypted index and returns the results that match the query to the users. However, directly using SE can leak access patterns that are vulnerable to statistical attacks, which can recover the plaintext of the datasets or the search token.

In the literature, recent works [24,25,27–29] have started to seal access pattern leakages by designing oblivious data structures. ObliviousMap [25] and Oblidb [28] propose the tree-based oblivious data structures by combining tree data structures and ORAM schemes. The POSUP [27] utilizes Intel SGX [32] and ORAM scheme [13] to optimize the efficiency of oblivious keyword search/update operation based on an inverted index. Unfortunately, neither of the existing schemes can support MMT queries. This is because of two challenging issues. First, unlike keyword search functions, the cloud server without pre-knowledge of the result information has to scan all the data in the dataset to make comparisons, which reduces the search performance. Second, existing systems [24,29] load the entire outsourced data within the trusted memory, called enclave created by Intel SGX, for each query to completely hide the access pattern, which can incur a significant delay.

In this paper, we propose a new oblivious data structure named OBTree to support oblivious MMT. It is worth noting to directly combine an index data structure and ORAM to hide access pattern leakage for MMT. We adopt the oblivious random access machine (ORAM) technique, trust hardware technology and an efficient index data structure for MMT in an efficient way, so that the execution time can be feasible for the practical application. The main contributions are as follows:

Optimize the Cloud Storage. Instead of directly packing the nodes into the ORAM blocks, we re-design the way of combing the tree-based index data structure with different node sizes with ORAM blocks. In particular, to save server-side storage, we "cut" a node to several splinters so that we no longer need to pad the ORAM blocks to the same size. In OBTree, each block of the ORAM stores a splinter. To construct a connection between the splinters and the ORAM blocks, we re-define the structure of each ORAM block by adding the next splinter's path identifier when storing the splinter.

Putting Hardware-Supported ORAM in Effect. In the search protocol, we achieve constant complexity bandwidth consumption in a complete access process. We design two map data structures within the enclave to minimize the bandwidth for the read operations. One is to map each node's first splinter identifier to the path where this splinter stores in ORAM. The other one is to map each splinter's position in BTree to the corresponding ORAM block's identifier. In addition, the search protocol on OBTree can achieve sub-linear search efficiency. Meanwhile, it hides access pattern leakages when accessing the untrusted memory.

2   Overview

2.1   System Model

The high-level design of our protocol is shown in Fig. 2. Our design involves three entities: the data owner, the data user, the untrusted SGX enabled cloud server.

- *Data owner/user:* data owner first constructs a secure channel with the enclave of the cloud server by remote attestation. Then the data owner shares a secret key for encryption/decryption with the enclave and the data user. The data user can send search tokens to the cloud server.
- *Cloud server:* cloud server is comprised of two parts, an enclave and an untrusted part. The enclave stores the ORAM controller which contains sensitive data structures for ORAM operation, and the untrusted part stores ORAM tree-based data structures. When receiving a search token, the cloud server can search on the encrypted index obliviously and return the results satisfying the queries to the data user.

2.2   Threat Model

We assure an SGX enabled system and consider a powerful adversary that can control the entire software stack except for the code inside the enclave. On the trusted side, it follows the client logic and ORAM controller logic that runs in the

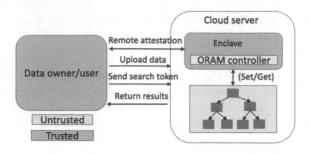

Fig. 2. High-level design

enclave but attempts to infer the plaintext information from the available data and the background knowledge about the dataset distribution. In particular, the adversary can monitor all interactions of the enclave with resources outside the enclave. We further assume that a secure channel is established between the remote party such as the data owner and the enclave of the cloud.

We do not consider denial of service (Dos) attacks and side channel attacks through cache access and time of process [8–10] which have been addressed by orthogonal studies like [11].

3   Building Blocks

3.1   Path ORAM

Oblivious Random Access Machine (ORAM) [12], a cryptographic primitive, hides access patterns on the untrusted cloud. ORAM shuffles the locations of blocks in memory and re-encrypts the data for each access. In our system, we use a tree-based ORAM scheme implemented by path ORAM [13]. There are three components in the Path ORAM scheme: a complete binary tree that stores data block, a position map which associates each block with a random leaf in the tree and a stash that temporarily stores some blocks at the trusted part such as the client side. Each access in a tree-based ORAM incurs a read and an eviction operations. The details of these two operations are deferred to [13]. In particular, ORAM guarantees that any two access patterns of the same length are computationally indistinguishable. Formally, ORAM's security is defined as follows:

ORAM Security [13]. Let $A := ((op_1, a_1, data_1), \cdots (op_q, a_q, data_q))$ denotes a data request sequence of length q, where each op_i denotes a $read(a_i)$ or a $write(a_i, data)$ operation and a_i denotes the address of the block being read or written, and $data_i$ denotes the data being written. Let $AP(A)$ denotes possibly randomized sequence of accesses to the untrusted storage given the sequence of data requests A. An ORAM scheme is secure if for any two data request sequences A and A' of the same length, their access patterns $AP(A)$ and $AP(A')$ are computationally indistinguishable by anyone but the client ORAM controller.

3.2   Intel SGX

Intel SGX [32] is a recent advance in hardware-based computer processor technology. It protects the security of applications from malicious operating systems and physical attacks by providing three main security properties: isolation, remote attestation and sealing. On SGX enabled cloud, memory can be divided into two parts: the trusted part and the untrusted part. On the one hand, the trusted part is called enclave and is backed by the encrypted and integrity-protected memory, known as EPC. Applications running on SGX enabled platform should define their sensitive part and put this part into the enclave. On the other hand, the untrusted part stores the ordinary code and data. Isolation enables that code within the enclave can access the entire virtual memory, while the code on the untrusted part can invoke the code within the enclave by the pre-defined interfaces. In the current SGX, the EPC size is limited to 128 MB. When the sensitive data and code are larger than the capability of the enclave, sealing makes to encrypt and authenticate the data within enclave. Remote attestation allows the client to verify the correct creation of an enclave on the cloud and can establish a secure channel between the client and the enclave. Using the secure channel, the external party can securely communicate with the enclave.

3.3   Bloom Filter and Bloom Tree

Bloom filter (BF) [31] is a space-efficient probabilistic data structure for fast set membership testing. BF is an array whose size is ζ and is initialized with 0 and associated with a set of k different hash functions $H = \left\{ h_i : \{0,1\}^* \rightarrow [\zeta]_{i=1}^k \right\}$. To add a keyword α to BF, we computer $h_i(\alpha), i\epsilon\{1, 2, \cdots k\}$ and need to set the element to be 1 at the position $h_i(\alpha)$. To check whether a keyword α is in the filter or not, we need to check if BF contains 1 at $h_i(\alpha)$ for $i\epsilon\{1, 2, \cdots k\}$. BF allows the false positive f to be tuned using the following formula:

$$f \approx \left(1 - e^{-\frac{km}{\zeta}} \right)^k,$$

where m is the number of elements encoded in BF.

Bloom tree (BTree) [7] is a balanced b-ary tree of height $log_b(n)$ whose node is BF and we let b equal 2 in our work. In the BTree, each leaf is associated with a group and a set of hash functions H, and each internal node is generated from its children and associated with two sets of hash functions H_j^i, where i denotes the node level and j denotes the j-th set of hash functions. We show an example in Fig. 3, as shown in [7]. Let $DB = \{g_1, g_2, ...g_4\}$ be a dataset which contains 4 groups. We encode DB to BTtree in a bottom-up fashion. First, we encode each keyword in g to a BF v, which is a leaf of the BTree. Second, for each internal node in whose right child or leaf child is v, we encode keywords in g to in. This process repeats until we construct the BF for all internal nodes. To check a keyword w, the search algorithm starts from the root of BTree and can return the right value with a certain probability in a top-down fashion when w is encoded into the BTree. Otherwise, BTree can return a value with a false-positive.

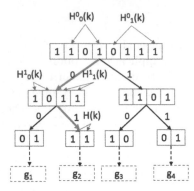

Fig. 3. An example of bloom tree

4   The Proposed Platform

4.1   Oblivious Data Structure

To do secure membership testing, we design an oblivious data structure, called OBTree which employs path ORAM, bloom tree and point technology in ODS [30]. To design an OBTree, there are three steps: construct a BTtree, divide BTree and construct OBTree. The detailed algorithm to construct OBTree is shown in Algorithm 1.

In the first step, we construct a BTree. Given multiple groups DB as input, we encode the elements in DB into BTree using the encoding algorithm in [7].

In the second step, we divide BTree and construct a hash table HT to store the divided BTree temporarily. Specifically, we first divide each node v_i of BTree into j splinters, where j is computed by the length of v_i and the pre-defined size of each splinter l. S_{ij} denotes the j-th splinter of v_i. Then, we use a string in the form of $path(v_i)|j|l$ as the key of S_{ij}, where $path(v_i)$ denotes a string concatenated with all tree branches beginning from the root of BTree to the current node v_i. If v_i is a root of BTree, $path(v_i)$ is an empty string denoted by & in Fig. 4. Otherwise, for an internal tree node, a tree branch to its left child is encoded as a character '0', and a tree branch to its right child is encoded as a character '1'. Finally, we use $path(v_i)|j|l$ as the key and the content of the splinter as the value and then store the key-value pair in HT temporarily.

In the third step, we package HT to OBTree as follows. To construct blocks of path ORAM inspired by the pointer trick in [30], we set each block in the form of $B := (\textbf{bID}, \textbf{DATA}, \textbf{NextPath})$, where **bID** is the block identifier, **DATA** contains the block data which corresponds to the value of HT and **NextPath** denotes the path of the next block in ORAM. Then, we store all constructed blocks into the path ORAM structure named PT. In addition, OBTree requires two hash tables named PID and H within the enclave. PID is to keep track of the first splinter's path in path ORAM. While H is to keep track of the splinters identifiers in path ORAM.

Algorithm 1: Build OBTree

Input: Groups DB, block size l, the number of leaves in ORAM L_0
Output: Oblivious data structure OBTree
create a BTree BT using DB
A is number of splinces that BTree divide
L is the hight of BTree
$IDS := 0, ..., A - 1$
$H \leftarrow \emptyset$, where H is a hashtable
$B \leftarrow \emptyset$
$PID \leftarrow \emptyset$, where PID is a hashtable
for *each node v_i in BT* **do**
   split v_i into m_i blocks of size l
   if *j=1* **then**
      $S_{ij}.pID \overset{r}{\leftarrow} \{0, 1, \cdots, L_0\}$
      $NID \leftarrow path(v_i)$
      $PID.add(pair(NID, S_{ij}.pID))$
   for *j=2,...,m_i* **do**
      $S_{ij}.pID \overset{r}{\leftarrow} \{0, 1, \cdots, L_0\}$
   for *j=1,..., m_i* **do**
      $S_{ij}.key \leftarrow path(v_i)|j|l$
      random select a number id form IDS
      $B_{ij}.ID \leftarrow id$
      $H.add(pair(S_{ij}.key, S_{ij}.pID))$
      $IDS.remove(id)$
      if *j=m_i& the length of m_i-th block;l* **then**
         pad the length of m_i-th splinter to l
      $B_{ij}.DATA \leftarrow j$-th splinter of v_i
      if *j=m_i* **then**
         $B_{ij}.NextPath \leftarrow null$
      else
         $B_{ij}.NextPath \leftarrow B_{ij+1}.pID$
      $B \leftarrow B \cup B_{ij}$
$PT \leftarrow BuildPathORAMTree(B, L_0)$
$OBTree \leftarrow (PID, H, PT)$

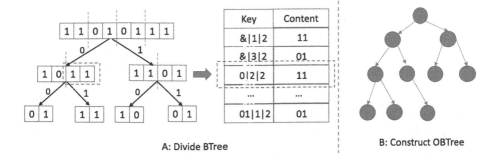

A: Divide BTree

B: Construct OBTree

Fig. 4. An illustration of constructing OBTree

In Algorithm 1, $BuildPathORAMTree(B, L_0)$ denotes to build an ORAM tree by inputting blocks B and the number of leaves in ORAM L_0.

Figure 4 shows how to construct an OBTree on the local client. For a dataset $DB = \{g_1, \cdots, g_4\}$, where g_i denotes a group which contains multiple keywords, we firstly construct a BTree as shown in Sect. 3.3. Then, we divide each node into splinters of the same size which equal the leaf size and stores these splinters in a hash table. For example, for the splinter "11", we let the key of this splinter be $0|2|2$, where '0' denotes the path of the splinter, the left '2' denotes the identifier of this splinter in node "1011", and the right '2' denotes the length of this splinter. Finally, we package this hash table into PT.

4.2   Search Protocols

We show the search protocol in Algorithm 2. Given a search token Tk, the search protocol searches over the OBTree in a top-down style. The overview of search protocol is as follows. First, the procedure in the enclave decrypts Tk as w. Then, given a node, the procedure within the enclave needs to recursively check whether its left child or the right node contains keyword w or not. If the length of the path is $L-1$, it denotes that the procedure reaches a leaf node and return this leaf's identifier id. The main difficulty is how to read the nodes of BTree obliviously into the enclave from the untrusted memory. We show the details in Algorithm 3. Specifically, the procedure of $searchObliviousNode$ in Algorithm 3 consists of the following three steps:

Algorithm 2: Search OBTree

Input: Search token Tk, BTree root size RZ, Block size l, BTree's level L
Output: group identifier id
Decrypt Tk as a keyword w
Get the root of BTree
Invoke, SearchObliviousNode

---

1. **Read a node.** The procedure within the enclave firstly looks up the hash table H to get the interested block identifier bID. Then the procedure determines where the path of the splinters stored in the OBTree. If the splinter is the first splinter of the node, the procedure gets the path identifier pid from the hash table PID. Otherwise, it gets pid from the $bNextPath$ stored in the block. Then it uses the ORAM read operation $ReadOram$ to get the corresponding block. Finally, the enclave splices all splinters to restore a whole node.
2. **Search the keyword.** When given the whole block, the procedure needs to tailor the block to the size of corresponding node in BTree, and loads the tailored block into a BF. Then, it tests whether the keyword w is belonging to the BF by using the procedure $search$ in [7].
3. **Determine the result.** When the length of the path is $L-1$, it denotes that a search reaches a leaf node and the current search is completed. Otherwise, it denotes that the search protocol need to recursively search the current

node's child and recursively sets $path \leftarrow path|0$ or $path \leftarrow path|1$. As shown in [7], false-positive errors may occur at any level of nodes in BTree. We can detect the false-positive errors by the number of results returned by the search protocol. If there is just one identifier, we can determine that the search procedure returns the right result. Otherwise, the false positive occurs. Note that we assume that the intersection of the groups is empty.

Algorithm 3: SearchObliviousNode

Input: keyword w, oblivious data structure $OBTree$, BTree root size RZ, BTree's level L, block size l, level of current node L_N, path of the current node $path$

Output: group identifier id

$Data \leftarrow \emptyset$

initialize a bloom filter BF $m_i \leftarrow RZ/2^{L_N}/l$

$pid \leftarrow null$

for $i=1,...,m_i$ **do**
  $bID \leftarrow H.get(path|i|l)$
  if $i=1$ **then**
    Get NID using path
    $pid \leftarrow PID.get[NID]$
  else
    $pid \leftarrow bNextPath$
  $(bData, bNextPath) \leftarrow ReadOram(pid, bID)$
  $Data \leftarrow Data|bData$

tailor the set $Data$ to be the length of 2^{L_N}

load the set $Data$ to BF

$flag := Search(BF, w)$

if *the length of path is* $L - 1 \& flag = 1$ **then**
  covert path to a number as a group identifier id
  return id

if *flag is left* **then**
  $path \leftarrow path\text{---}0$
  Invoke, SearchObliviousNode
else
  $path \leftarrow path\text{---}1$
  Invoke, SearchObliviousNode

---

5 Security and Complexity

Security. The main building block of OBTree is Path ORAM, and therefore, its security is mainly inherited from the security of Path ORAM. Given the same length accesses, as shown in Definition 1, path ORAM can hide access patterns. In our system, any search operations involved with the same number of ORAM accesses are indistinguishable. However, for a membership testing query, the tree traversal pattern consists of two kinds of cases. In the first case, the path reaches a leaf correctly. In the second case, there is more than one path satisfying the query in BTree, which means that false positives occur. Although OBTree can leak the path access pattern that denotes which path contains a given keyword, it hides each node's content and structure. We believe that path pattern leakage in OBTree is acceptable in many use cases.

Complexity. The search complexity of OBTree is sub-linear in the worst case. We compare OBTree with the related works in Table 1.

Table 1. Comparison with related works

Scheme	Query type	Index size	Search time	Construct time	Access pattern hiding
HardIDX [24]	*Range*	$\mathcal{O}(c \log n)$	$\mathcal{O}(x \log n)$	$\mathcal{O}(c \log n)$	×
POSUP [27]	*single* − *k*	$\mathcal{O}(c \log n)$	$\mathcal{O}(x \log n)$	$\mathcal{O}(c \log n)$	✓
OBTree	*MMT*	$\mathcal{O}(n \log n)$	$\mathcal{O}(x \log^2 n)$	$\mathcal{O}(n \log n)$	✓

† In this table, n denotes the number of documents, with x denotes the number of identifies in the search results, with *range* denotes range queries and *single* − *k* denotes keyword queries and with c denotes the maximum number of different keywords contained in all documents. The label ✓ shows that the protocol can prevent against access pattern leakages. Otherwise, it indicates ×.

6   Related Work

Searchabe encryption schemes [14–16] allow the untrusted cloud to conduct search operations on the encrypted dataset and recently are applied in blockchain system [2]. Song et al. [14] supports the first searchable encryption scheme for single plaintexts. To improve the performance, an encrypted inverted index data structure was proposed in [15]. Shen et al. [16] proposes a inner-product predicate encryption which has been made fully secure but has a linear search time. In contrast, schemes in [14] support private range queries with polylogarithmic search time by using the pseudorandom functions and symmetric encryption. Schemes in [18–20,33] support private boolean search with a sublinear search time. However, These encryption schemes can only search for keyword equality and leak much sensitive information such as access patterns.

SGX-based applications [22–24] use a trusted execution environment to operate the sensitive database in a hostile environment. In particular, EnclaveDB [22] provides an enclave protected database and HardIDX [24] and TrustedDB [23] design a read-only key-value index. But search protocols in [22–24] do not protect against statistical attacks. Oblivious computing [21,26–28] harness the trusted environment and ORAM technologies to hide access pattern leakage to defend statistical attacks. In particular, Oblidb [28] proposes an efficient scheme for generic oblivious access purposes by harnessing Intel SGX with Path ORAM. ZeroTrace [26] leverages Intel SGX with ORAM to enable oblivious memory primitives. However, these schemes can not support MMT on the encrypted dataset in an efficient way.

7 Conclusion

In this paper, we introduce a new oblivious index data structure, named OBTree, which enables oblivious multiple-set membership testing over encrypted dataset. OBTree is designed to carefully combine Intel SGX, compressed data structure and path ORAM. Compared with previous works, OBTree simultaneously enables a new search function of searchable encryption and hide access pattern leakage to protect against statistical attacks. In addition, OBTree is firstly working on the oblivious membership testing, and the search efficiency on it is sub-linear in the worst case.

References

1. Qi, S., Zheng, Y.: Crypt-DAC: cryptographically enforced dynamic access control in the cloud. IEEE Trans. Dependable Secure Comput. **18**(2), 765–779 (2019)
2. Qi, S., Lu, Y., Zheng, Y., Li, Y., Chen, X.: CPDS: enabling compressed and private data sharing for industrial IoT over blockchain. IEEE Trans. Ind. Inform. **17**(4), 2376–2387 (2021)
3. Qi, S., Zheng, Y., Li, M., Liu, Y., Qiu, J.: Scalable industry data access control in RFID-enabled supply chain. IEEE Trans. Network. **24**(6), 3551–3564 (2016)
4. Qi, S., Lu, Y., Wei, W., Chen, X.: Efficient data access control with fine-grained data protection in cloud-assisted IIoT. IEEE Internet Things J. **8**(4), 2886–2899 (2021)
5. Lu, Y., Qi, Y., Qi, S., Li, Y., Song, H., Liu, Y.: Say no to price discrimination: decentralized and automated incentives for price auditing in ride-hailing services. IEEE Trans. Mobile Comput. (2020). https://doi.org/10.1109/TMC.2020.3008315
6. Lu, Y., et al.: Secure deduplication-based storage systems with resistance to side-channel attacks via fog computing. IEEE Sens. J. (2021). https://doi.org/10.1109/JSEN.2021.3052782
7. Yoon, M., Son, J., Shin, S.: Bloom tree: a search tree based on bloom filters for multiple-set membership testing. In: INFOCOM, pp. 1429–1437 (2014)
8. Hähnel, M., Cui, W., Peinado, M.: High-resolution side channels for untrusted operating systems. In: USENIX Annual Technical Conference (ATC), pp. 299–312 (2017)
9. Lee, S., Shih, M.-W., Gera, P., Kim, T., Kim, H., Peinado, M.: Inferring fine-grained control flow inside SGX enclaves with branch shadowing. In: USENIX Security, pp. 557–574 (2017)
10. Brasser, F., Muller, U., Dmitrienko, A., Kostiainen, K., Capkun, S., Sadeghi, A.: Software grand exposure: SGX cache attacks are practical. In: WOOT (2017)
11. Shih, M.W., Lee, S., Kim, T., Peinado, M.: T-SGX: eradicating controlled channel attacks against enclave programs. In: NDSS (2017)
12. Goldreich, O., Ostrovsky, R.: Software protection and simulation on oblivious RAMs. J. ACM **43**(3), 431–473 (1996)
13. Stefanov, E., et al.: Path ORAM: an extremely simple oblivious RAM protocol. In: CCS, pp. 299–310 (2013)
14. Song, D.X., Wagner, D., Perrig, A.: Practical techniques for searches on encrypted data. In: S&P, pp. 44–55 (2000)
15. Curtmola, R., Garay, J., Kamara, S., Ostrovsky, R.: Searchable symmetric encryption: improved definitions and efficient constructions. In: CCS, pp. 79–88 (2006)

16. Shen, E., Shi, E., Waters, B.: Predicate privacy in encryption systems. In: Reingold, O. (ed.) TCC 2009. LNCS, vol. 5444, pp. 457–473. Springer, Heidelberg (2009). https://doi.org/10.1007/978-3-642-00457-5_27
17. Demertzis, I., Papadopoulos, S., Papapetrou, O., Deligiannakis, A., Garofalakis, M.: Practical private range search revisited. In: SIGMOD, pp. 185–198 (2016)
18. Cash, D., Jarecki, S., Jutla, C., Krawczyk, H., Roşu, M.-C., Steiner, M.: Highly-scalable searchable symmetric encryption with support for Boolean queries. In: Canetti, R., Garay, J.A. (eds.) CRYPTO 2013. LNCS, vol. 8042, pp. 353–373. Springer, Heidelberg (2013). https://doi.org/10.1007/978-3-642-40041-4_20
19. Pappas, V., Krell, F., Vo, B., Kolesnikov, V., Malkin, T., et al.: Blind seer: a scalable private DBMS. In: S&P, pp. 359–374 (2014)
20. Kamara, S., Moataz, T.: Boolean searchable symmetric encryption with worst-case sub-linear complexity. In: Coron, J.-S., Nielsen, J.B. (eds.) EUROCRYPT 2017. LNCS, vol. 10212, pp. 94–124. Springer, Cham (2017). https://doi.org/10.1007/978-3-319-56617-7_4
21. Jiang, Q., Qi, Y., Qi, S., Zhao, W., Lu, Y.: PBSX: a practical private Boolean search using Intel SGX. Inf. Sci. **521**, 174–194 (2020)
22. Priebe, C., Vaswani, K., Costa, M.: EnclaveDB: a secure database using SGX. In: S&P, pp. 264–278 (2018)
23. Bajaj, S., Sion, R.: TrustedDB: a trusted hardware based database with privacy and data confidentiality. In: SIGMOD, pp. 205–261 (2011)
24. Fuhry, B., Bahmani, R., Brasser, F., Hahn, F., Kerschbaum, F., Sadeghi, A.-R.: HardIDX: practical and secure index with SGX. In: Livraga, G., Zhu, S. (eds.) DBSec 2017. LNCS, vol. 10359, pp. 386–408. Springer, Cham (2017). https://doi.org/10.1007/978-3-319-61176-1_22
25. Roche, D.S., Aviv, A.J., Choi, S.G.: A practical oblivious map data structure with secure deletion and history independence. In: S&P, pp. 178–197 (2016)
26. Sasy, S., Gorbunov, S., Fletcher, C.W.: Zerotrace: oblivious memory primitives from intel SGX. IACR Cryptol. ePrint Arch. **2017**, 549 (2017)
27. Thang, H., Ozmen, M.O., Jang, Y., Yavuz, A.A.: Hardware-supported ORAM in effect: practical oblivious search and update on very large dataset. In: PETS, pp. 172–191 (2019)
28. Eskandarian, S., Zaharia, M.: ObliDB: oblivious query processing for secure databases. In: NDSS (2018)
29. Sun, W., Zhang, R., Lou, W., Hou, Y.T.: REARGUARD: secure keyword search using trusted hardware. In: INFOCOM, pp. 801–809 (2018)
30. Wang, X.S., Nayak, K., Liu, C., Chan, T., Shi, E., et al.: Oblivious data structures. In: CCS, pp. 215–226 (2014)
31. Bloom, B.: Space/time trade-offs in hash coding with allowable errors. Commun. ACM **13**(7), 422–426 (1970)
32. Costan, V., Devadas, S.: Intel SGX explained. IACR Cryptol. ePrint Arch. **2016**(86), 1–118 (2016)
33. Wang, H., Jiang, Q., He, H., Qi, Y., Yang, X.: Variable-length indistinguishable binary tree for keyword searching over encrypted data. In: Wang, G., Lin, X., Hendler, J., Song, W., Xu, Z., Liu, G. (eds.) WISA 2020. LNCS, vol. 12432, pp. 567–578. Springer, Cham (2020). https://doi.org/10.1007/978-3-030-60029-7_51

A Blockchain Architecture Design that Takes into Account Privacy Protection and Regulation

Xiangke Mao[✉], Xinhang Li, and Suwei Guo

BNRist, Department of Computer Science and Technology,
Institute of Internet Industry, Tsinghua University, Beijing, China
xiangkemao@tsinghua.edu.cn, xh-li20@mails.tsinghua.edu.cn

Abstract. The privacy protection technology on blockchain provides participants with good anonymity, but it also brings great difficulties to regulators. At the same time, for the permission chain, anonymity can not play the role of privacy protection. In this paper, a new framework is designed to meet the dual requirements of privacy protection and data supervision in the context of multi organizations and multi participants. The architecture can not only satisfy the privacy protection and supervision of transaction data, but also protect the data privacy under the condition of compliance with the supervision.

Keywords: Data privacy · Data supervision · Consortium blockchain · Smart contract

1 Introduction

In November 2008, Satoshi Nakamoto [1] published a white paper on Bitcoin, detailing how to build a decentralized cryptocurrency system. Later, in January 2009, it created the first block of Bitcoin-the genesis block, which marked the birth of the Bitcoin transaction system. In the past ten years, under the influence of Bitcoin, hundreds of cryptocurrencies have appeared [2–4]. Most of the underlying technologies of these digital currencies are based on blockchain technology. The blockchain has the characteristics of decentralization, openness and transparency, and tamper-proof. After more than ten years of development, the blockchain has surpassed the scope of use only in the field of cryptocurrency, and it is used in the Internet of Things [5,6], supply chain [7,8], and data sharing [9] etc. In 2013, the emergence of Ethereum [10] marked the arrival of the era of smart contracts. Its groundbreaking use of blockchain technology to implement smart contracts ensures that the contract can be automatically and effectively executed when the trigger conditions are met. In order to solve the needs of various enterprises and government departments for blockchain technology, the consortium blockchain technology has received great attention. Unlike

Supported by National Key R&D Program of China (2018YFB1404401).

C. Xing et al. (Eds.): WISA 2021, LNCS 12999, pp. 311–319, 2021.
https://doi.org/10.1007/978-3-030-87571-8_27

the public chain, the consortium chain has restrictions on the users that can be accessed, and can only be accessed when permission is obtained. This guarantees the security of the blockchain from the participant level. Hyperledger Fabric [11] is a representative of the consortium blockchain, which is widely used. In this paper, we mainly conduct research on the consortium blockchain.

Although the consortium chain confirms the identity of the joining node, it also uses the Hash algorithm to make the content on the blockchain unable to be tampered with. However, the data on the blockchain can be fully obtained and viewed by the participants of the consortium chain, which is prone to the issue of transaction privacy exposure. For some special application scenarios, it is not desirable for all participants to be able to see all transaction data. For example, in the scenario shown in Fig. 1. In the figure, there are three large organizations A, B, and C that trade with each other, and each organization contains three different participating members. Transactions between members within the organization are also required, but the transaction data between members within the organization is private transaction data, which can only be seen by participants inside the organization, and cannot be obtained by participants outside the organization. At the same time, transaction data between organizations can only be obtained by members of participating organizations, and not by members of other organizations that are not participating. At the same time, in the scenario shown in Fig. 1, in order to prevent fraudulent transactions within the organization or between organizations and deceive the review of the personnel of the regulatory agency, all transactions must be supervised. Similarly, since the regulator can view all transaction data, in order to prevent the relevant staff of the supervisory department from leaking the transaction data, it must be able to ensure the privacy of the data under the premise of satisfying the supervision.

In order to solve the above problems, this article proposes a new architecture based on the requirements of the scenario shown in Fig. 1 to solve three key issues (1) data privacy protection (2) transaction data supervision (3) how to regulate the blockchain under the premise of protecting privacy. In the following content, we will introduce the proposed new architecture in Sect. 2, describe related work in Sect. 3, and finally conclude the full text in Sect. 4.

2   Our Proposed Architecture

According to the existing blockchain architecture and the scenario in Fig. 1, we designed the architecture shown in Fig. 2. It consists of data layer, network layer, consensus layer, contract layer and application layer. Next, we will introduce each part of the architecture.

2.1   Data Layer

The data layer mainly includes block data and transaction data of blockchain. In order to avoid the excessive growth of data on the blockchain, the proposed architecture uses the combination of on chain storage and off chain storage. The

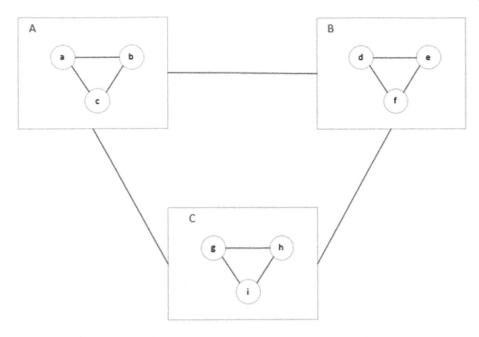

Fig. 1. An example of transaction network between multiple organizations and organization members.

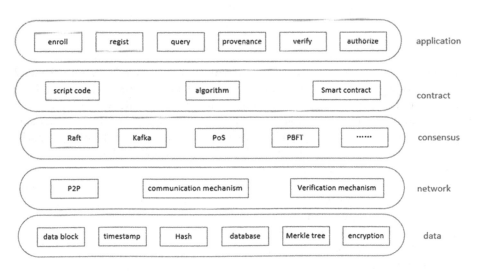

Fig. 2. The architecture of our designed blockchain.

hash value of transaction data is stored on the chain, while the database is used to store transaction data off chain. The data in the block is organized according to the structure of Merkle tree to facilitate the subsequent data authenticity and verification. According to the specific needs, the design of the new block is shown in Fig. 3.

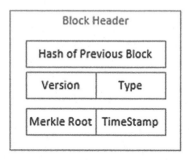

Fig. 3. The new block header.

Because the design blockchain is a consortium chain, there is no need for miners to mining, so there is no need for nonce. Because there are different types of blockchain in the designed architecture, the type of blockchain is added to facilitate the management of blockchain.

2.2 Network Layer

The network layer mainly realizes the information exchange between decentralized network nodes in the blockchain. The network layer includes distributed networking mechanism, data dissemination mechanism and data verification mechanism. Due to the use of completely P2P decentralized networking technology, each node in the network is equal in status and connected and interacted with each other in a flat topology structure. Each node will undertake the work of network routing, verification and dissemination of transaction information. In the proposed architecture, Gossip protocol is used to broadcast transaction data. Gossip can detect whether the nodes in the network are offline or online, synchronize the data between the nodes in the network, and when new nodes join, it can synchronize the data in time. In order to solve the three problems in Fig. 1, the blockchain connection network as shown in Fig. 4 is designed. In Fig. 4, there are three different types of roles: organization, organization member and regulation node.

1. Organization: an organization is composed of businesses or users engaged in the same type of activities. When users in different organizations conduct transactions, they will generate a new ledger.

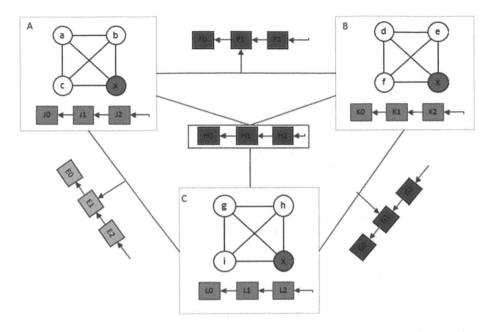

Fig. 4. The structure of blockchain network. (Color figure online)

2. Members of an organization: each member can belong to only one organization in the network.
3. Regulator: the regulator has the right to view and access all transaction data, but does not have the right to write. At the same time, its verification and data access are completed through the smart contract. The blockchain will record the executing process of the smart contract.

In the blockchain network shown in Fig. 4, there are also two different types of blockchain ledger, namely, the internal ledger and the inter organizational ledger.

1. The internal ledger of an organization is mainly used to record the transaction data between members of the organization and the transaction data queried and accessed by regulators. Only the members who are allowed can access the ledger. This ensures the privacy of transaction data within the organization.
2. The inter organization ledger mainly records the transaction data of inter organization members and the data accessed by regulators, so as to ensure the privacy of inter organization data transaction, and other organizations that are not involved will not be able to obtain the data.

When a new ledger is generated, a regulatory node (red) will be automatically generated. The regulatory node also uses the method of on chain storage and off chain storage. For the supervision node, there are only query and access functions. Moreover, when the supervisor node accesses the original transaction data, it can only access through the smart contract, that is, it can only perform the pre-set operation in the contract when it meets the pre-set conditions.

2.3 Consensus Layer

The consensus layer mainly encapsulates all kinds of consensus mechanism algorithms of network nodes. Consensus mechanism algorithm is the core technology of blockchain technology, which is responsible for allocating the task load of distributed network accounting nodes, so that highly dispersed nodes can reach a consensus on the effectiveness of blockchain data efficiently, affecting the security and reliability of the whole system. The designed consortium chain does not need mining, so it does not use the proof of work (POW) mechanism similar to that in bitcoin, but uses raft, Kafka and Practical Byzantine Fault Tolerance (PBFT) to reach a consensus, effectively ensuring the consistency of data.

2.4 Contract Layer

The contract layer encapsulates all kinds of script codes, algorithms and smart contracts of the blockchain, which is the basis for flexible programming and data operation of the blockchain system. We design data access, data verification, data storage and other contracts. The main purpose of designing the contract is to automatically write the user's record of access the original data or operation records into the blockchain, so as to reduce the operator's intervention in the data as much as possible, so as to better ensure the security of the data. Next, we describe the execution process of the smart contract by taking the regulator's verification and access to transaction data as an example.

1. The system authenticates the identity of the regulator. If the verification is passed, enter the second step. Otherwise, re-enter the authentication information.
2. Select data validation.
3. The system calls the smart contract of data verification. At this time, the contract will verify the operator's permissions. If you have permission, the execution contract will call other related contracts to return the verification results. If it does not have permission, it tells the operator that it does not have permission.
4. According to the verification result returned in the previous step, if the verification is passed, you can return to the second step to verify other transactions. If it fails, go to the next step.
5. For the failed data, the original data needs to be obtained, and the smart contract of original data access will be triggered.
6. Data access requires the signatures of multiple participants in the ledger to read the original data. At this time, the contract will collect the signatures of the ledger owner in the network. Once the pre-set signature number in the contract is obtained, the contract will read the original transaction data in the database and return it to the regulator.

In the above process, only when the transaction fails to pass the verification, the regulator has the authority to obtain the original transaction data. Moreover, when it accesses data, it also needs to get the signature of the data owner. This reduces the possibility of direct contact between regulators and data, and ensures the privacy and security of data in the regulatory environment.

2.5 Application Layer

The application layer encapsulates various application scenarios and cases of blockchain. It includes the interface of registration, authentication, query, provenance, verification and authorization, which facilitates the subsequent development.

3 Related Work

As the earliest blockchain system, bitcoin uses anonymity to hide the user's identity information in order to protect the user's privacy. But bitcoin network has three important characteristics (1) Transaction data is public and accessible to anyone (2) Using the UTXO model, the input and output addresses of each transaction can be obtained, so it is easy to build a transaction network, and then analyze the network. (3) multiple use of public key address. The payee and the payer are identified by public key address. If the same address is used many times, the user's identity information may be exposed. [12] pointed out the potential anonymous threat of bitcoin by analyzing the transaction data of bitcoin from January 3, 2009 to July 12, 2011. In order to solve the problem of bitcoin anonymity leakage, many solutions to improve the anonymity of bitcoin have been proposed [13,14], especially for some new digital currencies, which design different solutions to improve the anonymity of bitcoin. For example, in order to hide the original transaction information from the outside world, [15] use the method of combining multi transactions into one large transaction. In [2,16], Ring signature and address hiding technology are used to realize the anonymity of the sender and receiver of the transaction, and ring anonymity is used to realize the anonymity of the transaction itself.

Most of the above schemes are designed to hide the user's identity information in the network, which is not suitable for the scenario of consortium chain. Because in the consortium chain, the identity of each participant can be confirmed. As the most commonly used hyper ledger, it gives up the encryption and decryption to protect private data, and designs private data, pipeline and other mechanisms to ensure the privacy of data.

Data privacy and regulation interact. The higher the technology of privacy protection, the more difficult the regulation will be. Regulation mechanism is essential for the healthy development of blockchain. Without regulation, users may take advantage of privacy protection to engage in criminal activities. Data privacy protection is also indispensable. In order to serve the regulation, it is not advisable to give up data privacy completely. Therefore, how to protect the privacy of data under the condition of meeting the regulatory requirements is the problem of this paper. Inspired by the data privacy protection in Fabric, this paper integrates the regulatory mechanism and designs a new architecture to meet the requirements of the scenario in Fig. 1.

4   Conclusion

In this paper, a new block chain architecture is designed, which not only ensures data privacy, but also carries out data supervision under the situation of multi organization and multi participants. The architecture mainly includes data layer, network layer, consensus layer, contract layer and application layer. In the network layer, the regulation node is introduced to supervise the transaction data of the whole network. At the same time, two different types of ledger are designed to record the transaction data, which ensures the privacy of the data between organizations. In the contract layer, new contracts are designed, such as data verification and data access contracts, to control regulators' access to the original transaction data. In the future research, we will borrow the ideas of distribute system and database [18,19] to design a blockchain architecture that not only meets regulatory requirements but also ensures data privacy.

References

1. Nakamoto, S.: Bitcoin: a peer-to-peer electronic cash system. Decentralized Bus. Rev., 21260 (2008)
2. Noether, Shen: Ring SIgnature Confidential Transactions for Monero. IACR Cryptology ePrint Archive 2015, 1098 (2015)
3. Li, C., et al.: A decentralized blockchain with high throughput and fast confirmation. In: 2020 USENIX Annual Technical Conference (USENIXATC 20), pp. 515–528 2020
4. Miers, I., Garman, C., Green, M., Rubin, A.D.: Zerocoin: anonymous distributed E-cash from bitcoin. In: 2013 IEEE Symposium on Security and Privacy, pp. 397–411. IEEE (2013)
5. Latif, S., Idrees, Z., Ahmad, J., Zheng, L., Zou, Z.: A blockchain-based architecture for secure and trustworthy operations in the industrial Internet of Things. J. Ind. Inf. Integr. **21**, 100190 (2021)
6. Tseng, L., Wong, L., Otoum, S., Aloqaily, M., Othman, J.B.: Blockchain for managing heterogeneous internet of things: a perspective architecture. IEEE Netw. **34**(1), pp. 16–23 (2020)
7. Wamba, S.F., Queiroz, M.M., Trinchera, L.: Dynamics between blockchain adoption determinants and supply chain performance: an empirical investigation. Int. J. Prod. Econ. **229**, 107791 (2020)
8. Esmaeilian, B., Sarkis, J., Lewis, K., Behdad, S.: Blockchain for the future of sustainable supply chain management in Industry 4.0. Resour. Conserv. Recycl. **163**, 105064 (2020)
9. Qi, S., Youshui, L., Zheng, Y., Li, Y., Chen, X.: CPDS: enabling compressed and private data sharing for industrial internet of things over blockchain. IEEE Trans. Industr. Inf. **17**(4), 2376–2387 (2020)

10. Buterin, V.: A next-generation smart contract and decentralized application platform. White Paper **3**(37) (2014)
11. Cachin, C.: Architecture of the hyperledger blockchain fabric. In: Workshop on Distributed Cryptocurrencies and Consensus Ledgers, vol. 310, no. 4 (2016)
12. Reid, F., Harrigan, M.: An analysis of anonymity in the bitcoin system. In: Altshuler, Y., Elovici, Y., Cremers, A., Aharony, N., Pentland, A. (eds.) Security and Privacy in Social Networks, pp. 197–223. Springer, New York (2013). https://doi.org/10.1007/978-1-4614-4139-7_10
13. Bonneau, J., Narayanan, A., Miller, A., Clark, J., Kroll, J.A., Felten, E.W.: Mixcoin: anonymity for bitcoin with accountable mixes. In: Christin, N., Safavi-Naini, R. (eds.) FC 2014. LNCS, vol. 8437, pp. 486–504. Springer, Heidelberg (2014). https://doi.org/10.1007/978-3-662-45472-5_31
14. Ruffing, T., Moreno-Sanchez, P.: ValueShuffle: mixing confidential transactions for comprehensive transaction privacy in bitcoin. In: Brenner, M., et al. (eds.) FC 2017. LNCS, vol. 10323, pp. 133–154. Springer, Cham (2017). https://doi.org/10.1007/978-3-319-70278-0_8
15. Maurer, F.K., Neudecker, T., Florian, M.: Anonymous CoinJoin transactions with arbitrary values. In: 2017 IEEE Trustcom/BigDataSE/ICESS, pp. 522–529. IEEE (2017)
16. Heilman, E., Alshenibr, L., Baldimtsi, F., Scafuro, A., Goldberg, S.: TumbleBit: an untrusted bitcoin-compatible anonymous payment hub. In: Network and Distributed System Security Symposium (2017)
17. Möser, M., et al.: An empirical analysis of traceability in the Monero blockchain. arXiv preprint arXiv:1704.04299 (2017)
18. Zhao, X., Lei, Z., Zhang, G., Zhang, Y., Xing, C.: Blockchain and distributed system. In: Wang, G., Lin, X., Hendler, J., Song, W., Xu, Z., Liu, G. (eds.) WISA 2020. LNCS, vol. 12432, pp. 629–641. Springer, Cham (2020). https://doi.org/10.1007/978-3-030-60029-7_56
19. Ruan, P., et al.: Blockchains vs. distributed databases: dichotomy and fusion. In: Proceedings of the 2021 International Conference on Management of Data, pp. 1504–1517, June 2021

Knowledge Graph

Constructing Chinese Historical Literature Knowledge Graph Based on BERT

Qingyan Guo[1], Yang Sun[1], Guanzhong Liu[1], Zijun Wang[1], Zijing Ji[1], Yuxin Shen[1], and Xin Wang[1,2(✉)]

[1] College of Intelligence and Computing, Tianjin University, Tianjin, China
{qingyan,yangsun,3018216075,3018216240,jizijing,shenyuxin,
wangx}@tju.edu.cn
[2] Tianjin Key Laboratory of Cognitive Computing and Application,
Tianjin 300072, China

Abstract. Knowledge graph construction (KGC) aims to organize knowledge into a semantic network which can reveal relations between entities. Its basis is named entity recognition (NER) and relation extraction (RE) tasks. In recent years, KGC methods for Chinese have made great progress. However, most existing methods concentrate on modern Chinese and ignore the classical Chinese due to its complexity, making research in this field relatively lacking. In this paper, we construct a high-quality classical Chinese labeled dataset for NER and RE tasks. More specifically, we conduct a series of experiments to select an optimal NER model to strengthen the whole pipeline model for NER and RE tasks, augmenting our dataset iteratively and automatically. Additionally, we propose an improved RE model to better combine semantic entity information extracted by the NER model. Moreover, we construct a knowledge graph (KG) based on Chinese historical literature and design a visualization system with intuitive display and query functions.

Keywords: Named Entity Recognition · Knowledge graph · Relation extraction · BERT

1 Introduction

As a kind of structured knowledge, knowledge graph (KG) has attracted extensive attention in academic communities. KG, as a knowledge organization and representation method based on graph structure, composed of entities, relations, and semantic descriptions, is used for describing entities, abstract concepts, events and their relations in the real world. Specifically, KGs encode various entities and their relations into subject-predicate-object triples. With the rapid development of natural language processing (NLP), automatic knowledge mining from massive unstructured text has become an important trend. Compared with the traditional manual construction method, training deep learning models to extract knowledge from text automatically is necessary.

© Springer Nature Switzerland AG 2021
C. Xing et al. (Eds.): WISA 2021, LNCS 12999, pp. 323–334, 2021.
https://doi.org/10.1007/978-3-030-87571-8_28

Knowledge graph construction (KGC) aims to organize and visualize knowledge, which is based on tasks of Named Entity Recognition (NER) and Relation Extraction (RE). In recent years, a large number of KGs, such as FreeBase [1], YAGO [19], and Satori [6], have been constructed and show great potential to provide efficient services. However, most of the existing KGs in Chinese, including encyclopedia Zhishi.me [16], XLore [23], and CN-DBpedia [26], are based on modern Chinese. There are few studies of KGC on classical Chinese. As shown in Fig. 1, classical Chinese literature contains rich historical knowledge. However, the information has not been fully exploited. In addition, high-quality labeled datasets are rare due to the difficulty of grammars and semantics in classical Chinese, which hinders the process of KGC in classical Chinese.

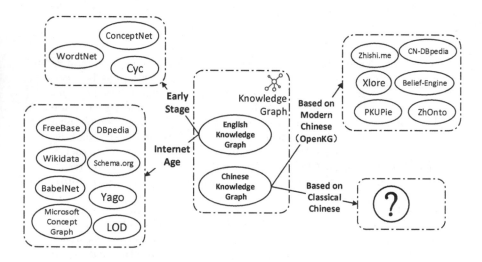

Fig. 1. State-of-the-art of KGs

To deal with the above issues, we use the pre-trained language model and pipeline model consisting of NER and RE to extract knowledge from large-scale unlabeled classical documents, and construct a high-quality KG based on classical Chinese literature.

The main contributions of this paper are as follows: (1) We propose an end-to-end pipeline-based knowledge extraction model based on BERT [5], which can be applied to knowledge extraction in Chinese classical text, and has achieved competitive results in Chinese Twenty-Four Histories. (2) To ensure the high-accuracy of relation predictions, we propose an improved RE model combined with entity type information predicted in NER phase. The experimental results show that the performance of the improved model is better than baseline models. (3) We construct a labeled dataset for NER and RE in classical Chinese (CCNR) with high-quality and conduct a series of experiments on CCNR. Extensive experiments show that the modified RoBERTa [15] and Guwen-BERT[1] (using

---

[1] GuwenBERT https://github.com/ethan-yt/guwenbert.

our RE model) perform better than original models without modification. (4) Based on Chinese historical literature, we construct a high-quality KG and design a visualization system to display CHL-KG and provide query functions.

2 Related Work

In this section, we introduce existing methods which are related to our research, including named entity recognition and relation extraction models.

2.1 Named Entity Recognition

Named entity recognition, the basis of NLP tasks, is a subtask of information extraction that seeks to locate and classify named entities mentioned in unstructured text into pre-defined categories, e.g., person names and organizations [24]. Yu et al. [27] integrated the BiLSTM-CRF model for classical Chinese NER and proved that BERT [5] has a strong ability in transfer learning.

Rule-Based Method. Rule-based methods can automatically discover and generate rules, and DL-Cotram proposed by Collins et al. [3] is the most representative method. However, the scalability of these methods is poor and cannot recognize unregistered words.

Statistics-Based Machine Learning Method. Based on probability, Hidden Markov Model (HMM) and Conditional Random Field (CRF) [12] improve the rule-based method and regard NER as a sequence labeling process.

Deep Learning-Based Method. Traditional deep learning-based NER methods concentrate on Convolutional Neural Networks (CNN) and Recurrent Neural Networks (RNN). Collobert et al. [4] proposed two network structures to complete NER tasks: the window method and sentence method. Li et al. [14] applied the RNN-based network to Chinese biomedical NER tasks. Additionally, Huang et al. [9] proposed a series of neural network models for sequence annotation, including LSTM [8], bidirectional LSTM (Bi-LSTM), LSTM with the CRF layer, and Bi-LSTM with the CRF layer, which perform well in various NLP tasks.

Recently, transformer-based end-to-end models have been widely applied. Vaswani et al. [21] proposed a transformer model for encoding and decoding intermediate language representation between different languages in order to solve the sequence to sequence problem. BERT proposed by Google [5] has achieved SOTA performance in 11 different NLP tasks and becomes a milestone in the history of NLP development. Compared with BERT, RoBERTa [15] dynamically adjusts the mask mechanism, improves the pre-training task, and uses a larger pre-training corpus. ALBert [13] also modifies the pre-training task of BERT, and reduces training parameters to accelerate the training process. ERNIE [20] proposed by Baidu fully combines the structured knowledge included in existing KGs, and takes the advantage of syntax and semantic information.

2.2 Relation Extraction

Relation extraction, also regarded as triple (subject-predicate-object) extraction, explores connections between entity pairs.

Joint RE Model. By extracting entities and their relations at the same time, the joint RE model avoids the accumulation of errors in the pipeline model. Katiyar et al. [10] first used the deep Bi-LSTM sequence annotation method for joint entity extraction, and further improved the accuracy of the model by adding constraints and relation optimization. Zheng et al. [29] proposed a new annotation strategy and relation annotation method. By learning the location information of entities, the type information of entity relations, the role information of entities, and triples were directly extracted by using the neural network.

Pipeline-Based RE Model. Hashimoto et al. [7] applied RNN to syntax tree analysis, and replaced word dependency matrix with additional features. Zeng et al. [28] proposed to use CNN for relation and feature extraction at word and sentence level. Aiming at specific entities, Wang et al. [22] added multi-layer attention mechanism to CNN architecture. Recently, pre-processing models, i.e., pre-training language models, have been widely proposed and used, e.g., BERT, Generative Pre-Training (GPT) [18], and Embeddings from Language Model (ELMo) [17]. Christopoulou et al. [2] applied BERT to relational representation and proposed a novel 'Matching the Blanks' pre-training task.

3 Methodology

The construction process of the Chinese historical literature KG (CHL-KG) is shown in Fig. 2. According to the entity and relation categories which we define in original unstructured text in classical Chinese, we annotate a portion of Chinese Twenty-Four Histories manually. Taking the initial labeled dataset of high quality as training set, we utilize the pipeline model based on BERT [5] to predict remaining entities and relations of unlabeled parts. The prediction results are manually modified and combined with the previous version to obtain the next version of dataset. Circulating several iterations, we import the final dataset to Neo4j² and construct a KG as well as a visualization platform.

3.1 Classical-Chinese-NER-RE-Dataset

To the best of our knowledge, CCNR (Classical-Chinese-NER-RE-Dataset) is the first open dataset for NER and RE tasks in classical Chinese. Considering the differences between classical Chinese and modern Chinese in terms of grammar, phrases and syntax, it is necessary to construct a labeled dataset by circulating

---

² https://neo4j.com/.

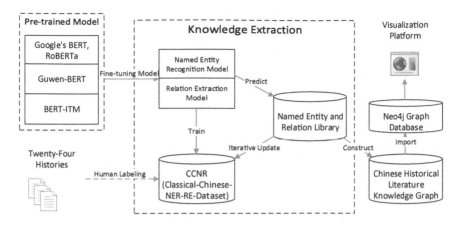

Fig. 2. Construction process of CHL-KG

manual annotation and model prediction processes for the neural network to extract knowledge.

Since the phrases, grammar and syntax of classical Chinese vary over time, we select four historical materials with a large span of eras from the Twenty-Four Histories as data sources to increase the applicability of the dataset, namely the Historical Records (史记), Records of the Three Kingdoms (三国志), Book of Northern Qi (北齐书), and Later History of the Five Dynasties (后五代史).

NER-Dataset. The dataset of NER task contains six entity categories: person (PER), location (LOC), organization (ORG), job (JOB), book (BOO), and war (WAR). The statistics of named entities are shown in Table 1. We utilize the 'BIO' labeling method to annotate entities and label each element as "B-X", "I-X" or "O". Among them, "B-X" means that the fragment in which this element is located belongs to type X and the element is at the beginning of this fragment. "I-X" means that the fragment in which this element is located belongs to type X and this element is in the middle of the fragment, and "O" means that this element does not belong to any type.

The data format is shown in Fig. 3. The example means that Huang Di (one of the legendary Chinese sovereigns [25]) was Shao Dian's (an ancient man) son, whose given name was Xuanyuan and family name was Gongsun. In this sentence, Huang Di, Shao Dian, Gong Sun, and Xuan Yuan all represent names of person and therefore are labeled as 'PER'.

RE-Dataset. The dataset of RE task contains 4,886 examples and includes 41 types of relations which can be divided into three categories: kinship, personal information, and social performance. We preprocess and standardize the dataset by deleting relation types consisting of less than 20 instances and complementing paired relationships. We mark each entity as the subject of the relation and other

Table 1. Entity statistics of each category

# PER	# LOC	# ORG	Total
13,472	4,149	2,100	
# JOB	# BOO	# WAR	23,362
3,178	157	6	

Fig. 3. Data format of the NER-dataset

entities within a certain distance as its corresponding objects. If there is a real relation between the pair, the relation type is marked as 'known', otherwise it is annotated as 'unknown' to strengthen the generalization ability of the model.

3.2   Named Entity Recognition Model

With extensive applications of deep neural network models in NLP, end-to-end deep learning models have gradually become the mainstream method. BERT learns from large-scale unlabeled corpus through unsupervised learning method, and focuses on word-level and sentence-level characteristics, enhancing the semantic representation of word vectors. In the NER phase, we use a model based on BERT, which is pre-trained by modern Chinese corpus. The model generates token, segment, and position embeddings for each word and feeds them into 12 transformer blocks with 768 hidden layers by default. Taking a sentence from Book of Northern Qi (北齐书) as an example input, the architecture of the NER model is shown in Fig. 4.

3.3   RE Model

Using the pre-trained language model as the encoder, we consider taking the entity category information as the input feature to equip the model with the ability to learn possible relations between entities more efficiently.

The architecture of the RE model is shown in Fig. 5. For the input sentence " 袁聿修字方德 " (Yuan Yuxiu's courtesy name is Fang De), the entity pair, "袁聿修 " (Yuan Yuxiu, an ancient man in China) and " 方德 " (Fang De, Yuan Yuxiu's courtesy name) can be predicted as 'PER' in the previous NER phase, which are taken as references in our RE model. Furthermore, we deliberately add the 'unknown' tag to facilitate the process of RE and strengthen the robustness of the model. That is, for two entities without any kind of relation, the model will tag their relationship as 'unknown'.

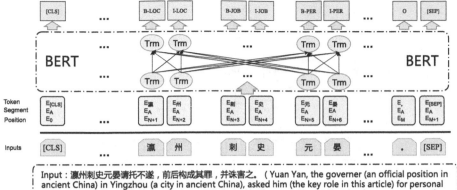

Fig. 4. Architecture of NER model based on BERT

Additionally, we utilize the encoder to obtain pooled-entity-level and sentence-level feature representations from entity pairs and the whole sentence. A linear transformation of the feature representation along with the encoder's hidden states is beneficial to capture the important hidden features. Then the representations and the corresponding entity type information are concatenated and then go through a fully connected network. Finally the relational classification results are given by Softmax function.

To sum up, we add the entity type labels to the model as additional inputs, promoting the RE model to better combine the semantic information offered by entity pairs predicted by the NER model.

4   Experiments

In this section, we evaluate the proposed models, and present performances on our CCNR dataset. We first introduce the experimental settings. Next, we select an optimal NER model and validate our RE model.

4.1   Experimental Settings

Datasets and Baselines. For NER phase, we evaluate four methods, including BERT-base [5], RoBERTa-base [15], Guwen-BERT, and BERT-ITM (an incremental training method base on BERT) on three datasets, including Modern Chinese, CCNR-3, and CCNR-6. Note that, CCNR containing 3 types and 6 types of entities are named CCNR-3 and CCNR-6, respectively.

Training. We set the proportion of linear warmup method as 10%, training batch size as 32, and use the Adam optimizer [11] with default parameters ($\beta_1 = 0.9, \beta_2 = 0.99, \varepsilon = 1e - 8$). We set the initial learning rate as $5e - 5$ for NER phase and $1e - 5$ for RE phase. And the weight decay is 0.01 for NER phase. Note that, RE phase use the Adam optimizer without weight decay.

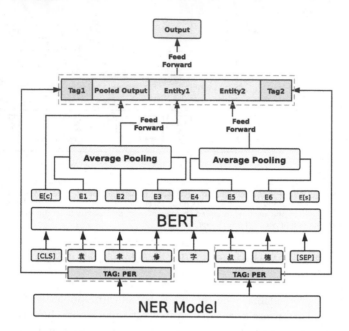

Fig. 5. Architecture of RE model

Implementation. Following the previous work BERT-base [5], we use a linear warmup method and optimize models with Adam optimizer. We use the precision, recall, and F1 as metrics, and conduct experiments on Tesla-V100 GPU.

Instance Processing. For both NER and RE phases, we assign the maximum length of each instance as 256 and mark punctuations as truncation positions to make the model fully consider contextual information.

4.2 NER Model Selection

In this part, we conduct extensive experiments to select an optimal NER model to strengthen the whole pipeline model.

Considering the number of entities labeled as 'BOO', 'WAR', and 'JOB' are less than other categories, we select 3 types of entities with the highest frequency in the whole dataset, i.e., 'PER', 'ORG', and 'LOC', as named entities for experiments. In addition to Guwen-BERT, we test BERT-base and RoBERTa-base, which are pre-trained by modern Chinese corpus, since the named entities in modern Chinese labeled dataset are the same as 3 entities selected above. The results are shown in Table 2.

We can observe that, Guwen-BERT outperforms RoBERTa-base and BERT-ITM on CCNR-3 dataset. Moreover, the performance of Guwen-BERT on CCNR-6 dataset is better than RoBERTa-base. The reason is that Guwen-BERT is pre-trained by a large-scale classical Chinese corpus, combines modern Chinese RoBERTa weights, and therefore transfers part of the language features of modern Chinese to classical Chinese. Thus, we choose Guwen-BERT as the NER model to extract entities and feed back to the RE model, thereby improving the whole pipeline process.

Table 2. Performances of NER models (%)

PTM	Dataset	F1	Precision	Recall
BERT-base [5]	Modern Chinese	43.95	35.96	57.60
RoBERTa-base [15]	Modern Chinese	45.17	37.38	57.67
RoBERTa-base [15]	CCNR-3	41.82	43.88	40.39
Guwen-BERT	CCNR-3	**53.18**	**52.28**	**61.08**
BERT-ITM	CCNR-3	42.11	42.54	42.32
BERT-ITM	CCNR-3	37.47	35.76	43.41
RoBERTa-base [15]	CCNR-6	36.47	34.99	39.87
Guwen-BERT	CCNR-6	36.78	35.34	40.59

4.3 RE Model Validation

In this part, we conduct extensive experiments to valid RE models. The semantic information, namely, the type and position embeddings of entity pairs are offered by the NER model selected above.

We use the original and improved RE models trained by CCNR to conduct RE experiments on the above models. The ablation results are shown in Table 3. We can observe that concatenating entity tags with pooled entity embeddings provides consistent improvement on all models, especially on RoBERTa-based models (RoBERTa and Guwen-BERT). We can also observe that adding tag information can be helpful to learn the semantic features of entity categories when the corpus is limited. In addition, Guwen-BERT and our improved RE method achieve the best performance, and BERT-ITM also achieves impressive performance. The main reason is that Guwen-BERT and BERT-ITM have a comprehensive representation of classical Chinese grammar and vocabulary since they are pre-trained by classical Chinese corpus.

Table 3. Performances of RE models (%)

PTM	RE model	F1
BERT-base	Original	56.65
BERT-base	With Tag (ours)	55.68
RoBERTa	Original	55.94
RoBERTa	With Tag (ours)	57.56
Guwen-BERT	Original	56.77
Guwen-BERT	With Tag (ours)	**59.35**
BERT-ITM	Original	57.28
BERT-ITM	With Tag (ours)	57.95

5   Chinese Historical Literature KG

Based on CCNR, we construct our named entity library and entity relation library used for KGC. Furthermore, we build a KG visualization platform through the Neo4j interface for educational purposes.

We utilize Neovis.js as the front-end to display CHL-KG and Django as the back-end framework. The platform shown in Fig. 6 provides three query functions, i.e., the natural language query, precise query, and professional query, to search CHL-KG and Fig. 6(a) shows the homepage of the platform. The natural language query shown in Fig. 6(b) takes natural language sentences as input, and matches the corresponding Cypher template at the back-end to search. The precise query shown in Fig. 6(c) uses the categories of entities and relations to return particular types of entities and relations, which provides a fuzzy query function. The professional query shown in Fig. 6(d), taking Cypher query language as input, is for users who are proficient in Graphical Query Language. These three types of interfaces are designed to meet the requirements of different end-users and help them to have an intuitive understanding of CHL-KG.

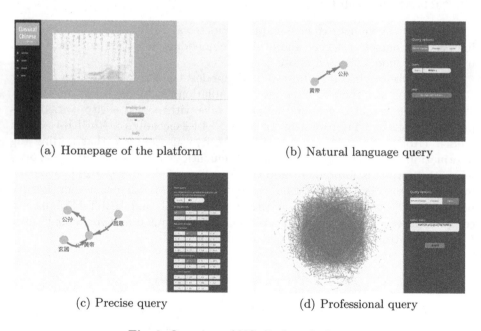

(a) Homepage of the platform   (b) Natural language query

(c) Precise query   (d) Professional query

Fig. 6. Overview of KG display platform

6   Conclusion

In this paper, we construct a high-quality classical Chinese annotation dataset for NER and RE, and enhance the pipeline model by selecting the best NER model to augment the labeled dataset automatically. The experimental results demonstrate that the RE model combined with semantic entity information performs better than basic models. Based on Chinese historical literature, we further construct a KG and develop a visualization platform to display CHL-KG.

Acknowledgement. This work is supported by the China Universities Industry, Education and Research Innovation Foundation Project (2019ITA03006), and the National Training Programs of Innovation and Entrepreneurship for Undergraduates (202010056117).

References

1. Bollacker, K., Evans, C., Paritosh, P., Sturge, T., Taylo, J.: Freebase: a collaboratively created graph database for structuring human knowledge. In: Proceedings of the 2008 ACM SIGMOD International Conference on Management of Data, pp. 1247–1250 (2008)
2. Christopoulou, F., Miwa, M., Ananiadou, S.: A walk-based model on entity graphs for relation extraction. arXiv preprint arXiv:1902.07023 (2019)
3. Collins, M., Singer, Y.: Unsupervised models for named entity classification. In: 1999 Joint SIGDAT Conference on Empirical Methods in Natural Language Processing and Very Large Corpora (1999)
4. Collobert, R., Weston, J., Bottou, L., Karlen, M., Kavukcuoglu, K., Kuksa, P.: Natural language processing (almost) from scratch. J. Mach. Learn. Res. **12**, 2493–2537 (2011)
5. Devlin, J., Chang, M.W., Lee, K., Toutanova, K.: BERT: pre-training of deep bidirectional transformers for language understanding. arXiv preprint arXiv:1810.04805 (2018)
6. Gao, Y., Liang, J., Ilan, B., Yakout, M., Mohamed, A.: Building a large-scale, accurate and fresh knowledge graph. In: KDD-2018, Tutorial, vol. 39, pp. 1939–1374 (2018)
7. Hashimoto, K., Miwa, M., Tsuruoka, Y., Chikayama, T.: Simple customization of recursive neural networks for semantic relation classification. In: Proceedings of the 2013 Conference on Empirical Methods in Natural Language Processing, pp. 1372–1376 (2013)
8. Hochreiter, S., Schmidhuber, J.: Long short-term memory. Neural Comput. **9**(8), 1735–1780 (1997)
9. Huang, Z., Xu, W., Yu, K.: Bidirectional LSTM-CRF models for sequence tagging. arXiv preprint arXiv:1508.01991 (2015)
10. Katiyar, A., Cardie, C.: Investigating LSTMs for joint extraction of opinion entities and relations. In: Proceedings of the 54th Annual Meeting of the Association for Computational Linguistics (Volume 1: Long Papers), pp. 919–929 (2016)
11. Kingma, D.P., Ba, J.: Adam: a method for stochastic optimization. arXiv preprint arXiv:1412.6980 (2014)
12. Lafferty, J., McCallum, A., Pereira, F.C.: Conditional random fields: probabilistic models for segmenting and labeling sequence data (2001)
13. Lan, Z., Chen, M., Goodman, S., Gimpel, K., Sharma, P., Soricut, R.: Albert: a lite BERT for self-supervised learning of language representations. arXiv preprint arXiv:1909.11942 (2019)
14. Li, J., et al.: WCP-RNN: a novel RNN-based approach for bio-NER in Chinese EMRs. J. Supercomput. **76**(3), 1450–1467 (2020)

15. Liu, Y., et al.: Roberta: a robustly optimized BERT pretraining approach. arXiv preprint arXiv:1907.11692 (2019)
16. Niu, X., Sun, X., Wang, H., Rong, S., Qi, G., Yu, Y.: Zhishi.me - weaving Chinese linking open data. In: Aroyo, L., et al. (eds.) ISWC 2011. LNCS, vol. 7032, pp. 205–220. Springer, Heidelberg (2011). https://doi.org/10.1007/978-3-642-25093-4_14
17. Peters, M.E., et al.: Deep contextualized word representations. arXiv preprint arXiv:1802.05365 (2018)
18. Radford, A., Narasimhan, K., Salimans, T., Sutskever, I.: Improving language understanding by generative pre-training (2018)
19. Suchanek, F.M., Kasneci, G., Weikum, G.: Yago: a core of semantic knowledge. In: Proceedings of the 16th International Conference on World Wide Web, pp. 697–706 (2007)
20. Sun, Y., et al.: ERNIE: enhanced representation through knowledge integration. arXiv preprint arXiv:1904.09223 (2019)
21. Vaswani, A., et al.: Attention is all you need. arXiv preprint arXiv:1706.03762 (2017)
22. Wang, L., Cao, Z., De Melo, G., Liu, Z.: Relation classification via multi-level attention CNNs. In: Proceedings of the 54th Annual Meeting of the Association for Computational Linguistics (Volume 1: Long Papers), pp. 1298–1307 (2016)
23. Wang, Z., et al.: XLore: a large-scale English-Chinese bilingual knowledge graph. In: International Semantic Web Conference (Posters & Demos), vol. 1035, pp. 121–124 (2013)
24. Wikipedia contributors: Named-entity recognition – Wikipedia, the free encyclopedia. https://en.wikipedia.org/w/index.php?title=Named-entity_recognition&oldid=959772078 (2020). Accessed 20 May 2021
25. Wikipedia contributors: Yellow emperor – Wikipedia, the free encyclopedia (2021). https://en.wikipedia.org/w/index.php?title=Yellow_Emperor&oldid=1038043350. Accessed 14 Aug 2021
26. Xu, B., et al.: CN-DBpedia: a never-ending Chinese knowledge extraction system. In: Benferhat, S., Tabia, K., Ali, M. (eds.) IEA/AIE 2017. LNCS (LNAI), vol. 10351, pp. 428–438. Springer, Cham (2017). https://doi.org/10.1007/978-3-319-60045-1_44
27. Yu, P., Wang, X.: BERT-based named entity recognition in Chinese twenty-four histories. In: Wang, G., Lin, X., Hendler, J., Song, W., Xu, Z., Liu, G. (eds.) WISA 2020. LNCS, vol. 12432, pp. 289–301. Springer, Cham (2020). https://doi.org/10.1007/978-3-030-60029-7_27
28. Zeng, D., Liu, K., Lai, S., Zhou, G., Zhao, J.: Relation classification via convolutional deep neural network. In: Proceedings of COLING 2014, The 25th International Conference on Computational Linguistics: Technical Papers, pp. 2335–2344 (2014)
29. Zheng, S., et al.: Joint learning of entity semantics and relation pattern for relation extraction. In: Frasconi, P., Landwehr, N., Manco, G., Vreeken, J. (eds.) ECML PKDD 2016. LNCS (LNAI), vol. 9851, pp. 443–458. Springer, Cham (2016). https://doi.org/10.1007/978-3-319-46128-1_28

Cost-Effective Memory Replay
for Continual Relation Extraction

Yunong Chen, Yanlong Wen$^{(\boxtimes)}$, and Haiwei Zhang

College of Computer Science, Nankai University, Tianjin, China
{chenyunong,wenyanlong,zhanghaiwei}@dbis.nankai.edu.cn

Abstract. Continual relation extraction incrementally learns to extract the relations between entities from unstructured text with a series of tasks. The state-of-the-art methods of continual relation extraction are based on memory replay, which allocate a fixed memory for each continuously coming task to store part of the training data and replay them in subsequent tasks. However, memory resources are usually limited in real scenarios. Existing methods haven't considered the limitation and use efficiency of memory. This paper introduces cost-effective memory replay (CEMR) based on existing methods, which stores as much training data as possible for each task and adopts effective strategies of samples selection and replacement. CEMR efficiently uses limited memory for memory replay and knowledge consolidation in the new tasks, which alleviates the catastrophic forgetting problem. Experiments are conducted on three different relation extraction datasets using multiple comparison methods. The final results show that CEMR outperforms the state-of-the-art methods, which proves the effectiveness of CEMR.

Keywords: Relation extraction · Continual learning · Nature language process

1 Introduction

The relation extraction aims to extract the relations between entity pairs from unstructured text. For example, with the sentence "Beijing is the capital of China", the relation "the capital of" between "Beijing" and "China" is expected to be extracted. Relation extraction is an important step in knowledge extraction. It is widely used in downstream tasks such as question answering and automatic knowledge completion.

Traditional relation extraction methods always train a model on the static dataset with fixed relations set. But in real-world scenes, data and new relations emerge incessantly. In this case, the model will forget the knowledge learned before, namely catastrophic forgetting [12], if it is simply retrained with the new training data. However, combining new data and previous data to train the model is impractical in storage and computation, especially when the amount of data for each task is large. Therefore, the traditional relation extraction methods

C. Xing et al. (Eds.): WISA 2021, LNCS 12999, pp. 335–346, 2021.
https://doi.org/10.1007/978-3-030-87571-8_29

are no longer applicable in this scenario. Continual learning methods are required to incrementally learn new knowledge from a series of tasks containing different sets of relations.

Recently, there are many kinds of research on continual learning and various continual learning methods are proposed. However, most of them are applied to image classification tasks such as digital recognition and object detection. Rare studies focus on the performance of natural language processing tasks. In fact, most continual learning methods cannot be directly applied to natural language processing tasks such as relation extraction to achieve good results. Some current studies suggest that memory replay should be used in the field of natural language processing [2,20]. Wang et al. [20] propose that even the simplest memory replay method has a better result on continual relation extraction than many other mainstream continual learning methods. The key of memory replay is to store a part of training data of learned tasks and replay them in the later tasks to consolidate knowledge and alleviate the catastrophic forgetting. However, existing continual relation extraction methods based on memory replay allocate a fixed size of memory for each task, ignoring the limited memory in the real scene. What's more, such a strategy can't fully use memory when learning few tasks.

In response to the above problems, this paper proposes CEMR(Cost-effective Memory Replay) method based on EMAR (Episodic Memory Activation and Reconsolidation) proposed by Han et al. [4]. CEMR efficiently uses memory by specific sample selection and replacement strategies. When learning a new task, CEMR utilizes stored training samples for memory replay and calculates the template of each relation to consolidate knowledge, effectively alleviating catastrophic forgetting. This paper compares different continual learning methods on several relation extraction datasets and analyzes the impact of memory size. The change of average accuracy and final result of CEMR is better than the state-of-the-art method, which proves the effectiveness of CEMR.

2   Related Work

The mainstream relation extraction methods are based on deep learning, which can be divided into supervised learning [9,13], unsupervised learning [8] and distantly supervised learning [11]. These traditional methods extract relations in the fixed relations set determined during training. But they can't detect and learn new relations out of the set, which may constantly emerge in the open domain. So continual learning methods are required in that case.

Continual learning can be divided into three categories: regularization, dynamic architectures, and memory replay [15]. The regularization method adds regularization terms to the loss function to limit the change of important parameters. For example, LwF(learning without forgetting) proposed by Li et al. [7] requires the new parameters can't deviate from the old parameters too much by knowledge distillation. Kirkpatrick et al. [6] proposed EWC(Elastic Weight Consolidation), which selectively slowing down learning on the parameters important for old tasks by adding L_2 regularization.

The dynamic structure method dynamically adjusts the structure of the neural network when learning a new task, such as increasing the number of neurons or layers of the network to improve the ability to learn new knowledge. Rusu et al. proposed PNN (progressive neural network) [18], which retains the old neural network and expands a sub neural network for training new tasks. In addition to expanding the new sub-network, DEN (dynamically expanding network) [22] also selectively shares part of the previous model structure for training. RCL (Reinforced Continual Learning) proposed by Xu et al. [21] uses reinforcement learning to find how to expand the optimal sub-network structure. However, the dynamic structure method needs to continuously expand the model structure for new tasks, which will lead to extremely complex network structures and huge parameter scales.

For memory replay, the main idea is to selectively store a few training samples for each incoming task and replay them in the subsequent tasks. The GEM (Gradient Episodic Memory) proposed by Lopez-Paz et al. [10] stores previous samples and requires that the training loss of these samples cannot increase. For the large calculation and storage of GEM, Chaudhry et al. [1] proposed A-GEM to improve. The iCaRL(Incremental Classifier and Representation Learning) proposed by Rebuffi et al. [17] uses the stored samples and training data to update model parameters and calculate feature vectors of relations. Sometimes because of memory restriction or privacy policies, it's impossible to directly store the training samples of previous tasks. So, Shin et al. [19] apply adversarial generation network to generate pseudo-data similar to actual training data to achieve memory replay.

Most of the methods mentioned above are proposed for image classification tasks. The earliest research on continual relation extraction is Wang et al. [20], they proved the effectiveness of memory replay and proposed EA-EMR (Embedding Aligned Episodic Memory Replay) to further improve the continual relation extraction. Based on it, Han et al. proposed the EMAR (Episodic Memory Activation and Reconsolidation) [4], which uses the relation template to consolidate knowledge. In addition to memory replay, Obamuyide et al. pointed out that meta-learning can improve continuous learning [14].

3   Cost-Effective Memory Replay

Existing continual relation extraction methods based on memory replay allocate a certain size of memory for each task to store training samples. However, in actual scenarios, data and relations emerge incessantly. Existing methods can't deal with it using limited memory resources. In addition, memory resources are not fully utilized if there are only few tasks. So, the paper proposes CEMR (Cost-effective Memory Replay) based on the EMAR [4], which is more in line with actual scenarios.

3.1 Problem Definition and Model Framework

Continual relation extraction aims to extract relations from a sequence of tasks containing different relation sets. The key is to alleviate catastrophic forgetting when learning new task. For k-th task T^k, it contains training data D^k_{train}, validate data D^k_{val}, test data D^k_{test} and relation set R_k. The training data $D^k_{train} = \{(x^1_k, y^1_k), (x^2_k, y^2_k), \cdots, (x^N_k, y^N_k)\}$, where x^i_k is input data, including nature language text and candidate relation set, and $y^N_k \in R_k$ is correct relation label. For all test data $\bigcup^k_{i=1} D^i_{test}$, the relation extraction model $f_\theta(\cdot)$ is expected to find the correct relation from seen relation set $\bigcup^k_{i=1} R_i$. To achieve that goal, storing all seen training data is impractical in both time and storage. Memory replay can be applied as an alternative. There is storage space $M = \{M_1, M_2, \cdots\}$, where $M_k = \{(x^1_{M_k}, y^1_{M_k}), (x^2_{M_k}, y^2_{M_k}), \cdots, (x^B_{M_k}, y^B_{M_k})\}$ represents the training samples stored in memory for task T^k. The number B of training samples depends on total size of memory and current number of tasks.

The framework of CEMR is shown in Fig. 1. There are three steps for learning new task T^k: (1) Optimizing loss function $l\left(f_\theta\left(x^i_k\right), y^i_k\right)$ with training data D^k_{train} and learning the relations in the task. (2) Selecting a certain amount of training samples from training data D^k_{train} and replacing some samples stored in memory M using a special strategy according to the number of tasks and the size of M. (3) Conducting memory replay using stored samples and calculating relation templates to consolidate previous knowledge.

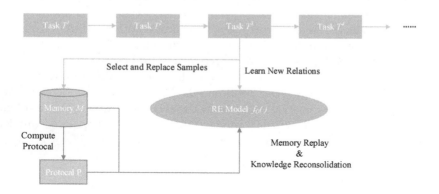

Fig. 1. The framework of CEMR

3.2 Relation Extraction Model

Relation extraction model $f_\theta(\cdot)$ is the basic part of CEMR. To get rid of the impact of the relation extraction model and compare with other continual learning methods fairly, CEMR chooses BiLSTM, same as other methods, to encode text and relation. The main process is simplified as Fig. 2. Sentence and relation are tokenized as words, represented by GloVe word embedding [16]. They are encoded by BiLSTM to get semantic feature vectors. Then, the similarity between sentence feature and relation feature is served as score. The relation which has the highest score is chosen as the final predicted relation. In fact, most continual learning methods are dependent on relation extraction models, so the relation extraction model used here can be replaced by any other proper model. Word embedding can also be trained by other models [3].

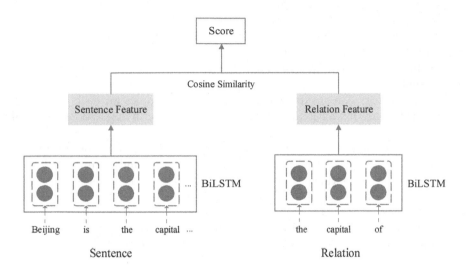

Fig. 2. The relation extraction model

3.3 Learning New Relation

A new task to learn has own dependent relation set, which may haven't been seen by model before, so model should learn them first. For new task T^k, it's training data $D_{train}^k = \{(x_k^1, y_k^1), (x_k^2, y_k^2), \cdots, (x_k^N, y_k^N)\}$. The model optimizes the loss function:

$$L(\theta) = -\sum_{i=1}^{N} \log \frac{\exp\left(g\left(f\left(x_k^i\right), r_{y_k^i}\right)\right)}{\sum_{j=1}^{|\tilde{R}_k|} \exp\left(g\left(f\left(x_k^i\right), r_j\right)\right)}$$

Where θ is the model parameters, $g(\cdot)$ is the function evaluating the similarity between vectors, such as cosine similarity, $\tilde{R}_k$ represent the seen relation set till task T^k.

3.4 Selecting and Replacing Samples

CEMR focus on how to effectively use limited memory resources. Before the training model, CEMR will determine the total memory size M according to the limited memory resources and the scale of tasks. The number of stored samples for each task depends on the total memory size M and learned task number k. Following the principle of making the best use of storage space, the memory size for task T^k is:

$$M_k = \min\left(\max\left(M_{left}, \frac{M}{k}\right), \left|D_{train}^k\right|\right)$$

where M_{left} is the available memory left, $\left|D_{train}^k\right|$ is the number of samples in training data. To ensure the model performing well on all tasks, memory M is averagely allocated for all seen tasks. The samples newly selected will replace the old samples when the memory is full. First, CEMR finds the tasks which have memory space M_i bigger than average space $\frac{M}{k}$. Then $M_i - \frac{M}{k}$ samples will be randomly selected to be replaced until there is enough space M_k for the new task. This strategy can ensure that the number of samples for each task in the memory is balanced, which can effectively alleviate overfitting and catastrophic forgetting.

When selecting samples from the training data D_{train}^k for storage, random selection is not an ideal way. The samples selected randomly cannot well cover the distribution of relations. To select more representative samples in the task, K-means clustering is used here. The embedding features of the text in training data are clustered into M_k classes using K-means. The sample closest to the center of each class is selected to be stored in the memory M, which contains more information and better covers the feature distribution of tasks.

3.5   Memory Replay and Knowledge Consolidation

Memory replay is an effective method to alleviate catastrophic forgetting. After learning new relations in task T^k, CEMR will take the samples stored in memory, $D_{replay}^k = \left\{\left(x_M^1, y_M^1\right), \left(x_M^2, y_M^2\right), \cdots, \left(x_M^{|\tilde{M}_k|}, y_M^{|\tilde{M}_k|}\right)\right\}$, as a complement to training data to reply, where $\tilde{M}_k = \bigcup_{i=1}^k M_i$ is the memory storing samples till task T^k. The loss function during the memory replay is:

$$L^M(\theta) = -\sum_{i=1}^{|\tilde{M}_k|} \log \frac{\exp\left(g\left(f\left(x_M^i\right), r_{y_M^i}\right)\right)}{\sum_{j=1}^{|\tilde{R}_k|} \exp\left(g\left(f\left(x_M^i\right), r_j\right)\right)}$$

If only conducting memory replay, overfitting may be occurred due to repeated training of same samples. So, CEMR also carry on knowledge consolidation via relation template. For each seen relation $r_i \in \tilde{M}_k$, the samples having same relation label in memory $P_i = \left\{x_{P_i}^1, x_{P_i}^2, \cdots, x_{P_i}^{|P_i|}\right\}$ will be selected. Then relation template is calculated as follow:

$$p_i = \frac{\sum_{j=1}^{|P_i|} f\left(x_j^{P_i}\right)}{|P_i|}$$

which is the average feature vector of that relation in the memory. To consolidate previous knowledge, optimizing model parameters as follows:

$$L^P(\theta) = -\sum_{i=1}^{|\tilde{R}_k|}\sum_{j=1}^{|P_i|} \log \frac{\exp\left(g\left(f\left(x_{P_i'}^j\right), p_i\right)\right)}{\sum_{l=1}^{|\tilde{R}_k|} \exp\left(g\left(f\left(x_{P_i}^j\right), p_l\right)\right)}$$

4   Experiment Result and Analysis

4.1   Dataset

FewRel [5] is a few-shot relation extraction dataset, including 80 relations and 56,000 instances. Referring to Wang et al. [20], FewRel is divided into 10 tasks via K-means clustering. Each task has its own different relation set and training set. Each training instance contains the correct relation label and 10 randomly selected candidate relation labels.

SimpleQuestions [23] is a knowledge-based question answering dataset containing 108,442 questions, which was later constructed by Yu et al. [24] as a relation extraction dataset consisting of 6,701 relations and 103,155 instances. Similarly, SimpleQuestions is divided into 20 tasks containing different relation sets referring to the Wang et al. [20].

TACRED [24] is a relation extraction dataset containing 42 relations and 21,784 instances. Similar to FewRel, the relations of TACRED are divided into 10 sets to construct 10 tasks according to Han et al. [4]. But there is a special relation type "n/a" (not available) in TACRED, which will be removed from the dataset.

4.2   Evaluation Metrics

The experiment mainly uses average accuracy and whole accuracy to evaluate the performance of the model in continual relation extraction. The average accuracy refers to the average of correct rates of all tasks that have been learned till task T^k. It expects a higher accuracy rate on each task with the same weight. So, it pays more attention to the solution of catastrophic forgetting. The formula for the average accuracy is as follows:

$$\text{ACC}_{\text{avg}} = \frac{1}{k}\sum_{i=1}^{k} acc_{f,i}$$

The whole accuracy refers to the correct rate on all test sets, including the test sets of unseen tasks. This metric takes both knowledge transfer and catastrophic forgetting into consideration. The formula for the whole accuracy is as follows:

$$\text{ACC}_{\text{whole}} = acc_{f,D_{test}}$$

4.3   Baseline

The experiment selects the following continual learning methods as baseline: (1) Origin method simply retrain the model using the training data of the new task, which will lead to catastrophic forgetting. (2) EWC [6] add a L_2 regularization into loss function to slow down the change of important parameters. (3) AGEM [1] stores samples in memory and requires the average loss of them not to increase in each training step. (4) EMR [20] stores samples in memory and simply replay them when training new tasks. (5) EAEMR appends embedding alignment model to EMR, which can alleviate the distortion of previous embedding space. (6) EMAR [4] is a state-of-the-art method that alleviates catastrophic forgetting via memory replay and knowledge consolidation.

4.4   Overall Result

Considering that other memory replay methods for comparison do not limit total memory, so the total storage space they ultimately use is taken as the storage limit of CEMR. For example, if baseline store 50 samples for each task, CEMR will set total storage to 500, 1000, and 500 for FewRel, SimpleQuestions, and TACRED respectively according to num of tasks. Each method runs 5 times with the same random seed. The average value of the final result is shown in Table 1. As can be seen from the table, compared to simply retraining the model for each new task, the continual learning method can improve the average accuracy and the whole accuracy. And in the continual relation extraction task, the method based on memory replay is better than the method based on regularization. The method described in this paper, CEMR, can use memory more efficiently and achieves the best results in most cases. CEMR can't perform better than EAEMR on SimpleQuestion, mainly because the dataset is relatively simple and the Origin method can also achieve good results.

Table 1. The average accuracy and whole accuracy of different methods on different dataset

Method	FewRel		SimpleQuestions		TACRED	
	ACC_{whole}	ACC_{avg}	ACC_{whole}	ACC_{avg}	ACC_{whole}	ACC_{avg}
Origin	0.181	0.277	0.588	0.408	0.148	0.099
EWC	0.213	0.234	0.626	0.587	0.072	0.154
AGEM	0.372	0.433	0.738	0.707	0.092	0.172
EMR	0.545	0.730	0.868	0.832	0.172	0.270
EAEMR	0.569	0.759	**0.876**	**0.833**	0.264	0.399
EMAR	0.656	0.769	0.848	0.815	0.415	0.518
CEMR	**0.663**	**0.787**	0.852	0.826	**0.492**	**0.588**

The change of average accuracy is shown in Fig. 3. As the number of learning tasks increases, the overall average accuracy decreases significantly, indicating the effect of catastrophic forgetting. However, the result on SimpleQuestion fluctuates greatly and the average accuracy on certain tasks even rises because of its simpleness. The average accuracy of CMER always maintains a high level during the process, which shows that CMER can effectively solve the problem of catastrophic forgetting.

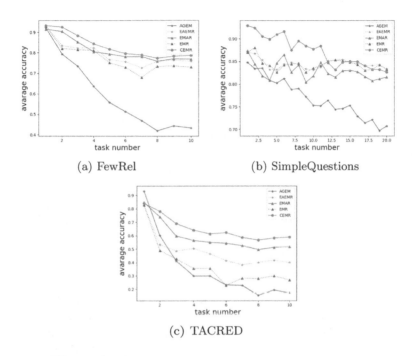

(a) FewRel   (b) SimpleQuestions

(c) TACRED

Fig. 3. The change of average accuracy on three datasets

The change of the whole accuracy is shown in Fig. 4. As the model learns more tasks, the whole accuracy obviously increases, which shows the effect of continual learning. The whole accuracy of CEMR is significantly better than other methods on TACRED, but is not prominent on FewRel and SimpleQuestion. It can be seen that the ability of knowledge transfer is related to the difficulty of the dataset.

4.5   The Effect of Memory Size

For the continual relation extraction methods based on memory replay, the size of the total memory space M has an important influence on the final result. Experiments tested the results on the dataset TACRED when allocating 10, 25, 50, and 100 samples for each task. The total memory space used by the CEMR set to 100, 250, 500, 1000 samples respectively because there are 10 tasks. The final results are shown in Table 2. With the growth of memory space,

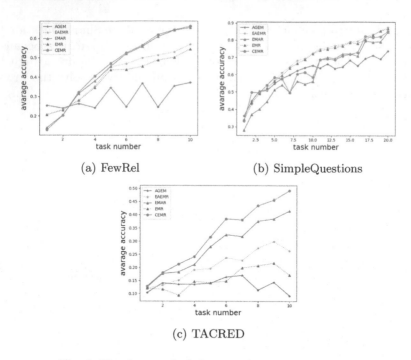

(a) FewRel   (b) SimpleQuestions

(c) TACRED

Fig. 4. The change of whole accuracy on three datasets

the performance of each method has improved, because each task can store more samples, which can better represent the data distribution of the original training set. On TACRED, CEMR has achieved a significantly better result in different memory sizes and can get the same result as other methods using smaller memory.

Table 2. The average accuracy and whole accuracy with different memory size on TACRED

	100		250		500		1000	
	ACC_{whole}	ACC_{avg}	ACC_{whole}	ACC_{avg}	ACC_{whole}	ACC_{avg}	ACC_{whole}	ACC_{avg}
AGEM	0.104	0.177	0.112	0.181	0.092	0.172	0.128	0.198
EMR	0.172	0.270	0.225	0.337	0.172	0.270	0.299	0.410
EAEMR	0.194	0.313	0.220	0.363	0.264	0.399	0.295	0.429
EMAR	0.338	0.419	0.366	0.476	0.415	0.518	0.488	0.578
CEMR	**0.346**	**0.450**	**0.419**	**0.518**	**0.492**	**0.588**	**0.619**	**0.673**

5   Conclusion and Future Work

This paper studied the continual relation extraction. On this task, many continual learning methods that perform well in the image classification are not applicable and memory replay is proved to be a good choice. Current continual relation extraction methods based on memory replay allocate a fixed size of memory space for each task, and only a small number of tasks are considered, which is not in line with the actual situation. Therefore, this paper proposes CEMR, which effectively uses memory to store as many samples as possible. It achieves a good result on three different relation extraction datasets. The catastrophic forgetting is greatly alleviated due to the memory replay and knowledge consolidation.

However, in this paper, the relation set and training data for each task are independent and different, which is not reasonable in actual scenarios. The overlap of relations and data between different tasks will bring new challenges to handle catastrophic forgetting. In the situation where memory sources are limited, sample selection and replacement are very important. There should be better strategies than clustering and average replacement, used by CEMR. In addition, CEMR uses a basic relation extraction model for a fair comparison. But it's possible that different relation extraction models are suitable for different continual learning methods, which also can be further studied in the future.

Acknowledgements. This research is supported by Chinese Scientific and Technical Innovation Project 2030 (No. 2018AAA0102100), National Natural Science Foundation of China (No. 62077031, 61772289, U1936206). We thank the reviewers for their constructive comments.

References

1. Chaudhry, A., Ranzato, M., Rohrbach, M., Elhoseiny, M.: Efficient lifelong learning with A-GEM. In: International Conference on Learning Representations (2018)
2. d'Autume, C.D.M., Ruder, S., Kong, L., Yogatama, D.: Episodic memory in lifelong language learning. In: NIPS, pp. 13132–13141 (2019)
3. Guo, C., Xie, L., Liu, G., Wang, X.: A text representation model based on convolutional neural network and variational auto encoder. In: Wang, G., Lin, X., Hendler, J., Song, W., Xu, Z., Liu, G. (eds.) WISA 2020. LNCS, vol. 12432, pp. 225–235. Springer, Cham (2020). https://doi.org/10.1007/978-3-030-60029-7_21
4. Han, X., et al.: Continual relation learning via episodic memory activation and reconsolidation. In: ACL, pp. 6429–6440 (2020)
5. Han, X., et al.: FewRel: a large-scale supervised few-shot relation classification dataset with state-of-the-art evaluation. In: EMNLP, pp. 4803–4809 (2018)
6. Kirkpatrick, J., et al.: Overcoming catastrophic forgetting in neural networks. Proc. Natl. Acad. Sci. **114**(13), 3521–3526 (2017)
7. Li, Z., Hoiem, D.: Learning without forgetting. IEEE Trans. Pattern Anal. Mach. Intell. **40**(12), 2935–2947 (2017)
8. Lin, H., Yan, J., Qu, M., Ren, X.: Learning dual retrieval module for semi-supervised relation extraction. In: The World Wide Web Conference, pp. 1073–1083 (2019)

9. Lin, Y., Shen, S., Liu, Z., Luan, H., Sun, M.: Neural relation extraction with selective attention over instances. In: ACL, vol. 1 (2016)
10. Lopez-Paz, D., Ranzato, M.: Gradient episodic memory for continual learning. Adv. Neural. Inf. Process. Syst. **30**, 6467–6476 (2017)
11. Marcheggiani, D., Titov, I.: Discrete-state variational autoencoders for joint discovery and factorization of relations. Trans. Assoc. Comput. Linguist. **4**, 231–244 (2016)
12. McCloskey, M., Cohen, N.J.: Catastrophic interference in connectionist networks: the sequential learning problem. Psychol. Learn. Motiv. **24**, 109–165 (1989)
13. Miwa, M., Bansal, M.: End-to-end relation extraction using LSTMs on sequences and tree structures. In: ACL, vol. 1 (2016)
14. Obamuyide, A., Vlachos, A.: Meta-learning improves lifelong relation extraction. In: Proceedings of the 4th Workshop on Representation Learning for NLP (RepL4NLP-2019), pp. 224–229 (2019)
15. Parisi, G.I., Kemker, R., Part, J.L., Kanan, C., Wermter, S.: Continual lifelong learning with neural networks: a review. Neural Netw. **113**, 54–71 (2019)
16. Pennington, J., Socher, R., Manning, C.D.: Glove: global vectors for word representation. In: EMNLP, pp. 1532–1543 (2014)
17. Rebuffi, S.A., Kolesnikov, A., Sperl, G., Lampert, C.H.: iCaRL: incremental classifier and representation learning. In: Proceedings of the IEEE Conference on Computer Vision and Pattern Recognition, pp. 2001–2010 (2017)
18. Rusu, A.A., et al.: Progressive neural networks. arXiv preprint arXiv:1606.04671 (2016)
19. Shin, H., Lee, J.K., Kim, J., Kim, J.: Continual learning with deep generative replay. In: NIPS, pp. 2994–3003 (2017)
20. Wang, H., Xiong, W., Yu, M., Guo, X., Chang, S., Wang, W.Y.: Sentence embedding alignment for lifelong relation extraction. In: Proceedings of NAACL-HLT, pp. 796–806 (2019)
21. Xu, J., Zhu, Z.: Reinforced continual learning. In: NIPS, pp. 907–916 (2018)
22. Yoon, J., Yang, E., Lee, J., Hwang, S.: Lifelong learning with dynamically expandable networks. In: International Conference on Learning Representations, ICLR (2018)
23. Yu, M., Yin, W., Hasan, K.S., dos Santos, C., Xiang, B., Zhou, B.: Improved neural relation detection for knowledge base question answering. In: ACL, vol. 1, pp. 571–581 (2017)
24. Zhang, Y., Zhong, V., Chen, D., Angeli, G., Manning, C.D.: Position-aware attention and supervised data improve slot filling. In: EMNLP, pp. 35–45 (2017)

Entity Alignment of Knowledge Graph by Joint Graph Attention and Translation Representation

Shixian Jiang, Tiezheng Nie[✉], Derong Shen, Yue Kou, and Ge Yu

Northeastern University, Shenyang 110169, China
{nietiezheng,shenderong,kouyue,yuge}@mail.neu.edu.cn

Abstract. Entity alignment is a crucial and challenging research task in the fields of knowledge graph and natural language processing. It aims to integrate knowledge graph information from different languages or different sources and apply it to subsequent knowledge graph construction and other downstream tasks. Currently, most entity alignment solutions are based on knowledge graph embedding, which align entities by mapping entities into low dimensional spaces. But it also face problems such as fail to make full use of the neighbor nodes and relation information of the graph, and the heterogeneity between the knowledge graphs. This paper proposes a method of combining graph attention and translation models. We use graph attention mechanism to convey information about neighboring nodes, and merge relation information into the entity representation, make full use of the internal information of the knowledge graph to learn a better entity representation. Then we combine the graph attention mechanism with knowledge translation representation model to constrain the consistency between different knowledge graphs and guarantee the accuracy of embedding vectors in low-dimensional space. The results of comparative experiments with other entity alignment methods show that the performance of our method is better than others, and the alignment effect is significantly improved.

Keywords: Entity alignment · Knowledge graph · Attention mechanism · Translation model · Cross-language

1 Introduction

With the explosive growth of Internet information, knowledge graphs of different languages in various fields emerge in an endless stream. Knowledge graph can accurately reflect the objective facts in real world, and can well express abstract meanings such as concepts and levels. At present, common knowledge graphs such as DBpedia, YAGO and Freebase, etc. However, the establishment and maintenance of different knowledge graphs are usually independent and are constructed in a monolingual environment, so they are used differently. When it comes to resources, there are problems such as insufficient knowledge versatility, and it needs to be fused to obtain a more comprehensive and unified knowledge graph.

© Springer Nature Switzerland AG 2021
C. Xing et al. (Eds.): WISA 2021, LNCS 12999, pp. 347–358, 2021.
https://doi.org/10.1007/978-3-030-87571-8_30

As a key step in constructing a high quality knowledge graph, the entity alignment task also has important applications in many fields. The entity alignment of cross-language knowledge graph can be used for many downstream natural language processing tasks, such as assisted machine translation, question answering system [1], information retrieval [2], recommendation system [3] and so on.

Entity alignment, also known as entity matching or entity resolution, is the process of judging whether two entities are the same object in the real world in multiple data sets. Figure 1 shows an example of entity alignment. Lines are used to connect pre aligned entities. The purpose of entity alignment task is to find new entity pairs in two knowledge graphs, which is represented by dotted lines in the graph.

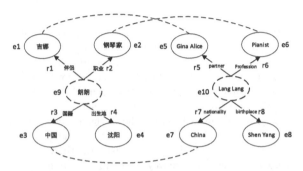

Fig. 1. An example of entity alignment

Traditional entity alignment methods usually use rule reasoning, artificial feature annotation, crowdsourcing [4] and other method. With the development of representation learning in recent years, many new embedding-based methods have emerged. Among them, the method based on the translation model regards the triple relation in the knowledge graph as the translation from the head entity to the tail entity, represented by TransE [5], TransH [6], TansD [7], etc. The representation learning method based on the translation model can capture the linear relation of entities in knowledge graph, making the boundaries of different entity vectors clear, but there are also limitations such as insufficient accuracy and high cost of obtaining seed sets.

There are also some researches using graph convolutional neural network GCN [8] for embedding representation of entities. By combining the information of neighboring entities around the central entity to enhance the vector representation of the entity. These methods are based on the assumption that similar entities are usually represented by similar fields, such as GCN-Align [9], but this type of method usually considers the surrounding neighbor structure and attribute information of the entity, and ignores the use of the relation between entities, and the convolution operation may affect the effect of entity alignment.

Based on the enlightenment of the above content, this paper proposes a new entity alignment method to solve the shortcomings of these existing solutions. The main contributions of this paper are as follows:

(1) An entity alignment method combining graph attention network and translation model is proposed (called GTEA),information of neighboring neighbor nodes is gathered through the attention mechanism, and the relation information is merged into the entity vector to obtain a more accurate vector representation, and then the translation model is used to lighten the impact of Knowledge graph heterogeneity, which makes the final entity representation more suitable for entity alignment.

(2) By making full use of graph information, we alleviate the dependence of entity alignment on seed set, and reduce the manpower and related costs of traditional methods to construct seed entity pairs, then the robustness of the model is improved.

(3) The feasibility and effectiveness of the key technologies proposed in this paper are verified through experiments.

The rest of this article is organized as follows. Section 2 reviews the work related to the entity alignment task. Section 3 defines the research question of this article. Section 4 presents the research method. Section 5 shows experimental results, and Sect. 6 is full text summary.

2 Related Work

In the field of entity alignment, methods based on traditional rules and similarity calculations have been deeply studied by researchers. However, these methods require a lot of manual participation and are therefore costly. Many methods use additional resources, such as owl attributes [10], entity descriptions [11], etc. These methods usually have a large number of complex parameters and are also limited by the availability of additional information about the knowledge graph.

In recent years, entity alignment methods based on embedding have received more attention. The representative method such as TransE model, according to the basic assumption of triple $h + r = t$, similar entities have similar distance to the embedding vector in the vector space. Based on the extension method, MtransE [12] learns entity embedding the mapping of vectors to their cross-language entity pairs completes the task of entity alignment. IPTransE [13] jointly encodes the entities and relations of different KGs into a unified low-dimensional space, and at the same time expands the seed set through iteration and parameter sharing. BootEA [14] tries to solve the problem of insufficient label data by using a bootstrapping strategy, and the method of editable alignment reduces the accumulation of errors in the propagation process. KDCoE [15] combine the structure of the knowledge graph with entity attributes and description information to further enhance the effect of entity alignment. MultiKE [16] model uses a multi-view combination method to align entities, and explores the effects of different combination strategies.

Due to the graph structure characteristics of knowledge graph, many researchers have also tried to use graph neural networks (GNN) to solve the entity alignment problem. Divided into spectral domain-based convolution method [17] and space-based convolution method [18], the convolution operation is performed on the topological graph and a single node respectively. The representative method GCN-Align uses the graph convolutional network GCN to learn the vector representation of attributes and structures, and then perform entity alignment based on these two vectors. R-GCN [19] uses a relation specific weight matrix to measure the influence of surrounding neighbors on the central entity, but it also adds many additional parameters, which makes it difficult to train on a large scale. Graph Attention Network (GAT) [20] can better realize the aggregation of surrounding neighbor information with weights through the attention mechanism. AliNet [21] expands the overlapping part of domain structure by introducing distant neighbors into attention mechanism to improve the accuracy of entity alignment.

Inspired by the above, in order to improve the accuracy of cross language knowledge mapping entity alignment problem, this paper proposes a relation fusion entity alignment method based on graph attention network and translation representation by combining graph attention network and translation model and fusing relation information into entity representation.

3 Problem Definition

Knowledge graph organizes the knowledge in real world by the form of triples. The triples can be expressed as $T = (h, r, t)$, where h represents the head entity, t represents the tail entity, and r represents the connection between the head and tail entities, multiple triples are related to each other, and loose unstructured knowledge is integrated through a graph structure. The task of cross-language entity alignment in this paper is to find more alignable entities through pre-aligned seed set data between the knowledge graphs of two different languages. The following is the formal definition of the task.

Definition 1 (Knowledge graph entity alignment): We express the knowledge graph as $KG = \{E, R, T\}$, $E = \{e_1, e_2, \ldots, e_n\}$, $R = \{r_1, r_2, \ldots, r_n\}$, $T = \{t_1, t_2, \ldots, t_n\}$. Respectively represent the entity set, relation set and triple set in the knowledge graph. For the cross-language entity alignment task, given two knowledge graphs expressed in different languages, $KG_1 = \{E_1, R_1, T_1\}$, $KG_2 = \{E_2, R_2, T_2\}$, the purpose is to find s other alignable entities between the two knowledge graphs based on the data in seed set.

Definition 2 (Seed Set): Extract some pre-aligned entity pairs from the two knowledge graphs as the seed set, expressed as $S = \{(e_1, e_2) | e_1 \in E_1, e_2 \in E_2\}$, where e_1 and e_2 are the entities in the two knowledge graph entities, which refer to the same things in real world.

Definition 3 (Set of neighbor entities): We call the associated entities around a central entity as neighbor entities, using the set $N_e = \{e_2 | (e_1, r, e_2 \in T)\} \cup \{e_2 | (e_2, r, e_1 \in T)\}$, where e_1 is center entity, T represents the triplet associated with the central entity.

4 Entity Alignment Model

4.1 Model Overview

In order to better capture the internal information of knowledge graph and alleviate the impact of the heterogeneity of the knowledge graph, we propose Graph Attention

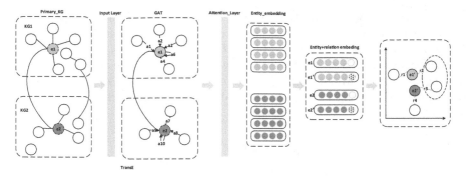

Fig. 2. Overview of GTEA

joint Translation Entity Alignment Model, called GTEA. Figure 2 shows the overall framework of the model.

In order to make better use of the surrounding domain information of the central entity of knowledge graph, firstly, the two original knowledge graphs are given, and the graph attention mechanism is used to capture the internal dependencies of the triples and the importance of neighbor nodes, and aggregate them through different attention coefficients. The surrounding node information is represented by the central entity, and then in order to make full use of the rich relation information in the knowledge graph, the relation representation of triples is merged into the connected entities to obtain a more semantically rich entity representation. Finally, the translation model is used to eliminate the influence of the heterogeneity of knowledge graph in the convolution process, then use distance based measurement functions to find new alignable entity pairs in a unified low dimensional vector space.

4.2 Graph Attention Layer

First, the graph attention network is used to embed the entities in the knowledge graph, and the initial input of the convolutional layer is node feature matrix, $X^{(l)} = \{x_1^{(l)}, x_2^{(l)}, \ldots, x_n^{(l)} | x_i^{(l)} \in \mathbb{R}^{d^{(l)}}\}$, Where $d^{(l)}$ represents the characteristics of the l layer, n represents the number of nodes in knowledge graph, and the hidden layer characteristics between networks are updated by the following feedforward network.

$$H^{(l+1)} = ReLU\left(A^{(l)}H^{(l)}W^{(l)}\right) \tag{1}$$

Where $H^{(l)} \in \mathbb{R}^{n \times d^{(l)}}$ represents the hidden layer state of upper layer, $W^{(l)}$ represents the specific trainable weight matrix of each layer. $ReLU(\cdot)$ represents the nonlinear activation function $ReLU(x) = Max(0, x)$, and $A^{(l)}$ represent the attention matrix of each layer, in which a single element is the attention coefficient of the corresponding node. Specifically, $a_{i,j}$ denotes attention coefficient entity e_i and entity e_j. The attention coefficient is calculated by the following formula:

$$a_{i,j} = \frac{\exp(\eta(q^T(wh_i \| wh_j)))}{\sum_{k \in N_i} \exp(\eta(q^T(wh_i \| wh_k)))} \tag{2}$$

Where h_i, h_j stands for e_i and e_j of hidden layer states. q, w are trainable parameters, $.^T$ represents transpose, and $\|$ represents the connection between vectors, η represents a nonlinear function. In this paper, we chooser *LeakyReLU* function, N_i stands for entity e_i is the index sets of neighbor nodes in knowledge graph. At the beginning of model training, this paper uses entity name to initialize the node feature matrix. According to a small number of pre-aligned seed sets, the model is trained by minimizing the following marginal loss function based on distance.

$$L_1 = \sum\nolimits_{(e_i,e_j)\in S} \sum\nolimits_{(e_i',e_j')\in S'} ReLU\left(d(e_i, e_j) - d(e_i', e_j') + \gamma\right) \tag{3}$$

Where $d(e_i, e_j) = \|e_i - e_j\|_{l1}$ represents the l_1 distance between the entity pair (e_i, e_j), S' represents the negative sample set, and $\gamma > 0$ is the marginal hyperparameter. Based on the basic assumption that the aligned entities should have similar domain structures, this paper selects the closest entity around the seed set entities as the negative sample sets.

4.3 Fusion Relation Information

After obtaining the vector representation of entity through graph attention network, the vector of entity has been fused with the information of surrounding neighbor nodes. Considering that the relation also contains part of the semantic information of the entity, the relation of the triples is merged in this step into the representation of the entity to obtain a more accurate feature vector of the entity. Given a specific relation r, first calculate the relation feature vector by the following way.

$$r = w_r(avg(\{e_h|e_h \in h_r\}\|avg(\{e_t|e_t \in t_r\})) \tag{4}$$

Among them, h_r and t_r respectively represent the head entity and tail entity set of the triple associated with the relation r. Firstly, the head and tail entity feature vectors are expressed as the mean, and then connected, and then linearly transformed by the learnable parameter w_r to obtain the relation vector representation of r.After the above steps, we have obtained the vector representation of entity and relation. For each entity e in set E, the representation of its relation is aggregated into a relation context vector, and then combined with the original trained entity representation to build a new entity feature vector.

$$e_{joint} = w_e\left(\sum\nolimits_{r\in R_e} r\|e\right) \tag{5}$$

R_e represents the set of relations associated with the entity e. First, all the relation vectors in R_e are summed, and the original entity vector e is connected with the summed relation vector to obtain the combined and optimized entity vector representation.

4.4 Translation Model

After getting the vector representation of the entity, considering the influence of the heterogeneity of the knowledge graph, we use the translation model to model the internal relation of the knowledge graph, so as to make the boundaries between entities clearer and improve the accuracy of alignment. In the experiment, we select the most classic TransE model as the translation model for application. For the triples $T = (e_h, r, e_t)$ in knowledge graph. In translation model, the relation r is regarded as the translation from head entity to tail entity, which satisfies the expression $e_h + r = e_t$ in the vector space. Therefore, the following function is used as the score function to measure the rationality of specific triplet, and $\| \cdot \|$ represents the L1 distance.

$$d = \|e_h + r - e_t\| \qquad (6)$$

For the entity alignment task, it is expected that the positive sample triples can have an absolutely low score function value, on the contrary, the negative sample triples are expected to have an absolutely high score. Based on this assumption, it can not only make the boundary between positive and negative triples clear in the vector space, but also reduce the drift phenomenon in the vector space of the positive and negative samples during the learning process of triple representation, making the model more stable, then we train model by the following marginal loss function.

$$L_2 = \sum_{(e_h, r, e_t \in t)} \sum_{e'_h, r, e'_t \in t')} \left[(d(e_h, r, e_t) - \gamma_1) + \left(d\left(e'_h, r, e'_t\right) - \gamma_2 \right) \right]_+ \qquad (7)$$

t represents the set of positive triples, and t' represents the set of negative triples. Negative triples are obtained by randomly replacing the head entity or tail entity, $[]_+$ represents $max(0, x)$, which is the maximum value not less than 0, and γ_1 and γ_2 are marginal parameters used to control the scores of positive and negative samples. We set $\gamma_2 > \gamma_1$ in the experiment.Finally, the model training is performed by combining the losses of above two parts, and the objective function is defined as:

$$L = L_2 + L_1 \qquad (8)$$

L_1 and L_2 are the loss of attention model and relational model, respectively. We use AdaGrad Optimizer to optimize the objective function in the experimental training.

5   Experiments

5.1   Dataset

According to previous research results, three large scale cross-language data sets are selected to evaluate the method in this paper. These data sets are extracted from the multilingual version of DBpedia, covering Chinese, Japanese, English, and French. Each data set contains knowledge graph information in two different languages and contains 15,000 reference entity pairs. Table 1 shows the detailed statistics of the data set.

Table 1. Dataset details

Datesets	Language	Entities	Reltion	Rel.trple
$DBP15K_{ZH-EN}$	Chinese	66,469	2,830	153,929
	English	98,125	2,317	237,674
$DBP15K_{FR-EN}$	Japanse	65,744	2,043	164,373
	English	95,680	2,096	233,319
$DBP15K_{JA-EN}$	French	66,858	1,379	192,191
	English	105,889	2,209	278,590

5.2 Parameter Settings

We have developed a method called GTEA by using tensorflow. Hist@K is used as the evaluation index of experiment, the higher Hist@K means the better performance of the model.

Hist@K: a metric to measure the proportion of correctly aligned entities in the top K candidate candidates. In the sorted list of candidate entities, traversal is made from the first to the K to see if the correct matching entity is found, and if so, Hist@K plus 1.

In the experiment, 30% of entity pairs are referenced as the seed set for training, and the rest as test set, the average value of model running 3 times independently is used as the final experimental result. The default parameters used in the experiment are set as follows: $\gamma = 1.0$, $\gamma_2 = 0.1$, $\gamma_1 = 0.05$, the dimensions of hidden layer and attention layer of the network is $d = 300$, using the Adagrad optimizer, setting the learning rate to 0.001 for model training.

5.3 Result Analysis

Comparison Method

In order to comprehensively evaluate the effect of the experiment, we chose the classic methods of entity alignment using graph neural network and translation model: MtransE, GCN_Align, IptransE were selected as comparative experiments.

MtransE: This method uses transe based translation method for the knowledge graph of each language to transform it into the corresponding embedding space, and then regards the alignment of cross-language entities as the topological transformation of the embedding space.

IptransE: This method considers the conversion of multi path complex relations on the basis of transe, and proposes an iterative alignment algorithm for expanding the seed set.

GCN_Align: This method uses graph convolutional network for the first time to model the equivalent relation between entities. GCN is selected to generate the embedding vector of the entity.

Entity Alignment Result

The experimental results of this paper and the comparison method are presented in

Tables 2, 3, and 4. The best experimental results in each data set are represented by boldface.

Table 2. The experimental results in $DBP15K_{ZH-EN}$

	$DBP15K_{ZH-EN}$		
	Hits@1	Hits@10	Hist@50
MTransE	30.65%	60.42%	70.53%
GCN-Align	36.48%	66.04%	76.88%
IPtransE	38.73%	70.56%	79.89%
GTEA	**59.58%**	**76.48%**	**85.21%**

Table 3. The experimental results in $DBP15K_{FR-EN}$

	$DBP15K_{FR-EN}$		
	Hits@1	Hits@10	Hist@50
MTransE	28.79%	56.23%	69.48%
GCN-Align	30.32%	63.51%	76.95%
IPtransE	37.95%	70.83%	79.63%
GTEA	**60.34%**	**75.59%**	**84.85%**

Table 4. The experimental results in $DBP15K_{JA-EN}$

	$DBP15K_{JA-EN}$		
	Hits@1	Hits@10	Hist@50
MTransE	26.82%	55.34%	70.29%
GCN-Align	32.41%	62.94%	77.05%
IPtransE	37.58%	67.63%	80.49%
GTEA	**66.23%**	**79.05%**	**88.90%**

It can be analyzed from the table that the entity alignment method adopted in this paper combines the advantages of graph neural network and translation model, and at the same time makes full use of the relation information between entities. Compared with those previous method, the experimental result has been improved, indicating the relation information can improve the accuracy of the graph's attention mechanism for entity representation. At the same time, the translation model helps to alleviate the impact of graph heterogeneity on entity representation, making full use of entity association relations to express richer semantics, and combining the two part can get a more accurate entity representation.

Seed Set Sensitivity

Since the seed set has always been the key point of model accuracy in the entity alignment task, in order to verify the effect of the seed set on entity alignment results in different methods, different proportions of prior seed sets are allocated to the data set for experiments, and the seed set proportions increase from 10% to 40%, with 10% as step size, as the training set, the larger the proportion of the seed set means more prior information.

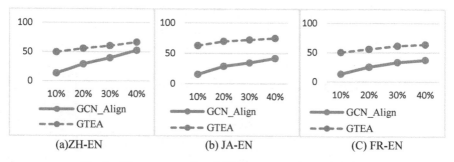

(a)ZH-EN   (b) JA-EN   (C) FR-EN

Fig. 3. Alignment results of different proportions of seed sets

Figure 3 shows the comparative experimental effect of GTEA and GCN_Align. The ratio of the pre-aligned seed set is used as the X axis, and the Hist@1 score reflecting the alignment accuracy is used as the Y axis.

The results show that as the proportion of the seed set increases, the results of all methods will get better. However, no matter what proportion of the seed set is used, the performance of GTEA is better than GCN_Align. At the same time, it can be seen that even when the proportion of the seed set is 10%, the accuracy of Hist@1 of the method in this paper can still reach 40%, but GCN-Align the result of will be greatly reduced, which reflects the robustness of GTEA, and at the same time proves that our method makes full use of the graph information to a certain extent, alleviating the high dependence of the entity alignment algorithm on the seed set.

Ablation Experiment

In order to further verify the effectiveness of integrating relational information and translation models in the graph attention network, two variants of GTEA were separated from GTEA for ablation studies, namely GTEA(NR) and GTEA(NT).

GTEA (NR): Keep the basic framework of the network unchanged, but don't incorporate the relation vector into the entity representation.

GTEA (NT): Delete the translation model that obtains the entity representation, and only use the attention network in the previous and merge the relation vector to obtain the final entity representation.

It can be concluded from Fig. 4 that GTEA and its variants will obtain better experimental results than GCN-Align, which reflects the effectiveness of the graph attention mechanism and translation model. At the same time, compared to GTEA (NT), the GTEA (NR) model only removes the process of fusing relational representations into entity representations, and improves the results more obviously. This shows that compared to fusing relational information in the graph attention network, it is more effective to improve the network structure and add the translation model.

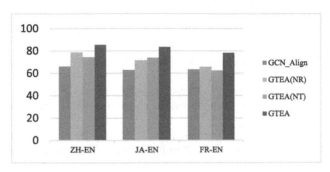

Fig. 4. The results of ablation experiment.

6 Conclusion

In this paper, we propose a knowledge graph entity alignment method, named GTEA. We use the graph attention mechanism to aggregate the information of neighboring nodes around the entity, and at the same time integrate the relation information into the entity representation to obtain a more accurate entity vector. Then through the translation model, the relation vector and entity vector are constrained in the low-dimensional space to make similar entities more distinguishable. The experimental results show that GTEA is superior in performance to the three benchmark methods.

Acknowledgments. This research is supported by National Natural Science Foundation of China (62072086, 62072084), Fundamental Research Funds for the Central Universities (N2116008, N180716010) and the National Defense Basic Scientific Research Program of China (JCKY2018205C012).

References

1. Zhang, Y., Dai, H., Kozareva, Z., Smola, A.J., Song, L.: Variational reasoning for question answering with knowledge graph (2017)
2. Chen, M., Tian, Y., Chang, K.-W., Skiena, S., Zaniolo, C.: Co-training embeddings of knowledge graphs and entity descriptions for cross-lingual entity alignment. IJCAI (2018)
3. Guo, Y., Yan, Z.: Collaborative filtering: graph neural network with attention. In: Wang, G., Lin, X., Hendler, J., Song, W., Xu, Z., Liu, G. (eds.) WISA 2020. LNCS, vol. 12432, pp. 428–438. Springer, Cham (2020). https://doi.org/10.1007/978-3-030-60029-7_39

4. Li, G., Zheng, Y., Fan, J., et al.: Crowdsourced data management: overview and challenges. In: The 2017 ACM International Conference, pp. 1711–1716. ACM, New York (2017)
5. Bordes, A., Usunier, N., Garcia-Duran, A., Weston, J., Yakhnenko, O.: Translating embeddings for modeling multi-relational data. In: NIPS, pp. 2787–2795 (2013)
6. Ji, G., He, S., Xu, L., et al.: Knowledge graph embedding via dynamic mapping matrix. In: Meeting of the Association for Computational Linguistics & the International Joint Conference on Natural Language Processing (2015)
7. Wang, Z., Zhang, J., Feng, J., Chen, Z.: Knowledge graph embedding by translatingon hyperplanes. In: Twenty-Eighth AAAI Conference on Artificial Intelligence, pp. 1112–1119 (2014)
8. Kipf, T.N., Welling, M.: Semi-supervised classification with graph convolutional networks (2017)
9. Wang, Z., et al.: Cross-lingual knowledge graph alignment via graph convolutional networks. In: Proceedings of the 2018 Conference on Empirical Methods in Natural Language Processing (2018)
10. Hu, W., Chen, J., Qu, Y.: A self-training approach for resolving object conference on the semantic web. In: Proceedings of the 20th International Conference on World Wide Web ACM (2011)
11. Yang, Y., Sun, Y., Tang, J., Ma, B., Li, J.: Entity matching across heterogeneous sources. In: Proceedings of the 21th ACMSIGKDD International Conference on Knowledge Discovery and Data Mining, pp.1395–1404 (2015)
12. Chen, M., Tian, Y., Yang, M.: Multilingual knowledge graph embeddings for cross-lingual knowledge alignment. In: Proceedings of the 26th International Joint Conference on Artificial Intelligence, pp. 1511–1517(2017)
13. Zhu, H., Xie, R., Liu, Z., Sun, M.: Iterative entity alignment via joint knowledge embeddings. In Proceedings of the 26th International Joint Conference on Artificial Intelligence, pp. 4258–4264. AAAI Press (2017)
14. Sun, Z., Hu, W., Zhang, Q., Qu, Y. Bootstrapping entity alignment with knowledge graph embedding. In: IJCAI, pp. 4396–4402 (2018)
15. Xiong, C., Power, R., Callan, J.P.: Explicit semantic ranking for academic search via knowledge graph embedding. In: WWW(2017)
16. Zhang, Q., Sun, Z., Hu, W., Chen, M., Guo, L., Qu, Y.: Multi-view knowledge graph embedding for entity alignment. IJCAI Int. Jt. Conf. Artif. Intell. **2019**, 5429–5435 (2019)
17. Bruna, J., Zaremba, W., Szlam, A., et al.: Spectral networks and locally connected networks on graphs. Comput. Sci. (2013)
18. James, A., Don, T.: Search-Convolutional Neural Networks. In: CoRR. abs/1511.02136, pp. 1–14 (2015)
19. Schlichtkrull, M., Kipf, T., Bloem, P.: Modeling relational data with graph convolutional networks. In: The Semantic Web, 2018, pp. p. 593—607. Springer International Publishing, Cham (2018). https://doi.org/10.1007/978-3-319-93417-4_38
20. Velickovic, P., Cucurull, G., Casanova, A., Romero, A., Lio, P., Bengio, Y.: Graph attention networks. In: Proceedings of the 6th International Conference on Learning Representations (2018)
21. Sun, Z., et al.: Knowledge graph alignment network with gated multi-hop neighborhood aggregation. AAAI (2019)

Incremental Validation of RDF Graphs

Yanling Wang, Xin Wang[✉], and Baozhu Liu

College of Intelligence and Computing, Tianjin University, Tianjin, China
{amandawang,wangx,liubaozhu}@tju.edu.cn

Abstract. With the scale of RDF graph data dramatically increasing, the demands for specifying constraints on RDF graphs and detecting violations of such constraints become urgent. Recently, the validation of RDF graphs has been attracting increasing research efforts, among which ShEx (Shape Expressions) and SHACL (Shapes Constraint Language) are widely used. However, incremental validation of RDF graphs still needs further study. To this end, we propose an incremental validation method of RDF graphs. Our method differentiates the types of constraint, adjusts the order of schema validation, and validates RDF graphs from a graph theory perspective as well as incrementally validates RDF graphs updates. The experimental results show that the incremental validation method proposed in this paper improves the efficiency of the validation of knowledge graphs updates.

Keywords: Incremental validation · RDF graph · SHACL

1 Introduction

With the proliferation of Knowledge Graphs (KG), the applications of KGs have a rapid growth in recent years [1,2]. Due to the flexible structure of *Resource Description Framework* (RDF) [3], RDF has become such a standard data model for knowledge graph. With the continuous increasing of the scale of RDF graph, new challenges have been brought to RDF graph data management. In order to provide correct data for downstream tasks, it is important to detect unqualified data from the massive RDF data. The valid RDF data should conform to a specified schema, therefore, the schema is essential for RDF data management. For instance, if a node n stands for a person, then n is valid only if there is an attribute of the name attached to n. Recent developments in knowledge graphs have heightened the need for RDF data validation in most applications. Although some attempts on RDF validation have been made, there is no unified standard for the validation currently.

The *Data Definition Language* (DDL) in SQL is a mature and widely-used schema for relational databases, and XML databases have similar schemas, e.g., XSD [4] and DTD [5]. In other words, both relational databases and XML databases have a well-established method to deal with validation and update issues. The schema can detect the unreasonable one when new data is added to the database. Incremental validation, as a partial validation method rather than

© Springer Nature Switzerland AG 2021
C. Xing et al. (Eds.): WISA 2021, LNCS 12999, pp. 359–371, 2021.
https://doi.org/10.1007/978-3-030-87571-8_31

complete validation of the database instance, attracts increasing research efforts in relational databases and XML databases. In practice, as new data is constantly being added to the RDF graph, RDF databases are expected to provide an efficient mechanism to handle data updates.

The Shapes Constraint Language (SHACL) [6], which was proposed by W3C in 2017, is a promising attempt on the RDF validation. In SHACL, constraints that restrict the property values of certain RDF nodes are grouped into a structure called *shape*, which can be viewed as a schema for RDF graphs. Figure 1(a) presents a SHACL schema of university RDF graph with two shapes called *Department* and *University*. The first one defines *Departments* which have at least one attribute of the name and is a sub-organization of at least one *University*, i.e., an instance of the next shape. The second shape shows that *Universities* have at least one attribute of the name. SHACL uses target declarations to indicate which nodes need to be validated, e.g., *sh:targetClass* in Fig. 1(a).

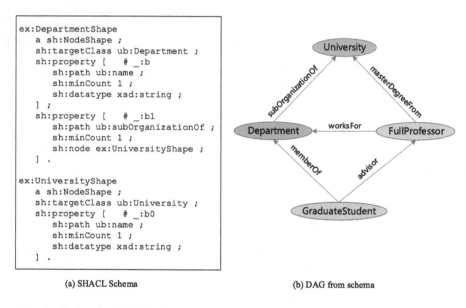

(a) SHACL Schema   (b) DAG from schema

Fig. 1. A simple SHACL shape schema and directed acyclic graph from shapes.

The most intuitive way to validate the updated graph is to validate the whole graph. However, the intuitive approach is not practical, while the validation efficiency will be significantly affected, since large-scale data will be involved and constant re-validation can result in resources waste. Therefore, a partial validation method will greatly improve the efficiency of RDF data update processing, which is the major objective of this work and has not been discussed in depth in previous researches.

Contributions. Based on the characteristics of RDF graphs and RDF shapes, this paper proposes an incremental validation algorithm for RDF graphs. Our contributions are as follows:

(1) We classify the constraints of each shape according to the constraint type, distinguish the constraints containing references (chained constraints) from other constraints (single constraints). In the validation process, the single constraint is first validated, and then the chained constraint is validated.

(2) We maintain a list of target nodes constrained by shapes in the schema. The list records the validation results of each node constrained by shapes, thereby reducing the time for chained validations.

(3) The reference of the shapes in the schema is constructed into a directed acyclic dependency graph, where shapes are validated according to the reverse order of the topological sort.

Organizations. The rest of this paper is organized as follows. In Sect. 2, we present the related work. The preliminaries are presented in Sect. 3, and in Sect. 4, we explain the method for validating graphs against shapes as well as the incremental validation and the algorithm. The experimental results are shown in Sect. 5. Finally, we conclude this paper in Sect. 6.

2   Related Work

Incremental validation has been successfully applied in the area of XML and semi-structured data management over decades. However, RDF graphs are more flexible than that of XML, which makes it difficult to apply incremental validation. Whereas, there are few research works on the validation of RDF graphs.

Two specifications for describing XML Documents schemas exist: Document Type Definition (DTD) and XML Schema Definition (XSD). The most classic method is to directly convert DTD documents into tree automata [7,8], which is not suitable for XML documents with too many elements. Neven et al. [9] distinguished the node schema from the path schema, transformed the node schema into automata, converted the XML document into sequences, and validated the node and path, respectively. Another use of automata is to abstract XML documents into trees and model schema into bottom-up tree automata, and execute incremental validation in trees [7,10].

With the proliferation of RDF graphs, the need for a schema that can describe and constrain RDF data becomes essential. ShEx [11] and SHACL are two schemas that describe and validate RDF graphs by shapes. Prud'hommeaux et al. [12] gave the definition of ShEx and implemented several validation tools as well as two extensions, GenX and GenJ, which can create ordered and closed content and XML/Json documents by leveraging the predictability of the graphs traversal, and provide a simple and declarative mapping from RDF data to XML and Json documents. However, ShEx does not provide a standard semantics for describing RDF constraints, while SHACL, which provides such standard semantics, leaves the recursion undefined in the W3C recommendation. Hence, Corman et al. [13] presented a semantic for the SHACL core constraint component which handles recursion. Since validating RDF graphs against SHACL constraints is NP-hard in the size of the graphs. They also provided an algorithm which finds an approximation to the validation problem by limiting the number of iterations. Andresel et al. [14] proposed a stricter, more constructive semantics for SHACL, based on Answer Set Programming (ASP). Furthermore, Corman et al. [15] investigated the problem of validating SHACL constraints over SPARQL endpoint and presented SHACL2SPARQL [16].

Although several attempts mentioned above have been made to tackle the RDF validation problem, there are still many problems worth exploring, one of which is the incremental validation of RDF graphs that studied in this paper.

3   Preliminaries

In this section, we introduce the definitions of relevant background knowledge. Due to the satisfactory scalability and flexibility of SHACL, this paper adopts SHACL schema for RDF graph validation. In addition, since SHACL has similar characteristics with other RDF graph schemas, makes it able to be transformed into other schemas and vice versa, it makes sense to use SHACL as a representative. Table 1 gives the notations used throughout this paper.

Table 1. List of Notations.

Notation	Description
$\top$	Boolean **true**
ϕ_s	The constraint belonging to a specific shape s
$\geq_n r.\phi$	At least n r-successors in G are valid against ϕ, and $n \in N^+$
$EQ(r_1, r_2)$	r_1-successors and r_2-successors of a node should coincide
$\sigma(v, s)$	An entity v belongs to the target nodes of s
$[\phi_s]^{G,v,\sigma}$	Evaluation of a node v of a graph G using an assignment σ
$u(v)$	The update to node v
G'	The RDF graph G after update
U	A set of $u(v)$ that makes G become G'

RDF is a graph-structured model based on three disjoint sets I, L and B, where I is the set of IRIs, L is the set of literals, and B is the set of blank nodes. In this paper, we represent an RDF graph as a directed graph.

Definition 1 (RDF Graph). *An RDF Graph G is a directed graph $G = (V, E)$, where*

– V is a finite set of nodes;

- E is the set of triples of the form $(v, p, v') \in \{I \cup B\} \times I \times \{I \cup B \cup L\}$, meaning that there is an edge in G from v to v' that labeled with property p.

An RDF schema is a set of RDF shapes that describe the structure of RDF graphs. RDF shapes can be visited from a particular node which is specified by some mechanism. The property value of a node as well as its neighbours can be restricted by constraints in RDF shapes. SHACL provides the components of "target" to defines constraints on RDF graphs. The notion of the "target" indicates for each shape which nodes have to conform to it.

For constraint expressions, this paper follows the abstract syntax introduced in [13], specifically, the constraints are defined by the following grammar:

$$\phi ::= \top \mid s \mid I \mid \phi_1 \wedge \phi_2 \mid \neg\phi \mid \geq_n r.\phi \mid EQ(r_1, r_2)$$

Definition 2 (RDF Schema). *An RDF schema S is a triple $(S, targ, def)$, where*

- *S is a set of shape names,*
- *$targ$ is a function that assigns target queries to each $s \in S$,*
- *def is a mapping from s to constraints ϕ.*

Example 1. The schema in Fig. 1 can be represented as follows.

S = {ex:DepartmentShape, ex:UniversityShape},
$targ$ (ex:DepartmentShape) =
    SELECT ?x WHERE { ?x a ub:Department},
$targ$ (ex:UniversityShape) =
    SELECT ?x WHERE { ?x a ub:University},
def (ex:DepartmentShape) = ($\geq_1$ub:name.xsd:string) $\wedge$
    ($\geq_1$ ub:subOrgainzationOf.ex:UniversityShape),
def (ex:UniversityShape) = ($\geq_1$ub:name.xsd:string) .

Definition 3 (Assignment). *An assignment σ for S and G is a mapping from S to V. If $\sigma(v, s)$ is true, then v is assigned to the shape s.*

Constraint Evaluation is grounded using assignments σ which map graph nodes to shape names [15].

Definition 4 (Validation). *A graph G is valid against a schema S if there is an assignment σ that makes $[\phi_s]^{G,v,\sigma}$ is true for s in S and v in V.*

4 Incremental Validation on RDF Graphs Against SHACL Shapes

Validating RDF graphs against the SHACL schemas is a prerequisite for incremental validation of RDF graphs, hence this section will cover the validation of RDF graphs first, followed by the incremental validation.

4.1   Single Validation and Chained Validation

The validation of most constraints in SHACL can be accomplished only by validating the neighboring data of the target node. For example, in Fig. 1, the type of the ub:name property of node v must be xsd:string. However, the validation of v and s may trigger the validation of v' and s', where v' is the successor to v and s' is the shape referenced by s. It can be deduced that validating v' may lead to more nodes being validated. Therefore, we separate constraints that contain references from other types of constraints and divide SHACL shape into two sub-shapes: *single-shape* without references, and *chained-shape* with references. A Validation against S can be converted to a single validation and a chained validation.

Then the RDF Schema S can be converted to S' as a tuple $(S,\ targ,\ defs,\ defc)$, where $defc$ and $defs$ are functions assign property constraint containing chained constraints and single constraints to each $s \in S$, respectively. Constraints in single-shape are properties that limit the local data of nodes and do not affect other nodes without taking chained-shape into account.

4.2   RDF Validation Against Schema

Since constraints with references are the reason for causing shape references, a single validation is an independent validation for each node. Therefore the validation process is the single validation first, followed by the chained validation.

Directed Acyclic Graphs Validation. The chained validation is to validate all nodes according to constraints with references on the premise of completing the single validation. As shown in Fig. 1(b), the shapes can be abstracted as DAG (Directed Acyclic Graph (DAG) refers to a directed graph that does not contain a cycle).

We assume that there are no cycles in S, which means the depth of the validation depends on the depth of the DAG that S abstracts into. When validating G, different access orders to chained-shapes result in different validation efficiencies, e.g., if GraduateStudent are validated first, all data needs to be loaded, whereas if the University is validated first, only this type of data needs to be loaded. Accordingly, we assume that the absence of constraints with references will not cause the loading of other data, hence if for each $s \in S$, $s \notin \phi_{s_n}$, s_n should be validated first, and the order of the validation is the reverse order of the topological sort.

Auxiliary Structure. In this paper, a validation of SHACL schema is divided into a single validation and a chained validation. During the process, temporary results need to be stored and accessed many times. Meanwhile, the chained validation requires that the results of the single validation can be obtained easily. Therefore, a list of target nodes constrained by the shape is maintained, where validation results for each s and v are stored.

The RDF graphs validation against schema S' is shown in Algorithm 1, in which we resort the shapes according to the inverse of topological sort. Each node and its corresponding shape, as well as the validation result are stored in the auxiliary structure CV.

Theorem 1. *The Algorithm 1 returns a correct result for the validation problem of a non-negative and non-recursive schema S over a graph G, and runs in $\mathcal{O}(|\mathcal{S}| \cdot |G|)$.*

Algorithm 1: RDF Validation Against Spilt Schema

Input: A schema $\mathcal{S}'$, an RDF graph G
Output: Auxiliary Structure CV

1   $S' \leftarrow$ ATopoSort (S)
2   $FNode \leftarrow \emptyset$, $CV \leftarrow \emptyset$, $res \leftarrow$ **false**
3   **for** $s \in S'$ **do**
4      $FNode \leftarrow targ(s)$ // $FNode$ **stores** v **in** V **assigned by** s
5   **for** $v \in Fnode$ **do**
6   **if** $[defs(s)]^{(G,v,\sigma)}$ *is* **true then**
7              $res \leftarrow [defc(s)]^{(G,v,\sigma)}$
8   **else**
9              $res \leftarrow$ **false**
10   set res to the value of v in CV
11 **return** CV

Proof. The correctness of the result directly comes from the validity of the assignment of each constraint. For DAG, a single validation can locally check whether each node satisfies the shape. For the chained validation, since the access order is the reverse of the topology sort of shapes, when validating the constraints with references, the references have been validated and the results have been stored in CV, therefore, each chained constraint can be validated in $\mathcal{O}(1)$. Then the time complexity is $\mathcal{O}(|\mathcal{S}| \cdot |G|)$.   □

4.3 Incremental Validation

We formally define a generic formulation of RDF graphs incremental validation problem, i.e., G is valid against $\mathcal{S}'$. We propose a method to validate the update and determine the legitimacy of the update by validating a part of G' instead of the entire graph.

Based on the auxiliary structure and validaiton method in Sect. 4.2, for the node $v \in V$, update $u(v)$, the incremental validation algorithm obtains the corresponding shape from the auxiliary structure.

Theorem 2. *The Algorithm 2 returns a correct result for the update to a graph G, and runs in $\mathcal{O}(|\mathcal{S}| \cdot |U|)$.*

Proof. What we consider is the atomic operation of the node, hence the validation of each update also follows the validation rules in Algorithm 1. The chained validation of each node can also be completed in $\mathcal{O}(1)$. Then the time for validating updates is $\mathcal{O}(|\mathcal{S}| \cdot |U|)$.   □

The incremental validation algorithm of RDF graphs is shown in Algorithm 2. When processing updates, IncValidation firstly sorts the shapes in S according to the inverse topological sort of the dependency graph. For each s, IncValidation gets the corresponding target nodes in U, and performs single validations and chained validations on each v respectively, and updates the value of conformance in CV according to whether v is already in G. For updates to nodes which are not constrained by any shapes, IncValidation accepts it by default.

Algorithm 2: RDF IncValidation

Input: A schema $\mathcal{S}'$, an RDF graph G, auxiliary structure CV, a set of updates U

Output: true if U is accepted, otherwise false

1   $flag \leftarrow$ **false**
2   $S' \leftarrow$ ATopoSort (S)
3   **for** $s \in S'$ **do**
4      $FNode \leftarrow targ(s)$ // $FNode$ stores v from U assigned by s
5   **for** $v \in FNode$ **do**
6         $g \leftarrow$ GetGraph (v) // Local graph of node v
7         $flag \leftarrow [defs(s)]^{(g,v,\sigma)}$
8   **if** $flag$ *is* **true then**
9            $s_r \leftarrow defc(s)$, $v' \leftarrow g.getObject(v,r)$
10           $flag \leftarrow CV.getValue(s_r, v')$
11   **if** $flag$ *is* **true then**
12   UpdateGraph $(u(v), G)$
13   **if** $v \notin V$ **then**
14   put v into s of CV and set the value true

15 **return** $flag$

---

5 Experimental Results

In this section, we evaluate the performance of our method. We conducted extensive experiments to verify the efficiency of our proposed algorithms.

5.1 Experimental Settings

Experimental Setup. Our experiments are performed on Windows 10 operating system with AMD A8-7100 Compute Cores 4C + Radeon R5 Compute Cores 4 G, and 1.8 GHz CPU.

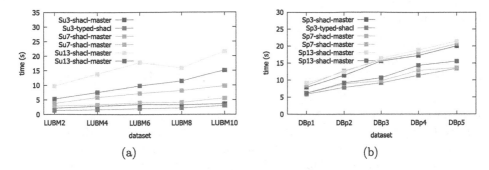

(a)   (b)

Fig. 2. Time for (a) validating LUBM against Su and (b) validating DBpedia against Sp.

Datasets. To the best of our knowledge, there is no benchmark to evaluate the performance of SHACL validators. Therefore, we generated our experimental datasets using the commonly recognized Lehigh University Benchmark (LUBM). This paper use LUBM to generate five datasets ranging in size from 20000 triples to 100000 triples with a span of 20000. In this paper, two updated datasets are generated for the LUBM datasets, which are 100 triples and 1000 triples, denoted as Lup100 and Lup1000, respectively. Furthermore, the latest version of DBpedia [17,18] (2020) is also used. We selected a subset of DBpedia data, involving related instances of 13 classes. We selected five datasets from 10000 to 50000 triples with a span of 10000 as the original datasets and denoted them as DBp1, · · · , DBp5. We also generated the updated datasets with the scale of 100 triples and 1000 triples respectively, which are represented as DBup100 and DBup1000.

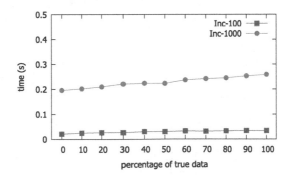

Fig. 3. Validating time for updating LUBM10 against Su13 with different ratio of true data.

Shape Schemas. We designed two sets of shapes for LUBM and DBpedia by observing the data pattern. They were designed to cover the shape references discussed in this paper. In order to evaluate the impact of different scale schemas on the validating efficiency, we divided each set of shapes into three inconsistent schemas, which we denoted as Su3, Su7, Su13, Sp3, Sp7 and Sp13.

5.2   Experimental Results

We compared the performance of the basic validation engine **SHACL-master** and optimized method **Typed-SHACL** about validating time on both LUBM and DBpedia. The incremental validation method **Inc-SHACL** is evaluated on update datasets mentioned in Sect. 5.1.

Efficiency of Validation Methods. This paper uses the traditional method **SHACL-master** and the improved method of adjusting the validation order **Atopo-shacl** to conduct experiments on the dataset. The experimental results are shown in Fig. 4. Accessing the shape according to the inverse topological sorting of the **DAG** can improve the efficiency of validation to a certain extent, and the improvement effect is obviously affected by the number of chained constraints.

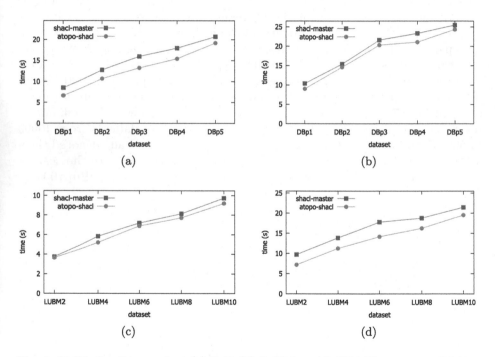

Fig. 4. Validation time against (a) Sp7, (b) Sp13 for with SHACL-master, validation time against (c) Su7, (d) Su13 and with Atopo-shacl.

It can be observed that the validation time of Typed-SHACL on both the benchmark datasets and the real-world datasets are less than that of SHACL-master in Fig. 2, indicating that Typed-SHACL is better than SHACL-master in the validation efficiency. In addition, it is obvious that the validation time on the benchmark datasets is less than on the real-world datasets. This is due to the fact that the data types in the benchmark datasets are uniformly distributed and the data content is relatively regular, while the data in the real-world datasets is randomly distributed.

Efficiency of Incremental Validation. The results are shown in Fig. 5. Several experiments are conducted to compare the efficiency of complete validation and incremental validation for updates. The efficiency of incremental validation is obviously higher than that of complete validation. Furthermore, it should be noted that the larger the update size, the more time it took for incremental validation.

We adjusted the amount of incorrect data in the updated data with a span of 10% to observe the effect of incorrect data in the updated data on validation time. The impact of updated data with different proportions of right and wrong on the validation efficiency is shown in Fig. 3. The experimental results show that, with the increase of the proportion of false data, the validation time will decrease. Since when the false data is validated, it is not necessary to fully validate all of its constraints in the shape, and the validation results are reported as long as a validation failure is found during the process.

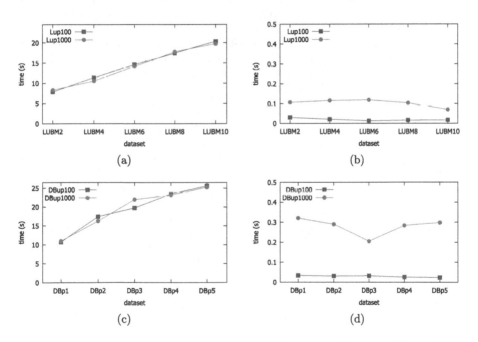

Fig. 5. Validation time against Sp13 for updating LUBM10 with (a) SHACL-master and (b) Typed-shacl, validation time against Su13 for updating Dbp5 with (c) SHACL-master and (d) Typed-shacl.

6   Conclusions and Future Work

In this paper, we propose an RDF graph validation method based on constraint type, and on this basis, we propose an incremental validation algorithm for RDF graph data. We conducted extensive experiments based on the benchmark datasets and the real-world datasets respectively to verify the validation efficiency of our method, which has obvious efficiency advantages when validating the data and the updated data. Since the work in this paper is not based on the complete SHACL semantics, we will work on implementing incremental validation based on the complete SHACL semantics in the future work.

Acknowledgments. This work is supported by National Natural Science Foundation of China (61972275).

References

1. Yun, H., He, Y., Lin, L., Pan, Z., Zhang, X.: Construction research and application of poverty alleviation knowledge graph. In: Ni, W., Wang, X., Song, W., Li, Y. (eds.) WISA 2019. LNCS, vol. 11817, pp. 430–442. Springer, Cham (2019). https://doi.org/10.1007/978-3-030-30952-7_42
2. Sheng, M., et al.: DSQA: a domain specific QA system for smart health based on knowledge graph. In: Wang, G., Lin, X., Hendler, J., Song, W., Xu, Z., Liu, G. (eds.) WISA 2020. LNCS, vol. 12432, pp. 215–222. Springer, Cham (2020). https://doi.org/10.1007/978-3-030-60029-7_20
3. World Wide Web Consortium, et al.: RDF 1.1 concepts and abstract syntax (2014)
4. Thompson, H.S., Mendelsohn, N., Beech, D., Maloney, M.: W3C xml schema definition language (xsd) 1.1 part 1: Structures. The World Wide Web Consortium (W3C), W3C Working Draft, December 3 (2009)
5. Wusteman, J.: Document type definition (DTD). Marcia J. Bates, 1640–1647 (2010)
6. Knublauch, H., Kontokostas, D.: Shapes constraint language (SHACL). W3C Candidate Recommendation **11**(8) (2017)
7. Bouchou, B., Alves, M.H.F.: Updates and incremental validation of XML documents. In: Lausen, G., Suciu, D. (eds.) DBPL 2003. LNCS, vol. 2921, pp. 216–232. Springer, Heidelberg (2004). https://doi.org/10.1007/978-3-540-24607-7_14
8. Bouchou, B., Alves, M.H.F., Laurent, D., Duarte, D.: Extending tree automata to model XML validation under element and attribute constraints. In: ICEIS, vol. 1, pp. 184–190 (2003)
9. Neven, F.: Automata, logic, and XML. In: Bradfield, J. (ed.) CSL 2002. LNCS, vol. 2471, pp. 2–26. Springer, Heidelberg (2002). https://doi.org/10.1007/3-540-45793-3_2
10. Papakonstantinou, Y., Vianu, V.: Incremental validation of XML documents. In: Calvanese, D., Lenzerini, M., Motwani, R. (eds.) ICDT 2003. LNCS, vol. 2572, pp. 47–63. Springer, Heidelberg (2003). https://doi.org/10.1007/3-540-36285-1_4
11. Boneva, I., Gayo, J.E.L., Hym, S., Prud'hommeau, E.G., Solbrig, H.R., Staworko, S.: Validating RDF with shape expressions. CoRR, abs/1404.1270 (2014)
12. Prud'hommeaux, E., Labra Gayo, J.E., Solbrig, H.: Shape expressions: an RDF validation and transformation language. In: Proceedings of the 10th International Conference on Semantic Systems, pp. 32–40 (2014)

13. Corman, J., Reutter, J.L., Savković, O.: Semantics and validation of recursive SHACL. In: Vrandečić, D., et al. (eds.) ISWC 2018. LNCS, vol. 11136, pp. 318–336. Springer, Cham (2018). https://doi.org/10.1007/978-3-030-00671-6_19

14. Andresel, M., Corman, J., Ortiz, M., Reutter, J.L., Simkus, M.: Stable model semantics for recursive SHACL. In: WWW 2020: The Web Conference 2020 (2020)

15. Corman, J., Florenzano, F., Reutter, J.L., Savković, O.: Validating SHACL Constraints over a SPARQL Endpoint. In: Ghidini, C., et al. (eds.) ISWC 2019. LNCS, vol. 11778, pp. 145–163. Springer, Cham (2019). https://doi.org/10.1007/978-3-030-30793-6_9

16. Corman, J., Florenzano, F., Reutter, J.L., Savkovic, O.: SHACL2SPARQL: validating a SPARQL endpoint against recursive SHACL constraints. In: ISWC Satellites, pp. 165–168 (2019)

17. Auer, S., Bizer, C., Kobilarov, G., Lehmann, J., Cyganiak, R., Ives, Z.: DBpedia: a nucleus for a web of open data. In: Aberer, K., et al. (eds.) ASWC/ISWC -2007. LNCS, vol. 4825, pp. 722–735. Springer, Heidelberg (2007). https://doi.org/10.1007/978-3-540-76298-0_52

18. Li, W., Chai, L., Yang, C., Wang, X.: An evolutionary analysis of DBpedia datasets. In: Meng, X., Li, R., Wang, K., Niu, B., Wang, X., Zhao, G. (eds.) WISA 2018. LNCS, vol. 11242, pp. 317–329. Springer, Cham (2018). https://doi.org/10.1007/978-3-030-02934-0_30

Unsupervised Entity Resolution Method Based on Random Forest

Wanying Xu, Chenchen Sun$^{(\boxtimes)}$, Lei Xu, Wenyu Chen, and Zhijiang Hou

Tianjin University of Technology, Tianjin 300384, China
hzj@tjut.edu.cn

Abstract. The task of entity resolution is to find records that describe the same entity in the real world, so as to solve the problem of data duplication. This paper proposes an unsupervised entity resolution method based on machine learning. This method first uses LSTM to convert records into vectors with semantic information. Next, we use the improved random forest method to map the records into the n-dimensional space to realize the partition operation of the records, and consider that the records in the same partition point to the same entity. Finally, we use an improved Affinity Propagation Clustering (AP) to cluster the partitions to determine whether the records in different partitions point to the same entity. Through experiments on real data sets, the effectiveness of the algorithm for solving entity resolution tasks is proved.

Keywords: Entity resolution · Random forest · LSTM · Affinity Propagation Clustering

1 Introduction

In the big data environment, the data generated on the Internet has exponentially increased. With the development of machine learning, people realize that data contains great value, and hope to use data mining and other technologies to learn valuable information from them. However, the quality of the data generated on the network platform is usually not high, and most of them have problems such as data inconsistency, incompleteness, and duplication. *Entity resolution* is a key technology to improve data quality, and it can effectively solve the problem of data duplication [1–4]. It plays a vital role in data cleaning, data integration and other fields, and is a guarantee for improving data quality.

Machine learning algorithms can be used to solve entity resolution problems [5–8]. This type of solution regards the entity resolution problem as classification problem, and classifies the record pairs into two categories: matching and non-matching. The basic idea is: firstly, establish an initial entity resolution model; secondly, use the labeled data set to train the model to learn appropriate matching functions, matching rules, and parameters; finally, apply these to the entity resolution model. However, in the real world, there are often no training examples that are known to match or not, so manual marking is required. This is a labor-consuming process that relies on professional knowledge, and

© Springer Nature Switzerland AG 2021
C. Xing et al. (Eds.): WISA 2021, LNCS 12999, pp. 372–382, 2021.
https://doi.org/10.1007/978-3-030-87571-8_32

the manually labeled training data cannot reach 100% accuracy. All these have affected the quality and efficiency of entity resolution.

For the above problems, this paper proposes an unsupervised entity resolution method (URF) based on random forest. First, use LSTM to model the records and convert the records into high-dimensional vectors containing semantic information. Then, by constructing an unsupervised random forest, the records are mapped into points in the n-dimensional space, and the positional relationship between the points reflects the similarity relationship between the records. According to the location of the points, we complete the partition operation of the records, and consider that the records in the same partition point to the same entity. Finally, we map the partitions in the n-dimensional space to a graph, and perform graph clustering through the improved AP clustering algorithm. Through this operation, we can judge whether the records in different partitions point to the same entity, and get the final result.

The organization structure of this paper is as follows: Sect. 2 introduces the related work of entity resolution task. Section 3 introduces the content of the unsupervised entity resolution model based on random forest proposed in this paper. Section 4 compares the work of this paper with existing work through experiments to verify the effectiveness of the method proposed in this paper. Section 5 summarizes the whole paper.

2   Related Work

The concept of entity resolution was formally proposed by [9] in 1959. After decades of development, entity resolution technology has been widely used in many fields such as artificial intelligence and databases. Since the mid-1990s, researchers have used machine learning algorithms to solve entity resolution tasks, thereby improving its quality and efficiency. Entity resolution methods based on machine learning can be divided into two categories: entity resolution methods based on classifiers and entity resolution methods based on probability graph models [10–12]. The entity resolution methods based on classifiers regards this problem as a classification problem, and finally records pairs are divided into two categories: matching and non-matching. Verykios et al. first proposed the use of decision tree to solve the problem of entity resolution in [13]. After that, more classifiers were applied to entity resolution tasks.

Decision tree is a common machine learning algorithm, which can use the parameters with strong discriminating ability as classification criteria to improve the performance of entity resolution [5]. Decision trees are used to solve the entity resolution problem in [6]. The basic idea is: First, select the optimal partition parameters for each attribute based on the CART algorithm. Then, according to the contribution of each attribute to the division, different weights are attached to different attributes, and finally a decision tree is generated.

Support vector machines (SVM) are also used to solve the problem of entity resolution [7]. A training model of SVM classifier is proposed in [8]. The model divides the entity resolution task into two phases. In the training phase, the similarity calculation function is learned through the training set, that is, dividing hyperplane is determined in the vector space. In the testing phase, the learned similarity calculation function is used for unlabeled entity pairs to measure their matching degree.

In addition to the above methods, entity resolution methods based on classifiers also include: entity resolution method based on active learning [14, 15] and entity resolution method based on genetic programming [16, 17].

3 Entity Resolution with Random Forest

3.1 Record Representation

If we want to solve the problem of entity resolution, we must first embed the records in the sample set into the vector space. That is, under the condition of retaining the semantic information, we represent the text data as a high-dimensional vector form.

Attribute Vector. The sample set contains n records, and each record contains m attributes. For each record, we segment it and preprocess each segment through the word2vec model [18]. Then, the N-dimensional word vector corresponding to each word can be obtained.

In the real world, records are composed of multiple attributes, and multiple attributes together determine an entity. We regard attributes as the smallest unit of entity resolution tasks. Different attributes have different contributions to entity resolution tasks, so we must distinguish the contribution of different attributes while turning records into vectors of uniform dimensions. Therefore, while converting the record into a sequence of word vectors, it is necessary to record the boundaries of each attribute.

Record Vector. We use the LSTM model to model the records [19]. Each LSTM unit includes a forget gate, an input gate and an output gate to achieve forgetting previous information and memorizing current information. The forget gate determines which of the previous information will be forgotten, and it updates the information through Eq. (1):

$$f_t = \sigma \left(W_f * [h_{t-1}, x_t] + b_f \right) \tag{1}$$

Among them, σ represents the sigmoid activation function, h_{t-1} is the state of the previous layer, and x_t is the current input. The input gate determines which current information will be memorized, Eq. (2) determines which information needs to be updated through the sigmoid activation function, and Eq. (3) determines the degree of information update through the tanh activation function.

$$i_t = \sigma (W_i * [h_{t-1}, x_t] + b_i) \tag{2}$$

$$\tilde{C}_t = \tanh(W_c * [h_{t-1}, x_t] + b_c) \tag{3}$$

$$C_t = f_t * C_{t-1} + i_t * \tilde{C}_t \tag{4}$$

The output gate determines which information will be output, Eq. (5) determines which parts will be output through the sigmoid activation function, and Eq. (6) realizes the processing of the current cell state through the tanh activation function.

$$o_t = \sigma (W_o * [h_{t-1}, x_t] + b_o) \tag{5}$$

$$h_t = o_t * \tanh(C_t) \tag{6}$$

We input word vectors into each unit of LSTM in sequence order, and output the last hidden layer of each attribute as the vectorized semantic expression of each attribute according to the division boundary of each attribute. Then we apply the vectorized semantic expression of each attribute to the attention mechanism layer, and apply Eq. (7) to the corresponding node:

$$u_j = \tanh(w_s h_j + b_s) \tag{7}$$

Among them, h_j is the output of the last hidden layer of each attribute, w_s and b_s are the weights and biases of the attention mechanism layer. Then use softmax to obtain the contribution rate of each attribute through Eq. (8).

$$\alpha_j = \frac{\exp(u_j^T u_s)}{\sum_j \exp(u_j^T u_s)} \tag{8}$$

Among them, u_j is the parameter that needs to be set. The contribution rate is the weight corresponding to each attribute. We use Eq. (9) to perform a weighted summation of the vectorized semantic expression and weight of each attribute to obtain the vectorized semantic expression of the entire record.

$$x_i = \sum_j \alpha_j h_j \tag{9}$$

3.2 Partition Based on Random Forest

By constructing a **random forest** [20, 21], we map the records to an n-dimensional space, so that the records correspond to the points in the space one by one. At the same time, the positional relationship of the points in the n-dimensional space represents the similarity relationship between the records. This avoids the pairwise comparison of record pairs when calculating the similarity, and reduces the time complexity of the algorithm.

One-Dimensional Partition by a Single Tree. Tree construction is a recursive process. We use all the records in the sample set to form the root node, and recursively split the records on each node. At each node, we will project the record x_i in a random direction, and convert the vector into a projection value. We define the calculation equation of the projection value p_i as (10)

$$p_i = x_i e^T \tag{10}$$

Among them, e is a random unit vector, generated from the normal distribution of each element in x_i. The median of the projection values of all records is used as the threshold to divide the records. The records with the projection value greater than the threshold constitute the right subtree, and the records with the projection value less than the threshold constitute the left subtree. Repeat the above division process on the left and right subtrees, until the height of the tree reaches a given threshold, stop the recursion, at this time we can get a tree.

Each tree partitions the records with different partitioning standards. The leaf nodes of tree T_j represent the partition of the j-th dimension in the n-dimensional space, and records that are assigned to the same leaf node are in the same partition of the j-th dimension. From this we can define the partition index I_i^j of the record, I_i^j represents the partition position of the record in the j-th dimension.

Multi-dimensional Partition by Random Forest. A single tree only partitions records with a single partition standard, which makes the partition result have some errors. So we need to randomly construct multiple trees to form a random forest. Each tree in the forest has a different partition criterion for records. Combining the partition results of multiple criterions can make the partition results lower in error, and improve the quality of entity resolution.

The random forest is composed of multiple trees, and the partitions formed by the leaf nodes of each tree divide the space in different dimensions. Each tree has different partition criteria for records, so the same record is located in a different partition on each tree. Each record is divided by each tree in the forest, and we can get the partition position of each dimension in the n-dimensional space. Combine the partition indexes of each dimension to get a vector, which we call the multi-dimensional partition index vector $IN(x_i)$. The multi-dimensional partition index vector represents the position of the record in the n-dimensional space, and we define it as (11):

$$IN(x_i) = \{I_i^1, I_i^2, ..., I_i^j\} \tag{11}$$

The multi-dimensional partition index vector contains the partition information of each record in each dimension, and the element I_i^j represents the partition index of the record x_i in the j-th dimension, that is, the leaf node position of the record x_i in the j-th tree. Synthesizing the partition of each dimension, we divide the n-dimensional space into different partitions. Each partition is called a multi-dimensional partition, and records with the same multi-dimensional partition index vector belong to the same multi-dimensional partition. At the same time, all records in the same multi-dimensional partition D_k contain the same partition index vector.

By defining multi-dimensional partitions, we realized the task of partitioning records in entity resolution. We call records that are in the same partition after being divided by multiple partition standards as highly similar records. By aggregating them into the same partition, a preliminary clustering operation is realized. For the records in the same multi-dimensional partition, we can think that they point to the same entity; for the records in different multi-dimensional partitions, we need to further cluster to determine whether they point to the same entity.

3.3 Partition Graph Based Record Clustering

The preliminary clustering process has aggregated records with great similarity into the same multi-dimensional partition, but records in adjacent multi-dimensional partitions may also point to the same entity. At this time, we need to perform another clustering operation.

Multi-dimensional Partition Graph. In order to facilitate the clustering operation, we map the distribution of the records in the n-dimensional space to a two-dimensional graph while preserving the similarity relationship between the records. The graph generated by this process is called the multi-dimensional partition graph $G(v, e)$. The node v in the graph represents the multi-dimensional partition, and the edge e represents the similarity between the multi-dimensional partitions.

Node Building. Each multi-dimensional partition contains multiple records, and these records contain the same multi-dimensional partition index vector. We need to correspond one-to-one between the multi-dimensional partitions in the n-dimensional space and the nodes in the graph. Therefore, in each multi-dimensional partition, we need to generate a record R_k, which is assigned to the node v_k in the graph as a representative. The representative record R_k, is described by the mean value μ_k and sample density ρ_k of all records in the multi-dimensional partition. We define the mean value and sample density as (12) and (13)

$$\mu_k = \frac{1}{N(D_k)} \sum_{x_i \in D_k} x_i \tag{12}$$

$$\rho_k = \frac{N(D_k)}{V(D_k)} \tag{13}$$

Among them, $N(D_k)$ represents the total number of records contained in the multi-dimensional partition D_k, and $V(D_k)$ represents the approximate value of the volume of the points distributed in the multi-dimensional partition. $V(D_k)$ can be calculated by Eq. (14)

$$V(D_k) = \prod_{n=1}^{N} \sigma_n^2 \tag{14}$$

Among them, σ_n^2 is the variance of the n-th element of all the record vectors in each multi-dimensional partition.

Edge Construction. We need to connect nodes with greater similarity. Now we define the concept that two nodes are adjacent. When the Hamming distance between the multi-dimensional partition index vectors of two nodes V_α and V_β is 1, that is, when the two vectors have only one element different, the two nodes are adjacent. We connect two adjacent nodes by adding an edge $E_{\alpha\beta}$. The weight $w_{\alpha\beta}$ of edge $E_{\alpha\beta}$ represents the degree of similarity between the records represented by two adjacent nodes, and we define it as (15):

$$w_{\alpha\beta} = \frac{N(D_\alpha) + N(D_\beta)}{V(D_{\alpha\beta})} \tag{15}$$

Graph Clustering with an Improved Affinity Propagation Algorithm. Multi-dimensional partition graphs contain similar relationships among records in different multi-dimensional partitions. We use *Affinity Propagation* (AP) algorithm for graph clustering to determine whether records in different multi-dimensional partitions point to the same entity [22].

First, the similarity matrix S is created from the similarity relationship between the nodes in the multi-dimensional partition graph. The elements in matrix $S(\alpha, \beta)$ represent the similarity between nodes α and β, and the calculation equation is (16)

$$S(\alpha, \beta) = \begin{cases} -\dfrac{1}{w_{\alpha\beta}}, \alpha \neq \beta \\ p(\beta), \alpha = \beta \end{cases} \tag{16}$$

Among them, P is the reference degree, which indicates the tendency of the data point to be selected as the cluster center. Next, we have to calculate the responsibility and availability between nodes, and repeatedly update and transmit these two messages. The responsibility $r(\alpha, \beta)$ represents the degree to which node β is suitable as the cluster center of node α, and its calculation equation is (17)

$$r(\alpha, \beta) = S(\alpha, \beta) - \max_{\gamma \neq \beta}\{a(\alpha, \gamma) + S(\alpha, \gamma)\} \tag{17}$$

Availability $a(\alpha, \beta)$ represents the attribution degree of node α selecting node β as the cluster center, and its calculation equation is (18) and (19).

$$a(\alpha, \beta) = \min\left\{0, r(\beta, \beta) + \sum_{\gamma \neq \alpha} \max\{0, r(\gamma, \beta)\}\right\} \tag{18}$$

$$a(\beta, \beta) = \sum_{\gamma \neq \alpha} \max\{0, r(\gamma, \beta)\} \tag{19}$$

In order to avoid shocks, we introduce a damping factor $\lambda \in [0, 1)$ when updating responsibility and availability, and combine the damping factor to perform a weighted summation between the newly calculated matrix and the original matrix. The update equations for responsibility and availability are (20) and (21):

$$r_{t+1}(\alpha, \beta) = \lambda \cdot r_t(\alpha, \beta) + (1 - \lambda) \cdot r_{t+1}(\alpha, \beta) \tag{20}$$

$$a_{t+1}(\alpha, \beta) = \lambda \cdot a_t(\alpha, \beta) + (1 - \lambda) \cdot a_{t+1}(\alpha, \beta) \tag{21}$$

Among them, t represents the previous iteration moment. $r(\alpha, \beta)$ and $a(\alpha, \beta)$ are continuously updated alternately, until the number of iterations reaches a set threshold or the cluster center does not change in several iterations, the iteration is terminated. Through the above clustering process, we can get to the cluster centers of all nodes, and the cluster center β corresponding to node α is determined by Eq. (22).

$$\phi = \arg \max\{a(\alpha, \beta), r(\alpha, \beta)\} \tag{22}$$

Through clustering, we can divide each node into the most suitable clustering center, so that the records with high similarity can be aggregated into clusters. The records contained in nodes belonging to the same cluster point to the same entity.

4   Experimental Results

4.1   Experimental Setup

Experimental Environment. The experiment uses Intel(R) Core(TM) i7-8565U CPU @ 1.80 GHz 1.99 GHz processor, 8.00 GB RAM and 64-bit operating system.

Experimental Data. We use three public data sets DBLP-ACM (DA), DBLP-Scholar (DS) and Amazon-GoogleProducts (AG) to conduct experiments. Each record in the data set contains four attributes.

Evaluation Indicators. In order to show the performance of the algorithm, we evaluate it with precision, recall and F-measure. F-measure is the main metric. Among them, precision indicates the proportion of record pairs that are actually matched among all the record pairs that are predicted to be matched. Recall indicates the proportion of record pairs that are judged to be matched by entity resolution among the truly matched record pairs. F-measure is the harmonic average of precision and recall.

4.2   Experimental Results and Analysis

Comparison with Traditional Machine Learning Methods. The traditional entity resolution method first uses the labeled data to train the machine learning model, and then completes the entity resolution task. Here we take the entity resolution method based on decision tree (Decision tree) and the entity resolution method based on support vector machine (SVM) as examples to compare with URF [6, 7].

The three bar charts in Fig. 1 respectively show the comparison of the algorithm performance of URF and traditional machine learning methods on three data sets. We can see from it that F-measure of URF on each data set is obviously higher than Decision tree method and SVM method. And there is a tradeoff between precision and recall. The unsupervised entity resolution method based on random forest proposed in our paper performs better in entity resolution tasks.

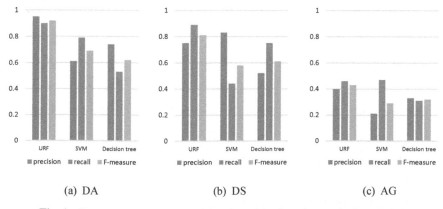

(a) DA	(b) DS	(c) AG

Fig. 1. Comparison between traditional machine learning methods and URF

Comparison with Traditional Clustering Algorithms. We compare the K-Means clustering algorithm with the AP clustering algorithm used in URF. Using the controlled variable method, except for replacing AP clustering with K-Means clustering, other parts of the model are consistent with this paper. Compare the newly generated algorithm (RF+K-Means) with URF, and the result is shown in Fig. 2.

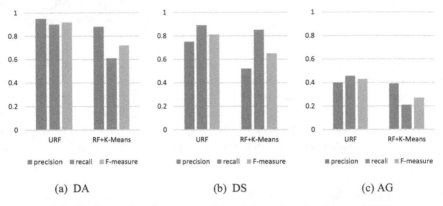

(a) DA   (b) DS   (c) AG

Fig. 2. Comparison between traditional clustering algorithm and URF

It can be seen from Fig. 2 that F-measure of URF is higher than the RF + K-Means algorithm on each data set. For the RF + K-Means algorithm, factors such as the number of clusters have a greater impact on the clustering results, but we do not know how many entities are included in the data set before entity resolution. However, URF does not need to specify the number of clusters and the initial cluster center, so URF is more effective.

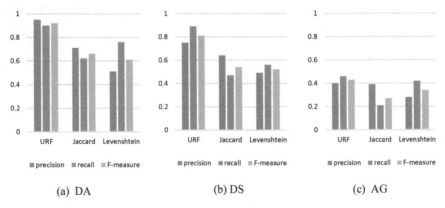

(a) DA   (b) DS   (c) AG

Fig. 3. Comparison between traditional similarity measurement and URF

Comparison with Traditional Similarity Measurement Methods. Traditional entity resolution algorithms mostly use field-based similarity algorithms or edit distance-based similarity algorithms. They only pay attention to the literal similarity, but ignore the semantic information, which reduces the quality of entity resolution. We replaced the similarity measurement method in this paper with the Jaccard similarity algorithm and the Levenshtein distance algorithm respectively, and compared with URF.

It can be seen from Fig. 3 that F-measure of URF in our paper is higher than the Jaccard algorithm and the Levenshtein algorithm on each data set. So URF pays full attention to semantic information when calculating similarity, and improves the quality of entity resolution.

5 Conclusions

This paper proposes an unsupervised entity resolution method based on random forest. First of all, we preprocess the records and convert them into vectors containing semantic information. Then, we partition the records by random forest and get the similarity relationship between the records. In the end, we use an improved AP clustering algorithm to cluster the partitions. For the method proposed in this paper, we conducted experiments on multiple data sets. Experimental results show that the entity resolution method proposed in this paper can improve the quality and efficiency of entity resolution. But the records in the real world are not as regular as those in the data set. In the next step, we are going to study how to perform entity resolution for incomplete or erroneous records.

Acknowledgements. This work is supported by the National Natural Science Foundation of China (Grant Nos. 62002262, 71804123), and the TJUT College Students' Innovative Entrepreneurial Training Plan Program (Grant No. 202010060040).

References

1. Yu, L., Nie, T., Shen, D., Kou, Y.: An approach for progressive set similarity join with GPU accelerating. In: Wang, G., Lin, X., Hendler, J., Song, W., Xu, Z., Liu, G. (eds.) WISA 2020. LNCS, vol. 12432, pp. 155–167. Springer, Cham (2020). https://doi.org/10.1007/978-3-030-60029-7_14
2. Naumann, F., Herschel, M.: An introduction to duplicate detection. Synth. Lect. Data Manage. **2**(1), 1–87 (2010)
3. Papadakis, G., Ioannou, E., Palpanas, T.: Entity resolution: past, present and yet-to-come. In: 23th International Conference on Extending Database Technology, pp. 647–650. Copenhagen, Denmark (2020)
4. Getoor, L., Machanavajjhala, A.: Entity resolution: theory, practice & open challenges. Proc. VLDB Endowm. **5**(12), 2018–2019 (2012)
5. Elfeky, M., Verykios V., Elmagarmid, A.: Tailor: a record linkage toolbox. In: 18th International Conference on Data Engineering, pp. 17–28 (2002)
6. Tejada, S., Knoblock, C., Minton, S.: Learning object identification rules for information integration. Inf. Syst. **26**(8), 607–633 (2001)
7. Bilenko, M., Mooney, R.: Adaptive duplicate detection using learnable string similarity measures. In: 9th International Conference on Knowledge Discovery and Data Mining, pp. 39–48 (2003)

8. Christen, P.: Automatic record linkage using seeded nearest neighbour and support vector machine classification. In: 14th ACM SIGKDD Conference on Knowledge Discovery and Data Mining, pp. 151–159 (2008)

9. Newcombe, H.B., Kennedy, J.M., Axford, S.J., James, A.P.: Automatic linkage of vital records. Science **338**1(130), 954–959 (1959)

10. Lu, C., Huang, G.Y.: Entity resolution in sparse encounter network using Markov logic network. IEEE Access **9**, 83055–83066 (2021)

11. Fisher, J., Christen, P., Wang, Q.: Active learning based entity resolution using markov logic. In: Bailey, J., Khan, L., Washio, T., Dobbie, G., Huang, J.Z., Wang, R. (eds.) PAKDD 2016. LNCS (LNAI), vol. 9652, pp. 338–349. Springer, Cham (2016). https://doi.org/10.1007/978-3-319-31750-2_27

12. Culotta, A., McCallum, A.: Joint deduplication of multiple record types in relational data. In: 14th International Conference on Information and Knowledge Management, pp. 257–258 (2005)

13. Verykios, V., Elmagarmid, A., Houstis, E.: Automating the approximate record-matching process. Inf. Sci. **126**(1–4), 83–98 (2000)

14. Arasu, A., Götz, M., Kaushik, R.: On active learning of record matching packages. In: Proceedings of the 2010 ACM SIGMOD International Conference on Management of Data, pp. 783–794 (2010)

15. Fu, Y.F., Zhu, X.Q., Li, B.: A survey on instance selection for active learning. Knowl. Inf. Syst. **35**(2), 249–283 (2013)

16. de Carvalho, M.G., Laender, A.H.F., Gonçalves, M.A., da Silva, A.S.: A genetic programming approach to record deduplication. IEEE Trans. Knowl. Data Eng. **24**(3), 399–412 (2012)

17. Isele, R., Bizer, C.: Active learning of expressive linkage rules using genetic programming. J. Web Semant. **23**, 2–15 (2013)

18. Sharma, A.K., Chaurasia, S., Srivastava, D.K.: Sentimental short sentences classification by using CNN deep learning model with fine tuned Word2Vec. Procedia Comput. Sci. **167**, 1139–1147 (2020)

19. Liu, G., Guo, J.B.: Bidirectional LSTM with attention mechanism and convolutional layer for text classification. Neurocomputing **337**, 325–338 (2019)

20. Flake, G.W., Tarjan, R.E., Tsioutsiouliklis, K.: Graph clustering and minimum cut trees. Int. Math. **1**(4), 385–408 (2003)

21. Perbet, F., Stenger, B., Maki, A.: Random forest clustering and application to video segmentation. In: Proceedings of the 2009 British Machine Vision Conference, pp. 1–10 (2009)

22. Bodenhofer, U., Kothmeier, A., Hochreiter, S.: APCluster: an R package for affinity propagation clustering. Bioinformatics **27**(17), 2463–2464 (2011)

Research and Application of Personalized Recommendation Based on Knowledge Graph

YuBin Wang[1], SiYao Gao[2(✉)], WeiPeng Li[3], TingXu Jiang[2], and SiYing Yu[2]

[1] State Grid Corporation of China, Beijing, China
yubin-wang@sgcc.com.cn
[2] College of Intelligence and Computing, Tianjin University, Tianjin, China
[3] State Grid Shandong Electric Power Company, Shandong, China

Abstract. Text data resources in the power domain have become increasingly abundant in recent years with the large scale popularization of information office in the power sector, but workers are facing an increasingly severe problem of data information overload. Since the concept of knowledge graph have been proposed, researchers have used professional datasets in various fields to construct corresponding knowledge graphs and proposed various knowledge graph completion algorithms to solve the problem of missing entity and relation links. In this paper, we introduce the knowledge graph as auxiliary information into the recommendation system of power domain. Our method uses translation-based models to learn the representations of users and items and applies them to optimize the recommender system. In addition, to address users diverse interests, we also build user profiles in our method to aggregate a users history with respect to candidate items. According to the characteristics of the data and the representativeness and universality of the data, extensive experiments are conducted on the Citeulike. We apply our approach to the power domain and construct the knowledge graph of the power domain dataset. The results validate the effectiveness of our approach on recommendation.

Keywords: Electric domain · Recommender system · User profile · Knowledge graph

1 Introduction

With the construction of informatization in the era of Big Data, online office has become an important part in people's daily work and many online office platforms have emerged [1]. At present, users mainly rely on searching in the file system based on the keywords. However, it is difficult to accurately filter the resources that meet the users' needs. Constructing a recommender system by analyzing users' historical browsing behaviors can more accurately recommend the file resources to users. The recommender system does not require users to

© Springer Nature Switzerland AG 2021
C. Xing et al. (Eds.): WISA 2021, LNCS 12999, pp. 383–390, 2021.
https://doi.org/10.1007/978-3-030-87571-8_33

search for required resources actively, but predicts users' behaviors by mining users' potential interests. Applying the recommender system to the recommendation of power office documents can meet users' needs and achieve personalized office.

User portrait is an important concept in recommender system, constructed by collecting the user's basic information and the user's historical browsing behavior information [2]. We use the knowledge graph to link the user portrait with the file information. We take into account the connection between the file resources, which can recommend the files that meet users' needs and interests. In the meanwhile, it dynamically recommends power file resources according to the user's official progress to achieve adaptive office resource recommendation.

In this paper, we start with the citeulike dataset and construct user portrait using knowledge graph technology. We take users and knowledge points as entities, and take the relations between the knowledge points learned by users as the relations, and combine user characteristics and learning resources to build a collaborative filtering recommender system for learning resources. Finally, the model with the best experimental effect is applied to the private power domain data collection. The manual evaluation method is used for evaluation, and the results are excellent.

2   Related Work

2.1   Knowledge Graph Embedding

Knowledge graphs (KGs) aim at semantically representing the world's truth in the form of machine-readable graphs composed of triplet facts [3]. In recent years, various knowledge graph embedding methods have been proposed, among which the translation-based models are simple and effective with good performances. Inspired by word2vec, given a triplet (h, r, t), TransE [4] learns vector embeddings h, r and t which satisfy $h + r \approx t$, but there are certain limitations in dealing with complex relation such as 1-to-N, N-to-1, and N-to-N. To solve this limitation of TransE, TransH [5] proposed a method for entities could express differently under different relations. TransR [6–8] treats an entity as a collection of multiple attributes. Different attributes correspond to different relations, and different relations have different semantic spaces. The TransD [9] can be viewed as a simplified version of the TransR model. The TransD uses two vectors to project the head entity and the tail entity into the sum of mapping matrix and reduces the complexity compared to the TransR model.

2.2   Knowledge Graphs for Recommendation

Inspired by the success of applying KG in a wide variety of tasks, researchers also tried to utilize KG to improve the performance of recommender systems [10–12]. Among various types of side information, knowledge graph (KG) usually contains much more fruitful facts and connections about items [13]. Wang

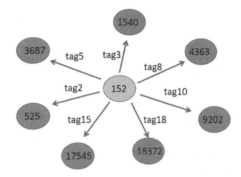

Fig. 1. User profile structure for user id 152. The figure shows the author 152 and the papers associated with him. The mark attached to each arrow indicates the relationship between the author and the paper

et al. [14] proposed DKN, which is a content-based click-through rate prediction recommendation model, with knowledge-aware convolutional neural network (KCNN) to process semantic representation and knowledge representation to form embeddings. Collaborative Knowledge base Embedding (CKE) obtains the structured information in the knowledge graph through TransR. It uses the denoising coding network to learn to obtain the text representation vector of the item and used the convolutional coding network to obtains the visual representation vector of the item. Ripple Network [15] simulates the propagation of water waves to represent the spread of user interests on the knowledge graph. The user's interests spread outward and gradually decays along the knowledge graph around the user's historical behavior entity. The Multi-task Learning for KG enhanced Recommendation (MKR) [16] model includes a recommendation part and a knowledge representation learning part. The recommendation part takes the feature representation of the item as input, and the predicted click rate as output. The knowledge representation learning part takes the head entity and relation as the input, and the predicted tail entity as the output. The model links the two through the cross-sharing unit (Fig. 1).

3   Knowledge Graph Representation Learning

Due to the limitation of the data in the dataset, this paper uses the paper labels as user interests feature labels to calculate the similarity between the semantic features of the user's historical learning resources and the semantic features of resource. This paper uses the translation model to represent the knowledge graph as entity vectors and relation vectors, and calculates the vectors of entities and relations by calculating the similarity in the vector space. Because the relationship of the knowledge graph is relatively complex, the training effect on TransR is better than that on TransE, we use TransR to calculate entity and relation vectors.

The translation model expects entities and relations to satisfy $h+t = r$, where H represents the head vector, R represents the relation vector, and T represents the tail vector. For each relation r, TransR set a projection matrix $M_r \in R_{k \times d}$, which may project entities from entity space to relation space. With the mapping matrix, we define the projected vectors of entities as $h_r = hM_r, t_r = tM_r$, the score function is correspondingly defined as:

$$f = \|h_r + r - t_r\|_{l_1/l_2} \tag{1}$$

We use TransR to obtain the vectors of user entities, label entities and paper entities, and use the cosine similarity to calculate the similarity of the entity vectors to represent the semantic similarity between user and resource.

$$similarity = \frac{user_i \cdot item_j}{|user_i| \cdot |item_j|} \tag{2}$$

$user_i$ represents the entity vector of user i and $item_j$ represents the entity vector of paper j.

In order to make the recommendation results more accurate, this paper considers the user profile, the reference relationships between papers, and the user's historical marked paper relation when calculating the vectors through the TransR model. User profile can express the relationship between user interests and learning resources The references of papers are closely related to the content of the paper, and the relationship between learning resources can be better considered by introducing the reference relationships of papers.

4   Recommendation Process

The specific process of recommending learning resources based on user interests and resource similarity is as follows:

1) For any learning resource, there is a list of knowledge point tags in the citeuilike-a dataset, which is used as the knowledge base of learning resources. For a certain learning resource, the knowledge base can be expressed as $K(item_i) = \{t_1, t_2, \cdots, t_n\}$, which represents the knowledge point lables. For the learning resource entities contained in the dataset, the dataset label is directly used as their knowledge base. For newly added learning resources, it is necessary to calculate the similarity with the existing knowledge point labels through character matching to obtain the learning resource labels, and build a resource knowledge base based on the matching labels. The knowledge base of learning resources is represented in Table 1.

2) According to the user's historical learning behavior, the knowledge point labels of the learning resources marked by the user can be used as the user labels to create a user profile. User profile can be expressed as $K(user) = \{K(item_1), K(item_2), \cdots, K(item_n)\}$. When the user marks a certain learning resource, all the knowledge point labels contained in the learning resource are

Table 1. Learning resource knowledge base

Learning resource	Corresponding knowledge base
$item_1$	$item_{11}, item_{12}, \cdots, item_{1a}$
$item_2$	$item_{21}, item_{22}, \cdots, item_{2b}$
$\cdots$	$\cdots$
$item_n$	$item_{n1}, item_{n2}, \cdots, item_{nc}$

added to the user profile. According to the user profile, the user's learning interest and cognitive level can be better understood. The recommender system can facilitate the prediction of the user's future learning behavior, and recommend appropriate learning resources for the user.

3) The translation model TransR is used to process user profile, learning resource knowledge bases and reference relationships among papers to obtain user vectors and learning resource vectors. We calculate the similarity between the user vectors and the learning resource vectors and select the top n resources with the highest similarity to recommend to the user. We construct a learning resource recommendation model based on the user profiles.

5 Experiment

5.1 Datasets

In this paper, we evaluate our methods with the publicly accessible dataset: citeulike-a. Citeulike-a dataset collects user and document information in citeulike and Google Scholar, including data that contains 5551 users and 16,980 papers. Citeulike-a includes user-marked paper information, paper title, abstract text information and citation information among papers. In practical applications, we use the power domain dataset, which contains personnel related information, the employees' browsing status of the documents, and the keywords of each document.

5.2 Experimental Results

In order to select the best knowledge representation model for recommendation, this article conducts experiments on TransE, TransR, and TransD under the conditions of K = 10, K = 20, K = 30, and K = 40 respectively, where K means the number of items recommended to user. Precision and recall rate (Recall) and F_1-score are used as the evaluation basis for evaluating the recommendation effect, and the performance is shown in Fig. 2. As shown in Fig. 2, the recommendation effect of TransR is better than other translation models. The accuracy generally decreases with the increase of K value. When K equals 10, the accuracy of TransR is much higher than TransE and TransD. The recall increases with

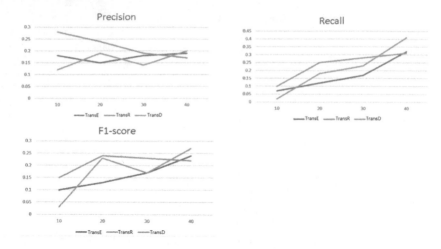

Fig. 2. The performance of the recommender system under different representation learning methods

the increase of K value. F_1-score combines the accuracy and recall, showing an overall upward trend. The dataset used in this paper contains a large number of 1-N, N-1, and N-N complex relations, so the recommendation effect of TransE is poor. The dataset in this article is small, compared with TransR, TransD reduces complexity and is more suitable for large-scale processing datasets, but TransR works better on small-scale datasets. Through the analysis and comparison of experimental results, TransR is finally selected to construct the recommender system.

The recommendation algorithm proposed in this paper is compared with the user-based collaborative filtering recommendation algorithm and the item-based collaborative filtering recommendation algorithm. The results are evaluated with Precision, Recall and F_1-score. The experimental results are shown in Fig. 3. It can be seen from the comparison experiment that the model we propose is better than the user-based collaborative filtering recommendation algorithm, and is similar to the item-based collaborative filtering recommendation algorithm. From the perspective of accuracy, *our method $\geq$ itemcf $>$ usercf*; From the perspective of recall, *our method $>$ itemcf $>$ usercf*, the recall of itemcf decreases with the increase of K value. From the perspective of F_1-score, *our method $\approx$ itemcf $>$ usercf*, since the user profile constructed in this paper mainly relies on the user's historical learning behavior, it lacks other characteristics and attribute information, so the recommendation effect will be closer to itemcf.

In the last stage of the research, we migrated the trained model that performed best on the citeulike to the State Grid electronic file dataset which is consistent with the actual generated scenario, and conducted simulation experiments using this dataset. In the experiment, the user profile is constructed according to the user's browsing history of the power grid files and the subject

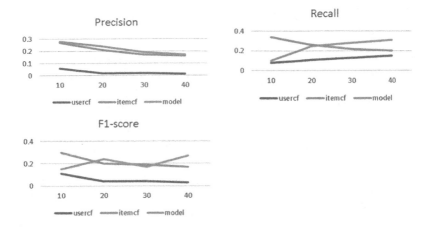

Fig. 3. Comparison of experimental performance

words of each power grid file. TransR processes the user profile, the power grid file resource knowledge base and the subject relation between the files to obtain the user embeddings and the power grid file resource embeddings. We calculate the similarity and select the top k resources with the highest similarity to recommend to users. We construct a learning resource recommendation model based on user profiles. The experimental results show that when the model is applied to the national grid electronic file dataset, it can still achieve excellent results. The accuracy of model recommendation can reach a satisfactory level.

6 Conclusion

In this paper, we adopt an adaptive learning resource recommendation algorithm to select appropriate learning resources for users, and migrate the model to the State Grid electronic file dataset consistent with the actual generation scenario for application. We test the recommendation effect of the recommender system on the citeulike and show that our method can outperform popular baselines. Finally, the model is applied to the dataset in the electric power domain and the effectiveness and robustness of the model are verified according to the results of manual evaluation. In future work, we plan to use more abundant datasets, combining with richer user characteristic information to build a more perfect user profile and make recommendations more accurately.

Acknowledgments. This work is supported by the Science and Technology Project of State Grid Corporation of China (Contract No.: SGSDWF00FCJS2000155).

References

1. Ma, K., Ilievski, F., Francis, J., et al.: Knowledge-driven data construction for zero-shot evaluation in commonsense question answering. In: 35th AAAI Conference on Artificial Intelligence (AAAI-21) (2020)

2. Gu, H., Wang, J., Wang, Z., Zhuang, B., Su, F.: Modeling of user portrait through social media. In: IEEE International Conference on Multimedia and Expo, pp. 1–6 (2018)
3. Li, J., et al.: Deep hybrid knowledge graph embedding for top-N recommendation. In: Workshop on Information Security Applications, pp. 59–70 (2020)
4. Bordes, A., Usunier, N., Garcia-Duran, A., Weston, J., Yakhnenko, O.: Translating embeddings for modeling multi-relational data. In: Annual Conference on Neural Information Processing Systems, pp. 2787–2795 (2013)
5. Wang, Z., Zhang, J., Feng, J., Chen, Z.: Knowledge graph embedding by translating on hyperplanes. In: AAAI Conference on Artificial Intelligence, pp. 1112–1119 (2014)
6. Lin, Y., Liu, Z., Sun, M., Liu, Y., Zhu, X.: Learning entity and relation embeddings for knowledge graph completion. In: AAAI Conference on Artificial Intelligence, pp. 2181–2187 (2015)
7. Wang, X., He, X., Cao, Y., Liu, M., Chua, T.S.: KGAT: knowledge graph attention network for recommendation. In: Proceedings of the 25th ACM SIGKDD International Conference on Knowledge Discovery & Data Mining (2019)
8. Trisedya, B.D., Qi, J., Zhang, R.: Entity alignment between knowledge graphs using attribute embeddings. In: Proceedings of the AAAI Conference on Artificial Intelligence, vol. 33 (2019)
9. Ji, G., He, S., Xu, L., Liu, K., Zhao, J.: Knowledge graph embedding via dynamic mapping matrix. In: Proceedings of the 53rd Annual Meeting of the Association for Computational Linguistics and the 7th International Joint Conference on Natural Language Processing, pp. 687–696 (2015)
10. Paulheim, H.: Knowledge graph refinement: a survey of approaches and evaluation methods. Semantic Web **8**, 489–508 (2017)
11. Zhao, B., Xu, Z., Tang, Y., Li, J., Liu, B., Tian, H.: Effective Knowledge-Aware Recommendation via Graph Convolutional Networks. In: Wang, G., Lin, X., Hendler, J., Song, W., Xu, Z., Liu, G. (eds.) WISA 2020. LNCS, vol. 12432, pp. 96–107. Springer, Cham (2020). https://doi.org/10.1007/978-3-030-60029-7_9
12. Shubha, C.A., et al.: Analysis and implementation of recommender system in E-commerce. In: Lecture Notes in Engineering and Computer Science: Proceedings of The World Congress on Engineering and Computer Science 2018, 23–25 October 2018, San Francisco, USA, pp. 143–148 (2018)
13. Khatib, K.A.l., et al.: End-to-end argumentation knowledge graph construction. In: AAAI Technical Track: Natural Language Processing, vol. 34, no. 05 (2020)
14. Wang, H., Zhang, F., Xie, X., Guo, M.: DKN: deep knowledge-aware network for news recommendation. In: Proceedings of the 2018 World Wide Web Conference, pp. 1835–1844 (2018)
15. Wang, H., et al.: RippleNet: propagating user preferences on the knowledge graph for recommender systems. In: Proceedings of the 27th ACM International Conference on Information and Knowledge Management, pp. 417–426 (2018)
16. Wang, H., et al.: Multi-task feature learning for knowledge graph enhanced recommendation. In: International World Wide Web Conferences, pp. 2000–2010 (2019)

Machine Learning

An X-Architecture SMT Algorithm
Based on Competitive Swarm Optimizer

Ruping Zhou[1], Genggeng Liu[1(✉)], Wenzhong Guo[1], and Xin Wang[2]

[1] College of Mathematics and Computer Science, Fuzhou University, Fuzhou, China
{200320100,liugenggeng,guowenzhong}@fzu.edu.cn
[2] College of Intelligence and Computing, Tianjin University, Tianjin, China
wangx@tju.edu.cn

Abstract. The X-architecture Steiner Minimum Tree (XSMT) is the best connection model of non-Manhattan multi-terminal nets in global routing, and it is an NP-hard problem. Particle Swarm Optimization (PSO), with its efficient searching ability and self-organizing ability, has become a powerful tool for constructing the XSMT. However, PSO is prone to fall into the local optimum due to its excessive exploitation intensity. To keep a smooth trade-off between exploitation and exploration capabilities of PSO, maintain the diversity of the population, and obtain a better solution, this paper proposes an XSMT algorithm based on Competitive Swarm Optimizer (called CSO-XSMT). The algorithm utilizes the methods of pairwise competition and roulette wheel selection to randomly select the learning objects of particles so as to enhance the exploration ability of the population and improve the algorithm performance. Meanwhile, to further reduce the wirelength of the Steiner tree, a refine strategy based on sharing edges is proposed, which adjusts the Steiner tree obtained by CSO to improve the quality of the final routing tree. Experimental results show that compared with other Steiner tree construction algorithms, the proposed algorithm has better wirelength optimization capability and superior stability.

Keywords: Competitive swarm optimizer · Roulette wheel selection · X-architecture Steiner minimum tree · Wirelength optimization · Refine strategy

1 Introduction

Chip design technology is a crucial factor to promote the development of Very Large Scale Integration (VLSI), as well as biochips [7–10]. In the physical design

This work was partially supported by the National Natural Science Foundation of China under Grants No. 61877010, No. 11501114, No. 11271002 and No. U1705262, State Key Laboratory of Computer Architecture (ICT,CAS) under Grant No. CARCHB202014, Fujian Natural Science Funds under Grant No. 2019J01243 and No. 2018J07005, Independent Innovation Fund between Tianjin University and Fuzhou University under Grant No. TF2021-8, and Fuzhou University under Grants No. GXRC-20060 and No. XRC-1544.

C. Xing et al. (Eds.): WISA 2021, LNCS 12999, pp. 393–404, 2021.
https://doi.org/10.1007/978-3-030-87571-8_34

of VLSI, the global routing is a key point. As the best model for multi-terminal nets in global routing, Steiner Minimum Tree (SMT) is a research hot spot, which introduces extra points to construct the minimum routing tree connecting the given pins.

The SMT problem can be classified into Manhattan and non-Manhattan structures according to the routing direction. At present, most studies are mainly based on Manhattan structure [2,6,14]. However, the Manhattan structure is limited by routing directions, so it is difficult to find a high-quality solution. To make full use of the routing resources and improve the routing quality, more and more scholars turn their attention to non-Manhattan structure. Compared with the traditional horizontal direction and vertical direction, the non-Manhattan structure has two more routing directions, namely $45^c irc$ and $135^c irc$, which can search the solution space of SMTs more adequately.

Precise algorithms [18,19], traditional heuristic algorithms [5,11], and Swarm Intelligence (SI) techniques [1,3,12,13,15–17] can be used to solve non-Manhattan SMT. However, the high complexity of precise algorithms makes it difficult to construct the SMT, which belongs to an NP-hard problem.

Aiming to overcome the shortage of precise algorithms, many works have applied heuristic strategy to the construction of non-Manhattan routing trees. Chiang et al. [5] proposed a heuristic algorithm with time complexity $O(n^3 \log_2 n)$ to solve the Y-architecture Steiner tree based on greedy strategy. Also based on greedy strategy, a heuristic algorithm of Rectilinear Steiner Minimum Tree (RSMT) with time complexity $O(n \log_2 n)$ is proposed in [11] and successfully applied to the construction of X-architecture Steiner tree by sacrificing a little time. Traditional heuristic algorithms are mostly based on greedy strategy, and accordingly, are prone to premature convergence.

Therefore, using evolutionary computing and intelligent decision-making, SI has gradually become a popular tool for solving VLSI routing problems. Arora et al. [1] used the Ant Colony Optimization algorithm to construct both the Manhattan and non-Manhattan Steiner trees. Liu et al. [12] utilized Discrete Differential Evolution and proposed several strategies to obtain excellent XSMT. An X-architecture Steiner Minimum Tree (XSMT) algorithm based on Particle Swarm Optimization (PSO) is designed in [13]. The algorithm can obtain various topologies of XSMT, which is beneficial to optimize the congestion in global routing. Based on adaptive PSO and hybrid transformation strategy, Liu et al. [16] proposed a unified algorithm for RSMT and XSMT constructions. In order to apply the PSO algorithm to different routing problems, a new discrete PSO with a multi-stage transformation strategy is proposed in [17], which can expand the search space of the algorithm. The social learning strategy based on the example pool that enables particles to learn from different excellent individuals is adopted in [3,15], thus optimizing the ability of PSO to construct the XSMT.

Although PSO has the characteristic of simple implementation and outstanding optimization ability, it still has a lot of room for improvement in convergence speed and convergence accuracy. Therefore, this paper designs and implements an XSMT algorithm based on Competitive Swarm Optimizer (CSO), called CSO-XSMT. The major contributions of this paper include:

(1) A learning mechanism based on pairwise competition and roulette wheel selection is proposed. In traditional PSO, particles are vulnerable to the influence of historical optimal positions and gradually fall into the local extremum, making it difficult to jump out. In this paper, there are no longer historical positions. The learning objects are selected randomly by pairwise competition and roulette wheel selection. It greatly enhances the diversity of the population and avoids particles falling into the local extremum.
(2) Considering the characteristics of Steiner tree as a discrete problem, mutation and crossover operators of genetic algorithm are combined into the particle update formula to realize the discretization of CSO and better construct the XSMT.
(3) A refine strategy considering sharing edges is designed to find the best local topology of all pins. By maximizing the length of sharing edges, the routing tree obtained by CSO is further optimized, and finally, the XSMT with higher quality is constructed.

2   Problem Formulation

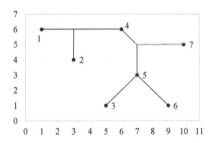

Fig. 1. An X-architecture Steiner tree connecting 7 pins.

The XSMT problem can be described as follows: Given the set of pins $P = \{p_1, p_2, p_3, ..., p_n\}$, each pin p_i is represented by the coordinate (x_i, y_i). The XSMT is constructed by adding extra Steiner points (called Pseudo-Steiner Point, PSP) to connect all given pins, in which the routing directions can be $45°$ and $135°$ apart from horizontal and vertical directions. The definitions of four choices of PSP are given in [3], namely Choice 0, Choice 1, Choice 2, and Choice 3. Taking a net with 7 pins as an example, Fig. 1 is an X-architecture Steiner Tree (XST) that connects those pins.

3 CSO-XSMT

3.1 CSO

PSO has the advantages of simple implementation and high search efficiency, and has been successfully applied to many optimization problems. However, in most cases, the increase of dimensions of optimization problems will significantly increase the number of local optimum, and then lead to premature convergence. To solve this problem, Cheng et al. [4] proposed CSO by eliminating *gbest* and *pbest*. Its learning mechanism of pairwise competition between different particles provides the driving force of the particle flight and balances the capabilities of exploration and exploitation of the population. The particle update formula of CSO is as follows:

$$V_{l,k}(t+1) = R_1(k,t)V_{l,k}(t)$$
$$+R_2(k,t)(X_{w,k}(t) - X_{l,k}(t)) \tag{1}$$
$$+\varphi R_3(k,t)(\overline{X}_k(t) - X_{l,k}(t))$$

$$X_{l,k}(t+1) = X_{l,k}(t) + V_{l,k}(t+1) \tag{2}$$

where $X_{w,k}(t)$ is the winner particle of kth competition in tth iteration, $X_{l,k}(t)$ is the loser particle, $\overline{X}_k(t)$ is the global mean position of the population, φ is the parameter that controls the influence of $\overline{X}_k(t)$, R_1, R_2 and R_3 are random numbers between $[0, 1]$.

To assimilate the merit of CSO and apply it to the discrete XSMT problem, a new learning strategy suitable for Steiner tree is designed in this paper. At the same time, due to the poor performance of the algorithm after discretization of $\overline{X}_k(t)$, this paper adopts roulette wheel selection as a new social learning strategy. Based on the learning mechanism, this paper improves the performance of PSO. The implementation of CSO-XSMT algorithm proposed in this paper will be described from four aspects: respectively particle encoding, fitness function, particle update formula, and refine strategy.

3.2 Particle Encoding

In CSO-XSMT, each particle represents a candidate Steiner tree and is encoded by the method of the edge-vertex encoding, which is more suitable for evolutionary algorithm. For a net with n pins, it is necessary to form a legal spanning tree with $n - 1$ edges. And each edge is represented by a substring $(E_i, E_j, pspc)$ where $pspc$ is the PSP choice of the edge (E_i, E_j) that connects pin i with pin j. It also requires a digit to represent the fitness of the particle. Therefore, a particle is encoded by a string with length $3 \times (n - 1) + 1$. The encoding of the routing tree in Fig. 1 is as follows:

$$\underline{5\ 7}\ \underline{2}\ \underline{5\ 4}\ \underline{0}\ \underline{2\ 4}\ \underline{1}\ \underline{5\ 3}\ \underline{1}\ \underline{2\ 1}\ \underline{1}\ \underline{5\ 6}\ \underline{0}\ \textbf{18.7279}$$

where the encoding length of the particle is 19, and the fitness of the particle is 18.7279 in bold. Take the third substring $(\underline{5\ 3\ 1})$ as an example, it indicates that PSP choice between pin 5 and pin 3 is choice 1.

3.3   Fitness Function

Since a particle in the population represents a candidate solution, that is, a candidate Steiner tree, the fitness function needs to reflect the quality of the corresponding Steiner tree. In this paper, the length of the Steiner tree is taken as the fitness of the particle, and the calculation formula is as follows:

$$fitness = L(T_x) = \sum_{c_i \subset T_x} l(e_i) \tag{3}$$

where $l(e_i)$ represents the length of the edge segment e_i of the routing tree T_x.

3.4   Particle Update Formula

The proposed algorithm based on the competition mechanism of CSO randomly selects two particles to compare their fitness values. In this way, CSO divides the population into losers and winners, and only updates losers while keeps winners unchanged. The update operation discussed below is only used for loser particles.

In CSO, every loser follows the following update formula:

$$X_{l,k}^t = SF_3(SF_2(SF_1(X_{l,k}^{t-1}, \omega), c_1), c_2) \tag{4}$$

where ω is inertia weight, c_1 and c_2 are acceleration factors, SF_1 is the inertia component, SF_2 and SF_3 are new individual cognition and social cognition. In CSO-XSMT, SF_1 is realized by mutation operation, which determines whether to keep the particle state of last iteration; SF_2 and SF_3 are realized by crossover operations, which carry out partial gene crossover with the winner of competition and the object selected by roulette respectively.

Inertia Component. The algorithm performs mutation operation by SF_1, which is expressed as follows:

$$W_{l,k}^t = SF_1(X_{l,k}^{t-1}, \omega) = \begin{cases} M_p(X_{l,k}^{t-1}), & r_1 < \omega \\ X_{l,k}^{t-1}, & \text{otherwise} \end{cases} \tag{5}$$

where $M_p()$ is the mutation operation, ω determines the probability of $M_p()$, r_1 is a random number between $[0, 1)$ and $W_{l,k}^t$ is the particle after mutation .

The mutation operation of CSO-XSMT is for PSP by using two-point mutation. If the generated random number $r_1 < \omega$, the algorithm will randomly select two edges to replace their PSP choices. Otherwise, keep the Steiner tree unchanged. As shown in Fig. 2, for a routing tree with 7 pins, the algorithm randomly selects two edges m_1 (3 2 1) and m_2 (2 5 3) of particle X_l to change their PSP choices. After mutation, the PSP choices of m_1 and m_2 are changed from Choice 1 and 3 to Choice 2 and 0, respectively.

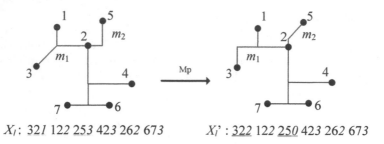

X_l: 32*1* 122 25*3* 423 262 673 X_l' : 3*22* 122 2*50* 423 262 673

Fig. 2. Mutation operation of CSO-XSMT.

Individual Cognition. In CSO algorithm, the loser no longer learn from its individual best position but is attracted by the winner. The algorithm performs crossover operation of individual cognitive component by SF_2, which is expressed as follows:

$$S_{l,k}^t = SF_2(W_{l,k}^t, c_1) = \begin{cases} C_p(W_{l,k}^t, X_{w,k}^t), & r_2 < c_1 \\ W_{l,k}^t, & \text{otherwise} \end{cases} \tag{6}$$

where $C_p()$ is the crossover operation, c_1 determines the probability that the loser crossovers with the winner, r_2 is a random number between $[0,1)$ and $S_{l,k}^t$ is the particle after crossover.

If the generated random number $r_2 < c_1$, crossover operation will be performed: The algorithm randomly determines the interval of the edges to be crossed, and finds the encoding on the interval of the winner particle. Then replaces the encoding on the interval of the loser with that of the winner. As shown in Fig. 3, X_l is the loser after mutation, and X_w is the winner with better fitness in pairwise competition. The algorithm selects the interval of the loser (that is, the red edges e_1, e_2 and e_3 in the figure), and the encoding corresponding to the winner is 1 2 *1* 2 5 *0* 4 2 *3* (that is, the blue edges in the figure). After replacing the encoding, the routing choices of edges e_1, e_2 and e_3 are changed from Choice 2, Choice 0, and Choice 3 to Choice 1, Choice 0, and Choice 3, respectively.

The crossover operation above enables the particle X_l to learn better genes from others. After repeated iterative learning, the particle can gradually approach the optimal position.

Social Cognition. The CSO-XSMT algorithm uses roulette wheel mechanism to randomly select the social learning object of the particle, and the probability of each particle is calculated according to the following formulas:

$$all\_fitness = \sum_{i=1}^{popsize} pop[i].fit\_value \tag{7}$$

$$prob[i] = \frac{poolVec[i].fit\_value}{all\_fitness} \tag{8}$$

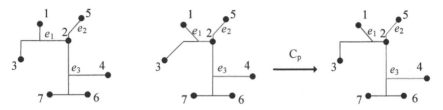

X_l: 322 <u>122</u> <u>250</u> <u>423</u> 262 673 X_w: 32<u>1</u> <u>12</u><u>1</u> <u>250</u> <u>423</u> 262 673 X_l' :322 <u>12</u><u>1</u> <u>250</u> <u>423</u> 262 673

Fig. 3. Crossover operation of CSO-XSMT.

where $pop[i].fit\_value$ is the fitness value of the ith particle, and $poolVec[i].fit\_value$ is the fitness value of the ith particle after sorting in ascending order of fitness. The formulas show that the better a particle is, the lower its probability of being selected.

The algorithm performs crossover operation of social cognitive component by SF_3, which is expressed as follows:

$$X_{l,k}^t = SF_3(S_{l,k}^t, c_2) = \begin{cases} C_p(S_{l,k}^t, X_{r,k}^t), & r_3 < c_2 \\ S_{l,k}^t, & \text{otherwise} \end{cases} \tag{9}$$

where $X_{r,k}^t$ is the particle selected by roulette wheel selection, c_2 determines the probability that $S_{l,k}^t$ crossover with $X_{r,k}^t$, r_2 is a random number between $[0,1)$ and $X_{l,k}^t$ is the particle obtained after all the above operations.

3.5   Refine Strategy

To further optimize XST obtained by CSO algorithm and get a routing tree with a shorter wirelength, this paper introduces a refine strategy based on sharing edges. For any pin p_i of XST, there is at least one optimal structure among all interconnection models. Figure 4 shows all local topologies of the 2-degree pin p and the one in the rectangle is the best local topology. Because pin p is connected with pin q and pin g, the interconnection wire has a sharing edge, which is the red bold edge in Fig. 4. It is easy to see that the longer the sharing edge is, the shorter the wirelength is. Therefore, our refine strategy aims to increasing the length of sharing edges as much as possible.

At first, the algorithm calculates the degree of each pin p_i and records the list of its adjacent pins. For each pin p_i, if its degree is greater than 1, the algorithm calculates the length of its sharing edges. Then sort each pin in descending order according to the length of sharing edges. The algorithm needs to enumerate all combinations of each pin by adjusting the PSP choices of its adjacent edges and select the combination with the shortest wirelength as the best local topology of the pin.

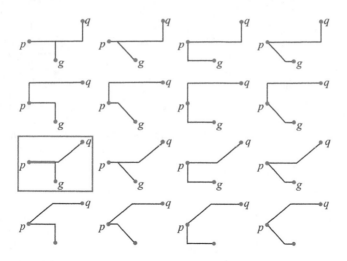

Fig. 4. 16 topologies with pin p of two degrees.

4   Experimental Results

In order to verify the effectiveness of the CSO-XSMT algorithm and related strategies, experiments are carried out on the benchmark circuit suite [20]. The population size is set to 100 and the number of iterations is 1000, while the other parameters are consistent with [16]. Considering the randomness of the algorithm, the average values of all experiments are obtained by independent run 10 times.

4.1   Validation of Learning Mechanism of CSO

In order to verify the effectiveness of the learning mechanism of our algorithm, this section compares CSO with DPSO (X) [10] on best wirelength, mean wirelength, and standard deviation. As shown in Table 1, CSO algorithm can obtain better wirelength than DPSO (X) on tests 4–10. Among them, CSO performs best on test 7, where the mean wirelength is optimized by 0.946% compared with DPSO (X). For the three indexes tested in the experiment, CSO can respectively achieve 0.308%, 0.406% and 38.24% optimization compared with DPSO (X) method. It is because the learning mechanism of CSO expands the scope of particle learning objects. DPSO (X) only updates the particle by learning from *gbest* and *pbest*, which can not guarantee the diversity of the population and balance exploitation and exploration well. In addition, the backward particles in DPSO (X) are far away from *gbest*, and one update may cause large position changes, so the stability of the algorithm is bad. The mechanism based on roulette wheel selection alleviates this situation, thus improving stability.

Table 1. Comparison between CSO and DPSO (X) algorithm

Test	Pins	Best wirelength			Mean wirelength			Standrad deviation		
		DPSO (X)	CSO	Imp (%)	DPSO (X)	CSO	Imp (%)	DPSO (X)	CSO	Imp (%)
1	8	16918	16918	0.000	16918	16918	0.000	0	0	0.00
2	9	18041	18041	0.000	18041	18041	0.000	0	0	0.00
3	10	19696	19696	0.000	19696	19696	0.000	0	0	0.00
4	20	32214	32193	0.065	32214	32197	0.053	10	6	40.00
5	50	48039	47959	0.167	48045	47965	0.167	74	7	90.54
6	70	56295	56277	0.032	56548	56313	0.416	138	41	70.29
7	100	68777	68334	0.644	69115	68461	0.946	180	80	55.56
8	410	141642	140496	0.809	141942	140672	0.895	186	120	35.48
9	500	154677	153487	0.769	155037	153635	0.904	147	90	38.78
10	1000	220824	219512	0.594	221211	219700	0.683	203	98	51.72
Average				**0.308**			**0.406**			**38.24**

4.2 Validation of Refine Strategy

In order to verify the effectiveness of the refine strategy, this section compares the best wirelength, mean wirelength, and standard deviation obtained by DPSO (X) and CSO-XSMT (CSO with refine strategy) in Table 2. We can find that the performance of CSO-XSMT is further improved, especially in the standard deviation. Compared with DPSO (X), the average optimization of our standard deviation is optimized by 51.83% on average. For test 5 with 50 pins, the standard deviation of ours is optimized to reach 93.24%. And for test 10 with the largest scale, our algorithm can reduce best wirelength by 2.646%, mean wirelength by 2.789% and standard deviation by 81.28% compared with DPSO (X). The experimental results in Table 2 demonstrate the effectiveness of refine strategy in reducing wirelength and stabilizing the algorithm.

Table 2. Comparison between CSO-XSMT and DPSO (X) algorithm

Test	Pins	Best wirelength			Mean wirelength			Standrad deviation		
		DPSO (X)	CSO-XSMT	Imp (%)	DPSO (X)	CSO-XSMT	Imp (%)	DPSO (X)	CSO-XSMT	Imp (%)
1	8	16918	16918	0.000	16918	16918	0.000	0	0	0.00
2	9	18041	18041	0.000	18041	18041	0.000	0	0	0.00
3	10	19696	19696	0.000	19696	19696	0.000	0	0	0.00
4	20	32214	32193	0.065	32214	32210	0.012	10	8	20.00
5	50	48039	47952	0.181	48045	47955	0.187	74	5	93.24
6	70	56295	56271	0.043	56548	56288	0.460	138	19	86.23
7	100	68777	68334	0.644	69115	68370	1.078	180	37	79.44
8	410	141642	138989	1.873	141942	139060	2.030	186	40	78.49
9	500	154677	151345	2.154	155037	151378	2.360	147	30	79.59
10	1000	220824	214982	2.646	221211	215047	2.786	203	38	81.28
Average				**0.761**			**0.891**			**51.83**

Table 3. Comparison between CSO-XSMT and other SMT algorithms

Test	Pins	Mean wirelength			Imp (%)	
		DPSO (R)	HTS-PSO (R)	Ours	DPSO (R)	HTS-PSO (R)
1	8	17931	17693	16918	5.649	4.38
2	9	20503	19816	18041	12.008	8.957
3	10	21910	21214	19696	10.105	7.156
4	20	35173	35153	32210	8.424	8.372
5	50	52682	52087	47955	8.973	7.933
6	70	61455	61155	56288	8.408	7.958
7	100	75997	74489	68370	10.036	8.215
8	410	159732	154210	139060	12.942	9.824
9	500	173352	166590	151378	12.676	9.131
10	1000	248624	240231	215047	13.505	10.483
Average					**10.273**	**8.241**

4.3 Comparison with Other SMT Algorithms

In order to further verify the effectiveness and performance of the proposed algorithm, this section compares the CSO-XSMT algorithm with other SMT algorithms, respectively DPSO (R) [14] and HTS-PSO (R) [16] in Table 3. It can be found that compares with DPSO (R) and HTS-PSO (R), ours reduces the mean wirelength by 10.273% and 8.241% on average. This means that our algorithm is superior to these two SMT algorithms in general. Especially in test 10, the wirelength of XSMT generated by DPSO (R) and HTS-PSO (R) are 248624 and 240231, while the wirelength generated by ours is only 215047, which is 13.505%, 10.483% shorter than those obtained by the other two algorithms. The results show that the more pins there are, the better the optimization of CSO-XSMT wirelength is. Experiments in this section prove that SMT based on X-architecture can obtain the shortest routing tree.

5 Conclusion

To optimize the wirelength of the XSMT in VLSI routing, this paper proposes the CSO-XSMT algorithm. Firstly, the algorithm greatly improves the traditional PSO. Through pairwise competition and roulette wheel selection, the algorithm can keep the diversity of the population in later iterations, thus overcoming the defects of PSO. Secondly, combining the mutation and crossover operations of genetic algorithm, and considering the characteristics of Steiner tree, the CSO is discretized to construct an ideal XST. Finally, a refine strategy based on sharing edges is designed to minimize the final routing tree by maximizing its sharing edges. Experimental results show that the CSO-XSMT algorithm proposed in this paper has excellent wirelength optimization capability and stability. In future work, we will further improve the performance of the algorithm to better apply it to various VLSI routing problems.

References

1. Arora, T., Moses, M.E.: Ant colony optimization for power efficient routing in manhattan and non-manhattan VLSI architectures. In: 2009 IEEE Swarm Intelligence Symposium, pp. 137–144 (2009). https://doi.org/10.1109/SIS.2009.4937856
2. Chen, G., Young, E.F.Y.: Salt: Provably good routing topology by a novel Steiner shallow-light tree algorithm. IEEE Trans. Comput. Aided Des. Integrated Circuits Syst. **39**(6), 1217–1230 (2020). https://doi.org/10.1109/TCAD.2019.2894653
3. Chen, X., Zhou, R., Liu, G., Wang, X.: SLPSO-based X-architecture Steiner minimum tree construction. In: Wang, G., Lin, X., Hendler, J., Song, W., Xu, Z., Liu, G. (eds.) WISA 2020. LNCS, vol. 12432, pp. 131–142. Springer, Cham (2020). https://doi.org/10.1007/978-3-030-60029-7_12
4. Cheng, R., Jin, Y.: A competitive swarm optimizer for large scale optimization. IEEE Trans. Cybern. **45**(2), 191–204 (2015). https://doi.org/10.1109/TCYB.2014.2322602
5. Chiang, C., Chiang, C.S.: Octilinear Steiner tree construction. In: The 2002 45th Midwest Symposium on Circuits and Systems. MWSCAS-2002, vol. 1, pp. I–603 (2002). https://doi.org/10.1109/MWSCAS.2002.1187293
6. Held, S., Rockel, B.: Exact algorithms for delay-bounded Steiner arborescences. In: Proceedings of the 55th Annual Design Automation Conference. DAC 2018. Association for Computing Machinery, New York, NY, USA (2018). https://doi.org/10.1145/3195970.3196048
7. Huang, X., Guo, W., Chen, Z., Li, B., Ho, T.Y., Schlichtmann, U.: Flow-based microfluidic biochips with distributed channel storage: Synthesis, physical design, and wash optimization. IEEE Trans. Comput. 1 (2021). https://doi.org/10.1109/TC.2021.3054689
8. Huang, X., Ho, T.Y., Chakrabarty, K., Guo, W.: Timing-driven flow-channel network construction for continuous-flow microfluidic biochips. IEEE Trans. Comput. Aided Des. Integrated Circuits Syst. **39**(6), 1314–1327 (2020). https://doi.org/10.1109/TCAD.2019.2912936
9. Huang, X., Ho, T.Y., Guo, W., Li, B., Chakrabarty, K., Schlichtmann, U.: Computer-aided design techniques for flow-based microfluidic lab-on-a-chip systems. ACM Comput. Surv. **54**(5), 1–29 (2021). https://doi.org/10.1145/3450504
10. Huang, X., Ho, T.Y., Guo, W., Li, B., Schlichtmann, U.: Minicontrol: synthesis of continuous-flow microfluidics with strictly constrained control ports. In: Proceedings of the 56th Annual Design Automation Conference 2019. DAC 2019. Association for Computing Machinery, New York, NY, USA (2019). https://doi.org/10.1145/3316781.3317864
11. Kahng, A., Mandoiu, I., Zelikovsky, A.: Highly scalable algorithms for rectilinear and octilinear Steiner trees. In: Proceedings of the ASP-DAC Asia and South Pacific Design Automation Conference, vol. 2003, pp. 827–833 (2003). https://doi.org/10.1109/ASPDAC.2003.1195132
12. Liu, G., Yang, L., Xu, S., Li, Z., Chen, C.H.: X-architecture Steiner minimal tree algorithm based on multi-strategy optimization discrete differential evolution. PeerJ Comput. Sci. **7**(6), e473 (2021). https://doi.org/10.7717/peerj-cs.473

13. Liu, G., Chen, G., Guo, W.: Dpso based octagonal Steiner tree algorithm for VLSI routing. In: 2012 IEEE Fifth International Conference on Advanced Computational Intelligence (ICACI), pp. 383–387 (2012). https://doi.org/10.1109/ICACI.2012.6463191

14. Liu, G., Chen, G., Guo, W., Chen, Z.: DPSO-based rectilinear Steiner minimal tree construction considering bend reduction. In: 2011 Seventh International Conference on Natural Computation, vol. 2, pp. 1161–1165 (2011). https://doi.org/10.1109/ICNC.2011.6022221

15. Liu, G., Chen, X., Zhou, R., Xu, S., Chen, Y.C., Chen, G.: Social learning discrete particle swarm optimization based two-stage x-routing for IC design under intelligent edge computing architecture. Appl. Soft Comput. **104**, 107215 (2021)

16. Liu, G., Chen, Z., Zhuang, Z., Guo, W., Chen, G.: A unified algorithm based on HTS and self-adapting PSO for the construction of octagonal and rectilinear SMT. Soft Comput. **24**(6), 3943–3961 (2019). https://doi.org/10.1007/s00500-019-04165-2

17. Liu, G., Zhu, W., Xu, S., Zhuang, Z., Chen, Y.-C., Chen, G.: Efficient VLSI routing algorithm employing novel discrete PSO and multi-stage transformation. J. Ambient Intell. Humanized Comput. **6**, 1–16 (2020). https://doi.org/10.1007/s12652-020-02659-8

18. Shang, S.P., Jing, T.: Steiner minimal trees in rectilinear and octilinear planes. Acta Mathematica Sin. Engl. Ser. **23**(9), 1577–1586 (2007). https://doi.org/10.1007/s10114-005-0910-0

19. Teig, S.L.: The X architecture: not your father's diagonal wiring. In: Proceedings of the 2002 International Workshop on System-Level Interconnect Prediction, pp. 33–37. SLIP 2002. Association for Computing Machinery, New York, NY, USA (2002). https://doi.org/10.1145/505348.505355

20. Warme, D., Winter, P., Zachariasen, M.: GeoSteiner software for computing Steiner trees (2003). http://geosteiner.net

A Novel Grayscale Image Steganography via Generative Adversarial Network

Zhihua Gan and Yuhao Zhong[✉]

School of Software, Henan University, Kaifeng 475001, China
gzh@henu.edu.cn

Abstract. Steganography is an effective technique in the field of information hiding that typically involves embedding secret information into an image to resist steganalysis detection. In recent years, several works on image steganography based on deep learning have been presented, but these works still have issues with steganographic image and revealed image quality, invisibility, and security. In this paper, a novel grayscale image steganography via generative adversarial network is proposed. To boost the invisibility of the model, we construct an encoding network, which is comprised of a secret image feature extraction module and an integration module that conceals a grayscale secret image into another color cover image of the same size. Moreover, considering the security of the model, adversarial training between the encoding-decoding network and the steganalyzer is used. As compared to state-of-the-art steganography models, experimental results show that our proposed steganography scheme not only has higher peak signal-to-noise ratio and structural similarity index but also better invisibility.

Keywords: Steganography · Deep learning · Generative adversarial network · Grayscale image

1 Introduction

Steganography is a fundamental technique for information hiding. In the early of steganography research, the researchers were mostly concerned with seeking ways to conceal secret information into images without causing human visual perception, such as S-UNIWARD [1], WOW [2], and HUGO [3]. The algorithms automatically identify the regions of the image that are appropriate for steganography. Even if it will mitigate the distortion created by the embedding of secret information, which affects the statistical features of the image, there is still no way to eliminate the traces of manipulation left in the image during the embedding of the information.

With the rapid rise of deep learning, multiple neural networks have appeared and made significant advances in the fields such as object recognition, image segmentation, and image processing. Volkhonskiy et al. [4] were the first to apply generative adversarial network (GAN [5]) to steganography, incorporating deep learning with conventional steganography algorithms. It embeds a secret message

© Springer Nature Switzerland AG 2021
C. Xing et al. (Eds.): WISA 2021, LNCS 12999, pp. 405–417, 2021.
https://doi.org/10.1007/978-3-030-87571-8_35

into images generated by GAN using traditional algorithms to boost steganography security. Baluja et al. [6] implemented the embedding of a color image into a color image of the same size by utilizing an encoding-decoding network. But it revealed secret images and steganographic images had poor visual quality. Atique et al. [7] instead conceal a grayscale image in a color image of the same size, and the steganographic image suffers from color distortion. Both Baluja's and Atique's models have one benefit, namely, they do not affect the understanding of the semantic information contained in the secret images, even though the secret image can not be completely reconstructed. Its apparent disadvantage is that the large amount of information embedded in the cover image, along with the lack of security considerations, results in its relatively poor resistance to steganalysis.

Based on the mentioned weaknesses, Zhang et al. [8] took into account the fact that the Y channel of an image includes only semantic information but no color information under the YUV channel, and embedded a grayscale image into the Y channel of the image to address the problem of image color distortion. In addition, to enhance the security of the model, an advanced steganalysis model, XuNet [9], was added as a discriminator for adversarial training. While its approach effectively eliminates image color distortion and enhances security, it still has issues with steganographic image quality, invisibility, and secret image reconstruction quality.

Inspired by the previous work, we propose a novel Grayscale Image Steganography via Generative Adversarial Network called GISGAN. It is capable of embedding a grayscale secret image into another color cover image of the same size. In comparison to ISGAN [8], the encoding network consists of a secret image feature extraction module and an integration module, which can fuse features from different layers to enhance the invisibility of the model. The contributions to our work are as follows:

1. A novel encoding network is proposed to enhance the invisibility of the secret information and the quality of the steganographic image, which is composed of a secret image feature extraction module for extracting secret images features and an integration module for concealing a secret image into another color cover image.
2. Given the security of the model, we incorporate a steganalyzer as a discriminator for adversarial training with the encoding-decoding network to resist the detection of the steganalysis model, thus strengthening the security of the model.
3. The experimental results on different datasets validate the effectiveness of our model. Compared with the previous work that was used to hide the grayscale secret images, our model performed better in image evaluation metrics PSNR as well as SSIM, and has better invisibility.

The rest of the paper is organized as follows. Section 2, discusses the previous related works of steganography based on Deep Learning. In Sect. 3, the details of the proposed steganography model and the loss function. In Sect. 4, describes the details of experiments, including datasets, parameter settings, and evaluation metrics. Finally, we analyze the experimental results as well as the conclusions in Sects. 5 and 6, respectively.

2   Related Works

In 2016, Volkhonskiy et al. [4] used the Deep Convolutional Generative Adversarial Network (DCGAN [10]) to propose the steganography network SGAN, which is made up of a generator G, a discriminator D, and a steganographic analyzer S. However, the SGAN training generator suffers from gradient disappearance as well as instability in training. Based on the SGAN, Shi et al. [11] have proposed SSGAN that is similar to the SGAN but uses Wasserstein GAN [12] to further improve the quality of steganographic images. SGAN and SSGAN utilize the GAN approach for image steganography. The security of the model has improved, but the generated images are not semantically realistic and are more easily to drawn attention than nature images.

Tang et al. [13] proposed an automatic steganographic distortion learning framework ASDL-GAN, wherein generator can transform the cover image into an embedded change probability matrix while the discriminator incorporates the XuNet [9]. Nevertheless, the performance of the model is inferior to that of S-UNIWARD [1], and it remains to depend on too much prior knowledge from traditional steganography algorithms. Hayes et al. [14] proposed SteGAN, the model for introducing an encoder-decoder network to steganography, but still the invisibility is weak. In 2017, Baluja et al. [6] first proposed an image steganography model, which implements embedding a color secret image into a color cover image of the same size, but reduces the security and the quality of the steganographic image. After that, Atique et al. [7] took another perspective by replacing the color secret image with a grayscale secret image. Unfortunately, these models suffer from poor image quality, color distortion and low security.

Based on Baluja and Atique, Zhang et al. [8] proposed the ISGAN model in 2019, which similarly implemented the steganography of grayscale images into color images, which embedded the grayscale image into the Y channel of the color image, avoiding the image color distortion, but still has problems with steganographic image quality and invisibility. Further enhancements have also been made recently around the previous work. Chen et al. [15] learned the thought of ISGAN to embed grayscale images into color images. In contrast, the model chooses the B channel under the RGB color space, in addition to adding the noise before the decoding network to simulate attacks to improve the robustness of the model. Li et al. [16] proposed a novel steganography model by combining traditional encryption algorithms [17] with deep learning, which provide a new thought for image steganography.

3   Our Method

The overall architecture of our model is shown in Fig. 1. In this section, we present the proposed steganography model from three aspects. Firstly, we describe our basic model, including the details of the encoding-decoding network. Then, the steganalyzer is implemented as a discriminator by adversarial training with encoding-decoding network to construct the GISGAN. Finally, we present the loss function that was used to train the network.

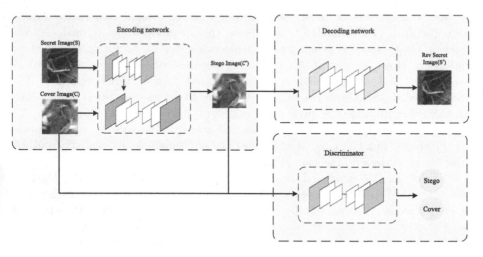

Fig. 1. The overall architecture of GISGAN. The encoding network conceals a grayscale secret image into another color cover image of the same size to obtain a steganographic image. The decoding network reveals the secret image from the steganographic image. The discriminator tries to distinguish whether the input image contains a secret image.

3.1   Basic Model

Our method is different from the previous steganographic networks in that most methods focus on the steganographic and have less attention to the secret image. We will pay more attention to the features of the secret image. By embedding the secret image into the cover image layer by layer to make it integrate better with the cover image to improve the quality and invisibility of the steganographic image. For this purpose, an end-to-end basic model is proposed to implement steganography and reconstruct secret images. It is divided into two parts: encoding network and decoding network.

Inspired by Atique [7], we construct an encoding network that includes a secret image feature extraction module and an integration module. Specifically, the secret and cover images are fed as inputs to two parallel branches of the secret image feature extraction module and the integration module to extract features, respectively. After each convolutional layer, the features extracted by the secret image feature extraction module are superimposed on the features extracted from the integration module. Compared with Atique, we perform a total of seven concat operations, and the feature maps are halved layer by layer to reduce the number of parameters in the model. Moreover, after concatenation, the features of the previous layer size are reused using residual blocks [18] to avoid information being lost. Several skip connections are used in the residual blocks to fuse shallow and deep features in the different convolution levels. The shallow features contain a lot of low-level efficient detail about the image outline and color, which is advantageous for steganographic images. Therefore, the residual block should be incorporated as the main module of the integration module. In comparison to directly concatenating the secret image with the cover image, our approach not only decreases the difference between the generated image and the original image, but it also enhances the invisibility of the model. The specific structure of which is shown in Fig. 2.

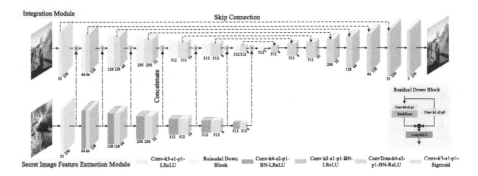

Fig. 2. The architecture of encoding network. Conv, ConvTran, BN, and LReLU represent convolution, transpose convolution, batch normalization, and LeakyReLU activation function, respectively. Besides, k, s, and p denote the kernel size, stride and padding of operators, respectively.

On the other hand, we design the decoding network to extract the secret image from the steganographic image. The decoding network is composed of a 6-layer fully convolutional network, and the input of the first layer is the steganographic image with size $3 \times 256 \times 256$. The network utilizes 3×3 convolution layers with stride of 1 and padding of 1, followed by batch normalization (BN) operation and ReLU activation function. But, the sigmoid activation function was applied after the last convolution layer. The specific decoding network is shown in Fig. 3.

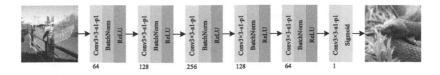

Fig. 3. The architecture of decoding network.

3.2   Discriminator

In view of the work of Baluja [6] and Atique [7], the steganography approaches based on deep learning used to embed a secret image into a cover image just focused on enhancing the visual quality of steganographic images and revealed secret images. It is well known that security is one of the most critical criteria for evaluating the quality of steganographic models. However, due to the large amount of information embedded in the cover image and the lack of protection, these steganographic models have low security. Subsequently, in consideration of the security of steganographic images, ISGAN [8] improved the security of the model by introducing an improved steganalysis model XuNet [9] as a discriminator for adversarial training. The steganalysis model can be applied as a binary discriminator to determine whether or not the cover image contains secret information. Based on this, the steganalysis model XuNet is used as a discriminator in our model. The difference is that we fine-tuned the discriminator to make it applicable to our model. The BN operation, LeakyReLU activation function, and the global average pooling are used after each convolutional layer. Certainly, we preserve the SPP [19] block, which can extract more features from different respective fields to improve the performance. The details of the discriminator are shown in Fig. 4.

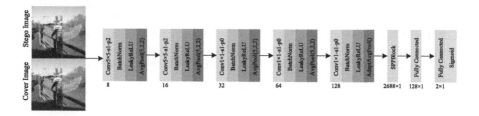

Fig. 4. The architecture of discriminator.

3.3   The Loss Function

In this paper, we use the mean square error (MSE) as the loss function of our training network to minimize the difference between the generated steganographic image c' and the original cover image c, as well as the difference between the revealed secret image s' and the original secret image s. This loss function has been used in the training of the proposed steganographic models based on encoder-decoder network, such as Deep Steganography [6] and ISGAN [8]. Thus, the basic model (encoding-decoding network) loss functions are as follows:

$$L(c, c') = \|c - c'\|_2^2 \tag{1}$$

$$L(s, s') = \|s - s'\|_2^2 \tag{2}$$

$$L_{basic} = L(c, c') + \alpha L(s, s') \tag{3}$$

where c, c', s, and s' represent the cover image, the steganographic image, the secret image, and the revealed secret image, in that order. Furthermore, the encoding-decoding network is regarded as the generator and the steganalyzer as the discriminator during adversarial training. The generator and discriminator play the min-max game [5]. It can be expressed as:

$$min_G max_D = E_{c \sim P(c)}[log D(c)] + E_{c \sim P(c), s \sim P(s)}[log(1 - D(En(c, s)))] \tag{4}$$

where $En(\cdot)$ and $D(\cdot)$ represents the encoding network and the discriminator, respectively. We use the binary cross-entropy loss as the possibility that the steganographic images can be detected as cover images class. The adversarial loss function is as follows:

$$L(c', y_t) = BinaryCrossEntropy(c', y_t) \tag{5}$$

where y_t represents the label of the cover image. The security of steganographic images is improved by adversarial loss.

The total loss function for GISGAN can be described as:

$$L_{total} = L(c, c') + \alpha L(s, s') + \beta L(c', y_t) \tag{6}$$

where α is a hyperparameter to trade off the quality of steganographic images and revealed secret images, and β is the weight the security of model.

4   Experiments Details

In this section, we'll introduce our experiment details. Firstly, we use ImageNet [20], Pascal VOC2012 [21], and LFW [22] as the training, validation, and test datasets of our experiments. The ImageNet contains more than 14 million images for various computer vision projects, such as image segmentation, target recognition, and target tracking. Pascal VOC 2012 is a dataset for object detection and semantic segmentation. The Labeled Faces in the Wild (LFW) contains more than 13000 face images belonging to different people collected from the website. From the datasets, we randomly choose 10k images to constitute 5k cover-secret image pairs as our training set, 4k images of the remaining part as our validation set, and 2k images to constitute 1k cover-secret images as our testing set.

The generator and discriminator are alternately trained in the training process of generative adversarial network. As a result, the encoding-decoding network is regarded as the generator and the Steganalyzer is regarded as the discriminator. During the encoding-decoding network training process, all parameters were initialized by the Xavier initialization [23], and the initial learning rate (lr) was set as 1e−3. And we set an initial learning rate of discriminator as 3e−5. The Adam optimization approach is used to update the parameters of the network automatically. The batch size was set as 8 limited by the computing power. After several experimental analysis, the parameters α, and β were eventually set as 0.75 and 0.005, respectively. The total number of training epochs is set as 150, and the size of all images we used is 256×256. The GPU is NVIDIA GeForce 2080 Ti, the experimental environment is Pytorch and the application in Python.

We used the peak signal to noise ratio (PSNR [24]) and structural similarity ratio (SSIM [24]) as metrics to measure the performance of the model. The PSNR measures the error between the respective pixels of two images to determine image quality, and a higher value means less image distortion after steganography. The SSIM analyses the luminance, contrast, and structure of a steganographic image to evaluate its quality. The higher the SSIM value, the greater the similarity between the image before and after steganography.

5 Results and Analysis

In this section, we conduct the comparison experiments with the previous work in two respects: (1) the quality of steganographic and revealed secret images, and (2) the security of the model.

5.1 The Quality of Steganographic and Revealed Secret Images

We evaluate the performance of the model in terms of steganographic image quality and the difference between the secret image and the revealed secret image. Table 1 shows the PSNR and SSIM value of steganographic images and revealed secret images on different datasets. The bolded figures represent the optimal values for that dataset. Better visual quality indicates that the cover image with secret information is more realistic and has a smaller difference from the original cover image.

As can be seen from Table 1, we can notice that the average PSNR values of steganographic images and revealed secret images generated by the basic model have improved by 3.78 and 4.72 compared with the ISGAN [8] on ImageNet, respectively. The average SSIM values of the steganographic images are almost identical to that of ISGAN. After adding the steganalyzer as a discriminator by adversarial training, the performance of our method is further improved. On the ImageNet datasets, the average PSNR of steganographic images generated by GISGAN increased by 0.68 from the basic model. In particular, on the LFW dataset, the average PSNR of steganographic images generated by our basic model and GISGAN is higher than ISGAN by 6.63 and 7.55. There is also a degree of improvement in our model on the Pascal VOC2012 dataset.

Table 1. The PSNR and SSIM value of steganographic images and revealed secret images.

Model	Cover Image	Secret Image	Stego-Cover PSNR (db)	Reveal-Secret PSNR (db)	Stego-Cover SSIM	Reveal-Secret SSIM
ISGAN	LFW	LFW	34.12	33.67	0.9660	0.9529
Basic model	LFW	LFW	40.75	38.95	0.9758	0.9810
GISGAN	LFW	LFW	**41.67**	**39.69**	**0.9821**	**0.9811**
ISGAN	ImageNet	ImageNet	34.28	33.05	0.9676	0.9554
Basic model	ImageNet	ImageNet	38.06	37.77	0.9624	0.9712
GISGAN	ImageNet	ImageNet	**38.74**	**37.90**	**0.9680**	**0.9713**
ISGAN	Pascal VOC2012	Pascal VOC2012	34.44	32.70	0.9635	0.9581
Basic model	Pascal VOC2012	Pascal VOC2012	38.28	37.62	0.9695	0.9715
GISGAN	Pascal VOC2012	Pascal VOC2012	**38.88**	**38.01**	**0.9713**	**0.9724**

It is also intuitively observed from Fig. 5 that the steganographic and the revealed secret images are superior to ISGAN. The steganographic images generated by our model remain free of color deviation from the revealed secret images, and the comparison with the original images shows no problems such as noise and distortion. Although from a visual perspective, the steganographic images generated by our model and ISGAN are not different from the original cover images, there is significant noise in the ISGAN as seen in the revealed secret images. Besides, the residual image obtained by subtracting the corresponding pixel values of the steganographic image and the cover image with different scales of enhancement is shown in the fifth to ninth columns of Fig. 5.

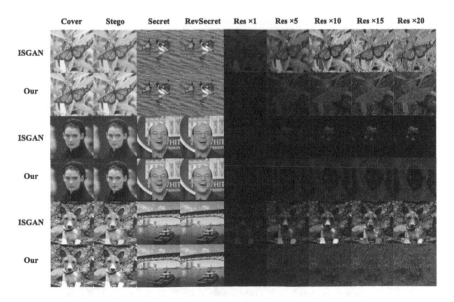

Fig. 5. Sample results on ImageNet, lfw and PASCAL-VOC2012.

It can be seen that the residual image of our method is close to dark, representing the smaller pixel difference between the steganographic image and the cover image, the smaller the distortion, and the better visual quality of the image.

On the other hand, the integrity of the revealed secret information is a crucial aspect when utilizing steganography for information transmission as well. Figure 6 illustrates the difference between the revealed secret image and the original secret image. The darker the residual image, the smaller the difference. The results indicate that neither model can fully reveal the secret image. Because the information in the secret image is redundant, it is acceptable. Still, the pixel differences between the revealed secret images and the corresponding secret images by our method are smaller compared to the ISGAN. It implies that our method can better accomplish secret information transmission.

5.2   The Security of Our Model

The security of the model is evaluated by two aspects. The first is the invisibility of the steganographic images, and the second is the ability of steganographic images to resist detection by the steganalysis model. Figure 5, the residual image is obtained by subtracting the corresponding pixel values of the steganographic images and the cover images with ×5, ×10, ×15, and ×20 enhancement. From the displayed figure, it is observed that our residual images are darker than ISGAN [8] with different multiplicative enhancements for different datasets, indicating that our steganographic images are more invisible than ISGAN. The results from our approach show that almost no detail was apparent in the residual images with ×5 enhancement. As the enhancement becomes larger, the information

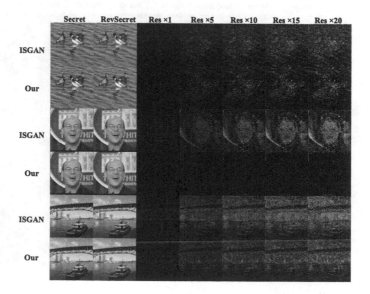

Fig. 6. The residual of secret images.

of the cover image gradually becomes clearer, but still, no information about the secret image can be visible under the ×20 enhancement condition, which indicates the strong security of secret images in the steganographic images.

The proposed model also experiments in terms of security against the steganalysis model. For this purpose, the well-trained basic model of ISGAN is used to generate 5000 steganographic images on ImageNet with their corresponding cover images to constitute the dataset. Then, the first layer structure of the advanced steganalyzer-SRNet [25] was modified to make it applicable to the detection of color images. Finally, we use SRNet trained for 30 epochs as our steganalysis tool. After the training, the well-trained SRNet was used to analyze 2000 steganographic images generated by the basic model and GISGAN. The LFW and PASCAL VOC2012 datasets were subjected to the same experiments. Table 2 displays the average SRNet accuracies for different models. The lower the accuracy of the detection, the better the security of the steganalysis resistance.

Table 2. The detection accuracy of steganalyzer

Steganalyzer	Dataset	Our basic model	GISGAN
SRNet	LFW	91.80%	40.02%
	ImageNet	94.48%	56.08%
	Pascal VOC 2012	90.72%	41.68%

On the ImageNet dataset, SRNet has a detection accuracy of only 56.08% for 2000 steganographic images generated by GISGAN, while it has the detection accuracy of 94.48% on the basic model. Compared with the basic model, the detection accuracy of GISGAN is also decreased by more than 40% on the LFW and Pascal VOC2012 datasets. Thus, it can prove that we have effectively improved the security of the model by adversarial training.

6   Conclusion

In this paper, we propose a novel grayscale image steganography via generative adversarial network called GISGAN consisting of three sub-networks: encoding network, decoding network, and discriminator. The encoding network is composed of the secret image feature extraction module and the integration module, which can fuse the features of secret images and cover image layer by layer to generate steganographic images to strengthen the invisibility of the model. While improving the quality of steganographic images, the security of the steganographic model is strengthened by adversarial training between the encoding-decoding network and the steganalyzer. The experimental results indicate that the proposed model generates steganographic images with less distortion, better invisibility, and higher revealed secret image quality. Despite the fact that there are still differences between the revealed secret images and the corresponding original secret images, we have made some progress in comparison to previous work.

References

1. Holub, V., Fridrich, J., Denemark, T.: Universal distortion function for steganography in an arbitrary domain. EURASIP J. Inf. Secur. **2014**(1), 1–13 (2014). https://doi.org/10.1186/1687-417X-2014-1
2. Holub, V., Fridrich, J.: Designing steganographic distortion using directional filters. In: 2012 IEEE International Workshop on Information Forensics and Security (WIFS), pp. 234–239. IEEE (2012)
3. Pevný, T., Filler, T., Bas, P.: Using high-dimensional image models to perform highly undetectable steganography. In: Böhme, R., Fong, P.W.L., Safavi-Naini, R. (eds.) IH 2010. LNCS, vol. 6387, pp. 161–177. Springer, Heidelberg (2010). https://doi.org/10.1007/978-3-642-16435-4_13
4. Volkhonskiy, D., Borisenko, B., Burnaev, E.: Generative adversarial networks for image steganography (2016)
5. Goodfellow, I.J., Pouget-Abadie, J., Mirza, M., Bing, X., Bengio, Y.: Generative adversarial nets. MIT Press (2014)
6. Baluja, S.: Hiding images in plain sight: deep steganography. In: Proceedings of the 31st International Conference on Neural Information Processing Systems, pp. 2066–2076 (2017)
7. Rahim, R., Nadeem, S., et al.: End-to-end trained CNN encoder-decoder networks for image steganography. In: Proceedings of the European Conference on Computer Vision (ECCV) Workshops (2018)
8. Zhang, R., Dong, S., Liu, J.: Invisible steganography via generative adversarial networks. Multimedia Tools Appl. **78**(7), 8559–8575 (2018). https://doi.org/10.1007/s11042-018-6951-z
9. Xu, G., Wu, H.Z., Shi, Y.Q.: Structural design of convolutional neural networks for steganalysis. IEEE Sig. Process. Lett. **23**(5), 708–712 (2016)
10. Radford, A., Metz, L., Chintala, S.: Unsupervised representation learning with deep convolutional generative adversarial networks. arXiv preprint arXiv:1511.06434 (2015)
11. Shi, H., Dong, J., Wang, W., Qian, Y., Zhang, X.: SSGAN: secure steganography based on generative adversarial networks. In: Zeng, B., Huang, Q., El Saddik, A., Li, H., Jiang, S., Fan, X. (eds.) PCM 2017. LNCS, vol. 10735, pp. 534–544. Springer, Cham (2018). https://doi.org/10.1007/978-3-319-77380-3_51
12. Arjovsky, M., Chintala, S., Bottou, L.: Wasserstein generative adversarial networks. In: International Conference on Machine Learning, pp. 214–223. PMLR (2017)
13. Tang, W., Tan, S., Li, B., Huang, J.: Automatic steganographic distortion learning using a generative adversarial network. IEEE Sig. Process. Lett. **24**(10), 1547–1551 (2017)
14. Hayes, J., Danezis, G.: Generating steganographic images via adversarial training. arXiv preprint arXiv:1703.00371 (2017)
15. Chen, B., Wang, J., Chen, Y., Jin, Z., Shim, H.J., Shi, Y.Q.: High-capacity robust image steganography via adversarial network. KSII Trans. Internet Inf. Syst. **14**(1), 366 (2020)

16. Li, Q., et al.: A novel grayscale image steganography scheme based on chaos encryption and generative adversarial networks. IEEE Access **8**, 168166–168176 (2020)
17. Qin, S., Tan, Z., Zhang, B., Zhou, F.: Evolutionary-based image encryption with DNA coding and chaotic systems. In: Wang, G., Lin, X., Hendler, J., Song, W., Xu, Z., Liu, G. (eds.) WISA 2020. LNCS, vol. 12432, pp. 592–604. Springer, Cham (2020). https://doi.org/10.1007/978-3-030-60029-7_53
18. He, K., Zhang, X., Ren, S., Sun, J.: Deep residual learning for image recognition. In: Proceedings of the IEEE Conference on Computer Vision and Pattern Recognition, pp. 770–778 (2016)
19. He, K., Zhang, X., Ren, S., Sun, J.: Spatial pyramid pooling in deep convolutional networks for visual recognition. IEEE Trans. Pattern Anal. Mach. Intell. **37**(9), 1904–1916 (2015)
20. Deng, J., Dong, W., Socher, R., Li, L.J., Li, K., Fei-Fei, L.: ImageNet: a large-scale hierarchical image database. In: 2009 IEEE Conference on Computer Vision and Pattern Recognition, pp. 248–255. IEEE (2009)
21. Everingham, M., Van Gool, L., Williams, C.K., Winn, J., Zisserman, A.: The pascal visual object classes (VOC) challenge. Int. J. Comput. Vis. **88**(2), 303–338 (2010)
22. Huang, G.B., Mattar, M., Berg, T., Learned-Miller, E.: Labeled faces in the wild: a database for studying face recognition in unconstrained environments. In: Workshop on Faces in 'Real-Life' Images: Detection, Alignment, and Recognition (2008)
23. Glorot, X., Bengio, Y.: Understanding the difficulty of training deep feedforward neural networks. In: Proceedings of the Thirteenth International Conference on Artificial Intelligence and Statistics, pp. 249–256. JMLR Workshop and Conference Proceedings (2010)
24. Wang, Z., Bovik, A.C., Sheikh, H.R., Simoncelli, E.P.: Image quality assessment: from error visibility to structural similarity. IEEE transactions on image processing **13**(4), 600–612 (2004)
25. Boroumand, M., Chen, M., Fridrich, J.: Deep residual network for steganalysis of digital images. IEEE Trans. Inf. Forensics Secur. **14**(5), 1181–1193 (2018)

Behavior Recognition Based on Two-Stream Temporal Relation-Time Pyramid Pooling Network (TTR-TPPN)

Mengxing Huang[1]($\boxtimes$), Zhenfeng Li[1], Yu Zhang[2], Yuchun Li[1], Xinze Li[1], and Siling Feng[1]

[1] State Key Laboratory of Marine Resource Utilization in South China Sea College of Information Science and Technology, Hainan University, Haikou, China
[2] School of Computer Science and Technology, Hainan University, Haikou 570288, China
https://www.scholat.com/huangmx09

Abstract. Nowadays, intelligent surveillance has received extensive attention from academia, business, and industry. Deep learning algorithms are widely used in the field of intelligent surveillance. Recently, most deep learning models are limited to a short-term behavior recognition in the entire video. In order to better identify human behavior in the video, we combined a Two-stream network and a Temporal Relation network (TRN) and added a time pyramid pooling operation. In this way, the Two-Stream Temporal Relation-Time Pyramid Pooling Network (TTR-TPPN) can be constructed. The relational pyramid pool network integrated the frame-level features in the video into video-level features. We applied the TTR-TPPN to the Internet public standard data set UCF101 and the self-made DW20 data set. It is found through experiments that this network has a higher recognition rate than other behavior recognition methods on both data sets, and it has better performance in long-term behavior recognition. Therefore, the TTR-TPPN enables it to recognize long-time sequence behavior and improves the accuracy of human behavior recognition.

Keywords: TTR-TPPN · Behavior recognition · Time pyramid pooling operation · TRN · Long-time sequence behavior

1 Introduction

The field of computer vision is a very important branch of artificial intelligence. It mainly studies how to make machines recognize and understand human actions in videos [1]. Human behavior detection has always been a hot spot in the field of computer vision. With the advancement of computer science and technology in recent years, behavior detection has become more and more integrated into our lives [2]. For example, it has played a huge role in smart surveillance, smart medical, smart home and other fields. With the popularity of surveillance, intelligent

© Springer Nature Switzerland AG 2021
C. Xing et al. (Eds.): WISA 2021, LNCS 12999, pp. 418–429, 2021.
https://doi.org/10.1007/978-3-030-87571-8_36

surveillance has made outstanding contributions to ensuring public safety. Traditional cameras can only display surveillance information and cannot respond to emergency situations in time. Therefore, it is necessary to perform online judgment manually, which consumes a lot of manpower and material resources. Deep learning can simulate the nerves of the human brain, learn image information autonomously from many training samples, and extend a multi-layer network structure to a certain neural network [3]. The intelligent surveillance based on deep learning aims at human visual perception and the ability to process images and the machine has the ability to automatically recognize human behavior. The intelligent monitoring system first automatically recognizes the targets contained in the video images collected by the camera, and then further analyzes, detects, locates and tracks and recognizes moving targets [4]. The behavior recognition of a machine is the computer's automatic processing, analysis and understanding of image information [5].

At present, traditional behavior recognition algorithms perform classification operations by extracting and describing behavior characteristics. Traditional behavior recognition methods can be divided into low-level behavior recognition methods and high-level behavior recognition methods [6]. Early human behavior recognition feature extraction methods include calculation methods based on geometric features of the human body [7], feature extraction methods of motion information, etc. [8]. With the advent of convolutional networks, deep learning is increasingly applied to behavior recognition model [9]. Huang et al. proposed an underwater multi-target detection method based on multi-scale convolution neural network (MSC-CNN) [10]. Chen L et al. proposed to introduce attention mechanism in the spatiotemporal dual-flow network model. This method can effectively add a weight to the static image features and the dense optical flow features between frames, pay attention to the beneficial regions in the feature information and then distinguish the feature information, so as to achieve more accurate behavior recognition [11]. The research on this algorithm has had a far-reaching impact on subsequent algorithms. Feichtenhofer et al. realized 3D convolution kernel fusion on the basis of space-time Two-stream volume and network [12]. Lan Z et al. further studied the spatio-temporal fusion of Two-stream networks and proposed linear coding on the coding layer [13]. Shi Y et al. added the trajectory to the Two-stream network and proposed a three-stream network that integrates the trajectory description stream, spatial flow, and temporal flow [14].

Existing behavior recognition models have relatively complex calculations and relatively large parameters, which makes the network difficult to train. The existing two-stream network behavior recognition model lacks the ability to learn long-term sequences. In order to better recognize the behavior in the video, we use the temporal reasoning network and the Two-stream model as the basis, and add the time pyramid pooling operation, which builds the ability to fuse the frame-level features in the video into video-level features. The Two-Stream Temporal Relation-Time Pyramid Pooling Network (TTR-TPPN) has the ability to recognize long sequence. This article applies the TTR-TPPN to the Internet public standard data set UCF101 and the self-made DW20 data set. It is found through experiments that the network has a higher recognition rate than other behavior recognition methods on two data sets, and it has a stronger robustness and generalization ability. In summary, our contributions are as follows:

1. In order to solve the problems of low accuracy of long-time sequence video recognition, we integrated the time-series reasoning network and the Two-stream network model. This model can maximize sparse sampling the video frames to the greatest extent to capture temporal relationships on scale.
2. We add the time pyramid pooling operation to the above model to obtain the TTR-TPPN model. Through the merging operation, the size and redundancy of the original input data can be reduced, thereby improving the recognition rate.
3. We apply TTR-TPPN to the Internet public data set UCF101 and the self-made data set DW20. The TTR-TPPN is higher than other network models in the accuracy of behavior recognition. The experimental results prove the effectiveness of our method and can be applied to the field of intelligent monitoring. The follow-up arrangements of this article are as follows: Sect. 2 describes the method of this experiment in detail. Section 3 is the results of the experiment, and the experimental results are discussed and analyzed, and the last section summarizes the work of this article.

2   Methods

We use Time Reasoning Network (TRN) to sample the frames. This model can learn and infer the timing dependence of frames in the video on multiple scales, and further improve the TRN. The TTR-TPPN contains multiple scale information passes and uses the time pyramid pooling method, which improves the accuracy of behavior recognition. The experimental data set adopts UCF101 standard data set and self-made DW20 data set.

2.1   TTR-TPPN Architecture

TTR-TPPN uses a timing network to sample frames sparsely, and then learns and infers its causality, which is more effective than dense sampling and convolution. We merged the Two-stream CN on the basis of the TRN framework structure model. By sparsely sampling the RGB image and the optical flow stack in the video, it is used as the input of spatial stream and temporal stream network to obtain Two-stream Temporal Relation Network (TTRN). After the last global pooling layer of the space network and the time network, frame-level features can be obtained by using the time pyramid pooling layer. It is further encoded into multiple time-scale video levels to form a new two-stream temporal inference pyramid network model. In order to capture the structure of human action time in the video, the structure from coarse to fine can be used. A compact video stage can be used to obtain models with multiple time scale information. In general, in the

TTR-TPPN model, the RGB image is used as the input of the spatial flow convolution network, and the optical flow image is used as the input of the temporal flow convolution network. The time pyramid pooling layer is added to the two networks. The frame-level feature encoding of the video is converted into video-level features, and the Two-stream network is merged to finally obtain the TTR-TPPN model. The TTR-TPPN architecture is shown in Fig. 1.

2.2   Principle of Time Pyramid Processing

In the field of digital image processing, data preprocessing is required, such as dimensionality reduction and redundancy reduction. The conventional method uses three methods: maximum pooling, average pooling, and global average pooling. However, their common problem is that they lack information extraction in the time dimension when extracting features. Therefore, we use the method of Time Pyramid Pooling (TPP) to extract the time series features of various scales in the input samples. Time pyramid pooling can ensure that behavioral features with inconsistent video duration can be mapped to video sequence features of the same duration. Assuming it is the time pyramid of the k layer, then there is $M = \sum_{i-1}^{k} 2^{i-1}$ time blocks are in this time pyramid. The samples in the input network can be divided into a total of n frames. Assuming that there are d filters at the end of the convolution operation of the network, then after the pooling operation of the pyramid pooling layer, an $d * (2i-1)$ -dimensional feature vector is generated, and the $d * (M)$ -dimensional feature vector is displayed in different time series frames. It is output after the time pyramid pooling operation. Through the aggregation and output operations through the aggregation function, the filter of the time pyramid pooling operation can be obtained. Average convergence and maximum convergence are two ways of convergence function.

The average convergence formula of the TTR-TPPN pooling operation is shown in formula 1 (S_i, S_j are the i-th and j-th $d * (M)$ -dimensional feature vectors, and $Q_{i \rightarrow j}$ represents the feature vector finally obtained through the average operation).

$$Q_{i \rightarrow j} = (S_i \oplus S_i + 1 \oplus \ldots \oplus S_j) / (j - i + 1) \tag{1}$$

The maximum convergence formula of TTR-TPPN pooling operation is shown in formula 2.

$$Q_{i \rightarrow j} = \max \{S_i, S_i + 1, \ldots, S_j\} \tag{2}$$

Assume that the number of videos in the input sample data set is N and contains N action classes. By using an end-to-end approach to further training the network, and then optimizing a series of parameters in the network, the video-level features can be obtained, which can be represented by P, and the output video-level feature prediction results are obtained by formula 3:

$$Y = \varphi (W_c P + b_c) \tag{3}$$

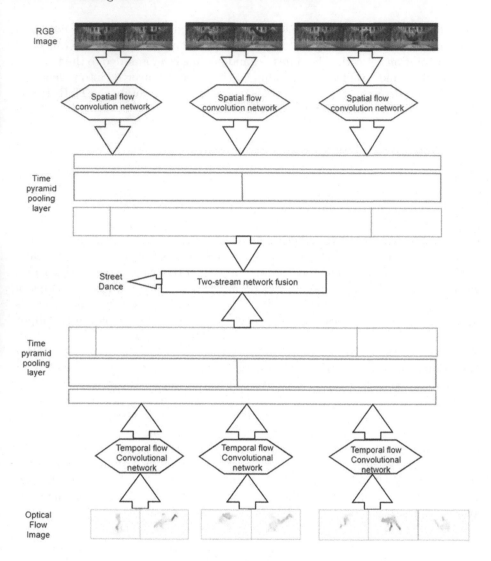

Fig. 1. TTR-TPPN architecture

In the formula 3, $\varphi()$ represents softmax operation, W_c represents the weight parameter of the fully connected operation, b_c represents the offset parameter. The calculation formula for the fusion of loss function and cross entropy is shown in formula 4.

$$L(W, b) = -\sum_{i=1}^{N} \log\left(Y\left(y_i\right)\right) \tag{4}$$

The specific time pyramid processing principle is shown in Fig. 2.

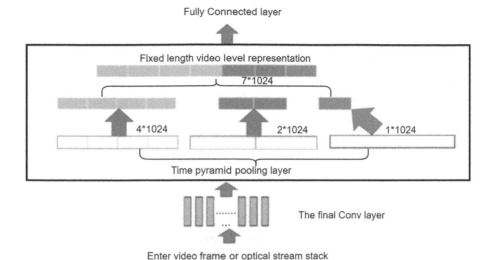

Fig. 2. Time pyramid processing flow

2.3 Principle of Time Pyramid Processing

We use Inception batch processing standardized deep convolution structure in the TTR-TPPN model to extract the features of the Two-stream model. In practice, we use cross-modal pre-training to reduce overfitting. We process the sample data set in a variety of ways to expand the number of samples and prevent overfitting of the spatial flow. Regularization is used in the training process to shorten the training time of the model. Table 1 shows the parameters and meanings used in network training.

(1) Cross-modal pre-training

When training a deep neural network, if the number of input samples is small, overfitting will occur. In TTR-TPPN, a large standard image data set is used to further complete the pre-training and initialize the parameter X1 in the spatial stream network. Because the surface features in the RGB image are inconsistent with the timing features in the optical flow diagram. Therefore, when the time flow network parameter X2 is initialized, the cross-modal pre-training method must be used to linearly change the value range of the optical flow field, so that the value X3 of the optical flow field is within the interval of 0 to 255. After that, the weight value X4 in the first convolutional layer of the space is averaged for the streaming network in the TTR-TPPN. Then calculating the number of channels X5 when the optical flow image is input into the network, which greatly reduces the overfitting phenomenon. The probability in the temporal flow network also improves the recognition accuracy of the time flow network.

Table 1. Parameters used for modal training

Parameter	Meaning
X1	Spatial flow network parameters
X2	Temporal flow network parameters
X3	Value range of optical flow field (0–255)
X4	The weight value of the convolutional layer of the first layer of spatial stream network
X5	Optical flow input network channel volume
$\bar{A}_1$	The mean of the first batch standardization layer of TTR-TPPN
$\overline{s_1}$	The variance of the first batch standardization layer of TTR-TPPN
A	Mean of all batch standardization layers
s	Variance of all batch standardization layers

(2) Data enhancement

In order to solve the overfitting caused by too few videos in the sample data set, we adopt the following methods: corner cropping, scale jitter, and horizontal flip to expand the sample data set. Corner cropping is to cut off the surrounding area or the center area of the screen, thereby reducing the problem of excessive focus on the center of the screen. Scale jitter is to unify the input size to 256×360. The next step is to take two numbers randomly from $\{168, 192, 224, 256\}$ as the height and width of the processed image, and then unified the processed video screen to 224×224. These processed samples are also used as sample data for training. Horizontal flip is to flip the picture to get a new data set.

(3) Regularization

In the model training process, the regularization method can shorten the training time of the model, and batch standardization can speed up the convergence speed of the network model. After the network parameter initialization operation in TTR-TPPN, the mean $\bar{A}_1$ and variance $\overline{s_1}$ parameters in the first batch normalization layer in the TTR-TPPN are fixed to remove the mean and variance parameters in all batch normalization layers in the TTR-TPPN. Because the surface features in the RGB image are inconsistent with the temporal connection features in the optical flow diagram, it is necessary to re-estimate the activation mean and variance parameters for the time flow network. The TTR-TPPN takes every 25 frames of input video as a stack for model training, and then specifies the temporal flow input window size of five consecutive optical flow fields as $224 \times 224 \times 10$. The network parameters are obtained through the small batch stochastic gradient descent algorithm. In actual operation, the batch value is 128. The gradient L2 norm in the model is specified as 40, the momentum term is 90%, and the learning rate is initialized to 0.0001. The learning rate will decrease to 1/10 of the original value if the test data and recognition accuracy no longer decrease sequentially. TTR-TPPN stipulates that the maximum number of iterations of the temporal flow network is 1200 and the random loss probability of the Dropout layer is positioned at 80%.

2.4 Run Environment

We use the Caffe framework in the experiment. The Two-stream network in TTR-TPPN uses NVIDIA GTX Titan X GPU to accelerate network training and testing.

3 Results and Discussion

We use UCF101 standard data and UT two-person interaction standard data and conduct training and testing to analyze the performance of TTR-TPPN. We use the method mentioned in the second part to operate on the UCF101 data set. For UT, we set the ratio of training set and test set to 70% and 30%. The Two-stream model in the TTR-TPPN is pre-trained using the Kinetic large data set, and the evaluation index of the experiment is the average prediction accuracy rate (Averr, hereinafter referred to as recognition rate). The calculation formula of the recognition rate is shown in formula 5. $\hat{y}_i$ represents the predicted value of behavior recognition, y_i represents the actual value of behavior recognition, n represents the total number of data samples.

$$Averr = \frac{1}{n} \sum_{i-1}^{n} \left| \frac{\hat{y}_i - y_i}{y_i} \right| \times 100\% \tag{5}$$

We initialize Two-stream network parameters to improve the model recognition rate. Table 2 shows the comparison of the recognition rate of TTR-TPPN, TRN, and LRCN on the UCF101 standard data and the UT two-person interaction standard data set. Table 2 shows that TTR-TPPN has the best performance.

Table 2. TTR-TPPN recognition rate

Method	UCF101	UT
LRCN	87.6%	89.3%
TRN	91.4%	92.7%
TTR-TPPN	94.8%	95.5%

3.1   The Influence of the Number of Time Pyramid Layers on the Experiment

When the number of TTR-TPPN layers is 1, it means that the pooling layer with the largest granularity collects the features in all frames. When the number of time pyramid levels increases, the recognition accuracy of the temporal flow model and the spatial flow model in the TTR-TPPN has been improved. When the number of layers of the temporal pyramid is 4, the recognition accuracy rate of the spatial flow model in the TTR-TPPN is unchanged, but the recognition rate of the time flow model in the TTR-TPPN begins to decrease. When the number of layers of the time pyramid is 4, it contains 15 time blocks $(1 + 2 + 4 + 8)$. When the number of layers of the time gold pagoda is 3, it contains 7 time blocks $(1 + 2 + 4 = 7)$. When the number of time pyramid levels is 4, the time block included is twice as many as the number of time pyramid levels is 3. Therefore, more calculations are required when the number of pyramid layers is 4. Therefore, we apply the pooling operation with 3 layers of time pyramid to the TTR-TPPN, so that the recognition accuracy of the TTR-TPPN reaches the highest. The changes of TTR-TPPN recognition accuracy rate to different pyramid levels are shown in Table 3.

Table 3. Changes in the numbers of pyramid layers at different times for the recognition accuracy of TTR-TPPN

The number of pyramid layers	Spatial stream model	Temporal stream model
TTR-TPPN (the first layer)	90.8%	93.6%
TTR-TPPN (the second layer)	91.7%	93.9%
TTR-TPPN (the third layer)	92.2%	94.9%
TTR-TPPN (the fourth layer)	92.2%	94.3%

3.2   Analysis of Two-Stream Network Convergence Ratio

As analyzed in the previous section, the Two-stream model has the highest recognition accuracy when the number of time pyramid levels is 3 in the TTR-TPPN. In this section, the spatial flow model and the time flow model in the TTR-TPPN are combined using a weighted average method to determine the corresponding label of the input experimental sample. When the number of time pyramid layers in the TTR-TPPN are 1 and 3 respectively, the change diagram of the Two-stream model after fusion with different weight ratios is shown in Fig. 3.

As shown in Fig. 3, when the weight ratios of the TTR-TPPN spatial flow model and the time flow model are 0.2 and 0.8 respectively, the recognition accuracy reaches the highest, which is 97.11%. It can be seen from the Fig. 3 that when the weight ratio of the spatial flow model in the TTR-TPPN is below 0.2,

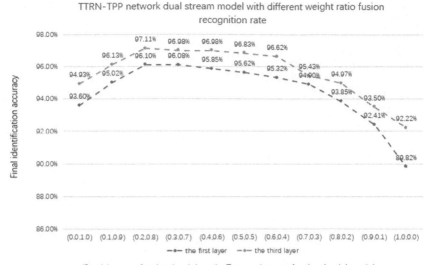

Fig. 3. TTR-TPPN Two-stream model with different weight radio fusion recognition rate

as the weight ratio of the spatial flow model increases, the recognition accuracy of the TTR-TPPN also increases. When the weight ratio of the spatial flow model in the TTR-TPPN is above 0.2, as the weight ratio of the spatial flow model increases, the recognition accuracy of the TTR-TPPN decreases. Therefore, it can be inferred that the influence of the spatial flow model on the final recognition accuracy in TTR-TPPN is less than that of the time flow model. Consequently, we set the number of time pyramid layers to be 3, and the weight ratios of the spatial flow model and the time flow model are 0.2 and 0.8 respectively as the final model of the experiment. We compare the final model with several other algorithms on the UCF101 data set. The comparison results are shown in Table 4.

It can be seen from Table 4 that the recognition accuracy of the TTR-TPPN is much higher than other traditional methods and deep learning methods when the number of layers of the time pyramid is 3 and the weight ratio of the spatial flow model to the time flow model is 0.2 and 0.8.

3.3 Analysis of Self-made Data Set Recognition Results

We use the TRN model and the TTR-TPPN model and use the self-made DW20 data set as the experimental sample data set for comparative testing. We use 70% of the videos in each category of 30 categories in the self-made data set DW20 as the training set, and the other 30% as the test set. When the self-made DW20 data set is not de-redundant, the TTR-TPPN has an improvement of more than 20% compared with the TRN network. It can be concluded that the TTR-TPPN has a strong ability to recognize long-time sequence behavior.

Table 4. Recognition rate of different behavior recognition methods on UCF101 dataset

Methods		Recognition of the long jump
Traditional methods	STIP	52.1%
	DT	82.7%
	IDT	83.9%
	Karpathy	62.7%
	Two-stream CN	81.1%
Deep learning methods	Simonyan	86.5%
	3DConvNet	87.3%
	LRCN	87.9%
	TRN	92.5%
	TTR-TPPN	97.1%

The accuracy of specific categories of the TTR-TPPN and TRN in the self-made DW20 data set is shown in Table 5.

Table 5. Accuracy of TRN and TTR-TPPN on the high jump and the long jump of the DW20 dataset

Network	Recognition of the high jump	Recognition of the long jump
TRN	64.1%	53.5%
TTR-TPPN	85.3%	74.9%

4   Conclusion

In order to improve the general recognition accuracy of some current behavior recognition methods, we merge the TRN with the Two-stream CN network. On this basis, a time pyramid pooling layer is added, and a TTR-TPPN with long-time sequence behavior recognition capability is constructed. Compared with other model networks, the recognition accuracy of this network is improved. In the actual training process of the TTR-TPPN, we use an end-to-end approach to encode the frame-level features in the input sample video into a compact video-level representation with multiple time scales. We use two kinds of experimental data sets: the standard data set UCF101 data set publicly available on the Internet and the self-made DW20 data set. Through experiments, it is concluded that the TTR-TPPN model has a higher recognition rate in these two data sets. Therefore, the TTR-TPPN has better performance than other methods and has stronger long-term sequence structure modeling capabilities, so it may be applied to actual scenarios.

Acknowledgements. Supported by the National Key Research and Development Program of China (Grant #: 2018YFB1404400), National Natural Science Foundation of China(Grant #: 62062030, Major Science and Technology Project of Haikou (Grant #: 2020-009), Project supported by the Education Department of Hainan Province (Grant #: Hnky2019-22).

References

1. Kupryanov, K., Gorodnichev, M.G.: Recognition of human behavior. In: 2021 Systems of Signals Generating and Processing in the Field of on Board Communications (2021)
2. Guan, S., Zhang, Y., Tian, Z.: Research on human behavior recognition based on deep neural network. In: Proceedings of the 3rd International Conference on Mechatronics Engineering and Information Technology (ICMEIT 2019) (2019)
3. Yu, Y.: Deep learning for image recognition. J. Jpn. Soc. Artif. Intell. **28**, 962–974 (2018)
4. Shao, Z., Cai, J., Wang, Z.: Smart monitoring cameras driven intelligent processing to big surveillance video data. IEEE Trans. Big Data 4(1), 105–116 (2018)
5. Zhang, X., Luo, L., Zhao, W., Guo, Z., Yue, J.: On combining multiscale deep learning features for the classification of hyperspectral remote sensing imagery. Int. J. Remote Sens. **36**(13–14), 3368–3379 (2015)
6. Ullah, M.M., Parizi, S.N., Laptev, I.: Improving bag-of-features action recognition with non-local cues. In: Proceedings - British Machine Vision Conference, BMVC 2010, Aberystwyth, UK, 31 August–3 September 2010 (2010)
7. Liu, L., Jiao, Y., Meng, F.: Key algorithm for human motion recognition in virtual reality video sequences based on hidden Markov model. IEEE Access **8**, 159705–159717 (2020)
8. Cai, W., Xia, S., Sun, R., Chen, H., Chen, W.: A micro-motion feature extraction method based on CORR-OMP. In: 2021 IEEE 4th International Conference on Electronics Technology (ICET) (2021)
9. Koohzadi, M., Charkari, N.M.: Survey on deep learning methods in human action recognition. IET Comput. Vis. **11**(8), 623–632 (2017)
10. Huang, S., Huang, M., Zhang, Yu., Li, M.: Under water object detection based on convolution neural network. In: Ni, W., Wang, X., Song, W., Li, Y. (eds.) WISA 2019. LNCS, vol. 11817, pp. 47–58. Springer, Cham (2019). https://doi.org/10.1007/978-3-030-30952-7_6
11. Chen, L., Liu, R., Zhou, D., Yang, X., Zhang, Q.: Fused behavior recognition model based on attention mechanism. Visual Comput. Ind. Biomed. Art **3**(1), 1–10 (2020). https://doi.org/10.1186/s42492-020-00045-x
12. Feichtenhofer, C., Pinz, A., Zisserman, A.: Convolutional two-stream network fusion for video action recognition. In: Computer Vision & Pattern Recognition (2016)
13. Lan, Z., Yi, Z., Hauptmann, A.G.: Deep local video feature for action recognition. In: 2017 IEEE Conference on Computer Vision and Pattern Recognition Workshops (CVPRW) (2017)
14. Shi, Y., Tian, Y., Wang, Y., Huang, T.: Sequential deep trajectory descriptor for action recognition with three-stream CNN. IEEE Trans. Multimedia **19**(7), 1510–1520 (2017)

CC-BRRT: A Path Planning Algorithm Based on Central Circle Sampling Bidirectional RRT

Wei Li[1,3], Menghan Ren[1,4], Yonglong Zhu[1,3], Yanyu Zhang[1,3], Sufang Zhou[2,3], and Yi Zhou[1,3(⊠)]

[1] School of Artificial Intelligence, Henan University, Kaifeng 475000, China
zhouyi@henu.edu.cn
[2] School of Computer and Information Engineering, Henan University, Kaifeng 475000, China
[3] International Joint Research Laboratory for Cooperative Vehicular Networks of Henan, Henan, China
[4] Eagle Drive Technology (Shenzhen) Co., Ltd., Shenzhen 518000, China

Abstract. Path planning through a bidirectional fast extended random tree algorithm cannot converge quickly, which does not meet the requirements of mobile robot path planning. To address this problem, an improved, central circle sampling bidirectional RRT (CC-BRRT) algorithm is proposed in this paper. The algorithm searches for the next sampling point by a central circle sampling strategy to reduce the numbers of searching nodes, and reduces the randomness by a target biasing strategy to speed up the convergence of the algorithm. For the obtained path, a sextic spline interpolation method is used to generate a smooth and executable path. Finally, experiments on mobile robot path planning are conducted both in simple with fewer obstacles and complex with more obstacles scenarios. The results show that the proposed CC-BRRT algorithm is superior to several other algorithms, with substantially fewer nodes sampled and a good smoothness and feasibility of the planned path.

Keywords: Path planning · Bidirectional RRT · Central circle sampling

1 Introduction

Path planning is the key technology that directs a mobile robot to reach a target location safely from a starting position under multiple constraints [1]. According to known information about the environment, the methods of path planning are divided into local path planning algorithms and global path planning algorithms. Local path planning involves a robot using sensors to extract information about the surrounding. The robot plans a safe path from the current path node to the next path node in a timely and efficient manner in an unknown environment. Global path planning involves the robot planning a collision-free path before moving from the starting position to the target position in a known environment.

In complex environments, how mobile robots move from the starting position to the target position efficiently and without collision has been the focus of researchers'

© Springer Nature Switzerland AG 2021
C. Xing et al. (Eds.): WISA 2021, LNCS 12999, pp. 430–441, 2021.
https://doi.org/10.1007/978-3-030-87571-8_37

attention [2]. Many methods are proposed and the methods are mainly divided into search-based methods and sampling methods. The search-based methods are represented by the A* algorithm [3, 4], which needs to discrete the state space and then search the path. The sampling-based methods are based on the RRT algorithm [5, 6], which needs to build a graph of random sampling in state space to search for paths. Compared with other traditional planning algorithms, the RRT algorithm of basic sampling converges faster and can effectively solve the problems of path planning in complex environment and high-dimensional environment [7]. However, the RRT algorithm also has some drawbacks. Because of the randomness of the algorithm, it often samples in the whole state space, which results in spending a lot of time and effort exploring invalid areas. And even some planned paths appear in a small range of right-angle changes, which do not satisfy the robot kinematic model, and not easy to walk along with for the robots [8].

In this paper, we propose an improved CC-BRRT algorithm to address the problems of slow convergence, low search efficiency and poor planning path of the basic RRT algorithm, which accelerates the convergence speed of the algorithm, reduces the running time and improves the quality of the planning path. The main contributions of this paper are as follows:

- A CC-BRRT algorithm is proposed, which aims at solving the efficiency problem of the algorithms and the smoothness problem of the planned path.
- We leverage a central circle sampling strategy to search for the next sampling point and reduce the randomness by a target biasing strategy, which speed up the convergence of the algorithm.
- We smooth the generated paths using sextic spline interpolation to obtain a smooth and executable path for kinematic robots.
- We perform the proposed algorithm and some common algorithms experimentally and compare their performance. The results indicate that the proposed algorithm is superior to other algorithms in searching time, the length, and smoothness of the final path.

2 Related Work

In recent years, many different improved RRT algorithms have been proposed. For example, Urmson et al. [9] proposed a target-biased RRT algorithm, which improved the computational efficiency because the target point will appear in the sampling point with a certain probability. Kuffner et al. [10] proposed a RRT-connect algorithm, which generated two search trees to search for the path and improved the search efficiency of the algorithm. Song et al. considered heuristic search strategy and shortened the search time by Gaussian sampling [11]. Karanan et al. achieved the convergence of the optimal solution by optimizing the selection of parent node, but needs to spend too much time to search for the path, which has less efficient [12].

With the development of artificial intelligent techniques, deep learning has also been applied to the RRT algorithm. The deep learning model has good performance in feature extraction and classification [13]. Pareekutty et al. put forward a RRT-HX algorithm, which used reinforcement learning to evaluate the generated tracks, and used the cost information of quality deviation to identify the low-cost paths. But the algorithm did not essentially reduce the randomness of RRT algorithm [14]. Zou *et al.* used reinforcement learning SARAS (λ) algorithm in the RRT algorithm to solve the problem of robot local path planning in the special circumstances [8].

The path planning algorithm based on the RRT algorithm produces an initial path trajectory that is not smooth. The path trajectory cannot be directly applied to the robot motion planning process. Liu *et al.* use cubic spline interpolation curves to design the motion trajectory [15].

On the problem of path smoothness, people generally use arcs or straight lines to smooth the paths. But this approach gives the robots more twists and turns, and even emergency stop or other situations, which easily cause damage to robot parts. The method of spline interpolation is used to fit smooth curves, which conforms to the robot kinematics, with incomparable advantages of arc or straight-line fitting.

3   Problem Formulation

The 2D state-space path planning system model of the mobile robot is shown in Fig. 1, where X is the state space, $X_{obs} \subset X$ is the obstacle space, and $X_{free} = X \backslash X_{obs}$ is the free space. $x_{start} \in X_{free}$ is the initial position and $x_{goal} \in X_{free}$ is the target location. $\sigma[0, 1] \to X_{free}$ denotes a feasible path, where $\sigma(0) = x_{start}, \sigma(1) = x_{goal}$ [16]. $\sum X_{free}$ denotes the set of all feasible paths. $\sigma : \sum X_{free} \to R \geq 0$, and then the optimal path planning problem can be defined as: finding a path σ^* such that the value of the cost function is minimized when connecting the initial state and the goal state. That is:

$$\sigma^* = \arg \min_{\sigma \in \sum} \{c(\sigma)|\sigma(0) = x_{start}, \quad \sigma(1) = x_{goal} \forall s \in [0, 1], \sigma(s) \in X_{free}\} \quad (1)$$

Fig. 1. System model

4 Basic Bidirectional RRT Algorithm

The search process of basic bidirectional RRT algorithm is shown in Fig. 2. Firstly, the starting point X_{start}, target point X_{goal}, obstacle point X_{obs} are initialized, respectively. The starting point X_{start} and the target point X_{goal} are treated as the root node of the random tree T_a and T_b. Secondly, the starting point X_{start} randomly generates a sampling point X_{rand} in state space, and then traverses the random tree T_a to find the node X_{near}^a closest to X_{rand}. Thirdly, the growth step of the random tree is extended towards the connection of X_{near}^a and X_{rand} to get the new node X_{new}^a. If there are no obstacles between X_{new}^a and X_{near}^a, X_{new}^a is added to the random tree T_a. The node closest to X_{new}^a in the random tree T_b is chosen as the next node X_{near}^b, and a certain steps are extended in that direction to get X_{new}^b. If X_{new}^b and X_{near}^b pass the collision detection, X_{new}^b will be added to the random tree T_b and the random tree will continue to grow in the direction of X_{new}^a. The search process is repeated until an obstacle is encountered or the nearest distance between the two random trees is less than the predefined threshold. If an obstacle is encountered, the two trees are exchanged and the sampling process above is repeated. If the distance between the two random trees is less than the connection threshold, the information of the connection point is returned and the path search process finishes.

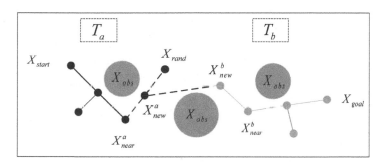

Fig. 2. Basic bidirectional RRT searching process

5 CC-BRRT Algorithm

In this section, several improvements to the bidirectional RRT algorithm are introduced first, and then the detailed steps and pseudocode of the CC-BRRT algorithm are described.

5.1 Goal Biased Strategy

The RRT algorithm performs path search by random sampling in the global state space, which wastes a lot of computing resources. In this paper, the sampling point X_{rand} is generated according to a random probability value p. If p is greater than the predefined probability value p_{goal}, the node generated by using the central circle sampling strategy is set as the sampling point of the extended tree sub-nodes. Otherwise, $X_{rand} = X_{goal}$.

5.2 Central Circle Sampling Strategy

The basic bidirectional RRT algorithm samples in the whole state space randomly, which helps to discover more feasible paths. But the number of the sampled nodes increases considerably, which increases the search time and reduces the search efficiency. To solve the above problems, this paper proposes a central circle sampling strategy based on the bidirectional RRT algorithm, as shown in Fig. 3. Make a random circle centered on the midpoint of the starting point and the target point, and set the sampling point X_{rand}, on the generated circle, which is calculated as:

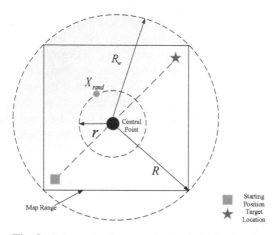

Fig. 3. Schematic diagram of central circle sampling

$$\begin{cases} r = R_w \cdot n^{\frac{m+1}{2}} \\ x_{rand} = x_{center} + r \cdot \cos(2\pi n), \ \ n \in (0,1) \\ y_{rand} = y_{center} + r \cdot \sin(2\pi n) \end{cases} \tag{2}$$

$$\begin{cases} x_{center} = (x_{start} + x_{goal})/2 \\ y_{center} = (y_{start} + y_{goal})/2 \end{cases} \tag{3}$$

where m is the central circle coefficient, R_w is the sampling radius of the outer circle, x_{start}, y_{start} are the abscissa and ordinate of the starting point, x_{goal}, y_{goal} are the abscissa and ordinate of the target point, x_{center}, y_{center} are the abscissa and ordinate of the central point and x_{rand}, y_{rand} are the abscissa and ordinate of X_{rand} generated on the central circle.

5.3 Sextic Spline Interpolation

In this paper, a method of sextic spline interpolation is proposed to smooth the planned path, which can satisfy the stability and safety of the mobile robot in the actual motion process [11].

Let there be $n + 1$ path nodes $a = x_0 < x_1 < \cdots < x_n = b$ on the interval $[a, b]$, given the function value of the node $f(x_i) = f_i, (i = 0, 1, 2 \cdots, n)$, if $h(x)$ satisfies the following conditions:

(1) $h(x) \in C^2$;

(2) $h(x_i)=f_i$;

(3) On each segment interval $[x_i, x_{i+1}]$, $h(x)$ is a sextic polynomial.

$h(x)$ is a sextic spline interpolation function.

The sextic spline interpolation function is a piecewise sixth degree polynomial in a partitioned interval $[x_0, x_n]$ that can be written as:

$$s(x) = \begin{cases} h_1(x) & x_0 \le x \le x_1 \\ \vdots & \vdots \\ h_m(x) & x_{m-1} \le x \le x_m \\ \vdots & \vdots \\ h_n(x) & x_{n-1} \le x \le x_n \end{cases} \tag{4}$$

$$h(x) = a_i x^6 + b_i x^5 + c_i x^4 + \quad \cdots g_i, \quad i = 0, 1, 2, \cdots 6$$

where $a_i, b_i, c_i, \cdots g_i$ is the coefficient to be determined, so $h(x)$ has a total of $7 \times n$ coefficients to determine.

5.4 Algorithm Description

This section describes the proposed CC-BRRT algorithm, and the specific idea of the algorithm and the pseudocode are as follows:

Step 1: Take the robot's current position as the starting point X_{start}, the destination position as the target point X_{goal} initialize a random tree for rapid exploration;

Step 2: Generate a random number p between 0 and 1. If p is less than p_{goal}, set a random sampling point in the tree T_a. Otherwise, perform the central circle sampling in the state space, and obtain the sampling point X_{rand};

Step 3: Through the central circle sampling point X_{rand}, find out the child node in the random tree T_a. That is, set the node closest to X_{rand} as the next node X_{near}^a;

Step 4: The new node X_{new}^a is generated by X_{near}^a in the direction of X_{rand} according to a certain step size S in (5). S is set 3. After do collision detection for the new node X_{new}^a and neighboring nodes X_{near}^a. If they pass collision detection, add the generated X_{new}^a to the random tree T_a, Otherwise, turn to *Step* 2.

$$\begin{cases} x_{new}^a = x_{near}^a + S \cdot \cos(\arctan(\frac{y_{rand}-y_{near}^a}{x_{rand}-x_{near}^a})) \\ y_{new}^a = x_{near}^a + S \cdot \sin(\arctan(\frac{y_{rand}-y_{near}^a}{x_{rand}-x_{near}^a})) \end{cases} \tag{5}$$

Step 5: If X_{near}^b in the extended tree T_b finds X_{new}^a in T_a and the distance between X_{near}^b and X_{new}^a is less than the threshold value for connecting the two trees, T_a and T_b will be connected and the planned path is found. T_b is extended in the same way as T_a in the state space until the path is found.

Step 6: Smooth the generated path by sextic spline interpolation.

Algorithm 1 : CC-BRRT

1: $V_a \leftarrow \{X_{start}\}$; $E_a \leftarrow \phi$;
2: $V_b \leftarrow \{X_{goal}\}$; $E_b \leftarrow \phi$;
3: $T_a \leftarrow (V_a, E_a)$; $T_b \leftarrow (V_b, E_b)$;
4: for k=1 to K do
5:     if $p > p_{goal}$ then
6:         $X_{rand} \leftarrow$ CC_Sample (x_{center} , y_{center});
7:     else
8:         $X_{rand} = X_{goal}$;
9:     end if
10:     if extend(T_a, X_{rand}) $\neq$ Trapped) then
11:         if (connect(T_b, X_{new})=Reached) then
12:             return path(T_a, T_b);
13:         end if
14:     end if
15:     swap(T_a, T_b);
16: end for
17: return path(T_a, T_b);

Function : connect

1: function connect ($T = (V, E), X$)
2:     repeat
3:         S $\leftarrow$ extend(T, X);
4:     until S $\neq$ Advanced;
5:     return S;
6: end function

Function: extend

1: function extend (T, X)
2:     $X_{near} \leftarrow$ Nearest (T, X_{rand});
3:     $X_{new} \leftarrow$ Steer(X_{near} , X_{rand});
4:     if isCollisionFree(X_{near}, X_{new}) then
5:         $V \leftarrow V \cup \{X_{new}\}$;
6:         $E \leftarrow E \cup \{X_{near}, X_{new}\}$;
7:         if ($X_{new} = X$) then
8:             return Reached;
9:         else
10:             return Advanced;
11:         end if
12:     end if
13:     return Trapped;
14: end function

6   Simulation and Results

6.1   Simulation Setup

To fully verify the effectiveness and feasibility in solving robot path planning problems, the CC-BRRT algorithm was experimentally compared with RRT-Connect [10], the goal-biased RRT (GRRT) algorithm [9], the goal-biased bidirectional RRT (GBRRT) algorithm [17].

All the simulations are performed using PyCharm2019. The simulation scenario is an area of 100 m × 100 m. The coordinates of the starting point are (0,0), and the coordinates of the target point are (100,100). The goal of the simulation is to find a collision-free, shortest distance and smooth effective path from the starting position to the target location.

6.2   Performance Analysis in a Simple with Fewer Obstacles Scenario

In this section, we perform the CC-BRRT and other algorithms in a simple with fewer obstacles scenario. As can be seen in Fig. 4(a), the GRRT algorithm is less efficient in planning paths. The extended trees are uniformly distributed in obstacle-free regions, and the final planned paths are more curved and have longer distance. From Fig. 4(b) and (c), it can be seen that the GBRRT and RRT-Connect algorithm has obvious twists and fluctuations in the multiple local areas. From Fig. 4(d), it can be seen that after using the CC-BRRT algorithm to search for the path, the number of extended trees is significantly reduced.

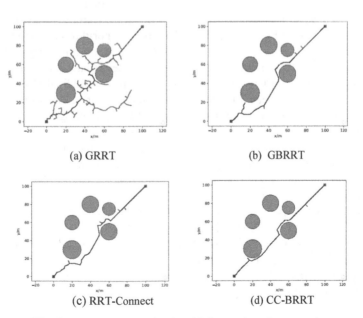

(a) GRRT   (b) GBRRT

(c) RRT-Connect   (d) CC-BRRT

Fig. 4. Simulation in a simple with fewer obstacles scenario

We compared these three algorithms after 30 iterations of path planning. The results listed in Table 1 are the average comparison results of the number of nodes, path length and time between CC-BRRT and the compared algorithms for 30 trials under the same environmental conditions. In terms of the number of nodes, it is reduced by 66.42% compared to the GRRT algorithm and by about 17.9% compared to the GBRRT algorithm. In terms of the path length, it also reduces the path length by a lot compared to other algorithms. The CC-BRRT algorithm reduces the path length by 13.49%, 7.6% than GRRT algorithm and GBRRT algorithm.

Table 1. Performance comparison of the algorithms in a simple scenario.

Algorithm	Number of nodes	Path length	Time/ms
GRRT	165.14	170.32	19.37
RRT-Connect	72.31	164.86	10.21
GBRRT	67.55	159.37	6.95
CC-BRRT	**55.44**	**147.34**	**5.78**

6.3 Performance Analysis in a Complex with More Obstacles Scenario

Figure 5 shows the simulation results of CC-BRRT and other algorithms in a complex with more obstacles scenario, and it can be directly seen from the figure that the paths are smoother and shorter. Figure 5(a) shows that GRRT algorithm sampling points cover the whole state space in the complex scenario, which makes the efficiency of the planned path seriously reduced. From Fig. 5(b) and (c), it can be seen that the overall performance from the GBRRT and RRT-Connect algorithm are okay, but the final planned paths are more tortuous and do not match the actual robot motion paths. From Fig. 5(d), it can be seen that CC-BRRT algorithm has significantly fewer sampling points, and the final planned path is smoother.

The results listed in Table 2 are the average results of the number of nodes, path length, and time obtained by running each algorithm independently for 30 trials under the same environmental conditions. It can be seen that under complex environmental conditions, the path length, the time spent on planning the path, and the number of nodes planned by CC-BRRT is better than the other algorithms. The CC-BRRT algorithm reduces the time by 72.6%, 11.7% than GRRT algorithm and GBRRT algorithm. It also reduces path length by 10.94% compared to GRRT algorithms. In terms of the number of nodes, it is reduced by 67.39% compared to the GRRT algorithm and by about 9.73% compared to the GBRRT algorithm. From simple to complex scenarios, the effectiveness and reliability of the improved algorithms in this paper are further verified.

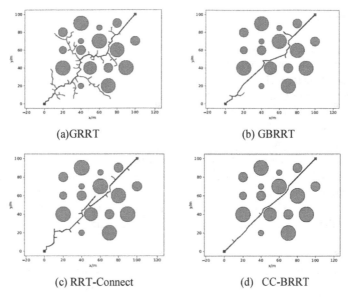

Fig. 5. Simulation in a complex with more obstacles scenario

Table 2. Performance comparison in complex scenarios with more obstacles.

Algorithm	Number of nodes	Path length	Time/ms
GRRT	170.13	165.24	28.14
RRT-Connect	68.25	158.26	14.86
GBRRT	61.45	154.59	8.74
CC-BRRT	**55.47**	**147.16**	**7.71**

Figure 6(a) and (b) indicate RRT, GRRT, RRT-Connect, GBRRT, and CC-BRRT, respectively, and it shows that the path planned by the CC-BRRT algorithm is smoother.

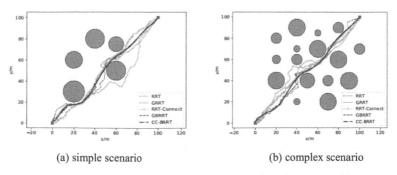

(a) simple scenario   (b) complex scenario

Fig. 6. The final path planning comparison

7 Conclusion

In this paper, a central circle bidirectional RRT (CC-BRRT) algorithm is proposed, which searches for the next sampling point by a central circle sampling strategy, and reduces the randomness by a target biasing strategy, and smooths the generated paths using sextic spline interpolation to obtain a smooth and executable path for kinematic robots. After performing in simple and complex scenarios, it is found that the CC-BRRT algorithm proposed in this paper reduces the blindness of sampling to some extent, shortens the path length, reduces the number of sampled nodes, and improves the smoothness of the path, compared with other path planning algorithm.

At present in the field of path planning, many researchers gather to apply the method of deep learning to the field of path planning, which has achieved great success. In the future works we will combine the algorithm with deep learning, and consider using the algorithm on the physical object for verification to test the actual performance of the algorithm.

Acknowledgment. This work was supported by National Natural Science Foundation of China (62071356), and Program for Science & Technology Development of Henan Province (202102310198, 212102210412).

References

1. Xiang, J.L., Wang, H.D., Ou, Y.: Algorithm of local path planning for unmanned surface vehicles based on improved Bi-RRT. Ship Build. China **061**(001), 157–166 (2020)
2. Zhang, W.B., Xiao, J.L.: Application of improved RRT in intelligent vehicle path planning under complicated environment. China J. Highway Transp. **34**(03), 225–234 (2020)
3. Ammar, A., Bennaceur, H., Châari, I., Koubâa, A., Alajlan, M.: Relaxed Dijkstra and A* with linear complexity for robot path planning problems in large-scale grid environments. Soft. Comput. **20**(10), 4149–4171 (2015). https://doi.org/10.1007/s00500-015-1750-1
4. Wang, D.J.: Indoor mobile-robot path planning based on an improved A* algorithm. J. Tsinghua Univ. Sci. Technol. **52**(08), 1085–1089 (2012)
5. Liu, C.J., Han, J.Q., Kang, A.: Dynamic path planning based on an improved RRT algorithm for roboCup robot. Robot **39**(01), 8–15 (2017). https://doi.org/10.13973/j.cnki.robot.2017.0008
6. Song, J.Z., Dai, B., Shan, E.: An improved RRT path planning algorithm. Acta Electron. Sin. **38**(2A), 225–228 (2010)
7. Tan, J.H., Pan, B., et al.: Robot path planning based on improved RRT*FN algorithm. Control Dec. **36**(08), 1834–1840 (2020)
8. Zou, Q.J., Liu, S.H., Hou, Y.P.: Rapidly-exploring random tree algorithm for path re-planning based on reinforcement learning under the peculiar environment. Control Theory Appl. **37**(08), 1737–1748 (2020)
9. Urmson, C., Simmons, R.: Approaches for heuristically biasing RRT growth. In: Proceedings 2003 IEEE/RSJ International Conference on Intelligent Robots and Systems (IROS 2003), vol. 2, pp. 1178–1183 (2003). https://doi.org/10.1109/IROS.2003.1248805
10. Kuffner, J.J., LaValle, S.M.: RRT-Connect: an efficient approach to single-query path planning. In: IEEE International Conference on Robotics and Automation, vol.2, pp. 995–1001(2000). https://doi.org/10.1109/ROBOT.2000.844730

11. Song, X.L., Zhou, N., Huang, Z.Y.: An improved RRT algorithm of local path planning for vehicle collision avoidance. J. Hunan Univ. Nat. Sci. **44**(4), 30–37 (2017)
12. Karaman, S., Frazzoli, E.: Incremental sampling-based algorithms for optimal motion planning. In: Robotics: Science and Systems (2010). https://arxiv.org/abs/1005.0416
13. Gao, M., et al.: Online anomaly detection via incremental tensor decomposition. In: Ni, W., Wang, X., Song, W., Li, Y. (eds.) WISA 2019. LNCS, vol. 11817, pp. 3–14. Springer, Cham (2019). https://doi.org/10.1007/978-3-030-30952-7_1
14. Pareektty, N., James, F., Ravindran, B.: RRT-HX: RRT with heuristic extend operations for motion planning in robotic systems. In: ASME International Design Engineering Technical Conferences & Computers & Information in Engineering Conference (2016). https://doi.org/10.1115/DETC2016-60547
15. Liu, J.S., Ji, H.Y., Li, Y.: Robot path planning based on improved bat algorithm and cubic spline interpolation. Acta Automatica Sinica (2020).
16. Mashayekhi, R., Idris, M., Anisi, M., et al.: Informed RRT*-Connect: an asymptotically optimal single-query path planning method. IEEE Access **8**, 19842–19852 (2020). https://doi.org/10.1109/ACCESS.2020.2969316
17. Li, X.W., Yu, H.S.: RRT robot path planning algorithm based on bidirectional growth improvement. Mod. Comput. **21**(4), 28–31 (2019)

Rule Reduction for EBRB Classification Based on Clustering

Longjiang Chen(ID), Yanggeng Fu(✉)(ID), Nannan Chen(ID), Jifeng Ye(ID),
and Genggeng Liu(ID)

College of Mathematics and Computer Science, Fuzhou University, Fuzhou, China
{fu,liugenggeng}@fzu.edu.cn

Abstract. The extended belief rule base (EBRB) system has been successfully applied to classification problems in various fields. However, the existing EBRB generation method converts all data into extended belief rules, which leads to the large scale of rule base and affects the efficiency and accuracy of subsequent inference. In view of this, this paper proposes an EBRB rule reduction method based on the adaptive K-means clustering algorithm (RC-EBRB). In the rule generation process, the K-means clustering algorithm is applied to obtain the rule cluster centers, which are used to generate new rules. In the end, these new rules form a reduced EBRB. Moreover, in order to determine the initial cluster centers and the number of clusters in the K-means clustering algorithm, the algorithm idea of K-means++ is introduced and a reduction granularity adjustment algorithm with threshold is proposed, respectively. Finally, four datasets on commonly used classification datasets from UCI are used to verify the performance of the proposed method. The experimental results are compared with the existing EBRB methods and the traditional machine learning methods, which prove the effectiveness of the method.

Keywords: Extended belief rule base · Rule reduction · Data-driven · K-means clustering algorithm

1 Introduction

Classification problem is significant in a wide range of fields, and many methods have been proposed to solve the problem, such as rule-based system [1], neural network [2], etc. In order to model the incompleteness, uncertainty and ambiguity of information in actual problems, Yang et al. [3] proposed a belief rule base (BRB) inference methodology using the evidence reasoning (ER), which is called RIMER. This method combines traditional IF-THEN rules with decision theory [4], fuzzy set theory [5], Dempster-Shafer (D-S) evidence theory [6,7] and other theories, so that it has the ability to deal with incomplete, uncertain, and fuzzy information.

Supported by the Natural Science Foundation of Fujian Province, China (2019J01647) and the National Natural Science Foundation of China (61773123).

The expert system based on RIMER is called the BRB system, which uses BRB as a carrier to express knowledge and uses ER algorithm to realize knowledge reasoning. Since the number of rules in BRB is exponentially related to antecedent attributes and their referential values, the scale of BRB is prone to "combination explosion". In order to better deal with this problem, Liu et al. [8] proposed a data-driven extended belief rule base (EBRB) system. On the basis of the traditional BRB system, the EBRB system extends the antecedent attribute part of the rule and introduces the belief distribution form, making the rule base have a stronger ability to express vague and uncertain information. The EBRB system has been successfully applied in many professional fields. Chen et al. [9] proposed an EBRB system based on an improved rule activation rate to solve the problem of non-activation of rules and balance the consistency and completeness of activated rules. Fang et al. [10] analysed the imbalance of dataset that existed in the generated EBRB and proposed a balance adjusting approach to eliminate the influence of imbalance in EBRB system. To solve the problem of low reasoning efficiency caused by the disorderly storage of rules, Yang et al. [11] proposed a multi-attribute search framework of KD tree and BK tree for EBRB to deal with different attribute dimensions and improve the reasoning performance. Lin et al. [12] proposed an EBRB system based on the VP tree and MVP tree index structure and realized the automatic selection of index parameters. In order to reduce the scale of EBRB, Yang et al. [13] introduced data envelopment analysis (DEA) to evaluate the effectiveness of each rule in EBRB and proposed a rule reduction method based on DEA.

The scale of the rule base will affect its inference efficiency, so it is necessary to start from the perspective of rule reduction to essentially improve the rule retrieval efficiency of EBRB. Therefore, this paper proposes an EBRB rule reduction method based on rule clustering analysis (RC-EBRB). Firstly, the K-means clustering algorithm is introduced into EBRB. By constructing the feature vectors of the rules in EBRB, the clustering analysis of the rules is realized, and the obtained cluster centers are used to construct a new EBRB, thereby a new EBRB rule reduction algorithm is proposed. Secondly, this paper refers to the algorithm idea of K-means++ [14] to determine the initial rule cluster centers. Thirdly, a reduction granularity adjustment algorithm with threshold is proposed, which improves the traditional elbow method to determine the reduction granularity, so as to better weigh the inference accuracy and efficiency of the reduced EBRB. Finally, through several experiments on the datasets from UCI [15], the performance of the proposed RC-EBRB is analyzed. And by comparing with some existing work of EBRB and traditional machine learning methods, the effectiveness of RC-EBRB is proved.

2   Overview of EBRB System

2.1   Representation of EBRB

The EBRB system is an extension of the BRB system. It embeds the belief structure into the entire antecedent attributes based on the BRB system to better deal with the ambiguity, incompleteness, and uncertainty of information. Generally, the kth extended belief rule in EBRB can be expressed as:

$$R_k : IF \ U_1 \ is \ \{(A_{11}, \alpha_{11}^k), (A_{12}, \alpha_{12}^k), ..., (A_{1J_1}, \alpha_{1J_1}^k)\} \wedge ... \wedge$$

$$U_T \ is \ \{(A_{T1}, \alpha_{T1}^k), (A_{T2}, \alpha_{T2}^k), ..., (A_{TJ_T}, \alpha_{TJ_T}^k)\}$$

$$THEN \ \{(D_1, \beta_1^k), (D_2, \beta_2^k), ..., (D_N, \beta_N^k)\}$$

$$with \ a \ rule \ weight \ \theta_k \ and \ attribute \ weights \ \{\delta_1, \delta_2, ..., \delta_T\} \tag{1}$$

$$s.t. \ \sum_{j=1}^{N} \beta_j^k \le 1, \quad \sum_{j=1}^{J_i} \alpha_{ij}^k \le 1, \ \forall i \in \{1, ..., T\}$$

where k is the index of rule, U_i denotes the ith antecedent attribute, and α_{ij}^k is the belief degree to which the ith antecedent attribute is evaluated to be the jth referential value A_{ij}^k. J_i is the number of referential values of the ith antecedent attribute and T denotes the number of antecedent attributes. The consequent term of rule $\{(D_n, \beta_n^k), n = 1, .., N\}$ is the belief distribution of consequent attribute, and β_n^k denotes the belief degree to which the nth referential value D_n is believed to be the consequent in the kth rule. θ_k is the rule weight of kth rule and δ_i denotes the attribute weight of ith attribute. If $\sum_{j=1}^{N} \beta_j^k = 1$, the kth rule is called complete, otherwise the rule is incomplete.

2.2   Construction of EBRB

Different from the construction of rule base in BRB system, rules in EBRB system can be generated based on given data. Suppose there are L pieces of data, and each piece of data has T antecedent attributes and a consequent attribute, denoted as $\{(x_1^k, x_2^k, ..., x_T^k; y^k) | k = 1, 2, ..., L\}$. Then the rule generation steps of the EBRB system are given below:

1) Determine the referential values. Use the experience of domain experts [3] or fuzzy membership function [1] to determine the referential values of each antecedent attribute and the utility values of the consequent attribute.
2) Obtain the belief distribution. Using the referential values of antecedent attributes and the utility values of the consequent attribute determined in the previous step, the input X and the consequent y of the training data are respectively transformed into the corresponding belief distribution form. For the input part of the kth data $X^k = (x_1^k, x_2^k, ..., x_T^k)$, let γ_{ij} denote the value corresponding to the referential value A_{ij} and ensure that $\gamma_{i(j+1)} > \gamma_{ij}(j = 1, 2, ..., J_i - 1)$, then the belief distribution form of the antecedent attributes can be obtained by Eqs. 2–4.

$$E(x_i^k) = \{(A_{ij}, \alpha_{ij}^k), j = 1, 2, ..., J_i\} \tag{2}$$

$$\alpha_{ij}^k = \frac{\gamma_{i(j+1)} - x_i^k}{\gamma_{i(j+1)} - \gamma_{ij}}, \ \alpha_{i(j+1)}^k = 1 - \alpha_{ij}^k, \quad \gamma_{ij} \le x_i^k \le \gamma_{i(j+1)} \tag{3}$$

$$\alpha_{it}^k = 0, \quad t = 1, 2, ..., J_i \text{ and } t \neq j, j+1 \tag{4}$$

Similarly, according to the consequent value y^k of the kth data, the belief distribution form of the consequent attribute of the rule can also be calculated.

2.3 Reasoning of EBRB

After the rules of the EBRB system are generated, EBRB reasoning can be carried out. The reasoning processes are as follows.

1) Convert the input data into belief distribution. Given a T-dimensional input data $X = (x_1, x_2, ..., x_T)$, according to Eqs. 2–4, the corresponding belief distribution form can be obtained:

$$E(x_i) = \{(A_{ij}, \alpha_{ij}), j = 1, 2, ..., J_i\}, \ i \in \{1, ..., T\} \tag{5}$$

2) Calculate the activation weight of each rule. First, the distance between the input data and the ith rule can be calculated by Euclidean distance.

$$d_i^k = \sqrt{\sum_{j=1}^{J_i} (\alpha_{ij} - \alpha_{ij}^k)^2} \tag{6}$$

Then, the individual matching degree between the kth rule and the input regarding the ith antecedent attribute is calculated as follows:

$$S_i^k = \exp\left(-\frac{(d_i^k)^2}{2\sigma^2}\right), \ \sigma > 0 \tag{7}$$

where σ is a parameter used to control the activation rate of the rule. And the activation weight of kth rule can be obtained by Eq. 8:

$$\omega_k = \frac{\theta_k \prod_{i=1}^{T_k} (S_i^k)^{\overline{\delta}_i}}{\sum_{l=1}^{L} \left[\theta_l \prod_{i=1}^{T_l} (S_i^l)^{\overline{\delta}_i} \right]}, \quad \overline{\delta}_i = \frac{\delta_i}{\max\limits_{j=1,2,...,T_k} \{\delta_j\}} \tag{8}$$

Remark 1. In the traditional EBRB proposed by Liu et al. [8], the following equation is used to calculate the individual matching degree:

$$S_i^k = 1 - d_i^k \tag{9}$$

It can be inferred that $S_i^k \in [1 - \sqrt{2}, 1]$, which may lead to counterintuitive individual matching degrees. Therefore, the calculation equation of individual matching degree based on Gaussian kernel function in [9] is introduced, namely Eq. 7, which makes the individual matching degrees belong to $(0, 1]$.

3) Integrate the activated rules using ER. All rules whose activation weight is greater than the threshold e are activated and aggregated by the analytical evidential reasoning algorithm.

$$\hat{\beta}_j = \frac{\mu \times \left[\prod_{k=1}^{L} (\omega_k \beta_j^k + 1 - \omega_k \sum_{n=1}^{N} \beta_n^k) - \prod_{k=1}^{L} (1 - \omega_k \sum_{n=1}^{N} \beta_n^k) \right]}{1 - \mu \times \left[\prod_{k=1}^{L} (1 - \omega_k) \right]} \tag{10}$$

where

$$\mu = \left[\sum_{j=1}^{N} \prod_{k=1}^{L} (\omega_k \beta_j^k + 1 - \omega_k \sum_{n=1}^{N} \beta_n^k) - (N-1) \prod_{k=1}^{L} (1 - \omega_k \sum_{n=1}^{N} \beta_n^k) \right]^{-1} \tag{11}$$

4) Convert the obtained belief distribution into the reasoning result. For classification problems, the final conclusion is determined by:

$$f(x) = D_j, \quad j = \underset{n=1,\ldots,N}{\arg\max} \hat{\beta}_n \tag{12}$$

3   EBRB System Based on Rule Clustering

This section will introduce the design ideas of the rule clustering algorithm together with the method of determining the initial rule cluster centers. Finally, the reduction granularity adjustment algorithm with threshold is proposed.

3.1   Rule Clustering Algorithm

In order to cluster the rules, the feature vectors of the rules need to be defined first. Assuming that an EBRB rule has T attributes, and each attribute has J referential values, then the vector composed of the antecedent attribute beliefs has $T \times J$ dimensions, and most of the belief values are 0. The vector dimension is high and the distribution is sparse, which is not conducive to rule clustering. So this paper improves the compression method proposed by [11], therefore the belief values of the kth rule can be mapped to a T-dimensional vector, which is denoted as $Z^k = \{z_1^k, z_2^k, \ldots, z_T^k\}$ and can be calculated by Eq. 13.

$$z_i^k = \sum_{j=1}^{J_i} (\alpha_{ij}^k \gamma_{ij}) + \frac{(\gamma_{i1} + \gamma_{iJ_i})}{2} (1 - \sum_{j=1}^{J_i} \alpha_{ij}^k) \tag{13}$$

With the feature vector of the rule as the basis, the clustering algorithm can be applied to the rule clustering. The clustering algorithm used here is K-means clustering algorithm. K-means clustering algorithm is an iterative clustering analysis algorithm, which can classify and organize the rules in some similar aspects, summarize the rule set, and get the representative rule cluster centers.

Algorithm 1: Rule clustering algorithm

    Input   : Origin EBRB *ruleSet*, Number of clusters k, Maximum Iterations
            maxIter
    Output: Reduced EBRB *rcRuleSet*
1 *centers* = select k rules in *ruleSet*
2 *converged* = $false$
3 *iter* = 0
4 **while** *converged is false and iter* < *maxIter* **do**
5   **for** *rule in ruleSet* **do**
6   *index* = $\underset{i}{\text{argmin}}$ (GetDistence(*rule, centers*[i])), $i \in [1, k]$
7   *clusters*[*index*].append(*rule*)
8   *newCenters* = means of *clusters*
9   **if** *centers equal to newCenters* **then**
10   *converged* = true
11   *centers* = *newCenters*
12   *iter* = *iter* + 1
13 *rcRuleSet* = generate rules with *centers*
14 **return** *rcRuleSet*

The steps of rule clustering algorithm based on K-means are as follows. Firstly, k rules are selected as the initial cluster centers by specific selection algorithm. Secondly, the distance between each rule and each cluster center is calculated by using the feature vector of the rule, and each rule is assigned to the nearest cluster center. A cluster set refers to a cluster center and the rules assigned to this cluster center. Finally, after each assignment, the center of each cluster set will be recalculated by taking the average of all feature vectors of rules in the cluster set as the new cluster center. The above steps will continue to be repeated until the cluster centers no longer change or the specified number of iterations is reached. The details of the algorithm are shown in Algorithm 1.

Figure 1 shows an example of rule distribution and clustering projected on the two-dimensional space by using the feature vectors of the rules. Three cluster centers can be obtained by using the K-means clustering algorithm. And three corresponding rules, which form the reduced EBRB, can be obtained according to the given rule generation method. It can be seen that the reduced EBRB can obtain representative rules and reduce the scale of the rule base effectively.

3.2 Initial Cluster Centers Selection

The selection of initial cluster centers will affect the performance of rule clustering. Generally, it is easier to converge to more representative cluster centers by selecting rules that are farther apart. So this section refers to the algorithm ideas of K-means++ [14] to select the initial cluster centers. The algorithm steps are as follows:

1) Randomly select the feature vector of a rule from the rule set as the first initial cluster center.

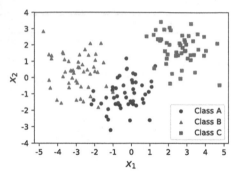

Fig. 1. Example of rule distribution and clustering

2) Firstly calculate the shortest distance between the feature vector of ith rule and the current cluster center (that is, the distance from the nearest cluster center), denoted by $D(Z_i)$. Secondly calculate the probability $P(Z_i)$ that the feature vector of ith rule is selected as the next cluster center, as shown in Eq. 14.

$$P(Z_i) = \frac{D(Z_i)^2}{\sum_{j=1,\ldots,n} D(Z_j)^2} \tag{14}$$

Finally, choose one new rule at random as a new cluster center, where a rule $rule_i$ is chosen with probability proportional to $P(Z_i)$.
3) Repeat step 2 until k cluster centers are selected.

3.3   Reduction Granularity Adjustment Algorithm with Threshold

The reduction granularity of the rule clustering algorithm depends on the value of k in the K-means clustering algorithm. This section will improve the traditional elbow method and propose a reduction granularity adjustment algorithm with threshold for rule clustering.

The core indicator of the elbow method is the sum of the squared errors (SSE), as shown in Eq. 15:

$$SSE = \sum_{i=1}^{k} \sum_{p \in C_i} |p - M_i|^2 \tag{15}$$

where C_i is the i-th cluster, p is the feature vector corresponding to the rules in C_i, M_i is the cluster center of C_i, and SSE is the clustering error of all rules, representing the clustering effect.

The core idea of the traditional elbow method is that the division of rules will be more refined with the increase of the number of clusters k, so the aggregation degree of each cluster will gradually improve, and the SSE will naturally become smaller. When the value of k is less than the real number of clusters, the increase

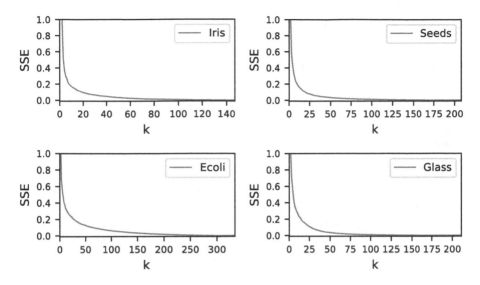

Fig. 2. Examples of the elbow method on different datasets

of k will greatly increase the aggregation degree of each cluster, so the decrease of SSE will be great. When k is close to the real number of clusters, the increase of aggregation degree will slow down, so the decrease of SSE will also tend to be flat. SSE tends to converge as k continues to increase. It can be seen that the relationship between SSE and k is the shape of an elbow, as shown in Fig. 2, and the value of k at the elbow will be closest to the real number of clusters.

Although the traditional elbow method is simple and effective, it is necessary to manually draw a two-dimensional line graph of SSE and k, and judge the approximate value range of k through manual observation. And when applied to rule clustering, although the elbow can obtain a better reduction effect, the lack of precision is relatively severe. Therefore, this paper introduces a threshold mechanism based on the elbow method to replace manual observation. When the SSE value is less than the threshold, the corresponding value of k is used as the final number of rule cluster centers. By adjusting the threshold to make it appear on the right side of the elbow, it is possible to better balance the inference accuracy and inference efficiency of the reduced rule base.

4   Experiments

In order to verify the effectiveness of the proposed method, this paper uses the four commonly used classification datasets Iris, Seeds, Ecoli, and Glass from UCI [15] for experiments. The experimental environment is Intel(R) Core i7-6700@ 3.40 GHz, 16 GB RAM, Windows 10 operating system. The algorithm is implemented in Python 3.6 environment.

4.1 Influence of Reduction Granularity on System Reasoning

For verifying the impact of different reduction granularity on classification accuracy, this section controls the reduction granularity of EBRB by setting different numbers of clusters. In order to ensure the reliability of the experimental results, each set of data is subjected to 100 rounds of 5-fold cross-validation experiments, and the average result of the experiments is used as the final evaluation basis.

As shown in Fig. 3, the abscissa represents the number of rules (the number of cluster centers), and the ordinate represents the classification accuracy. As the number of rules increases, the accuracy gradually improves and converges. Among the three datasets of Iris, Seeds, and Ecoli, when the number of rules exceeds 10, the accuracy fluctuation is basically stable, while the Glass dataset has always shown an upward trend and has not converged. This is mainly due to the respective distribution characteristics of the rule bases corresponding to different datasets. The reason why the rule base of the first three datasets can converge quickly is mainly that the rule base itself is relatively high in redundancy, and it can reflect the distribution of the overall rule base through a small number of cluster centers. By summarizing a few representative cluster centers, a better rule reduction effect can be achieved, so the convergence effect is obvious. But the rule base corresponding to Glass has more categories and higher attribute dimensions, so it is hard to extract and summarize redundant rules and achieve better rule reduction effects. Thus it can be seen that the proposed method can summarize the distribution of the rule base to a certain extent, and reduce the rules. The reduction effect of the rules is affected by whether the distribution of the original rule base has good aggregation characteristics.

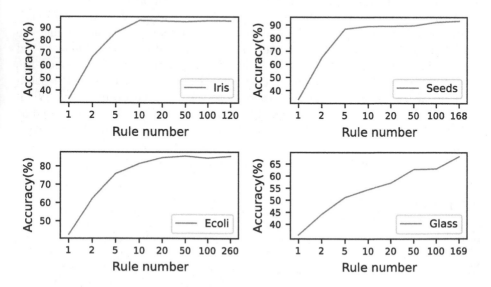

Fig. 3. Experimental results under different reduction granularity

4.2   Comparison with Other Methods

This section firstly compares the method proposed in this paper with the DEA-EBRB method proposed by Yang et al. [13]. DEA-EBRB is an EBRB rule reduction method based on data envelopment analysis. The proposed method is an EBRB rule reduction method based on clustering analysis, which is called RC-EBRB here. In order to make the comparison result more convincing, the method of controlling variables is then used to compare the reduction effects of the two algorithms under the premise that the classification accuracy of RC-EBRB is the same as or slightly better than that of DEA-EBRB. Note that the number of rules in RC-EBRB here is determined by gradually increasing until the afore-mentioned premise is met, and the accuracy of RC-EBRB is the average result of 100 rounds of 5-fold cross-validation. Table 1 shows the comparative experimental results of these two EBRB rule reduction algorithms. The indicator "number of rules" in Table 1 represents the number of rules in reduced EBRB and that in original EBRB.

Table 1. Comparison with DEA-EBRB in reduction effect

Datasets	Aspects	DEA-EBRB	RC-EBRB
Iris	Accuracy (%)	95.4	96.33
	Number of rules	92/120	**8/120**
Seeds	Accuracy (%)	91.67	91.81
	Number of rules	155/168	**80/168**
Ecoli	Accuracy (%)	83.33	83.35
	Number of rules	210/269	**11/269**
Glass	Accuracy (%)	69.44	69.45
	Number of rules	**155/171**	**155/171**

It can be seen that the reduction effect of the RC-EBRB method is significantly better than that of the DEA-EBRB method on Iris, Ecoli, and Seeds. The num of rules in the reduced EBRB has decreased by 84 on Iris, 75 on Seeds, and 199 on Ecoli. Therefore, compared with DEA-EBRB, RC-EBRB achieves a better rule reduction effect by refining a more representative rule set.

In order to further verify the effectiveness of the proposed RC-EBRB method, this paper compares the RC-EBRB method with some existing EBRB methods and traditional machine learning methods. The reduction granularity adjustment algorithm with threshold is used to obtain the optimal number of cluster centers under the four datasets of Iris, Seeds, Ecoli, and Glass to be 51, 49, 77, and 56 respectively.

Table 2 shows the corresponding experimental results. It can be seen that RC-EBRB ranks first on Iris and Ecoli in accuracy. Although the performance is average on Seeds and Glass, the number of rules on Seeds is only 49, and that on Glass is only 56, which is much lower than other EBRB methods. From the experimental results, it can be concluded that RC-EBRB can not only reduce the scale of the rule base but also achieve ideal inference accuracy. Compared with other EBRB methods including DEA-EBRB, it can better balance reasoning efficiency and reasoning accuracy.

Table 2. Comparison with other methods in classification accuracy

	Iris (%)	Seeds (%)	Ecoli (%)	Glass (%)
KNN [16]	85.17	**92.38**	81.27	61.21
EFRBCS [17]	93.00	82.38	77.79	61.38
BRBCS [1]	93.67	87.00	82.89	68.57
BPNN [18]	91.59	–	75.65	64.58
C4.5 [19]	**96.00**	–	80.24	66.82
EBRB [8]	95.13	91.10	80.27	68.81
DEA-EBRB [13]	95.40	91.67	83.33	**69.44**
RC-EBRB (this paper)	**96.00**	90.86	**85.88**	66.20

5   Conclusion

Since the scale of EBRB will affect its inference efficiency, this paper proposes a rule reduction method based on the rule clustering method. The clustering method used in this paper is the K-means clustering algorithm, which is applied to rule clustering by constructing feature vectors of the rules. The algorithm idea of K-means++ is referred to initialize the cluster centers, so that the initial cluster centers are relatively scattered, and improve the performance of rule reduction. And this paper adopts an improved elbow method to realize the reduction granularity adjustment algorithm with threshold. Finally, this paper compares the proposed method with some existing EBRB methods and traditional machine learning methods. From the experimental results, it can be seen that the rule clustering algorithm can mine the aggregation distribution rules between the internal rules of the original EBRB, summarize more suitable rules, achieve an ideal effect of rule reduction, and improve the reasoning effect of EBRB. For future research, we will investigate how to improve the reasoning performance of EBRB under different clustering methods.

References

1. Jiao, L., Pan, Q., Denoeux, T., Liang, Y., Feng, X.: Belief rule-based classification system: extension of FRBCS in belief functions framework. Inf. Sci. **309**, 26–49 (2015). https://doi.org/10.1016/j.ins.2015.03.005

2. Xue, L., Huang, W., Wang, J.: Ranking-based fuzzy min-max classification neural network. In: Wang, G., Lin, X., Hendler, J., Song, W., Xu, Z., Liu, G. (eds.) WISA 2020. LNCS, vol. 12432, pp. 352–364. Springer, Cham (2020). https://doi.org/10. 1007/978-3-030-60029-7_33

3. Yang, J.B., Liu, J., Wang, J., Sii, H.S., Wang, H.W.: Belief rule-base inference methodology using the evidential reasoning approach-RIMER. IEEE Trans. Syst. Man Cybern. Part A: Syst. Hum. **36**(2), 266–285 (2006). https://doi.org/10.1109/ TSMCA.2005.851270

4. Hwang, C.L., Yoon, K.: Methods for multiple attribute decision making. In: Multiple Attribute Decision Making, pp. 58–191. Springer, Heidelberg (1981). https:// doi.org/10.1007/978-3-642-48318-9_3

5. Zadeh, L.A., Klir, G.J., Yuan, B.: Fuzzy Sets, Fuzzy Logic, and Fuzzy Systems: Selected Papers, vol. 6. World Scientific, Singapore (1996). https://doi.org/10. 1142/2895

6. Dempster, A.P.: A generalization of Bayesian inference. J. R. Stat. Soc. Ser. B (Methodol.) **30**(2), 205–232 (1968). https://doi.org/10.1111/j.2517-6161.1968. tb00722.x

7. Shafer, G.: A Mathematical Theory of Evidence, vol. 42. Princeton University Press, Princeton (1976). https://doi.org/10.2307/j.ctv10vm1qb

8. Liu, J., Martinez, L., Calzada, A., Wang, H.: A novel belief rule base representation, generation and its inference methodology. Knowl.-Based Syst. **53**, 129–141 (2013). https://doi.org/10.1016/j.knosys.2013.08.019

9. Chen, N.N., Gong, X.T., Fu, Y.G.: Extended belief rule-based reasoning method based on an improved rule activation rate (in Chinese). CAAI Trans. Intell. Syst. **14**(6), 1179–1188 (2019). https://doi.org/10.11992/tis.201906046

10. Fang, W., Gong, X., Liu, G., Wu, Y., Fu, Y.: A balance adjusting approach of extended belief-rule-based system for imbalanced classification problem. IEEE Access **8**, 41201–41212 (2020). https://doi.org/10.1109/ACCESS.2020.2976708

11. Yang, L.H., Wang, Y.M., Su, Q., Fu, Y.G., Chin, K.S.: Multi-attribute search framework for optimizing extended belief rule-based systems. Inf. Sci. **370**, 159–183 (2016). https://doi.org/10.1016/j.ins.2016.07.067

12. Lin, Y.Q., Fu, Y.G., Su, Q., Wang, Y.M., Gong, X.T.: A rule activation method for extended belief rule base with VP-tree and MVP-tree. J. Intell. Fuzzy Syst. **33**(6), 3695–3705 (2017). https://doi.org/10.3233/JIFS-17521

13. Yang, L.H., Wang, Y.M., Lan, Y.X., Chen, L., Fu, Y.G.: A data envelopment analysis (DEA)-based method for rule reduction in extended belief-rule-based systems. Knowl.-Based Syst. **123**, 174–187 (2017). https://doi.org/10.1016/j.knosys.2017. 02.021

14. Vassilvitskii, S., Arthur, D.: K-means++: the advantages of careful seeding. In: Proceedings of the Eighteenth Annual ACM-SIAM Symposium on Discrete Algorithms, pp. 1027–1035 (2006). http://ilpubs.stanford.edu:8090/778/

15. Dua, D., Graff, C.: UCI machine learning repository (2017). http://archive.ics.uci. edu/ml

16. Derrac, J., Chiclana, F., García, S., Herrera, F.: Evolutionary fuzzy k-nearest neighbors algorithm using interval-valued fuzzy sets. Inf. Sci. **329**, 144–163 (2016). https://doi.org/10.1016/j.ins.2015.09.007

17. Cordón, O., Del Jesus, M.J., Herrera, F.: A proposal on reasoning methods in fuzzy rule-based classification systems. Int. J. Approximate Reasoning **20**(1), 21–45 (1999). https://doi.org/10.1016/S0888-613X(00)88942-2

18. Bhardwaj, A., Tiwari, A., Bhardwaj, H., Bhardwaj, A.: A genetically optimized neural network model for multi-class classification. Expert Syst. Appl. **60**, 211–221 (2016). https://doi.org/10.1016/j.eswa.2016.04.036
19. Abellán, J., Baker, R.M., Coolen, F.P., Crossman, R.J., Masegosa, A.R.: Classification with decision trees from a nonparametric predictive inference perspective. Comput. Stat. Data Anal. **71**, 789–802 (2014). https://doi.org/10.1016/j.csda.2013.02.009

Image Noise Recognition Algorithm
Based on Improved DenseNet

Mengxing Huang[1,2]([⊠]), Lirong Zeng[1], Yu Zhang[2], Yuchun Li[1], Zehao Ni[1],
Di Wu[1], and Siling Feng[1]

[1] State Key Laboratory of Marine Resource Utilization in South China Sea College
of Information Sc ience and Technology, Hainan University, Haikou, China
[2] School of Computer science and Technology, Hainan University,
Haikou 570288, China
`https://www.scholat.com/huangmx09`

Abstract. In the process of image generation or transmission, the image
quality is often degraded due to the interference and influence of speckle
noise, which will adversely affect the subsequent image processing. However, most of the existing Optical Coherence Tomography (OCT) image
denoising methods usually only use part of the prior information of the
OCT image, and ignore the changes in the texture and structure of the
OCT image. There is often a problem that the network is too deep and
the calculation complexity is too large. Aiming at this shortcoming, this
paper proposes an improved deep convolutional neural network image
denoising algorithm based on the original DenseNet algorithm idea. First,
by constructing a speckle noise image data set, a series of preprocessing is
performed on the input image data set, then the visual statistical feature
map of the image is extracted, and finally the DenseNet network structure is improved for network training. The experimental results show
that the improved DenseNet has better performance no matter compared with BM3D, which is recognized as the best denoising algorithm
in the field of image denoising, or compared with DnCNN, an advanced
image denoising algorithm in the field of deep learning.

Keywords: OCT image · Speckle noise · Image noise recognition ·
DenseNet network

1 Introduction

With the rapid development of information technology, people's demand for
images is increasing, and the requirements for image quality are also increasing [1]. OCT as a new type of optical imaging technology, has been widely used
in the medical field, especially in ophthalmology, which has played an important
diagnostic role. OCT imaging technology is based on the principle of interference detection of coherent beams, which is inevitably contaminated by speckle
noise [2]. In order to solve the problem of OCT, this paper proposes an image
noise recognition algorithm based on improved DenseNet [3].

© Springer Nature Switzerland AG 2021
C. Xing et al. (Eds.): WISA 2021, LNCS 12999, pp. 455–467, 2021.
https://doi.org/10.1007/978-3-030-87571-8_39

Domestic research on image noise recognition is roughly concentrated in the spatial domain, time domain and machine learning domains. Yang Zhiguo et al. judged whether the noise in the image is white noise based on the power spectral density of white noise is a constant, and the autocorrelation function is the principle of impulse response [4]. Wang Tiantian et al. After performing wavelet decomposition according to the noise image, the noise characteristics are mainly concentrated In the high frequency part of each scale of wavelet, the Gaussian noise and salt and pepper noise in the image are judged by fitting the histogram signal-to-noise ratio curve of the wavelet high-frequency subband coefficients [5].

In recent years, foreign research on this problem has been mainly based on extracting filtered images, and some algorithms have introduced deep learning. Vasuki et al. used Lee filtering instead of homomorphic filtering to extract the kurtosis and skewness of the residual image after image filtering, and used a multi-layer feedforward neural network classifier to judge Gaussian noise, speckle noise and salt and pepper noise [6]. Karibasappa et al. still use Wiener filtering, homomorphic filtering and median filtering after residual image kurtosis and skewness statistical characteristics, using probabilistic neural networks to distinguish Gaussian noise, uniform noise, speckle noise and salt and pepper noise in the image. Finally, fuzzy logic is used to estimate the parameters of Gaussian noise [7].

A large number of researchers have discussed the nature of denoising. However, it is quite challenging to propose a method suitable for removing the types of noise present in the image. When the nature of noise degradation is not clear, its recognition in the image is an important step in the interpretation system based on visual information [8]. In the face of most real-life images, there is no prior knowledge about the nature of the noise in the image. But it is reported in many related documents that the nature of the noise is known. This makes it necessary to estimate the statistical parameters of noise. More importantly, we must first determine the nature of the noise to the noise source, which can be achieved through the method of identity verification, otherwise, the ideal denoising result will not be obtained [9].

In response to the above problems, we propose an improved DenseNet network structure to identify image noise by preprocessing the input image data set and combining the advantages of feature fusion [10]. The main contributions of this article are as follows:

(1) We have different statistical characteristics according to various noises, so the statistical characteristics of the image will also have different results, and the convolutional neural network is used for training. Visual features are classified to identify different types of noise.
(2) By performing a series of preprocessing on the noise data set, we can extract the visual statistical feature map of the noise image and suppress the generation of other irrelevant features.

(3) Our improved DenseNet structure can effectively reduce the complexity of the network structure, reduce the problem of excessive network parameters, and improve the training effect.

We compare the proposed method with the traditional image noise recognition algorithm DenseNet on OCT ophthalmic images. The experimental results verify the effectiveness of this method.

2 Related Work

2.1 Dense Block

The network structure of DenseNet is mainly composed of Dense block and transition layer. The nonlinear composite function $H_n()$ in the dense block of DenseNet uses the structure of batch normalized BN layer + linear rectification function ReLU layer + 3×3 convolutional layer, and each composite function in the dense block outputs the number of features of k As shown in the figure, the nth layer has a total of $k_0 + k^*(n-1)$ and uses a dense block plus a transition layer structure. The transition layer module includes a 1×1 convolutional layer and a feature map, where k_0 represents the number of channels in the input layer, and k is also called the growth rate of the network.

In order to reduce the size of the feature map, network 2×2 average pooling layer, two adjacent dense blocks are connected by a transition layer, and the pooling operation is used to reduce the size of the feature map and compress the model. Figure 1 shows a DenseNet network structure containing 3 dense blocks:

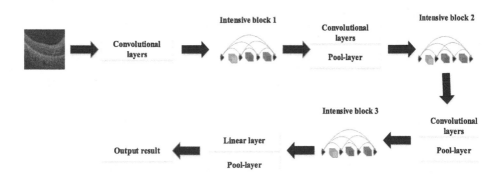

Fig. 1. Structure of DenseNet with three dense blocks.

Table 1 shows the different network structures trained by DenseNet on the ImageNet dataset. The network adopts the structure of 4 dense blocks plus three transition layers. The image size of the ImageNet dataset is 224×224. Before entering the first dense block, it is necessary to pass 2k 7×7 initial convolutional layers with a step size of 2 and a 3×3 maximum pooling layer with a step size of 2. In order to reduce the number of input feature maps, a 1×1 convolutional layer is placed before each 3×3 convolutional layer as a bottleneck layer to improve computational efficiency. And in order to maintain the size of the feature map, the 3×3 convolution kernels in the convolution layer in the dense block all use zero padding for convolution operations. After the last dense block is connected a classification layer composed of a global average pooling layer and a fully connected softmax layer.

Table 1. Structures of DenseNet with different growth rates

Layers	Output Size	DenseNet-121	DenseNet-169	DenseNet-201	DenseNet-264
Convolution	112×112	7×7 conv, stride 2			
Pooling	56×56	3×3 max pool, stride 2			
Dense Block 1	56×56	$\begin{pmatrix} 1 \times 1 \\ 3 \times 3 \end{pmatrix} \times 6$	$\begin{pmatrix} 1 \times 1 \\ 3 \times 3 \end{pmatrix} \times 6$	$\begin{pmatrix} 1 \times 1 \\ 3 \times 3 \end{pmatrix} \times 6$	$\begin{pmatrix} 1 \times 1 \\ 3 \times 3 \end{pmatrix} \times 6$
Transition Layer 1	56×56	1×1 conv			
	28×28	2×2 average pool, stride 2			
Dense Block 2	28×28	$\begin{pmatrix} 1 \times 1 \\ 3 \times 3 \end{pmatrix} \times 12$	$\begin{pmatrix} 1 \times 1 \\ 3 \times 3 \end{pmatrix} \times 12$	$\begin{pmatrix} 1 \times 1 \\ 3 \times 3 \end{pmatrix} \times 12$	$\begin{pmatrix} 1 \times 1 \\ 3 \times 3 \end{pmatrix} \times 12$
Transition Layer 2	28×28	1×1 conv			
	14×14	2×2 average pool, stride 2			
Dense Block 3	14×14	$\begin{pmatrix} 1 \times 1 \\ 3 \times 3 \end{pmatrix} \times 24$	$\begin{pmatrix} 1 \times 1 \\ 3 \times 3 \end{pmatrix} \times 32$	$\begin{pmatrix} 1 \times 1 \\ 3 \times 3 \end{pmatrix} \times 48$	$\begin{pmatrix} 1 \times 1 \\ 3 \times 3 \end{pmatrix} \times 64$
Transition Layer 3	14×14	1×1 conv			
	7×7	2×2 average pool, stride 2			
Dense Block 4	7×7	$\begin{pmatrix} 1 \times 1 \\ 3 \times 3 \end{pmatrix} \times 16$	$\begin{pmatrix} 1 \times 1 \\ 3 \times 3 \end{pmatrix} \times 32$	$\begin{pmatrix} 1 \times 1 \\ 3 \times 3 \end{pmatrix} \times 32$	$\begin{pmatrix} 1 \times 1 \\ 3 \times 3 \end{pmatrix} \times 48$
Classification Layer	1×1	7×7 global average pool			
		1000D fully - connected, softmax			

2.2   Experimental Data

Because there is currently no data set for noisy images, the paper first builds a noisy image data set through the OCT data set [11]. Select 150 images in the OCT image data set, convert the IMG format to BMP, adjust the size to $224 \times 224 \times 3$ and add the parameters of 0.5% (density) salt and pepper noise; 1% salt and pepper noise; $\mu = 0.005, \sigma = 0.1$ Gaussian noise; $\mu = 0.01, \sigma = 0.1$ Gaussian noise; $\mu = 0.01, \sigma = 0.1$ Gaussian noise & 0.5% salt and pepper noise; $\mu = 0.01, \sigma = 0.1$ Gaussian noise & 1% salt and pepper noise; $\mu = 0.005, \sigma = 0.1$ Gaussian noise & 0.5% salt and pepper noise; $\mu = 0.005, \sigma = 0.1$ Gaussian noise & 1% salt and pepper noise [12] (Fig. 2).

2.3   Image Preprocessing

The image preprocessing process is as follows:

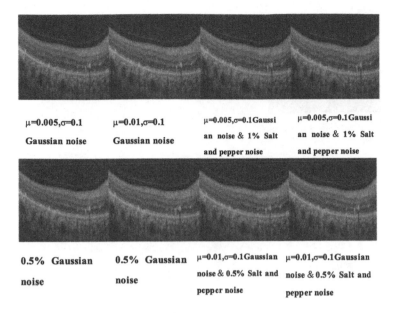

Fig. 2. Part of the noise image data set.

(1) Draw a gray histogram of the noise image.
(2) Fit the frequency information at different gray values of the gray histogram into a curve to improve the training speed and highlight the features that need training.
(3) The frequency range of the gray value in the normalized gray value histogram is from 0 to 1. The image shows the proportion of the gray value frequency corresponding to each gray value.
(4) In order to highlight the characteristics of various noises, the y-axis interval is adjusted again to 0 to 0.1.
(5) Remove the coordinate axis and the coordinate axis scale to reduce other unnecessary features and improve the training speed.
(6) In order to be consistent with the input layer size of DenseNet, reset the image size to $224 \times 224 \times 3$. Figure 3 shows a diagram of the image preprocessing process:

3   Improved DenseNet Network

Due to the excessive amount of network parameters, the training time becomes longer. In response to this problem, the Dense block unit is improved and 3 Dense blocks are added, as shown in Fig. 4. By splitting a 3×3 convolution in the original Dense block into a 1×3 convolution and a 3×1 convolution in series, the purpose of this can be to improve the parameters that need to be trained. The original 3×3 convolution The kernel needs to train a total of 9 parameters.

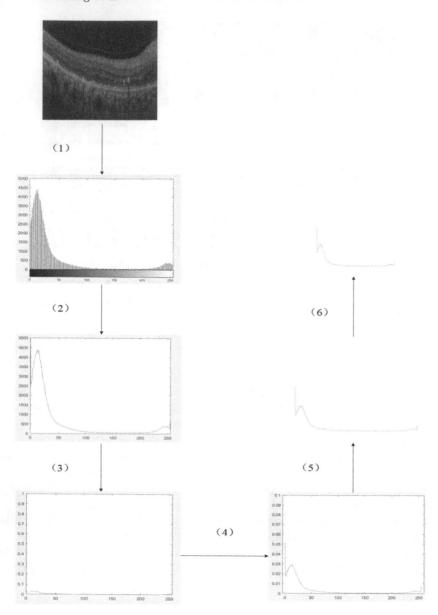

Fig. 3. Image preprocessing.

After splitting into a 1×3 convolution and a 3×1 convolution, there are a total of 6 parameters that need to be trained. That is to say, after the improvement, the parameters that need to be trained for a Dense block unit block are reduced by 3, which is 1/3 of the original. However, the entire network parameters are mainly concentrated in the 2nd to 16th layers. Therefore, after the improvement, the amount of parameters that the entire network needs to be trained is also 1/3 of the original, thereby improving the detection accuracy.

This article uses 4 dense block network structures. The first dense block contains 8 1 × 3 and 3 × 1 convolution operations. The input of the eighth substructure is the output result of the previous layer, the number of output channels in each layer is 8, assuming that the growth rate is a small integer 8. In the traditional DenseNet network, each 3 × 3 convolutional layer is followed by a 1 × 1 convolutional layer, which can reduce the number of feature maps input to the convolutional layer, thereby increasing the operating speed of the network. The calculation efficiency is also improved accordingly. After comparison, it is found that this design is very effective for the DenseNet network, so the one-dimensional convolutional layer will be retained in the network of this article.

From the structure diagram, we can see that every two Dense blocks contains a Transition Layer. In Transition Layer, the parameter reduction takes a value of 0 to 1, which represents how many times the output is reduced. In this article, the reduction is set to 0.5, so that the number of channels will be reduced by half when it is passed to the next dense block. The Dense Block module immediately passes through the Transition Layer, which is a 1 × 1 convolution, 2 × 2 average pooling, and BN. The next module structure is similar to the previous one, the only difference is Because the size of the output feature map is getting smaller and smaller, the depth of the network is gradually deepening.

Prove the theoretical feasibility by formula (1):

$$output = \frac{input - k + 2padding}{s} + 1 \tag{1}$$

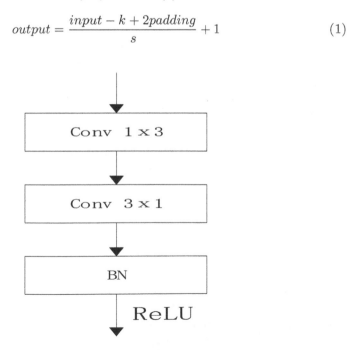

Fig. 4. Improved Unit block structure diagram.

In formula (1): output is the side length of the output feature map; input is the side length of the image input; k is the side length of the convolution kernel; padding is the side length of the input image plus 0; s is the convolution step length. Equation (1) is the calculation method of the output feature map size after convolution. Suppose the input image size is N × N, the convolution step length in DenseNet is all 1, so sets = 1. Without considering padding, the output feature map size after 3 × 3 convolution is $(N-2) \times (N-2)$; if the image is first convolved 1 × 3 and then the output feature map size is $N \times (N-2)$, and then after 3 × 1 convolution, the output feature map size is $(N-2) \times (N-2)$. Through comparison, it can be found that the final output feature map size of the two methods is the same, therefore, the improved Dense block unit block is theoretically feasible.

4   Algorithm Simulation

4.1   Simulation Environment

Experiments were conducted under the Linux system for network training using the Deep Learning Toolbox deep learning framework with hardware configured with Intel Xeon CPU, memory 16 G, graphics card of 11 GB NVIDIA Tesla V100 PCIe GPU.

4.2   Simulation Results

In the experiment, 4 Dense blocks are used. There are two convolutional layers inside each Dense block module. The convolutional layers are 1 × 3 and 3 × 1 convolutional layers, and because Dense block is followed by 1 × 1 With the convolutional layer of 1, the parameters are also simplified. The experimental results are shown in Table 2.

Table 2. Experimental results

Recognition methods	Recognition rate	Parameter	Loss function
DenseNet	79.87	839754	0.85
Improved DenseNet	85.28	475838	0.72

From Table 2, it is easy to know that the recognition rate of classical DenseNet is only 79.87, but the improved recognition rate is obviously improved, which is 85.28. The results show that the improved structure can obviously improve the recognition rate. The number of parameters decreased from 839754 to 475838, which obviously shortened the running time and improved the running efficiency.

Comparing the original dense block and the improved dense block structure, the experimental results are shown in Fig. 5. It can be seen that when the training

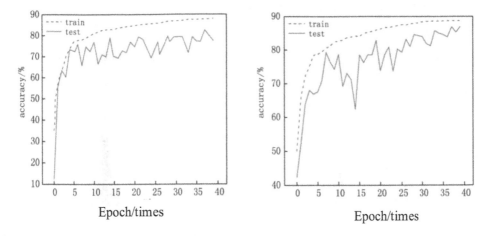

Fig. 5. Comparison of traditional DenseNet and improved DenseNet experimental results.

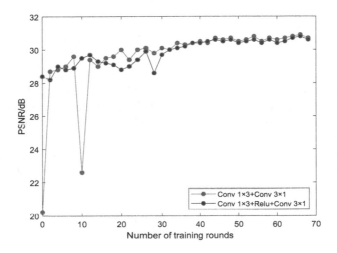

Fig. 6. Dense block Structure ReLU or not.

gradually stabilizes, the PSNRs of the two are basically the same, which means that although the improved dense block reduces the amount of training network parameters, the image denoising effect has not been reduced. The denoising effect of the original DenseNet image is basically the same, which proves that this method is not only theoretically feasible but also experimentally feasible. In order to verify whether there is an activation function ReLU between the convolutional layer 1×3 and the convolutional layer 3×1 in the improved dense block structure, the effect of the denoising effect on the image is carried out, Conv1×3 +ReLU+ Conv3×1 and Conv1×3 + Conv3×1, the experimental results are shown in Fig. 5. It can be seen from Fig. 6 that Conv1×3 + Conv3×1 has a better effect. This is because

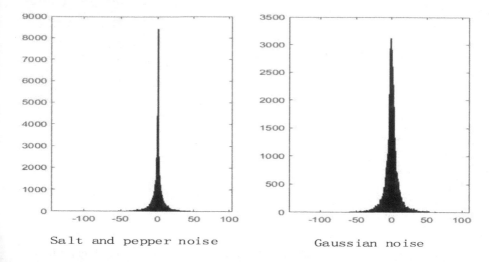

Fig. 7. Intensive connection mechanism of the DenseNet network.

when there is a ReLU activation function, some neurons that are in negative values output by Conv1 × 3 will be inactivated, resulting in some information. If there is no ReLU activation function, some negative values output by Conv1 × 3 may make this part of the negative value become positive after Conv3 × 1, so as to be retained, and this part of the information can be retained at the same time. Therefore, Conv1 × 3 + Conv3 × 1 has a better denoising effect. Select the contrast algorithm of three image noise recognition algorithms.

The first comparison algorithm is an improved image noise recognition algorithm based on wavelet domain high frequency subband coefficients. The image noise recognition algorithm based on wavelet domain high frequency subband coefficients judges the salt and pepper noise and Gaussian noise in the image according to the characteristics of the energy distribution in the wavelet domain high frequency subband coefficient histogram. As shown in Fig. 6, after wavelet decomposition and HH subband coefficient histogram extraction, the HH subband coefficients containing salt and pepper noise are concentrated near 0, the distribution of HH subband coefficients containing gaussian noise tends to gaussian distribution (Fig. 7).

In order to use objective evaluation indicators, the algorithm is improved: first, the noisy image is decomposed by db1 wavelet, and then the HH histogram of the high frequency subband coefficients is extracted, and finally the extracted image is trained using the same DenseNet network structure.

The second contrast algorithm is an improved image noise recognition algorithm based on image gray histogram. The image noise recognition algorithm based on the image gray histogram manually intercepts the gray uniform area of the noise image, draws its histogram, and judges the noise type according to the characteristics of the histogram. In order to use objective evaluation indicators, improve the algorithm: first manually intercept the 20 × 20 gray uniform area in the noise data set, then draw the gray histogram of the uniform area, and finally use the same DenseNet network structure to train the gray histogram.

The third contrast algorithm is an image noise recognition algorithm based on convolution neural network. By using the same DenseNet network structure, the gray histogram of noisy image is trained, and the preprocessing operation of gray histogram except modified size is not carried out.

Global accuracy, macro accuracy, macro recall, and macro F1 scores of the three contrast algorithms are shown in Table 3:

Table 3. Accuracy, macro-precision, macro-recall, and macro-F1 score of the comparison algorithms

Directory	Paper Algorithm	Comparison Algorithm 1	Comparison Algorithm 2	Comparison Algorithm 3
Accuracy	0.914	0.490	0.477	0.801
Macro Accuracy	0.918	0.493	0.471	0.806
Macro Call Rate	0.915	0.489	0.439	0.801
Macro F1 score	0.917	0.490	0.454	0.801

Table 3 shows that the accuracy of contrast algorithm 1 is 49%. The algorithm can distinguish salt and pepper noise from other noises with gaussian noise, but the intensity of noise can not be classified well. To some extent, the algorithm can distinguish salt and pepper noise from other noises with gaussian noise but can not classify gaussian noise and mixed noise accurately.

The accuracy of contrast algorithm 2 is 47.7%. Although contrast algorithm 2 can not be classified well between 1% salt and pepper noise and 0.5% salt and pepper noise directory, it can be distinguished well between salt and pepper noise and other noise with gaussian noise. At the same time, the contrast algorithm 2 can not classify the noise intensity very well. To some extent, it shows that the algorithm can distinguish salt and pepper noise from other noise with gaussian noise under the condition of low intensity noise, but can not carry out accurate mixed noise classification and noise intensity classification.

The accuracy of contrast algorithm 3 is 80.1%. With low-intensity noise, the accuracy of contrast algorithm 3 is much higher than that of the other two methods. The accuracy is 100% when identifying low intensity salt and pepper noise. The accuracy of the Gaussian noise is 90%, When the Gaussian noise of $\mu = 0.005$ is identified, the mixed noise of $\mu = 0.005$ Gaussian noise &0.5% salt and pepper noise is misjudged three times, When identifying $\mu = 0.01$ Gaussian noise, it is misjudged as $\mu = 0.01$ Gaussian noise &0.5% salt and pepper noise 3 times; The accuracy of the four mixed noises is further reduced. Different from the noise type recognition algorithm based on visual statistical features, The contrast algorithm 3 only extracts the gray histogram of the noise image, and reset the gray histogram size to $224 \times 224 \times 3$, There is no other preprocessing operation. Compared with the DenseNet-based image noise recognition algorithm, To some extent, it shows that image preprocessing can improve the ability of model recognition.

5   Conclusion

This paper first introduces the noise model of OCT image, Gaussian noise, salt and pepper noise, Gaussian noise and salt and pepper noise mixed noise, and then introduces the image noise recognition algorithm based on improved DenseNet. By improving the network structure of the DenseNet, the training parameters are input for network training. Finally, the experimental results are analyzed and compared with the other 3 image noise recognition algorithms. The results show that the DenseNet based image noise recognition is superior to other algorithms to some extent.

Acknowledgements. Supported by Hainan Provincial Natural Science Foundation of China(Grant #: 2019CXTD400), National Key Research and Development Program of China (Grant #: 2018YFB1404400), National Natural Science Foundation of China(Grant #: 62062030, Major Science and Technology Project of Haikou (Grant #: 2020-009), Project supported by the Education Department of Hainan Province (Grant #: Hnky2019-22).

References

1. Dang, N., Prasath, S., Nguyen, V.S., Minh, H.L.: An adaptive image inpainting method based on the modified Mumford-Shah model and multiscale parameter estimation. Comput. Opt. **43**(2), 251–257 (2019)
2. Yao, X., Ji, K., Liu, G., Shi, W., Gao, W.: Blood flow imaging by optical coherence tomography based on speckle variance and doppler algorithm. Laser Optoelectron. Prog. **54**(3), 031702 (2017)
3. Zhang, J., Ding, S., Zhang, N.: An overview on probability undirected graphs and their applications in image processing. Neurocomputing **321**, 156–168 (2018)
4. Fang, L., Li, S., Cunefare, D., Farsiu, S.: Segmentation based sparse reconstruction of optical coherence tomography images. IEEE Trans. Med. Imaging (2017)
5. Xing, Y., Xu, J., Tan, J., Li, D., Zha, W.: Deep CNN for removal of salt and pepper noise. Image Process. IET (2019)
6. Karibasappa, K.G., Karibasappa, K.: AI based automated identification and estimation of noise in digital images. In: El-Alfy, E.-S.M., Thampi, S.M., Takagi, H., Piramuthu, S., Hanne, T. (eds.) Advances in Intelligent Informatics. AISC, vol. 320, pp. 49–60. Springer, Cham (2015). https://doi.org/10.1007/978-3-319-11218-3_6
7. Chang, X., Shi, W., Zhang, F.: Signed network embedding based on noise contrastive estimation and deep learning. In: Ni, W., Wang, X., Song, W., Li, Y. (eds.) Web Information Systems and Applications, pp. 40–46. Springer, Cham (2019). https://doi.org/10.1007/978-3-030-30952-7_5
8. Lin, C., Li, Y., Feng, S., Huang, M.: A two-stage algorithm for the detection and removal of random-valued impulse noise based on local similarity. IEEE Access **99**, 1 (2020)

9. Huo, F., Zhang, W., Wang, Q., Ren, W.: Two-stage image denoising algorithm based on noise localization. Multimedia Tools Appl. **2** (2021)
10. Vasuki, P., Bhavana, C., Roomi, S., Deebikaa, E.L.: Automatic noise identification in images using moments and neural network. In: International Conference on Machine Vision & Image Processing (2013)
11. Amini, Z., Rabbani, H.: Optical coherence tomography image denoising using Gaussianization transform. J. Biomed. Opt. **22** (2017)
12. Fu, B., Zhao, X., Li, Y., Wang, X., Ren, Y.: A convolutional neural networks denoising approach for salt and pepper noise. Multimedia Tools Appl. **78**(21), 30707–30721 (2018). https://doi.org/10.1007/s11042-018-6521-4

Particle Picking Method for Cryo Electron Tomography Image Based on Active Learning

Mingjie Mo, Fang Kong, and Qing Liu[✉]

School of Information, Renmin University of China, Beijing 100872, China
{2014201884,2017104078,qliu}@ruc.edu.cn

Abstract. Cryo electron tomography (cryo-ET) is an important tool to obtain macromolecular complex assembly and in situ macromolecular structure. The purpose of cryo-ET processing is to obtain the position of particles of interest from the reconstructed tomogram through particle picking, and then classify and average them to obtain a high-resolution three-dimensional structure. Particle picking is critical to obtaining high-resolution structures. Therefore, we focuses on the characteristics of cryo-ET image. The combination of region proposals and classifiers are used to achieve object detection tasks on special images with large size, low signal-to-noise ratio, and low contrast. We redesigned class labels based on granular and non-granular features in the image. Active learning strategies were used to achieve effective particle picking. In the experimental part of this article, a double-output resnet3d network was designed as a classifier based on an active learning strategy for the cryo-ET image dataset of ribosomes of saccharomyces cerevisiae, and achieved good results in particle picking. The accuracy rate can reach the above 81.9%.

Keywords: Deep learning · Active learning · Cryo-ET · Object detection · Particle picking

1 Introduction

Since the 20th century, with the rapid development of life science, people can not only observe life activities from the macro level of organs and tissues, but also obtain the structure of biological macromolecules from the micro perspective to reveal the material basis of life activities.

In the 1950s, Watson and Crick discovered DNA double helix structure [10], and structural biology was born. In the 1960s, X-ray crystal diffraction was used to analyze the structure of the first globulin [5]. In the 1980s, the cryo electron microscope was born.

Single particle analysis technology is a common data analysis method in current cryo electron microscopy technology [2]. Protein macromolecules in images are called particles. Single particle analysis technology requires collecting enough, usually several hundred thousand single protein 2D projections to reconstruct the image in Fourier space, so as to calculate the real 3D structure.

© Springer Nature Switzerland AG 2021
C. Xing et al. (Eds.): WISA 2021, LNCS 12999, pp. 468–479, 2021.
https://doi.org/10.1007/978-3-030-87571-8_40

The cryo electron microscopy technology relies on making the purified single protein into frozen samples, using high-voltage electron beam to penetrate the samples, and using the direct electron detector to collect the signal for imaging at the camera, so as to obtain the gray image with low signal-to-noise ratio and low contrast. There are three problems in the image of cryo electron microscope: The cryo-ET image is formed when the sample is penetrated by the electron beam, and the electrical signal intensity is converted into the light and dark of the image, whose signal-to-noise ratio is very low; The size of the cryo-ET image is large, the objects are small and the number of objects is large; The work of labeling is time-consuming and labor-consuming, so the number of reliable labels is very small. In view of the problem of image processing of cryo-ET, this paper combines the traditional image processing methods and deep learning methods, uses the method of combining region proposal and image classification task to select particles, and uses active learning method to train the model for the problem of lack of real annotation, so as to explore how to carry out accurate and efficient particle selection and improve the effect of the model.

The rest of the paper is organized as follows: Sect. 2 reviews the related work; In Sect. 3, the method used in this experiment is introduced; Sect. 4 introduces the experimental work in detail and the experimental results were compared; We summarize the main work on the particle selection of cryo electron tomography images in Sect. 5.

2   Related Work

Particle selection is a typical object detection task. Unlike face target detection [1], there are two main methods of particle selection in freeze electron microscope images, one is semi-automatic, the other is fully automatic.

Semi automatic method mainly relies on template matching, which usually takes the existing structure as the template, or a small number of good particles selected manually as the template. This method traverses all possible proposal regions on the image, and outputs the regions highly correlated with the template. The semi-automatic method based on template matching has been widely used in the processing software of cryo electron microscope, including EMAN, RELION, CRYOSPARC [6] and so on. The main function of IMOD software package is automatic tomographic image reconstruction [4]. RELION [8] is a widely used image processing software for cryo electron microscopy. Its particle selection method depends on manual selection or template matching method, which is time-consuming and the accuracy is easily affected by personal differences.

The automatic method follows the idea of object detection, which mainly adopts the method based on region proposal plus classifier. Applepicker [3] designed a method to extract region candidates. This method selects the particles according to the statistical characteristics and the traditional idea of object detection instead of preprocessing the image. However, the sliding window method will cause a lot of time cost. At the same time, only the feature vector composed of two statistical features is used for SVM classification after the data dimension is increased. It is difficult to mine more deep-seated features in the particle region, which affects the classification effect. In 2016, a DeepPicker [9] method for particle selection using CNN method was proposed, which still follows the sliding window from the process and uses trained CNN to judge in turn. A special point of this method is the training of CNN model, which uses the data of several known structures to train together, and obtains the influence of using different kinds of known structures to make training sets on the recognition ability of the model on new particles. Although this paper claims that the pre training model can be used directly, it often needs to manually select some samples according to the new particles to fine-tune the model. In addition, the use of sliding window to provide region candidates will still bring a lot of time overhead. R-cnn series or Yolo [7] methods based on proposal box position regression have also been used in particle selection of cryo electron microscope images. These methods can achieve a certain effect on the plane image, but significantly depend on the training set and annotation.

3   Model

In this section, we will introduce a particle selection algorithm based on active learning and region proposal combined with image classification.

Directly following the typical idea of object detection, we select all proposal areas on the image according to particle size at first, then we classify all proposal areas, and keep the proposal areas which containing particles. The overall process of particle selection is shown in Fig. 1.

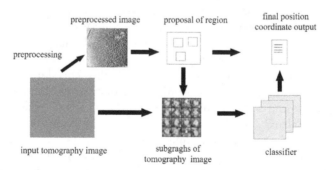

Fig. 1. Process of particle selection

3.1   Binarized Based Region Proposal

The Binarized Based Region Proposal (BBRP) is described in Algorithm 1:

Algorithm 1. Region-Proposal

Input: I: original image; d_{max}, d_{min}: upper and lower bounds of the preset area diameter; $iter$: preset iteration number; T: preset initial threshold; $\triangle T$: preset threshold reduction;

Output: O: List of particle regions;

1: **function** REGION-PROPOSAL($I, d_{max}, d_{min}, iter, T, \triangle T$)
2:     $O\{\}$=NULL;
3:     $s_{max}, s_{min} = calc_{area}(d_{max}, d_{min})$;
4:     **for** i in range($iter$) **do**
5:         $I' = Binarized(I, T)$;
6:         $Clusters=regionprops(L)$;
7:         **for** $item$ in $Clusters$ **do**
8:             **if** $size\_fit(item)\&\&!\ overlap(item, O) > d_{min}$ **then**
9:                 $O.insert(item)$;
10:            **end if**
11:        **end for**
12:        $T = T - \triangle T$
13:    **end for**
14: **end function**

In Algorithm 1, All the detected regions containing particles are considered as members of the positive class, and each element of O{} is a coordinate (x, y, z) of the positive class. The $Binarized()$ algorithm in BBRP relies on the assumption that the contrast between the particle region and the background noise region in the image is significantly different, so it can use a specific pixel value as the threshold to binarize the image. In binary image, the region with pixel value of 1 represents particle region, while the region with pixel value of 0 represents non particle region.

In BBRP, due to the trend that the average value of local particle pixels and the average value of local noise pixels change together, when the final proposal region set is provided, the particle and noise in different regions can be distinguished by using the changed pixel value as the threshold value and expanding the output set iteratively, Each iteration adds the particle region obtained from the current round to the final particle proposal list. Details of the iterative method are described as follow:

1. Select the iteration initial pixel value X. The initial threshold x can be selected as 250.
2. After getting the binary image according to the pixel threshold, the connected region labeling algorithm is implemented on the binary image to get all the connected region information. The label image L obtained by BBRP algorithm is used to obtain the size of all connected regions $area$ and the geometric center coordinates (x, y, z). The label ID of the connected region is combined to form a triplet set $Clusters$ which contains statistical information of all connected regions.
3. In each iteration, $size\_fit(item)$ judges whether the size of the connection area is beyond the range of particle size. $overlap(item, 0)$ judges whether the centroid coordinates of $item$ overlap the existing positions in the output

set O. If the distances are not less than the minimum preset diameter of a particle, the connected region to be added is considered as a new particle region.

According to the experimental results, BBRP algorithm can get good effect after three iterations.

3.2   Two-Output Resnet3d Network

Network Structure. According to the results of the region proposal algorithm, resnet18 is used as the template. According to the active learning strategy and data characteristics, the two-output resnet3d network is reconstructed, which includes two modules, the object module (OBM) and the loss prediction module (LPM), corresponding to two outputs. The function of the OBM is to output the classification results of the image, and the function of the LPM is to output the predicted value of the loss function of the image in the OBM. Figure 2 shows the structure of two-output resnet3d network.

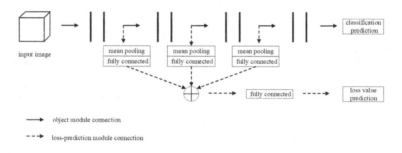

Fig. 2. Structure of two-output resnet3d network

The LPM uses the output of the first three residual modules of the OBM as the input after being activated by relu. After an average pooling, the output is directly input to the full connection layer to get the intermediate result. The three intermediate results are spliced and mapped to the final output value through the full connection layer again. This value represents the predicted value of the loss function.

The OBM is the standard residual network module. The structure of the OBM is shown in Fig. 3. In the figure, the solid line connection represents the connection with the same feature dimension, and the dotted line connection represents the connection with different feature dimensions. The size of the feature graph is halved by pooling method, and then the dimension of the feature is increased by using padding method, so as to ensure the complexity of the features before and after connection. The output of the network is the subgraph of the preprocessed image based on the region proposal method.

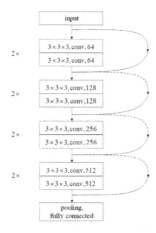

Fig. 3. The structure of the object module

Loss Function. The loss function of the network adopts cross entropy. In the OBM, the size of the loss function value reflects the accuracy of the network model to predict a sample. From the perspective of active learning, the network should choose more samples with large loss function value to learn instead of those with small loss function value.

However, in the unlabeled sample pool, because of lacking label information, it is impossible to obtain the loss function value directly. It is necessary to adopt the design of LPM. In the training process of labeled samples, the LPM predicts the loss function value of the network for these samples.

Algorithm 2 is the training algorithm of two-output resnet3d network.

We need to divide a batch for training into two subsets of the same size each time. For each pair of samples A and B of the two subsets, If compared with sample B, sample A is more difficult to learn for the network at this time, sample A has higher loss function value than sample B in the OBM. The optimization goal of the LPM is to make the output value of sample A in the LPM higher than that of sample B, and return the LPM label according to the loss function value of each pair of samples. The equation is as follow:

$$L(A, B) = \max\left(0, elt * \left(loss_{pred\ A} - loss_{pred\ B}\right)\right) \tag{1}$$

loss $_{pred\ A}$ and loss $_{pred\ B}$ represent the loss prediction values of sample A and B output by the LPM respectively. *elt* is a coefficient, and its calculation formula is as follows:

$$elt = \begin{cases} -1, & \text{if } loss_A > loss_B \\ 1, & \text{else} \end{cases} \tag{2}$$

$loss_A$ and $loss_B$ are the real loss values of samples a and B in the network OBM. According to the above two formulas, for a pair of samples A and B, the LPM is labeled as follows:

Algorithm 2. Training algorithm of two-output resnet3d network

---

Input: *model*: model of network; X_train, Y_train: training set; L: size of training set; *epochs, batch_size*: training parameters;

Output: *model*: model of network;

1: **function** TRAIN($model, X\_train, Y\_train, L, epochs, batch\_size$)
2:     $Shuffle(X\_train, Y\_train)$;
3:     **for** $current\_ep$ in range($epochs$) **do**
4:         **for** i in range($int(L/batch\_size)$) **do**
5:             $loss\_label = zeros((batch\_size, 1))$;
6:         **end for**
7:     **end for**
8:     **for** ei in range($batch\_size$) **do**
9:         $t1 = model.evaluate(X\_train[i * batch\_size + ei], [Y\_train[i * batch\_size + ei]])$;
10:         $eval[ei] = t1[0]$ \\ Using array $eval[]$ to store the real loss value of samples;
11:         $outscore, loss\_pred[ei] = model.predict(X\_train[i * batch\_size + ei])$;
12:         $group\_num = int(batch\_size/2)$;
13:         **for** $groupi$ in range($group_num$) **do**
14:             $loss\_label = Calc\_Loss\_label(eval, loss\_pred)$;
15:         **end for**
16:         $model.train\_on\_batch()$;
17:     **end for**
18: **end function**

$$\begin{cases} loss_{labelA} = - \ \text{elt} \ * \ L(A, \ B) + loss_{\text{pred} \ A} \\ loss_{labelB} = \ \text{elt} \ * \ L(A, \ B) + loss_{\text{pred} \ A} \end{cases} \tag{3}$$

If the LPM correctly predicts the relative size of the loss function values of the two samples, then the loss function values of the two samples in the LPM are all 0, and the label returned to the LPM is equal to the predicted value of the loss function of the model. Otherwise, the label in the loss function prediction module of the two samples is the predicted value of the other party's loss function.

3.3 Model Training Combined with Active Learning Strategy

Training Process. As the network model is divided into OBM and LPM, two active learning strategies can be generated according to the output of these two modules. The model training process combined with active learning strategy is shown in Fig. 4.

The difference between the two strategies is the sample selection of the query function Q. The central part of model learning process is the updating process of labeled sample pool and unlabeled sample pool. The construction of labeled sample pool and unlabeled sample pool can be divided into two cases: initial construction and iterative update.

The same number of positive and negative samples are randomly selected as the initial labeled sample pool, on which the initial classifier model C is trained.

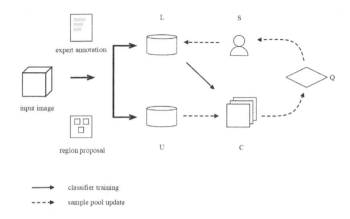

Fig. 4. Model training process based on active learning strategy

---

Algorithm 3. Iterative algorithm of sample pool

Input: L: labeled sample pool; U: unlabeled sample pool; C: classifier model; K: he number of samples to be selected;
Output: L: labeled sample pool; U: unlabeled sample pool; C: classifier model after iteration;
1: **function** UPDATE(L, U, C, Q)
2: $\quad samples = Q(C, U, K)$;
3: $\quad Update\_pool(L, samples, U)$;
4: $\quad C = Train(C, L)$;
5: **end function**

The iteration process of the Sample pool is shown in Algorithm 3.

The design of query function Q is different under two strategies. The following two sections will introduce it respectively.

Active Learning Strategy Based on Loss Prediction Module. Under the active learning strategy designed according to the output of the network LPM, the sample with the largest prediction value of the loss function is selected as the sample to be labeled. Thus, the design of the query function based on the active learning strategy designed by the LPM is shown in Algorithm 4.

Active Learning Strategy Based on Object Module Output. The samples with an output score of about 0.5 mean that the model has not been able to distinguish its features accurately. Therefore, in the active learning strategy, the samples that are difficult to be accurately classified by these models should become the samples that need to be labeled. The design of the query function based on the active learning strategy designed by the OBM is shown in Algorithm 5.

Algorithm 4. Q_{loss}

Input: U: unlabeled sample pool; C: classifier model; K: he number of samples to be selected;

Output: *samples*: labeled sample pool;

1: **function** Q_LOSS(C, U, K)
2:    *scores* = $C(U)$;
3:    *scores_sort* = $argsort(-scores)$;
4:    \\ All scores are sorted in descending order, and *argsort* returns the original array index value corresponding to each element after descending order.
5:    *samples* = *scores*[*scores_sort*[$0 : K + 1$]];
6: **end function**

Algorithm 5. Q_{obj}

Input: U: unlabeled sample pool; C: classifier model; K: he number of samples to be selected;

Output: *samples*: labeled sample pool;

1: **function** Q_OBJ(C, U, K)
2:    *scores* = $C(U)$;
3:    *scores_sort* = $argsort(abs(scores - 0.5))$;
4:    \\Sort all scores according to the distance of 0.5, and *argsort* returns the original array index value corresponding to each element after ascending sorting.
5:    *samples* = *scores*[*scores_sort*[$0 : K + 1$]];
6: **end function**

---

4   Experiment

4.1   Experimental Data and Environment

The experimental environment is Ubuntu 16.04 operating system, 11 GB memory GPU. All the processing programs use Python for experiments, and the deep learning part uses Keras framework. The image data used in this experiment is 80S ribosome of Saccharomyces cerevisiae, which can be obtained from EMPIAR library, No. EMPIAR-10045. It contains 7 reconstructed cryo-ET images and annotation files of particle positions in the images.

The original data set consists of 7 large-scale cryo-ET images and the annotation file of the particle center. Each image is less than 90G. We use the particles in the first five images as classifier training. After that, all proposal region images are extracted and saved from the preprocessed feature images according to the particle positions provided by the region proposal method. All these candidate images constitute the unlabeled sample pool in the active learning strategy. The last two images are used to test the whole particle selection process. Due to the lack of publicly available datasets in this domain, only a single dataset was used for the experiments.

4.2   Experimental Results and Analysis

Evaluation Index of Particle Selection. The results of particle selection are mainly evaluated by precision, recall and F1 score, and the calculation formula is as follows:

$$Precision = \frac{P_t}{P_t + N_f} \quad Recall = \frac{P_t}{P_t + P_f} \tag{4}$$

P_t is the number of correctly detected particles. P_f is the number of wrongly detected particles. N_f is the number of wrongly detected particles in the background image. If the deviation between the selected position and the real marked position is less than $1/4$ of the particle diameter, the selected position is considered to be correct.

Comparative Experiment. In order to analyze the influence of active learning strategy in the process of particle selection, the following four groups of experiments were carried out and the results were compared: (a)Only regional proposal results (BBRP) are used as the baseline; (b)Based on the region proposal method, the model without active learning strategy training is used as the classifier to generate the result (BBRP + C); (c)Based on the region proposal method, the model trained by the active learning strategy based on the loss function design is used as the classifier to generate the result (BBRP + C_{loss}); (d)Based on the region proposal method, the model trained by the active learning strategy based on the network output design is used as the classifier to generate the result (BBRP +C_{obj}). For the output of the classifier, 0.5 is used as the threshold value. If it is higher than 0.5, the output is considered to contain particles. Otherwise, it is considered not to contain particles, or the deviation of particle center is too far. The comparison of experimental results is shown in Table 1.

Table 1. Results of different experimental strategies.

Experimental method	Precision	Recall	F1-score
BBRP	0.4	0.83	0.54
BBRP+C	0.808	0.701	0.751
BBRP+C_{obj}	0.794	0.726	0.758
BBRP+C_{loss}	0.783	0.759	0.771

From the Table 1 we find that the comprehensive effect of particle selection has been improved after using the region proposal method and splicing classifier training, among which the BBRP+C_{loss} strategy can obtain the largest F1-score improvement. Therefore, it proves that the active learning strategy based on the LPM has the greatest effect on improving the classifier performance and the final particle selection effect.

In addition, according to the final total number of samples and the number of positive samples in the labeled sample pool when different active learning strategies are used for model training, and different thresholds are selected as the basis for output evaluation, the results are summarized as shown in Table 2.

Table 2. Comparison of detailed data under different training strategies.

Experimental method	Number of samples	Number of positive samples	Output threshold	Precision	Recall	F1-score
BBRP+C	200	100	0.2	0.789	0.759	0.77
			0.5	0.808	0.701	0.751
			0.8	0.817	0.603	0.694
BBRP+C_{obj}	300	156	0.2	0.775	0.774	0.774
			0.5	0.794	0.726	0.758
			0.8	0.819	0.663	0.733
BBRP+C_{loss}	300	170	0.2	0.74	0.8	0.769
			0.5	0.783	0.759	0.771
			0.8	0.802	0.7	0.748

From Table 2 we can find:

1. No matter which output threshold is used for truncation, BBRP+C_{loss} method can obtain the highest recall rate under the same threshold condition, which means that the active learning strategy based on the design of network loss prediction output module can retain the recall rate obtained by the region proposal method to the greatest extent. BBRP+C_{loss} method can ensure high recall rate and improve accuracy if the region proposal method uses a more comprehensive method.
2. Except for using 0.2 as the output threshold, BBRP+C_{loss} method has the highest F1-score under the same threshold, which means BBRP+C_{loss} has the best effect on improving the classifier and the result of particle selection task.
3. No matter which active learning strategy is used, the final effect of particle selection is improved. Compared with BBRP+C_{loss} method, BBRP+C_{obj} tends to obtain higher precision rate under the same output threshold. For particle selection task, the recall rate ensures that "enough comprehensive" particles can be found, which makes it possible to make more comprehensive analysis of samples. In the case of particle selection task, it is more important to ensure the recall rate, which means the active learning strategy based on LPM is better.

5   Conclusion

According to the characteristics of cryo-ET image, we try to use image threshold segmentation method for region proposal and residual neural network for further feature extraction of proposal region, so as to get the classification results for particle selection. At the same time, combined with the active learning strategy, the efficient use of tagging can effectively reduce the training cost and improve the performance of the model. It is a successful application of the object detection method in this large-scale, low signal-to-noise ratio data, and realizes the effective particle selection.

References

1. Chen, R., Jin, Yu., Xu, L.: A classroom student counting system based on improved context-based face detector. In: Wang, G., Lin, X., Hendler, J., Song, W., Xu, Z., Liu, G. (eds.) WISA 2020. LNCS, vol. 12432, pp. 326–332. Springer, Cham (2020). https://doi.org/10.1007/978-3-030-60029-7_30
2. Cheng, Y., Grigorieff, N., Penczek, P., Walz, T.: A primer to single-particle cryo-electron microscopy. Cell **161**(3), 438–449 (2015)
3. Heimowitz, A., Andén, J., Singer, A.: Apple picker: automatic particle picking, a low-effort Cryo-EM framework. J. Struct. Biol. **204**(2), 215–227 (2018)
4. Mastronarde, D.N., Held, S.R.: Automated tilt series alignment and tomographic reconstruction in IMOD. J. Struct. Biol., S1047847716301526 (2016)
5. Perutz, M.F., Kendrew, J.C., Watson, H.C.: Structure and function of haemoglobin: II. Some relations between polypeptide chain configuration and amino acid sequence. J. Mol. Biol. **13**(3), 669–678 (1965)
6. Punjani, A., Rubinstein, J.L., Fleet, D.J., Brubaker, M.A.: cryoSPARC: algorithms for rapid unsupervised Cryo-EM structure determination. Nat. Methods **14**(3), 290–296 (2017)
7. Redmon, J., Divvala, S., Girshick, R., Farhadi, A.: You only look once: unified, real-time object detection. IEEE (2016)
8. Scheres, S.: RELION: implementation of a Bayesian approach to Cryo-EM structure determination. J. Struct. Biol. **180**(3), 519–530 (2012)
9. Wang, F., et al.: DeepPicker: a deep learning approach for fully automated particle picking in Cryo-EM. J. Struct. Biol. **195**(3), 325–336 (2016)
10. Watson, J.D., Crick, F.: Molecular structure of nucleic acids: a structure for deoxyribose nucleic acid. Nature **248**(4), 623–624 (1953)

Application of Neural Network in Oracle Bone Inscriptions

Mengting Liu[1,2,3], Feng Gao[1,2,3], Bang Li[1,2,3], and Guoying Liu[1,2,3(✉)]

[1] School of Computer and Information Engineering, Anyang Normal University, Anyang 455000, Henan, China
guoying.liu@aynu.edu.cn
[2] Key Laboratory of Oracle Bone Inscriptions Information Processing, Ministry of Education of China, Anyang 455000, Henan, China
[3] Henan Key Laboratory of Oracle Bone Inscriptions Information Processing, Anyang 455000, Henan, China

Abstract. The recognition of Oracle Bone Inscriptions(OBIs) is of great significance to archaeology, history, and linguistics. To realize the fast and accurate retrieval of images for large-scale OBIs datasets and break through the limitations of current conventional retrieval methods, this paper proposes a convolutional neural network for OBIs recognition. The model is designed according to the characteristics of OBIs. The experimental results show that the improved network can better extract the features of OBIs characters, and the recognition rate reaches 84.45%, which is 13.74% higher than the network before the improvement.

Keywords: Convolutional neural network · Oracle bone inscriptions · Image recognition

1 Introduction

Oracle Bone Inscriptions(OBIs) refer to the characters carved on tortoise shells and animal bones for divination in ancient times. In the process of OBIs examination and interpretation, a large number of marked OBIs databases are needed, and manual marking takes time and effort, so the identification of OBIs characters is increasingly important.

At present, many scholars have used different methods to recognize OBIs. Zhou et al. proposed a method to identify the shape of OBIs using graph theory [2] and stroke characteristics. Li et al. proposed a method to identify the shape of the nail using the principle of graph features [3]. Li et al. proposed the method of graph isomorphism to identify the shape of OBIs [4]. Liu et al. proposed to use support vector machine classification technology to study the identification of OBIs [5]. Gu et al. proposed an identification method based on topological registration [6] and a method using fractal geometry [7] to identify the shape of the nail. Lu et al. [8] proposed a classification method characterized by Fourier descriptors of curvature histograms. However, the background noise of OBIs on oracle rubbings is large and there are many variant characters, which brings some difficulties to the image recognition of OBIs. The design and extraction of

C. Xing et al. (Eds.): WISA 2021, LNCS 12999, pp. 480–488, 2021.
https://doi.org/10.1007/978-3-030-87571-8_41

image features by the above methods depending on the experience of researchers and the features of current images, which affects the generality of the algorithm.The deep learning algorithm based on the convolutional neural network [9–12] has continuously made breakthroughs in large-scale competitions in the field of image classification and recognition and even exceeded the recognition accuracy of human beings in some fields. Bao et al. [13] proposed a method of name entity recognition in aircraft design field based on deep learning. It is possible to recognize OBIs characters using the powerful feature extraction capability of deep learning.

The remainder of this paper is organized as follows: Sect. 2 introduces the source and making method of the dataset used in this experiment; Sect. 3 introduces the proposed architecture of OBIs recognition; Sect, 4 provides the experimental results and analysis; Sect. 5 discusses some limitations and concludes the paper.

2 OBIs Dataset

All the experiments in this paper are from Jia Gu Wen He Ji, Jia Gu Wen He Ji Bu Bian, Dong Jing Da Xue Dong Yang Wen Hua Yan Jiu Suo Cang Jia Gu Wen Zi, Yin Xu Hua Yuan Zhuang Dong Di Jia Gu, Huai Te Shi Suo Cang Jia Gu Wen Ji, Rui Dian Si De Ge Er Mo Yuan Dong Gu Wu Bo Wu Guan Cang Jia Gu Wen Zi, Su De Mei Ri Suo Jian Jia Gu Ji, Tian Li Da Xue Fu Shu Tian Li Can Kao Guan Cang Jia Gu Wen Zi, Xiao Tun Nan DI Suo Cang Jia Gu, Ying Guo Suo Cang Jia Gu Ji, which are composed of OBIs characters and appear on all OBIs rubbings and are determined by experts.

These ten books were scanned electronically with a scanner (resolution 600 lb). As shown in Fig. 1, it is a page of images scanned from Jia Gu Wen He Ji Bu Bian. Then, clipping the scanned image. Figure 2 shows one of the rubbings cut from Fig. 1. Finally, cutting out OBIs characters from a single cut rubbings image. Figure 3 shows the cutting process of a single OBIs. Based on OBIs dictionary tool, the clipping function is added to it, which makes the dataset construction easier.

Fig. 1. A page in Jia Gu Wen He Ji Bu Bian.   **Fig. 2.** OBI rubbings exhibition.

There are 163 classes of OBIs characters, and 300 original pictures are selected from each category to form OBIs character data set OBIs163, it consists of a training set and a test set. Each class of the training set contains 250 pictures of original OBIs characters, and each class of the test set contains 50 pictures so that the number of pictures in the training set and the test set is 5:1. Figure 4 shows the sample data of OBIs character "San".

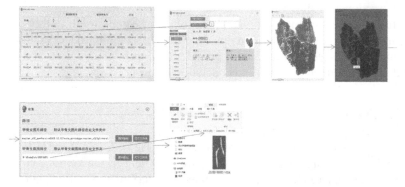

Fig. 3. Cutting process of OBIs.

Fig. 4. Dataset sample of OBIs "San".

3   Methodology

3.1   Architecture

Since alexnet directly refreshed the recognition rate of ImageNet in 2012, the network depth of subsequent models such as vgg, resnet and densenet has gradually increased, and it has been proved that the performance on the ImageNet dataset is getting better

Table 1. Structure of alexnet, vgg, resnet, densenet

Model	alexnet	vgg19	resnet18	densenet
Year	2012	2014	2015	2017
Input Size	224×224	224×224	224×224	224×224
Depth	8	19	18	121
Kernel	11,5,3	3	7,3,1	3,1

and better. In this paper, four classical convolutional neural networks, alexnet, vgg19, resnet18, and densenet, are experimented on OBIs dataset, their network model structure is shown in Table 1.

The trained recognition network is used to recognize the data in the OBI-CNN test set. The accuracy of the top five predictive recognition (Top-5) is alexnet > resnet18 > vgg19 > densenet. Therefore, this paper will take convolutional neural network as the basic theory and alexnet network architecture as the guiding ideology to realize an OBIs recognition network which is named OBI-CNN, the network model structure as shown in Fig. 5.

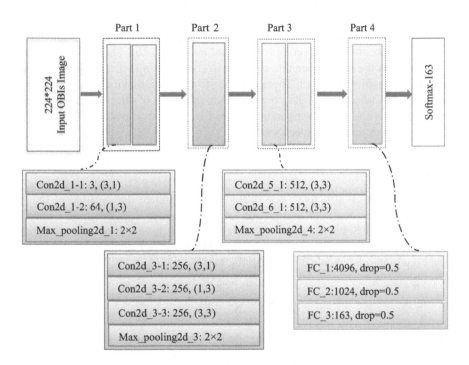

Fig. 5. The architecture of the OBI-CNN.

Part 1 includes two convolution layers and one pooling layer, the convolution kernel size of the first convolution layer is 3×1, the convolution kernel size of the second convolution layer is 1×3, and the step size is set to 1, which is activated by Relu function, and the maximum pooling output is used, and the step size is set to 2.

Part 2 includes three convolution layers and one pooling layer. The convolution kernel sizes of the first and second convolution layers are still 3×1 and 1×3, and the convolution kernel size of the third convolution layer is 3×3. BN batch normalization is added to improve the training effect and speed up the training, and maximum pooling is adopted.

Part 3 includes two convolution layers and one pooling layer. The convolution kernel is 3 × 3, the moving step is 1, and batch normalization processing is added to improve the training accuracy of the model, and the maximum pooling processing is used.

Part 4 is the full connection of the network, which has three full connection layers, mainly mapping the learned features to the sample space, and finally connecting with the Softmax classifier to realize the final OBIs character recognition.

It can be seen from these four parts that the OBI-CNN is designed according to the characteristics of OBIs. The convolution kernels of the first six convolution layers are all strip convolution kernels. Considering the font characteristics of OBIs characters, they are depicted by sharp tools, which are mostly stripped instead of square Chinese characters. When the network performs strip convolution, two convolution superpositions are performed. However, it should be noted that the superposition of 3 × 1 convolution kernel and 1 × 3 convolution kernel is not equal to a 3 × 3 convolution kernel. Experiments show that the recognition rate of the strip convolution kernel is higher than that of the square convolution kernel. Therefore, the strip convolution not only increases the depth of the network, which makes the data characteristics of Oracle characters image more obvious but also has fewer parameters than the square convolution kernel, which makes the network operation speed faster.

3.2 Loss Function

The essence of model structure optimization is to minimize the iteration of the loss function, and OBI-CNN uses Softmax to the cross-entropy loss function. Generally, in neural networks, Softmax, as the output layer of the classification task, outputs the probabilities of selecting several categories, and the sum of probabilities of all categories is 1, and the function formula is shown in Eq. (1):

$$S_i = \frac{e^{Z_i}}{\sum_k e^{Z_k}} \tag{1}$$

S_i represents the output of the i neuron, and k is the number of output nodes, so the output of neuron is shown in Eq. (2):

$$Z_i = \sum_j \omega_{ij} x_{ij} + b \tag{2}$$

ω_{ij} is the j weight of the i neuron, and b is the offset, which indicates the i output of the network. Adding a softmax function to the output becomes Eq. (3):

$$A_i = \frac{e^{Z_i}}{\sum_k e^{Z_k}} \tag{3}$$

A_i represents the i output value of Softmax.

The loss function is used to calculate the inconsistency between the predicted value and the true value of the network model. In this paper, the cross-entropy function is used. The derivative result is simple and easy to calculate. The function formula is shown in Eq. (4):

$$C = -\sum_i y_i \ln A_i \qquad (4)$$

In which y_i represents the true classification result.

4 Experments

4.1 Implementation Details

In the training stage, we train the proposed model with batch size 64 with tensors resized at 224*224 on GPU for 40 epochs, and the initial learning rate is set to 0.01. We use a weight decay of 5e-4 and a Nesterov momentum of 0.9, All the networks are optimized by using stochastic gradient descent(SGD), we adopt the weight initialization introduced by [14].

4.2 Evaluation Index

In this experiment, the accuracy of the test set-top(k) is used to evaluate the model. The calculation formula is shown in Eq. (5):

$$top(k) = \frac{r}{a} \qquad (5)$$

In the formula, the value of k is 1,2,... n, n is the number of tree species images, a represents the total number of model test set images, and r represents the number of images correctly tested in the first k results predicted by the model.

4.3 Performance Evaluation

Figure 6 is the curve of recognition rate and loss value of different network models changing with epoch. It can be seen that there is no serious oscillation phenomenon in the training process of each network, and all of them have converged. Among them, the OBI-CNN network has the best performance because it has the highest recognition rate and the smallest loss value.

Figure 7 is the training time diagram of different networks. In the deep learning model, the more layers of the network, the more computation, and the longer the training time.However, the training time of OBI-CNN proposed in this paper is 5 h and 19 min, which is lower than that of alexnet, but the network depth is 19, which is greater than that of it.

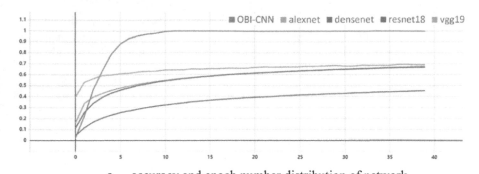

a.   accuracy and epoch number distribution of network

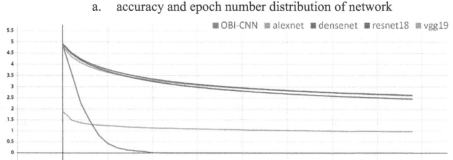

b.   loss and epoch number distribution of network

Fig. 6. Accuracy and loss of epoch number distribution of model.

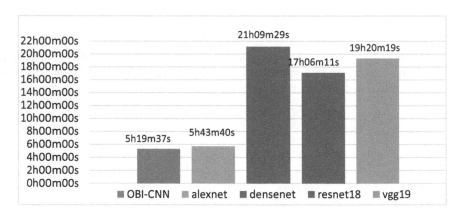

Fig. 7. Network training time column chart.

The recognition results of different networks on the test set are shown in Table 2. It can be seen that the recognition accuracy of OBI-CNN designed in this paper in Top-1 and Top-5 is higher than that of the other four classical convolutional neural networks. Among them, the Top-5 recognition accuracy of OBI-CNN reached 84.45%, which was higher than other networks and 13.74% higher than the second-ranked alexnet network. The OBI-CNN network is optimized and improved based on alexnet network, which increases the network depth without increasing the training time and improves the recognition accuracy, so the result is satisfactory.

Table 2. Test results of different networks

Model	Top-1 accuracy(%)	Top-5 accuracy(%)
Alexne	46.94	70.71
vgg19	31.04	56.71
resnet18	30.83	55.17
Densenet	28.12	46.75
OBI-CNN	65.64	84.45

5 Conclusion

In this paper, we propose a network of OBIs recognition named OBI-CNN. It changes the square convolution kernel into a strip convolution kernel according to the characteristics of OBIs and superimposes the feature images obtained from two strip convolution layers, which makes the extracted OBIs feature more remarkable. The experimental results fully prove that the OBI-CNN proposed in this paper greatly shortens the training time of the algorithm, improves the correct rate of OBIs163 classification, and speeds up the speed of image classification at the same time.

Acknowledgment. The authors acknowledge the supports from the National Natural Science Foundation of China (No. U1804153, 61806007), the Oracle Bone Inscriptions Research and Application Special Project of Ministry of Education and National Language Committee of China (No. YWZ-J010, YWZ-J023), the National Social Science Fund Major Entrusted Project of China (No. 16@ZH017A3), the Program for Changjiang Scholars and Innovative Research Team in University (No. 2017PT35), the Development Projects of Henan Province Science and Technology(No. 202102310562).

References

1. LeCun, Y., Boser, B., Denker, J.S., et al.: Backpropagation applied to handwritten zip code recognition. Neural Comput. **1**(4), 541–551 (1989)
2. Krizhevsky, A., Sutskever, I., Hinton, G.E.: Imagenet classification with deep convolutional neural networks. Adv. Neural. Inf. Process. Syst. **25**, 1097–1105 (2012)
3. Simonyan, K., Zisserman, A.: Very deep convolutional networks for large-scale image recognition. arXiv preprint arXiv:1409.1556 (2014)
4. Szegedy, C., Liu, W., Jia, Y., et al.: Going deeper with convolutions. In: Proceedings of the IEEE Conference on Computer Vision and Pattern Recognition, pp. 1–9 (2015)
5. He, K., Zhang, X., Ren, S., et al.: Deep residual learning for image recognition, pp. 770–778 (2015)

6. Gu, J., Wang, Z., Kuen, J., et al.: Recent advances in convolutional neural networks. Pattern Recogn. **77**, 354–377 (2018)
7. Zhou, X.L., Hua, X.C., Li, F.: A method of Jia Gu Wen recognition based on a two-level classification. In: Proceedings of 3rd International Conference on Document Analysis and Recognition. (2), 833–836 IEEE (1995)
8. Jiang, M., Deng, B., Liao, P., et al.: Construction on word-base of oracle-bone inscriptions and its intelligent repository. Comput. Eng. Appl. **40**(4), 45–47 (2004)
9. Shao-tong, G., Xiao-hu, M., Yi-ming, Y.: Topological frame based input method coding of Jiaguwcn. J. Chin. Inf. Proc. **22**(4), 123–128 (2008)
10. Qingsheng, Y.Y., Aimin, W.: Recognition of inscriptions on bones or tortoise shells based on graph isomorphism. Comput. Eng. Appl. **47**(8), 112–114 (2011)
11. Gao Feng, W., Qinia, L., et al.: Recognition of fuzzy inscription character based on component for bones or tortoise shells. Sci. Technol. Eng. **14**(30), 67–70 (2014)
12. Guo, J., Wang, C., Roman-Rangel, E., et al.: Building hierarchical representations for oracle character and sketch recognition. IEEE Trans. Image Process. **25**(1), 104–118 (2016)
13. Bao, Y., An, Y., Cheng, Z., Jiao, R., Zhu, C., Leng, F., Wang, S., Wu, P., Yu, G.: Named entity recognition in aircraft design field based on deep learning. In: Wang, G., Lin, X., Hendler, J., Song, W., Xu, Z., Liu, G. (eds.) WISA 2020. LNCS, vol. 12432, pp. 333–340. Springer, Cham (2020). https://doi.org/10.1007/978-3-030-60029-7_31
14. Dress, A., Grünewald, S., Zeng, Z.: A cognitive network for oracle bone characters related to animals. Int. J. Mod. Phys. B **30**(4), 1630001 (2016)

Improving Text Summarization Using Feature Extraction Approach Based on Pointer-generator with Coverage

Yongchao Chen[1], Xin He[2(✉)], Guanghui Wang[2], and Junyang Yu[1]

[1] Henan Provincial Engineering Research Center of Intelligent Data Processing, Kaifeng, China
[2] Henan International Joint Laboratory of Intelligent Network Theory and Key Technology, Henan University, Kaifeng, China
hxsyjkf@foxmail.com

Abstract. The sequence-to-sequence models have been widely used in the abstractive summarization tasks. However, the existing models have inaccuracy issues due to the insufficient understanding of overall semantics, the emergence of Out Of Vocabulary (OOV) words and the duplication of generated summarizations. In order to solve the above issues, in this paper, we propose an automatic text summarization model through combining the feature extraction, pointer-generator and coverage mechanism. First, in order to obtain input text with the multiple semantics, a feature extraction method is proposed to extract hidden features in the source text. Then, the pointer-generator (PGN) is used to select words from the source text or vocabulary to form a summarization, which solves the problem of OOV words in the generated summarizations. Finally, the coverage mechanism is applied to reduce the duplication of generated summarizations. The experimental results show that the model achieves better performance on the real-world CNN/Daily Mail summarization dataset.

Keywords: Abstractive summarization · Sequence-to-sequence · Feature extraction · Coverage mechanism

1 Introduction

With the rapid development of the Internet, the information society is filled with a large amount of text data. New media has become main channels to obtain information. However, the obtained information is often accompanied with redundant and worthless text. Automatic text summarization [1] is proposed to process information.

The Sequence-to-Sequence (Seq2Seq) model was used to text summarization and achieved remarkable results. The source content is expressed in more concise text. The automatic summarization based on Seq2Seq still has the following issues: the insufficient understanding of overall semantics, the emergence of OOV words and the duplication of generated summarizations.

The low global relevance of the generated summarization is due to the lack of multiple semantic features during training. The relationship between adjacent words is not considered when encoding. The size of the vocabulary and the generated words may not be in the source text, resulting in OOV words in the generated abstract [2]. In the test phase, the attention mechanism may reuse the reference from the previous phase [3], which causes the problem of duplication.

In order to address the above problems, we propose an automatic text summarization model (SumCNN-PC) based on the SumCNN (Summarization Convolutional Neural Network) method, pointer-generator, and coverage mechanism. We use SumCNN for feature extraction before encoding to improve the global relevance of the generated summarization. The pointer-generator and coverage mechanisms are added during training to solve the problem of OOV words and repetitions in the abstract. The main contributions of this paper are as follows:

- This paper proposes a text summarization model that combining the feature extraction, pointer-generator and coverage mechanism. The model is divided into two parts. The first part extract features from the source text. The second part inputs the result of feature extraction into the pointer-generator network to complete summary generation.
- We propose a feature extraction method suitable for text summarization. In the decoding process, pointer network and coverage mechanism are introduced. The pointer is used to deal with OOV words, and the coverage mechanism solves the problem of duplication in the abstracts.
- The experimental results show that the model SumCNN-PC has achieved better performance using the CNN/Daily dataset, in terms of the indicators of ROUGE-1, ROUGE-2, and ROUGE-L.

2   Related Work

The Seq2Seq model was first proposed by Cho *et al.* [4] and Sutskever *et al.* [5] in 2014. The Seq2Seq model combined with the attention mechanism was first used in the field of text summarization by Rush *et al.* [6]. This model generates summarization on the basis of understanding the source text. [7] also adopted the seq2seq model to achieve cross-language summarization.

Because of the obvious limitation of the dictionary size, OOV will appear after the summary is generated. Gu *et al.* [2] proposed the COPYNEY model to solve this problem. In addition, See *et al.* [8] proposed the PGN network model. It has outstanding effects after adding the coverage mechanism. Xie *et al.* [9] uses a combination of extractive and abstractive, and also introduced PGN and coverage mechanisms. [10] uses reinforcement learning as a method to optimize the objective function.

In recent years, Convolutional Neural Network (CNN) has achieved obvious success in natural language processing (NLP). More contextual features can be extracted from the text, so that the model training effect is better.

Based on the above research, we have made improvements on the source CNN to make it a feature extraction method suitable for text summarization, and is called Sum-CNN. SumCNN with pointer-generator network, and introduce a coverage mechanism to propose a text summarization model based on sequence-to-sequence model.

3   Our Models

In this section, we present a text summarization model combining the feature extraction, PGN and coverage mechanism, which framework is shown in Fig. 1.

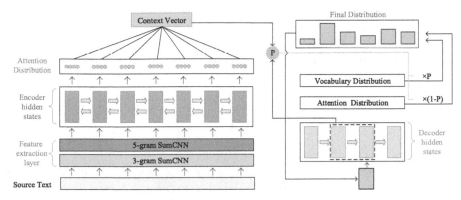

Fig. 1. SumCNN-PC model.

3.1   SumCNN

In order to solve the problem of low global relevance in the summarization, we introduce the CNN into baseline model. We improved the CNN into a method suitable for text summarization, which is called SumCNN. Figure 2 is SumCNN model.

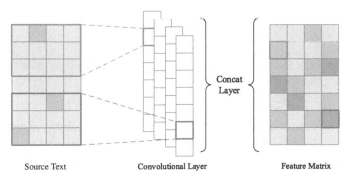

Fig. 2. SumCNN.

SumCNN contains two convolutional layers and two splicing layers. The convolutional layer is used for feature extraction. The splicing layer is used to splice the feature vectors. The number of convolution kernels is set to the size of the word vector.

3.2 Seq2Seq Attentional Model

Our Seq2Seq is similar to the model proposed by Nallapati [11]. The difference is that the encoder uses Bi-LSTM and the decoder uses LSTM.

In the encoding stage, the encoder be used to generate the hidden layer state h_i:

$$h_i = \left[\overrightarrow{h}_i; \overleftarrow{h}_i \right] \tag{1}$$

where h_i is formed by splicing $\overrightarrow{h}_i$ and $\overleftarrow{h}_i$, $\overrightarrow{h}_i$ represents the forward hidden layer state, and $\overleftarrow{h}_i$ represents the reverse hidden layer state. The generating formulas are:

$$\overrightarrow{h}_i = LSTM\left(\overrightarrow{h}_{i-1}, x_i\right) \tag{2}$$

$$\overleftarrow{h}_i = LSTM\left(\overleftarrow{h}_{i-1}, x_i\right) \tag{3}$$

where x_i represents the ith word vector, $i \in [1, m]$.

At the tth step of decoding, the decoder accepts the hidden state of the previous moment and the previous word embedding. It is used to generate current decoder hidden layer state s_t. The calculation process of attention distribution a^t is:

$$e_i^t = v^T \tanh(W_h h_i + W_s s_t + b_{attn}) \tag{4}$$

$$a^t = \frac{\exp\left(e_i^t\right)}{\sum\limits_{j=1}^{m} \exp\left(e_i^t\right)} \tag{5}$$

where m is the length of the input sequence, and v, W_h, W_s, b_{attn} are all trainable parameters. A weighted sum of the hidden layer state of the encoder with attention weight is generated, which is called the context vector C_t:

$$C_t = \sum\limits_{i=1}^{m} a_i^t h_i \tag{6}$$

The context vector at that moment and the state of the hidden layer of the decoder are used to generate the vocabulary distribution P_{vocab}:

$$P_{vocab} = softmax\left(V'(V[s_t; C_t] + b) + b'\right) \tag{7}$$

where V, V', b, b' are learnable parameters, P_{vocab} is the probability distribution of all words in the vocabulary. Finally, predict the final distribution of word w:

$$P(w) = P_{vocab}(w) \tag{8}$$

The loss function at time t is set as:

$$loss_t = -logP\left(w_t^*\right) \tag{9}$$

where w_t^* represents the target vocabulary at time t. The entire loss is:

$$loss = \frac{1}{T} \sum\limits_{t=0}^{T} loss_t \tag{10}$$

3.3 Pointer Generation Network

PGN network incorporates a pointer on Seq2Seq model. We set a pointer P_{gen}, and $P_{gen} \in [0, 1]$. The pointer is calculated by the C_t, h_i and the s_t:

$$p_{gen} = \sigma\left(W_C^T C_t + W_h^T h_i + W_s^T s_t + b_{ptr}\right) \tag{11}$$

where $W_C^T, \cdot W_h^T, \cdot W_s^T, \cdot b_{ptr}$ are all parameters that can be trained. P_{gen} is used as a selector to determine whether the generated word is generated from the vocabulary:

$$P(y) = p_{gen}P_{vocab}(w) + \left(1 - p_{gen}\right) \sum_{i:w_i=w} a_i^t \tag{12}$$

When generating the target vocabulary, if the generated word at the current moment is OOV, $P_{gen} = 0$. If the generated word w is not in the source text, then $\sum_{i:w_i=w} a_i^t$ is zero.

3.4 Coverage Mechanism

In order to solve repetition problem, the coverage mechanism was introduced.
First, we add up all the attention distributions before time t:

$$A^t = \sum_{t'=0}^{t-1} a^{t'} \tag{13}$$

When $t = 0$, A^0 is a zero vector, the formula (4) is updated to:

$$e_i^t = v^T \tanh\left(W_h h_i + W_s s_t + W_A A^t + b_{attn}\right) \tag{14}$$

A loss is additionally defined for the repetition of the summary:

$$covloss_t = \sum_i \min\left(a_i^t, c_i^t\right) \tag{15}$$

then the loss function is redefined as:

$$loss = -logP\left(w_t^*\right) + \lambda covloss_t \tag{16}$$

4 Experimental Analysis

4.1 Dateset

The CNN/Daily dataset is used as our evaluation dataset. This dataset was originally collected from about 1 million pieces of news data from CNN and Daily Mail as machine reading comprehension corpus [12]. It contains 286817 training pairs, 13368 verification pairs and 11487 test pairs.

4.2 Evaluation Metric

The evaluation index of model is the same as that of Nallapati *et al.* [11], using the standard ROUGE metric [13] to evaluate trained model. We calculate the F1 value in ROUGE-1, ROUGE-2, and ROUGE-L. It is to measure the"similarity" between the summarization generated by our model and the standard summarization.

4.3 Parameter Setting

Because the PGN network has the advantage of processing OOV words, the size of our vocabulary is set to 50000. The dimension of word vector is set to 128. The dimensions of the hidden layer vector is set to 256. The initial learning rate is set to 0.15, and the initial accumulator is set to 0.1. The two-layer convolution kernel size of the convolution layer of SumCNN is set to 3 and 5 respectively. The number of convolution kernels is set to 128. The convolution step size is set to 1.

During the experiment, we train on an NVIDIA Tesla P100 GPU. The batch size is set to 16. Only the 400 words at the beginning of the article were intercepted as the source text. The length of the generated summarization is limited to 100.

4.4 Experimental Results

We compare our proposed method with other predominant models. They are mentioned in related work.

Table 1. Comparison of experimental results

Model	ROUGE F1 scores		
	1	2	L
Seq2Seq+Attention	31.33	11.81	28.83
Abstractive model	35.46	13.30	32.65
Pointer-generator	36.44	15.66	33.42
Pointer-generator+coverage	39.53	17.28	36.38
WordNet+Dual-attn+PGN+Cov	39.32	17.15	36.02
ML+RL+Intra-attn	39.87	15.82	34.41
SumCNN-PGN (Ours)	37.96	16.58	34.97
SumCNN-PC (Ours)	**39.67**	**17.46**	**36.39**

As shown in the experimental results shown in Table 1. Compared to baseline, the effect of our models has been greatly improved. SumCNN-PGN is improved on the pointer-generator. According to the results, compared to the pointer-generator, the ROUGE-1 score of SumCNN-PGN increased by 1.52% points. The ROUGE-2 score increased by 0.82% points. The ROUGE-1 score increased by 1.55% points. After the coverage mechanism was introduced on the basis of the SumCNN-PGN model, ROUGE score has been improved again. Compared to pointer-generator + coverage, ROUGE-1, ROUGE-2, ROUGE-1 have been upgraded.

Comparing SumCNN-PC with the WordNet + Dual-attn + PGN + Cov, ROUGE-1 score is 0.35 percentage points higher. The ROUGE-2 score is 0.31 percentage points higher. The ROUGE-1 score is 0.94 percentage points higher. Compared with the ML + RL + Intra-attn, although our ROUGE-1 and ROUGE-1 scores are not prominent, it is 1.71 higher than the ROUGE-2 of the model using reinforcement learning.

Figure 3 shows a sample summary generated by the SumCNN-PC. It can be seen from Fig. 3 that the summarization generated by the model has high readability. Without repetition and OOV words in the sentence.

Source Text:
no need to tell padraig harrington what a difference a year makes . twelve months ago he stood on the same spot on the same practice ground at the same tournament , the houston open , looking as crestfallen as i have ever seen him . now look at him . there 's an obvious contentment as he hits a few shots and breaks off to chat with his long-time caddie , ronan flood . he 's taking in the surroundings with the grateful bearing of a man handed an unlikely lifeline . in short , here 's a three-time major champion who 's got his mojo back . three time major winner padraig harrington has rediscovered his form ahead of the masters . the irishman practices during the pro-am before the start of the houston open in texas . the small miracles that occur in this game . a month ago harrington had fallen outside the world 's top 300 when he accepted an invitation to play in the honda classic in florida . he ended up winning his first tournament in america since the 2008 uspga championship .

Reference summarization:
padraig harrington had fallen out of top 300 before winning honda classic .irishman returns to the masters after failing to qualify last year .the three times major winner tips dustin johnson to challenge rory mcilroy at augusta .

SumCNN-PC (Ours) summarization:
padraig harrington stood on the same spot on the same practice ground .there 's an obvious contentment as he hits a few shots and breaks off to chat with his long-time caddie , ronan flood .three time major winner padraig harrington had fallen outside the world 's top 300 when he accepted an invitation to play in the honda classic .

Fig. 3. Examples of summaries generated by SumCNN model

---

5   Conclusion

In this paper, aiming at the problems of low global semantic relevance, OOV and repetition of summarization, a automatic text summarization model is proposed. This model combines the attention mechanism with the proposed SumCNN method. Then, the PGN and the coverage mechanism was added. Finally, better results was achieved on the CNN/Daily dataset. In the next step, we plan to improve the versatility of the model to language and improve it into a model that is suitable for Chinese datasets.

Acknowledgements. This work was supported in part by the Key Technologies Research and Development Program of Henan under Grant (212102210078), the Elite Postgraduate Students Program of Henan University (SYL20060174), and Major public welfare projects in Henan under Grant (201300210400).

References

1. Gambhir, M., Gupta, V.: Recent automatic text summarization techniques: a survey. Artif. Intell. Rev. **47**(1), 1–66 (2016). https://doi.org/10.1007/s10462-016-9475-9
2. Gu, J., Lu, Z., Li, H., Li, V.O.: Incorporating copying mechanism in sequence-to-sequence learning. arXiv preprint arXiv:1603.06393 (2016)
3. Tu, Z., Lu, Z., Liu, Y., Liu, X., Li, H.: Modeling coverage for neural machine translation. arXiv preprint arXiv:1601.04811 (2016)
4. Cho, K., Van Merriënboer, B., Gulcehre, C., Bahdanau, D., Bougares, F., Schwenk, H., Bengio, Y.: Learning phrase representations using RNN encoder-decoder for statistical machine translation. arXiv preprint arXiv:1406.1078 (2014)
5. Sutskever, I., Vinyals, O., Le, Q.V.: Sequence to sequence learning with neural networks. arXiv preprint arXiv:1409.3215 (2014)
6. Rush, A.M., Chopra, S., Weston, J.: A neural attention model for abstractive sentence summarization. arXiv preprint arXiv:1509.00685 (2015)
7. Yang, F., Cui, R., Yi, Z., Zhao, Y.: Cross-language generative automatic summarization based on attention mechanism. In: Wang, G., Lin, X., Hendler, J., Song, W., Xu, Z., Liu, G. (eds.) WISA 2020. LNCS, vol. 12432, pp. 236–247. Springer, Cham (2020). https://doi.org/10.1007/978-3-030-60029-7_22
8. See, A., Liu, P.J., Manning, C.D.: Get to the point: summarization with pointer-generator networks. arXiv preprint arXiv:1704.04368 (2017)
9. Xie, N., Li, S., Ren, H., Zhai, Q.: Abstractive summarization improved by WordNet-based extractive sentences. In: Zhang, M., Ng, V., Zhao, D., Li, S., Zan, H. (eds.) NLPCC 2018. LNCS (LNAI), vol. 11108, pp. 404–415. Springer, Cham (2018). https://doi.org/10.1007/978-3-319-99495-6_34
10. Paulus, R., Xiong, C., Socher, R.: A deep reinforced model for abstractive summarization. arXiv preprint arXiv:1705.04304 (2017)
11. Nallapati, R., Zhou, B., Gulcehre, C., Xiang,B., et al.: Abstractive text summarization using sequence-to-sequence rnns and beyond. arXiv preprint arXiv:1602.06023 (2016)
12. Hermann, K.M., Kočiský, T., Grefenstette, E., et al.: Teaching machines to read and comprehend. arXiv preprint arXiv:1506.03340 (2015)
13. Lin, C.Y.: Rouge: a package for automatic evaluation of summaries. In: Text Summarization Branches Out, pp. 74–81 (2004)

Mixed Multi-channel Graph Convolution Network on Complex Relation Graph

Chengzong Li[1], Rui Zhai[1], Fang Zuo[2,3(✉)], Junyang Yu[1], and Libo Zhang[1]

[1] Henan Provincial Engineering Research Center of Intelligent Data Processing, Kaifeng, China
[2] Henan International Joint Laboratory of Intelligent Network Theory and Key Technology, Henan University, Kaifeng, China
zuofang@henu.edu.cn
[3] Subject Innovation and Intelligence Introduction Base of Henan Higher Educational Institution-Software Engineering Intelligent Information Processing Innovation and Intelligence Introduction, Base of Henan University, Kaifeng, China

Abstract. Graph Convolutional Networks (GCNs) have attracted increasing attention on network representation learning because they can generate node embeddings by aggregating and transforming information within node neighborhoods. However, the classification effect of GCNs is far from optimal on complex relationship graphs. The main reason is that classification rules on relationship graphs rely on node features rather than structure. most of the existing GCNs do not preserve the feature similarity of the node pair, which may lead to the loss of critical information. To address these issues, this paper proposes a node classification network based on semi-supervised learning: Mif-GCN. Specifically, we propose a mixed graph convolution module to adaptively integrate the adjacency matrices of the initial graph and the feature graph in order to explore their hidden information. In addition, we use an attention mechanism to adaptively extract embeddings from the initial graph, feature graph, and mixed graph. The ultimate goal of the model is to extract relevant information for improving classification accuracy. We validate the effectiveness of Mif-GCN on multiple datasets, including paper citation networks and relational networks. The experimental results outperform the existing methods, which further explore the classification rules.

Keywords: Graph convolutional networks · Semi-supervised classification · Deep learning

1 Introduction

Graphs are common data structure that interpret the interrelationship of objects, and real-world data is increasingly represented as graph data. The purpose of graph analysis is to extract a large amount of hidden information from data, thus deriving many important tasks. The purpose of semi-supervised node classification task is to use a small number of labeled nodes to predict the labels of a large number of unlabeled nodes.

In recent years, graph neural networks have shown remarkable results in node representation learning, and GCN [9] being one of the most popular models. Existing GCNs

© Springer Nature Switzerland AG 2021
C. Xing et al. (Eds.): WISA 2021, LNCS 12999, pp. 497–504, 2021.
https://doi.org/10.1007/978-3-030-87571-8_43

models often follow the message-passing manner to obtain the embedding of all nodes by aggregating the node features within the node neighborhood. However, recent studies have provided some new insights into the mechanisms of GCNs. Li et al. [5] prove that the convolution operation of GCNs is actually a kind of Laplacian smoothing, and explores the multi-layer structure. Wu et al. [6] show that graph structure as low-pass-type filter on node features. Zheng et al. [12] utilize the Louvain-variant algorithm and jump connection to obtain different granularity of node features and graph structure information for adequate feature propagation.

Most of the current GCNs rely on the structural information of graph, which can seriously affect the model performance when the structure of graph is not optimal. The reason may be that most of the current GCNs rely on the structural information of the graph to classify. In this work, we retain graph structure information and feature similarity, consider the influence of the mixed information of the two on the results, and introduce the attention mechanism for classification.

2   Related Work

The first representative work of applying deep learning models to graph structure data is network embedding, which learns fixed-length representations for each node by constraining the proximity of nodes, such as DeepWalk [1], Node2Vec [2]. After that, the research defines graph convolution in the spectral space based on convolution theorem. ChebNet [3] parameterize the convolution kernel in the spectral approach, which avoids the characteristic decomposition of the Laplace matrix and greatly reduces the complexity. Kipf [9] simplifie the ChebNet parameters and proposes a first-order graph convolution neural network. SGC [4] put the first-order graph convolution neural introduces the first-order approximation of Chebyshev polynomials, which further simplifies the mechanism. With the increasing importance of attention mechanism, it has a considerable impact in various fields. GAT [7] leverage the attention mechanism to learn the relative weights between two connected nodes. GNNA [14] improve the embeddedness of users and projects by introducing attention mechanisms and high-level connectivity.

Some experiments show that the graph neural network is inefficient when the graph structure is noisy and incomplete. kNN-GCN [8] take the k-nearest neighbor graph created from the node features as input. DEMO-Net [13] formulate the feature aggregation into a multi-task learning problem according to nodes' degree values. AM-GCN [10] use consistency constraints and disparity constraints to obtain information and improves the ability to fuse network topology and node features. SimP-GCN [11] make use of self-supervised methods for node classification and explore the different effects of GCNs on assortative and disassortative graphs. Hence, it motivates us to develop a new approach that combines graph structure and feature similarity on the node classification task.

3   The Proposed Model

The key point is that Mif-GCN license node features to propagate in both structure space and feature space, because we consider that the classification rules of datasets depend on graph structure or feature similarity. We propose a mixed graph convolution module to adaptively balance the mixed information from initial graph and feature graph, which may contain hidden information. This module is designed to explore depth-related information.

3.1 Multi-Channel Convolution Module

Feature Graph Construction. We first calculate the cosine similarity between the features of each node pair and construct the similarity matrix $S \in \mathbb{R}^{n \times n}$. Then, we select predetermined number of nodes to establish the connection relationship, in order to obtain a k-nearest neighbor (kNN) graph $G_f = (A_f, X)$.

The cosine similarity can be calculated from the cosine of the angle between two vectors:

$$S_{jk} = \frac{x_j \cdot x_k}{|x_j| \, |x_k|} \tag{1}$$

Where X_j and X_k are the feature vectors of nodes j and k Fig. 1.

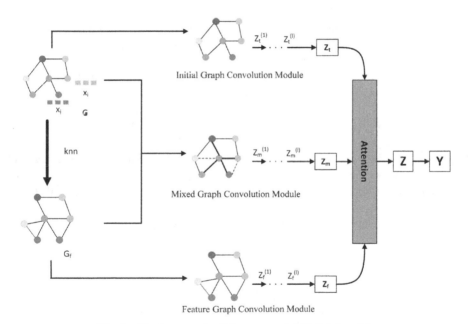

Fig. 1. The framework of the proposed Mif-GCN model.

Initial Graph Convolution Module and Feature Graph Convolution Module. In this module, node features are propagated in structure space and feature space separately so that we can distinguish correlations. Both modules use the typical GCN to get the node representation.

Then with the initial graph $G_i = (\mathbf{A}_i, \mathbf{X})$, the l-th layer output $\mathbf{Z}_i^{(l)}$ can be represented as:

$$\mathbf{Z}_i^{(l)} = ReLU\left(\tilde{\mathbf{D}}_i^{-\frac{1}{2}}\tilde{\mathbf{A}}_i\tilde{\mathbf{D}}_i^{-\frac{1}{2}}\mathbf{Z}_i^{(l-1)}\mathbf{W}_i^{(l)}\right) \tag{2}$$

Where $\mathbf{W}_i^{(l)}$ is the weight matrix of the l-th layer, $\tilde{\mathbf{A}}_i$ is the adjacency matrix of the initial graph of the added self-loop and $\tilde{\mathbf{D}}_i$ is the diagonal matrix of $\tilde{\mathbf{A}}_i$. We have $\tilde{\mathbf{A}}_i = \mathbf{A}_i + \mathbf{I}, \mathbf{A}_i$ is the adjacency matrix of the initial graph, $\mathbf{I}$ is the identity matrix. $ReLU$ is the ReLU activation function and $\mathbf{Z}_i^{(0)} = \mathbf{X}$.

Then with the feature graph $G_f = (\mathbf{A}_f, \mathbf{X})$, the l-th layer output $\mathbf{Z}_f^{(l)}$ can be represented as:

$$\mathbf{Z}_f^{(l)} = ReLU\left(\tilde{\mathbf{D}}_f^{-\frac{1}{2}}\tilde{\mathbf{A}}_f\tilde{\mathbf{D}}_f^{-\frac{1}{2}}\mathbf{Z}_f^{(l-1)}\mathbf{W}_f^{(l)}\right) \tag{3}$$

Where $\mathbf{A}_f$ is the adjacency matrix of the feature graph, $\mathbf{Z}_f^{(0)} = \mathbf{X}$, other parameters are the same as Eq. (2).

Mixed Graph Convolution Module. This module can mix information from the initial and feature graphs. We define a goal vector to balance the degree of contribution from the two graphs. Then, this module gets the node embedding.

Then with the initial graph and the feature graph, the l-th layer output of the mixed propagation matrix $\mathbf{P}^{(l)}$ can be represented as:

$$\mathbf{P}^{(l)} = \mathbf{g}^{(l)} * \tilde{\mathbf{D}}_i^{-\frac{1}{2}}\tilde{\mathbf{A}}_i\tilde{\mathbf{D}}_i^{-\frac{1}{2}} + \left(1 - \mathbf{g}^{(l)}\right) * \tilde{\mathbf{D}}_f^{-\frac{1}{2}}\tilde{\mathbf{A}}_f\tilde{\mathbf{D}}_f^{-\frac{1}{2}} \tag{4}$$

Where $\mathbf{g}^{(l)} \in \mathbb{R}^n$ is the goal vector that balances the information from the initial and feature graphs, $'*'$ denotes the operation of multiplying the j-th element of a vector with the j-th row of a matrix. $\mathbf{g}^{(l)}$ can combine information from two graphs to different degrees.

Then the l-th layer output of goal vector $\mathbf{g}^{(l)}$ can be represented as:

$$\mathbf{g}^{(l)} = \sigma\left(\mathbf{Z}_i^{(l-1)}\mathbf{W}_g^{(l)} + \mathbf{b}_g^{(l)}\right) \tag{5}$$

Where $\mathbf{Z}_i^{(l-1)} \in \mathbb{R}^{n \times d^{(l-1)}}$ is the input hidden representation of the previous layer, and $\mathbf{Z}_i^{(0)} = \mathbf{X}$. $\mathbf{W}_g^{(l)} \in \mathbb{R}^{d^{(l-1)} \times 1}$ and $\mathbf{b}_g^{(l)}$ are parameters in order to calculate $\mathbf{g}^{(l)}$. σ denotes the sigmoid activation function.

Then with the mixed propagation matrix $\mathbf{P}^{(l)}$, the l-th layer output $\mathbf{Z}_m^{(l)}$ can be represented as:

$$\mathbf{Z}_m^{(l)} = \sigma\left(\mathbf{P}^{(l)}\mathbf{Z}_m^{(l-1)}\mathbf{W}_m^{(l)}\right) \tag{6}$$

Where $\mathbf{P}^{(l)}$ denotes the mixed propagation matrix and $\mathbf{Z}_m^{(0)} = \mathbf{X}, \mathbf{W}_m^{(l)}$ is the parameter, here σ denotes the sigmoid activation function.

3.2 Attention Mechanism

We use the attention mechanism to combine to get the final node embedding. The ultimate goal is to explore the most relevant information in the three embeddings. $\mathbf{Z}_I, \mathbf{Z}_F, \mathbf{Z}_M$ contain the embedding of n nodes,$\alpha_i, \alpha_f, \alpha_m$ represent the respective attention values.

For node j,$\mathbf{z}_F^j \in \mathbb{R}^{1 \times h}$ denotes the embedding of node j in $\mathbf{Z}_F$. Then use a weight vector $\vec{\mathbf{a}} \in \mathbb{R}^{h' \times 1}$ to obtain the attention values ω_F^j:

$$\omega_F^j = \vec{\mathbf{a}}^T \cdot tanh\left(\mathbf{W} \cdot \mathbf{z}_F^j\right) \tag{7}$$

Where $\mathbf{W} \in \mathbb{R}^{h' \times h}$ are parameters. Similarly, we get the attention values ω_I^j and ω_M^j for node j in $\mathbf{Z}_I$ and $\mathbf{Z}_M$.

Then we use the *softmax* activation function to obtain α_F^j:

$$\alpha_F^j = softmax\left(\omega_F^j\right) = \frac{exp(\omega_F^j)}{exp(\omega_I^j) + exp(\omega_F^j) + exp(\omega_M^j)} \tag{8}$$

Where α_F^j is the percentage of embedding. Similarly, we can get α_I^j and α_M^j. For all n nodes, there are $\alpha_I = diag\left(\alpha_I^1, \alpha_I^2, \cdots, \alpha_I^n\right), \alpha_F = diag\left(\alpha_F^1, \alpha_F^2, \cdots, \alpha_F^n\right)$ and $\alpha_M = diag\left(\alpha_M^1, \alpha_M^2, \cdots, \alpha_M^n\right)$.

Finally, we combine the three embeddings to obtain $\mathbf{Z}$:

$$\mathbf{Z} = (\alpha_I \cdot \mathbf{Z}_I + \alpha_F \cdot \mathbf{Z}_F + \alpha_M \cdot \mathbf{Z}_M) \tag{9}$$

3.3 Objective Function

Our model has shown good results without adding other loss functions. We use the cross-entropy error as our overall objective function.

Then with the final embedding $\mathbf{Z}$, We can get the class predictions for n nodes as $\hat{\mathbf{Y}}$:

$$\hat{\mathbf{Y}} = softmax(\mathbf{W} \cdot \mathbf{Z} + \mathbf{b}) \tag{10}$$

Where *softmax* is a commonly used activation function for classification.

If the training set is L, for each $l \in L$, $\mathbf{Y}_l$ denotes the real label and $\hat{\mathbf{Y}}_l$ denotes the predicted label. Then, our classification loss can be expressed as:

$$\mathcal{L}_{class} = -\sum_{l \in L} \sum_{j=1}^{C} \mathbf{Y}_{lj} \ln \hat{\mathbf{Y}}_{lj} \tag{11}$$

4 Experiments

4.1 Experimental Settings

Datasets. We selecte a representative dataset and four complex relational datasets for our experiments, which include paper citation networks (Citeseer and ACM) and social networks (UAI2010, BlogCatalog and Flickr). The statistics of these datasets are shown in Table 1.

Table 1. Dataset statistics.

Datesets	Nodes	Edges	Classes	Features	Training	Test
Citeseer	3327	4732	6	3703	120	1000
ACM	3025	13128	3	1870	60	1000
UAI2010	3067	28311	19	4973	380	1000
BlogCatalog	5196	171743	6	8189	120	1000
Flickr	7575	239738	9	12047	180	1000

Parameters Setting. To validate our experiments, we choose 20 nodes per class as the training set and 1000 nodes as the test set. We set hidden layer units $nhid1 \in \{512, 768\}$ and $nhid2 \in \{32, 128, 256\}$. We set learning rate $0.0002 \sim 0.0005$, dropout rate is 0.5, weight decay $\in \{5e - 3, 5e - 4\}$ and $k \in \{2 \ldots 8\}$. For other baseline methods, we refer to the default parameter settings in the authors' implementation. For all experiments, we run 10 times and recorded the average results. And we use accuracy (ACC) and macro F1 score (F1) to evaluate the performance of the model.

4.2 Node Classification

In this subsection, we discuss the classification effects of each model on different datasets. The experimental results are shown in Table 2.

Table 2. Node classification accuracy (%). The best performance is highlighted in bold.

Datesets	Metrics	GCN	kNN-GCN	GAT	DEMO-Net	AM-GCN	If-GCN	Mif-GCN
Citeseer	ACC	70.50	61.40	72.40	69.40	**73.10**	71.48	72.96
	F1	67.62	58.96	68.34	67.42	68.42	67.16	**68.64**
ACM	ACC	87.95	78.53	87.25	83.29	90.40	90.28	**90.70**
	F1	87.87	78.14	87.38	83.17	90.43	90.20	**90.64**
UAI2010	ACC	50.12	66.09	58.64	25.41	70.10	69.24	**70.98**
	F1	32.90	50.45	41.48	19.82	55.61	55.30	**56.13**
BlogCatalog	ACC	69.48	76.54	64.32	61.09	81.98	81.82	**83.00**
	F1	68.53	75.59	63.43	58.79	81.36	81.20	**82.38**
Flickr	ACC	42.20	71.28	38.54	55.26	75.26	74.98	**76.94**
	F1	39.98	71.49	37.10	54.13	74.63	74.32	**76.56**

In the paper citation networks, the existing methods have achieved relatively excellent results, which shows the consistency of our method. In the social relationship networks, our method, kNN-GCN and AM-GCN have all achieved good results. This phenomenon may be that they retain the feature similarity between nodes. Other models showed varying degrees of fluctuation and even failure. The reason may be that they depend on graph structure for classification. The importance of node similarity for GCN is further confirmed.

The experimental results show that our model Mif-GCN outperforms other comparison baselines. Our model removes the mixed graph convolution module as If-GCN. Comparing If-GCN and Mif-GCN, the results show the effectiveness of our model. We visualize the attention trend graph. Among the three types of social networks, the feature graph convolution module and the mixed graph convolution module are more valued. They make the classification results biased toward the favorable side.

5 Conclusion

We propose Mif-GCN, a framework that simultaneously preserves graph structure and features similarity. We consider classification rules on different datasets including simple and complex graphs, our method is more generally applicable to networks with complex node relationships.

Acknowledgement. This work is supported by the Henan Province Science and Technology R&D Project(212102210078); the Major Public Welfare Project of Henan Province(201300210400); Key Scientific Research Project of Universities in Henan Province under Grant No.19A520016; Henan Higher Education Teaching Reform Research and Practice Project(Graduate Education) under Grant No.2019SJGLX080Y.

References

1. Perozzi, B., Al-Rfou, R., Skiena, S.: Deepwalk: online learning of social representations. In: Proceedings of the 20th ACM SIGKDD International Conference on Knowledge Discovery and Data Mining, pp. 701–710 (2014)
2. Grover, A., Leskovec, J.: Node2vec: scalable feature learning for networks. In: Proceedings of the 22nd ACM SIGKDD International Conference on Knowledge Discovery and Data Mining, pp. 855–864 (2016)
3. Defferrard, M., Bresson, X., Vandergheynst, P.: Convolutional neural networks on graphs with fast localized spectral filtering. arXiv preprint arXiv:1606.09375 (2016)
4. Klicpera, J., Bojchevski, A., Günnemann, S.: Predict then propagate: graph neural networks meet personalized pagerank. arXiv preprint arXiv:1810.05997 (2018)
5. Li, Q., Han, Z., Wu, X.M.: Deeper insights into graph convolutional networks for semi-supervised learning. In: Proceedings of the AAAI Conference on Artificial Intelligence. vol. 32 (2018)
6. Wu, F., Souza, A., Zhang, T., Fifty, C., Yu, T., Weinberger, K.: Simplifying graph convolutional networks. In: International Conference on Machine Learning, pp. 6861–6871. PMLR (2019)
7. Veličković, P., Cucurull, G., Casanova, A., Romero, A., Lio, P., Bengio, Y.: Graph attention networks. arXiv preprint arXiv:1710.10903 (2017)
8. Franceschi, L., Niepert, M., Pontil, M., He, X.: Learning discrete structures for graph neural networks. In: International Conference on Machine Learning, pp. 1972–1982. PMLR (2019)

9. Kipf, T.N., Welling, M.: Semi-supervised classification with graph convolutional networks. arXiv preprint arXiv:1609.02907 (2016)
10. Wang, X., Zhu, M., Bo, D., Cui, P., Shi, C., Pei, J.: Am-gcn: adaptive multi-channel graph convolutional networks. In: Proceedings of the 26th ACM SIGKDD International Conference on Knowledge Discovery and Data Mining, pp. 1243–1253 (2020)
11. Jin, W., Derr, T., Wang, Y., Ma, Y., Liu, Z., Tang, J.: Node Similarity Preserving Graph Convolutional Networks. In: Proceedings of the 14th ACM International Conference on Web Search and Data Mining, pp. 148–156 (2021)
12. Zheng, W., Qian, F., Zhao, S., Zhang, Y.: Multi-granularity graph wavelet neural networks for semi-supervised node classification, Neurocomputing (2020)
13. Wu, J., He, J., Xu, J.: Net: degree-specific graph neural networks for node and graph classification. In: Proceedings of the 25th ACM SIGKDD International Conference on Knowledge Discovery and Data Mining, pp. 406–415 (2019)
14. Guo, Y., Yan, Z.: Collaborative filtering: graph neural network with attention. In: Wang, G., Lin, X., Hendler, J., Song, W., Xu, Z., Liu, G. (eds.) WISA 2020. LNCS, vol. 12432, pp. 428–438. Springer, Cham (2020). https://doi.org/10.1007/978-3-030-60029-7_39

Natural Language Processing

Chinese Administrative Penalty Event Extraction for Due Diligence in Financial Markets

Ying Li[1], Yuting Chen[1], Jun Wang[2], Ruixuan Li[1(✉)], and Yuhua Li[1]

[1] School of Computer Science and Technology, Huazhong University of Science
and Technology, Wuhan, China
{li_ying,yuting_chen,rxli,idcliyuhua}@hust.edu.cn
[2] Nanjing Wudaozhixin Information Technology Co., Ltd, Nanjing, China
jwang@iwudao.tech

Abstract. Discovering major penalties against investment targets is critical to due diligence and risk control in financial markets, but extracting and evaluating penalty events from millions of announcements released by thousands of administrative agencies at all levels is also very challenging. In this paper, we developed the first system for Chinese administrative penalty event extraction, which was formulated as question answering tasks, and constructed corresponding benchmark data for training and evaluation. We also applied the Retrospective Reader mechanism to replace hand-crafted heuristic rules for no-answer scenarios and devised new strategies for generating informative questions. Empirical results on the benchmark demonstrated that our method outperforms the previous state-of-the-art methods. Recently, our system has been successfully deployed in a due diligence project for a top tier Chinese investment bank.

Keywords: Penalty event extraction · Due diligence · Machine reading comprehension

1 Introduction

Discovering major negative events of investment targets through due diligence is critical and challenging for risk control in the financial market. Especially, thousands of different administrative agencies released millions of penalties of various types which are scattering around the Web and are highly time-consuming and costly to be processed and analyzed by hand due to diverse formats and specifications. Therefore, we need to develop an automated system to extract penalty events from massive text documents to produce valuable structured information such as penalty types, litigants and amounts, and further evaluate their significance and check if there is any non-compliance with concealment of negative information disclosure.

Currently, existing Chinese financial event extraction systems [18,22] mainly deal with public company announcements with relatively standardized content

© Springer Nature Switzerland AG 2021
C. Xing et al. (Eds.): WISA 2021, LNCS 12999, pp. 507–518, 2021.
https://doi.org/10.1007/978-3-030-87571-8_44

and format in line with regulations of the China Securities Regulatory Commission (CSRC). Penalty announcements instead come from a wide range of administrators, regulators and government departments at ministry, provincial, municipal and county levels, so they are full of irregularity on formats and specifications and more challenging for corresponding event extraction.

Traditional event extraction systems typically rely heavily on entity recognition as a preprocessing/concurrent step [1,3,10,14,16,20], causing the well-known problem of error propagation [6]. Some recent research based on machine reading comprehension (MRC) [6,9,13] try to avoid this issue by formulating event extraction as a question answering (QA) task, which extracts the event arguments in an end-to-end manner. At the same time, it can help extract event arguments by offering informative priori knowledge with well-constructed questions, thus achieving state-of-the-art (SOTA) performance [6]. When extracting specific events, not all arguments may appear in the context, so the MRC model needs to have the ability to handle the no-answer situations [15]. For example, in the example shown in Fig. 1 there are two events: a *Fine* event triggered by '*fined*' and a *Confiscation* event triggered by '*confiscated*'. Neither the *Penalty Date* of the *Fine* event schema nor the *Penalty Date* of the *Confiscation* event schema appears in the sentence. Therefore, the *NULL* answer should be returned when asked about these arguments. However, the current methods mainly deal with the no-answer cases through hand-crafted heuristic rules [6,9], which are prone to errors due to the variability and ambiguity of natural language [19].

To address the above issues, we developed the first automated system for Chinese administrative penalty event extraction for due diligence and constructed a corresponding benchmark based on real-world data. We were also the first to apply the Retrospective Reader model for the penalty event extraction task, introducing an answer verification mechanism to replace the previous hand-crafted heuristic rules for no-answer scenarios, thus better handling the case where some event-specific arguments do not exist. Also, we devised new strategies for generating more informative questions and investigated their impact on model performance. We conducted experiments on real-word data and demonstrated the effectiveness of our method by outperforming the previous STOA methods. And our system has been deployed in a due diligence project for a top tier Chinese investment bank.

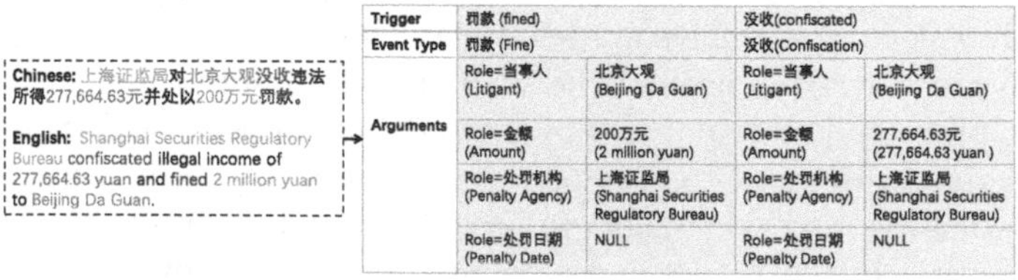

Fig. 1. A typical example of Chinese administrative penalty event

2   Related Work

Some recent work including DCFEE and Doc2EDAG also studied on Chinese event extraction in financial domain [18,22], but they mainly worked on official announcements of Chinese listed companies and extracted events such as *Equity Freeze*, *Equity Repurchase*, *Equity Underweight*, *Equity Overweight* and *Equity Pledge*, and they framed the event argument extraction as a traditional sequence labelling problem [18,22].

MRC has become a unified framework successfully applied to a variety of NLP applications [7,11,12]. Du et al. [6], Liu et al. [13] and Li et al. [9] formulated event extraction as MRC tasks, and proposed serval end-to-end systems overcoming error propagation in traditional methods. Du et al. [6] proposed the BERT-QA-trigger system with three question generation strategies, and Liu et al. [13] used machine translation methods to design more natural questions to extract event arguments, and Li et al. [9] formulated event extraction as multi-turn question answering. They all worked on the English texts from the 2005 Automatic Content Extraction (ACE) data set, in which Liu et al. [13] and Du et al. [6] made some assumptions on this dataset, such as the trigger words are all single tokens, which obviously does not match many real scenarios, especially for Chinese. Moreover, in the event argument extraction phase, all three of them used hand-crafted heuristic rules for the no-answer scenarios, which may lead to judgment errors as mentioned in the introduction.

In recent years, evaluation tasks related to Chinese financial event extraction have also been organized at the China Conference on Knowledge Graph and Semantic Computing (CCKS). For example, a financial event and argument extraction evaluation was released in 2020, which includes nine event types: *Bankruptcy Liquidation*, *Equity Freeze*, *Equity Underweight*, *Equity Overweight*, *Equity Pledge*, *Asset Loss*, *Accident*, *Leader Death*, and *External Indemnity* from official announcements of Chinese listed companies. The PIEE [17] model achieved the first place on this evaluation task using a MRC-based method, which generated questions containing only argument roles and event types and also used hand-crafted heuristic rules to cope with the no-answer scenarios.

In addition to hand-crafted heuristic rules, there are some other methods to identify unanswerable questions. For example, Liu et al. [5] appended an empty word token to the context and added a simple classification layer to the reader. Hu et al. [8] used two types of auxiliary loss, independent span loss to predict plausible answers and independent no-answer loss to decide answerability of the question. Back et al. [2] developed an attention-based satisfaction score to compare question embeddings with the candidate answer embeddings. However, these studies only stacked the verifier module in a simple way or just jointly learned answer positions and non-answer losses, and have been shown to be suboptimal [21].

Instead of working on English texts of ACE2005 or official announcements of Chinese listed companies, we focused on Chinese penalty event extraction for due diligence, which involves more data diversity and irregularity due to a large number of different data sources. It is also very critical to detect no-answer scenarios and further enhance the MRC methods for event extraction tasks. Different from all the above methods dealing with no-answer scenarios, our system exploited *Retrospective Reader* which adopts a two-stage humanoid design based on a comprehensive survey over existing answer verification solutions [21].

3   System Framework

Figure 2 illustrates the framework of our system of Chinese Penalty Event Extraction for Due Diligence. We will briefly introduce each component in detail as follows.

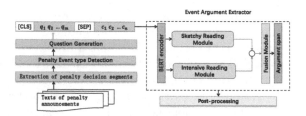

Fig. 2. The framework of the system of Chinese Penalty Event Extraction for Due Diligence.

3.1   Extraction of Penalty Decision Segments

The penalty announcements can be naturally parsed into a series of informative segments including *litigant profile, violation fact, penalty basis* and *penalty decision*, and penalty events can be located in the *penalty decision* segments. Generally, the textual features for each segment are distinct and clear, and with BERT and only limited labelled segment data we can train a robust sentence classifier to identify to which segment a sentence belongs. The input to the classifier may include context information of a current sentence such as its previous sentence and/or its next sentence.

3.2   Penalty Event Type Detection

Liu et al. [13] found that with MRC-based methods it is difficult to construct suitable questions for extracting trigger words, so we also skipped trigger detection and directly formulated corresponding penalty event type detection as a classification task. We denote a *penalty decision* segment as $S = \{s_1, s_2, s_3, \ldots, s_N\}$ where N represents the number of tokens in the segment. The token [CLS] is placed at the start of the segment. The input sequence of the segment $C = \{[CLS], s_1, s_2, s_3, \ldots, s_N\}$ is encoded using BERT, and the entire sequence can be represented with the encoding of [CLS]. For each individual penalty event type, we trained an individual binary classifier to determine whether the segment contains the event type.

3.3 Question Generation

Given a specific event type, the question generator aims to generate informative questions for extracting arguments. Du et al. [6] found that questions incorporating different semantic information had quite different effects on the event extraction results with the ACE2005 dataset, and we also designed and explored three template-based question generation methods with our dataset of penalty event extraction as shown in Table 1.

Template 1: We generated simple questions for extracting event arguments by just selecting the interrogative words based on corresponding argument roles.

Template 2: To encode the event type information, we added 'in [event type]' at the beginning of the questions generated by Template 1.

Template 3: In addition to event types and argument roles, the occurrence patterns of event arguments were encoded into the question as additional prior knowledge.

Table 1. Real examples of questions for extracting arguments of event type Recovery.

Argument	Template 1	Template 2	Template 3
Amount	金额是多少? (How much is the amount?)	在追缴中,金额是多少? (How much is the amount of the recovery event?)	找到追缴中的金额,该参数主要以元或美元结尾(Find the amount of recovery events, usually ends in yuan or dollar.)
Litigant	当事人是谁? (Who is the litigant?)	在追缴中,当事人是谁? (Who is the litigant of the recovery event?)	找到追缴中的当事人,该参数主要是人名或公司名(Find out the litigant of recovery events, usually a person or company)
Penalty Date	处罚日期是什么时候? (When is the penalty date?)	在追缴中,处罚日期是什么时候? (When is the penalty date of the recovery event?)	找到追缴中的处罚日期,该参数主要包含年月日(Find out the penalty date of recovery events, usually contains the date)
Penalty Agency	处罚机构是什么? (What is the penalty agency?)	在追缴中,处罚机构是什么? (What is the penalty agency of the recovery event?)	找到追缴中的处罚机构,该参数主要是机构名(Find out the penalty agency of recovery events,usually the agency name)

3.4 Event Argument Extractor

Inspired by Zhang et al. [21], we converted argument extraction task into the span-based MRC task, which can be represented as a triplet $\langle C, Q, A \rangle$, where C is the context, and Q is a question generated by question generator, in which a span is an argument A. Our method is supposed to not only predict the start and end positions in the context C and extract span as argument A but also return a NULL when the question is unanswerable which means the argument does not exist in the context. Our argument extractor is composed of three parallel modules as shown in Fig. 3: a sketchy reading module, an intensive reading module and a fusion module. The sketchy reading makes a coarse judgment about the answerability, whether the question is answerable; and then the intensive reading predicts candidate answers and gains confidence in answerability, and finally the fusion module combines the two stages of judgment scores with the candidate answers to arrive at the final argument.

We concatenate question and the context as the input sequences of BERT [4]. Let $H = \{h_1, h_2, h_3 \ldots, h_{m+n+2}\}$ be the last-layer hidden states in BERT of the input sequence of length $m + n + 2$, where m denotes the length of the question and n denotes the length of segments.

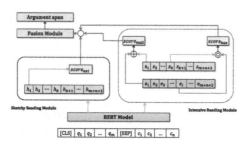

Fig. 3. The architecture of event argument extractor

Sketchy Reading Module. After reading, the sketchy reader will make a preliminary judgment, whether the question is answerable given the context. This indicates whether the argument exists in the context. We passes the [CLS] - the pooled first token representation $h_1 \in H$ to a fully connection layer to get classification logits $\hat{y}_i$ composed of answerable ($logit_{ans}$) and unanswerable ($logit_{na}$) elements with training objective of cross entropy, then we calculate the difference as the preliminary verification score [21]:

$$score_{ext} = logit_{na} - logit_{ans}. \tag{1}$$

Intensive Reading Module. The objective of the intensive reader is to verify the answerability, produce candidate argument spans. Inspired by previous works [13], given the joint representation H of question Q and context C, we feed H as the input of a linear layer with $SoftMax$ operation to obtain 2 probability vectors containing the start and end positions of the answer over every position in the input sequence: $S = \{s_1, s_2, s_3 \ldots, s_{m+n+2}\}$ and $E = \{e_1, e_2, e_3 \ldots, e_{m+n+2}\}$. Then we calculate the $has\text{-}answer$ score $score_{has}$ and the $no\text{-}answer$ score $score_{null}$:

$$score_{has} = max\,(s_k + e_l), 1 < k \le l \le m + n + 2, \tag{2}$$

$$score_{null} = s_1 + e_1. \tag{3}$$

The training objectives for both answer span prediction and answerable probability calculation are defined as cross entropy loss.

Fusion Module. Rear verification is required in the fusion module. Specifically, rear verification is the combination of predicted probabilities of sketchy reading module and intensive reading module, which is an aggregated verification for final answer [12].

The final no-answer score $score_{final}$ will be calculated as:

$$score_{final} = \beta_1 (score_{null} - score_{has}) + \beta_2 score_{ext}, \qquad (4)$$

where β_1 and β_2 are weights. An answerable threshold δ is set and determined according to the development set. The model predicts the answer span that gives the has-answer score if the final score is above the threshold δ, and null string otherwise.

3.5   Post-processing

For arguments unanswerable in *penalty decision* segments, we can try to search and locate answers which may be available in other related locations. Specifically, the *Litigant* argument may be found in *litigant profile* segments, and the *Penalty Agency* argument may be obtained from the information of source websites where original penalty announcements were crawled.

4   Experiments

4.1   Dataset and Evaluation Metrics

Previously, there was no data set for training and evaluating the administrative penalty event extraction system yet, so we constructed the first benchmark of Chinese administrative penalty for Due Diligence by ourselves. We randomly sampled a set of penalty decision segments extracted from real-world announcements released by different agencies, and labelled event types and arguments using the open source annotation tool *Doccano*[1] as shown in Fig. 4. We finally obtained 11,335 event instances and split into three parts: 9335 as the training set, 2000 as the validation set, and 2000 as the testing set. As shown in Table 2, the dataset contains 8 types of penalty events.

We adopted the same evaluation criteria used in Li et al. [10]: An event argument is correctly identified (ID) if its offsets and event type match those of any of the reference argument mentions in the document; and it is correctly classified if its semantic role is also correct. Since we did not detect specific trigger words and directly classified the event types, we skipped the evaluation on trigger detection as well. We also adopted *precision* (P), *recall* (R), and *F1-score* (F1) as evaluation metrics to ensure comparability.

---

[1] https://github.com/doccano/doccano.

Table 2. Statistics of the data set, including the distribution of various event types on the training set, validation set, and test set

Event Type	Train	Dev	Test	Total
Payment of investigation fees (缴付调查费)	25	3	3	31
Fine And Confiscation (罚款没收)	22	6	7	35
Recovery(追缴)	59	5	7	71
Total Fines(罚款合计)	230	29	33	292
Confiscation(没收)	1607	135	155	1897
Warning(警告)	2088	219	223	2530
Order(责令)	2084	232	242	2558
Fine(罚款)	9025	065	983	10973
All Events	15140	1594	1653	18387

Fig. 4. An example of the labeling system doccano.

4.2 Experimental Results and Analyses

We exploited the pre-trained Chinese BERT-base model in all our experiments. Other hyper-parameters were tuned with the validating set. Specifically, the epoch number was set as 5. The dropout rate was set as 0.5. The initial learning rate was set as 2e-6 for event detection, 6e-6 for event argument extraction. The answer prediction threshold δ was set as 0.5, and $\beta_1 = \beta_2 = 0.5$ in this work.

At first, we investigated how the question generation strategies affect the performance of event extraction. We conducted the experiments and evaluated argument extraction with 3 different templates as shown in Table 3. As expected, the template 3 outperforms the template 1 and template 2, and the template 2 achieves better performance than the template 1. The results showed that encoding more informative prior knowledge into questions is helpful. The template 3 was selected as the default setting for all the following experiments.

Table 3. Comparison of different question generation strategies

	Argument identification			Argument ID + Classification		
	P	R	F1	P	R	F1
Template 1	94.49	90.09	92.23	90.28	89.66	89.97
Template 2	95.36	91.03	93.14	93.89	90.84	92.34
Template 3	95.16	92.07	**93.58**	95.09	92.04	**93.55**

In order to verify the effectiveness of our system[2], we compare our method to a few prior competitive models: DCFEE [18], Doc2EDAG [22], BERTEE [13],

---

[2] For fair evaluation, the post-processing is not involved.

DYGIE++[16], XF-Event-Extraction and BERT-QA-Trigger [6]. Both DCFEE and Doc2EDAG framed event extraction tasks as sequence labelling, and they used convolutional neural networks(CNN) and Transformer as the basic component, respectively. BERTEE [13], which adopts BERT representations but uses classification strategy for event extraction. DYGIE++[16] is a BERT-based framework that models text spans and captures within-sentence and cross-sentence context. XF-Event-Extraction[3] is the first place model of iFLY-TEKA.I.2020 Event Extraction Challenge[4], which assumes that there is only one event in an input sequence. BERT-QA-Trigger [6] is the first work on transforming event extraction tasks into MRC tasks.

As shown in Table 4, the experimental results show that our method attains state-of-the-art performance and outperforms all baselines by considerable margins in argument extraction. Specifically, our method outperforms BERTEE with 4.91% in argument extraction, which indicates that the improvements are mainly from the task reformulation as MRC, rather than introducing the BERT encoder. Compared to the BERT-QA-trigger model, our method is 2.09% higher in the argument extraction task, which shows that introducing a verification mechanism responding to no-answer cases does improve performance.

Table 4. Argument extraction results on the test set.

	Argument Identification			Argument ID + Classification		
	P	R	F1	P	R	F1
DCFEE	74.65	69.78	72.13	54.28	62.67	58.18
Doc2EDAG	91.86	83.81	87.65	90.59	83.62	86.96
BERTEE	91.96	85.54	88.64	91.96	85.54	88.64
DYGIE++	82.43	73.47	77.69	82.33	73.39	77.60
XF-Event-Extraction	68.24	58.12	62.16	67.85	58.07	62.00
BERT-QA-trigger	92.52	90.42	91.46	92.52	90.42	91.46
Our Model	95.16	92.07	**93.58**	95.09	92.04	**93.55**

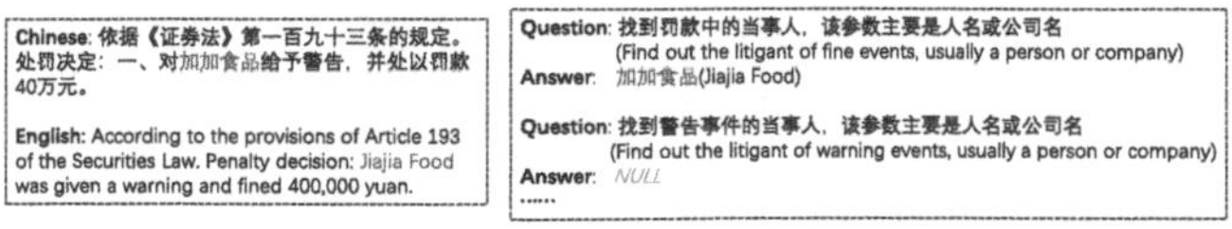

Chinese: 依据《证券法》第一百九十三条的规定。处罚决定: 一、对加加食品给予警告，并处以罚款40万元。

English: According to the provisions of Article 193 of the Securities Law. Penalty decision: Jiajia Food was given a warning and fined 400,000 yuan.

Question: 找到罚款中的当事人，该参数主要是人名或公司名
(Find out the litigant of fine events, usually a person or company)
Answer: 加加食品(Jiajia Food)

Question: 找到警告事件的当事人，该参数主要是人名或公司名
(Find out the litigant of warning events, usually a person or company)
Answer: NULL
......

Fig. 5. An example of a *Fine* event and a *Warning* event sharing the same span as an argument.

---

[3] https://github.com/WuHuRestaurant/xf_event_extraction2020Top1.
[4] http://challenge.xfyun.cn/topic/info?type=hotspot.

In Table 5, we studied the argument extraction results on different event types and selected the four event types, as they account for major proportion of the data set and perform more consistently. From the results, we found that our model achieves the optimal performance compared to other models on all event types except *Warning* events. The BERTEE model performs better than the MRC-based model on *Warning* events, probably because *Warning* events and *Fine* events often share the same span as arguments. As shown in Fig. 5, '*Jiajia Food*' play the role of *Litigant* in both *Fine* and *Warning* event. The BERTEE model uses two separate classifiers to identify *Fine-Litigant* and *Warning-Litigant*, respectively. However, the MRC-based approach shares model parameters to calculate start index and end index in the argument extraction phase. Therefore, when the number of *Fine* events far exceeds the number of *Warning* events, BERTEE can extract both arguments independently, while the MRC-based model tends to ignore the *Warning-Litigant* because the questions we designed for *Fine-Litigant* and *Warning-Litigant* is not differentiated enough. Also, our model does not perform as well as the BERT-QA-trigger model on *Warning* events. Based on the fact that no-answer cases in *Warning* events is very rare, we believe that the answerable threshold δ set for the whole dataset is too high for Warning events, which may lead to incorrect prediction results. During the system deployment in a due diligence project for a top tier Chinese investment bank, the end-users told us that *Warning* events generally do not involve monetary punishments and are less important compared with *Fine* events. So fortunately the issues on *Warning* events will not really cause problems for evaluating total significance of discovered penalties in real applications.

Table 5. The argument extraction results of the five main event types.

Model	Fine	Order	Warning	Confiscation
DCFEE	59.32	42.27	46.38	54.92
Doc2EDAG	87.02	83.05	70.81	85.71
BERTEE	90.26	91.52	**84.57**	88.40
DYGIE++	83.06	60.95	69.61	68.15
XF-Event-Extraction	63.89	64.95	73.08	46.10
BERT-QA-trigger	90.58	90.62	83.27	91.82
Our Model	**95.26**	**94.27**	75.50	**94.61**

5 Conclusion and Future Work

In this paper, we developed the first automated system for Chinese administrative penalty event extraction using the MRC framework, which has been successfully deployed in a due diligence project for a top-tier Chinese investment bank. We applied the Retrospective Reader mechanism to replace hand-crafted heuristic rules for no-answer scenarios, and investigated how question generation strategies affect the argument extraction performance. We also constructed the first benchmark data for training and evaluating the Chinese administrative penalty event extraction tasks, and our experimental results on the benchmark demonstrated that our method outperforms the previous state-of-the-art methods. In the future, we will extend our work to other financial events such as investment events to enable efficiently capturing comprehensive profiles of various investment Institutions.

Acknowledgements. This work is supported by the National Key Research and Development Program of China under grants 2016QY01W0202, National Natural Science Foundation of China under grants U1836204, U1936108, and Major Projects of the National Social Science Foundation under grant 16ZDA092.

References

1. Ahn, D.: The stages of event extraction. In: Proceedings of the Workshop on Annotating and Reasoning about Time and Events, pp. 1–8 (2006)
2. Back, S., Chinthakindi, S.C., Kedia, A., Lee, H., Choo, J.: NeurQuRi: neural question requirement inspector for answerability prediction in machine reading comprehension. In: International Conference on Learning Representations (2019)
3. Chen, Y., Xu, L., Liu, K., Zeng, D., Zhao, J.: Event extraction via dynamic multi-pooling convolutional neural networks. In: Proceedings of the 53rd Annual Meeting of the Association for Computational Linguistics and the 7th International Joint Conference on Natural Language Processing (Volume 1: Long Papers), pp. 167–176. Association for Computational Linguistics, Beijing, July 2015
4. Devlin, J., Chang, M.W., Lee, K., Toutanova, K.: BERT: pre-training of deep bidirectional transformers for language understanding. In: Proceedings of the 2019 Conference of the North American Chapter of the Association for Computational Linguistics: Human Language Technologies, Volume 1 (Long and Short Papers), pp. 4171–4186. Association for Computational Linguistics, Minneapolis, June 2019
5. Dhingra, B., Liu, H., Yang, Z., Cohen, W., Salakhutdinov, R.: Gated-attention readers for text comprehension. In: Proceedings of the 55th Annual Meeting of the Association for Computational Linguistics (Volume 1: Long Papers), pp. 1832–1846. Association for Computational Linguistics, Vancouver, July 2017
6. Du, X., Cardie, C.: Event extraction by answering (almost) natural questions. In: Proceedings of the 2020 Conference on Empirical Methods in Natural Language Processing (EMNLP), pp. 671–683. Association for Computational Linguistics, Online, November 2020
7. Gao, S., Sethi, A., Agarwal, S., Chung, T., Hakkani-Tür, D.: Dialog state tracking: a neural reading comprehension approach. In: Nakamura, S., et al. (eds.) SIGdial, pp. 264–273. Association for Computational Linguistics (2019)
8. Hu, M., Wei, F., Peng, Y., Huang, Z., Yang, N., Li, D.: Read + verify: machine reading comprehension with unanswerable questions. In: Proceedings of the AAAI Conference on Artificial Intelligence, vol. 33, no. (01), pp. 6529–6537 (2019)
9. Li, F., et al.: Event extraction as multi-turn question answering. In: Proceedings of the 2020 Conference on Empirical Methods in Natural Language Processing: Findings, pp. 829–838 (2020)

10. Li, Q., Ji, H., Huang, L.: Joint event extraction via structured prediction with global features. In: Proceedings of the 51st Annual Meeting of the Association for Computational Linguistics (Volume 1: Long Papers), pp. 73–82. Association for Computational Linguistics, Sofia, August 2013

11. Li, X., Feng, J., Meng, Y., Han, Q., Wu, F., Li, J.: A unified MRC framework for named entity recognition. In: Proceedings of the 58th Annual Meeting of the Association for Computational Linguistics, pp. 5849–5859. Association for Computational Linguistics, July 2020

12. Li, X., et al.: Entity-relation extraction as multi-turn question answering. In: Proceedings of the 57th Annual Meeting of the Association for Computational Linguistics, pp. 1340–1350. Association for Computational Linguistics, Florence, July 2019

13. Liu, J., Chen, Y., Liu, K., Bi, W., Liu, X.: Event extraction as machine reading comprehension. In: Proceedings of the 2020 Conference on Empirical Methods in Natural Language Processing (EMNLP), pp. 1641–1651 (2020)

14. Nguyen, T.H., Cho, K., Grishman, R.: Joint event extraction via recurrent neural networks. In: Proceedings of the 2016 Conference of the North American Chapter of the Association for Computational Linguistics: Human Language Technologies, pp. 300–309 (2016)

15. Rajpurkar, P., Jia, R., Liang, P.: Know what you don't know: unanswerable questions for SQuAD. In: Proceedings of the 56th Annual Meeting of the Association for Computational Linguistics (Volume 2: Short Papers), pp. 784–789. Association for Computational Linguistics, Melbourne, July 2018

16. Wadden, D., Wennberg, U., Luan, Y., Hajishirzi, H.: Entity, relation, and event extraction with contextualized span representations. In: Proceedings of the 2019 Conference on Empirical Methods in Natural Language Processing and the 9th International Joint Conference on Natural Language Processing (EMNLP-IJCNLP), pp. 5784–5789. Association for Computational Linguistics, Hong Kong, November 2019

17. Wang, H., et al.: A prior information enhanced extraction framework for document-level financial event extraction (2020)

18. Yang, H., Chen, Y., Liu, K., Xiao, Y., Zhao, J.: DCFEE: a document-level Chinese financial event extraction system based on automatically labeled training data. In: Proceedings of ACL 2018, System Demonstrations, pp. 50–55 (2018)

19. Yu, B., Mengge, X., Zhang, Z., Liu, T., Yubin, W., Wang, B.: Learning to prune dependency trees with rethinking for neural relation extraction. In: Proceedings of the 28th International Conference on Computational Linguistics, pp. 3842–3852. International Committee on Computational Linguistics, Barcelona, December 2020

20. Yu, J., et al.: Extraction and portrait of knowledge points for open learning resources. In: Wang, G., Lin, X., Hendler, J., Song, W., Xu, Z., Liu, G. (eds.) WISA 2020. LNCS, vol. 12432, pp. 46–56. Springer, Cham (2020). https://doi.org/10.1007/978-3-030-60029-7_5

21. Zhang, Z., Yang, J., Zhao, H.: Retrospective reader for machine reading comprehension. arXiv preprint arXiv:2001.09694 (2020)

22. Zheng, S., Cao, W., Xu, W., Bian, J.: Doc2EDAG: an end-to-end document-level framework for Chinese financial event extraction. In: Proceedings of the 2019 Conference on Empirical Methods in Natural Language Processing and the 9th International Joint Conference on Natural Language Processing (EMNLP-IJCNLP), pp. 337–346. Association for Computational Linguistics, Hong Kong, November 2019

CoEmoCause: A Chinese Fine-Grained Emotional Cause Extraction Dataset

Zhuojin Liu, Zhongxin Jin, Chaodi Wei, Xiangju Li, and Shi Feng[✉]

School of Computer Science and Engineering, Northeastern University,
Shenyang, China
fengshi@cse.neu.edu.cn

Abstract. Emotion Cause Extraction (ECE) is an emerging hot topic in the field of sentiment analysis. The purpose of the ECE task is to extract the causes of emotions in the text according to the text and given emotions. Emotional Cause Pair Extraction (ECPE) is a brand-new research problem on ECE, whose main purpose is to extract the emotion clauses and emotion cause clauses in texts at the same time. At present, the ECPE task has received extensive attentions from both academia and industry communities. The existing ECPE researches are mainly carried out on online news report corpus. However, this kind of corpus has several limitations, such as the small scale and rough tagging granularity. In this paper, we construct an emotion cause extraction dataset containing 5,195 COVID-19 pandemic-related discussion posts collected from Sina Weibo, and leverage 10 state-of-the-art models to perform ECE and ECPE tasks on this new dataset. We find that the performances of most existing models on our newly constructed dataset decrease dramatically compared with the reported results in the online news dataset. We further analyze the causes of this phenomenon.

Keywords: Emotion cause pair extraction · Emotion cause clause extraction · Deep learning · Dataset construction

1 Introduction

In recent years, detecting the emotional causes in social media is a hot re- search topic in the NLP field [13,14], which has attracted extensive attention from both academia and industry communities. This task can help companies find users' opinions on products, so as to make targeted improvements. Moreover, this emerging task can also help government departments find people's opinions on policies and social phenomena, so as to better formulate policies.

At present, two important research directions of this task are (1) Emotion Cause Extraction (ECE), that is, extracting the cause of a certain emotion clause from the document according to the given emotion; (2) Emotional Cause Pair Extraction (ECPE), which goes further, extracting both emotion clause and emotion cause clause (emotion-cause pair) from text simultaneously where both

© Springer Nature Switzerland AG 2021
C. Xing et al. (Eds.): WISA 2021, LNCS 12999, pp. 519–530, 2021.
https://doi.org/10.1007/978-3-030-87571-8_45

emotion and cause are unknown. Figure 1 shows an example of these two research problems.

In this example sentence, clause 3 (c_3) is the emotion clause, which contains the emotion word "scared". The reason for this emotion is clause 1 (c_1) and clause 2 (c_2). In the ECE task, the text and emotion clauses are given to the model, hoping to find the cause clause that causes the emotion clause. In Fig. 1, the sentence and emotion clause (c_3) are provided to the model, and we expect the ECE model can find two cause clauses, namely c_1 and c_2.

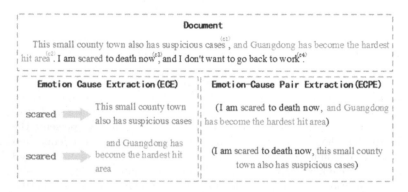

Fig. 1. An example of ECE and ECPE tasks. The example sentence has four clauses that are annotated by superscript labels.

In the ECPE task, only the sentence text is provided to the model, and this task aims to automatically find the emotional cause pair (ECP) in the text, namely both the emotion clause and the emotion cause clause. In the above example, we hope to find the emotion pair ($c_3 - c_1$) and ($c_3 - c_2$).

In the previous literature, the tasks of ECE and ECPE were generally carried out on a Chinese dataset from Sina News and an English dataset [6,8]. The scale of these two data sets is relatively small, with only about 2000 documents, and each document has at least one emotion-cause pair. All these limitations may affect the training of the model and cause the over-fitting problem. In addition, a microblog dataset with multi-user (retweet) structure was released [1]. However, this dataset only contains multi-user microblogs, but has limited single-user (original) microblogs.

To tackle these challenges, we construct a brand-new dataset, CoEmoCause, for emotion cause analysis based on the discussions of the COVID-19 epidemic on Weibo. The dataset has a scale of 5,195 posts, and all the data are obtained from Sina Weibo, the largest microblog platform in China. More than 95% of texts in CoEmoCause contain 0–3 emotion-cause pairs, and a few have more than 3 emotion-cause pairs. Compared with the previous datasets, the size of our dataset is twice the size of the Chinese online news dataset, which enables the model to be better generalized and effectively reduces the possibility of over-fitting. In addition, since our dataset is collected from public social media platforms, the expression of text in the dataset is more colloquial, and it contains more semantic information than the above two datasets based on online news texts, which have much more formal expressions. For example, on social platforms, people will use sarcasm, using positive words to express their negative feelings, which rarely appears in online news articles. Moreover, because the dataset constructed in our paper is based on the microblog discussion data of the COVID-19 epidemic, it has certain guiding significance for the follow-up public opinion monitoring of the pandemic situation.

In this paper, we will introduce our dataset, and then evaluate 10 state-of-the-art models for ECE and ECPE tasks on our newly constructed dataset, and compare the performance with their performance on the existing datasets.

We can summarize our contributions as follow:

- We constructed a large-scale fine-grained dataset for emotional cause analysis, and the dataset is released online.[1]
- We implemented 10 state of the art models for ECE and ECPE tasks, and analyzed the reasons for the decrease in algorithm performance.

The rest of the paper is organized as follows: Sect. 2 summarizes the related works in the field of context-based sentiment analysis tasks; Sect. 3 describes the proposed context-aware sentiment classification algorithm; Sect. 4 describes the experimental results on the benchmark dataset; Sect. 5 gives the conclusions.

2 Related Work

Lee S.Y.M. et al. [10] first proposed the ECE task to extract the causes behind a given emotion expression in the text, the authors also provided a corpus and proposed a rule-based solution. After that, Gui et al. [8] redefined the ECE task from word level to clause level binary classification problem and utilized an event-driven multi-kernel support vector machine to extract emotional causes. The authors also constructed a corpus based on Sina News, and this dataset gradually became the benchmark dataset of ECE and ECPE tasks. Subsequently, the method based on the neural network began to appear. Gui et al. [7] considered the ECE question as a question answering problem. Li et al. [11,12] took context into account and added attention mechanism to the model. Xia et al. [17] applied transformer in ECE tasks for the first time. Ding et al. [2] took the relative position and global label between emotion and cause into account. Hu et al. [9] better obtained the structural information between clauses through graph neural network.

ECE task aims to find the cause on the basis of known emotions, which cannot be well applied to real scenes. In order to solve this problem, Xia et al. [16] further proposed a new task, Emotion Cause Pair Extraction, that is, directly extracting the emotion-cause pair from the text. This paper proposed a two-step method, which first extracted emotions and causes, and then combined them into emotion-cause pair and filtered out possible ECP. However, the second step in this two-step

---

[1] https://github.com/neuChatbotDS/CoEmoCause-Dataset.

method may suffer from the error of the first step, and results in affecting the over-all performance. Therefore, the subsequent paper proposes a series of end-to-end methods. Ding et al. [3] represented the relationship between clauses through a 2D structure and extracted ECP from it. Wei et al. [15] improved the performance by modeling the relationship between clauses; Fan et al. [5] performed ECPE by predicting the sequence of actions on the graph. Ding et al. [4] extracted the corresponding causes and emotions from a sliding window with emotion as the pivot and cause as the pivot, and finally combined these two predictions to obtain the final emotion-cause pairs.

We select 5 ECE models and 5 ECPE models to test their performance on our dataset, and analyze their performance.

3   Problem Definition

First, we define the notation as follows: Let D be a document in the dataset, and D is divided into n clauses, then D can be expressed as $D = \{c_1, c_2, ..., c_n\}$, where c_i represents the i-th clause in D. In addition, we define c^e as emotional clause and c^c as cause clause.

Based on the above definition, we formally define ECE and ECPE tasks as follows: In the ECE task, for a document D, our purpose is to identify whether clause $c_i (i = 1 ... n)$ stimulates the emotion expression in a given document. Therefore, for each clause, the task can be defined as a binary classification task:

$$f(c_i, c^e) = \begin{cases} Y \ if \ c_i \ contains \ the \ emotion \ cause \ resulting \ c^e \\ N \ otherwise \end{cases} \tag{1}$$

where c^e is the emotion clause in D, and only one exists in each document; c_i is an arbitrary clause in the document.

The goal of ECPE is to extract the set of emotion-cause pairs. i.e. extract

$$P = \{..., (c^e, c^c), ...\} \tag{2}$$

where c^c is the cause of c^e.

4   CoEmoCause Dataset

4.1   Dataset Annotation

Our dataset comes from the epidemic dataset of the SMP2020 microblog emotion classification competition.[2] There are 7,317 entries of data in the original dataset. In this paper, each document is annotated by one person, all of the annotators are undergraduates majoring in computer science and fully trained. Each paragraph

---

[2] https://smp2020.aconf.cn/smp.html#4.

of text is marked with a topic, emotion category, emotion clauses, and cause clauses. In addition, we also add information such as the length of the text, the beginning and end positions of emotions, and cause in the corresponding clauses. After removing the texts whose topic have nothing to do with the epidemic, there are 5,195 documents left in the dataset.

In our dataset, each document is divided into clauses according to punctuation marks. The splitting rule is formulated based on our experience and observation. For example, the sentence before and after the retweet symbol "//" in microblog is divided into two clauses. When encounters multiple continuous punctuation marks, such as "I haven't gone to Wuhan yet!!!! Hold on brother", the clause will not be split until the continuous punctuation marks are over.

Besides emotion cause annotation, the Weibo posts are also labeled with six topic categories, namely, virus source and transmission, pandemic data report, prevention and control measures, response to the pandemic, deeds of people and others. There are nine emotion labels: respect, support, anger, happiness, surprise, disgust, sadness, fear, and anticipation. There may be situations where one emotional word corresponds to multiple causes (Fig. 2) and one cause triggers multiple emotions (Fig. 3).

In Fig. 2, the causes of emotion "frustrating" are "infected because of taking care of her" and "watched her partner pass away". In Fig. 3, the cause of emotion "damn it" and "[anger]" is "ate wild animals".

Compared with the previous datasets, our constructed CoEmoCause dataset is mainly different in the following aspects: 1) The language is more colloquial, some emotions are more obvious and some are subtler. 2) Some emotional expressions have no obvious emotional words, which requires the model to have a deeper understanding of semantics. 3) Some emotions are expressed in sarcasm, which leads to positive emotional words, while the expressed emotions are negative. 4) Often use microblog emoji to express emotions. In addition, expressions will be combined with the third point. For example, [smiling] expressions have evolved into a way to express negative emotions instead of being happy. In our dataset, the emoji provided by microblogs have been transformed into the form of "[emoji name]".

4.2 Dataset Statistics

CoEmoCause is the corpus we constructed. CN denotes the dataset based on Sina News released by [8], and also the benchmark dataset for ECE and ECPE tasks. EN denotes the English dataset, which is constructed from novel text and released by NTCIR-13 ECA task [6] (Fig. 4).

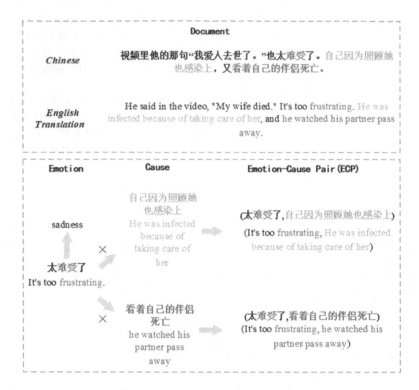

Fig. 2. Example 1 of an annotated post in CoEmoCause dataset.

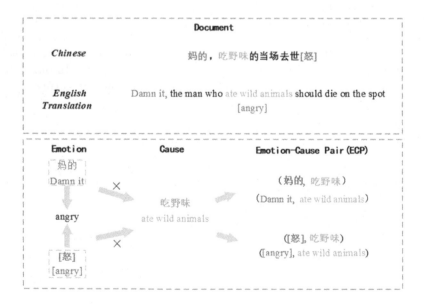

Fig. 3. Example 2 of an annotated post in CoEmoCause dataset.

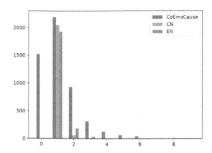

Fig. 4. The comparison of number of emotion cause pairs in the three datasets.

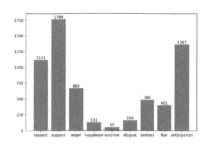

Fig. 5. The statistics of each emotion category in CoEmoCause dataset.

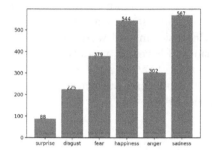

Fig. 6. The statistics of each emotion category in CN dataset.

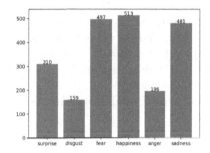

Fig. 7. The statistics of each emotion category in EN dataset.

Because the documents in the CoEmoCause dataset are all selected from microblog texts related to the COVID-19 pandemic, the number of emotions is more unbalanced than the other two datasets. Specifically, there are a large number of texts supporting and paying respects to doctors going to Wuhan and hope that the epidemic will pass as soon as possible, which leads to a large number of emotions of respect, support and anticipation. Dissatisfaction with government policies, fear of COVID-19, and sadness over this disaster caused by the epidemic are reflected in anger, fear and sadness emotions, with a medium number of the dataset. Finally, some positive responses to good news, the aversion to the people who caused the epidemics and some scattered documents constitute happiness disgust and surprise. Compared with other datasets, the imbalance among emotion categories in CoEmoCause is more serious, as shown in Fig. 5, 6, and 7.

In Table 1, it can be seen that the average clause length of our dataset is only 67% of the CN dataset under the condition that the average document length is not much different from that of the CN dataset. This is also consistent with the impression that people tend to write more short sentences on social media, while in news reports, writers tend to write well-organized long sentences. At the same time, in our dataset, the average number of emotion-cause pairs per

Table 1. The statistics of the three datasets

Dataset	CoEmoCause	CN	EN
Average document length	65.71	64.74	316.62
Number of ECP per document	1.19	1.03	1.14
Average emotion length	3.12	2.3	8.06
Average cause length	6.39	9.47	41.74
Average emotion-cause distance (clause)	0.92	1	1

text reaches 115% of the CN dataset, which shows that people tend to express more emotions on social media than formal texts. 9

In addition, 29.2% of documents in our dataset have no emotion-cause pair. In contrast, each document in the CN and EN datasets has at least one emotion-cause pair. When we perform ECE and ECPE tasks in real-world scenes, the obtained text probably does not contain emotions. Therefore, models trained on CN and EN data sets are likely not to adapt to this situation and tend to find at least one emotion-cause pair in each document. Therefore, our data set is very helpful for the model to be applied to real-world scenes.

5   Experiment

5.1   Models

We conduct experiments using ten state-of-the-art ECE and ECPE models on the CoEmoCause dataset. The comparison models are as follows.

MANN [11] designs a multi-attention mechanism to capture the mutual influences between the emotion clause and each candidate clause, and then generates the representations for the above two clauses separately. And these representations are fed into a convolutional neural network to model the emotion cause clause.

CANN [12] encodes the clauses with a co-attention based bidirectional LSTM into high-level input representations, which are further fed into a convolutional layer for emotion cause analysis.

RTHN [17] is composed of a lower word-level encoder based on RNNs, and an upper clause-level encoder based on Transformer. RTHN encodes the relative position and global prediction information into a Transformer that can capture the causality between clauses and make RTHN more efficient.

FSS-GCN [9] narrows focus from global structure to local structure by continuously injecting structural constricts into networks, and constructs inter-clause dependency by parsing a document into graph structure based on the syntactic dependency.

PAEDGL [2] proposes a model based on the neural network architecture to encode text content, relative position and global label in a unified and end-to-end fashion, and introduces a relative position augmented embedding learning algorithm, and transforms the task from an independent prediction problem to a reordered prediction problem.

ECPE [16] first performs individual emotion extraction and cause extraction via multi-task learning, and then conduct emotion-cause pairing and filtering.

ECPE-2D [3] represents the emotion-cause pairs in the forms of a square matrix, then a 2D Transformer framework is applied to capture the interactions of different emotion-cause pairs. Finally, valid emotion-cause pairs are extracted by conducting a binary classification.

Trans-ECPE [5] transforms the ECPE problem into a procedure of directed graph construction. The directed graph is constructed by incrementally creating the labeled edges according to the causal relationship between the connected nodes, through a sequence of defined actions 10.

Rank-CP [15] ranks clause pair candidates in a document, and proposes a one-step approach to perform end-to-end extraction.

ECPE-MLL [4] transforms the ECPE task into an emotion-pivot cause extraction problem and a cause-pivot emotion extraction problem, and gets the final emotion-cause pairs based on the combination of these two predictions.

5.2   Experiment Results and Discussions

The experimental results are shown in the following Table 2.

Table 2. The experiment results for the ten models on the two tasks. In this table, CEC represents the CoEmoCause dataset; Re denotes the performance reduction of CEC compared with CN dataset.

Method	Model	Precision			Recall			F1		
		CEC	CN	Re	CEC	CN	Re	CEC	CN	Re
ECE	MANN	.774	.784	−.01	.519	.759	−.2394	.621	.771	−.1497
	CANN	.646	.772	−.1262	.597	.689	−.0925	.620	.727	−.1067
	RTHN	.749	.770	−.0208	.634	.733	−.099	.686	.750	−.0638
	FSS-GCN	.802	.786	.0156	.649	.757	−.1085	.717	.771	−.0543
	PAEDGL	.801	.762	.0392	.584	.691	−.1072	.675	.724	−.0496
ECPE	ECPE	.609	.690	−.0816	.340	.514	−.1734	.435	.590	−.1548
	ECPE-2D	.599	.692	−.0991	.515	.590	−.0729	.553	.636	−.0835
	TransECPE	.544	.737	−.1938	.469	.631	−.162	.503	.680	−.1766
	Rank CP	.397	.661	−.2645	.241	.670	−.4287	.300	.655	−.3548
	ECPE-MLL	.851	.730	.1213	.620	.642	−.0216	.716	.682	.0345

We further analyzed the reasons for the decline of the model performance. In the ECPE task, the inter-ce cross-road method of the ECPE-2D model has the smallest performance decline. The inter-ce model uses the same clause vector representation as to the ECPE model, but on this basis, the inter-ce model combines the emotion specified vector representation and the cause specified vector representation to form a two-dimensional matrix, and applies the attention mechanism on this matrix to extract emotion-cause pairs.

The performance of these two models in the original dataset and our dataset is shown as follows:

Table 3. The performance comparison of CoEmoCause and CN datasets

Model	Precision		Recall		F1	
	ECPE	ECPE-2D	ECPE	ECPE-2D	ECPE	ECPE-2D
CoEmoCause	0.6086	0.5989	0.3401	0.5149	0.4353	0.5533
CN	0.6902	0.6922	0.5135	0.5904	0.5901	0.6356
Resuction	−0.082	−0.093	−0.173	−0.076	−0.155	−0.082

The two models use the same method to extract the vector representation of clauses, the only difference is that ECPE directly extracts emotions and reasons by using vector representation of clauses, and makes Cartesian product of emotion set and cause set. The result of the Cartesian product is used to obtain all emotion-cause pairs, and then a filter is applied to filter out the emotion-cause pair which is not a real emotion-cause pair. ECPE-2D combines emotion-specific clause vectors and cause-specific clause vectors to form a two-dimensional matrix. On this basis, a transformer is applied to extract information and an attention mechanism is used to extract inter-clause relationships. Finally, the information is integrated together to get emotion-cause pairs. However, in the original paper, the difference of F1 performance between the two models is only 0.0455, while the difference of performance on the CoEmoCause dataset is 0.1180, which is more than twice that of the original dataset. By using a transformer and attention mechanism to extract semantic information and considering the relationship between clauses, ECPE-2D effectively improves the prediction accuracy.

The difference in experimental results between the two models shows that our dataset requires the model to have a deeper understanding of semantics and to use higher-level structures to extract the information contained in the text, and transformer and attention mechanism can effectively extract the semantic information from the central emotion-cause pair and surrounding emotion-cause pairs.

6 Conclusion

In this paper, we provided a new corpus for emotional cause extraction and emotional-cause pair extraction. The corpus has the characteristics of large document scale, colloquial language, and rich semantics. We used 10 state-of-the-art

ECE and ECPE models to train on the dataset, and found that the performance of all models decreased significantly. Combined with the dataset, we further analyzed the reasons for these declines.

Acknowledgments. The work was supported by the National Natural Science Foundation of China (61872074, 61772122), the Fundamental Research Funds for the Central Universities (N180716010, N2124001), and the National Training Program of Innovation and Entrepreneurship for Undergraduates (S202110145126).

References

1. Cheng, X., Chen, Y., Cheng, B., Li, S., Zhou, G.: An emotion cause corpus for Chinese microblogs with multiple-user structures. ACM Trans. Asian Low Resour. Lang. Inf. Process. **17**(1), 6:1–6:19 (2017)
2. Ding, Z., He, H., Zhang, M., Xia, R.: From independent prediction to reordered prediction: integrating relative position and global label information to emotion cause identification. In: AAAI, pp. 6343–6350 (2019)
3. Ding, Z., Xia, R., Yu, J.: ECPE-2D: emotion-cause pair extraction based on joint two-dimensional representation, interaction and prediction. In: Jurafsky, D., Chai, J., Schluter, N., Tetreault, J.R. (eds.) ACL, pp. 3161–3170 (2020)
4. Ding, Z., Xia, R., Yu, J.: End-to-end emotion-cause pair extraction based on sliding window multi-label learning. In: Webber, B., Cohn, T., He, Y., Liu, Y. (eds.) EMNLP, pp. 3574–3583 (2020)
5. Fan, C., Yuan, C., Du, J., Gui, L., Yang, M., Xu, R.: Transition-based directed graph construction for emotion-cause pair extraction. In: Jurafsky, D., Chai, J., Schluter, N., Tetreault, J.R. (eds.) ACL, pp. 3707–3717 (2020)
6. Gao, Q.: Overview of NTCIR-13 ECA task (2017)
7. Gui, L., Hu, J., He, Y., Xu, R., Lu, Q., Du, J.: A question answering approach to emotion cause extraction. CoRR arXiv:1708.05482 (2017)
8. Gui, L., Wu, D., Xu, R., Lu, Q., Zhou, Y.: Event-driven emotion cause extraction with corpus construction. In: EMNLP, pp. 1639–1649 (2016)
9. Hu, G., Lu, G., Zhao, Y.: FSS-GCN: a graph convolutional networks with fusion of semantic and structure for emotion cause analysis. Knowl. Based Syst. **212**, 106584 (2021)
10. Lee, S.Y.M., Chen, Y., Huang, C.R.: A text-driven rule-based system for emotion cause detection. In: Proceedings of the NAACL HLT 2010 Workshop on Computational Approaches to Analysis and Generation of Emotion in Text, CAAGET 2010, pp. 45–53. Association for Computational Linguistics, New York (2010)
11. Li, X., Feng, S., Wang, D., Zhang, Y.: Context-aware emotion cause analysis with multi-attention-based neural network. Knowl. Based Syst. **174**, 205–218 (2019)
12. Li, X., Song, K., Feng, S., Wang, D., Zhang, Y.: A co-attention neural network model for emotion cause analysis with emotional context awareness. In: Riloff, E., Chiang, D., Hockenmaier, J., Tsujii, J. (eds.) EMNLP, pp. 4752–4757 (2018)
13. Liang, Y., Xu, L., Huang, T.: Sentiment tendency analysis of NPC&CPPCC in German news. In: Ni, W., Wang, X., Song, W., Li, Y. (eds.) WISA 2019. LNCS, vol. 11817, pp. 298–308. Springer, Cham (2019). https://doi.org/10.1007/978-3-030-30952-7_30

14. Liu, W., Zhang, M.: Semi-supervised sentiment classification method based on Weibo social relationship. In: Ni, W., Wang, X., Song, W., Li, Y. (eds.) WISA 2019. LNCS, vol. 11817, pp. 480–491. Springer, Cham (2019). https://doi.org/10.1007/978-3-030-30952-7_47
15. Wei, P., Zhao, J., Mao, W.: Effective inter-clause modeling for end-to-end emotion-cause pair extraction. In: Jurafsky, D., Chai, J., Schluter, N., Tetreault, J.R. (eds.) ACL, pp. 3171–3181 (2020)
16. Xia, R., Ding, Z.: Emotion-cause pair extraction: a new task to emotion analysis in texts. In: Korhonen, A., Traum, D.R., Màrquez, L. (eds.) ACL, pp. 1003–1012. Association for Computational Linguistics (2019)
17. Xia, R., Zhang, M., Ding, Z.: RTHN: a RNN-transformer hierarchical network for emotion cause extraction. In: Kraus, S. (ed.) IJCAI, pp. 5285–5291 (2019)

Fusing Various Document Representations for Comparative Text Identification from Product Reviews

Jing Liu[1,2(✉)], Xiaoying Wang[2], and Lihua Huang[1]

[1] School of Management, Fudan University, 200433 Shanghai, People's Republic of China
[2] School of Management Science and Engineering, Tianjin University of Finance and Economics, 300222 Tianjin, People's Republic of China

Abstract. Identifying reviews that convey comparative opinions is a preliminary and important task for comparative opinion mining. However, prior studies are time- and cost-consuming, present weak generalization capability and weak robustness, and are semantic-unaware. To address these issues and considering that document representations are differentially informative for different samples, we propose a novel neural network architecture named DRFA (Document Representation Fusion with Attention) for identifying comparative text from large amounts of product reviews on E-commerce platforms. Our developed method adopts the attention mechanism to fuse various representations derived by applying LSTM, CNN, and BERT. A series of experiments verified the effectiveness of our proposed DRFA in terms of improving predictive performance and enhancing robustness. This study has implications for both researchers and practitioners.

Keywords: Representation fusion · Deep neural network · Attention · Comparative text identification

1 Introduction

The popularity of E-commerce platforms has witnessed a rapid increase in the amount of product reviews, since people tend to exchange views, experiences and opinions on E-commerce platforms [1]. Product reviews are valuable for both consumers and manufactures. On one hand, in 2020, 87% of consumers would read online reviews when they make a purchase [2]. On the other hand, manufacturers can mine consumers' opinions regarding their products or services, thereby clarifying product positioning, estimating core competencies of products, and improving product quality [1]. Compared with traditional information collection channels (e.g., annual reports, questionnaires, and interviews), mining product reviews is more credible and timely [3]. Comparison, a common expression form on E-commerce platforms, can convey information regarding commonalities and differences between products [4]. Comparative opinion mining, i.e., mining opinions expressed through comparing different entities, has been widely used for competitor analysis [5], product strengths and weaknesses analysis [6], and product-level customer preference analysis [7], et al.

© Springer Nature Switzerland AG 2021
C. Xing et al. (Eds.): WISA 2021, LNCS 12999, pp. 531–543, 2021.
https://doi.org/10.1007/978-3-030-87571-8_46

However, conducting comparative opinion mining is a non-trivial task. One of main challenges is the low value density of web big data, which means that massive product reviews do not convey comparative opinions [1]. Therefore, a preliminary and important subtask for comparative opinion mining is to effectively find relevant product reviews (i.e., comparative text identification), for which there is still no widely recognized techniques [8]. Existing methods generally combine Class Sequence Rule (CSR)-generated rules and manually predefined keywords as features, and feed them into machine learning algorithms, such as Support Vector Machines (SVM) and Naive Bayes (NB) [5, 6]. These conventional comparative text recognition methods cannot be liberated from labor-intensive work, and present weak robustness and generalization capability when applied to new datasets. Moreover, they are conducted at the lexical and syntactic level, rather than at the semantic level, thus incapable to identify implicit comparative reviews which are without obvious comparison keywords or sequence patterns.

Comparative text recognition can be regarded as a text classification task, in which deep neural networks (DNNs) have made outstanding contributions [9]. In recent years, Convolutional Neural Networks (CNN) [10], Recurrent Neural Networks (RNN), and Bidirectional Encoder Representation from Transformers (BERT) [11] are the most popular deep learning architectures that can extract discriminative information automatically from raw data for addressing natural language processing (NLP) tasks. Moreover, Long Short-term Memory (LSTM) [12] mitigates the gradient vanishing problem of conventional RNN by developing the gating mechanism. Due to the variety of comparative expression patterns and the respective structure of each DNN, we argue that there is no deep learning model capable to effectively recognize all styles of comparative expressions since different deep learning methods are expert at dealing with different types of comparative expressions. This is supported by our preliminary experimental results. Several comparative reviews and their identification results by BERT, CNN, and LSTM individually are shown in Table 1. The CNN model, which is good at capturing local information, is expert at recognizing reviews with obvious comparative keywords (e.g., "more convenient"); the LSTM model, which is adept at capturing sequence dependency, performs better for comparative reviews that rely on the context; the BERT model, which is based on the transformer structure, is expert at addressing comparative reviews that require global semantic understandings [11]. Therefore, for the comparative text recognition task, the intuition underlying this study is twofold: (1) It is observed that LSTM, CNN, and BERT-derived document representations are differentially informative for different samples. (2) Different DNNs focus on capturing different facets of information, therefore may complement with each other.

Therefore, in this paper, we develop a novel neural network architecture named DRFA (Document Representation Fusion with Attention) for identifying product reviews that convey comparative opinions on E-commerce platforms. Firstly, we conduct automatic feature learning by applying LSTM, CNN, and BERT, respectively. Secondly, to fully explore the complementary nature of various document representations and achieve sample-dependent importance adjustment for these document representations, we incorporate attention mechanism to fuse them and assign weights adaptively. The main contributions of this study are as follows:

Table 1. Several comparative reviews and their identification results.

Comparative reviews	Identification results		
	BERT	CNN	LSTM
有home键的手机使用起来更方便了 Mobile phones with home buttons are more convenient to use	0	1	0
拿着iPhone8看着新手机iPhonexr, 感觉还是xr比较好用, 但是实体 使用还是iphone 8比较舒服 When holding the iPhone 8 and looking at the new mobile phone iPhone xr, I thought that the xr is easier to use, but the actual use of iPhone 8 is more comfortable	0	0	1
非常nice, 第一次用iPhone, 果然最吸引人的地方是流畅的系统, 安 卓系统给予不了的感觉, 高颜值的机身和舒适的屏幕, 总给人的感 觉是完美的 Very nice, it is the first time that I use iPhone. It turns out that the most attractive part is the smooth system. The Android system can't give you that feeling. The high-value body and comfortable screen always give people a perfect feeling	1	0	0

(1) A novel deep neural network-based framework is developed for comparative text identification. The adoption of deep learning can conduct feature learning automatically, effectively capture different types of discriminative information including semantic information, and hence enhancing the identification capability.

(2) We explore the complementary nature of document representations derived from different DNNs by incorporating the attention mechanism. The fusion of various document representations and incorporation of attention mechanism can enhance the predictive capability and the robustness of comparative text recognition models.

(3) To evaluate the generalization capability of the proposed approach, we conducted extensive experiments. The empirical results demonstrate the effectiveness of our proposed approach.

2 Literature Review

2.1 Comparative Opinion Mining

Prior studies on comparative opinion recognition mainly include two subtasks: identification of comparative text and extraction of the comparative relation. For the first subtask, prior work often uses the combination of CSR rules [13] and machine learning [14, 15]. For the second tasks, comparative relation extraction aims to extract comparative subject, comparative object, comparative content, and comparative result, or assess the competitiveness of an enterprise [16]. In this Section, we focus on reviewing studies on comparative text identification, i.e., the first subtask. Jindal et al. [13] pioneered the line of research, proposed a recognition method based on CSR and the supervised learning. For Chinese comparative text identification, Huang et al. [17] defined various comparative types, compared a keyword-based classifier and a pattern-based method. Wang et al. [15] adopted the improved PrefixSpan algorithm [18] to extract sequence patterns, and feed CSR-derived rules together with some manual rules, into the SVM algorithm. Panchenko et al. [8] demonstrated that combining CSR, keywords, and machine learning has achieved the best performance. They pointed out that the future direction of comparative text recognition should be towards the construction of neural network models with contextualized representations.

Conventional comparative text identification methods heavily rely on CSR-derived sequence patterns and pre-defined keywords that play an essential role for obtaining satisfactory performance. These methods have three main limitations: first, extensive labor work is required to build the keyword list. Second, they cannot guarantee to cover all rules and cannot be directly applied to other domains, which limits the generalization ability and robustness of models. Third, with these methods, the identification is performed using lexical information and simple syntactic information, e.g., part-of-speech (POS), thus being incapable to identify implicit comparative text due to unawareness of semantic information.

2.2 Text Classification Based on Deep Learning

Text classification methods have experienced the process from shallow to deep learning [19]. Among DNNs, RNN is an intuitive architecture for processing sequence text data due to its strong capability of capturing long-range dependency. For example, Dieng et al. [20] proposed a TopicRNN model, in which local and global dependencies are captured by RNN and latent topics, respectively. Many studies advance the text classification research using LSTM, an improvement of conventional RNN. For example, Tai et al. [9] used the tree structure to formulate sequence, and enable the whole subtree contributing little on the result be forgotten. Although CNN was initially proposed to address images, its effectiveness on text classification has been verified by prior studies. Kim et al. [21] proposed a CNN architecture with one convolution layer. Johnson et al. [22] trained a semi-supervised CNN model using labeled and unlabeled samples. Many studies have pointed out that CNN and LSTM have different performance when dealing with different types of tasks [23]. Yin et al. [24] compared the differences between CNN and LSTM for sentiment classification, observing that CNN is suitable for classifying samples whose sentiment were determined by certain keywords, while RNN is suitable for predicting samples whose sentiment were determined by long-range semantic dependency. The proposal of BERT [11], a transformer-based method, has boosted various NLP tasks. Su et al. [25] employed a hierarchical LSTM to receive sentence and word embeddings obtained from BERT. However, the complementary nature of different document representations supported by our preliminary experiments is overlooked and worth exploring.

3 Proposed Model

In this research, our goal is to establish a feature fusion-based comparative text recognition framework (as shown in Fig. 1), which uses the attention mechanism to adaptively

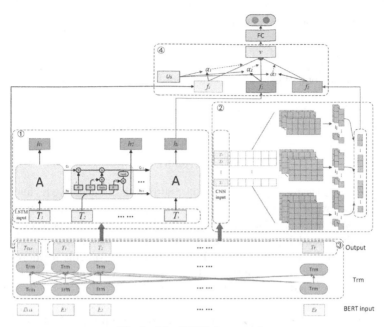

Fig. 1. The DRFA framework.

weight document representations obtained by LSTM, CNN, and Bert. As a result, we obtain the fused text representation.

Most comparative reviews follow certain sequential patterns, which is the reason why the CSR is the dominated method in previous studies. RNN is the conventional sequence modeling architecture, which considers the sequence dependencies of input data and can model fixed sequence patterns for identifying comparative text. Beyond that, LSTM is proposed to capture and preserve the long-term dependencies, achieving promising results in various sequence modeling tasks [26]. Therefore, we adopt LSTM, whose structure is shown in the first dashed box in Fig. 1. The LSTM unit has three gates to protect and control the unit status and information flow. The *forget gate* determines which part of information should be discarded based on the previous state h_{t-1} and the input vector x_t:

$$f_t = \sigma\left(W_f x_t + U_f h_{t-1} + b_f\right) \tag{1}$$

where W_f and U_f are learnable weight matrices, and b_f is the bias vector, σ is the sigmoid activate function. The following W_0, U_0, b_0 denote similar meanings.

The second gate is the input gate (i_t), which controls which part of information (g_t) can added to the current layer.

$$i_t = \sigma(W_i x_t + U_i h_{t-1} + b_i) \tag{2}$$

$$g_t = \tan h(W_g x_t + U_g h_{t-1} + b_g) \tag{3}$$

We can subsequently update states of cell memories as follows:

$$c_t = f_t \odot c_{t-1} + i_t \odot g_t \tag{4}$$

The third gate is the output gate. Using c_t and output gate, we can output the long-term memory:

$$o_t = \sigma(W_o x_t + U_o h_{t-1} + b_o) \tag{5}$$

$$h_t = o_t \odot \tanh c_t \tag{6}$$

In addition to sequence patterns, there also exist numerous comparative reviews that can be identified by a word or certain key phrases. For example, the comparative review " 华为荣耀4可属千元机之最 (Huawei Honor 4 could be the best in phones whose price is around one thousand RMB)" can be recognized correctly based on the discriminative word "最 (the best)". Compared with LSTM, CNN is more suitable to address above-mentioned comparative types, mainly attributed to its outstanding ability of capturing local features. In this study, the CNN framework is shown in the second dashed box in Fig. 1.

This study adopts the traditional CNN framework proposed by Kim et al. [21]. A review can be represented as $X_{1:t} = x_1 \oplus x_2 \oplus ... \oplus x_t$, where t is the length of each review (with padding where necessary), $x_i \in R^d$ denotes the d-dimensional word representation corresponding to the i th word in the review, and $\oplus$ represents the concatenation operator. A filter $w \in R^{h \times d}$ is involved in each convolution operation, aiming at generating a feature using a word matrix $X_{i:i+h-1}$ derived from a window of (i.e., h) words. The generated feature is calculated as:

$$c_i = f(w \cdot x_{i:i+h-1} + b) \tag{7}$$

where $b \in R$ is a bias term, f represents a nonlinear activate function. When applied to all possible windows of words in the review, the filter can generate a $(t-h+1)$-dimensional feature map:

$$c = \begin{bmatrix} c_1, c_2, ..., c_{t-h+1} \end{bmatrix} \tag{8}$$

Subsequently, a max-pooling operation is applied on the above-mentioned feature map, which means the maximum value $\hat{c} = \max\{c\}$ is taken as the feature finally derived by the filter. Suppose k filters are implemented, the document representation obtained by applying CNN is $\hat{c} = \begin{bmatrix} \hat{c}_1, \hat{c}_2, ..., \hat{c}_k \end{bmatrix}$.

There also exist many reviews that need global semantic understandings to be correctly classified. Pre-trained language models can effectively capture global semantics and significantly improve the performance of NLP tasks [19]. One of popular pre-trained models is BERT, which is designed to pre-train bidirectional representations using unlabeled text considering both right and left context in every layer [11]. Multi-layer bidirectional Transformers constitute the architecture of BERT. Transformer uses self-attention instead of traditional RNN to calculate the relationships between all word pairs in a sequence, which can be performed in parallel [27]. Two steps are involved in BERT for downstream tasks: pre-training and fine-tuning. The pre-training of BERT is guided by two supervised tasks: The Masked Language Model (MLM) task and Next Sentence Prediction (NSP). To learn a task-specific model, a fully connected layer is added to the pre-trained BERT model, jointly fine-tuning all parameters. The input representations of BERT are the sum of corresponding token, segment and position embedding. We use the [CLS] output as the document representation derived by BERT, and the BERT word vectors are fed into the LSTM and CNN architectures as inputs [28].

To take advantage of different DNNs' merits and adapt to varied importance of each representation for different samples. In this paper, we incorporate the attention layer (as shown in the fourth dashed box in Fig. 1) to adaptively fuse different document representations. Let $f_1, f_2,$ and f_3 be the *CLS* vector generated by BERT([*CLS*]), the long-term information generated by LSTM (h_t), and the document representation generated by CNN ($\hat{c}$), respectively. To implement the attention mechanism, inspired by the work [29], we introduce a context vector u_s, which is randomly initiated and learned during the training phase. The normalized importance of each feature type and fused text vector are respectively calculated as:

$$\alpha_i = \frac{\exp(f_i^T u_s)}{\sum_{i=1}^{3} \exp(f_i^T u_s)} \tag{9}$$

$$v = \sum_{i=1}^{3} \alpha_i f \tag{10}$$

Minimizing the cross-entropy loss function is the objective function of our model. The formula is as follows:

$$L = -\sum_{i=1}^{N} y^{(i)} \log \hat{y}^{(l)} \tag{11}$$

where $\hat{y}^{(i)}$ is the possible probability under each label category, $y^{(i)}$ is the true label of the sample, and N is the number of samples.

4   Experimental Design

We implemented our experiments on a dataset derived from customer reviews regarding mobile phones on JD.com (a well-known E-commerce platform in China, https://www.jd.com/). The collected information is composed of review content, review post time,

Table 2. The detailed information of involved product reviews.

Detailed categories	Detailed Information	Detailed categories	Detailed Information
order time span	2017/02/05 – 2019/10/10		
# reviews	132,696	# products	106
# product brands	12	# product sub-categories	1011

star rating, product attributes, URL, page title from which corresponding product name can be extracted. The detailed information is shown in Table 2.

As a preliminary work of labeling, we reviewed the prior work on the category division of comparative sentences. Jindal and Liu [14] concluded four types of comparatives: Non-Equal Gradable, Equative, Superlative, and Non-Gradable. In this study, we focused on classifying customer reviews into two categories: comparative and non-comparative reviews. All reviews belong to abovementioned comparative categories were labeled as positive, and other reviews were labeled as negative. We randomly labeled 20,000 reviews, among which the number of comparative reviews (positive class) and that of non-comparative reviews (negative class) are 3,284 and 16,716, respectively. To address the imbalance issue, we performed oversampling on the training and validation set, while kept the original distribution on the test set. In our experiments, the ratio of training set, validation set, and test set is 8:1:1.

Online reviews in Chinese generally contain large numbers of non-normal characters, gibberish, and emoticons. Pre-processing implemented in this study included converting traditional Chinese characters to simplified Chinese characters for uniform formatting, converting half-angle symbols to full-angle symbols, and removing meaningless symbols and emoticons.

In order to evaluate the effectiveness of our DRFA approach for identifying comparative reviews, we adopt the F-Score and accuracy (ACC), which have been widely used in previous literature on text classification [13].

In this study, we conducted a series of experiments to verify the effectiveness of the proposed DRFA method. The first experiment compared our method with the state-of-the-art approach (i.e., CSR + SVM) in the previous related research on comparative text identification. Experiment 2 was designed to verify the effectiveness of fusing various representations by comparing our method with implementations using a single representation, i.e., BERT, CNN, and LSTM, respectively. Experiment 3 was intended to investigate whether our attention-based fusion strategy can deliver an improved performance over other fusion strategies, i.e., model-level fusion and feature-level fusion with direct concatenation (referred as "feature-concatenation"). We performed 10-fold cross-validation for all experiments.

The experiments were performed on a Linux server with GPU, and implemented in Python with PyTorch 1.3. After parameter-tuning, the learning rate, batch size, pad size, epoch were set to 1e–5, 32, 120, and 20, respectively. For BERT, we adopted the $BERT_{BASE}$ model. For CNN, we adopted three types of filters with the size of 3×768, 4×768, and 5×768. Each type is composed of 256 filters.

5   Experiment Results and Discussion

Table 3 reports the experimental results. The values following " $\pm$ " represent standard deviations. Our proposed method delivers the highest performance values and the lowest standard deviations. Paired t-tests were performed to compare the performance of DRFA against other methods. As shown in Table 3, deep representation learning-based methods outperform the traditional comparative text identification approach, i.e., CSR + SVM. Especially, our DRFA method obtains a significant F-Score improvement of

19.14% over CSR + SVM (with p-values < 0.01). The results support our hypothesis that using automatic feature learning via DNNs can facilitate enhanced comparative text identification capabilities. Moreover, when compared with using single representation, fusing various documents representations significantly improves the F-Score (from 78.07%, 78.20%, 77.38% to 81.76%) and reduces the standard deviations from 1.81, 1.81, 2.08 to 0.84. The results suggest exploring the complementary nature of different representations can contribute to the enhanced identification capability and robustness. With respect to different fusion strategies, our attention-based fusion strategy delivers an improved F-Score and presents enhanced robustness. As shown in Table 3, the performance of directly concatenating various representations is unstable, mainly attributed to the high dimensionality.

Table 3. Experimental results.

	Model	F-Score (%)	ACC (%)
Traditional method	CSR + SVM	$62.62 \pm 1.53^{**}$	$86.47 \pm 1.04^{**}$
Using single representation	Fine-tuned BERT	$78.07 + 1.81^{**}$	$91.68 \pm 0.90^{**}$
	CNN	$78.20 \pm 1.81^{**}$	$91.58 \pm 0.99^{**}$
	LSTM	$77.38 \pm 2.08^{**}$	$91.30 \pm 0.86^{**}$
Using different fusion strategies	Model-level	$78.88 \pm 1.64^{**}$	$91.86 \pm 0.71^{**}$
	Feature-concatenation	$79.23 \pm 2.59^{*}$	$92.36 \pm 1.41^{*}$
	DRFA method	$\mathbf{81.76 \pm 0.84}$	$\mathbf{93.63 \pm 0.40}$

*p-values significant at a $= 0.05$.
**p-values significant at a $= 0.01$.

We calculate the Pearson correlation of attention weights assigned to different representations. As shown in Fig. 2, the attention weight of BERT is strongly negatively correlated with that of LSTM ($r = -0.97$) and CNN ($r = -0.83$). The *substitution relation* supports our hypothesis that the importance of each representation should be adaptively updated for different samples, for which we incorporate attention mechanism in our framework. The attention weight of LSTM and that of CNN present a strong positive correlation ($r = 0.67$), indicating that representations of LSTM and CNN are simultaneously utilized and work together to classify a sample. This *complementary relation* verifies the necessity of fusing different representations.

Figure 3 shows the average attention weights for identifying comparative text and non-comparative text. As shown in Fig. 3, CNN-based representation obtains the highest importance for predicting both positive and negative samples. For identifying comparative text, the final feature vector mainly depends on the CNN (49.97%) and LSTM (32.31%)-derived representations, while the weight of the representation provided by BERT accounts for 17.73%. This finding suggests that many comparative reviews can be classified by automatically learned keywords or long-term sequence patterns, while the overall semantic understandings are essential supplements.

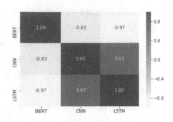

Fig. 2. Pearson correlation results

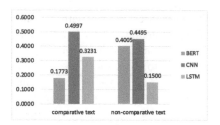

Fig. 3. Average attention weights of representations

Based on the analysis of our results, most comparative reviews need to consider all three document representations, but weight distribution varies significantly across different reviews. Table 4 presents attention weight distributions of several comparative reviews. For the first example, CNN-based representation is assigned the highest weight (65.53%), because the most discriminative information lies in the local expressions "比较", and CNN is expert at capturing local information. For the second example, it is very short and contains the sequence pattern "…比…快…". CNN and LSTM based representations obtain approximately equal importance, i.e., 46.28% and 43.87%, respectively, while BERT derives an unimportant representation with a low attention weight. However, for the third example which is characterized as semantic-aware, BERT's weight is relatively high (i.e., 47.75%).

Table 4. The distribution of attention weights of several comparative reviews.

	Content	BERT	CNN	LSTM
1	拿着iPhone 8和新手机iPhone xr, 感觉还是xr比较 好用,但是实体使用还是8比较舒服	13.64%	65.53%	20.83%
	Holding the iPhone 8 and the new mobile phone iPhone xr, I feel that the xr is easier to use, but physically using iPhone 8 is more comfortable			
2	8要比我以前的7 pus快很多	9.84%	46.28%	43.87%
	8 is much faster than my previously used 7 plus			
3	在v10和play之间犹豫了很久,都是1699,最后还是选了play. 虽然v10配置和相机都完爆play,但是我还是喜欢play的屏幕够大,颜值担当	47.75%	40.33%	11.92%
	I hesitated for a long time between V10 and play because both of them are priced as 1699. Finally, I chose play. Although the configuration and camera of V10 are better than that of play, I prefer play's screen, which is big enough and good-looking			

In order to clearly demonstrate the effectiveness of our proposed method, we performed t-SNE [30] to visualize the BERT, CNN, LSTM-based, and the fused representation, respectively. As shown in Fig. 4, our proposed DRFA-based representations of positive and negative reviews are easier to be classified.

Fig. 4. Visualization of single and the fused representation

6 Conclusion

In this paper, we propose a novel neural network architecture named DRFA for identifying comparative reviews from large numbers of product reviews on E-commerce platforms. Firstly, we obtain different document representations by performing LSTM, CNN, and BERT, respectively. Secondly, we adopt the attention mechanism to adaptively update the weight distribution of these representations. The experimental results demonstrated that our proposed DRFA approach can deliver significantly improved performance and enhance the robustness of the model across different test datasets. Based on the result analysis, we obtain the following conclusions for comparative text identification: (1) There exists a substitution relation between BERT and the other two methods. They are expert at recognizing different types of comparative reviews. (2) There exists a complementary relation between CNN and LSTM. Simultaneously utilizing them contributions to the enhanced performance. (3) For comparative review identification, CNN and LSTM contribute most while BERT plays a supplementing role, while BERT and CNN contribute more for non-comparative review recognition.

This study has implications for both practitioners and researchers. On one hand, this study can benefit stakeholders. For example, recognizing reviews that convey comparative opinions can lay a foundation for subsequent comparative relation mining, thereby identifying product strengths and weaknesses for product manufacturers, and providing decision making support for customers. On the other hand, our study takes a step forward in fusing multi-source information by regarding representations by different DNNs as different sources, and incorporating the attention mechanism to fuse them. The DRFA approach is scalable since it is not limited by the number of representations.

In the future, we would like to apply our approach to other text classification tasks, in order to further verify its generalization capability. Moreover, it would be interesting to investigate the strategy of preventing the incorporation of redundant information. In addition, we also plan to try other fusion strategies, such as gate mechanism.

Acknowledgments. This work is partially supported by the National Natural Science Foundation of China (Nos. 71701142), and China Postdoctoral Science Foundation (No. 2018M640346).

References

1. Varathan, K.D., Giachanou, A., Crestani, F.: Comparative opinion mining, Rev. (2017)
2. https://www.brightlocal.com/learn/local-consumer-review-survey/
3. Kuruzovich, J., et al.: Marketspace or marketplace? online information search and channel outcomes in auto retailing. Inf. Syst. Res. **19**(2), 182–201 (2008)
4. Jin, J., Ji, P., Gu, R.: Identifying comparative customer requirements from product online reviews for competitor analysis. Eng. Appl. Artif. Intell. **49**, 61–73 (2016)
5. Song, G., et al.: Identifying competitors through comparative relation mining of online reviews in the restaurant industry. Int. J. Hospitality Manage. **71** (2018)
6. Wang, H., Wang, W.: Product weakness finder: an opinion-aware system through sentiment analysis. Ind. Manag. Data Syst. **114**(8), 1301–1320 (2014)
7. Jin, J., Ji, P., Yan, S.: Comparison of series products from customer online concerns for competitive intelligence. J. Ambient Intell. Humanized Comput. (2018)
8. Panchenko, A., et al.: Categorizing comparative sentences. (2018)
9. Tai, K.S., Socher, R., Manning, C.: Improved semantic representations from tree-structured long short-term memory networks. Comput. Sci. **5**(1), 36 (2015)
10. LeCun, Y., et al.: Gradient-based learning applied to document recognition. Proc. IEEE **86**(11), 2278–2324 (1998)
11. Devlin, J., et al.: BERT: Pre-training of deep bidirectional transformers for language understanding. (2018)
12. Hochreiter, S., Schmidhuber, J.: Long short-term memory. Neural Comput. **9**(8), 1735–1780 (1997)
13. Jindal, N., Bing, L.: Identifying comparative sentences in text documents. In: SIGIR 2006: Proceedings of the 29th Annual International ACM SIGIR Conference on Research and Development in Information Retrieval, Seattle, Washington, USA, pp. 6–11 (2006)
14. Jindal, N., Liu, B.: Mining comparative sentences and relations, vol. 2. (2006)
15. Wang, W., et al.: Exploiting machine learning for comparative sentences extraction. Int. J. Hybrid Inf. Technol. **8**(3), 347–354 (2015)
16. Chang, Y., Li, Y., Chen, C., Cao, B., Li, Z.: An enterprise competitiveness assessment method based on ensemble learning. In: Ni, W., Wang, X., Song, W., Li, Y. (eds.) WISA 2019. LNCS, vol. 11817, pp. 79–84. Springer, Cham (2019). https://doi.org/10.1007/978-3-030-30952-7_9
17. Huang, X., et al.: Learning to identify comparative sentences in chinese text. (2008)
18. Liu, B.: Web Data Mining: Exploring Hyperlinks, Contents, and Usage Data. Springer-Verlag, New York (2007)
19. Li, Q., et al.: A survey on text classification: from shallow to deep learning. (2020)
20. Dieng, A.B., et al.: TopicRNN: a recurrent neural network with long-range semantic dependency. (2016)
21. Kim, Y.: Convolutional neural networks for sentence classification. Eprint Arxiv (2014)
22. Johnson, R., Tong, Z.: Semi-supervised convolutional neural networks for text categorization via region embedding. Adv. Neural Inf. Process. Syst. **28**, 919–927 (2015)
23. Vu, N.T., et al.: Combining recurrent and convolutional neural networks for relation classification. In: NAACL 2016. (2016)
24. Yin, W., et al.: Comparative study of CNN and RNN for natural language processing. (2017)
25. Su, J., et al.: BERT-hLSTMs: BERT and hierarchical LSTMs for visual storytelling. Comput. Speech Lang. **67**, 101169 (2021)

26. Bahdanau, D., Cho, K., Bengio, Y.: Neural machine translation by jointly learning to align and translate. Comput. Sci. (2014)
27. Vaswani, A., et al.: Attention Is All You Need. arXiv (2017)
28. Wan, T., Wang, W., Zhou, H.: Research on information extraction of municipal solid waste crisis using BERT-LSTM-CRF. (2020)
29. Yang, Z., et al.: Hierarchical attention networks for document classification. In: Proceedings of the 2016 Conference of the North American Chapter of the Association for Computational Linguistics: Human Language Technologies. (2016)
30. Laurens, V.D.M., Hinton, G.: Visualizing data using t-SNE. J. Mach. Learn. Res. 9(2605), 2579–2605 (2008)

The Code Generation Method Based on Gated Attention and InterAction-LSTM

Yuxuan Wang and Junhua Wu[✉]

School of Computer Science and Technology, Nanjing Tech University,
Nanjing 211816, China
wujh@njtech.edu.cn

Abstract. Code generation is an important research field of software engineering, aiming to reduce development costs and improve program quality. Nowadays, more and more researchers intend to implement code generation by natural language understanding. In this paper, we propose a generation method to convert natural language descriptions to the program code based on deep learning. We use an encoder-decoder model with gated attention mechanism. Here, the decoder is an InterAction-LSTM. The gated attention combines the previous decoding cell state with source representations to improve the limitation of invariant source representations. The decoder makes the information interact each other before putting them into the gate of the LSTM. The code generation is verified on two datasets, Conala and Django. Compared with known models, our model outperforms the baselines both in Accuracy and Bleu.

Keywords: Code generation · Encoder-decoder · Gated attention · InterAction-LSTM

1 Introduction

With the popularization of software and the emergence of open source communities, a large number of free programs and code are got easily. After studying 420 million lines of code, Gabel et al. [1] find code duplication always occurs. It means that artificial implementation of the same code will prolongs the software development cycle. Xia et al. [2] show that developers spend more than half of their time in understanding programs, and poor readability and maintainability also greatly affect the efficiency and quality of development. Automatic code generation has gotten much attention in academia and industry as it reduces the effort of developers while greatly improves readability and maintainability.

At early stage of code generation, researchers made various attempts based on grammar rules [3–5]. These researches use typically rules defined by people and only applicable to generate domain-specific logic languages. In 2012, Hindel et al. [6] pointed out that programming code is also repetitive just as the representation of natural language theoretically. This property could be exploited to solve the problem of code generation. It is a hypothesis as the foundation of code generation by transmitting natural language descriptions to executable

© Springer Nature Switzerland AG 2021
C. Xing et al. (Eds.): WISA 2021, LNCS 12999, pp. 544–555, 2021.
https://doi.org/10.1007/978-3-030-87571-8_47

code. Ling et al. [7] were inspired by Mandal et al. [8] to consider code generation as the process of sequence modeling, and proposed the latent predictor networks (LPN) model on code generation by using Sequence to Sequence (Seq2Seq). However, they did not consider the structure of code. Compared to natural languages, code often contains rich structural information.

In order to solve the problem that Ling et al. ignored, most of researchers focused on the abstract syntax tree (AST) [9]. Combining with AST, Yin et al. [10] proposed a syntactic neural network model (SNM) which used actions to model the construction process of the syntax tree firstly. After that, some researchers proposed a series of models, such as the abstract syntax networks model (ASN) [11], the retrieval based neural code generation model (ReCode) [12], a transition-based abstract syntax parser (Tranx) [13], the best list of code pieces reranked with Tranx (Tranx+Rerank) [14], and multiple input sources mixed with Tranx (Tranx+API) [15]. Besides using popular recursive neural network (RNN), researchers also proposed the GanCode model [16] and the Tree-Gen model [17]. The former is based on generative adversarial network (GAN), and the latter is based on transformer framework so that the long-dependency problem about sentence can be solved.

In these methods, most of models only use the soft attention mechanism. The invariance of source representations in soft attention will lead to great similarity of adjacent contexts. Meantime, the input of the LSTM is independent before going into each gate. This may result in some errors in generation of syntax trees. To address the problems above, this paper proposes NL2Code (Natural Language Description to Code), a neural network model with the Gated attention (GATT) and InterAction-LSTM (IA-LSTM). GATT combines the previous decoding cell state with source representations to improve the limitation of invariant source representations. The decoder uses the IA-LSTM to implement the further interaction of information before it is input to each gate. This will reduce the probability of syntax tree errors.

2   The Syntactic Rule Generation with Encoder-Decoder Framework

Given a segment of natural language descriptions, our purpose is to generate corresponding code that is matched with the description. Dong et al. [18] proposed a method that uses the rules of the abstract syntax tree to make the code generation. AST represents the code as a syntactic tree structure, where each node corresponds to a piece of code, so that the result generated should conform to the syntax rules of the programming language.

In our NL2Code model, after the source information is encoded, the decoder will generate an AST step by step. Finally a complete AST is converted into the specific code representation (e.g. python). While the first word in the input sequence is "if", NL2Code will generate the trunk of the syntax tree according to conditional statement rules (see Table 1). We have two types of actions to build an AST: ApplyRule and GetToken. ApplyRule applies syntax rules to non-leaf nodes, and GetToken populates variable terminal nodes by appending terminal tokens.

The probability of generating a syntax tree can be decomposed into the probability of generating action sequences and is shown as Eq. (1):

$$p(a \mid x) = \prod_{t=1}^{T} p\left(y_{(act,t)} \mid y < t, x\right) \tag{1}$$

where x is the natural language description and $y_{(act,t)}$ is the action taken at the time t, and $y < t = y_{(act,1)}, y_{(act,2)}, ..., y_{(act,t-1)}$. T is the number of the whole action sequence resulting in AST a.

Table 1. Examples of Python syntax rules.

Rule	Explanation	
Call → expr[func] expr*[args] keyword*[keywords]	Function call rules	func: function to call
	args: parameter list	keywords: keyword list
If → expr[test] stmt*[body] stmt* [orelse]	Conditional statement rules	test: conditional expression
	body: statement in If clause	orelse: elif or else statement

As the illustration of Fig. 1, the action ApplyRule is taken at $t_3 \rightarrow t_4$ and the rule is call → expr[func] (Table 1). The syntax tree is traversed by top-to-bottom and left-to-right rule started from root. The left node is selected first at t_3. The action GetToken fills leaf nodes with values ("str(open)") at t_6 (shown in the dashed box). Actions will be triggered at different time to get different values, and finally the entire syntax tree will be filled and generated.

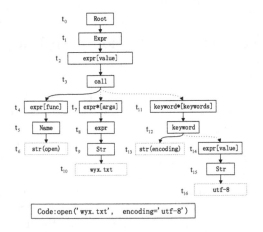

Fig. 1. The Abstract Syntax Tree (AST) for Python.

3   NL2Code Model

NL2Code uses a bidirectional GRU encoder, a gated attention (GATT) and an IA-LSTM decoder. The input sequence is encoded into source representations. GATT mixes the previous cell state of decoder into the source representations to improve its invariance. The IA-LSTM is used to carry out information interaction before it is input to each gate. This information is independent during the decoding process (see Fig. 2).

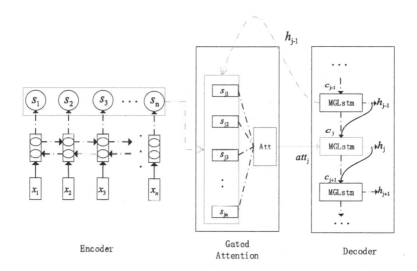

Fig. 2. Workflow of NL2Code.

3.1   Bidirectional GRU

Gated recurrent unit (GRU) is the variant form of recurrent neural network except long short-term memory (LSTM). Compare to the LSTM, GRU converges faster. Moreover, it presents a simpler structure with about 1/3 reduction of parameters to the same goal. GRU is assigned the capability to the preserving and discarding information by reset gate and update gate. These gates ensure the model with memory property. The input vector at time t is x_t and the hidden state at time $t-1$ is s_{t-1}. The forward propagation equation of the neural network of GRU is defined as Eq. (2):

$$s_t = f_{gru}\left(x_t, s_{t-1}\right) \tag{2}$$

The forgetfulness of the recurrent neural network usually makes the information loss, so the training of the forward and backward GRU network has got researchers' attention. Bidirectional GRU can capture the association of each unit between sequences more completely and provide more input sequences to the model. The input sequence $X = (x_1, x_2, ..., x_n)$ will be mapped into forward hidden states $(\overrightarrow{s_1}, \overrightarrow{s_2} \ldots \ldots \cdot \overrightarrow{s_n})$ and backward hidden states $(\overleftarrow{s_1}, \overleftarrow{s_2} \ldots \ldots \cdot \overleftarrow{s_n})$. The overall hidden state is obtained by splicing the forward and backward hidden states at that time, which is shown as Eq. (3):

$$s_t = \left[\overrightarrow{s_t}; \overleftarrow{s_t} \right] \tag{3}$$

The information is encoded into source representations S of the same length as the input sequence. S represents the overall source representations of input sentences and n represents the length of the input sequences, which is shown as Eq. (4):

$$S = (s_1, s_2, ..., s_n) \tag{4}$$

3.2   Gated Attention (GATT)

The attention mechanism is a bridge coupled tightly between encoder and decoder, which is very important in code generation. In 2015, Bahdanau et al. [19] applied the attention mechanism to the field of natural language processing, defined as follows:

$$Att_j = ATT\left(S, c_{j-1}\right) \tag{5}$$

Equation (5) is also called soft attention and is widely used in automatic summarization [20], machine translation [21] and other fields. However, this attention presents the problem of great similarity between adjacent context vectors Att_j. The lack of discrimination will affect the accuracy of code generated.

It should be noted whatever the attention weights change, the source representations S does not change. This invariance will lead to overhigh similarity between Att_j, so the gated attention (GATT) is introduced in this paper.

GATT consists of two layers: GAT layer and ATT layer. Before S is input to the ATT layer, the cell state c_{j-1} in decoding is introduced into the GAT layer. The invariance is broken by segmenting S. It will bring some difference between code in different time of decoder.

GAT layer: The GAT layer controls the semantic matching through one-way GRU, and the equation is as follows:

$$S_j^g = GAT\left(S, c_{j-1}\right) \tag{6}$$

Equation (6) can be split into Eq. (7)–Eq. (10). The cell state c_{j-1} and the tth source representation s_t, are fed into the activation function to calculate the reset gate r_{jt} and the update gate z_{jt} (Eq. (7), Eq. (8)). The reset gate ignores the information from the previous time step and the update gate preserves that. By reset gate and update gate, match degree between the input and words generated can be measured.

$$r_{jt} = \sigma\left(w_{rc}c_{j-1} + w_{rs}s_t + b_r\right) \tag{7}$$

$$z_{jt} = \sigma\left(w_{zc}c_{j-1} + w_{zs}s_t + b_z\right) \tag{8}$$

The new hidden state $\tilde{s_{jt}}$ is calculated with the cell state c_{j-1}, the code s_t and the reset gate result r_{jt} by function tanh (Eq. (9)). If r_{jt} is 0, the new hidden state $\tilde{s_{jt}}$ only depends on c_{j-1}, in other words, the previous hidden state is completely ignored:

$$s_{jt} = \tanh\left(w_s c_{j-1} + w_h \left(r_{jt} * s_t\right) + b_s\right) \qquad (9)$$

$(1 - z_{jt}) * s_t$ represent the code should be forgotten which is deleted from the previous hidden state. $z_{jt} * \tilde{s_{jt}}$ represents assigns weights to code, and finally the tth code of the source representation s_{jt}^g is obtained as Eq. (10):

$$s_{jt}^g = (1 - z_{jt}) * s_t + z_{jt} * \tilde{s_{jt}} \qquad (10)$$

Finally, each s_{jt}^g will be combined to get the source representations S_j^g corresponding to the decoder at the time j. Among them, σ, $tanh$, w_* and b_* represent the sigmod function, the hyperbolic tangent function, the weight matrix and its offset. $*$ represents the multiplication of corresponding elements of the matrix.

ATT layer: The input of the ATT layer is provided by S_j^g. It should be noted that this layer does not focus on the source representation S, only on the code that is subdivided by the GAT layer. The ATT layer attention equation is similar to the soft attention i.e. Eq. (11):

$$Att_j = GATT\left(S_j^g, c_{j-1}\right) = ATT\left(GAT\left(S, c_{j-1}\right), c_{j-1}\right) \qquad (11)$$

Specifically, The computing to ATT(.) is the same way in Eq. (5) and Eq. (11). Here, att_j represents the context vector corresponding to the decoder at the time j. After GATT being calculated, adjacent context vectors are sufficiently distinguished. It will provide enough difference information for generating AST nodes and support the code generation well.

3.3   InterAction-LSTM (IA-LSTM)

The decode is the process of generating AST by syntax rules. Empty nodes of the AST are filled with the obtained information. It is noted that the context vector att_j and hidden states of each gate in the original LSTM are independent, which may lead to the missing of information and affect the quality of the generated code. Therefore, our model uses an IA-LSTM to generate an AST that make an interaction between att_j and h_{j-1} before inputting them to each gate (see Fig. 3).

This process will initialize the input att_j and h_{j-1} with att_j^{-1} and h_{j-1}^0. att_j^i and h_{j-1}^i interaction calculating is according to the parity of parameter i. M and H are auxiliary matrixes defined with the size of $hiddensize * inputsize$.

$$att_j^i = 2\,\text{sig mod}\left(M \cdot h_{j-1}^{i-1}\right) * att_j^{i-2}, i \in (1, 3, 5 \ldots n) \qquad (12)$$

$$h_{j-1}^i = 2\,\text{sig mod}\left(H \cdot att_j^{i-1}\right) * h_{j-1}^{i-2}, i \in (2, 4, 6 \ldots n) \qquad (13)$$

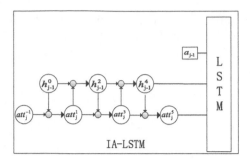

Fig. 3. The Structure of IA-LSTM.

By the Eq. (12) and Eq. (13), we know that h_{j-1}^2 is calculated by att_j^1 and h_{j-1}^0, while att_j^1 is computed out by att_j^{-1} and h_{j-1}^0. Thus, the final vector is obtained alternately and input to each gate of LSTM for calculating. As the value of sigmoid function ranges from 0 to 1 ($[0,1]$), calculating by many times will lead values of att_j and h_{j-1} to become lower. So the stability of values is ensured by multiplying with the constant 2.

$$h_j = f_{lstm} \left(\left[a_{j-1} : att_j^i : c_{j-1} \right], h_{j-1}^i \right) \tag{14}$$

$$p_{(act,j)} = \text{soft max} \left(h_j \right) \tag{15}$$

The hidden state h_j of the decoding layer at the time j is obtained by computing the context vector att_j, the hidden state h_{j-1}, the cell state c_{j-1}, and the action vector a_{j-1} at the time $j-1$ (Eq. (14)). We also define two matrices R and G to aid the action vector a_j. Each row of these auxiliary matrix corresponds to an action ApplyRule or GetToken. Finally, the hidden state h_j is as the input of the softmax function to predict the probability of the action $p_{(act,j)}$ at the time j (Eq. (15)). The IA-LSTM makes the interaction between independent vectors att_j and h_{j-1} before inputting them to each gate. The work aims to strengthen the relation of information and reduce the possibility of generating wrong branches for the abstract syntax tree.

4 Experiments and Results

4.1 Datasets

Django is the collection with 18,805 code lines of Python extracted from the Django web framework by Oda et al. [22], each line with a natural language description manually annotated. Conala is the dataset extracted from Stack Overflow by Yin et al. [23] with 598,237 examples. Considering the information redundancy of Conala, this paper uses 2879 examples manually annotated by Yin et al. 70% of the code in Django is variable assignment, function definition,

Table 2. Examples of Django and CoNaLa.

	Django	Conala
Describe:	remove 'backend' key from params dictionary and put it into back	How to convert a list of multiple integers into a single integer
Code:	back = params.pop ('backend')	r = int(' ',join(map(str,x)))

etc. In contrast, descriptions to the same code in Conala are composed of natural language required by programmers with different intents, which are more complex (see Table 2).

The model in this paper is based on an NVIDIA GeForce RTX 3080 graphics card and Ubuntu 20.04 system, and trained on the GPU using the Pytorch framework. The operating environment of Python 2.7 is created for Django with epoch of 50 and batch of 10, and the operating environment of Python 3.6 is created for CoNaLa with epoch of 80 and batch of 64. Both datasets use word embedding set to 128. The hidden-size of the encoder and decoder is 256. The initial learning rate lr is 0.001, and the dropout is 0.3. Both datasets use Adam as the optimizer. Compared to greedy-search, the model of this paper uses beam-search and sets beam size as 10.

4.2 Metrics

We use accuracy of exact match and bilingual evaluation understudy (Bleu) score to measure the match degree between code piece and natural language descriptions [24]. Accuracy is referred to the percentage of natural language that is correctly parsed into the same word, which reflects the match degree intuitively. However, considering the complex code structure, accuracy is not enough to evaluate the model, so Bleu is also an important metric. The traditional Bleu only considers whether the translation is adequate, but the Bleu improved is considered as percentage of correct translation to n successive words by n-gram, and the fluency of the sentence is also taken into account. This paper sets n as 4.

4.3 Experimental Evaluation

We compared our model to others about code generation which work on these two datasets. The eight models are: (1) natural language programing with automatic code generation (NMT) [8]; (2) latent predictor networks (LPN) [7]; (3) syntactic neural network model (SNM) [10]; (4) transition-based abstract syntax parser (Tranx) [13]; (5) generative adversarial network (GanCode) [16]; (6) retrieval based neural code generation (ReCode) [12]; (7) a multi-source representations generation model (Tranx+API) [15]; (8) a code reranking model (Tranx+Rerank) [14].

According to Table 3, our model NL2Code outperforms the other 7 models in terms of accuracy and Bleu scores in the Django. NL2Code improves accuracy

Table 3. Results of comparison experiment.

DateSets	Django		Conala
Models	Accuracy	Bleu	Bleu
NMT	45.1%	63.40	–
LPN	62.3%	77.60	–
SNM	71.6%	84.50	–
Tranx	73.7%	–	24.35
GanCode	69.7%	–	–
ReCode	72.8%	84.70	–
Tranx+API	–	–	25.90
Tranx+Rerank	80.2%	–	30.11
NL2Code (our model)	81.5%	85.50	29.89

by 1.3% compared to Tranx+Rerank, 7.8% compared to Tranx, and at least 8.7% compared to the other 5 models. In terms of the Bleu score, ReCode, SNM and NL2Code are relatively close, but the Bleu score of our model is still at least 0.8 higher than others. Considering the complexity of the Conala we used the same metrics as the other 3 models and only compared the Bleu scores of the Conala. The NL2Code model is 5.54 higher compared to Tranx and 3.99 higher compared to Tranx+API. Unfortunately, NL2Code is 0.22 lower compared to Tranx+Rerank. But NL2Code performs better in the Django, so it does not affect the validity of the NL2Code overall.

It should be noted that "–" means the given model has no score in Django or Conala. And our model NL2Code has better accuracy and Bleu scores than known models, which indicates gated attention and IA-LSTM playing an important roles. Compared to the models using soft attention such as SNM, Tranx, etc., gated attention breaks the invariance of source representations to differentiate the adjacent vectors, which improves the Accuracy and Bleu scores. When IA-LSTM is used as the decoder unit, the information interacts before it is input to each gate, which strengthens the connection between the information. It will improve the accuracy and Bleu scores compared to other models originated LSTM.

4.4   Component Evaluation

We designed some ablation experiments for our model to analyze the contribution of each component in Table 4. NO GATT means replacing GATT with soft attention and only using IALSTM in NL2Code. NO IALSTM means replacing IALSTM with BILSTM in NL2Code and only using GATT. NO Both means, instead of using GATT and IALSTM, that NL2Code uses soft attention and BILSTM (see Table 4).

Table 4. Effectiveness of each component.

DateSets	Django		Conala
Models	Accuracy	Bleu	Bleu
NO GATT	78.6%	80.65	25.92
NO IALSTM	80.4%	84.20	27.85
NO Both	77.3%	77.80	24.50
NL2Code (our model)	81.5%	85.50	29.89

We observe that Bleu scores in Conala are still much lower than that in Django because of complexity of the Conala data set, even in the ablation experiment. Compared to the NL2Code, NO GATT, NO IALSTM, and NO Both show decreases in accuracy and Blue scores, especially NO Both. Compared to NL2Code, NO Both has a 4.2% decrease in accuracy, a 7.7 decrease in Bleu score in the Django, and a 5.39 decrease in Bleu score in the Conala. By comparing to NO GATT and NO IALSTM, we find that NO IALSTM performs better both in accuracy and Blue score, which indicates that GATT attention plays an important role. However, it is indisputable that although NO GATT replaces GATT attention with soft attention, it still outperforms the NO Both, so the role of IALSTM still cannot be ignored.

5 Conclusion

In this paper, we propose a code generation method NL2Code which can convert natural language descriptions into code snippets. The model uses gated attention mechanism and IALSTM decoder to predict code generation, which breaks the invariance of source representations and improves the interaction between decoded information. Compared to other eight models, our model has better performance both in Accuracy and Bleu scores. However, NL2Code only learns some implicit syntactic information from pair corpus without taking the structural information of natural language description. In the future, we will try to make use of the structure of natural language to aid code generation more effectively.

References

1. Gabel, M., Su, Z.: A study of the uniqueness of source code. In: SIGSOFT FSE, pp. 147–156 (2010)
2. Xia, X., Bao, L., Lo, D., Xing, Z., Hassan, A.E., Li, S.: Measuring program comprehension: a large-scale field study with professionals. IEEE Trans. Softw. Eng. **44**, 951–976 (2017)
3. Zettlemoyer, S.L., Collins, M.: Learning to map sentences to logical form: structured classification with probabilistic categorial grammars. In: Uncertainty in Artificial Intelligence, pp. 658–666 (2005)

 4. HeYue: The design and research of automatic code generation in UML language. Ph.D. thesis, North China Electric Power University (2017)
 5. Raghothaman, M., Wei, Y., Hamadi, Y.: SWIM: synthesizing what I mean: code search and idiomatic snippet synthesis. In: ICSE, pp. 357–367 (2016)
 6. Hindle, A., Barr, T.E., Su, Z., Gabel, M., Devanbu, P.: On the naturalness of software. In: Proceedings of the 6th India Software Engineering Conference, pp. 61–61 (2013)
 7. Ling, W., et al.: Latent predictor networks for code generation. In: ACL, pp. 599–609 (2016)
 8. Mandal, S., Naskar, K.S.: Natural language programing with automatic code generation towards solving addition-subtraction word problems. In: ICON, pp. 146–154 (2017)
 9. Joshi, K.A., Levy, S.L., Takahashi, M.: Tree adjunct grammars. J. Comput. Syst. Sci. **10**, 136–163 (1975)
10. Yin, P., Neubig, G.: A syntactic neural model for general-purpose code generation. In: ACL, pp. 440–450 (2017)
11. Rabinovich, M., Stern, M., Klein, D.: Abstract syntax networks for code generation and semantic parsing. In: Meeting of the Association for Computational Linguistics, pp. 1139–1149 (2017)
12. Hayati, A.S., Olivier, R., Avvaru, P., Yin, P., Tomasic, A., Neubig, G.: Retrieval-based neural code generation. In: EMNLP, pp. 925–930 (2018)
13. Yin, P., Neubig, G.: TRANX: a transition-based neural abstract syntax parser for semantic parsing and code generation. In: EMNLP (Demonstration), pp. 7–12 (2018)
14. Yin, P., Neubig, G.: Reranking for neural semantic parsing. In: ACL, vol. 1, pp. 4553–4559 (2019)
15. Shi, M.: Research on code generation method using multi-source information. Ph.D. thesis, Nanjing Normal University (2020)
16. Zhu, Y., Zhang, Y., Yang, H., Wang, F.: GANCoder: an automatic natural language-to-programming language translation approach based on GAN. In: Tang, J., Kan, M.-Y., Zhao, D., Li, S., Zan, H. (eds.) NLPCC 2019. LNCS (LNAI), vol. 11839, pp. 529–539. Springer, Cham (2019). https://doi.org/10.1007/978-3-030-32236-6_48
17. Sun, Z., Zhu, Q., Xiong, Y., Sun, Y., Mou, L., Zhang, L.: TreeGen: a tree-based transformer architecture for code generation. In: National Conference on Artificial Intelligence, pp. 8984–8991 (2020)
18. Dong, L., Lapata, M.: Language to logical form with neural attention. In: Meeting of the Association for Computational Linguistics, pp. 33–43 (2016)
19. Chorowski, J., Bahdanau, D., Serdyuk, D., Cho, K., Bengio, Y.: Attention-based models for speech recognition. In: Annual Conference on Neural Information Processing Systems, pp. 735–750 (2015)
20. Yang, F., Cui, R., Yi, Z., Zhao, Y.: Cross-language generative automatic summarization based on attention mechanism. In: Wang, G., Lin, X., Hendler, J., Song, W., Xu, Z., Liu, G. (eds.) WISA 2020. LNCS, vol. 12432, pp. 236–247. Springer, Cham (2020). https://doi.org/10.1007/978-3-030-60029-7_22
21. Wu, S., Zhang, D., Yang, N., Li, M., Zhou, M.: Sequence-to-dependency neural machine translation. In: ACL, pp. 698–707 (2017)
22. Oda, Y., et al.: Learning to generate pseudo-code from source code using statistical machine translation (t). In: Automated Software Engineering, pp. 574–584 (2015)

23. Yin, P., Deng, B., Chen, E., Vasilescu, B., Neubig, G.: Learning to mine aligned code and natural language pairs from stack overflow. In: ICSE 2018: 40th International Conference on Software Engineering, Gothenburg, Sweden, May 2018, pp. 476–486 (2018)
24. Papineni, K., Roukos, S., Ward, T., Zhu, W.J.: BLEU: a method for automatic evaluation of machine translation. In: ACL 2002: Proceedings of the 40th Annual Meeting on Association for Computational Linguistics, pp. 311–318 (2002)

Named Entity Recognition of BERT-BiLSTM-CRF Combined with Self-attention

Lei Xu, Shuang Li, Yuchen Wang, and Lizhen Xu$^{(\boxtimes)}$

Department of Computer Science and Engineering, Southeast University, 21189 Nanjing, China
lzxu@seu.edu.cn

Abstract. Neural network models are often used in Chinese named entity recognition methods, but this model has the problem of too single character vector representation in the training process, and cannot handle the ambiguity of words well. Therefore, this paper uses the pre-trained model BERT (bidirectional encoder representation from transformers) as the embedding layer of the model, adds a self-attention layer, and proposes a Chinese named entity recognition research method based on the Bert-BiLSTM-CRF model combined with self-attention. The semantic vector of the word is enriched according to the context information of the word, and the word vector sequence output by BERT is sent as input to the BiLSTM-CRF model for training. Experimental results show that this method has achieved 94.80%, 95.44% and 95.12% in accuracy, recall and F1 value on the task of Chinese named entity recognition. Compared with other traditional methods, the effect of this method is significantly improved.

Keywords: Named entity recognition · BERT · Self-attention

1 Introduction

With the rapid development of artificial intelligence, natural language processing has become a hot research direction, and has important applications in intelligent question answering systems, machine translation, recommendation systems, and knowledge graphs. Named entity recognition is a basic task in natural language processing, and its purpose is to identify entities with specific meanings and types from text [1]. In this digital age where the amount of data is rapidly increasing, it is a crucial step to quickly filter useful information from massive data and accurately obtain named entities.

The methods of named entity recognition include rules and dictionaries [2, 3], statistics-based methods, rules and statistics combined methods [4], and deep learning methods [5]. Rule-based methods can achieve better recognition results on specific corpus, but the recognition effect relies on the formulation of manual rules, which is greatly restricted by domains. When the domains differ greatly, the formulated rules usually cannot be transplanted. The method based on deep learning is to solve the data sparseness and semantic problems of named entity recognition. Commonly used are models combining LSTM and CRF [6], convolutional neural networks [7], and so on.

C. Xing et al. (Eds.): WISA 2021, LNCS 12999, pp. 556–564, 2021.
https://doi.org/10.1007/978-3-030-87571-8_48

2   Related Work

In recent years, deep learning-based models have been widely used in NER tasks, and neural network models are used to learn the feature information of text. This is an end-to-end model that does not rely on artificial features, treat entity recognition as a task of sequence labeling [8]. Hamilton [9] et al. put LSTM neural network for the task of named entity recognition for the first time; Collobert et al. proposed the use of CNN-CRF neural network model structure [10]; Lample et al. used character-level word representation, and then used LSTM and CRF to extract entities [11]; Huang et al. used the BiLSTM-CRF model to identify English entities in the CONNLL2003 corpus, and its F1 value reached 88.83% [12]. Peng Yu [13] used BERT-based to realize the named entity recognition task in the twenty-four history of China.

Using the BiLSTM model can solve the problem of strong dependence on manual annotation features and domain knowledge in current machine learning. However, in the output result of the BiLSTM model, the labeling results of the words are scattered, which makes the recognition result unable to form a reasonable labeling sequence. CRF can better pay attention to the prediction results and optimize the labeling sequence. The combination of BiLSTM model and CRF model can effectively improve the accuracy of named entity recognition.

3   Proposed Model

3.1   BERT Embedding Layer

In 2018, the Google artificial intelligence team proposed the BERT pre-training language model [14], which refreshed the best results in 11 natural language processing tasks. Using the two-way transformer neural network as an encoder, the prediction of each word can refer to the text information in both front and rear directions.

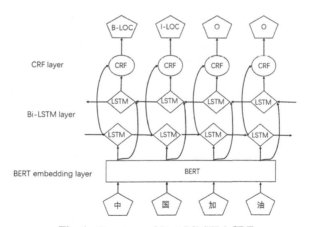

Fig. 1. Structure of Bert-BiLSTM-CRF

The sentence to be recognized is first input to the Bert module layer for pre-training, and then the output word vector sequence is input to the BiLSTM module for semantic coding, and the output result of the BiLSTM layer is sent to the CRF layer to calculate the optimal tag sequence. Compared with other traditional deep learning named entity recognition methods, the introduction of the Bert pre-training language model is the main difference. The Bert-BiLSTM-CRF model is learned on a large amount of corpus. It can calculate the vector representation of a word according to the context information of the word. The structure of BERT-BiLSTM-CRF is shown in Fig. 1.

The advantages of BERT are obvious in its achievements. The disadvantage is that it consumes huge resources in training. This method solves the problem of large consumption of computing resources to a certain extent. In the application of the BERT pre-training model, the method of fixed parameters is adopted. The internal parameters of the BERT are not updated during the training process, only the other parts of the overall model except the BERT are trained.

3.2 BiLSTM Layer

In NER tasks, recurrent neural networks (RNN) are often used to deal with this kind of sequence labeling problem, but when the length of the sequence is too long, it is difficult to learn the long-term dependence features of the sequence. Long and short-term memory neural network (LSTM) has made improvements to traditional RNN, introduced the memory unit and threshold mechanism to capture long-distance information and solved the problem of gradient disappearance. The unit structure is shown in Fig. 2.

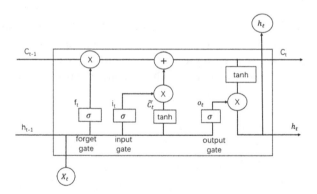

Fig. 2. LSTM unit structure

The output expression of the hidden layer of the LSTM network is shown in formula (1)-formula (5).

$$i_t = \sigma(W_{ix}x_t + W_{ih}h_{t-1} + b_i) \tag{1}$$

$$f_t = \sigma\left(W_{fx}x_t + W_{fh}h_{t-1} + b_f\right) \tag{2}$$

$$o_t = \sigma(W_{ox}x_t + W_{oh}h_{t-1} + b_o) \tag{3}$$

$$\tilde{c}_t = \tanh(W_{cx}x_t + W_{ch}h_{t-1} + b_c) \tag{4}$$

$$h_t = o_t \otimes \tanh(c_t) \tag{5}$$

Among them, W and b respectively represent the weight matrix and the bias vector connecting the two layers, σ is the sigmoid activation function, $\otimes$ is the dot multiplication operation, x_t is the input vector, i_t, f_t and o_t respectively represents the input gate, forget gate and output gate at time t, $\tilde{c}_t$ represents the value at time t, then h_t is the output at time t.

BiLSTM is a combination of forward LSTM and backward LSTM. It calculates the input sequence in order and reverse order to obtain two different hidden layer representations, and then obtains the final hidden layer feature representation by vector splicing. BiLSTM neural network can well capture bidirectional semantic information, learn context relationships, it has become a mainstream model in sequence labeling tasks.

3.3 CRF Layer

The conditional random field model is to calculate the conditional probability distribution P(Y|X) of random variable sequence $Y = (Y_1, Y_2, \ldots, Y_n)$ under the condition that the given random variable sequence $X = (X_1, X_2, \ldots, X_n)$.

$$P(Y_i|X, Y_i, \ldots, Y_n) = P(Y_i|X, Y_{i-1}, Y_{i+1}) \tag{6}$$

X represents the input observation sequence; Y represents the corresponding state sequence. In 2001, Lafferty et al. proposed a linear chain conditional random field model. The linear chain conditional random field is one of the algorithms widely used in sequence labeling tasks. CRF can obtain the occurrence probabilities of various label sequences and reduce the occurrence of label sequences that do not meet the restriction relationship. Let P be the weight matrix output by the decoding layer, and then the evaluation score S (x, y) can be obtained:

$$S(x, y) = \sum_{i=0}^{n} M_{y_i, y_{i+1}} + \sum_{i=0}^{n} P_{i, y_i} \tag{7}$$

M is the transition matrix, $M_{y_i, y_{i+1}}$ indicates the probability of transferring from the y_i label to the y_{i+1} label, P_{i, y_i} represents the probability of the i-th word being marked as y_i, and n is the length of the sequence. Finally, the maximum likelihood method is used to solve the maximum posterior probability $P(y|x)$, and the loss function value of the model is obtained.

$$\log P(y|x) = S(x, y) - \sum_{i=0}^{n} S(x, y_i) \tag{8}$$

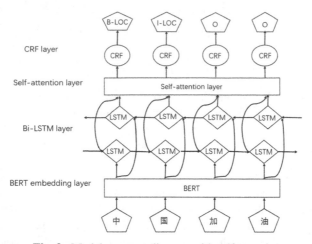

Fig. 3. Model structure diagram with self-attention

3.4 Self-attention Layer

In order to compensate for the loss of model named entity recognition accuracy caused by BERT not doing fine-tuning training, this article tries to add a self-attention layer to the model. The increased self-attention layer model is shown in Fig. 3.

Self-attention is often calculated by zooming and reducing the dot product attention. The calculation formula is shown in (9).

$$Attention(Q, K, V) = SoftMax\left(\frac{QK^T}{\sqrt{d_k}}\right)V \tag{9}$$

In this formula, Q, K and V are the three matrices obtained from the same input and different parameters. Firstly, the matrix multiplication of Q and K is calculated. In order to prevent the multiplication result from being too large, divide by $\sqrt{d_k}$, and finally normalize the result to probability distribution by soft-Max operation, and multiply by matrix V to get the final result. The self-attention structure is shown in Fig. 4.

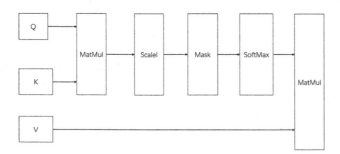

Fig. 4. Self-attention structure

4   Experiment and Result Analysis

4.1   Experimental Data Set and Data Annotation

This paper uses a data set to test the named entity recognition effect of the proposed model. The data set is the SIGHAN 2006 competition MSRA data set. The detailed information of the data set is shown in Table 1. The MSRA data set does not contain the validation set.

Table 1. Statistics of data set.

Data set	Type	Train	Validation	Test
MSRA	Sentence	46364		4365
	word	2169879	—	172601
			—	

The mainstream labeling modes of named entities are divided into two types: BIO labeling and BIOES labeling. The data set used in this experiment is labeled with BIO, label B is the first word of the named entity, label I is the non-first word of the named entity, and label O is the non-named entity. The MSRA data set contains 3 types of labels in three categories: place name (LOC), organization name (ORG), and person name (PER).

4.2   Experimental Environment and Parameter Settings

The experimental model in this article is built with TensorFlow. The experimental hardware and software environment configuration is shown in Table 2.

Table 2. Experimental setting.

Environment	Configuration
RAM	16 GB
GPU	NVIDIA GeForce RTX 2070
Python version	Python 3.7
TensorFlow version	TensorFlow 1.13.0

In order to ensure the fairness of the experiment, in each experimental model, except for some special experimental parameter settings, try to maintain the consistency of the parameters. The detailed parameters are as follows: the maximum length of a single sentence is limited to 128 words, the LSTM hidden layer dimension is 200, the dropout is 0.5, the Adam optimizer is used, the training learning rate is 0.001, the batch size of the MSRA data set is 128. In the BERT fine-tuning experiment, the training learning rate is $5 * 10^{-5}$, and the batch size is 16.

4.3 Experimental Results and Analysis

This article proposes a named entity recognition model with an additional self-attention layer based on the BERT-BiLSTM-CRF model with fixed BERT parameters. And selected 4 deep neural network models related to the proposed model to compare with the model proposed in this paper. The 4 contrast neural network models are as follows.

1) BiLSTM-CRF, the most commonly used neural network named entity recognition model at this stage, consists of a two-way long and short-term memory network layer and a conditional random field layer.
2) BiLSTM-self-attention-CRF model, a self-attention layer without pre-training model is added to the BiLSTM-CRF model.
3) BERT model, directly use the data set to adjust the parameters in the BERT model.
4) The BERT-BiLSTM-CRF model consists of a BERT embedding layer, a two-way long short-term memory network layer and a conditional random field layer.

The experiments in this article all use precision, recall, and F1 value as the evaluation criteria for model accuracy. The calculation formulas are as follows:

$$P = \frac{TP}{TP + FP} * 100\% \tag{10}$$

$$R = \frac{TP}{TP + FN} * 100\% \tag{11}$$

$$F1 = \frac{2 * P * R}{P + R} * 100\% \tag{12}$$

In these formulas: TP indicates the number of named entities correctly identified; FP indicates the number of misidentified named entities; FN indicates that the number of entities that the name has not been identified. P is the precision rate; R is the recall rate. Table 3 shows the experimental results of 5 models on the MSRA data set.

Table 3. Experimental comparison results on MSRA

Model	P/%	R/%	F1/%
BiLSTM-CRF	89.88	87.93	88.89
BiLSTM-self-attention-CRF	90.52	87.44	88.95
BERT	94.91	94.72	94.82
BERT-BiLSTM-CRF	94.48	94.48	94.48
BERT-BiLSTM-self-attention-CRF	94.80	95.44	95.12

From the comparison results in Table 3, it can be seen that the addition of the BERT pre-training model has significantly improved the accuracy of model named entity recognition, with an average improvement rate of 5.55%. However, the addition of the self-attention layer has a limited effect on the accuracy of the model.

In summary, the Bert-BiLSTM-CRF model combined with self-attention proposed in this paper can achieve better recognition results than traditional neural network-based models.

5 Conclusion

Aiming at the problems of Chinese named entity recognition based on the traditional character vector representation being too singular and unable to deal with the ambiguity features of words well, this paper proposes the Bert-BiLSTM-CRF model combined with self-attention. The Bert model uses the bidirectional transformer as the encoder by jointly adjusting the context in all layers. After comparative experiments, the overall recognition effect of the model proposed in this paper is better than other models in terms of person, place, and organization names.

References

1. Liu, L., Dongbo, W.: Summary of research on named entity recognition acta. Information **37**(3), 329–340 (2018)
2. Farmakiotou, D., Karkaletsis, V., Koutsias, J., Sigletos, G., Spyropoulos, C.D., Stamatopoulos, P.: Rule-based named entity recognition for Greek financial texts. In: Proceedings of the Workshop on Computational lexicography and Multimedia Dictionaries COMLEX 2000, pp. 75–78 (2000)
3. Chandel, A., Nagesh, P.C., Sarawagi, S.: Efficient batch top-k search for dictionary-based entity recognition. In: 22nd International Conference on Data Engineering ICDE'06, p. 28. IEEE (2006)
4. Zheng gao, P.: Research on the recognition of Chinese named entity based on rules and statistics. Inf. Sci. **30**(5), 708–712 (2012)
5. Yao, L., Liu, H., Liu, Y., Li, X., Anwar, M.W.: Biomedical named entity recognition based on deep neutral network. Int. J. Hybrid Inf. Technol. **8**(8), 279–288 (2015)
6. Huang, Z., Xu, W., Yu, K.: Bidirectional LSTM-CRF models for sequence tagging. arXiv preprint arXiv:1508.01991 (2015)
7. Shen, Y., Yun, H., Lipton, Z.C., Kronrod, Y., Anandkumar, A.: Deep active learning for named entity recognition. arXiv preprint arXiv:1707.05928 (2017)
8. Wang, L., et al.: Segment-level Chinese named entity recognition based on neural network. J. Chin. Inf. Proc. **32**(3), 84–100 (2018)
9. Hammerton, J.: Named entity recognition with long shortterm memory. In: Seventh Conference on Natural Language Learning at Hlt-naacl. Association for Computational Linguistics (2003)
10. Collobert, R., Weston, J., Bottou, L., et al.: Natural language processing almost from scratch. J. Mach. Learn. Res. **12**(1), 2493–2537 (2011)
11. Lample, G., Ballesteros, M., Subramanian, S., et al: Neural architectures for named entity recognition (2016)
12. Huang, Z., Xu, W., Yu, K.: Bidirectional LSTM-CRF models for sequence tagging. Comput. Sci. (2015)

13. Yu, P., Wang, X.: BERT-based named entity recognition in chinese twenty-four histories. In: Wang, G., Lin, X., Hendler, J., Song, W., Xu, Z., Liu, G. (eds.) WISA 2020. LNCS, vol. 12432, pp. 289–301. Springer, Cham (2020). https://doi.org/10.1007/978-3-030-60029-7_27
14. Devlin, J., Chang, M., Lee, K., et al: BERT: pre-training of deep bidirectional transformers for language understanding computation and language, (10), 1810–4805 (2018)

Recommendation

A Method of MOBA Game Lineup Recommendation Based on NSGA-II

Mengwei Li[1], Jia Tian[2], Wei Liu[2], Kangwei Li[1], Zhaozhao Xu[1], Tiezheng Nie[1], Derong Shen[1(⊠)], and Yue Kou[1]

[1] School of Computer Science and Engineering, Northeastern University, Shenyang 110004, China
{nietiezheng,shenderong,kouyue}@cse.neu.edu.cn
[2] Beijing System Design Institute of the Electro-Mechanic Engineering, Beijing 100854, China

Abstract. The existing research on MOBA game lineup recommendation mainly takes the heroes with high winning rate in the historical game as the recommendation results, without comprehensively considering the synergy relationship and the counter relationship between heroes. In the recommended results, there will be a situation where a hero has great ability, but the synergy between the heroes is not good, or the recommended hero is counter by the enemy. This paper summarizes the problem as a multi-objective optimization problem, and proposes a MOBA game lineup recommendation method base on NSGA-II. Firstly, based on the performance of the heroes in historical games, the ability of these heroes is quantified. At the same time, according to different heroes as teammates or opponents, the synergy relationship and the counter relationship between heroes are evaluated. Then, NSGA-II is used to solve the multi-objective optimization problem, and the recommendation result is obtained. Finally, the results are evaluated based on prediction models. The method proposed in this paper integrates the hero's ability, the synergy relationship and the counter relationship between heroes, which has strong rationality. Experimental results show that the method has high accuracy, and can provide a new idea for solving this kind of problems.

Keywords: Lineup recommendation · NSGA-II · MOBA · The synergy relationship and the counter relationship

1 Introduction

With the development of the electronic game industry, MOBA (Multiplayer Online Battle Arena) games occupy an important position in the electronic sports market. The most popular MOBA games include DOTA2, League of Legends, and Glory of Kings. A MOBA match consists of a fight between two teams of five players, both sides attempt to destroy one another's fortress, and each player controls a single character known as a hero. Each hero has an array of unique powers which are accessed by leveling up. In order to win the game, heroes must gain experience to level up and gold to purchase helpful items by killing enemy heroes and minions. The choice of heroes plays a large role in determining the match outcome. Each hero has different strengths and weaknesses, and

© Springer Nature Switzerland AG 2021
C. Xing et al. (Eds.): WISA 2021, LNCS 12999, pp. 567–579, 2021.
https://doi.org/10.1007/978-3-030-87571-8_49

these strengths and weaknesses can be used to complement other heroes on the same team or counter the heroes on the opposing team. Therefore, the rationality of the lineup has a key influence on the outcome.

There are two types of research on lineup recommendation, one is to recommend the lineup based on the winning rate of the game, and then evaluate the recommended lineup based on the outcome prediction model. For example, [1] artificially adds specific heroes to optimize the prediction model. [2] uses the outcome prediction model to evaluate the lineup generated by MCTS. The other one is to use the objective function to obtain the best lineup. For example, [3] defines the objective function and uses genetic algorithm to optimize the obtained lineup. [4] solves the problem of recommendation through multi-objective optimization. [15] proposes an explainable recommendation method to solve the recommendation problem.

Analyzing the existing works on the lineup recommendation in MOBA games, the main deficiencies are as follows: 1) Existing research only considers the winning rate of heroes, without quantitatively evaluating the synergy relationship and the counter relationship between heroes. 2) The lineup is recommended completely based on the objective function and the hero's performance in the history game, without considering the lineup of the enemy.

Based on the above description, this paper uses a multi-objective optimization method to recommend the best lineup. It not only introduces the quantified value of relationship between heroes, but also considers the influence of the enemy on the recommended result. The main contributions are as follows:

1) We propose a model for quantifying hero's ability and the relationship between heroes. It can improve the accuracy of the lineup recommendation effectively.
2) We propose a lineup recommendation method based on NSGA-II by integrating the ability of the hero, the relationship of heroes, and the specific enemy lineup, which can improve the winning rate of the lineup recommended.
3) We propose a model to evaluate the recommended lineup, and the results show that our lineup recommendation method has good performance and rationality.

This paper will take the DOTA2 game as an example in the follow-up. Section 2 introduces related work on the lineup recommendation; Sect. 3 introduces concepts; Sect. 4 introduces lineup recommendation method; Sect. 5 is an experimental part; Sect. 6 is the summary of this paper.

2 Related Work

In the related research of lineup recommendation, it mainly includes two parts: the lineup recommendation model and the relationship between heroes. The research results of these two parts are respectively introduced below.

2.1 Lineup Recommendation Model

[1] proposed a logistic regression prediction model for MOBA game, which optimizes the model by manually adding hero. However, the method mainly relies on human

judgment. [2] made the best choice based on the winning rate of the game, and the Monte Carlo Tree Search (MCTS) algorithm is used to simulate the next chosen hero. [5] used the discriminative neural network to simulate the lineup. This method believed that the lineup generated by the generator is the closest to the data in the real game.

In recent years, some scholars focus on lineup recommendation by using evolutionary algorithms, such as [3] used genetic algorithms to generate a subset of heroes, and defined the reward function for different strategies. Each strategy is evaluated and ranked by the reward function generated by the genetic algorithm. Similarly, [4] used a fuzzy metric-based method to evaluate the candidate lineup and the enemy lineup, then used a multi-objective optimization method to calculate the lineup. However, this method ignored the relationship between heroes, which has great limitations.

2.2   The Relationship Between Heroes

The relationship between heroes plays an important role in the lineup recommendation. [6] used PCA to analyze the synergy and counter among heroes, and the results showed that the accuracy of PCA was lower than that of logistic regression. In order to build a balanced lineup, this method used the k-means algorithm to find the best clustering to synthesize the relationship between heroes, but the result was not ideal. To solve this problem, [7] used logistic regression to verify the importance of lineup. The best results cannot be obtained because the model is difficult to evaluate the synergy relationship and the counter relationship between heroes. Similarly, [8] introduced a new hero vector representation method, which simply regarded the hero winning rate as hero vector.

[9] simply estimated the synergy relationship and the counter relationship between heroes, taking the winning rate of two heroes as teammates to evaluate the synergy relationship, and the winning rate as the opponents to evaluate the counter relationship. However, this method does not consider the mutual synergy relationship and the counter relationship between heroes. [10] used LSTM for lineup recommendation, in which, a game was regarded as a sentence, and different heroes were seen as words, then the CBOW model is used to generate hero vectors, and the correlation between vectors is presented as the synergy relationship between heroes, and the log5 formula is used to estimate the counter relationship between heroes. The disadvantage of the work is that the precondition is too high, which is difficult to realize in practical applications.

In summary, the existing work neither quantify the synergy relationship and counter relationship between heroes nor apply it to the lineup recommendation model. In this paper, the multi-objective optimization method is used to recommend the best lineup, which not only introduces the quantitative value of synergy index and counter index, but also considers the influence of enemy lineup on the recommended lineup.

3   Preliminaries

3.1   Hero Ability

The heroes in MOBA games have some inherent attributes, such as strength, agility, intelligence, health, mana, etc. These factors have a greater impact on the hero's ability and are directly related to the outcome of the game.

In this paper, according to the method in RSR (Ran-Sum Ratio) [13], we divide the inherent attributes into the following two categories:

(1) High-quality attributes: the lowest value indicates the worst outcome.
(2) Low-quality attributes: the highest value indicates the worst outcome.

After categorizing the hero's attributes, the RSR is used to obtain the evaluation result of different heroes. $ability_{pi}$ represents hero i's high-quality attribute ability, and $ability_{ni}$ represents hero i's low-quality attribute ability.

3.2 Synergy Relationship and Counter Relationship Between Heroes

The relationship between heroes is particularly important in lineup but it is difficult to define them directly from hero attributes.

Therefore, this paper proposes a method to evaluate the relationship between heroes for improving the accuracy of lineup recommendations. [10] used the CBOW model to obtain hero vectors to estimate the synergy relationship between heroes. However, this method takes the co-occurrence frequency of heroes in the game as the standard to measure the synergy relationship between heroes, which has great limitations. Based on the performance of each hero in the historical game, this paper proposes a relationship evaluation model to quantify the synergy relationship and the counter relationship.

3.3 Lineup Recommendation

In MOBA game, a reasonable lineup plays a decisive role in the outcome of the game, while the hero's ability and the relationship between heroes are necessary for constructing the lineup. Under the condition of given enemy lineup, this paper proposes a method based on NSGA-II to obtain reasonable lineup, so as to counter the enemy and win the game.

4   Lineup Recommendation Based on NSGA-II

The framework of the lineup recommendation method in this paper is shown in Fig. 1. Firstly, the RSR is used to evaluate the ability of heroes. At the same time, the relationship between heroes is evaluated by the evaluation method of relationship. Then a reasonable objective function is designed to solve the problem of recommendation. Finally, the lineup recommendation result is evaluated.

4.1   Estimating Hero's Ability Based on RSR

In current research, the commonly methods to quantify the hero's ability including topsis (distance between superior and inferior) [11], fuzzy integral [12], and RSR [13]. RSR is a non-parametric comprehensive evaluation method which is widely used in multi-index comprehensive evaluation. Its basic principle is to obtain dimensionless statistics RSR through rank transformation in a matrix of m rows and n columns. RSR is convenient

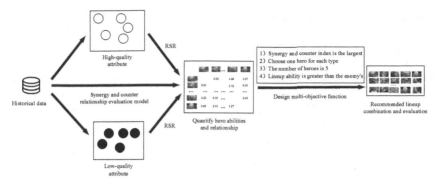

Fig. 1. The lineup recommendation framework

to integrate other methods. Compared with several other commonly methods, RSR is more suitable for evaluating hero's ability.

For the MOBA games, we divide the strength, agility, intelligence, health, mana, kills, assists, and gold, damage and positive complement into high-quality attributes, while death and denials into low-quality attributes.

First, an $m * n$ matrix was created: m means the number of the hero and n means the number of the attributes. the original attributes matrix of $m * n$ is formed:

$$X = \begin{pmatrix} x_{11} & \cdots & x_{1n} \\ \vdots & \ddots & \vdots \\ x_{m1} & \cdots & x_{mn} \end{pmatrix}$$

Where X_{ij} represents the value of the j-th attributes of the i-th hero.

Second, rank conversion was performed: suppose all attributes were valued as high-quality attributes, therefore the attributes were sorted in ascending order in the rank conversion process.

Third, the ranks R_{ij} were summed over rows, giving each column or attribute a score, and then summed over these columns. the rank matrix of $m * n$ is formed:

$$R = \begin{pmatrix} R_{11} & \cdots & R_{1n} \\ \vdots & \ddots & \vdots \\ R_{m1} & \cdots & R_{mn} \end{pmatrix}$$

Where R_{ij} represents the value of the j-th attribute of the i-th hero.

the RSR values are calculated according to Eq. (1). This paper regards RSR_i as the ability of hero i.

$$RSR_i = \frac{1}{m * n} \sum_{j=1}^{n} R_{ij} \tag{1}$$

4.2 Estimating the Relationship Between Heroes

The synergy index and the counter index are difficult to obtain from hero's attributes. [9] regard the winning rate of heroes as the synergy (counter) relationship between heroes when the heroes are teammates (opponents). This method ignores the mutual relationship between heroes, and fewer matches can be obtained by this method. The quantitative results lack practical significance. Therefore, this paper proposes an evaluation method of relationship between heroes, which is defined as follows:

$$p_I^i = \frac{count_w^i}{count^i} \tag{2}$$

$$P_{counter}(i|j) = \frac{count_{jw}^i}{count_j^i} - p_I^i \tag{3}$$

$$p_{synergy}(i, j) = \frac{count_{i,j_w}}{count_{i,j}} - p_I^i \tag{4}$$

$count^i$ means the number of games that hero i participated in, $count_w^i$ means the number of games that hero i won. For hero i and j, we express $P_{counter}(i|j)$ as the counter index of hero j against i, where $count_j^i$ means that hero i and hero j are selected as opponents in the same game, $count_{j_w}^i$ means the number of matches where hero i defeated hero j.

We denote $p_{synergy}(i, j)$ as the synergy index of heroes i and j, where $count_{i,j}$ means the number of matches where hero i and hero j are selected as teammates in the same game, $count_{i,j_w}$ represents the number of matches obtained by hero i and hero j together. Based on the above Equation, the synergy relationship matrix and the and counter relationship matrix can be obtained.

4.3 Lineup Recommendation Base on NSGA-II

The lineup recommendation problem can be regarded as a multi-objective problem. This paper proposes a lineup recommendation method base on NSGA-II. The following rules should be met in the process of multi-objective optimization:

1) Make sure that the synergy index and the counter index between heroes is maximized; 2) Make sure only one hero of each type is selected; 3) Make sure that the number of heroes selected is 5; 4) Make sure that the selected lineup is more capable than the enemy.

Base on rule 1), the relationship is embedded into the Eq. (5) and Eq. (6) after dimensionality reduction through principal component analysis, which respectively indicate the largest synergy index and the largest counter index between the selected heroes.

$$\max(synergy) = \sum_{i=1}^{5} synergy_i \tag{5}$$

$$\max(counter) = \sum_{i=1}^{5} counter_i \tag{6}$$

Base on rule 2), cluster analysis is performed using k-means to obtain the hero's category. Heroes in MOBA game are roughly divided into five positions: confrontation, jungle, mid, ADC, and support according to the roles they play in the game. Therefore, we group all heroes into the above five categories to design the objective function, which means each category is required to select a hero. Assuming that $x_1 \ldots x_i$ is the first category, it satisfies the solution of the following objective function.

$$\sum_{j=1}^{i} x_j = 1 \tag{7}$$

Base on rule 3), each team in MOBA game selects 5 heroes. We design Eq. (8) base this mechanism.

$$\sum_{i=1}^{117} x_i = 5 \tag{8}$$

Base on rule 4), select the hero subset whose ability is greater than that of the enemy as the recommended lineup, we design Eq. (9).

$$\sum_{i=1}^{5} pow_i > pow_{enemy} \tag{9}$$

Where, pow_i represents the sum of the selected hero's high-quality attribute ability and low-quality attribute ability, and the right side represents the ability of the enemy lineup. According to this objective function, the ability of the recommended lineup must be greater than that of the enemy lineup.

4.4 Lineup Evaluation Prediction Models

This paper evaluates the recommended lineup based on outcome prediction models built based on the historical information of the game. We calculated the average value of hero's attributes at the beginning of the game. Hero's attributes refer to the basic attributes of heroes calculated from the real data, including 11 attributes such as strength, agility, intelligence, armor, and average damage etc.

The attribute vector of hero is shown in Table 1. We trained three outcome prediction models based on the attribute vectors of ten heroes from the historical game data, using random forest, Xgboost, and GBDT as the prediction models. The accuracy of the prediction model is regarded as the evaluation metric of the recommendation model.

Table 1. Attribute vector of hero.

Attribute 1	Attribute 2	Attribute 3	...	Attribute 11
Strength	Agility	Armor	...	Damage

5 Experiment

5.1 Experiment Datasets

The data used in this experiment comes from the professional league of DOTA2. We use OpenDota's api to obtain the ID and video download address of the professional game, and then use the DOTA2 api to obtain the basic information of the above ID game. The video is downloaded using IDM, and the data of 5,000 games are shared as the data set of this paper.

The inherent attributes of the hero are divided into high-quality attributes and low-quality attributes, and the original attribute matrix are obtained. Then calculate the above attributes of different heroes in the game they participate in, and obtain the average value of different attributes at the end of the game. fea_i represents the final statistics of the abilities, such as strength and agility, which can be obtained according to Eq. (10). In order to eliminate the influence of the match duration on attributes, we design Eq. (11). att_{end} means the hero i's attributes at the end of the game. $times$ means the number of matches the hero has participated in. t means the match duration.

$$fea_i = \frac{att_i}{times} \tag{10}$$

$$att_i = \frac{att_{end}}{t} \tag{11}$$

Hero selection is coded with 0/1 to indicate. There are 117 heroes participating in the data set. Therefore, this feature can be represented by a vector which size is 117. The initial value is 0. When the hero is selected, the hero's corresponding bit is 1, the rest are 0, and the final output is a 0/1 vector, of which a total of five bits are 1, and the rest are 0.

5.2 Evaluation Metrics

This paper uses *Precision* (Eq. 12) and hit rate as the evaluation metric to measure the experimental results.

Precision is often used to measure the accuracy of prediction model. This paper uses the outcome prediction model to evaluate the result of the lineup recommendation. Therefore, it is reasonable and effective to use the accuracy as one of the evaluation metrics of the experimental results.

$$Precision = \left(\frac{TP}{TP + FP} \right) \tag{11}$$

With reference to the evaluation of the results in [5], this paper uses hit rate as another evaluation metric. The hit rate is the ratio of matches predicted in the test set.

5.3 Hero's Ability

The attributes of heroes are calculated from historical game, and RSR is used to evaluate the hero's ability based on attributes matrix. The result is shown in Table 2. The evaluation result show that the larger the RSR is, the better the ability.

Through the comparison of the RSR evaluation result, the difference between the pros and cons of the heroes is obtained. For Example, the high-quality RSR of hero3 is poorer than other heroes, while low-quality RSR of hero3 is better than other heroes.

Table 2. RSR evaluation result.

Hero id	High-quality score	Low-quality score
1	0.706	0.727
2	0.670	0.743
3	0.483	0.836
...	...	...
117	0.552	0.760

In order to verify the effectiveness of the RSR, this paper uses the TOPSIS [11] to evaluate the hero's ability, and the results are shown in Table 3.

Table 3. TOPSIS evaluation result.

Hero id	High-quality score	low-quality score
1	81	78
2	66	80
3	12	115
...	...	...
117	30	96

Compared with the RSR evaluation results, the trends of different heroes are similar. For example, the high-quality score of hero3 is poorer than other heroes, while low-quality score of hero3 is better than other heroes. It shows that that both evaluation methods are effective. However, these methods evaluate hero's ability from different perspectives. RSR uses a rank matrix to quantify and has a strong comprehensive ability, while TOPSIS is based on the distance to the worst solution and the optimal solution to evaluate hero's ability. This paper compares the accuracy of the evaluation methods of RSR and TOPSIS in the lineup recommendation model, and the results are shown in Fig. 2.

From the comparison of different evaluation methods in Fig. 2, it can be seen that. when GBDT is used as the lineup evaluation model, the average accuracy is the highest. Compared with TOPSIS, RSR is more stable as the evaluation method recommended by the lineup. RSR evaluates the hero's ability from the perspective of the rank matrix, which is more comprehensive. The results show that RSR as a hero's ability evaluation method is relatively stable.

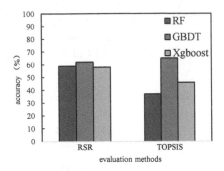

Fig. 2. Comparison of different evaluation methods

5.4 The Relationship Between Heroes

Table 4 shows the synergy between heroes. The diagonal element indicates the counter index of hero i. The synergy index of hero 1 to hero 2 is 3.04, and the synergy index of hero 2 to hero 1 is 0.32. This is due to the fact that different heroes are used as references, and heroes have different synergy indexes.

Table 4. Synergy between heroes.

	Hero 1	Hero 2	...	Hero 117
hero 1		3.04	...	1.37
hero 2	0.32		...	0.33
...	...	...	...	...
hero 117	1.06	3.38	...	

Similar to Table 4 and Table 5 shows the counter relationship between heroes. A negative counter index indicates that hero i has a weak counter relationship with hero j.

Dotamax [14] provides queries about the synergy relationships and the counter relationships between heroes, but the website does not provide detailed calculation methods. In order to verify the effectiveness of the quantified model proposed in this paper, this paper obtains the data from Dotamax [14], and compares the impact of the different method on the prediction results. The results are as follows as shown in Fig. 3.

From Fig. 3 that the two different calculation methods have certain effects. Among them, the synergy relationship and the counter relationship obtained by Dotamax [14] can achieve higher accuracy when using GBDT prediction, but the quantification method proposed in this paper is more stable.

Table 5. Counter between heroes.

	hero 1	hero 2	...	hero 117
hero 1		−6.53	...	−0.72
hero 2	0.13		...	−1.46
...	...	...	...	...
hero 117	−2.63	−5.25	...	

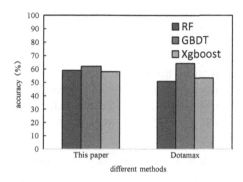

Fig. 3. Comparison of different calculation methods

5.5 The Evaluation Model

The inherent attributes of each hero are taken as the hero vector. This paper uses the current mainstream classification methods for comparative experiments, including RF (random forest), GBDT, Xgboost, SVM (support vector machine), and LR (logistic regression). Use 80% of the data set as the training set, and the remaining part as the test set. The training results are shown in Table 6.

Table 6. Comparison of different evaluation model.

Model	RF	GBDT	Xgboost	SVM	LR
Accuracy	54.85%	52.95%	53.64%	52.75%	51.97%

This paper uses the lineup information of both teams as the prediction data. Among the above models, RF, GBDT, and Xgboost have the highest accuracy as lineup recommendation evaluation models. The accuracy of the model is stable at 53%-55%. Due to the randomness of the evaluation model of the lineup recommendation result, this paper integrates three models to evaluate the lineup recommendation result.

5.6   Lineup Recommendation Method

Based on NSGA-II algorithm, objective function is designed to solve the lineup recommendation problem. After obtaining the recommended result, assuming that the recommended lineup can win, and predicting the outcome. Use the ratio of the number of positive samples to the data set as the accuracy. The results are shown in Table 7.

Table 7. Accuracy of recommended model.

Model	RF	GBDT	Xgboost
Accuracy	59%	62%	58%

In order to verify the hit rate of this method, recommend five lineups for each enemy in the test set. This paper chooses to hit 1 hero, hit 2 heroes, and hit 3 heroes to verify. Hit 1 hero means that one of the lineups actually selected in the test set and the recommend result have one hero in common. The experimental results are shown in Table 8.

Table 8. Hit rate of recommended model.

Number of heroes	1	2	3
Hit rate	45.6%	10%	2%

From Table 8 that the method proposed in this paper has better practical application scenarios, with better reference and higher accuracy. With reference to the method of [4], this paper adds a MOEA/D comparison to verify the effectiveness of the method we proposed. The results are shown in Fig. 4. The two different multi-objective optimization methods in Fig. 4. can obtain relatively stable and effective recommendation results in the three prediction models, but NSGA-II is better than MOEA/D.

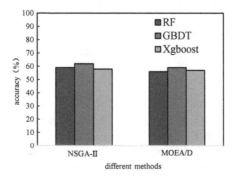

Fig. 4. Comparison of different multi-objective optimization methods

6   Conclusion and Outlook

This paper uses DOTA2 as the experimental data set, and proposes a method to solve the lineup recommendation in MOBA games. We integrate multiple factors such as hero abilities, synergy relationship and counter relationship between heroes, and enemy lineup, and recommends a reasonable lineup. Experimental result shows that the method has a certain degree of rationality and reference, and can provide reference for professional player and coaches to develop training plans. Follow-up will in-depth study of the serialization recommendation of the MOBA game lineup.

Acknowledgment. This work is supported by the National Natural Science Foundation of China (62072084, 62072086), the National Defense Basic Scientific Research Program of China (JCKY2018205C012) and the Fundamental Research Funds for the C-entral Universities (N2116008).

References

1. Song, K., Zhang, T., Ma, C.: Predicting the winning side of DotA2. In: Sl: sn. (2015)
2. Chen, Z., Xu, Y., et al.: The art of drafting: a team-oriented hero recommendation system for multiplayer online battle arena games. In: ACM Conference, pp. 200–208 (2018)
3. Costa, L.M., Souza, A.C.C., Souza, F.C.M.: An approach for team composition in league of legends using genetic algorithm. In: IEEE, pp. 52–61 (2019)
4. Wang, L., Zeng, Y., Chen, B., Pan, Y., Cao, L.: Team recommendation using order-based fuzzy integral and NSGA-II in starcraft. In: IEEE Access, pp. 59559–59570 (2020)
5. Gourdeau, D., Archambault, L.: Discriminative neural network for hero selection in professional heroes of the storm and DOTA 2. In: IEEE Transactions on Games (2020)
6. Agarwala, A., Pearce, M.: Learning Dota 2 team compositions. In: Sl: sn. (2014)
7. Conley, K., Perry, D.: How does he saw me? a recommendation engine for picking heroes in dota 2. In: Np, nd Web, p. 7 (2013)
8. Wang, N., Li, L., Xiao, L., Yang, Zhou, Y.: Outcome prediction of dota2 using machine learning methods. In: Proceedings of 2018 International Conference on Mathematics and Artificial Intelligence, pp. 61–67 (2018). https://doi.org/10.1145/3208788.3208800
9. Kinkade, N., Lim, K.: DOTA2 win prediction. In: Univ Calif, pp. 1–13 (2015)
10. Xu, C.: Research of Dota2 lineup recommendation and winning percentage prediction system based on deep learning. In: Henan University (2019)
11. He, M., Ma, X., Jin, Y.: Station importance evaluation in dynamic bike-sharing rebalancing optimization using an entropy-based TOPSIS approach. In: IEEE Access, pp. 38119–38131 (2021). https://doi.org/10.1109/ACCESS.2021.3063881
12. Lian, L., Ouyang, T., Ma, F., Liu, J.: Fuzzy integral sliding mode control based on microbial fuel cell. In: Complexity, pp. 1–8 (2021). https://doi.org/10.1155/2021/6670039
13. Yusuf, D., Rew, M., Davis, G., et al.: An alternative method for the evaluation of docking performance: rsr vs rmsd. J. Chem. Inf. Model. **48**(7), 1411–1422 (2008). https://doi.org/10.1021/ci800084x
14. Dotamax Homepage. www.dotamax.com. Accessed 25 July 2021
15. Wang, H., Kou, Y., Shen, D., Nie, T.: An explainable recommendation method based on multi-timeslice graph embedding. In: Wang, G., Lin, X., Hendler, J., Song, W., Xu, Z., Liu, G. (eds.) WISA 2020. LNCS, vol. 12432, pp. 84–95. Springer, Cham (2020). https://doi.org/10.1007/978-3-030-60029-7_8

Explainable Recommendation via Neural Rating Regression and Fine-Grained Sentiment Perception

Ziyu Yin[1], Yue Kou[1(✉)], Guangqi Wang[2], Derong Shen[1], and Tiezheng Nie[1]

[1] Northeastern University, Shenyang 110004, China
{kouyue,shenderong,nietiezheng}@cse.neu.edu.cn
[2] Liaoning Provincial Higher and Secondary Education Enrollment Examination
Committee Office, Shenyang 110031, China

Abstract. Compared with traditional recommendation systems, explainable recommendation systems have certain advantages in terms of system transparency, result credibility, and user satisfaction. However, the existing text explanation generation methods are often limited by predefined templates which limit the ability of text expression. The freestyle text generation methods have stronger expressive ability, but are less controllable and ignore fine-grained sentiment perception from user comments. In this paper, a Dual Learning-based Explainable Recommendation Model (called DLER) is proposed, which uses a dual learning mechanism to perform rating prediction and explanation generation respectively. The parameters can be adjusted iteratively via the collaborative promotion between the two phases. A rating prediction algorithm based on neural rating regression and an explanation generation algorithm based on fine-grained sentiment perception are respectively proposed. On the one hand, the ratings are predicted via MLP. On the other hand, users' fine-grained sentiments are perceived by analyzing comments, which will be used for GRU-based explanation generation. The experiments demonstrate the effectiveness and the efficiency of our proposed method in comparison with traditional methods.

Keywords: Explainable recommendation · Rating prediction · Explanation generation · Neural rating regression · Fine-grained sentiment perception

1 Introduction

With the development of information technology and the Internet, people have gradually moved from the era of information shortage to the era of information overload. As an information consumer, it is very difficult to find information

Supported by National Natural Science Foundation of China (62072084, 62072086),
Fundamental Research Funds for the Central Universities (N2116008, N180716010).

that people are interested in from the massive amount of information. Recommendation system is an important tool to solve this problem. On one hand, it can help users find valuable information, and on the other hand, it can show the information to users who are interested in it, so as to achieve a win-win situation for information consumers and information producers. Many effective recommendation algorithms have been proposed, such as user/item based collaborative filtering [1,2], matrix decomposition [3,4] and deep neural network [5,6], to improve the accuracy of recommendation. An explainable recommendation system has more applicable value, and its application scenarios are diverse. For example, in e-commerce sites, such as Amazon, Taobao and other shopping sites, considering explainable recommendations can facilitate users to make better choices to meet their needs.

Users often publish their own opinions, preferences, and sentimental tendencies about items in the form of text comments. By analyzing these comments, the accuracy and explainability of recommendations can be improved. Therefore, some research work began to use the rich online text comments to propose a series of text explanation generation methods, including: template-based text explanation generation methods [7], free-style text explanation generation methods [8] and a method of compromise between the two [9]. The above text explanation generation method can understand the user's psychology by analyzing their comments while giving the recommendation result, and give the recommendation reason in the form of text. However, the above methods still have shortcomings:

First of all, template-based text explanation generation methods need to manually define sentence templates in advance. The cost of creating such templates is high, and these methods are often limited to pre-defined templates and limit text expression capabilities. For example, suppose that the template is defined as: "You may be interested in a feature, and this product performs well on the feature" to construct an explanation. In the template above, all the project features are described as "being well". However, the user's attention to different features of items is different, and this method cannot reflect the differences between the features of different items. Secondly, the free-style text explanation generation methods mostly aim at text readability and fluency, but lack controllability, and it is difficult to guarantee the quality of the generated explanation. In addition, although some research work combines the above two methods to support text generation guided by a certain feature, it often ignores the perception of fine-grained sentiments in user comments. For example, the focus is on increasing the variety of explanatory sentences [9].

Aiming at the above limitation, the contributions of this paper are as follows:

First, a Dual Learning-based Explainable Recommendation Model (called DLER) is proposed. Different from traditional recommendation models, this model uses a dual learning mechanism that can predict ratings and generate explanation at the same time, and use the synergy between the two to iteratively adjust parameters to improve the accuracy of rating predictions and generate high-quality explanation.

Second, a rating prediction algorithm based on neural rating regression and an explanation generation algorithm based on fine-grained sentiment perception are respectively proposed. The multi-layer perception is used for rating prediction. The user's fine-grained sentiments are perceived through the analysis of comments, and the explanation text is learned and generated based on GRU, while ensuring the expression ability and quality of the explanation text.

Thirdly, experiments have verified the feasibility and effectiveness of the key technologies proposed in the paper.

The rest of this paper is organized as follows. Section 2 reviews related work. Section 3 proposes our DLER model. Section 4 proposes an explainable recommendation algorithm based on DLER. Section 5 shows the experimental results and Sect. 6 concludes.

2 Related Work

This section first introduces the relevant work of recommender system, and then compares the method proposed in this paper with the existing methods.

Collaborative filtering recommendation algorithm [10] is an algorithm that has been widely studied and applied in recent years. Model-based collaborative filtering algorithms (such as matrix decomposition) can reduce the dimensions of high-order sparse matrices and effectively obtain the potential eigenmatrices of users and projects. Porteous et al. [11] added auxiliary information to Bayesian matrix decomposition to improve the accuracy of recommendation results. Chen et al. [7] proposed the attribute-based collaborative filtering, which was modeled from the perspective of personalization and explainability to implement alternative recommendations. The traditional collaborative filtering algorithm described above essentially initializes an embedded vector for the user and the project, and then optimizes the model with interactive information. However, the two are asynchronous and do not encode the interaction information into the embedded vectors, so these embedded vectors are not precise enough.

Deep neural network (DNN) has a strong feature learning ability, which can represent the relevant information of users and projects by learning the structure of deep network. Fan et al. [12] uses deep neural network to learn the representation of each user from social relations, and integrates it into probability matrix factorization for rating prediction. Kim et al. [13] proposed a convolutional matrix decomposition model, which used convolutional neural network to obtain contextual features of text and integrate them into the probabilistic matrix decomposition to further improve the accuracy of rating prediction. Some studies use deep neural network model to model user behavior sequence to predict user preferences [14,15]. In terms of explainable recommendation models, there are template-based explainable recommendation models, NLP-based explainable recommendation models, and explainable recommendation models that combine templates and NLP.

The method proposed in this paper differs from the above techniques in these aspects:

First, the explainable recommendation model based on dual learning and the explainable recommendation algorithm based on DLER proposed in this paper can effectively predict user preferences and generate comments, and further explain to improve the explainability and transparency of the model. Second, our method perceives the user's fine-grained sentiments and generates explanation by analyzing the user's comments, which improves the accuracy of explanation. Whereas the previous methods usually use a text summary [16,17], the first sentence [18], or the comment title [19], which may generate an explanation irrelevant to the recommended item.

3   DLER: The Proposed Model

In order to improve the accuracy of rating prediction and support explainable recommendation, a Dual Learning-based Explainable Recommendation Model (called DLER) is proposed. This section first introduces the basic idea of DLER model, then describes its components.

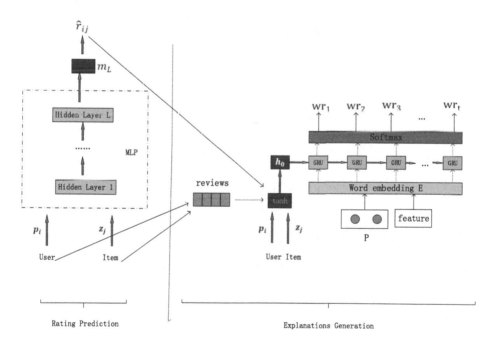

Fig. 1. The basic idea of DLER

3.1   Model Overview

The basic idea of DLER model is given in Fig. 1. The model consists of two parts: rating prediction and explanation generation. The goal of the rating prediction submodule of the DLER model is to give a user u_i and an item v_j, perform rating prediction, and the obtained $\hat{r}_{ij}$ is used as an input of the explanation generation submodule. Based on a feature f_{uv} of the user's item of interest, our model can perceive the user's fine-grained sentiments to train the template, and generate explanatory sentences as an explanation of the recommended results. The input feature f_{uv} can be any feature that we want to generate for the explanation and discussion. In summary, the training data includes users, items, ratings, features, and explanation sentences, however only users, items, and features are needed in the testing phase.

Next, the rating prediction method is introduced, the user latent vector and the item latent vector implicitly express the user's preferences and item characteristics. As shown in Fig. 2, after obtaining the user's final latent vector $\mathbf{p}_i$ and the item latent vector $\mathbf{z}_j$, first connect them, and then enter multi-layer perceptron to predict the rating and get the user's predicted rating for the item:

$$\hat{r}_{ij} = MLP[\mathbf{p}_i, \mathbf{z}_j] \tag{1}$$

The rating prediction task aims to find a function f: $u_i, v_j \rightarrow \hat{r}_{ij}$ which is used to minimize the difference between the predicted rating and the true value. As for the explanation generation task, it aims to find a function g: u_i, v_j ,$\hat{r}_{ij}, f_{uv} \rightarrow c_{uv}$, which is used to minimize the difference between the explanation sentence and the real comment.

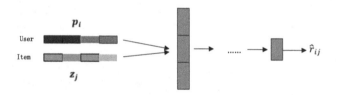

Fig. 2. Rating prediction

3.2   Neural Rating Regression

Traditionally, rating prediction task is achieved by the inner product between the user and item latent factors, or by adding bias items to the corresponding users and items respectively, but its bilinear nature may make it difficult to model complex user-item interactions. In recent research on representation learning in some fields, such as computer vision and natural language processing, the knowledge base is completed, indicating that non-linear transformation will enhance the representation ability. Therefore, we employ multi-layer perceptron (MLP) with L hidden layers to capture the interactions between users and items, and map the user latent vector and the item latent vector to the real rating, as shown in the left part of Fig. 1. Formally, given the IDs of user u_i and item v_j, we can obtain their latent vectors $\mathbf{p}_i$ and $\mathbf{z}_j$ (also called embeddings), which can predict the user's preference for the item, and then the recommendation module is defined as:

$$\begin{cases} m_1 = \sigma(W_1[\mathbf{p}_i, \mathbf{z}_j] + b_1) \\ m_2 = \sigma(W_2 m_1 + b_2) \\ \quad \ldots\ldots \\ m_L = \sigma(W_L m_{L-1} + b_L) \end{cases} \qquad \hat{r}_{ij} = w_r m_{L-1} + b_r \qquad (2)$$

where $[\mathbf{p}_i, \mathbf{z}_j]$ denotes the concatenation of vectors, $\sigma(\cdot)$ is the sigmoid activation function, $W_x \in R^{2d \times 2d}$ and $b_x \in R^{2d}$ are weight matrices and bias vectors in the hidden layers, while $w_r \in R^{2d}$ and $b_r \in R$ respectively correspond to the weight and bias parameters of the final linear layer.

$$h_{cl} = tanh(W_{uv}[\mathbf{p}_i, \mathbf{z}_j] + b_h) \qquad (3)$$

In order to minimize the difference between the real rating and the predicted rating value and improve the high accuracy of the recommendation model, we adopt the mean square error loss as its objective function:

$$L_r = \frac{1}{|\mathbf{X}|} \sum_{u,i \in \mathbf{X}} (r_{ij} - \hat{r}_{ij})^2 \qquad (4)$$

where $\mathbf{X}$ is the training data set. In this way, the randomly initialized latent vectors $\mathbf{p}_i$ and $\mathbf{z}_j$ can be updated through backpropagation.

For personalized recommendations, we can predict the rating of each item that the user hasn't noticed, and recommend the top-ranked items.

3.3 Explanation Generation Based on Fine-Grain Sentiment Perception

In this subsection, we designed a model for the explanation and generation task of given preference data and fine-grained sentimental characteristics of users. After obtaining $\mathbf{p}_i$ and $\mathbf{z}_j$ from user matrix $\mathbf{U}$ and item matrix $\mathbf{V}$, we should design a strategy to "translate" these two latent vectors into a fluent sequence of words. Recently, gated recurrent neural networks such as long short-term memory (LSTM) and gated recurrent units (GRU) have shown high abilities in tasks related to text generation. Taking into account the advantages of GRU performance, parameters and computational efficiency, we use GRU as the basic model to generate the explanation of module control (as shown on the right side of Fig. 1). Each input record is associated with a user ID, an item ID, and a rating r_{ij}. We use the user u_i's representation $\mathbf{p}_i$ and the item v_j's representation $\mathbf{z}_j$ as the input of the encoder, so that the decoded word sequence can be personalized to different user-item pair. In addition, we also consider the prediction rating $\hat{r}_{ij}$ and item characteristics to strengthen the control of the sentiment and accuracy of the generated explanation. We first map the predicted rating $\hat{r}_{ij}$ obtained in the neural rating regression stage into a vector, and then map the user potential vector $\mathbf{p}_i$ and item potential vector $\mathbf{z}_j$ is mapped to a hidden space:

Suppose h_{cl} is the output of the last hidden layer. We add the final generation layer to map h_{cl} to a vector $\hat{c}$ of size $|\mathbf{V}|$, where $\mathbf{V}$ is the vocabulary in the comment and explanation sentence, perceiving the user's fine-grained sentiment.

$$\hat{c} = softmax(W_{hc}h_{cl} + b_c) \tag{5}$$

where softmax($\cdot$) represents the softmax function, and $W_{hc} \in R^{|\mathbf{V}| \times d}$ and $b_c \in R^{|\mathbf{V}|}$ are model parameters. h_0 is used as the initial hidden state of the decoder.

$$h_0 = g(u_i, v_j, \hat{c}, \hat{r}_{ij}) \tag{6}$$

The hidden state of other time steps can be calculated by repeatedly feeding the input word q_{t-1} of the $t-1$ layer and the given feature f_{uv} and the user's fine-grained sentimental representation of the item to the decoder.

$$h_t = g(q_{t-1}, f_{uv}, h_{t-1}) \tag{7}$$

where g($\cdot$) is the gated recursive unit (GRU).

$$s_t = softmax(W_v h_t + b_v) \tag{8}$$

where softmax($\cdot$) represents the softmax function, $W_v \in R^{d \times |\mathbf{V}|}$ and $h_t \in R^{|\mathbf{V}|}$ are model parameters, and Eq. 8 represents the words generated before time t to predict the words generated at the current time. At time step t, the decoder receives the hidden vector h_t and maps it to a vector of $|\mathbf{V}|$, where $\mathbf{V}$ is the vocabulary of the words in the data set. This vector can be regarded as the distribution on the probability vocabulary, from which the word wr_t with the greatest probability is sampled.

In order to train the explanation generation module, we use the widely used cross-entropy loss as our objective function, and calculate the loss of each user-item pair in the training set.

$$L_{eg} = \frac{1}{|\mathbf{X}|} \sum_{u,i \in \mathbf{X}} \frac{1}{|\mathbf{P}_{uv}|} \sum_{t=1}^{|\mathbf{P}_{uv}|} -log s_t \tag{9}$$

where $\mathbf{P}_{uv}$ is the user's real comment on the item, and $|\mathbf{P}_{uv}|$ is the word length of the word group.

4   DLER-Based Explainable Recommendation Algorithm

Based on the DLER model, this section proposes an explainable recommendation algorithm. The specific steps of the DLER-based interpretable recommendation algorithm are as follows (as shown in Algorithm 1). In the training phase, the rating function $\hat{r}_{ij} = f(u_i, v_j, c_{uv})$ and the explanation generating function $\hat{c} = g(u_i, v_j, \hat{r}_{ij}, f_{uv})$ are modeled. In the test phase, for each candidate user item pair (u_i, v_j) representation to predict the user's preferences. Then, the predicted rating will be approximated to the user's preference, perceive the user's fine-grained sentiments on the item, and be fed to the explanation generation function for generating reviews during testing. Test phase: First, the test set is used as input to construct the DLER model. Then we use the trained model parameters to calculate the rating as the final rating prediction output and text explanation.

Algorithm 1: Explainable Recommendation Algorithm Based on DLER

Input: user id u, item id v, $\Pi = \phi$, and explanation generation model **G**.
Output: candidate rating and explanation.

1   Initialize all trainable parameters of preference prediction θ and review generation φ.
2   Precompute the marginal preference distribution and the marginal review distribution with GRU.
3   **for** Each user **do**
4   Get a mini-batch of m user-item pairs, and corresponding preferences and reviews;
5   **for** Each pair of input (u, v, C_{uv}, r_{uv}) in the mini-batch **do**
6   Compute the objective function of preference prediction and compute the conditional loss in Eq. 4;
7   **End for**
8   Optimize θ, φ to minimize loss;
9   Until Convergence of parameters.
10   Initialize the parameter set.
11   **for** Each pair of input(u, v, C_{uv}, f_{uv}) in the mini-batch **do**
12   Compute the real review representation h_{uv};
13   Compute the approximated review representation s_t;
14   Update the loss in Eq. 9;
15   **End for**
16   Optimize the parameter set;
17   Until Convergence of model parameters.
18   **End for**

In order to learn the parameters, a loss function needs to be specified for optimization. The recommendation task we are concerned about is rating prediction and explanation generation, so the loss function is expressed as Eq. 4 and Eq. 9. In this paper, the gradient descent method is used to propagate and update the parameters in the negative gradient direction through neural network back propagation to reduce the value of the loss function. When the early stop condition is met (the error has not fallen for 5 times), the iteration is stopped. In order to avoid the problem of neural network overfitting, we use dropout to randomly discard some neurons during the training process to make the model more general.

5   Experiments

In the experiments, we use two real data sets in different fields, namely, Amazon and Yelp data sets. Both Amazon and Yelp are websites that can provide product rating and review services. Users can use a rating of 1 to 5 to rate the products on the website. Since the two datasets are very large, operations to delete users and items with fewer than 20 interactions were performed, resulting in two subsets AZ-1 and YELP-1. Each comment record in our data sets includes user ID, item ID, overall rating from 1 to 5, and text comments. Table 1 shows the key characteristics of the three processed data sets. The statistics of the two data sets are shown in Table 1.

Table 1. Statistics of the experimental datasets.

Dataset	User	Items	Reviews	Features
AZ-1	7506	7360	441783	5399
YELP-1	27147	20266	1293247	7340

For the rating prediction task of the recommendation model, the root mean square error (RMSE) and mean absolute error (MAE) are usually used to evaluate the prediction accuracy. RMSE and MAE measure the quality of recommendation by calculating the deviation between the predicted value and the true value. The smaller the value, the higher the accuracy of the recommendation algorithm.

For explainability, the generated explanation is evaluated from two perspectives: the relevance to the real sentence and the degree of personalization. For the first perspective, we use ROUGE in the text summary to evaluate the textual similarity between the generated explanation and the basic facts, and use the accuracy, recall and F1 of ROUGE-1 to measure the sentences generated at different granularities. The greater the ROUGE rating, the closer the generated text is to reality. For the second perspective, the degree of personalization, we use two indicators to evaluate: the ratio of unique sentences, the feature coverage ratio in all of the generated sentence:

(1) **Unique Sentence Ratio (USR).** We present this metric to calculate how many distinct sentences are generated:

$$USR = \frac{|S|}{N} \tag{10}$$

where S denotes the set of generated unique explanation, and N is the number of total explanation.

(2) **Feature Coverage Ratio (FCR).** We adopt this metric to measure how many different features are shown in the generated explanation:

$$FCR = \frac{N_f}{|F|} \tag{11}$$

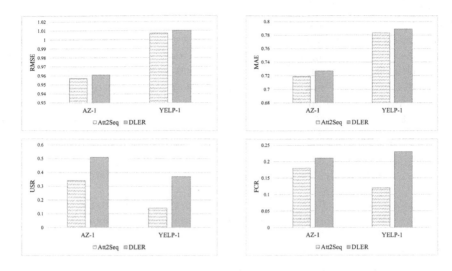

Fig. 3. Performance comparison between DLER algorithm and other algorithms

where F is the collection of all features in the dataset, and N_f is the number of distinct features shown in the generated explanation.

To evaluate algorithmic performances, the algorithms are compared as follows:

- **NRT**: Neural Rating and Tips generation [19]. This model predicts a rating based on user ID and item ID, and formulates the explanation generation problem as a text summarization task.
- **Att2Seq**: Attribute-to-sequence [20] adopts two-layer LSTM to decode the encoded representations to generate a textual review.

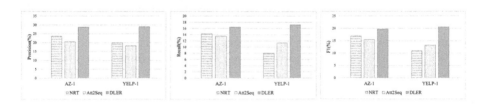

Fig. 4. Comparisons of different methods on precision, recall and F1

For all neural generative models, namely NRT, Att2Seq, and DLER, we set the maximum length of the generated text to 15. Experimental results of different recommendation algorithms. As far as RMSE and MAE are concerned, the accuracy of the method of rating prediction tasks is not much different (see Fig. 3); in terms of interpreting sentence performance, we use two indicators,

USR and FCR (see Fig. 3), analyzing the degree of personalization of all neural generation methods, it can be seen that compared with other recommendation algorithms, the DLER algorithm performs better on the USR and FCR evaluation indicators. Our DLER model consistently outperforms all baselines on the two datasets (see Fig. 4). Table 2 shows some examples of reviews generated based on Amazon's baseline model. The real cases come from three pairs of user projects. From the results, we can observe that the explanation generated by DLER are more fluent and have fewer grammatical errors.

Table 2. Generated explanation based on Amazon dataset.

Case 1	Rating: 1; Review: I read the description for this product which said it would be compatible with my macbook and upon receiving it. It did not work at all.
NRT	The case is good.
Att2Seq	This case was easy for my ipad air to the tablet.
DLER	The product received is **not compatible** with my tablet, it is not recommended to buy.
Case 2	Rating: 5; Review: The case is sturdy the keyboard is perfect for my kids to do their studies with and type on the very easy to use.
NRT	The product is good
Att2Seq	The buttons are comfortable and accurate.
DLER	The quality of this product is **very good**, I **like** it very much.

6   Conclusions

In this paper, a Dual Learning-based Explainable Recommendation Model (DLER) is proposed. In addition, a rating prediction algorithm based on neural rating regression and an explanation generation algorithm based on fine-grained sentimental perception are proposed to improve the accuracy of rating prediction and the sentimental matching between explanation sentences and users. Compared with existing methods, this algorithm has better results. In the next step, we will further consider the improvement of the sub-module of interpreting generated sentences. While ensuring the accuracy of the sentiment of users interpreting sentences, we will increase the diversity of interpreting sentences.

References

1. Resnick, P., Iacovou, N., Suchak, M., Bergstrom, P., Riedl, J.: GroupLens: an open architecture for collaborative filtering of netnews. In: Proceedings of the 1994 ACM Conference on Computer Supported Cooperative Work, pp. 175–186 (1994)
2. Sarwar, B., Karypis, G., Konstan, J., Riedl, J.: Item-based collaborative filtering recommendation algorithms. In: Proceedings of the 10th International Conference on World Wide Web, pp. 285–295 (2001)

3. Koren, Y.: Factorization meets the neighborhood: a multifaceted collaborative filtering model. In: Proceedings of the 14th ACM SIGKDD International Conference on Knowledge Discovery and Data Mining, pp. 426–434 (2008)
4. Koren, Y., Bell, R., Volinsky, C.: Matrix factorization techniques for recommender systems. Computer **42**(8), 30–37 (2009)
5. Zhang, S., Yao, L., Sun, A., Tay, Y.: Deep learning based recommender system: a survey and new perspectives. ACM Comput. Surv. (CSUR) **52**(1), 1–38 (2019)
6. Zheng, L., Noroozi, V., Yu, P.S.: Joint deep modeling of users and items using reviews for recommendation. In: Proceedings of the Tenth ACM International Conference on Web Search and Data Mining, pp. 425–434 (2017)
7. Chen, T., Yin, H., Ye, G., Huang, Z., Wang, Y., Wang, M.: Try this instead: personalized and interpretable substitute recommendation. In: Proceedings of the 43rd International ACM SIGIR Conference on Research and Development in Information Retrieval, pp. 891–900 (2020)
8. Chen, H., Chen, X., Shi, S., Zhang, Y.: Generate natural language explanations for recommendation. arXiv preprint arXiv:2101.03392 (2021)
9. Li, L., Zhang, Y., Chen, L.: Generate neural template explanations for recommendation. In: Proceedings of the 29th ACM International Conference on Information & Knowledge Management, pp. 755–764 (2020)
10. Wu, Y., Ester, M.: FLAME: a probabilistic model combining aspect based opinion mining and collaborative filtering. In: Proceedings of the Eighth ACM International Conference on Web Search and Data Mining, pp. 199–208 (2015)
11. Porteous, I., Asuncion, A., Welling, M.: Bayesian matrix factorization with side information and Dirichlet process mixtures. In: Proceedings of the AAAI Conference on Artificial Intelligence, vol. 24 (2010)
12. Fan, W., Li, Q., Cheng, M.: Deep modeling of social relations for recommendation. In: Proceedings of the AAAI Conference on Artificial Intelligence, vol. 32 (2018)
13. Kim, D., Park, C., Oh, J., Lee, S., Yu, H.: Convolutional matrix factorization for document context-aware recommendation. In: Proceedings of the 10th ACM Conference on Recommender Systems, pp. 233–240 (2016)
14. Kang, W.C., McAuley, J.: Self-attentive sequential recommendation. In: 2018 IEEE International Conference on Data Mining (ICDM), pp. 197–206. IEEE (2018)
15. Wang, H., Kou, Y., Shen, D., Nie, T.: An explainable recommendation method based on multi-timeslice graph embedding. In: Wang, G., Lin, X., Hendler, J., Song, W., Xu, Z., Liu, G. (eds.) WISA 2020. LNCS, vol. 12432, pp. 84–95. Springer, Cham (2020). https://doi.org/10.1007/978-3-030-60029-7_8
16. Truong, Q.T., Lauw, H.: Multimodal review generation for recommender systems. In: The World Wide Web Conference, pp. 1864–1874 (2019)
17. Wang, Z., Zhang, Y.: Opinion recommendation using neural memory model. arXiv preprint arXiv:1702.01517 (2017)
18. Chen, Z., et al.: Co-attentive multi-task learning for explainable recommendation. In: IJCAI, pp. 2137–2143 (2019)
19. Li, P., Wang, Z., Ren, Z., Bing, L., Lam, W.: Neural rating regression with abstractive tips generation for recommendation. In: Proceedings of the 40th International ACM SIGIR conference on Research and Development in Information Retrieval, pp. 345–354 (2017)
20. Dong, L., Huang, S., Wei, F., Lapata, M., Zhou, M., Xu, K.: Learning to generate product reviews from attributes. In: Proceedings of the 15th Conference of the European Chapter of the Association for Computational Linguistics: Volume 1, Long Papers, pp. 623–632 (2017)

Friend Relationships Recommendation Algorithm in Online Education Platform

Jingda Kang, Juntao Zhang, Wei Song, and Xiandi Yang[✉]

School of Computer Science, Wuhan University, Wuhan, China
{jingdakang,juntaozhang,songwei,xiandiy}@whu.edu.cn

Abstract. With the development and application of online education, mining friend relationships between learners can improve interactive communication, enhance collaborative learning, and motivate learners to make mutual progress. However, existing methods only recommend well-known members through the number of likes and fans, which fail to consider the hidden interest points and content topics of members in the text. To address this problem, we propose an Evaluation Latent Dirichlet Allocation (EvaluationLDA) algorithm to recommend suitable friends for learners in online education. The EvaluationLDA algorithm clusters learners with similar learning interests to obtain Top-N friend recommendation sequences based on constructing learner document datasets, calculating learner similarity, and modeling the friend topic. We conduct experiments to demonstrate the effectiveness of our EvaluationLDA algorithm. The result shows that our EvaluationLDA algorithm can effectively recommend Top-N friend sequences in the online education platform.

Keywords: Online education · Friend recommendation · EvaluationLDA algorithm

1 Introduction

With the development and successful application of Massive Open Online Courses (MOOCs), online education has become a new education trend that people pursue lifelong learning. Especially, affected by the global COVID-19 epidemic in 2020, hundreds of millions of people have participated in online learning that further expands the social needs of online education. The interaction between learners can promote learners' communication, resource sharing, collaborative learning, and even establish precious friendships in the online learning process. Recent studies have focused on the social dimension of learning in MOOCs and then analyzed discussion forums of MOOCs to understand the patterns of student interactions [11,15]. Zou et al. explored social presence and its relationship with learners prestige in the learner network of a MOOC [18]. Therefore, mining learners' knowledge status, interaction behaviors, and friend relationships are meaningful and valuable for promoting their efficient learning.

Despite the popularity of MOOCs, studies have found that a lot of MOOCs had low completion rates that were less than 10% [6,8]. For the past few years,

C. Xing et al. (Eds.): WISA 2021, LNCS 12999, pp. 592–604, 2021.
https://doi.org/10.1007/978-3-030-87571-8_51

many researchers have been trying to find potential factors that may explain the low engagement and low completion rates in MOOCs. Some works find that the social factor plays an important role in learners' engagement and performance in MOOCs through stimulating higher-order thinking skills [1,13]. Therefore, the online education platform is a place for learning and a world for people to communicate, interact, and collaborate. The problem about how to recommend high-quality friends (friends with similar interests) for users more accurately has become a popular research spot of scholars in the field of the social network, and online education platform also needs a similar friend recommendation mechanism to search for learners with similar interests. At present, mainstream online education platforms such as edX, Coursera, and XuetangX provide learners with the function of actively following others and cannot recommend the best friends intelligently for learners through the system. On the other hand, learners cannot directly obtain the social network relationship established by following and forum likes to understand the learner group relationship [5]. Therefore, the personalized friend recommendation system based on the friend recommendation algorithm is missing in the current research.

To solve the problems raised above, we design a friend recommendation algorithm called the Evaluation Latent Dirichlet Allocation algorithm (EvaluationLDA), which achieves clustering of learners with similar topics and obtains initialized friend relationship recommendations with similar interests via judging the clustering results according to the cosine similarity. Then, we use the learner's comprehensive evaluation results in the online education forum as the new dimension of cosine similarity and recalculate the friend similarity to obtain the Top-N friend recommendation sequence. Finally, we conduct experiments to demonstrate the effectiveness of our EvaluationLDA algorithm. The experiment result shows that our EvaluationLDA algorithm can effectively recommend Top-N friend sequences.

2   Related Work

Social network structure refers to the social relationship networks formed between the individuals in the network system. Friend recommendation in the social network can be regarded as the problem of the link prediction method that only considers network topology structure, users are treated as network nodes, and the edges between nodes are friends [10]. Existing friend recommendation algorithms include friend recommendations based on link prediction [9,16], friend recommendations based on information content [12], etc. The link prediction method is applied to calculate the possibility of generating a new link after a while to predict the possibility of two users become friends, and we recommend these users as friends with a high probability [16]. Li et al. merged the Jaccard Coefficient, the Adamic-Adar, and user features to construct an algorithm to calculate the probability of a new link between two nodes [9]. Tang et al. constructed the user model by considering four aspects: user profile, the content information that user posted, the link relationship, and the interaction relationship between users, and presented a friend recommendation approach based on the similarity of the microblog user model [12]. At present, most online education platforms do not include the personalized friend recommendation functions and cannot recommend suitable friends for learners by mining learners' interests. In this paper, we design and implement a friend recommendation algorithm in an online education platform to recommend suitable friends for learners.

3   Proposed Algorithm

In this section, we first introduce the Jaccard Coefficient and Jaccard-Score algorithm based on ternary closure theory. Then we describe the detail of our EvaluationLDA algorithm.

3.1   Jaccard-Score Algorithm Based on Social Network

Ternary Closure Theory. Definition of a ternary closure: a ternary closure in a social network where two people have mutual friends implies they are highly likely to become friends in the future. As shown in Fig. 1, x, y and k_1 and x, y, k_2 form ternary closures respectively. In real life, mutual friends can serve as bonds between strangers to understand each other, form trust, and become friends. Therefore, the number of mutual friends between two learners determines the strength of the relationship between them and can use for friend recommendations.

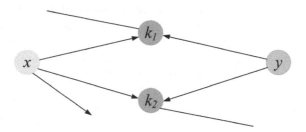

Fig. 1. Example of a ternary closure

Jaccard Coefficient. The Jaccard Coefficient calculates [7] the similarity between users according to the number of mutual friends. Its idea is to form a social network based on the user's follow relationship, where nodes represent users, and the edges between nodes represent the follow relationships between users, such as follower or following. The more common neighbors two nodes have in a social network, the greater the similarity between these two nodes, and the greater the likelihood that they will become friends. The equation of the Jaccard Coefficient is as follows:

$$Jaccard(x, y) = \frac{|Neighbour(x) \cap Neighbour(y)|}{|Neighbour(x) \cup Neighbour(y)|} \tag{1}$$

where Neighbour(x) represents the neighbor node of node x. However, the disadvantage of this formula is that it fails to consider the different attributes of mutual friends.

Jaccard-Score Algorithm. To solve the defect of the Jaccard Coefficient, we need to set different weights for each mutual friend according to the number of mutual friends. According to the features of the experimental dataset, we decide to take the inverse square root of the number of friends of each mutual friend for weighting processing. The Jaccard-Score equation expresses as follows:

$$Jaccard - Score(x, y) = \sum_{k \in Neighbour(x) \cap Neighbour(y)} \frac{1}{\sqrt{|Neighbour(k)|}} \quad (2)$$

The flowchart of the Jaccard-Score algorithm is shown in Fig. 2.

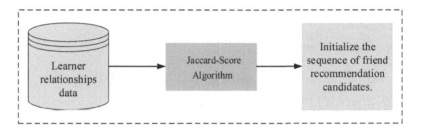

Fig. 2. Flow chart of Jaccard-Score algorithm

Where Neighbour(x) is the neighbor node of node x. The Eq. (2) calculates the similarity score between learners according to the weight of their mutual friends. The Eq. (2) calculates the similarity score between learners according to the weight of their mutual friends, and it had applied in social networks. For example, Facebook's friend recommendation module uses the variant of this equation. After calculating according to the equation, we sort the learner scores in descending order to obtain the initial friend recommendation candidate sequence.

3.2   EvaluationLDA Algorithm

The architecture graph of the EvaluationLDA algorithm is shown in Fig. 3. The learner clustering process is described based on the EvaluationLDA algorithm as follows.

Learner Document Preprocessing. Firstly, we extract the text data of posts from the dataset and construct an integrated document set for each learner as the initially processed data set. Then, we implement data cleansing through word segmentation, POS tagging, and deletion of stop words with Jieba tools. Finally, we combine all words into a unified document set containing all the data sets for each learner.

Parameters Determination. According to existing research experience [14], we set the two parameters α and β of the LDA model 50/K (K is the number of topics) and 0.01. This setting can ensure that the clustering results have a

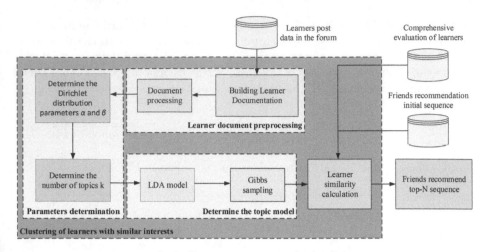

Fig. 3. The architecture graph of the EvaluationLDA algorithm

certain degree of difference and achieve rapid convergence. The size of K determines the range of topic categories and the complexity of training time in the model training process. We estimate the number of the most reasonable topics K according to the degree of confusion index. The lower the degree of confusion index, the better the clustering effect. The formula for calculating the degree of confusion is given by:

$$perplexity(D) = \exp(-\frac{\sum \log p(w)}{\sum_{d=1}^{M} N_d})p(w) = p(z|d) * p(w|z) \tag{3}$$

with

$$p(w) = p(z|d) * p(w|z) \tag{4}$$

where M is the total number of test set documents. N_d represents the set of all words in number d document, without removing duplication. p(w) represents the probability of each word in the test set [17]. p(z|d) represents probability of each theme in each document. p(w|z) represents probability of each word in the dictionary under each topic.

Determine the Topic Model. We implement the EvaluationLDA topic model based on the standard LDA topic model [2], and the LDA model has a wide range of applications, such as semantic Web services discovery. The input is the learner document after data preprocessing, that is, the document content which is used to uniquely identify the learning interest of each learner. The document-topic probability distribution obtained after the standard LDA model training is the learner-topic probability distribution. The standard LDA model is shown in Fig. 4.

In Fig. 4, m represents a collection of documents $m \in \{1, 2, ..., M\}$. Each document corresponds to a unique learner. α and β are two parameters satisfying the Dirichlet distribution, and both represent the control parameters at the corpus level. The α vector parameter is used to generate the topic distribution θ_m of the m-*th* document, and finally generate the topic number $Z_{m,n}$ of the n-*th* word in the m-*th* document. During this process, the model first extracts a document-topic dice θ_m, then throw the dice to generate the topic number $Z_{m,n}$ for the n-*th* word in the document. φ_k represents the k-*th* topic word distribution vector. The β vector parameter is used to generate the n-*th* word $W_{m,n}$ of the m-*th* document from K topic vectors, that is, each topic β vector corresponds to the probability distribution matrix $p|z(w)$. In this process, the model selects and throws the dice $k = Z_{m,n}$ from K kinds of topic-word dice within $\{\phi_1, \phi_2, ..., \phi_k\}$, and then generates the word $W_{m,n}$.

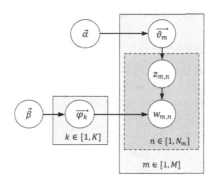

Fig. 4. The standard LDA model

In fact, every time the model generates a word in a document, it has to throw two dices. The first time it throws a document-topic dice to get the topic, and the second time it throws a topic-word dice to get the word. In order to generate the word in the document, it has to throw two dices to get the final result. A composed of N word document will dice $2N$ times. In order to speed up the training time and improve efficiency, the model can equivalently adjust the order of the $2N$ dice: first N times throwing the document-topic dice to get all of the topics in corpus, then according to the topic of each word number, second N times throwing topic-word dice to produce N words. The equivalent adjustment equation is given by:

$$p(w, z|\alpha, \beta) = p(w|z, \beta)p(z|\alpha) = \prod_{k=1}^{K} \frac{\Delta(\phi_K + \beta)}{\beta} \prod_{m=1}^{M} \frac{\Delta(\theta_m + \alpha)}{\alpha} \qquad (5)$$

We implement Gibbs Sampling to solve the obtained probability distribution matrix, that is, the joint probability of document-topic distribution θ and the joint probability of topic-word distribution are sampled to obtain the joint probability of document-topic distribution and the joint probability of topic-word distribution. Hence, according to the one-to-one correspondence between learner and document, The joint probability distribution of learner-topic $p(w, z)$ is obtained.

The Gibbs Sampling process of our EvaluationLDA topic model is as follows: (1) Randomly select a topic for each feature word z_i in the learner document as the initial state of the Markov chain. (2) Implement posterior probability equation to iteratively generate a new Markov chain state for learner documents until the Markov chain converges to the posterior probability distribution of the parameters to be estimated. (3) Estimate the parameter of θ and ϕ to get the document-topic probability distribution and topic-word probability distribution, then record the value of the feature word Z_i. Through mathematical deduction, the posterior distribution of topics in M is given by:

$$p(\theta_m|z_{-i}, w_{-i}) = Dir(\theta_m|n_{m,-i} + \alpha) \tag{6}$$

The posterior probability distribution of feature words in the k-*th* topic is given by:

$$p(\phi_k|z_{-i}, w_{-i}) = Dir(\phi_k|n_{k,-i} + \beta) \tag{7}$$

Both equation (6) and formula (7) represent the n-*th* word randomly selected in the initial Markov chain (representing the n-*th* word in the m-*th* document). The estimates of parameters θ and ϕ can be derived as follows:

$$\theta_{mk} = \frac{n_{m,-i} + \alpha_k}{\sum_{k=1}^{K}(n_{m,-i} + \alpha_k)} \tag{8}$$

$$\phi_{kz} = \frac{n_{k,-i} + \beta_z}{\sum_{z=1}^{Z}(n_{k,-i} + \beta_z)} \tag{9}$$

Thus, the conditional probability distribution of the output of EvaluationLDA model is successfully obtained through Gibbs Sampling, and the document-topic probability distribution θ and topic-word probability distribution are obtained, which quantifies the learner's interest preference and gets the learner-topic preference matrix.

Learner Similarity Calculation. Next, we convert the obtained document-topic probability distribution correspondingly to learner-topic probability distribution, and get learner-topic preference matrix. In order to combine with learner comprehensive evaluation, we process the learner-topic preference matrix with an additional dimension to indicate the comprehensive evaluation scores of learners in the online education platform, as shown in matrix:

$$\begin{matrix} U_1 \\ \cdots \\ U_M \end{matrix} \begin{bmatrix} z_{11} & \cdots & z_{1K} & E_1 \\ \vdots & \ddots & \vdots & \vdots \\ z_{M1} & \cdots & z_{MK} & E_M \end{bmatrix}$$

$E_1...E_M$ are the comprehensive evaluation scores which mark the overall performance of M learners in the online education platform. It can be considered that the comprehensive evaluation of learners' platform is also a new topic, which can be divided into learners' 'topic distribution' as the basis of learner clustering and friend recommendation. From the perspective of sociology, the level of learners should be taken as one of the criteria for clustering. This method can make the clustering calculation of learners more accurate and the recommendation result more accurate.

According to approach explained above, this step essentially calculates the document similarity. After obtaining the joint probability distribution of learner and topic obtained by EvaluationLDA, plus the learner evaluation score obtained by comprehensive evaluation module in platform, we implement cosine similarity method to calculate the similarity between target learner and other learners, and groups with similar learning interests of target learner are found for clustering. Top-N learners were selected as the original friend recommendation sequence after descending sorting.

Friends Recommendation Top-N Sequence. According to the original Top-N learner friend recommendation sequence, combined with the initial Top-N friend recommendation sequence obtained by Jaccard-Score algorithm, we sum the similarity of the learners in the two sequences to obtain the new similarity sequence. The equation is given by:

$$R = xE + yJ \tag{10}$$

In the equation, x and y stands for the weight parameter of learner similarity in the original friend recommendation sequence and the initial friend recommendation sequence, respectively. Until now we have got the unsorted final friend recommendation sequence.

Then we sort the new sequence in descending order, and fetch Top-N learners as the final friend recommendation sequence. Thus, the friend recommendation sequence output by EvaluationLDA algorithm is obtained, and the friend recommendation process for the target learner is successfully completed. The relevant process is as described in Fig. 3.

4   Experiments

4.1   Datasets and Evaluation Metrics

Datasets. We collect the Sina Weibo dataset to conduct experiments to verify the effectiveness of our algorithm. The Weibo website is an open social platform that contains a large amount of user information and text data. Its forum resources involve education, technology, games, and many other fields, and it is suitable for researching social network relationship mining algorithms. The detailed statistics of this dataset are shown in Table 1.

Table 1. Statistics of the Sina Weibo dataset

Dataset	Learners number	Posts number	Following relationship
Weibo	1000	5000	142368

Evaluation Metrics. To evaluate the effectiveness of the EvaluationLD algorithm, we use multiple evaluation metrics for verification, such as Precision, Recall, and F1 measure. The larger the value of these evaluation metrics, the better the performance of the algorithm.

4.2   Experimental Results and Analysis

We obtain the Top-N recommended friend sequence based on the Weibo dataset. When determining the parameters used in the EvaluationLDA model, it is necessary to select the most appropriate parameter K (the number of topics) according to the degree of confusion. According to the number of topics K, we calculate the corresponding confusion degree, as shown in Fig. 5.

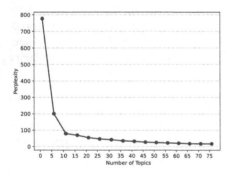

Fig. 5. The degree of confusion index varies with the number of topics

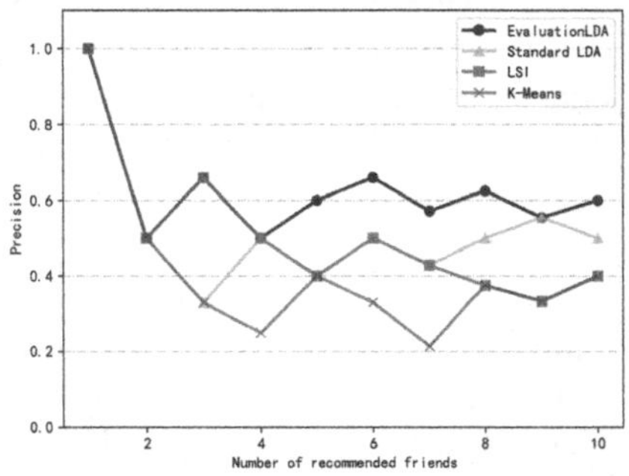

Fig. 6. Precision calculation results

We can see from Fig. 5 that when the number of topics is greater than 10, the degree of confusion begins to converge and eventually becomes stable with the increase of topics. Therefore, we decide to set the number of topics as $K = 10$ in our experiments.

According to the existing research and empirical data, the EvaluationLDA model parameter α is determined to be $50/K$, β is determined to be 0.01, to ensure that the clustering results have a certain degree of differentiation. The topic-word matrix and learner-topic matrix are obtained by experiments on weibo dataset. The topic-word matrix result is shown in Table 2:

According to Table 2, we can observe that the clustering results of the topic meet the basic expectation. The 1–5 topic categories belong to education, economy, technological products, policy, and games, which meet the standard of fundamental classification effect. The results of some learner's topic distribution matrix are shown in Table 3.

After the learner-topic distribution table is obtained, according to the learner similarity calculation stage in the EvaluationLDA algorithm, we add the item of

Table 2. Topic-word matrix table

Topic number	1	2	3	4	5
Feature words	exam	fund	function	China	player
	Information	stock	shooting	new energy	game
	situation	company	edit	market	Samsung
	publish	manager	Sony	activity	activity
	change	market	hd	support	reward
	provide	manage	lens	investment	race
	English	risk	company	industry	card

Table 3. Learner-topic matrix table

Learner ID	Topic 1	Topic 2	Topic 3	Topic 4	Topic 5
0001	0.25159374	0.001	0.00	0.00	0.73445415
0002	0.58133951	0.02123515	0.00	0.02	0.00
0003	0.73381263	0.00	0.00	0.00	0.26280597
0004	0.11558618	0.03563155	0.02684684	0.001	0.00
...	...	...	...	...	...
0999	0.00	0.00	0.00	0.082537815	0.20322542
1000	0.051224303	0.0122249	0.00	0.00	0.00

learner comprehensive evaluation score of the topic, and calculate the cosine similarity between target learner and other learners, then sort the result in descending order. We take the target learner ID of 0001 as an example and calculate the Top-5 learners in descending order of similarity, the result as shown in Table 4. We sum the initial recommendation friend sequence obtained by the Jaccard-Score algorithm to get the final ranking of learner similarity. Then we select the final recommended sequence of the Top-N friends according to the needs and get the result of friend recommendation.

Table 4. Learn similarity calculation

Ranking	Learner ID	Cosine similarity
1	0176	0.9622584
2	0003	0.9456756
3	0027	0.9430328
4	0657	0.9220919
5	0426	0.9174225

To better evaluate the performance of the EvaluationLDA, we use the standard LDA, LSI(Latent Semantic Indexing) algorithm [4], and K-Means [3] clustering algorithm as comparative experiments to verify our algorithms recommendations of friends on the Weibo dataset. 70% of the data set is taken as the training set for training, and the other 30% is taken as the test set to recommend friends for the users in the test set and verify the recommendation result. We set the number of recommended friends as 1–10, and calculate the accuracy rate, recall rate, and F1 measure of all methods, respectively. Specific experimental results are shown in Figs. 6, 7 and 8.

According to the analysis of the experimental results, with the increase of the number of friends, the precision rate of the all algorithms gradually decreases, the recall rate gradually increases, and the F1 measure also gradually increases as a whole. That is to say, when Top-10 friends are recommended to learners, the effect is better than that of N less than 10, because there are more topic categories in the Weibo data set. It can be seen from the experimental results that the evaluation index of the EvaluationLDA algorithm is generally better than that of the comparative algorithms. Because we considered the comprehensive evaluation score of the learner in the online education platform and the learner's actual follower relationship, and cluster the categories more accurately, which reflects in the results of the learner similarity calculation. Overall, the experimental results are in line with expectations, and the EvaluationLDA is better than current text cluster-based algorithms.

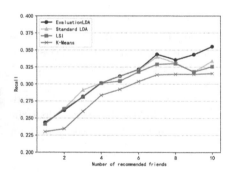

Fig. 7. Recall calculation results   **Fig. 8.** F1 Measure calculation results

5   Conclusion

In this paper, we study the problem of friend recommendation in the online education platform. To implement learner friend recommendation on an online education platform, we propose an EvaluationLDA friend recommendation algorithm. This algorithm clusters the content of learners' posts in the forum system and increases the comprehensive evaluation of learners. Then, the text content understands the learners' real interest direction and combines the user evaluation module to calculate the sequence of recommended friends. Finally, the final recommended friend is completed from the initial recommended friend candidate sequence.

In the future, we provide different friend recommendation methods in different ways according to learner needs, rather than the recommended method based on the same topic model clustering for all learners.

Acknowledgements. This work is partially supported by National Natural Science Foundation of China Nos. U1811263, 62072349, National Key Research and Development Project of China No. 2020YFC1522602. We also thank anonymous reviewers for their helpful reports.

References

1. Al-Rahmi, W.M., Alias, N., Othman, M.S., Marín, V.I., Tur, G.: A model of factors affecting learning performance through the use of social media in Malaysian higher education. Comput. Educ. **121**, 59–72 (2018)
2. Blei, D.M., Ng, A.Y., Jordan, M.I.: Latent Dirichlet allocation. J. Mach. Learn. Res. **3**, 993–1022 (2003)
3. Bradley, P.S., Fayyad, U.M.: Refining initial points for k-means clustering. In: Proceedings of the Fifteenth International Conference on Machine Learning ICML, pp. 91–99 (1998)
4. Deerwester, S.C., Dumais, S.T., Landauer, T.K., Furnas, G.W., Harshman, R.A.: Indexing by latent semantic analysis. J. Am. Soc. Inf. Sci. **41**(6), 391–407 (1990)
5. Dou, B., Li, S., Zhang, S.: Social network analysis based on structure. Chin. J. Comput. **35**(04), 741–753 (2012)
6. Gütl, C., Rizzardini, R.H., Chang, V., Morales, M.: Attrition in MOOC: lessons learned from drop-out students. In: Uden, L., Sinclair, J., Tao, Y.-H., Liberona, D. (eds.) LTEC 2014. CCIS, vol. 446, pp. 37–48. Springer, Cham (2014). https://doi.org/10.1007/978-3-319-10671-7_4
7. Jaccard, P.: The distribution of the flora in the alpine zone 1. New Phytol. **11**(2), 37–50 (1912)
8. Joo, Y.J., So, H., Kim, N.H.: Examination of relationships among students' self-determination, technology acceptance, satisfaction, and continuance intention to use K-MOOCs. Comput. Educ. **122**, 260–272 (2018)
9. Li, B., Chen, Z., Huang, R., Zheng, X., Xu, C., Zhou, Q., Long, Z.: Link prediction friends recommendation algorithm for online social networks named JAFLink. J. Chin. Comput. Syst. **38**(08), 1741–1745 (2017)
10. Lü, L., Zhou, T.: Link prediction in complex networks: a survey. Physica A Stat. Mech. Appl. **390**(6), 1150–1170 (2011)
11. Poquet, O., Dawson, S.: Untangling MOOC learner networks. In: Proceedings of the Sixth International Conference on Learning Analytics & Knowledge, LAK 2016, pp. 208–212 (2016)
12. Tang, F., Zhang, B., Zheng, J., Gu, Y.: Friend recommendation based on the similarity of micro-blog user model. In: 2013 IEEE International Conference on Green Computing and Communications and IEEE Internet of Things and IEEE Cyber, Physical and Social Computing, pp. 2200–2204 (2013)

13. Wang, W., Guo, L., He, L., Wu, Y.J.: Effects of social-interactive engagement on the dropout ratio in online learning: insights from MOOC. Behav. Inf. Technol. **38**(6), 621–636 (2019)
14. Wei, X., Croft, W.B.: LDA-based document models for ad-hoc retrieval. In: Proceedings of the 29th Annual International ACM SIGIR Conference on Research and Development in Information Retrieval, pp. 178–185 (2006)
15. Wise, A.F., Cui, Y.: Learning communities in the crowd: characteristics of content related interactions and social relationships in MOOC discussion forums. Comput. Educ. **122**, 221–242 (2018)
16. Xie, F., Chen, Z., Shang, J., Feng, X., Li, J.: A link prediction approach for item recommendation with complex number. Knowl. Based Syst. **81**, 148–158 (2015)
17. Zhao, H., Chen, J., Xu, L.: Semantic web service discovery based on LDA clustering. In: Ni, W., Wang, X., Song, W., Li, Y. (eds.) WISA 2019. LNCS, vol. 11817, pp. 239–250. Springer, Cham (2019). https://doi.org/10.1007/978-3-030-30952-7_25
18. Zou, W., Hu, X., Pan, Z., Li, C., Cai, Y., Liu, M.: Exploring the relationship between social presence and learners' prestige in MOOC discussion forums using automated content analysis and social network analysis. Comput. Hum. Behav. **115**, 106582 (2021)

RHE: Relation and Heterogeneousness Enhanced Issue Participants Recommendation

Huiyu Jiang$^{(\boxtimes)}$, Liang Wang, Xianping Tao, and Hao Hu

State Key Laboratory for Novel Software Technology, Nanjing University, Nanjing 210023, People's Republic of China
jhy@smail.nju.edu.cn, {wl,txp,myou}@nju.edu.cn

Abstract. Open source software (OSS) platform users frequently join issue discussions in various repositories, and establish numerous co-talk (i.e. cross-issue reference and discussion) relationships between issues both within or cross repositories. In this work, we collect and analyze issue discussion data from GitHub to study the unique features of co-talk relationships, and discover that many participants play a versatile role during issue discussions across repositories. Based on the discovery, we enhance existing bug triaging technologies with the Relation and Heterogeneousness Enhance (RHE) method to include potential participants with cross-repository co-talk histories. RHE integrates co-talk relationship embedding and heterogeneous graph embedding for complex OSS communities. We conduct experiments with real-world data collected from GitHub to show the effectiveness and usefulness of RHE. The results suggest that RHE achieves an improved performance comparing to the baseline approaches.

Keywords: OSS · Co-talk · Heterogeneous graph · Embedding

1 Introduction

During the development of open source software (OSS), more and more software repositories are maintained in OSS community (e.g. GitHub), and everyone can submit an issue in a public repository. These issues can help developers find software vulnerabilities, but it is often too much to be reviewed in time by the maintainers manually. Besides, manual reviewing bug reports and choosing participants is heavy and tedious, and it will worsen in a large software repository.

The same problem is also faced by bug tracking systems (another type of OSS community that often maintains single repository. e.g. Mozilla). And in bug tracking system, bug triaging [3,8,17,18] treats this problem as a classifier from bug report text to developer or fixer. And existing work on bug triaging only focuses on recommending potential fixers from developers *inside* the repository. But in GitHub, this problem becomes more complicated because everyone can join in issues' discussion or submit some extra information about this issue, and someone may be experts but do not maintain the repository.

© Springer Nature Switzerland AG 2021
C. Xing et al. (Eds.): WISA 2021, LNCS 12999, pp. 605–616, 2021.
https://doi.org/10.1007/978-3-030-87571-8_52

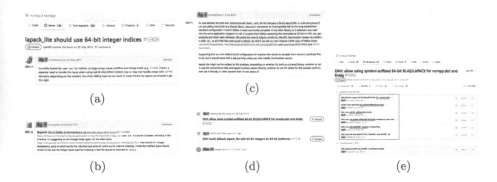

Fig. 1. *numpy/numpy#5906* fixing process.

Considering the following example, Fig. 1(a) issue *numpy/numpy#5906*, *dev-n* is the one of the maintainer of *numpy/numpy*, and he report array indices crash problem in this issue, and *dev-n* suggested using 64-bit instead of 32-bit integer indices. Then, *dev-a* doesn't support this, and listed code fragment links showing in Fig. 1(b), but they don't have a good solution. *Dev-t*, a *julia* developer, submitted a comment for this issue to describe a solution from same issues of *JuliaLang/julia* and *xianyi/OpenBLAS*. In the *JuliaLang/julia*, they append a suffix '64_' to BLAS and lapack symbols to avoid this problem and enhance security, showing in Fig. 1(c). Finally, Fig. 1(d) and 1(e), pull-request *numpy/numpy#15012* shows the suffix '64_' are appended to variables and macros. In other words, *dev-t*'s comments including other repositories experience are adopted in this issue, but he does not maintain *numpy/numpy*.

Developers (e.g. *dev-t*) break the barrier between with repositories by issue link, and these links between issues are named co-talk relationship in this paper. In the 833 maintainers of our experiments, 15.4% maintain multiple repositories, and up to 54.9% of them participated issues of various repositories. We study the problem of issue participants recommendation in this paper. To solve this, two sub-problems need to be solved:

GitHub issues contain a lot of natural language and co-talk relationship information, like the example above. First problem is how to combine the two pieces of information reasonably. There are many types of entities and have more relationships in OSS community. It should be modeled and expressed reasonably, and combine with the problem above.

In this work, we propose RHE (**R**elation and **H**eterogeneousness **E**nhanced) method that include: 1) embed text and co-talk relationships of issues with relational-doc2vec model, 2) embed issues and developers with OSS community heterogeneous graph, and 3) recommend issue participants by distance metric.

In summary, this paper makes the following contributions:

- We build a database to study co-talk relationship in GitHub;
- We take into account the participants that are not maintainers but potentially helpful in tackling the issues when performing recommendations;

– We propose the RHE method that includes the co-talk relationship and heterogeneous structure embedding for the recommendation problem.

2 Related Work

Bug Triaging: Čubranić, D. et al. [3] proposed a text categorization approach, using Naïve Bayes classifier, to tackle bug triaging as a text classification problem in bug tracking systems. Jeong, G. et al. [8] show that 37%–44% of bug reports are "tossed" (or reassigned) to other developers, the length of tossing sequence is closely related to the time-to-correction for a bug. Thus, their experiments show that the tossing graph is effect to improving on Naïve Bayes and Bayesian Network for bug triaging. Xia, X. et al. [18] proposed a new framework which maps the words in the bug reports to their corresponding topics, and a model named MTM which extends Latent Dirichlet Allocation for bug triaging. Their approach considers products and component information (metadata) in bug tracking systems, and achieved good results in their experiments. Synthesize the point of the above approaches, Xi, S. et al. [17] proposed ITRIAGE to integrate the features from textual content, metadata, and tossing sequence of bug report in bug tracking systems.

Pull-Request Reviewers Recommendation: Yu, Y. et al. [19] analyze social relations between contributors and reviewers in GitHub, and propose a novel approach by mining project's comment networks to tackle the challenge of assigning reviewers to pull-requests. Rahman, M.M. et al. [15] emphasize the effect of external libraries and specialized technologies in pull-request, built the relation ships between pull-request and some reviewers by similar project experience or similar specialized technology experience, and proposed CoRReCT. Hannebauer, C. et al. [7] compare eight algorithms, showing that the algorithms based on review expertise achieves better recommendations results.

Cross-Repository Bug Tracking: Ma, W. et al. [12] study the process of fixing cross-project correlated bugs. In their survey responses, 60.5% of the surveyed downstream developers believe that tracking the root cause of cross-project bugs is difficult, and developers have to take longer time(more than 1 day) to find bug root. This paper provides an important factual basis for our study. Tan, S.H. et al. [16] show that the similar bugs across different Android apps, and the authors find 34 new bugs in 5 Android apps using their collaborative method, their main idea is that similar bugs or issues can be appeared on different apps or projects. Their work is also a good practice, but their method don't have enough flexibility and scalability to expand to OSS communities.

3 Preliminary Study

We present our preliminary study on co-talk issues, starting with the introduction to the co-talk database used in this work. First, we define concepts related

Table 1. The preliminary statistic result of co-talk/normal issues

Item	Normal	Co-talk inner	Co-talk outer
#issues	**4,607,782**	1,877,050	694,531
Avg. consuming time	125 d	129 d	**155 d**
Avg. #participants	2.43	2.84	**4.55**
Avg. #comments	4.27	5.68	**10.10**

to 'co-talk' as follows (notations used: letters in Frak font including $\mathfrak{r}$, $\mathfrak{i}$, and $\mathfrak{d}$ mean repository, issue, and developer, respectively):

- **Co-talk Issue:** Given two issues $i_{\mathfrak{r}_1}$ and $i_{\mathfrak{r}_2}$ in repository $\mathfrak{r}_1$ and $\mathfrak{r}_2$ ($\mathfrak{r}_1$ and $\mathfrak{r}_2$ can be the same), when a user submitted a comment including a reference link that links to issue $i_{\mathfrak{r}_2}$ in the issue $i_{\mathfrak{r}_1}$, we then call the two issues has established a co-talk relationship $\langle i_{\mathfrak{r}_1}, i_{\mathfrak{r}_2} \rangle$, and are both co-talk issues.
- **Co-talk outer issue:** Satisfies the above condition and $\mathfrak{r}_1 \neq \mathfrak{r}_2$.
- **Co-talk inner issue:** Satisfies the first condition and $\mathfrak{r}_1 = \mathfrak{r}_2$.
- **Normal issue:** A issue that is not co-talk issue is called normal issue.

3.1   GitHub Co-talk Database

The following main ideas should be considered for GitHub co-talk database. Intuitively, the number of stars is a witness to the thriving of the repository. And a survey [2] shows that 73% of the participant does consider the number of stars before using or contributing to a repository. Under the guidance above, we collect all issues of 3,058 large repositories as our co-talk database.

1. Enough number of large repositories to represent GitHub community;
2. Repositories that fork from another are eliminated (e.g. *microsoft/clang-1*);
3. Repositories that don't maintain a software project (e.g. *996icu/996.ICU*) or is a mirror from another system (e.g. *v8/v8*) or is archived to read-only (e.g. *npm/npm*) are excluded;
4. Repositories that use non-GitHub as bug/issue tracking system (e.g. *apple/swift*) are don't considered.

3.2   Co-talk/Normal Issue

The co-talk relationship is so important for this paper, to understand the differences between co-talk/normal issues, some statistic results will be showed in this part. And we made some statistics as shown in Table 1. Where *consuming time* is the time interval from an issue open to the issue close.

   Although the number of co-talk issues is less than normal issues, the other statistical results is converse obviously in Table 1. And Ma, W. et al. [12] also showed the cross-project correlated bug is difficult than within-project bug and

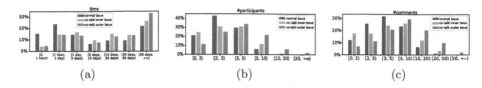

(a)   (b)   (c)

Fig. 2. Co-talk/normal issue statistic details. (a) A large percentage (above 50%) of normal issues will be closed within 5 d. (b) In normal issues, only 2 developers (including issue submitter) are needed to solve up to 60% issues. (c) Up to 70% issues of normal issues only have few comments from open to close.

have to take long time(more than a day) to find cross-project bug root. In our scenes, the co-talk outer issue we defined is also showing its difficulty to tackle, consuming more time, and take more developers' intelligence.

More details of Table 1 showing in Fig. 2, three sub-figures reveal the co-talk issue is more difficult and more complex than normal issue generally. And that is why are we pay more attention to co-talk relationships.

4   Problem and Method

4.1   Problem Definition

All repositories are represented by a set $\mathbf{R}$; all issues in a repository $\mathfrak{r}$ make up issue set $\mathbf{I}_\mathfrak{r}$, so that all issues in GitHub can be represented by a set $\mathbf{I} = \bigcup_{\mathfrak{r} \in \mathbf{R}} \mathbf{I}_\mathfrak{r}$; and the $\mathbf{D}$ denotes the set of $|\mathbf{D}|$ developers who maintain a certain repository above or participated a certain issues discussion. In formally, $\mathbf{R} = \{\mathfrak{r}_1, \mathfrak{r}_2, \cdots, \mathfrak{r}_{|\mathbf{R}|}\}$, $\mathbf{I}_{\mathfrak{r}_j} = \{i_{\mathfrak{r}_j \# 1}, i_{\mathfrak{r}_j \# 2}, \cdots, i_{\mathfrak{r}_j \# |\mathbf{I}_{\mathfrak{r}_j}|}\}$ and $\mathbf{D} = \{\mathfrak{d}_1, \mathfrak{d}_2, \cdots, \mathfrak{d}_{|\mathbf{D}|}\}$.

Furthermore, we use $\langle \mathfrak{d}, \mathfrak{r} \rangle$ to describe a **maintain** relationship, and the $\mathbf{M}$ is a set including all $\langle \mathfrak{d}, \mathfrak{r} \rangle$. $\langle \mathfrak{d}, i_{\mathfrak{r} \# k} \rangle$ represent $\mathfrak{d}$ submitted one or more comments in issue $i_{\mathfrak{r} \# k}$, the $\mathbf{P}$ is the set of all **participate** relationships. **Co-talk** relationship is represented by $\langle i_{\mathfrak{r}_j \# k}, i_{\mathfrak{r}_l \# m} \rangle$, $\mathbf{X}$ is complete works of co-talk relationship.

In a certain issue $i_{\mathfrak{r}_j \# k}$, it must contain a title and a summary, may contain some comments. The text information can be used as the issue's attributes. In addition, a toy example of OSS community shown on Fig. 3(a).

Given a heterogeneous graph $G = (V, E)$ (which $V = \mathbf{R} \cup \mathbf{D} \cup \mathbf{I}$, and $E = \mathbf{M} \cup \mathbf{P} \cup \mathbf{X}$) including all information of issue text, and a new issue belong to an existing repository $i_{\mathfrak{r} \# k} \notin \mathbf{I}, \mathfrak{r} \in \mathbf{R}$, find the $\mathbf{D}_{i_{\mathfrak{r} \# k}} \subseteq \mathbf{D}$ to satisfy $\forall \mathfrak{d} \in \mathbf{D}_{i_{\mathfrak{r} \# k}} (\langle \mathfrak{d}, i_{\mathfrak{r} \# k} \rangle \in \mathbf{P}_g) \wedge \forall \langle \mathfrak{d}, i_{\mathfrak{r} \# k} \rangle \in \mathbf{P}_g (\mathfrak{d} \in \mathbf{D}_{i_{\mathfrak{r} \# k}})$. Which $\mathbf{P}_g$ is ground truth participate relationships set for issue $i_{\mathfrak{r} \# k}$. To simplify the problem, we only predict the score of each developer and rank the developers accordingly. In this way, we can get $\mathbf{D}_{i_{\mathfrak{r} \# k}}$ in any top-n.

4.2   Method Framework

We propose RHE for issue participants recommendation problem, showing in Fig. 4. In our framework, phase 1 builds a semantic model to spread the semantic

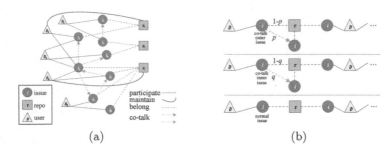

(a)   (b)

Fig. 3. (a) OSS community heterogeneous graph including three entities (*issue, repository*, and *user*), four relationships (*participate, maintain, co-talk*, and *belong*). Note, the *belong* relationship is implied in the symbol of issue (e.g. issue $i_{r_j\#k}$ belong to r_j). (b) Metapath for co-talk relationship of OSS community.

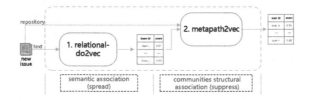

Fig. 4. RHE framework. Phase 1 is building a natural language embedding model with issue reference relationships. And the goal is to get issues that are text meaning similar to input issue. Phase 2 constructing a heterogeneous graph embedding model. The goal is to synthesize the results of phase 1 and to select candidates.

to another issue, regardless of the repository or submitter. And phase 2 mines potential relationships between new issue, semantic similar issues, and users; and to suppress scope according to the heterogeneous graph structure, resulting in suitable developers.

Relational-Doc2vec. We propose relational-doc2vec model based on doc2vec [9]. Differences from other documentation embedding methods, relational-doc2vec can embed the relationship information between documentations. It is worth noting that the co-talk relationship is explained as relational between issues. In addition, we do not limit the scope of repositories of output issues to spread the semantic connect to all over the OSS communities.

Figure 5(a) is an example to describe the main idea of relational-doc2vec. In training stage, we update co-talk embedding vector 2–4 twice in Fig. 5(a), representing the shared semantic embedding of issue 2 and issue 4. In relational-doc2vec model showing in Fig. 5(b), our object is to minimize the average log probability following:

$$-\frac{1}{T}\sum_{t=k}^{T-k}\log p(w_t|d_i; c_{i\sim j}, \cdots, c_{m\sim i}; w_{t-k}, \cdots, w_{t+k}), \tag{1}$$

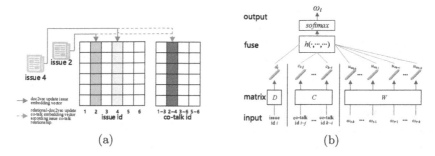

Fig. 5. (a) The main idea of relational-doc2vec model. (b) The matrix D and W is same to paper [9]; the matrix C is relational matrices, it can embed public semantics of two relational documentations. The d_i is an embedding vector from i-th column of matrix D, $c_{i \sim j}$ from matrix C, and u_{w_i} from matrix W. The fuse function can be simply defined as the average function.

where k is window size, T is length of the documentation d_i. The $p(\cdot | \cdot; \cdots; \cdots)$ is defined by following softmax function:

$$p(w_t | d_i; c_{i \sim j}, \cdots, c_{m \sim i}; w_{t-k}, \cdots, w_{t+k}) = \frac{e^{u_{w_t}^{\top} v_h}}{\sum_w e^{u_w^{\top} v_h}}, \tag{2}$$

where u_{w_t} is the embedding vector of word w_t, $\sum_w$ will enumerate all the words from vocabulary, and the $v_h = h(d_i; c_{i \sim j}, \cdots, c_{m \sim i}; w_{t-k}, \cdots, w_{t+k})$, $h(\cdot; \cdots; \cdots)$ is the fuse function (average in experiments) in Fig. 5(b), the matrices D, C, W can be regarded as three parameters of this function.

In practice, the denominator of Eq. (2) is too complicated to calculation because the vocabulary is very large [13] in general (about 3×10^5 in our experiments), and every parameter update will cause recalculation, combined with Negative Sampling [13]. Our objective function of word w_t is then formalized as:

$$\mathcal{O}(w_t) = -\log \sigma(u_{w_t}^{\top} v_h) - \sum_{j=1}^{N} \mathbb{E}_{w_j \sim P_n(w)} \log \sigma(-u_{w_j}^{\top} v_h), \tag{3}$$

where N is negative samples for each data [13], and $\sigma(\cdot)$ is sigmoid function.

Training with issue text and co-talk relationships, we get two matrices D, C to describe issue documentation semantic embedding and issue reference relationship embedding, respectively. We combine these two embedding vectors with fuse function $fuse(\cdot; \cdots)$, and calculate issue report embedding vectors. The fuse function is defined by the following equation (example in Fig. 5(b)).

$$fuse(d_i; \underbrace{c_{i \sim j}, \cdots, c_{m \sim i}}_{n}) = \begin{cases} \frac{1}{2}(d_i + avg(c_{i \sim j}, \cdots, c_{m \sim i})), & (n > 0) \\ d_i, & (n = 0) \end{cases}, \tag{4}$$

where $avg(\cdots)$ is average function for vectors.

In addition, according to the original doc2vec [9], we can keep the matrix W and adjust a new issue vector to infer the new issue text's embedding vector. In this paper, we use **cosine distance** to measure semantic similarity between two vectors. Thus, for an input issue, we can get the top-n semantically similar issues and its similarity from train set.

Metapath2Vec. To better embed different entities of OSS community, Metapath2Vec [4] is a more scalable and practical choice. We design the metapath showing in Fig. 3(b), including co-talk outer/inner and normal issue strategies. In addition, we count the historical co-talk relationship in each repository and filter out these irrelevant issues in the result of phase 1 first.

In the metapath, we focus on the relationships between developer and issue and building cross repositories relationship by enhancing random walk between co-talk outer issue, building relationship between developer and similar issue by enhancing co-talk inner issue. In our experiments, we set $p = 0.6$ and $q = 0.6$.

In Fig. 4, phase 2 receives data from phase 1 and new issue's repository information. We denote the input as $\langle \mathfrak{r}, [\langle \mathfrak{i}_1, s_1 \rangle, \langle \mathfrak{i}_2, s_2 \rangle, \cdots, \langle \mathfrak{i}_n, s_n \rangle] \rangle$, where $\mathfrak{r}$ is the repository of new issue, $\mathfrak{i}_j$ is sorted by similarity score s_j, s_j satisfies $s_1 \geqslant s_2 \geqslant \cdots \geqslant s_n$. n is the number of phase 1 outputs (50 in experiments).

We combine all developer **normalized embedding vector** as a matrix $V \in \mathbb{R}^{d \times |\mathbf{D}|}$ first, where d (300 in experiments) is graph embedding vector dimension. And we construct a matrix $U \in \mathbb{R}^{(n+1) \times d}$, the first row is repository $\mathfrak{r}$ **normalized embedding vector**, remaining rows is constructed by n issues. More importantly, we define vector $s = [\alpha s_1, s_1, s_2, \cdots, s_n]^\top$ as coefficient vector, where α (1.0 in experiments) is a constant for repository importance.

We need to calculate all cosine similarity for each pair $\langle \mathfrak{i}, \mathfrak{d} \rangle$, so a cosine similarity matrix C is defined by equation $C = s \cdot (U \times V)$ and $C \in \mathbb{R}^{(n+1) \times |\mathbf{D}|}$. The entry $(C)_{jk}$ of C is the distance score between issue $\mathfrak{i}_j$ and $\mathfrak{d}_k$ or between repository $\mathfrak{r}$ and $\mathfrak{d}_k$ when $j = 0$. We can sort all entries of matrix C, extract the developers ordered, remove duplication developers, and get the ordered recommendation candidate developers list finally.

5   Experiments

5.1   Dataset

We choose 40 repositories, divided into eight categories, to be included in our dataset as shown in Table 2. For these repositories, we collect all issues from 2018-01-01T00:00:00Z to 2020-01-01T00:00:00Z and split the training and testing sets by time 2019-07-01T00:00:00Z. After preprocessing, the dataset includes 146,653 training issues and 47,285 testing issues. And there are 10,341 users in our train set. In Table 2, we enumerate the distribution of issues in each repository. The train set contains 27,129 co-talk outer issues, 36,845 co-talk inner issues, and 82,679 normal issues. The test set contains 6,999 co-talk outer issues, 12,040 co-talk inner issues, and 28,246 normal issues, respectively.

Table 2. The repositories of dataset group by category.

Category	Repositories	Train	Test	Category	Repositories	Train	Test
VSC	microsoft/vscode	23,626	7,044	GO	golang/go	7,800	2,709
	Dart-Code/Dart-Code	450	175	RUST	rust-lang/rust	8,037	2,685
	flutter/flutter	10,090	5,067		rust-lang/rfcs	231	49
	dart-lang/sdk	3,207	1,086		rust-lang/cargo	1,377	410
	OmniSharp/omnisharp-vscode	696	172		servo/servo	1,513	638
TS	microsoft/TypeScript	6,101	1,733		web-platform-tests/wpt	2,268	509
	xtermjs/xterm.js	354	89	K8S	kubernetes/kubernetes	17,175	5,763
	electron/electron	4,896	1,752		openshift/origin	2,977	497
	DefinitelyTyped/DefinitelyTyped	9,177	2,975		etcd-io/etcd	907	242
	angular/angular	7,587	2,249		openshift/openshift-ansible	2,392	180
	angular/angular-cli	4,024	1,022		google/cadvisor	211	61
TF	tensorflow/tensorflow	8,261	3,079	SCI	scipy/scipy	1,339	548
	bazelbuild/bazel	2,159	449		scikit-learn/scikit-learn	2,463	979
	keras-team/keras	1,518	108		numpy/numpy	2,177	838
DOCK	moby/moby	2,095	445		jupyter/notebook	614	77
	docker/cli	686	124		ipython/ipython	353	89
	opencontainers/runc	247	55		astropy/astropy	1,198	566
	docker/swarmkit	216	33		pandas-dev/pandas	5,490	1,912
	docker/compose	712	126		dask/dask	1,124	429
	moby/libnetwork	172	24		pydata/xarray	733	300

5.2   Experiments Setup

We compare the proposed RHE method with the following three baseline approaches modified from RHE to show the effectiveness of co-talk relationships, relational-doc2vec, and OSS community heterogeneous structure in issue participant recommendation.

- RHE_{no_cotalk}: We prohibit it from accessing any co-talk relationship of dataset. Thus, the relational-doc2vec of phase 1 will be simplified to the original doc2vec, and metapath of phase 2 does not include any co-talk path.
- RHE_{no_rd2v}: We replace relational-doc2vec with doc2vec in phase 1.
- RHE_{no_hete}: The Metapath2Vec will be replaced by simple DeepWalk as graph embedding method. We do not set another experiment to replace this with Node2Vec [6], because DeepWalk is similar to Node2Vec when $p = 1$ and $q = 1$.

Hyperparameters: In phase 1, we set training epoch as 50, negative samples 5, window size 7, embedding vector size 300. And output 50 similar issues in train set for each input issue. Phase 2, the training epoch is 50, 6 times random walk start with each node, walk length 20, embedding vector size also 300, output 1,000 ordered users as recommendation result.

5.3   Preprocessing

Natural Language: First, we only consider English issue title and issue body. Second, we identify and remove comment quote and code segment (code block

Table 3. Results outlook.

Metric	Method			
	RHE_{no_cotalk}	RHE_{no_rd2v}	RHE_{no_hete}	RHE
MRR	0.3028	0.3015	0.3085	**0.3176**
Hits@1	0.4374	0.4371	**0.4387**	0.4142
Hits@5	0.6315	0.6221	0.6280	**0.6745**
Hits@15	0.7797	0.7762	0.7657	**0.8363**
Hits@30	0.8430	0.8409	0.8560	**0.8952**

> 100 chars) by issue's HTML. Then, We use NLTK [1] to tokenize, stem (Snow-ballStemmer [14]), and lemmatize (WordNetLemmatizer [5]) all of our issue doc-uments, and stop words (e.g. of, to) also be removed in this step. Finally, we attach issue body to issue title with a special separator '\n\n' to form issue documentation.

Data Cleaning: We eliminated short title issue (less than 20 characteristics or fewer than 4 words) and short body issue (less than 50 characteristics or fewer than 10 words) from the dataset, a total of 90,434 issues. Besides, we can not predict any developer who only participant issue of test set. Thus, we can only predicate developer who appears in train set.

5.4 Results

Metrics: According to the evaluation metrics of the recommendation question, we select Mean Reciprocal Rank (MRR) [10, 20] and Hits@1, 5, 15, 30 [11] as the evaluation metrics. Note, we use MRR for every ground truth developer instead of issue, and our experiments only output 1,000 ordered developers. Thus developer who is contained in the result will obtain rank score 0 in MRR.

Results: The experiment results are showing in Table 3. Obviously, RHE with co-talk relationship embedding and heterogeneous graph embedding achieve the highest MRR and mostly Hits@n. RHE_{no_cotalk} and RHE_{no_rd2v} scores are simi-lar in MRR or Hit@n, because these two methods are same before phase 2, and Metapath2Vec cannot show all the effects without co-talk relationship document embedding.

To better understand each method's details, we split the test set according to the issue type, and the results are shown in Table 4. It is reasonable; RHE_{no_hete} achieves the best results in Hits@1 when we focus on normal issue only. But RHE_{no_hete} performs poorly in all issues because co-talk relationship is extra information for co-talk issue but not for normal issue, and normal issue is more likely to find suitable candidates in their repository. No surprise, RHE_{no_hete} is also similar to RHE_{no_rd2v} in the three types of issue.

Table 4. Results detail group by issue type.

Metric	Method											
	RHE_{no_cotalk}			RHE_{no_rd2v}			RHE_{no_hete}			RHE		
	Outer	Inner	Normal	Outer	Inner	Normal	Outer	Inner	Normal	Outer	Inner	Normal
MRR	0.2352	0.2794	0.3415	0.2329	0.2776	0.3410	0.2425	0.2968	0.3401	**0.2495**	**0.3065**	**0.3498**
Hits@1	0.4121	0.4221	**0.4495**	0.4131	0.4232	0.4484	**0.4259**	**0.4423**	0.4402	0.3937	0.4301	0.4124
Hits@5	0.6117	0.6302	0.6364	0.6133	0.6192	0.6253	0.6292	0.6544	0.6166	**0.6810**	**0.6936**	**0.6648**
Hits@15	0.7708	0.7617	0.7891	0.7708	0.7628	0.7829	0.7663	0.7882	0.7561	**0.8349**	**0.8404**	0.8349
Hits@30	0.8449	0.8328	0.8468	0.8501	0.8311	0.8428	0.8634	0.8656	0.8502	**0.8973**	**0.8949**	0.8948

Table 3 suggests that RHE_{no_hete} gets the highest score of the Hits@1, as well as in columns of co-talk outer and inner of Table 4. Under this setting, relational-doc2vec shows its outstanding characteristics in co-talk relationship embedding aspect. But the excellent score only appears in Hits@1, which also illustrates the importance of heterogeneous embedding in OSS community. The proposed RHE method achieves the best score in general. The problem of issue participants recommendation is a multi-label prediction problem; Hits@1 is often not the most important aspect to achieve better results.

6   Conclusion

In this paper, we propose the RHE method to model co-talk issues in OSS community, and show its effectiveness in issue participants recommendation. RHE spreads semantic associations to the whole OSS community by relational document embedding, and suppress irrelevant issues and developers by heterogeneous graph embedding. We conduct real-world experiments to show the effectiveness of the proposed method and perform in-depth analysis. We also build a database including co-talk and normal issues, pull-requests, and comments to facilitate future research in related topics. As for our future work, we plan to improve RHE's adaptability and scalability when applied to large-scale OSS communities.

Acknowledgement. This work is supported by the National Key R&D Program of China under Grant No. 2018AAA0102302, and the Collaborative Innovation Center of Novel Software Technology and Industrialization.

References

1. Bird, S.: NLTK: the natural language toolkit. In: COLING/ACL 2006, Sydney, Australia, pp. 69–72. Association for Computational Linguistics, July 2006. https://doi.org/10.3115/1225403.1225421
2. Borges, H., Tulio Valente, M.: What's in a GitHub Star? Understanding repository starring practices in a social coding platform. J. Syst. Softw. **146**, 112–129 (2018). https://doi.org/10.1016/j.jss.2018.09.016
3. Čubranić, D.: Automatic bug triage using text categorization. In: SEKE 2004, pp. 92–97. KSI Press (2004)

4. Dong, Y., Chawla, N.V., Swami, A.: Metapath2vec: Scalable representation learning for heterogeneous networks. In: KDD 2017, Halifax, NS, Canada, pp. 135–144, KDD 2017. Association for Computing Machinery, August 2017. https://doi.org/10.1145/3097983.3098036

5. Fellbaum, C.: WordNet: An Electronic Lexical Database. The MIT Press, Cambridge (1998). https://doi.org/10.7551/mitpress/7287.001.0001

6. Grover, A., Leskovec, J.: Node2vec: scalable feature learning for networks. In: KDD 2016, San Francisco, California, USA, pp. 855–864. ACM Press (2016). https://doi.org/10.1145/2939672.2939754

7. Hannebauer, C., Patalas, M., Stünkel, S., Gruhn, V.: Automatically recommending code reviewers based on their expertise: an empirical comparison. In: ASE 2016, Singapore, Singapore, pp. 99–110. ACM Press (2016). https://doi.org/10.1145/2970276.2970306

8. Jeong, G., Kim, S., Zimmermann, T.: Improving bug triage with bug tossing graphs. In: ESEC/FSE 2009, Amsterdam, The Netherlands, p. 111. ACM Press (2009). https://doi.org/10.1145/1595696.1595715

9. Le, Q., Mikolov, T.: Distributed representations of sentences and documents. In: ICML 2014, Beijing, China, pp. 1188–1196. PMLR (2014)

10. Li, J., Xu, Z., Tang, Y., Zhao, B., Tian, H.: Deep hybrid knowledge graph embedding for Top-N recommendation. In: Wang, G., Lin, X., Hendler, J., Song, W., Xu, Z., Liu, G. (eds.) WISA 2020. LNCS, vol. 12432, pp. 59–70. Springer, Cham (2020). https://doi.org/10.1007/978-3-030-60029-7_6

11. Lin, Y., Liu, Z., Sun, M., Liu, Y., Zhu, X.: Learning Entity and Relation Embeddings for Knowledge Graph Completion. In: AAAI 2015 (Feb 2015)

12. Ma, W., Chen, L., Zhang, X., Zhou, Y., Xu, B.: How do developers fix cross-project correlated bugs? A case study on the GitHub scientific Python ecosystem. ICSE **2017**, 381–392 (2017). https://doi.org/10.1109/ICSE.2017.42

13. Mikolov, T., Chen, K., Corrado, G., Dean, J.: Efficient estimation of word representations in vector space. arXiv:1301.3781 [cs], September 2013

14. Porter, M.: An algorithm for suffix stripping. Prog. Electron. Libr. Inf. Syst. **14**(3), 130–137 (1980). https://doi.org/10.1108/eb046814

15. Rahman, M.M., Roy, C.K., Collins, J.A.: CoRReCT: code reviewer recommendation in GitHub based on cross-project and technology experience. In: ICSE 2016, Austin, Texas, pp. 222–231. ACM Press (2016). https://doi.org/10.1145/2889160.2889244

16. Tan, S.H., Ziqiang, L.: Collaborative bug finding for Android apps. In: ICSE 2020, Seoul, Republic of Korea, pp. 1335–1347. ACM Press (2020). https://doi.org/10.1145/3377811.3380349

17. Xi, S.-Q., Yao, Y., Xiao, X.-S., Xu, F., Lv, J.: Bug triaging based on tossing sequence modeling. J. Comput. Sci. Technol. **34**(5), 942–956 (2019). https://doi.org/10.1007/s11390-019-1953-5

18. Xia, X., Lo, D., Ding, Y., Al-Kofahi, J.M., Nguyen, T.N., Wang, X.: Improving automated bug triaging with specialized topic model. IEEE Trans. Softw. Eng. **43**(3), 272–297 (2017). https://doi.org/10.1109/TSE.2016.2576454

19. Yu, Y., Wang, H., Yin, G., Wang, T.: Reviewer recommendation for pull-requests in GitHub: What can we learn from code review and bug assignment? Inf. Softw. Technol. **74**, 204–218 (2016). https://doi.org/10.1016/j.infsof.2016.01.004

20. Zhang, Y., Wu, Y., Wang, T., Wang, H.: iLinker: a novel approach for issue knowledge acquisition in GitHub projects. World Wide Web **23**(3), 1589–1619 (2020). https://doi.org/10.1007/s11280-019-00770-1

Cold Start Recommendation Algorithm Based on Latent Factor Prediction

Wenan Tan[1,2(✉)], Xin Zhou[1], Xiao Zhang[1], Xiaojuan Cai[1], and Weinan Niu[1]

[1] College of Computer Science and Technology,
Nanjing University of Aeronautics and Astronautics, Jiangsu, Nanjing 211106, China
wtan@foxmail.com, zzzhouz@163.com
[2] School of Computer and Information Engineering,
Shanghai Polytechnic University, Shanghai 201209, China

Abstract. At present, recommendation algorithms have been widely used in online shopping, commercial services and entertainment. Although the recommended methods are endless, they are generally recommended based on the user's historical behavior or project information. Since the emergence of collaborative filtering in the 20th century, it become a very popular method in recommendation. But on the issue of cold start, collaborative filtering is not effective due to the lack of information. In this paper, we focus on solving the user side cold start problem. We propose a cold start recommendation algorithm based on latent factor prediction by combining collaborative filtering algorithm and matrix factorization. When it is just an ordinary user, we predict the rating by matrix factorization, and during the training iteration, we propose an improved optimization algorithm based on Alternate Least Squares (ALS). When the user is a new user, we predicted the latent factor by feature prediction. And then, the final outcome is calculated. The experiment uses the movielens data set. The experiment results show that our proposed recommendation algorithm has approximately 20.4% improvement over the baseline cold start algorithm. In addition, this paper design experiment to validate the proposed optimization algorithm effect, and the experiment results prove that the proposed algorithm has good precision and scalability.

Keywords: Recommendation algorithm · Alternate least square method · Matrix factorization · Cold-start · Collaborative filtering

1 Introduction

Since the collaborative filtering recommendation system was proposed in the 1990s, personalized recommendation services that provided by learning users' historical behaviors have played a good role in solving the problem of data overload. Collaborative filtering is based on the selection of some people with similar interests to give a list of services that you may be interested in, and then filter out the items/services you are interested in from the recommended list [1]. Collaborative filtering recommendation algorithms are mainly divided into item-based collaborative filtering (item-CF), user-based collaborative filtering (user-CF) and model-based collaborative filtering algorithm. Traditional

C. Xing et al. (Eds.): WISA 2021, LNCS 12999, pp. 617–624, 2021.
https://doi.org/10.1007/978-3-030-87571-8_53

recommendation algorithm based on matrix factorization (MF) decompose the rating matrix R into U (user feature matrix) and V (item feature matrix).

These algorithms have their own advantages and disadvantages. For MF, when the user needs to be recommended is a new user, the user latent factor is 0. To solve the user cold start problem, this paper proposes a cold start solved method, which combines the calculate similarity methods of user-CF and dimensionality reduction of MF. Besides, this paper also proposes an improved ALS optimization algorithm.

The rest of this paper is organized into following sections. Section 2 reviews some related work. Section 3 shows the details of our proposed algorithm. Section 4 presents the experimental results. Finally, Sect. 5 concludes this paper and points out the future work.

2   Related Work

MF focuses on factorizing the rating matrix into low-dimension user latent vectors and item latent vectors. Y Chen [2] et al. propose a novel approach called Multifaceted Factorization Models which incorporate a great variety of features in social networks. In response to the large sparsity and poor scalability of traditional collaborative filtering algorithms, Jinchao Guo [3] et al. proposed a collaborative filtering algorithm based on improved singular value decomposition (SVD) algorithm and K-means clustering algorithm. Wenan Tan [4] et al. proposed a dual-neighbor hybrid recommendation algorithm, which improves the recommendation accuracy and effectively solves the cold start problem. Uroš Ocepek [5] propose a novel approach for alleviating the cold start problem by imputing missing values into the input matrix. This approach combines local learning, attribute selection, and value aggregation into a single approach. Y Guo [6] et.al. present a general method named ANCF (Attention Neural network Collaborative Filtering), it captures collaborative filtering signals, refines the embedding of users and items and introduce an attention mechanism. And finally, they inject user-item collaboration signals into the embedding process. Jian Wei [7] propose two recommendation models to solve the cold start problems for new items, which are based on a framework of tightly coupled CF approach and deep learning neural network. Blerina Lika [8] et al. incorporates classification methods in a pure CF system while the use of demographic data help for the identification of other users with similar behavior. Xue H. J [9] propose a new factorization model that combines multi-view visual information with the implicit feedback data for restaurant prediction and ranking. Wang H [10] propose a Multi-Timeslice Graph Embedding (MTGE) model based on deep neural networks, which can be used to model users' behavior sequences and predict their interest based on the historical behavior. X Zhang [11] propose a new factorization model that combines multi-view visual information with the implicit feedback data for restaurant prediction and ranking.

3   Proposed Recommendation Model

This section will introduce a recommended algorithm based on matrix factorization which we proposed to solve the cold start problem with a new ALS optimization algorithm, called nALSCS. The algorithm proposed in this section mainly addresses the user cold start problem on the user side, that is, how to effectively perform personalized recommendations for users when there is a user who has not rated any items. We use u to represent users, $\mathbb{C} = \{c1, c2, \ldots cM\}$ to represent user attributes, i to represent items, and r to represent rating.

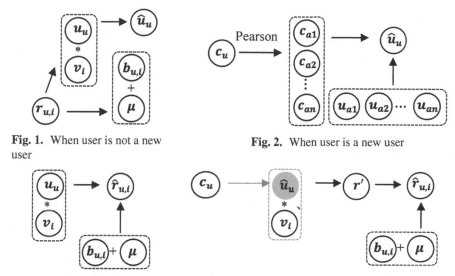

Fig. 1. When user is not a new user

Fig. 2. When user is a new user

Fig. 3. The graphical of rating prediction method of new user ordinary user and new user

First, we build a model to decompose the high-dimensional matrix into two low-dimensional feature matrices, called feature factorization. The source of our idea is that users have their own preference characteristics. For example, if Lily likes watching cartoons (preference characteristics), then he is likely to be interested in "Toy Story". Therefore, we only need to calculate the user's preference characteristic u_u, and then we can use this preference characteristic to infer the type that the user likes.

Secondly, we calculate the similarity between users according to the user attributes $\mathbb{C}$ when user is a new user. And we predict the user's latent features through the user's similar users (neighbors). We call this step feature prediction. The neighbors of the new user are obtained by calculating the similarity of the new user. Calculating the latent feature of the new user through combining the preference of the neighbors and the similarity with the new user.

Finally, we combine the feature factorization and feature prediction to rating predictions, which we call collaborative prediction. In this part, we predict user feature and combine item features to rating prediction.

$$r_{u,i} = u_u v_i^T + b_{u,i} + \mu \tag{1}$$

Among them, u_u refers to the latent factor vector of user u, v_i^T refers to the latent factor vector of item i, $b_{u,i}$ refers to the paranoia of user u relative to item i, and μ is the average rating.

The loss calculation method in the experimental algorithm uses square error loss (L2 loss):

$$L' = \frac{1}{2}\sum_{u=1}^{n}\sum_{i=1}^{m}\left\|r_{u,i} - \left(u_u v_i^T + b_{u,i} + \mu\right)\right\|^2 + \frac{\lambda}{2}\left(\sum_{u=1}^{n}\sum_{i=1}^{m}\|u_u\|^2 + \|b_{u,i}\|^2 + \|v_i\|^2\right) \quad (2)$$

Where λ is the regularization coefficient, and $\|u_u\|^2 + \|b_{u,i}\|^2 + \|v_i\|^2$ is the regularization term. The introduction of regularization can prevent over-fitting.

The alternate least squares method updates the features by judging the similarity between the entire prediction matrix and the rating matrix. The drawback is that the rating matrix needs to be set to an initial value for unrated items during calculation, which is generally set to 0. In the optimization process, the set initial value 0 is used as the real value for calculation, which results in huge errors when calculating the evaluation index. Therefore, when the user does not rate the item, we set $t_{u,i}$ to [0, 1], and set $t_{u,i}$ to [0, 1] if the user has rated it.

$$t_{u,i} = \begin{cases} [1,0] \; r_{u,i} > 0 \\ [0,1] \; r_{u,i} = 0 \end{cases} \quad (3)$$

$$L = min_{u*,v*,b*}t_{u,i} \cdot \left[L', 0\right]^T \quad (4)$$

According to the above regularized squared error function, a newALS optimization method is used to iteratively learn the model parameters. The algorithm loops through all ratings in the training set iteratively and updates each parameter according to the associated gradient until system converges.

$$\begin{cases} u_u = \left(V^TV + \lambda E\right)^{-1}\left(V^TR_i - V^TB_i\right) \\ v_i = \left(U^TU + \lambda E\right)^{-1}\left(U^TR_u - U^TB_u\right) \; r_{u,i} > 0 \\ b_{u,i} = \frac{1}{\lambda+1}\left(r_{u,i} - u_u^T v_i\right) \end{cases} \quad (5)$$

3.1 Feature Prediction

As show in Fig. 1, when a user has historical records in the rating matrix, that is, when the user is not a new user. We obtain the user's latent factor feature through iterative optimization. When the user is not a new user, the way that we predict the rating is as followed

$$r_{u,i} = u_u v_i^T + b_{u,i} + \mu \quad (6)$$

As show in Fig. 2, when a user has never rated any item. That is, when the user is a new user, then the user preference characteristics obtained through the matrix decomposition model are not representative, then we need to predict the user's preference characteristics, here we use the user's attributes to calculate the user's similarity. The similarity between. The calculation method of person similarity is as follows

$$s_{u,a} = \frac{\sum_{i\in I_{u,a}}(c_{u,i} - \overline{c}_u)(c_{a,i} - \overline{c}_a)}{\sqrt{\sum_{i\in I_{u,a}}(c_{u,i} - \overline{c}_u)^2}\sqrt{\sum_{i\in I_{u,a}}(c_{a,i} - \overline{c}_a)^2}} \quad (7)$$

$$\hat{u}_u = \frac{1}{\sum_{a \in \mathcal{N}_u} s_{u,a}} \sum_{a \in \mathcal{N}_u} s_{u,a} u_a \qquad (8)$$

Where $s_{u,a}$ is the similarity between user u and user a, c_u is the attribute value of user u, $\overline{c}_u$ and $\overline{c}_a$ are the average values of the attributes of user u and user a, and $\mathcal{N}_u$ is the list of similar users of user u.

3.2 Collaborative Prediction

As show in Fig. 3, when predict the rating, we adopt different calculation methods for different users. For ordinary users, we use the decomposed user characteristics and item characteristics to calculate.

$$\hat{r}_{u,i} = u_u v_i + b_{u,i} + \mu \qquad (9)$$

For new users, because the user characteristics obtained by decomposition cannot be used, we use the user attributes to predict the new user preference characteristics by using the similarity, calculate it, and then calculate the rating

$$\hat{r}_{u,i} = \hat{u}_u v_i + b_{u,i} + \mu \qquad (10)$$

Among them, $\hat{u}_u$ is the user latent factor feature obtained through feature prediction.

4 Experimental Design and Result Analysis

4.1 Evaluation Metrics and Experimental Settings

The data set used in the experiment is the ratings provided by MovieLens-1M [12] about the user's rating of the movie. In the rating set, each user's rating data for a movie includes each user's id, each movie's id, and user rating. The more interested in a movie, the higher the user's rating of this movie. The MovieLens-1M dataset is a fairly sparse rating matrix, with 93.7% of the data not being rated.

In this experiment, two indicators, mean absolute error and root mean square error, are mainly used for evaluation.

Mean Absolute Error (MAE): MAE can better reflect the actual situation of the predicted value error. The formula is as follows.

$$MAE = \frac{1}{\mathcal{N}_p} \sum_{u,i} |\hat{r}_{ui} - r_{u,i}| \qquad (11)$$

Root Mean Square Error [13] (RMSE): RMSE is used to measure the deviation between the observations and politics. Shows the degree of dispersion of the data. The formula is as follows.

$$RSME = \sqrt{\frac{1}{\mathcal{N}_p} \sum_{u,i} (\hat{r}_{ui} - r_{u,i})^2} \qquad (12)$$

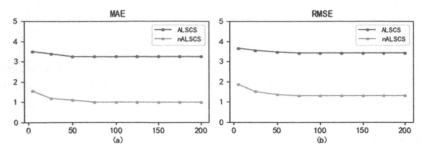

Fig. 4. ALSCS and nALSCS performance in MAE and RMSE

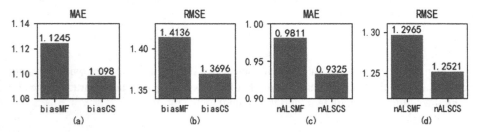

Fig. 5. The performance of model using latent factor prediction on cold start

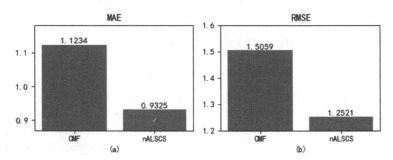

Fig. 6. The performance of the proposed model and CMF

4.2 Results and Analysis

First, we compared our proposed algorithm (ALSCS) with the recommendation algorithm using nALS (nALSCS).

We compare two algorithms from two metrics of MAE and REME on the same data set. The performance is show in Fig. 4.

Figure 4(a)(b) shows the changes of the two algorithms in MAE and RMSE when the number of iterations increases from 0 to 200. The result show that the nALSCS has a great improvement.

In order to verify the improvement effect of the method which use the latent factor feature prediction on the cold start problem, we compared the matrix factorization model biasMF) with the MF using feature prediction (biasCS). Figure 5(a)(b) compare the biasMF and biasCS in MAE and RMSE metrics of performance in the same cold start data sets. According to figure, biasCS have increased approximately 2.3% in MAE and 3.1% in RMSE. And then, we compare MF using nALS (nALSMF) with our proposed algorithm (nALSCS). The performance of the model on the cold start data set is show in Fig. 5(c)(d). The Fig. 5(c)(d) compare the nALSMF and nALSCS in MAE and RMSE metrics of performance in the same cold start data sets. According to figure, nALSCS have increased approximately 4.9% in MAE and 3.4% in RMSE.

Besides, this paper compares nALSCS with the CMF [14]. Figure 6(a)(b) shows that the nALSCS has a 16.99% improvement on MAE and 20.4% improvement on RMSE.

Table 1. Time consumption of recommendation

Algorithms	ALSMF	ALSCS
100k	25.8	0.2
1M	355.1	4.2

This paper experiment on the data set of movielens100K and 1M. The result is shown in Table 1, the unit of the result is second. The nALSCS proposed in this paper establishes an offline model, and determine the user type to rating prediction. In this way, nALSCS greatly reduces the operation time. And the data shows that the larger the data set, the more time shorter, which fully demonstrates the scalability of the algorithm.

5 Summary

This paper proposes a recommendation model for cold start problem on user side. This algorithm adopts a three-phase approach in order to solve the cold start problem and increase the accuracy of ratings prediction. For new user, we present a method which called feature prediction to predict the user latent factor. In optimization process, we present a new alternate least square method to improve the accuracy metric. Our experimental results show the performance of the proposed techniques. Our algorithm has about 20% improvement compared to the baseline algorithm for solving the cold start problem.

Acknowledgments. The paper is supported in part by the National Natural Science Foundation of China under Grant No. 61672022, No. 61272036, No. U1904186, and Key Disciplines of Computer Science and Technology of Shanghai Polytechnic University under Grant No. XXKZD1604.

References

1. Peña, F.J., O'Reilly-Morgan, D.: Combining rating and review data by initializing latent factor models with topic models for top-n recommendation. In: Fourteenth ACM Conference on Recommender Systems, vol. 438. Association for Computing Machinery, Virtual Event (2020)

2. Chen, Y., Liu, Z., Ji, D.: Context-aware ensemble of multifaceted factorization models for recommendation prediction in social networks (2021)
3. Jinchao, G., Jigang, Y.: Collaborative filtering algorithm based on improved SVD algorithm and bipartite K-means clustering algorithm. J. Light Ind. **35**(4), 88–95 (2020)
4. Tan, W., Qin, X., Wang, Q.: A hybrid collaborative filtering recommendation algorithm using double neighbour selection. In: Tang, Y., Zu, Q., Rodríguez García, J.G. (eds.) HCC 2018. LNCS, vol. 11354, pp. 416–427. Springer, Cham (2019). https://doi.org/10.1007/978-3-030-15127-0_42
5. Ocepek, U., Rugelj, J., Bosnić, Z.: Improving matrix factorization recommendations for examples in cold start. Expert Syst. Appl. **42**, 6784 (2015)
6. Guo., Y., Yan, Z: Recommended system: attentive neural collaborative filtering. In: IEEE Access, vol. 8, pp. 125953–125960 (2020)
7. Wei, J., He, J., Chen, K.: Collaborative filtering and deep learning based recommendation system for cold start items. Expert Syst. Appl. **69**, 29 (2016)
8. Lika, B., Kolomvatsos, K., Hadjiefthymiades, S.: Facing the cold start problem in recommender systems. Expert Syst. Appl. **41**, 2065 (2014)
9. Xue, H.J., Dai, X., Zhang, J.: Deep matrix factorization models for recommender systems. In: Twenty-Sixth International Joint Conference on Artificial Intelligence (2017)
10. Wang, H., Kou, Y., Shen, D., Nie, T.: An explainable recommendation method based on multi-timeslice graph embedding. In: Wang, G., Lin, X., Hendler, J., Song, W., Xu, Z., Liu, G. (eds.) WISA 2020. LNCS, vol. 12432, pp. 84–95. Springer, Cham (2020). https://doi.org/10.1007/978-3-030-60029-7_8
11. Zhang, X., Luo, H., Chen, B., Guo, G.: Multi-view visual Bayesian personalized ranking for restaurant recommendation. Appl. Intell. **50**(9), 2901–2915 (2020). https://doi.org/10.1007/s10489-020-01703-6
12. Zhang, F., Yuan, N.J., Lian, D.: Collaborative knowledge base embedding for recommender systems. In: The 22nd ACM SIGKDD International Conference on Knowledge Discovery and Data Mining, vol. 353. Association for Computing Machinery (2016)
13. Guo, G., Zhang, J., Smith, N.: A novel recommendation model regularized with user trust and item ratings. IEEE Trans. Knowl. Data Eng. **28**, 1607 (2016)
14. Yong, G.W., Yu, W.C.: Hybrid recommendation algorithm combining content and matrix factorization. Comput. Appl. Res. **037**, 1359 (2020)

Robust Graph Collaborative Filtering Algorithm Based on Hierarchical Attention

Ping Feng[1,2](✉), Yang Qian[2], Xiaohan Liu[2], Guoliang Li[2], and Jian Zhao[2]

[1] Jilin University, Changchun 130012, Jilin, China
[2] Changchun University, Changchun 130022, Jilin, China
fengping@ccu.edu.cn

Abstract. Learning the embedding representation of users and items is the core of the collaborative filtering algorithm. In recent years, the graph neural network (GNN) has been applied to the recommendation field due to its excellent performance. However, in the process of GNN iteratively aggregating neighbor information, the occasional noise in the graph structure will transmit errors to neighbor nodes along with the aggregation process, which will worsen the embedding representation of other nodes. Noise information is ubiquitous in real life. Therefore, while mining high-order collaborative information in depth, improving the robustness of the GNN model is also an important factor that needs to be considered in the recommendation task. Based on the above problems, this paper proposes a robust graph collaborative filtering algorithm based on hierarchical attention, which includes node-level and graph-level attention. Node-level attention performs preference aggregation and occasional noise information filtering on different neighbor nodes by learning the attention coefficients of different neighbor nodes; graph-level attention performs fusion and secondary filtering of occasional noise information on different deep graph embeddings by learning the attention coefficients of different dimensional nodes. The node-level and graph-level attention can fully realize the noise reduction of the graph structure in the deep propagation process. While ensuring that the nodes encode high-order collaborative information, it minimizes the noise information carried. Extensive experimental results on three data sets show that the recommendation algorithm is better than the existing mainstream recommendation algorithm in all evaluation indicators.

Keywords: Graph neural network · High-order collaborative information · Occasional noise · Hierarchical attention

1 Introduction

Personalized recommendations have been widely used in various areas of people's lives such as e-commerce platforms, social media and short video websites. Recently, because of the excellent performance of graph neural network, it has also been applied to the recommendation field [1]. However, most work does not take into account the robustness of GNN technology. The majority of GNN-based methods are very sensitive to the quality of the graph structure [2]. Incorrect graph structure information is continuously

© Springer Nature Switzerland AG 2021
C. Xing et al. (Eds.): WISA 2021, LNCS 12999, pp. 625–632, 2021.
https://doi.org/10.1007/978-3-030-87571-8_54

iteratively propagated through multi-layer GNN, which will affect the embedding representation of neighboring nodes, and ultimately affect the recommendation result. How to reduce the noise of incorrect graph structure information while mining high-order collaborative information is particularly important.

To address the above problems, this paper proposes a robust graph collaborative filtering algorithm based on hierarchical attention, called HAR-GCF. These two levels of attention act as information aggregators and filters at the same time, to carry out high-level aggregation of information, and to filter incorrect graph structure information at the same time.

2   Related Work

The attention mechanism was widely used in various fields of deep learning [3], including the field of recommendation systems. Many works [4, 5] have further improved the effect of the model by improving the weight distribution problem of the attention mechanism, but these models still do not consider high-level collaborative filtering information.

MF [6] is a widely used recommendation algorithm model, but it can only extract direct interaction information between users and items, and is not sufficient for extracting collaborative information. The recent emergence of graph neural networks is able to handle graph structures very well. Compared to traditional neural networks, graph neural networks have more powerful information extraction capabilities. GCMC [7] uses graph convolution to model users and item one-hop neighbors. Based on this, NGCF [8] modeled the multi-hop neighbor information through the overlay graph neural network, and extracted high-order connectivity information in the graph structure. LightGCN [9] proposes that nonlinear feature conversion and nonlinear activation have little effect on collaborative filtering tasks. On the basis of exploring multi-hop neighbor information aggregation, some works [10, 11] use auxiliary information to further improve the effect of recommendation.

3   Methodology

We propose a robust graph collaborative filtering algorithm model based on hierarchical attention, as shown in Fig. 1. The architecture of the model includes an embedding layer, a node-level attention layer, a graph-level attention layer, and a prediction layer.

3.1   Embedding Layer

The embedding layer maps the user and item IDs to d-dimensional vectors respectively. We construct a user (an item) lookup table $T_u \in \mathbb{R}^{U \times E}$ ($T_i \in \mathbb{R}^{I \times E}$), where E denotes the dimension of embedding. I (U) denotes the number of items (users). e_u and e_i denote the embedding vector of users and items, respectively.

$$T_u = [e_1, \ldots, e_U], T_i = [e_1, \ldots, e_I] \tag{1}$$

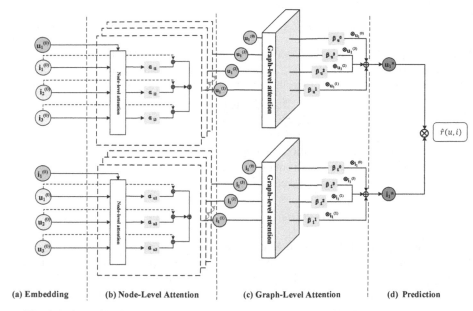

(a) Embedding   **(b) Node-Level Attention**   **(c) Graph-Level Attention**   **(d) Prediction**

Fig. 1. Hierarchical attention-based robust graph collaborative filtering model framework

3.2 Node-level Attention Layer

In the recommendation task, for a given user, the contribution of each neighboring item to the user's preference is different. This paper introduces node-level attention, learns the attention scores of different neighbor nodes, automatically optimizes the aggregation to complete the node representation, and at the same time denoises the graph structure from the node level.

One-Hop Neighbor Information Aggregation

A user's preferences are related to the items he has interacted with. By aggregating single-hop neighbor information, the representation update of the central node is completed. The aggregation process is defined as follows:

$$e_u^{new} = Agg(e_u, \{e_k, \forall k \in I(u)\}) \tag{2}$$

Similar to LightGCN, the non-linear feature conversion and non-linear activation function are removed here. e_u^{new} denotes the updated user embedding representation. $I(u)$ denotes the set of node u and all its neighbors. $Agg(\cdot)$ denotes information aggregation function, which completes the update of the user u vector by passing all neighbor information. Many explicit information aggregation functions are not enough to describe the information aggregation process. A node-level attention mechanism is defined here. Through an end-to-end architecture, the attention weight of neighbor nodes is implicitly captured. Define $Agg(\cdot)$ as follows:

$$Agg(e_u, \{e_k, \forall k \in I(u)\}) = attn_{node}(e_u, \{e_k, \forall k \in I(u)\}) \tag{3}$$

$$Agg(e_u, \{e_k, \forall k \in I(u)\}) = \sum\nolimits_{k \in I(u)} \alpha_{uk} e_k \tag{4}$$

Where $attn_{node}(\cdot)$ denotes the node-level attention aggregation filter function. α_{uk} denotes the attention coefficient of neighboring nodes to node u in the process of information aggregation. The node-level attention mechanism can be defined as follows:

$$\alpha_{ui} = \frac{\exp(\sigma(W_n^T[e_u||e_i]))}{\sum_{k \in I(u)} \exp(\sigma(W_n^T[e_u||e_k]))} \tag{5}$$

Where σ represents the activation function $LeakyReLU$, $||$ denotes the connection operation, W_n^T denotes the weight matrix in the node-level attention layer. The whole aggregation process can be expressed as:

$$e_u^{new} = \sum\nolimits_{k \in I(u)} \alpha_{uk} e_k, \; e_i^{new} = \sum\nolimits_{k \in I(i)} \alpha_{ik} e_k \tag{6}$$

Multi-hop Neighbor Information Aggregation
We extend the information aggregation approach from a single layer to multiple layers, enabling the aggregation of multi-hop neighborhood information to obtain a higher-order embedding representation of the node. The following will derive the first-order formula to higher-order, as follows:

$$e_u^{(h)} = \sum\nolimits_{k \in I(u)} \alpha_{uk}^{(h-1)} e_k^{(h-1)}, \; e_i^{(h)} = \sum\nolimits_{k \in I(i)} \alpha_{ik}^{(h-1)} e_k^{(h-1)} \tag{7}$$

Where h denotes the number of superimposed node-level attention layers.

3.3 Graph-Level Attention Layer

Different higher-order graph representations are obtained by stacking multiple node-level attention layers. Different levels of graph structures represent user preferences from different dimensions. Taking the node embedding of the group h graph structure as input, the graph-level attention layer can be expressed as follows:

$$e_u^* = attn_{graph}\left(e_u^{(0)}, e_u^{(1)}, \dots, e_u^{(h)}\right), e_i^* = attn_{graph}(e_i^{(0)}, e_i^{(1)}, \dots, e_i^{(h)}) \tag{8}$$

$$e_u^* = \sum\nolimits_{h=0}^{H} \beta_u^h e_u^{(h)}, \; e_i^* = \sum\nolimits_{h=0}^{H} \beta_i^h e_i^{(h)} \tag{9}$$

Where $attn_{graph}(\cdot)$ denotes the graph-level attention aggregation filter function. β^h denotes the importance of different $e^{(h)}$ to the prediction result in the process of information aggregation. We implement the graph-level attention mechanism through a single-layer MLP, as follows:

$$a_u^h = h^T Relu\left(W_g e_u^{(h)} + b\right) \tag{10}$$

$$\beta_u^h = \frac{\exp(a_u^h)}{\sum_{h=0}^{H} \exp(a_u^h)} \tag{11}$$

Where W_g denotes the weight matrix of hidden layer, b denotes the bias parameter of the hidden layer, h denotes the attention vector at the graph level, a_u^h denotes the attention score. e_u^* and e_i^* are the final user and item embedding representation.

3.4 Prediction Layer

Through the information aggregation and filtering of the node-level attention layer and the graph-level attention layer, the final node encodes high-level collaborative information while minimizing the noise information carried. The prediction layer uses the inner product operation to obtain the final prediction value.

$$\hat{r}(u, i) = e_u^{*T} e_i^*$$

(12)

3.5 Model Learning

The implicit feedback is considered here, and Bayesian personalized ranking [12] is used as the objective function. The specific objective function is as follows:

$$loss_{HAR-GCF} = \sum_{(u,i,j)\in S} -\ln\sigma\left(\hat{r}(u, i) - \hat{r}(u, j)\right) + \lambda||\Theta||_2^2$$

(13)

Where $S = \{(u, i, j)\}$ denotes the training data set. $\hat{r}(u, i)$ denotes the positive sample prediction score, $\hat{r}(u, j)$ denotes the negative sample prediction score. σ denotes the sigmoid activation function. Θ denotes the parameters included in the model. λ denotes the regularization parameter.

4 Experiments

4.1 Dataset and Evaluation Metrics

In order to verify the effectiveness of the model in this paper, Gowalla, Yelp2018 and MovieLens are selected as the datasets. The statistical information of the data set is shown in Table 1.

The evaluation metrics are recall@K and ndcg@K. K is set to 20 by default. We present the average metrics for all users in the test set.

Table 1. Information statistics of each dataset

Dataset	#Users	#Items	#Interaction	Density
Gowalla	29858	40981	1027370	0.00084
Yelp2018	31668	38048	1561406	0.00130
MovieLens1M	6040	3706	1000209	0.04468

4.2 Baselines

To verify the effectiveness of the algorithm proposed in this paper, we compare the algorithm in this paper with other state-of-the-art models. They are MF, GC-MC, NGCF and LightGCN which are mentioned in related work.

4.3 Parameter Settings

Same as LightGCN, the dimension of the embedded vector is set to 64 for all models. The batch size is set to 4096 for speed, and the above two parameters remain unchanged. The learning rate is adjusted between {0.1, 0.01, 0.001, 0.0001}, the optimal value is 5e-4, 1e-4, 1e-4 on Gowalla, Yelp2018 and MovieLens. L2 regularization parameter is tuned between {0.1, 0.01, 0.001, 0.0001, 0.00001}, the optimal value is 1e-4 in most case. In addition, we adopt early termination strategy and dropout strategy to prevent the occurrence of overfitting problems.

4.4 Overall Performance Comparison

As shown in Table 2, the performance of this algorithm is compared with other baselines algorithms.

Table 2. Performance comparison.

	Gowalla		Yelp2018		MovieLens	
	Recall	NDCG	Recall	NDCG	Recall	NDCG
MF	0.1388	0.1201	0.0495	0.0365	0.1189	0.1951
GC-MC	0.1413	0.1245	0.0536	0.0426	0.1158	0.1924
NGCF	0.1540	0.1335	0.0574	0.0467	0.1224	0.1982
LightGCN	0.1801	0.1512	0.0642	0.0524	0.1359	0.2195
HAR-GCF	**0.1895**	**0.1585**	**0.0686**	**0.0565**	**0.1412**	**0.2274**
%Improv	5.2%	4.8%	6.9%	7.8%	3.9%	3.5%

The HAR-GCF proposed in this paper is superior to other baselines on the three datasets. It can be seen that HAR-GCF works better on the sparse Gowalla and Yelp2018 datasets. The above shows the efficiency of the HAR-GCF algorithm. The dual attention mechanism can extract enough high-order collaborative information, and realize the noise reduction of the graph structure.

4.5 Model Parameter Analysis

Effect of Node-level Attention Layer Numbers

In order to study the influence of different number of node-level attention layers on model performance, we conducted related experiments. Table 3 shows the performance of different number of node-level attention layers (from 1 to 4).

Table 3. Effect of node-level attention layer numbers.

	Gowalla		Yelp2018		MovieLens	
	Recall	NDCG	Recall	NDCG	Recall	NDCG
1-layer	0.1852	0.1568	0.0619	0.0511	0.1398	0.2250
2-layer	0.1888	0.1575	0.0664	0.0549	0.1409	0.2263
3-layer	**0.1895**	**0.1585**	0.0679	0.0558	**0.1412**	**0.2274**
4-layer	0.1892	0.1584	**0.0686**	**0.0565**	0.1411	0.2272

- In general, as the number of layers increases, the model can mine more high-level collaboration information, and the performance of the model gradually improves. But too many layers will introduce more noise.
- On the Gowalla and Yelp2018 datasets, as the number of layers increases, the result of the model increases more than that of MovieLens, which shows that multi-layer node attention layer is more effective for sparse data.

Influence of Different Attention Mechanisms

In order to verify the influence of different attention mechanisms on the final model results, ablation experiments were performed on different attention mechanisms. There are three variants of HAR-GCF. HAR-node means that only node-level attention is used. HAR-graph means only graph-level attention is used. HAR-non means both node-level attention and graph attention are removed. And the results are shown in Fig. 2:

In general, the HAR-node and HAR-graph models with only one attention mechanism have better results than the HAR-non model without the attention mechanism, the results are not as good as the HAR-GCF using both attention mechanisms at the same time.

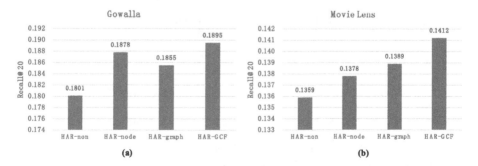

Fig. 2. Effect of different attention

5 Conclusion

In order to solve the problem of recommendation result decline caused by iterative propagation of occasional noise, this paper proposes a robust graph collaborative filtering algorithm based on hierarchical attention. Extensive experiments have proved that the node-level attention mechanism and graph-level attention mechanism are effective, and their recommendation results are better than mainstream recommendation algorithms. In the future, the research of this article will focus on the use of auxiliary information to improve the recommendation effect.

Acknowledgments. This work is supported by the Scientific and technological research projects (No.2021LY505L16).

References

1. Wu, Z., Pan, S., Chen, F., Long, G., Zhang, C., Philip, S.Y.: A comprehensive survey on graph neural networks. IEEE Trans. Neural Netw. Learn. Syst. **32**(1), 4–24 (2020)
2. Zhu, Y., Xu, W., Zhang, J., Liu, Q., Wu, S., Wang, L.: Deep graph structure learning for robust representations: a survey. arXiv preprint arXiv:2103.03036 (2021)
3. Chaudhari, S., Mithal, V., Polatkan, G., Ramanath, R.: An attentive survey of attention models. arXiv preprint arXiv:1904.02874 (2019)
4. Ebesu, T., Shen, B., Fang, Y.: Collaborative memory network for recommendation systems. In: The 41st International ACM SIGIR Conference on Research & Development in Information Retrieval, pp. 515–524 (2018)
5. Xiao, J., Ye, H., He, X., Zhang, H., Wu, F., Chua, T.S.: Attentional factorization machines: learning the weight of feature interactions via attention networks. arXiv preprint arXiv:1708.04617 (2017)
6. Koren, Y., Bell, R., Volinsky, C.: Matrix factorization techniques for recommender systems. Computer **42**(8), 30–37 (2009)
7. Berg, R.V.D., Kipf, T.N., Welling, M.: Graph convolutional matrix completion. arXiv preprint arXiv:1706.02263 (2017)
8. Wang, X., He, X., Wang, M., Feng, F., Chua, T.S.: Neural graph collaborative filtering. In: Proceedings of the 42nd International ACM SIGIR Conference on Research and Development in Information Retrieval, pp. 165–174 (2019)
9. He, X., Deng, K., Wang, X., Li, Y., Zhang, Y., Wang, M.: Lightgcn: simplifying and powering graph convolution network for recommendation. In: Proceedings of the 43rd International ACM SIGIR Conference on Research and Development in Information Retrieval, pp. 639–648 (2020)
10. Wang, X., He, X., Cao, Y., Liu, M., Chua, T.S.: KGAT: knowledge graph attention network for recommendation. In: Proceedings of the 25th ACM SIGKDD International Conference on Knowledge Discovery & Data Mining, pp. 950–958 (2019)
11. Li, J., Xu, Z., Tang, Y., Zhao, B., Tian, H.: Deep hybrid knowledge graph embedding for top-n recommendation. In: Wang, G., Lin, X., Hendler, J., Song, W., Xu, Z., Liu, G. (eds.) Web Information Systems and Applications, pp. 59–70. Springer International Publishing, Cham (2020)
12. Rendle, S., Freudenthaler, C., Gantner, Z., Schmidt-Thieme, L.: BPR: Bayesian personalized ranking from implicit feedback. arXiv preprint arXiv:1205.2618 (2012)

Architecture and Systems

An Adaptive Sharing Framework for Efficient Multi-source Shortest Path Computation

Xinyi Liu[1], Zhigang Wang[1(✉)], Ning Wang[1], Xiangtan Li[1], Bo Zhang[2],
Jun Qiao[3], Zhiqiang Wei[1], and Jie Nie[1]

[1] Faculty of Information Science and Engineering, Ocean University of China,
Qindao, China
{wangzhigang,wangning8687,weizhiqiang,niejie}@ouc.edu.cn,
liuxinyi5699@stu.ouc.edu.cn
[2] China Mobile System Integration Co., Ltd., Xi'an, China
[3] Shenyang Hangsheng Technology Co., Ltd., Shenyang, China

Abstract. Many real-world applications need to know the distances
between some or all pairs of vertices in a large graph, but the existing distributed single-source shortest path (SSSP) solution can not efficiently handle the problem of such multi-source computation. While,
distributed multi-source shortest path (MSSP) has not attracted enough
attention. This paper thereby proposes a distributed multi-source shortest path algorithm by scheduling source vertices in different batches and
sharing graph traversal operations in the same batch for efficiency (B-
MSSP), which effectively solves the performance bottleneck of MSSP on
big graphs. To correctly run multi source vertices in the same batch, a
message cluster mechanism is designed to effectively manage the communication. A further analysis reveals that the overall performance of
B-MSSP is affected by the batch size. Thus, an adaptive cost-benefit
model is proposed to quickly and easily select a proper batch size for
better performance. Extensive experiments over lots of real graphs validate the efficiency and effectiveness of our proposals.

Keywords: Multi-source shortest path · Sharing graph traversal ·
Distributed computation

1 Introduction

Many applications in real life need to know the distances between some or all
pairs of vertices in a given graph [1–3]. For example, given a social network, a
lot of knowledge can be discovered and then used for community detection and
marketing analysis. Thus, the network characteristics, such as intermediateness
and closeness, are greatly important and they usually depend on the evaluation
of the distance between any two vertices [4]. Another example is to perform
the friend recommendation or find close neighbors for some selected vertices

© Springer Nature Switzerland AG 2021
C. Xing et al. (Eds.): WISA 2021, LNCS 12999, pp. 635–646, 2021.
https://doi.org/10.1007/978-3-030-87571-8_55

(sources) where distances between sources and other vertices are usually used as the metrics [3]. The two examples mentioned above actually can be regarded as the typical multi-source shortest path (MSSP for short) computation. However, the volume of graph data is rapidly growing, which poses great performance challenges even for the traditional SSSP computation which just evaluates the distances between a specified source vertex and any other ones, let alone the MSSP variant where each vertex can possibly become a source.

To tackle the challenging MSSP problem over large graphs, there are three basic solutions. (1) Traditional Floyd Warshall algorithm and its distributed variants [5,6]: The main idea is to sequentially broadcast the distances from every vertex as source to others and then all vertices update their distances in parallel or a distributed manner [5,6]. That means the distance between any pair of vertices must be computed as a whole, which cannot efficiently support the scenario where only some of vertices are selected as sources. Also,the number of iterations is equal with the number of vertices, which is so large especially for large graphs. (2) S-SSSP: Besides Floyd warshall, a straightforward solution is to sequentially submit a series of SSSP jobs with varied source vertices. Clearly, the flexibility can be guaranteed because only cared vertices will be selected as sources for SSSP computation. Also, all existing optimization techniques can be directly utilized since SSSP as a classic problem has been researched in last decades [7–11]. While, the disadvantage of S-SSSP is that the sequentially submitted jobs execute traversals completely independently, which yields many redundant operations since these traversals are performed on the same graph. (3) B-MSSP: Another solution is to schedule all source vertices in different batches and within one batch multiple sources (MSSP) are computed concurrently in a single pass of graph traversal. Sources computed in the same batch of course can share related data access, computation and communication operations during the traversal process, leading to better performance.

On the other hand, to solve the performance bottleneck over large graphs, many efforts have been devoted into distributed solutions, among of which Hadoop is a general open source big data processing platform and has good scalability by dynamically adding new physical machines into the Hadoop cluster. Till now, some researchers have studied the SSSP algorithm or the A*-based variant using Hadoop [7,8] or other cloud computing environments [9–11]. However, the distributed MSSP has not attracted enough attention. We are also aware some recent works focusing on concurrent multi-source computation problems on special parallel devices, like multi-source breadth-first-search using shared memory on a single physical machine with multi cores [1,3] or Graphic Processing Unit (GPU) [12]. Compared with Hadoop, such devices have comparable computation efficiency but the storage scalability is very limited. The most recently published work CGraph [13] studies the temporal and spatial locality when running different kinds of graph algorithms (like PageRank and SSSP) on a graph. It designs a data-centric mechanism to improve locality to reduce redundant operations. However, this paper only focuses in our setting, we focus on the same kind of shortest path computation problems, and hence the locality is naturally good.

This paper thereby proposes a distributed implementation of MSSP (and hence B-MSSP) computation based on Hadoop. Since existing multi-source computation works use shared memory, they do not care about communication operations. While, for Hadoop deployed on a cluster with distributed memories, it is greatly important to efficiently manage messages among machines, especially when messages are generated by vertices traversed from different sources. We thereby design a simple but efficient communication mechanism to cluster messages. We further find that the overall performance of our B-MSSP computation is affected by the granularity of concurrent sharing, i.e., how many source vertices are concurrently computed in a single graph traversal (batch size). Intuitively, when the granularity is too small, B-MSSP will degrade into S-SSSP. On the contrary, if the granularity is too large, some additional communication overhead and resource contention will be caused and cannot be ignored. To achieve the best performance, one solution is to manually run MSSP computation multiple times by varying the batch size. That is clearly time-consuming and not friendly for end-users. Instead, we design an adaptive cost-benefit model, which automatically optimizes the concurrent granularity by analyzing some important yet easily-collected metrics. Note that our solution is designed based on Hadoop, but it can be easily extended to other big data processing platforms like Spark [14], Apache Flink [15], Pregel and its open source implementations [16,17].

2 Multi-source Shortest Path Computation Framework

2.1 Implementation of Single Source Shortest Path

The traditional solution about shortest path, i.e., Dijkstra, is a centralized algorithm and cannot be executed in the distributed manner. Therefore, Yang Ling et al. propose a new distributed implementation based on Hadoop. The main idea is to execute a series of Hadoop jobs to simulate the iterative computation of SSSP, where each job can be further divided into two phases Map and Reduce. In Map, each vertex transmits its graph topology and if it is active, its distance value is also broadcast to out-neighbors along outgoing edges. In Reduce, a vertex receives possible messages from its in-neighbors and then updates its value and is marked as active if necessary.

2.2 Implementation of Multi-source Shortest Path

S-SSSP: Multi-source Shortest Path Implementation Based on Serial Submission. It simulates multi source vertices one by one using the implementation of SSSP given in Sect. 2.1. However, in the big data era, a graph may have millions or even tens of millions of vertices. The scheme of S-SSSP is not suitable when the number of source vertices is large. First, it is not very friendly for end-users to manually organize the serial submission of so many source vertices. Second, for different SSSP computation, graph data structure is repeatedly transferred and stored, leading to redundant network and disk I/O costs. Finally, S-SSSP will generate a large amount of jobs but starting a MapReduce job is not free, which yields expensive job management overheads.

B-MSSP: Multi-source Concurrent Shortest Path Implementation Based on Batched Submission by Sharing Graph Traversal. In order to solve the disadvantages of S-SSSP, this paper studies the B-MSSP, a distributed implementation of MSSP computation based on Hadoop. The dataflow of B-MSSP is similar to that of SSSP. The difference is mainly focused on the data structure. The former needs to expand the structures of value ($dist$), active flag ($state$) and messages to support the update and propagation of the distance values of multiple source vertices. The new structure is shown as below:

$$\langle vertexId, \{dist\,[] : state\,[]\} : edges\,(dstId : weight)\rangle$$
$$\langle vertexId, \{dist[]\}\rangle$$

For example, if source vertex x is updated at some iteration, then, $dist[x]$ is the distance from the xth source vertex to the current vertex, and $state[x]$ represents the related state. In the following, we give more details about the Map and Reduce function design. Since the dataflow is similar to SSSP, now we only focus on the difference.

Map phase: Now the $dist$ and $state$ fields of the graph structure and construction message are expanded into arrays. The size of the array is obviously determined by the number of source vertices. In the first iteration, the algorithm performs breadth-first traversal of all source vertices. In subsequent iterations, map functions first traverse the $state$ value of each source vertex and if at least one is set as 1, the current vertex should construct the distance message and then broadcast it. The initmap module in Fig. 1 shows the output of the map function in the first iteration. Now assume that we have two sources v1 and v2. The source vertex v1 constructs two distance messages $\langle 2, \{10 : INF\}\rangle$ and $\langle 4, \{5 : INF\}\rangle$ for its out-neighbors v2 and v4. Similarly, source vertex v2 also constructs two distance messages $\langle 3, \{INF : 1\}\rangle$ and $\langle 4, \{INF : 2\}\rangle$. As can be seen from the figure, in the message structure, only the source related $dist$ value is set and others are invalid. In fact, these invalid values incur additional network and disk I/O costs which will be introduced in Sect. 3. Reduce phase: the Reduce function integrates the map output and compares the distance message with the $dist[x]$ value embedded in the static graph structure. If the former is shorter, the latter will be updated and then the related $state$ value is set as 1. MSSP converges until the number of active vertices in the current iteration step is zero. And after that, we should schedule the next batch of source vertices for MSSP computation. Figure 1 shows the entire Reduce phase. Distance message $\langle 2, \{10 : INF\}\rangle$ is integrated with graph structure $\langle 2, \{INF : 0 : 0 : 0 \quad 3 : 1 : 4 : 2\}\rangle$.

Compared with S-SSSP, B-MSSP schedules multiple source vertices in a batch to execute the shortest path computation so as to share the graph traversal operations and hence reduce redundant network and disk I/O costs. Figure 2(a) schedules source vertices v1 and v2 in the same batch, which is equivalent to the collection of two SSSP computation shown in Fig. 3(b) and (c) on a graph but now the graph structure is traversed only once. And the simple yet efficient message cluster communication mechanism in B-MSSP compacts the messages in a single array which will be sent by initializing the network connection only once, instead of n times in S-SSSP. At this time, the number of transmissions of graph data is the maximum number of iteration steps in B-MSSP, rather than n multiplied by the maximum number in S-SSSP. That saves a lot of communication overhead and maintenance time of MapReduce tasks.

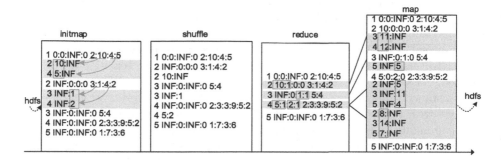

Fig. 1. Outputs results of the Map stage in MSSP

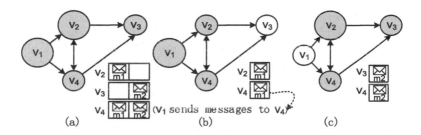

Fig. 2. Comparison of B-MSSP and S-SSSP

3 Adaptive Optimization for the Batch Size in B-MSSP

3.1 Analysis of the Impact Batch Size

The key advantage of B-MSSP's benefits is to share the graph traversal operations for source vertices scheduled in the same batch. Clearly, the number of source vertices in a batch (that is, the concurrent sharing granularity) will affect the overall performance. When the sharing granularity is too small, the sharing revenue will decrease and at the worst case it degenerates into S-SSSP. When the sharing granularity is too large, although the repeated transmission and traversal of the graph structure will be effectively reduced, other problems such as redundant communication and resource contention as analyzed later will happen. These problems greatly degrade the performance of B-MSSP.

Redundant Communication. The message cluster communication mechanism integrates the *dist* messages of multiple source vertices and packages them in a single network transmission, which saves the cost of frequently building network connection but also generates additional data transmission cost. As shown in Fig. 2(b), for SSSP, source vertex v1 starts to perform shortest path computation. Such a message will be generated when a vertex is updated (active). As shown in Fig. 2(a), for MSSP, two source vertices v1 and v2 are scheduled together, and the structure of message is: $\langle verteId, \{dist[1] : dist[2]\}\rangle$. However, the traversal of the two sources are not exactly synchronized. When source vertex v2 traverses to vertex v3, the traversal from source vertex v1 has not yet reached. As a result, vertex v2 sends a message with invalid INF in $dist[1]$ and valid distance value in $dist[2]$, and later, the traversal from vertex v1 generates a message with valid $dist[1]$ and invalid $dist[2]$. Such invalid values lead to redundant communication overheads. A bitmap index can specify the valid values and hence avoid sending invalid values. However, our tests show that such a design cannot work well as expected because of the frequent and not-free bit operations. Usually, the probability of transferring invalid values directly increase when more source vertices are scheduled in a batch.

Resource Contention. When MapReduce works, data need to be read/written between memory and disk. When the number of source vertices scheduled in a batch is too large, it is most likely that the amount of transferred data can exceed the limited bus bandwidth between memory and disk, which will cause data congestion during transmission and hence increase the blocking waiting time. Similarly, the shuffle phase will transfer many graph data and message data, which consumes a large amount of network resources. When the data transmission volume exceeds the network bandwidth, the waiting time will also significantly increase. Therefore, the larger the graph sharing granularity of B-MSSP is, the heavier the contention degree of bus and network bandwidth resources is, which generates negative impact on overall performance.

3.2   Optimizing Batch Size Based on Cost-Benefit Analysis

Till now, we know the shared granularity affects the execution efficiency of the B-MSSP computation. However, the optimal sharing granularity depends on many factors, such as the number of edges, the number of vertices, and the average out-degree of the graph. Therefore, it is difficult to give an one-fit-all fixed and optimal shared granularity value. One solution is brute force enumeration (manually testing multiple times to find an optimal granularity value) for a given input graph and cloud computing cluster, but it is too complicated and time-consuming. As a result, it is urgent to design an automatic calculation model of shared granularity based on cost-benefit analysis by comparing S-SSSP and B-MSSP. This model can quickly estimate a proper value of the optimal concurrency granularity based on simple and easy-to-obtain statistical data in the following. Table 1 summarizes important symbols used in our analysis and the related explanations.

Table 1. Description of important symbols

Symbol	Note
ζ	Sharing granularity (batch size)
Δ	Time of a hadoop job (no message is generated)
$f(x)$	The bytes of invalid messages generated by x source vertices
ε	Pure computational time for a single source vertex
λ	The average out of the data set
β	The number of edges of the data set/84000000
$|V_i|$	The number of active vertices in the i-th iteration
$|V|$	The total number of vertices in the data set
$\varUpsilon$	Number of machines in the cluster
τ	The number of bytes occupied by a given data type
φ	The number of iteration steps in an SSSP job
$g(m,\zeta)$	Total pure computing time
m	The number of vertices, If $(m\%\zeta==0)z=0$, else $z=1$
α	0.0006

The actual benefit of B-MSSP is mainly determined by two aspects: benefit and additional cost.

The benefit is the time saved compared to S-SSSP, including avoiding repeated transmission and repeated traversal of graph structure and MapReduce maintenance time. That can be computed by three factors: the number of iteration steps per MSSP in a batch of source vertices, the time of a Hadoop job without message transmission, and the average number of saved SSSP-based graph traversals. Thus, we need to execute a SSSP randomly to get the number of iteration steps. A Hadoop job without messages actually is one in which only the graph structure is transferred in one iteration. Accordingly, we can get MapReduce job maintenance time and graph transfer time of an iterative step.

The cost is mainly incurred by transferring invalid message values in the message array as analyzed in Sect. 3.1, and the performance degradation caused by resource contention. To estimate the volume of invalid message values, we can run a SSSP computation randomly selected source vertex to record the number of active vertices per iteration. Accordingly, we can obtain the average number of invalid bytes of each message array in each iteration step under different granularity values by using the mathematical expectation of binomial distribution $E(X) = NP$. Then we can calculate the number of redundant bytes according to the average value, and then calculate the transmission cost of redundant bytes. Since note that in each batch, the pure calculation time generated by another $(batchsize - 1)$ source vertices is also an additional when comparing MSSP and SSSP. Therefore, the cost part of the model also needs to take into account the extra pure computation time generated by multiple sources. Specifically, we can run SSSP on a stand-alone machine to estimate the pure computation time (now no network message is generated), and then multiply its runtime by the number of saved SSSP traversals as the total pure computation time.

To sum up, we can get the calculation Formula (1) of the cost-benefit model:

$$y(m) = \left(\frac{m}{\zeta}\right)(\zeta - 1)^* \Delta + z * (m\%\zeta - 1)^* \Delta - \left[\left(\frac{m}{\zeta}\right) f(\zeta) + f(m\%\zeta) + g(m,\zeta)\right]\alpha \quad (1)$$

Here, y is the actual benefit. Δ is the MapReduce job maintenance time and graph transfer time calculated by SSSP. $\left(\frac{m}{\zeta}\right)(\zeta - 1)$ and $z * (m\%\zeta - 1)$ are the total number of iterations saved, which is multiplied by Δ to indicate the benefit. Since the number of source vertices is not necessarily divisible by the shared granularity, we also should specify $m\%\zeta$ as a part of the benefit. Similarly, when analyzing the additional cost, we should first calculate the transmission overhead of the number of invalid bytes when the source vertices are full of ζ, where (m/ζ) is the number of batches executed. $f(\zeta)$ represents the number of invalid bytes generated by ζ source vertices; and then calculate the cost when the sharing granularity is not enough for ζ (i.e., $f(m\%\zeta)$). While, the result of $f(x)$ can be computed based on Formula (2).

$$f(x) = \sum_{j=1}^{3} \frac{\sum_{i=1}^{\varphi}\left(1 - \frac{|V_i|}{|V|}\right) * \left(x^\wedge(x * \beta) * |V_i|^* \lambda^* \tau\right)}{1024^* \Upsilon^* V_j} \quad (2)$$

Here, j=3 and V_j in Formula (2) indicate that the number of invalid bytes obtained above needs to be divided by the disk read/write rate and the network transfer rate to get the final time cost. $1 - \left|\frac{V_i}{V}\right|$ is the probability that a vertex is not active in the i-th iteration ($|V_i|$ and $|V|$ respectively stand for the number of active vertices at the i-th iteration and the whole vertices). Using the mathematical expectation of the binomial distribution, the average number of inactive vertices in each iteration can be obtained. However, as the sharing granularity n increases, the resource usage of cpu, memory, and disk increases rapidly, leading to resource competition and performance degradation. Therefore, the n in $E = np$ can be modified to: $n^{(n*\beta)}$ to indicate the impact of resource contention based on our tests. Based on expected value, the out-degree of active vertex and τ, we can get the number of invalid message bytes, so as to calculate the additional communication overhead in all iteration steps that i from 1 to φ. The purpose of dividing the summation by 1024 in the formula is to unify the calculation unit. Sharing granularity is also affected by the number of edges in the data set. Thus, for β, the number of edges in the data set is divided by a constant 840,000,000 to reflect the magnitude of the number of vertices. At last, the cost part should be multiplied by 0.0006 accordingly. $(g(m,\zeta))$ represents the pure computation time, which is the number of source vertices multiplied by the pure computation time of a single source vertex. The formula is as follows:

$$g(m, \zeta) = \frac{\varepsilon}{\Upsilon} * (m - m/\zeta - z) \quad (3)$$

where, Υ is the number of machines in the cluster, and the communication overhead is equally shared by the machines in the cluster.

4 Experimental Evaluation

4.1 Experimental Environment and Data Set

The experiments are run on a cluster of 5 physical machines (1 as the master node and 4 as the computing nodes). The master node runs 64 GB of memory; slave nodes' memory is 32 GB. Hadoop-0.20.2 is deployed on each machine. The specific conditions of the data set used in the experiments are shown in Table 2.

Table 2. Graph datasets

Data set	Number of vertices	Number of edges
Livej	4,847,571	69,028,541
Orkut	3,072,599	223,609,329
Hollywood	2,180,759	228,985,632

4.2 Overall Performance

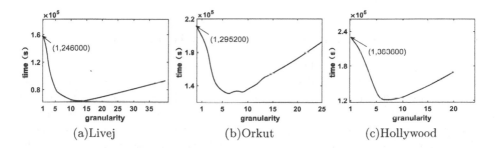

(a)Livej   (b)Orkut   (c)Hollywood

Fig. 3. Overall performance of B-MSSP

Figure 3 shows the overall performance of B-MSSP over the three data sets. We manually vary the batch size to observe the variation of the total runtime. Note that B-MSSP with batch size 1 is actually the counterpart objective S-SSSP. We can get the runtime of S-SSSP by multiplying the running time of SSSP by the number of source vertices (we randomly select 300 vertices). It can be seen that as the batch size increases, the total execution time first decreases and then increases. This validates our analysis in Sect. 3.1 that when the sharing granularity exceeds a certain threshold, B-MSSP generates a large amount of invalid message values and heavy resource contention. The growth rate of benefits brought from batch processing is less than the growth rate of additional costs. Particularly, in the best case, B-MSSP can improve the performance up to 73.7% compared with S-SSSP.

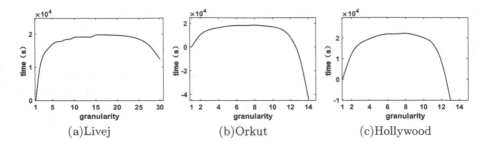

Fig. 4. Results of the cost-benefit model

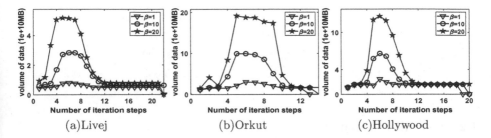

Fig. 5. The total number of read and written bytes

4.3 Effectiveness Analysis About the Cost-Benefit Model

Figure 4 is performed to validate the effectiveness of the adaptive mechanism of computing optimized batch size over the three data sets. The x-axis indicates the shared granularity. while the y-axis stands for the estimated actual benefit (i.e., the output of our model in Eq. (3)). The increase of y-axis actually means that the actual execution time of B-MSSP is being decreased. Comparing Figs. 3 and 4, the x-axis value corresponding to the highest point of the curve in the latter figure is very close to the x-axis value corresponding to the point with the shortest total execution time in the former figure. Such an x-axis value is indeed the optimal sharing granularity. And the comparison result validates that our cost-benefit model can very nicely capture the proper batch size to guide the schedule of source vertices in B-MSSP.

Figure 5 demonstrates the total bytes of data read and written from/to local disk or Hdfs per iteration, by running B-MSSP on all of three datasets and varying the batch size from 1, 10, to 20. Clearly, the communication volume increases significantly and directly with the increase of batch size, which of course consumes a lot of data transferring bus bandwidth and hence becomes a performance bottleneck.

Figure 6 shows the network traffic with the same setting used in Fig. 6. We compute this metric by collecting HDFS_BYTES_READ, HDFS_BYTES_WRITTEN and Reduce shuffle bytes reported by Hadoop. We observe the similar increase

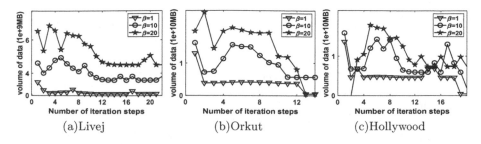

Fig. 6. The network traffic of B-MSSP

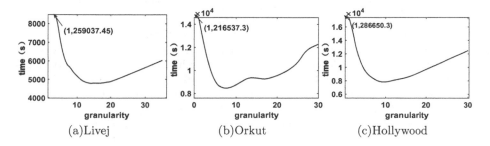

Fig. 7. Scalability of B-MSSP

trend when varying the batch size, which competes the limited network bandwidth and increases waiting time for data transmission.

Figure 7 shows an extended experiment for these three data sets, extending from four to eight states.

5   Conclusion

SSSP computation is an important underlying problem and the multi-source variant also has been widely used in many real-world applications. However, the current multi-source variant problem lacks a distributed solution that can effectively deal with big graphs. Thus, this paper proposes a B-MSSP based on graph traversal sharing, and also designs an automatic calculation model of shared granularity by analyzing benefits and costs of the proposed solution.

Acknowledgments. This work was supported by the National Natural Science Foundation of China (61902366 and 61902365), the Fundamental Research Funds for the Central Universities (202042008), the Project funded by China Postdoctoral Science Foundation (2020T130623, 2019M652474 and 2019M652473), the Projects funded by Postdoctoral Creative Foundation in Shandong Province and Postdoctoral Application&Research Foundation in Qingdao City, the Qingdao Independent Innovation Major Project (20-3-2-12-xx).

References

1. Then, M., et al.: The more the merrier: efficient multi-source graph traversal. Proc. VLDB Endow. **8**(4), 449–460 (2014)
2. Then, M., Günnemann, S., Kemper, A., Neumann, T.: Efficient batched distance, closeness and betweenness centrality computation in unweighted and weighted graphs. Datenbank-Spektrum **17**(2), 169–182 (2017)
3. Kaufmann, M., Then, M., Kemper, A., Neumann, T.: Parallel array-based single- and multi-source breadth first searches on large dense graphs. In: EDBT, pp. 1–12 (2017)
4. Shen, D.: Lower bounds on rate of convergence of matrix products in all pairs shortest path of social network. arXiv preprint arXiv:2006.13412 (2020)
5. Kang, S.J., Lee, S.Y., Lee, K.M.: Performance comparison of OpenMP, MPI, and MapReduce in practical problems. In: Advances in Multimedia 2015 (2015)
6. Pradhan, A., Mahinthakumar, G.: Finding all-pairs shortest path for a large-scale transportation network using parallel Floyd-Warshall and parallel Dijkstra algorithms. J. Comput. Civil Eng. **27**(3), 263–273 (2013)
7. Yang, L., Li, R., Tang, Z.: Research on the single source shortest path algorithm using MapReduce. Microcomput. Inf. **27**(12), 97–99 (2011). (in Chinese)
8. Adoni, W.Y.H., Nahhal, T., Aghezzaf, B., Elbyed, A.: The MapReduce-based approach to improve the shortest path computation in large-scale road networks: the case of A* algorithm. J. Big Data **5**(1), 1–24 (2018)
9. Gao, Y., Yao, L., Yu, J.: Research of query verification algorithm on body sensing data in cloud computing environment. In: Ni, W., Wang, X., Song, W., Li, Y. (eds.) WISA 2019. LNCS, vol. 11817, pp. 176–188. Springer, Cham (2019). https://doi.org/10.1007/978-3-030-30952-7_20
10. Arfat, Y., Suma, S., Mehmood, R., Albeshri, A.: Parallel shortest path big data graph computations of US road network using apache spark: survey, architecture, and evaluation. In: Mehmood, R., See, S., Katib, I., Chlamtac, I. (eds.) Smart Infrastructure and Applications. EICC, pp. 185–214. Springer, Cham (2020). https://doi.org/10.1007/978-3-030-13705-2_8
11. Malewicz, G., et al.: Pregel: a system for large-scale graph processing. In: Proceedings of the 2010 ACM SIGMOD, pp. 135–146 (2010)
12. Liu, H., Huang, H.H., Hu, Y.: IBFS: concurrent breadth-first search on GPUs. In: the 2016 International Conference on Management of Data, pp. 403–416 (2016)
13. Zhang, Y., et al.: CGraph: a correlations-aware approach for efficient concurrent iterative graph processing. In: 2018 Annual Technical Conference, pp. 441–452 (2018)
14. https://spark.apache.org/
15. https://flink.apache.org/
16. http://giraph.apache.org/
17. https://hama.apache.org/

Fabric Defect Target Detection Algorithm Based on YOLOv4 Improvement

Ying Wang[1], Zhengyang Hao[1], Fang Zuo[1(✉)], and Zixiang Su[2]

[1] Henan University, Kaifeng 475001, China
{wangying,zuofang}@henu.edu.cn
[2] Central China Normal University, Wuhan 430079, China

Abstract. Fabric defect detection is a key part of product quality assessment in the textile industry. It is important to achieve fast, accurate and efficient detection of fabric defects to improve productivity in the textile industry. For the problems of varying scales, irregular shapes and many small objects, an improved YOLOv4 object detection algorithm for fabric defects is proposed. Firstly, in order to improve the detection accuracy of small objects, the RFB module is introduced and fused with shallow features, which can obtain receptive fields of different scales to improve the features extracted from the backbone network. Secondly, the introduction of spatial and channel attention mechanisms can enhance fused features, allowing the network to focus on useful information. Experimental results show that the mean average precision of the improved YOLOv4 object detection algorithm in fabric defect map detection is 71.89%. The improved algorithm can accurately improve the accuracy of fabric defect positioning.

Keywords: Fabric defect detection · YOLOv4 · RFB · Attention mechanisms

1 Introduction

Fabric defect detection is an important part of a textile company's production. The accuracy of fabric detection determines the economic efficiency of the company. At this stage, most textile companies use manual detection, which requires workers to stand in front of the detection machine for a long time and carefully observe the defective state of the cloth in motion through uninterrupted detection. However, the detection speed of workers was unable to comply with the basic online detection requirements in practice, and the manual detection was inefficient.

Defect detection of fabric can be considered as an object detection problem. With the growing development of artificial intelligence technology and the expansion of application fields, deep learning methods [1] represented by deep convolutional neural networks have shown better performance than traditional methods in a large number of practical application scenarios in the industrial field [2]. The existing fabric detection algorithm has problems such as low efficiency of

© Springer Nature Switzerland AG 2021
C. Xing et al. (Eds.): WISA 2021, LNCS 12999, pp. 647–658, 2021.
https://doi.org/10.1007/978-3-030-87571-8_56

defect recognition and low recognition accuracy. Therefore, the paper proposed an improved YOLOv4 [3] algorithm to help detect the tiny fabric defect. The contributions to our work are as follows:

1. Aiming at the problem of variable scale, irregular shape and the existence of more small objects, a Fabric defect object detection algorithm based on YOLOv4 improvement is proposed.
2. The RFB module is introduced to increase the sensing field of shallow feature maps to improve the detection accuracy of small objects. It fuses the output of the RFB moudle on the second remaining block in the backbone with the output of the 8-fold downsampling of the original network and incorporates the feature pyramid structure.
3. By adding an attention module to each of the three branches at the end of the feature fusion network, the weights of the channel features, and spatial features of the feature map are allocated to increase the weight of useful features while suppressing the weight of invalid features to improve the accuracy of object detection.

The rest of the paper is organized as follows. Section 2, discusses the previous related works of fabric defect object. In Sect. 3, the details of the proposed model. In Sect. 4, describes the details of experiments, including datasets, parameter settings, and evaluation metrics. Finally, we present our conclusions in Sect. 5.

2 Related Works

The traditional methods of fabric defect detection are structure-based analysis, model-based analysis, and spectrum-based analysis [4], which are used frequently. KarleKar et al. [5] proposed a combined wavelet transform and morphology method to extract defect features by detecting cloth texture information. Jia et al. [6] automatically segmented the grid patterned fabric with repeating patterns by morphological processing and then detected the defect information by Gabor filter. Chao Deng et al. [7] proposed a method based on edge detection to automatically detect fabric defects using morphological processing and discrete cosine transform.

Deep learning models have powerful feature representation capabilities, and deep learning methods have a significant advantage over detectors with artificially set features in the target detection problem. These deep learning methods accomplish two tasks when performing target detection: localization and classification of the detected frames. Deep learning-based target detection methods are mainly divided into two-stage target detection algorithms and one-stage target detection algorithms. The two-stage target detection algorithm is used to achieve detection of targets in two stages: First, the suggested regions are extracted from the images input to the network, then classification and position regression are performed for each suggested region, and finally target detection is achieved. Typical representatives of such algorithms are R-CNN [8], SPP-Net [9], Fast R-CNN [10] and Faster R-CNN [11]. At the same time, single-stage target detection algorithms do not need the regional proposal stage, and directly generate the class probability and position coordinate values of the object, and directly get the final detection results after a single detection, so they have faster detection speed, and more typical algorithms such as YOLO [12], SSD [13], YOLOv2 [14], YOLOv3 [15], RetinaNet [16], YOLOv4. YOLOv4 is based on the original YOLO target detection architecture and uses the best optimization strategies in the field of convolutional neural networks in recent years, proposed by Alexey Bochkovskiy in 2020. The network structure of YOLOv4 is shown in Fig. 1. Our paper propose an improved YOLOv4 method which exploits the RFB module and attention mechanism to enhance the semantics of features.

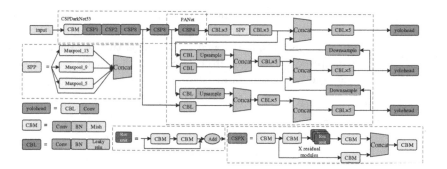

Fig. 1. The network structure of YOLOv4.

3 The Proposed Model

3.1 Attention Mechanism

Attention mechanism can effectively improve neural network performance. In recent years, the convolutional attention mechanism CBAM [17] has emerged on the basis of the SE-Net [18] model. Unlike SE-Net, the CBAM attention network contains two modules: channel attention and spatial attention. The structure of CBAM is shown in Fig. 2.

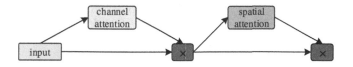

Fig. 2. The structure of CBAM.

The channel attention mechanism first performs a pooling operation on the incoming feature maps to retain the contextual information. Then the features are summed by the MLP structure, and then the input features required by the

Spatial attention module are generated by the sigmoid activation operation. The structure of the channel attention module is illustrated in Fig. 3.

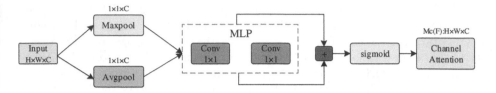

Fig. 3. The structure of channel attention module.

The construction of the spatial attention is demonstrated in Fig. 4. We first perform the average pooling and maximum pooling of one channel dimension separately to obtain two channels, and concatenate these two channels together. Then, the feature maps are obtained after a convolution with a convolution kernel of 7 and a sigmoid activation function.

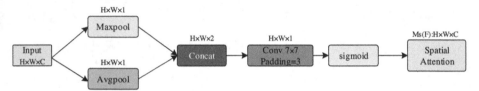

Fig. 4. The structure of spatial attention model.

The calculation method of attention module is shown in (1) and (2). A feature map F is given as input, where $M_c(F)$ denotes the weight of the channel attention map and $M_s(F)$ denotes the weight of the spatial attention map. $\otimes$ indicates factor-by-element division. In the multiplication process, F_c represents the channel attention value. F_r is the final output of the attention weights.

$$F_c = F \otimes M_c(F) \tag{1}$$

$$F_r = F_c \otimes M_r(F_c) \tag{2}$$

Equation (3) and (4) respectively show the process of channel attention and spatial attention. $\oplus$ indicates the collocation operation. $f^{n \times n}$ denotes the convolution operation with convolution kernel of n. σ denotes the sigmoid activation function.

$$M_c(F) = \sigma(f_1^{1 \times 1}(f_0^{1 \times 1}(AvgPool(F)) + f_1^{1 \times 1}(f_0^{1 \times 1}(MaxPool(F))) \tag{3}$$

$$M_s(F_c) = \sigma(f^{7 \times 7}([AvgPool(F_c) \oplus MaxPool(F_c)]) \tag{4}$$

Equation (3) shows the calculation of channel attention. First use the avg pooling and maxpooling operations to the feature mapping. Then the concatenation operation is performed after two more convolution operations with a convolution kernel of 1. Finally, the channel attention map is obtained after σ operate. Equation (4) shows the calculation of spatial attention. It operates in a similar way to channeled attention.

3.2 RFB Convolution Module

The RFB module uses different convolution kernels, a multi-branch convolution is designed, a pooling operation, and a wormhole convolution (dilation convolution) to control the eccentricity of the perceptual field, which forms the generated features after the final step of the reshaping operation. Figure 5 shows the flow of the RFB module.

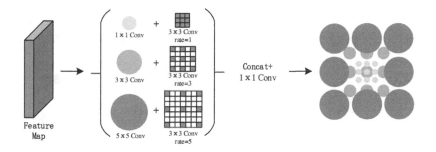

Fig. 5. Flow of the RFB module.

Figure 6 shows the structure of the RFB module, first forming a three-branch structure through 1×1, 3×3 and 5×5 convolution kernels, and then introducing dilation rate = 1, dilation rate = 2 and dilation rate = 3 in each branch. Finally, the outputs of the three branches are connected together for the purpose of fusing different features.

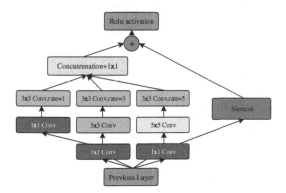

Fig. 6. The structure of RFB module.

3.3   The Proposed Model

This paper adds a CBAM module to each of the three branches at the end of the YOLOv4 feature fusion network. It allows the network to focus more on useful information. YOLOv4 is better for large object detection, but there is still leakage for small object objects in practical scenarios. The RFB module is introduced to increase the sensing field of shallow feature maps to improve the detection accuracy of small objects. It fuses the output of the RFB module on the second remaining block in the backbone with the output of the 8-fold downsampling of the original network and incorporates the feature pyramid. Algorithm 1 shows the flow process steps of the improved YOLOv4 defect detection method. Figure 7 shows the improved network structure of YOLOv4.

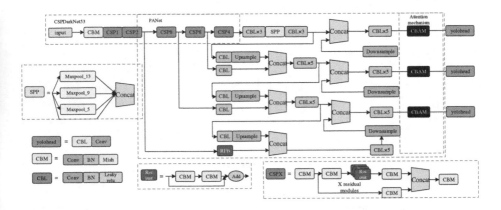

Fig. 7. The improved network structure of YOLOv4.

Algorithm 1. The improved YOLOv4 defect detection method

Require: The fabric images $T_1, T_2, ..., T_n$

1: The fabric image T_1 is divided into $S \times S$ girds, and each gird is provided for k priori boxes in different sizes

2: Improved YOLOv4 network was used to extract target and predict boundary frame coordinates (x, y, w, h), boundary box confidence (c) and class condition probability (Pr)

3: bounding box with confidence (c) below the threshold indicates that there are no targets in the bounding box, and the bounding boxes with confidence higher than the threshold are reserved

4: Select the category with the highest category conditional probability as the target category

5: Non-maximum suppression (NMS [19]) was used to remove redundant bounding boxes

6: The target category and the predicted box are output in the fabric images T_1

3.4   Loss Function

The YOLOv4 model loss function is composed of a weighted sum of multi-part losses, including the bounding box regression loss (L_{ciou}), confidence loss (L_{conf}), and classification loss (L_{class}). Equation (5) defines the total loss function.

$$L_{loss} = L_{ciou} + L_{conf} + L_{class} \tag{5}$$

In the bounding box regression loss (L_{ciou}), b and b^{gt} denote the centroids of the prediction box and the real box, respectively, p^2 denotes the Euclidean distance between the two points, c denotes the diagonal distance that can contain the minimum closed area of both the prediction box and the real box, a is the parameter used as a trade-off, and v is the consistency parameter that measures the aspect ratio. Compared with the traditional mean square error loss function, L_{ciou} effectively avoids the problem of being sensitive to the scale of the object and can better focus on the relationship between the position of the prediction frame and the actual frame. Equation (6), (7), (8), and (9) shows the loss (L_{ciou}) function.

$$L_{ciou} = \sum_{i=0}^{S^2} \sum_{j=0}^{B} I_{i+j}^{obj} \left[1 - IoU + \frac{p^2(b, b^{gt})}{c^2} \right] + av \tag{6}$$

$$IoU = \frac{|A \bigcap B|}{|A \bigcup B|} \tag{7}$$

$$a = \frac{v}{1 - IoU + v} \tag{8}$$

$$v = \frac{4}{\pi^2} \left(\arctan \frac{w^{gt}}{h^{gt}} - \arctan \frac{w}{h} \right)^2 \tag{9}$$

The confidence loss (L_{conf}) contains two parts, with and without the object in the prediction frame. The formula incorporates cross entropy, $\hat{C}_i^j$ denotes the probability value that the prediction frame contains the true object, C_i^j denotes the true value, and λ_{noobj} is a weight value to control the amount of contribution lost when there is no object. Equation (10) shows the loss(L_{conf}) function.

$$L_{conf} = \sum_{i=0}^{S^2} \sum_{j=0}^{B} I_{i+j}^{obj} \left[\hat{C}_i^j \log C_i^j + \left(1 - \hat{C}_i^j \right) log \left(1 - C_i^j \right) \right] - \lambda_{noobj}$$
$$\sum_{i=0}^{S^2} \sum_{j=0}^{B} I_{i+j}^{noobj} \left[\hat{C}_i^j \log C_i^j + \left(1 - \hat{C}_i^j \right) log \left(1 - C_i^j \right) \right] \tag{10}$$

The classification loss (L_{class}) is based on the grid horizontal and vertical coordinates, P_i^j denotes the probability that the $[i]\,[j]$ prior box belongs to category C. When the $[i]\,[j]$ prior box is responsible for detecting a certain object, the bounding box generated by this prior box will only go to calculate the classification loss function. Equation (11) shows the loss(L_{class}) function.

$$L_{class} = -\sum_{i=0}^{S^2} I_{i+j}^{obj} \sum_{C \in classes} \left[\hat{P}_i^j \log \left(P_i^j \right) + \left(1 - \hat{P}_i^j \right) \log \left(1 - P_i^j \right) \right] \quad (11)$$

4   Experiment

The dataset used in this paper is from a textile workshop in Foshan City, Guangdong Province. The dataset contains 3946 defective images with a resolution of 1000 × 1000. The object size of the dataset is mostly small. Figure 8 shows the samples of fabric defects. Figure 9 shows the shape distribution of fabric defects. Table 1 shows the sample distribution of the data set.

Fig. 8. Samples of fabric defects.

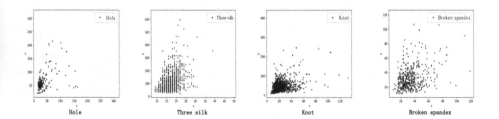

Fig. 9. Shape distribution of fabric defects.

Table 1. Samples distribution of the data set

Category	Hole	Three silk	Knot	Broken spandex	Total
Number	548	1762	2444	1044	5798

4.1 Evaluation Metrics

In this paper, mAP50 is used as an evaluation metric in the experiments. mAP50 is the mean value of AP50 (average precision) of the detection results for all categories. The AP50 value is the enclosed area of the accuracy and recall curves for an IoU threshold of 0.5. Equation (12) shows the calculation of precision and recall values. Equation (13) shows the calculation of AP50 and mAP50.

$$
\begin{aligned}
precision &= \frac{TP}{TP + FP} \\
recall &= \frac{TP}{TP + FN}
\end{aligned}
\tag{12}
$$

TP (True Positives) indicates the number of objects actually detected in the data set. FP (False Positives) indicates the number of objects detected by the detection model in error. FN (False Negatives) indicates the number of objects missed by the detection model.

$$
\begin{aligned}
AP50 &= \int_0^1 P_r dr, IoU >= 0.5 \\
mAP50 &= \frac{\sum_{i=1}^n AP50_i}{N}
\end{aligned}
\tag{13}
$$

Times: The time spent to detect each image. Use this metric to evaluate the detection speed of the target detection network.

4.2 Experimental Environment and Configuration

In this paper, the training and testing sets are divided in a split ratio from 8:2. The GPU is NVIDIA GeForce P100 and the experimental environment is Pytorch. The input size is $608 \times 608 \times 3$. The first stage freezes the first 249 network layers of the model with an initial learning rate of 0.001, 25 epochs of training, and a batchsize set to 4. The second stage trains all network layers with an initial learning rate of 0.0001, 25 epochs of training, and a batchsize set to 2. The value of gamma is 0.92. The learning rate decays to one gamma fraction of the original every other generation.

4.3 Test Result and Analysis

In order to evaluate the algorithm performance, the improved YOLOv4 which incorporates attention mechanism and multi-feature fusion, proposed in this paper were compared with YOLOv3, YOLOv4, Faster R-CNN and RetinaNet in various aspects. The same training set and test set were used for algorithm comparison experiment. The same learning rate and the epochs of training were used for contrastive algorithms. The AP50, mAP50 and times results of different models on the data set were shown in Table 2.

Table 2. Different network detection results

Model	Input size	AP50				mAP50	Times
		Hole	Three silk	Knot	Broken spandex		
Faster R-CNN	800 * 800	79.80%	50.50%	48.80%	27.10%	51.55%	0.1 s
RetinaNet	800 * 800	82.40%	62.60%	57.80%	57.90%	65.17%	0.08 s
YOLOv3	608 * 608	81.72%	72.34%	54.39%	45.48%	63.48%	0.03 s
YOLOv4	608 * 608	82.48%	74.75%	60.55%	51.36%	66.90%	0.03 s
Improved YOLOv4	608 * 608	86.20%	76.62%	58.26%	66.51%	71.89%	0.04 s

By comparing the mAP data obtained by different algorithms in Table 2, it is found that the improved YOLOv4 object detection algorithm based on attention mechanism and multi-feature fusion proposed in this paper has good performance in detection accuracy. Compared with the two-stage classical algorithm Fasters-RCNN, the proposed algorithm has a 20.34% improvement. In addition, the proposed algorithm improves 6.72%, 8.41% and 4.99% compared with the classical single-stage networks Retinanet, YOLOv3 and YOLOv4. Among them, the improved YOLOv4 algorithm enhances the AP values of holes by 3.72%, three silk by 1.87%, and broken spandex by 15.15% compared with the YOLOv4 algorithm.

In order to reflect the performance of the improved object detection algorithm more intuitively. The YOLOv4-based object detection model (the original algorithm) is firstly used for detection, and then the improved YOLOv4 algorithm model is put into detection. Figure 10 shows the detection results of YOLOv4 and improved YOLOv4. Columns (a) and (c) are the detection results of the YOLOv4 algorithm. Columns (b) and (d) show the detection results of the improved YOLOv4 algorithm. In the detection picture, label 1 represents the hole, label 3 represents the three silk, label 4 represents the knot and label 17 represents the broken spandex. From the experimental results, it is obvious that YOLOv4 has some missed detection objects and the confidence level is very low. The improved YOLOv4 algorithm in test images with good detection and high confidence. In summary, the improved YOLOv4 algorithm proposed in this paper shows a high detection rate. Although the speed is a little slow compared to the original algorithm, it still meets the requirements of fabric defect detection.

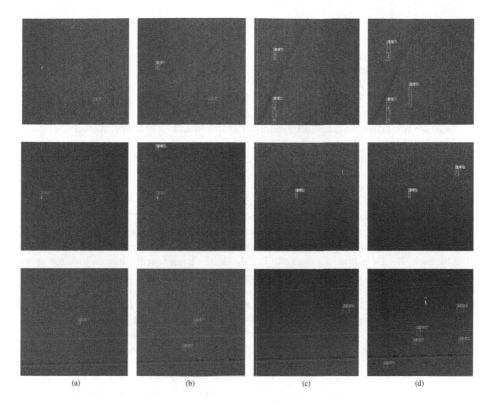

(a)   (b)   (c)   (d)

Fig. 10. The detection results of YOLOv4 and improved YOLOv4.

5   Conclusions

In this paper, we firstly add a top-down feature fusion structure by introducing the second residual block of the backbone feature extraction network to extract local information and improve the utilization of local information, and secondly introduce an attention mechanism module to focus the network on useful information. These improvements based on the YOLOv4 algorithm make the target detection with higher accuracy, and the next step will be considered to design the network structure by itself to construct a lightweight network with superior performance, so as to improve the detection accuracy and real-time performance of the model.

References

1. Zhang, H., Wang, K.F., Wang, F.Y.: Advances and perspectives on applications of deep learning in visual object detection. Zidonghua Xuebao/Acta Automatica Sinica **43**(8), 1289–1305 (2017)

2. Chen, R., Jin, Yu., Xu, L.: A classroom student counting system based on improved context-based face detector. In: Wang, G., Lin, X., Hendler, J., Song, W., Xu, Z., Liu, G. (eds.) WISA 2020. LNCS, vol. 12432, pp. 326–332. Springer, Cham (2020). https://doi.org/10.1007/978-3-030-60029-7_30

3. Bochkovskiy, A., Wang, C.Y., Liao, H.Y.M.: YOLOv4: optimal speed and accuracy of object detection. arXiv:2004.10934 (2020)

4. Yapi, D., Allili, M.S., et al.: Automatic fabric defect detection using learning-based local textural distributions in the contourlet domain. IEEE Trans. Autom. Sci. Eng. **15**(3), 1014–1026 (2017)

5. Vaibhav, M., Karlekar, V., Bhangale, K., et al.: Fabric defect detection using wavelet filter. In: 2015 International Conference on Computing Communication Control and Automation. IEEE (2015)

6. Jia, L., Chen, C., Liang, J., et al.: Fabric defect inspection based on lattice segmentation and Gabor filtering. Neurocomputing **238**(MAY17), 84–102 (2017)

7. Deng, C., Liu, Y.: Defect detection of twill cloth based on edge detection. Meas. Control Technol. **37**(12), 110–113 (2018)

8. Girshick, R., Donahue, J., Darrell, T., Malik, J.: Rich feature hierarchies for accurate object detection and semantic segmentation. In: Proceedings of the IEEE Conference on Computer Vision and Pattern Recognition, pp. 580–587, June 2014

9. He, K., Zhang, X., Ren, S., et al.: Spatial pyramid pooling in deep convolutional networks for visual recognition. IEEE Trans. Pattern Anal. Mach. Intell. **37**(9), 1904–16 (2014)

10. Girshick, R.: Fast R-CNN. In: Proceedings of the IEEE International Conference on Computer Vision (ICCV), pp. 1440–1448, December 2015

11. Ren, S., He, K., Girshick, R., et al.: Faster R-CNN: towards real-time object detection with region proposal networks. In: Advances in Neural Information Processing Systems, pp. 91–99 (2015)

12. Redmon, J., Divvala, S., Girshick, R., et al.: You only look once: unified, real-time object detection. In: Proceedings of the IEEE Conference on Computer Vision and Pattern Recognition, pp. 779–788 (2015)

13. Liu, W., et al.: SSD: single shot MultiBox detector. In: Leibe, B., Matas, J., Sebe, N., Welling, M. (eds.) ECCV 2016. LNCS, vol. 9905, pp. 21–37. Springer, Cham (2016). https://doi.org/10.1007/978-3-319-46448-0_2

14. Redmon, J., Farhadi, A.: YOLO9000: better, faster, stronger. In: Proceedings of the IEEE Conference on Computer Vision and Pattern Recognition, pp. 7263–7271 (2017)

15. Redmon, J., Farhadi, A.: YOLOv3: an incremental improvement. arXiv preprint arXiv:1804.02767 (2018)

16. Lin, T.Y., Goyal, P., Girshick, R., et al.: Focal loss for dense object detection. IEEE Trans. Pattern Anal. Mach. Intell. **(99)**, 2999–3007 (2017)

17. Woo, S., Park, J., Lee, J., Kweon, I.S.: CBAM: convolutional block attention module. In: Proceedings of the European Conference on Computer Vision (ECCV), pp. 3–19 (2018)

18. Jie, H., Li, S., Gang, S.: Squeeze-and-excitation networks. In: 2018 IEEE/CVF Conference on Computer Vision and Pattern Recognition (CVPR). IEEE (2018)

19. Neubeck, A., Gool, L.: Efficient non-maximum suppression. In: International Conference on Pattern Recognition (2006)

Design of General Aircraft Health Management System

Xiaoming Xie[1], Tengfei Zhang[2(✉)], Qingyu Zhu[3], and Guigang Zhang[2]

[1] Harbin Aircraft Industry Group Co., LTD, Harbin, China
[2] Institute of Automation, Chinese Academy of Sciences, Beijing, China
`tenfei.zhang@ia.ac.cn`
[3] China Aero-polytechnology Establishment, Beijing, China

Abstract. This paper studies the basic architecture of general aircraft health management, and analyzes the key technologies of general aircraft health management system, including data management technology and database technology, fault diagnosis technology, health assessment technology. Finally, a general aircraft health management software is developed and verified.

Keywords: General aircraft · Health management · Fault diagnosis · Health assessment

1   Introduction

General aircraft health management technology refers to the use of the obtained general aircraft state parameters, with the help of various traditional and modern mathematical methods to evaluate the health state of general aircraft and its subsystems, predict the development trend of performance state and possible faults, provide scientific and appropriate troubleshooting and maintenance suggestions, and predict the remaining service life [1–6]. The research on health management technology of general aircraft needs to pay attention to the consistency with international standards. According to the OSA CBM international standard, the functional hierarchy of the aircraft health management system is divided into the following seven parts, as shown in Fig. 1.

The capabilities of general aircraft health management include fault detection, fault isolation and performance monitoring; Prediction ability of key systems and components; Remaining service life prediction capability; Fault selective reporting capability; Ability to assist decision making and resource management; Performance degradation trend tracking capability; Component life tracking capability. From the perspective of signal processing, health management system can be divided into the following six levels of information structure:

The signal processing layer receives and processes data information from sensors and control systems and converts them into information needed by health management system.

© Springer Nature Switzerland AG 2021
C. Xing et al. (Eds.): WISA 2021, LNCS 12999, pp. 659–667, 2021.
https://doi.org/10.1007/978-3-030-87571-8_57

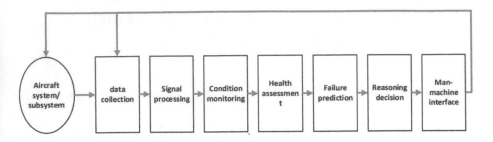

Fig. 1. Aircraft health management OSA-CBM international standard structure diagram.

The state monitoring layer receives data information from each sensor and control system after signal processing, detects and feeds back the results of the subsystem and its components according to the received information, and outputs the status and performance information data about the subsystem and its components [5–8].

The health assessment layer forecasts and evaluates the health status of the designated components and subsystems according to the system status information data. The status information data of the system may be directly from the sensor, or from the information after the status monitoring or other health assessment components [9–12].

The prediction layer comprehensively uses the data information of the first three layers and the given standard configuration status information to evaluate the current and future health status of components and subsystems, including the residual life prediction, etc. The standard configuration status information may be the default or generated by the estimation of the operating cycle.

This part of decision support layer includes the operation and support system. Comprehensive analysis of data information from the status monitoring layer, health assessment layer and prediction layer, evaluation and planning of the operation capability of the system, subsystem and components, management of maintenance resources (such as measures and suggestions for generating component replacement and system maintenance), and monitoring the effectiveness and performance of the system in the process [13–16].

Display and interface layer is the interface between health management system and user, which can be connected by handheld devices, computers, etc. Through the display interface, the system can be directly maintained and managed. It also includes interfaces for signal exchange and transmission with other control systems or health management systems.

2   Key Technologies of General Aircraft Health Management

General aircraft health management not only involves many disciplines, but also includes many technologies, such as sensor technology, data management technology, various modeling methods, complex algorithms, information fusion technology, prediction technology, system integration, etc. This research introduces some key technologies of general aircraft health management

2.1 Data Management Technology and Database Technology

Data management is the basis of general aircraft health management. The objects of its health management may come from different companies and different types of aircraft. The logic structure and data types of general aircraft health management system are various. Diagnosis and prediction need a lot of data, and the requirements of real-time and safety of data are relatively high. In order to improve the efficiency and universality of general aircraft health management system and realize cross platform application, it is very important to unify the management of general aircraft data. Therefore, database technology is the core of data cross platform application and unified management. Data from different platforms will be converted into a unified format through data conversion, and stored in a unified database from different working areas. In this scheme, the data is stored in HBase, which is a distributed, column oriented open source database and suitable for unstructured data storage.

2.2 Fault Diagnosis Technology

After data cleaning and preprocessing, the intelligent learning and analysis of fault diagnosis model is started, which is divided into the following stages.

Data dimension increasing stage: investigate the characteristics, advantages and disadvantages of potential algorithms that can be used for data dimension increasing, analyze and compare, and select the appropriate algorithm for general aircraft fault diagnosis data dimension increasing. Aiming at the parameters of general aircraft and the faults or abnormal conditions that need to be monitored, the algorithm that is more suitable for monitoring such faults or abnormal conditions is developed, and the algorithm is packaged and tested. A new parameter monitoring algorithm model is proposed, and the model is tested comprehensively.

Baseline learning stage: the purpose of baseline learning is to extract the baseline model of general aircraft from the normal operation data. Due to individual differences and degradation in the operation process of general aircraft, multidimensional sensor data are scattered in multi-dimensional space even under the same working condition, The purpose of baseline learning is to find the normal configuration of relevant sensor parameters from many general aircraft data. The configuration should be corresponding to the working conditions, and the normal baseline of each working condition can be obtained finally. The baseline is interpretable, and it is the clustering center and boundary of the normal state of each working condition.

Clustering analysis: the initial general aircraft data often have no fault tag, so it needs to adopt unsupervised learning algorithm, that is clustering algorithm to analyze the data. Firstly, the data is divided into a single working condition data set according to the working condition discrimination standard, and then the data is clustered by clustering analysis algorithm. Finally, the data sets containing faults in each working condition are automatically clustered into several classes, and then each class is marked according to the expert experience, including normal class and various fault mode classes.

Classification algorithm training: according to the data marked by clustering analysis steps, the supervised classification learning algorithm is used to carry out classification model learning. This study intends to use multi-dimensional data fusion, integrated learning strategy, and ultimately select the optimal sensor data combination and algorithm parameter settings. The construction of data set is very important, and the selection of different dimension subsets can determine which sensor data combination is the most effective for classification. In addition, the data set is divided into training sample set, test sample set and verification sample set. The training sample set and test sample set are used to train and adjust the algorithm parameters, and the verification sample set is used to evaluate the generalization ability of the model [17,18].

Finally, after off-line learning of a large number of historical data, the data-driven model library of fault diagnosis is formed, that is, all kinds of information and trained classification algorithm model. Then real-time fault diagnosis can be carried out.

In the real-time fault diagnosis stage, the real-time data also need to be cleaned and preprocessed to form a multi-dimensional data set for subsequent analysis. Firstly, anomaly detection is carried out. Firstly, the corresponding working conditions are identified, and then compared with the baseline of the corresponding working conditions. When the baseline boundary is exceeded, the abnormal alarm is made and the fault diagnosis process is triggered. Fault diagnosis can be realized based on fault clustering similarity analysis, or by trained classification algorithm. When the real-time data goes beyond the historical knowledge model, new fault knowledge is generated, which triggers the model updating process.

2.3   Performance Evaluation Technology

a. Comprehensive evaluation method of rough set

The structure of general aircraft is complex and there are many monitoring parameters. The importance of each parameter in performance evaluation is unknown. Therefore, it is necessary to determine the weight of each parameter by special calculation method. The attribute weight discrimination method of rough set can be used to determine the comprehensive weight coefficient corresponding to each evaluation factor, and then the factor parameters are weighted, and finally the weight of each parameter in the performance evaluation is obtained. This method effectively overcomes the shortcomings of the traditional method of partial subjectivity, makes the final parameter weight results more objective, and improves the effectiveness and accuracy of the system comprehensive evaluation.

b. Analytic hierarchy process evaluation technology

General aircraft health assessment is a multi-objective decision-making problem, which needs the comprehensive assessment method in system engineering theory.

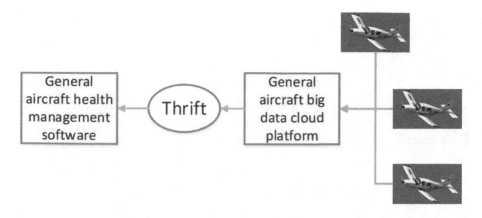

Fig. 2. General aircraft health management data acquisition process

Analytic hierarchy process (AHP) is a simple, flexible, multi criteria and multi-objective hierarchical weight decision-making analysis method which combines qualitative and quantitative analysis. On the basis of in-depth analysis of the problem, a complex problem is stratified according to certain principles, and the hierarchical structure model is established according to the degree of interaction between the factors and the subordinate relationship. The system analysis problem is transformed into the ranking problem of the relative importance weights of the lowest level (such as indicator level) and the highest level (such as target level), and finally provides the basis for the decision-making scheme of the management system.

3   Software Platform

The operation data of general aircraft is stored in the HBase database of general aircraft big data cloud platform. The general aircraft health management software is developed based on QT. It accesses the operation data of general aircraft in the HBase database through thrift to manage the health of general aircraft. General aircraft health management data acquisition process as shown in Fig. 2.

The interface of general aircraft health management software is divided into system main interface, system menu, data receiving interface, network configuration interface, data reading interface, condition monitoring information management interface, fault diagnosis interface, health assessment interface, health management interface, etc.

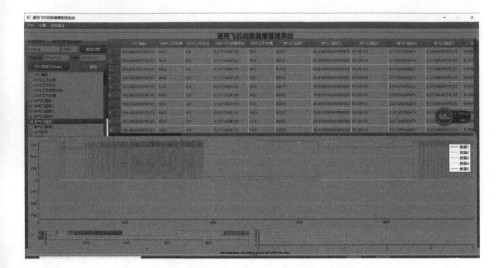

Fig. 3. Data loading and graphic display interface

3.1   Data Loading and Graphic Display

Data loading and graphic display interface are used to load and display data. Data flydata is flight control data and vibdata is vibration data. Parameters can be selected for visualization, and the software can display the time domain diagram and frequency domain diagram for vibration data. Data loading and graphic display interface as shown in Fig. 3.

3.2   Parameter Setting Interface

The main function of parameter setting interface is to set the condition monitoring threshold and time of each engine system and flight control system monitoring parameters. After setting the information, save the new information to the database. As shown in Fig. 4.

On the left side of the parameter setting interface is a tree control, which is used to display the important parameter names in the engine system and flight control system, in which the engine system includes (intake pressure, speed, fuel flow, cylinder head temperature, exhaust temperature, oil pressure, oil temperature, fuel pressure). The parameter setting interface can set the minimum, maximum and standard values of key parameters.

3.3   Health Management Interface

The health management interface of the software is shown in Fig. 5.

The software fault diagnosis module makes real-time diagnosis through the trained fault diagnosis model. The fault diagnosis results are displayed in the list

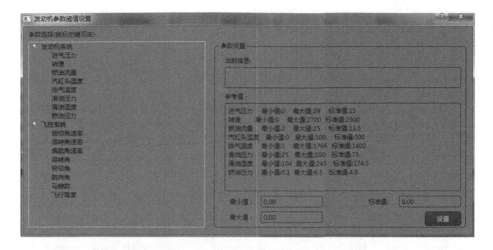

Fig. 4. Parameter setting interface

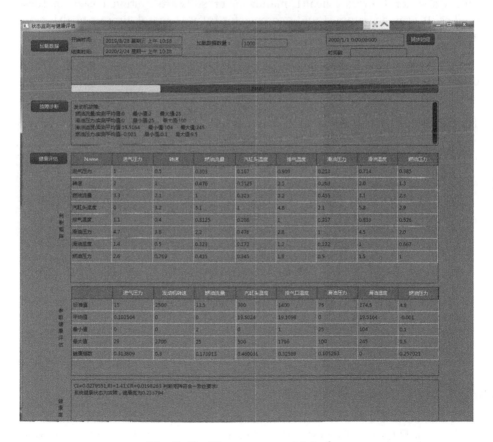

Fig. 5. Health management interface

box of fault diagnosis results. Generally, whether there is a fault or not is displayed. If there is a fault, the relevant information of the parameters causing the fault is listed. The health assessment function will display the judgment matrix of important parameters, the health assessment information of parameters and the final health level calculated according to the general aircraft operation parameters.

This paper studies the architecture of general aircraft health management system, analyzes the key technologies of general aircraft health management, and builds a general aircraft health management software system based on these studies. It provides support and reference for future general aircraft health management.

4   Conclusion

This paper studies the architecture of general aircraft health management system, analyzes the key technologies of general aircraft health management, and builds a general aircraft health management software system based on these studies. It provides support and reference for future general aircraft health management.

Acknowledgment. This work was supported by Aviation Science Funding (2020 00020M0002).

References

1. Ogaji, S.O.T., Singh, R.: Advanced engine diagnostics using artificial neural networks. Appl. Soft Comput. **3**(3), 259–271 (2003)
2. Wang, Z., Zarader, J.L., Argentieri, S.: A novel aircraft engine fault diagnostic and prognostic system based on SVM. In: International Conference on Condition Monitoring and Diagnosis, 2012, pp. 723–728. IEEE (2012)
3. Kobayashi, T., Simon, D.L.: Integration of on-line and off-line diagnostic algorithms for aircraft engine health management. NASA, TM-2007-214980 (2007)
4. Heider, R.: Improving the quality of technical data for developing case based reasoning diagnostic software for aircraft maintenance. In: Proceedings 13th International Conference on Data Engineering, 584–584 (1997)
5. Akaike, H.: Fitting autoregressive models for prediction. Ann. Inst. Stat. Math. **21**(1), 243–247 (1969). https://doi.org/10.1007/BF02532251
6. Said, S.E., Dickey, D.A.: Testing for unit roots in autoregressive-moving average models of unknown order. Biometrika **71**(3), 599–607 (1984)
7. Lütkepohl, H.: Vector autoregressive models. In: Lovric, M. (ed.) International Encyclopedia of Statistical Science, pp. 1645–1647. Springer, Heidelberg (2011). https://doi.org/10.1007/978-3-642-04898-2_609
8. Johansen S.: Estimation and hypothesis testing of cointegration vectors in Gaussian vector autoregressive models. Econometrica: J. Econ. Soc., 1551–1580 (1991)
9. Osborn, D.R.: Exact and approximate maximum likelihood estimators for vector moving average processes. J. R. Stat. Soc. Ser. B (Methodol.) **39**, 114–118 (1977)

10. Tsay, R.S.: Testing and modeling threshold autoregressive processes. J. Am. Stat. Assoc. **84**(405), 231–240 (1989)
11. Baillie, R.T., Bollerslev, T., Mikkelsen, H.O.: Fractionally integrated generalized autoregressive conditional heteroskedasticity. J. Econom. **74**(1), 3–30 (1996)
12. Koskela, T., Lehtokangas, M., Saarinen, J., et al.: Time series prediction with multilayer perceptron, FIR and Elman neural networks. In: Citeseer, pp. 491–496 (1996)
13. Gardner, M.W., Dorling, S.R.: Artificial neural networks (the multilayer perceptron)-a review of applications in the atmospheric sciences. Atmos. Environ. **32**(14–15), 2627–2636 (1998)
14. Leung, H., Lo, T., Wang, S.: Prediction of noisy chaotic time series using an optimal radial basis function neural network. IEEE Trans. Neural Netw. **12**(5), 1163–1172 (2001)
15. Zhang, J., Xiao, X.: Predicting chaotic time series using recurrent neural network. Chin. Phys. Lett. **17**(2), 88 (2000)
16. Han, M., Xi, J., Xu, S., et al.: Prediction of chaotic time series based on the recurrent predictor neural network. IEEE Trans. Sig. Process. **52**(12), 3409–3416 (2004)
17. Ren, G., Zhang, G., Wang, J.: Application of data mining in aeronautical information system. Comput. Digit. Eng. **48**(12), 2826–2829+2856 (2020)
18. Chen, H., Dong, Y., Gu, Q., Liu, Y.: An end-to-end deep neural network for truth discovery. In: Wang, G., Lin, X., Hendler, J., Song, W., Xu, Z., Liu, G. (eds.) WISA 2020. LNCS, vol. 12432, pp. 377–387. Springer, Cham (2020). https://doi.org/10.1007/978-3-030-60029-7_35

Graph-Encoder and Multi-decoders Solution Framework with Multi-attention

Hui Cai, Tiancheng Zhang$^{(\boxtimes)}$, Xianghui Sun, Minghe Yu, and Ge Yu

School of Computer Science and Engineering, Northeastern University,
Shenyang 110169, China
{tczhang,yuminghe,yuge}@mail.neu.edu.cn

Abstract. The goal of math word problem (MWP) is to design an automatic solution model to answer the mathematical questions given in the text. In recent years, many neural network models have achieved good results. However, most models do not consider the unique language features of mathematical problems in feature extraction, and ignore the structural information of mathematical problems and equations in the design of encoder and decoder. In order to solve these problems, we propose a graph-encoder and multi-decoders solution framework with multi-attention. On the one hand, we introduce the multi-attention mechanism in the encoding part to capture multiple features, and construct graph to express quantitative information. On the other hand, in the decoding part, we combine tree-structured and sequence-structured decoders. The experimental results on the Math23K dataset show that our model outperforms the existing state-of-the-art methods.

Keywords: Natural language processing · Math word problem · Automatic solution

1 Introduction

With the advent of the wave of education informatization, the traditional centralized school education is changing to online education, personalized education, and smart education. In this development process, how to effectively provide students with correct answers to questions is also a problem that must be solved. Compared with traditional question answering, machine answering is more intelligent and efficient.

In the field of machine answering, the automatic answering of primary school math application problems has always been one of the key research problems. This is very challenging and requires accurate natural language understanding to connect natural language text and mathematical expressions. Table 1 shows

This work is supported by National Natural Science Foundation under Grant (Nos. U1811261, 61902055), China Postdoctoral Science Foundation funded project (2019 M651134), the Fundamental Research Funds for the Central Universities (N180716010, N2117001).

C. Xing et al. (Eds.): WISA 2021, LNCS 12999, pp. 668–679, 2021.
https://doi.org/10.1007/978-3-030-87571-8_58

an example of math word problem. The input is a text description of the mathematical problem. Our goal is to extract the relevant quantity, map the problem into an expression, and return the final solution result.

Table 1. A math word problem.

Problem: From A to B, if you ride a bicycle 16 km per hour, you can reach it in 4 h. If it only takes 2 h by car, how many kilometers per hour the car travels?
Solution Expression: $x = 16 \times 4 \div 2$
Answer: 32

The research on math word problem has a long history [1], and with the continuous development of natural language processing technology, this problem has attracted the attention of more and more researchers. Early statistical learning methods [2–4] extract templates and features from the problem, and generate corresponding expressions based on these templates and features. This method has two main disadvantages: First, it requires additional annotation overhead, which makes it cannot handle large-scale data sets; second, these methods are essentially based on a set of predefined templates, which are weak in scalability and methods are not robust under diversified data. In recent years, many end-to-end models [5] have been developed to directly generate mathematical expressions from the question text. These models have the ability to generate new expressions that do not exist in the training set, such as [6–8].

Based on the previous research on the math word problems, and considering the characteristics of MWP, we propose a solution framework based on graph-encoder and multi-decoders, with a multi-attention mechanism. On the one hand, using the deep topological information of the graph structure, the dependency analysis information and numerical comparison information are integrated into the model in the form of graphs, which enriches the quantitative representation; On the other hand, considering the diversity of equation expressions and the limitations of a single decoder, we designed an effective multi-decoders architecture to improve the output accuracy. In addition, in order to better represent the global information of the problem, we adopt the multi-attention mechanism, which can enhance the performance of the model by acquiring various types of MWP features. The main contributions of this paper are summarized as follows:

(1) We propose a solution framework in which the encoder incorporates graph structure information, and the decoding part combines multiple types of decoders.
(2) We introduce the multi-attention mechanism in the model, which can not only capture the characteristics related to the quantity and the problem, but also consider the global information. It improves the performance of the model.

(3) Our experiments on the dataset Math23K have proved the effectiveness of our model. Compared with the existing state-of-the-art method, our model has improved accuracy by 1.2%.

2 Related Work

The existing solutions for math word problems can be divided into three categories: rule-based, statistic-based and deep learning-based methods [1]. Early rule-based methods and statistical-based methods are not flexible enough, such as [9,10], which requires a huge amount of manpower to design suitable features and expression templates. At present, the automatic solution technology of math word problems based on deep learning is becoming the mainstream research direction. Because it fits the law of text characteristics through learning [11], it improves the accuracy of the solution through continuous research on text representation and construction of more effective models. This method is more flexible and has a strong generalization ability.

Wang et al. [6] first tried to apply deep learning technology to math word problem solving, using the seq2seq model to map the problem text into mathematical expressions, and the encoding and decoding sides were recurrent neural networks (RNN). Robaidek B et al. [12] explored several data-driven technologies to solve MWP problems, including retrieval, classification, and generative models, and evaluated the seq2seq model with LSTM and CNN as the encoder and decoder, respectively. It turns out that a well-tuned classifier is better than the generation and retrieval model in performance, and semantic understanding and common sense knowledge are necessary for the success of the solution. Considering the uniqueness of expression trees, Wang et al. [7] proposed an equation normalization method to standardize repeated equations, and proposed model integration techniques for three mainstream seq2seq models to improve solution performance. Wang et al. [13] first tried to use deep reinforcement learning to solve math word problems, and proposed a DQN-based solver. The author designed the important components of DQN: state, action, reward function, and a two-layer forward neural network to approximate the Q value function. Wang et al. [8] proposed a template-based solver. The author first apply the seq2seq model to predict a tree-structured template, with the number as leaf nodes and internal nodes as unknown operators, and then use the designed recurrent neural network to infer internal operators. Chiang et al. [14] used BiLSTM on the encoding side to encode the semantics of the problem, and used stacks on the decoding side to generate correlation equations. Liu et al. [15] proposed a tree structure decoding method to generate an abstract tree of expressions in a top-down manner. Xie et al. [16] proposed a tree-structured neural model to generate expression trees in a goal-driven manner. The model first identifies and encodes its goal to achieve, and then the goal gets decomposed into sub-goals combined by an operator in a top-down recursive way. The whole process is repeated until the goal is simple enough to be realized by a known quantity as leaf node. However, this tree-based solution method cannot capture the sequential relationship between quantities. Considering that graphs are intuitive mathematical expressions of objects and relationships between objects, Zhang et al. [17]

proposed a novel Graph-tree architecture. This model combines the advantages of graph-based encoders and tree-based decoders, and can effectively express the order relationship between quantities. It is also the most advanced model currently.

3 Model

The overall framework of our model is shown in Fig. 1. On the encoding side, every word in the question text is encoded as a contextual representation. We first use a bidirectional long short-term memory (LSTM) network [18] to extract the problem representation (Sect. 3.1.1). At the same time, we construct the parse graph and comparison graph (Sect. 3.1.2). Then, we use the multi-attention mechanism to consider different types of features of questions, which are global features, quantity-related features, quantity-pair features, and question-related features (Sect. 3.1.3). Then we use graph convolution networks (GCNs) [19] to perform graph learning on the constructed graph (Sect. 3.1.4). On the decoding side, it aims to build an equation that can solve the given problem. We combine tree-structured decoder (Sect. 3.2.1) and sequence-structured decoder (Sect. 3.2.2) to select the final generation result according to the generation probability of different decoders.

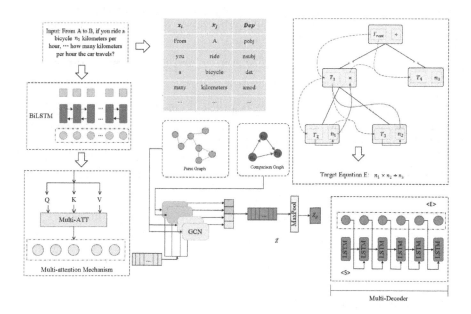

Fig. 1. Overall solution framework.

3.1 Encoder

3.1.1 Problem Representation Extraction

In order to capture the representation of each word, we use the BiLSTM neural network to learn the hidden state sequences of the input text. For the given problem sequence $P = \{x_1, x_2, \cdots, x_n\}$, we encode each word x_i into a vector h_i^p:

$$h_i^p = \text{BiLSTM}(e(x_i), h_{i-1}^p) \in R^{n \times 2d} \tag{1}$$

where $e(x_i)$ is the word embedding for word x_i. n and d are the size of the input sequence and the dimension of hidden vectors. We define the final vector representation h_i^p as the concatenation of the forward and backward hidden states:

$$h_i^p = \left[\overrightarrow{h_i^p} : \overleftarrow{h_i^p} \right] \tag{2}$$

where $\overrightarrow{h_i^p}$ and $\overleftarrow{h_i^p}$ are the BiLSTM hidden states in the forward and backward order, respectively. : represents the connection operation.

3.1.2 Graph Construction

In this part, we construct the parse graph and comparison graph, which helps to find the correct operator between the two quantities. By implementing dependency parsing [20], we can easily get the dependency relationship between word pairs in the text. Therefore, we define the parse graph to be constructed as follows:

- **Parse Graph (G_p):** For word pairs $x_i, x_j \in P$, there is an edge $e_{ij} = (x_i, x_j)$, if the pair has dependency relationship.

Negative numbers are rare in MWP and will affect the generation of correct answers. Therefore, in order to prevent this situation, the comparison information of the quantity also plays an important role in the model. Many models often ignore this situation, resulting in insufficient accuracy of the model. We constrain small numbers to subtract large numbers by constructing a comparison graph, which is defined as follows:

- **Comparison Graph (G_c):** First, we denote a series of quantities in the problem as $n_p = \{n_1, n_2, \cdots, n_l\}$. For two quantities $n_i, n_j \in n_p$, a directed edge $e_{ij} = (n_i, n_j)$ pointing from n_i to n_j will be added to the comparison graph if $n_i > n_j$.

3.1.3 Multi-attention Mechanism

Inspired by GROUP-ATT [21], we introduce a multi-attention mechanism including four different features in the encoding process to capture more text features. First, we separate the text into quantity spans and question spans by commas and periods, then we pass H^p through three different linear layers to get Q, K, V. According to the previous division, we feed Q, K, V to multi-attention, which mainly includes the following four types of attention:

1. **Global Attention** (O_g): O_g is calculated based on the entire input sequence by a self-attention computation module (SDPA) [22]. Q_g, K_g and V_g are set to H^p.
2. **Quantity-Related Attention** (O_c): Q_c is calculated based on the current quantity span by SDPA. In other words, Q_c, K_c and V_c are all derived from the quantity span where the current number is located, and have nothing to do with other spans.
3. **Quantity-Pair Attention** (O_p): The calculation of O_p consists of two parts: 1) Attention for quantity span: Q_p is derived from the i-th quantity span, and its corresponding K_p, V_q are derived from the j-th quantity span($i \neq j$). 2) Attention between quantities and question: Q_p is generated by question span, and its corresponding K_p, V_p are generated from quantity span.
4. **Question-Related Attention** (O_q): The calculation of O_q also includes two parts: 1) Attention from quantity span: Q_q is originated from quantity span, and its corresponding K_q, V_q are from question span. 2) Attention for question span: for Q_q related to the question span, its corresponding K_q, V_q are extracted from related quantity span.

Finally, we concatenate and project$\{O_g, O_c, O_p, O_q\}$to get the output of the multi-attention O:

$$O = \text{Concat}\left[O_g, O_c, O_p, O_q\right] \tag{3}$$

3.1.4   Graph Learning

We utilizes graph convolution networks (GCNs) to learn node features. For the constructed multiple graphs, we set up K-head graph convolution. For the constructed multiple graphs, we set up K-head graph convolution, which is to use K convolution networks separately and finally connect them. More specifically, a single GCN has its parameter $W_{gk} \in R^{d \times d_k}$, where $d_k = d/K$.

The inputs are the multiple graphs $\{G_m\}_{m=1}^{M}, G_m \in \{G_p, G_c\}$ and the feature matrix F(in the beginning, F is set as O), where M is the number of graphs and G_m is the m-th graph, and then we define learning of GCN as follow:

$$\text{GCN}(G_m, F) = \text{GConv}_2(G_m, \text{GConv}_1(G_m, F)) \tag{4}$$

Here, GCN contains two different graph convolution operations:

$$\text{GConv}(G_m, F) = \text{ReLU}(G_m F^T W_{gk}) \tag{5}$$

We implement parallel learning for GCN, and generate d_k-dimensional output for each GCN. The output values are concatenated, resulting in the final values:

$$\begin{aligned} \overline{Z} &= \overset{M}{\underset{m}{||}} \text{GCN}(G_m, O) \\ Z_g &= \text{MaxPool}(\overline{Z}) \end{aligned} \tag{6}$$

Here, $||$ represents the connection operation of K-head GCN. $\overline{Z}$ denotes the final encoder vectors of each word, and Z_g denotes the global vector of text descriptions for next decoding.

3.2   Decoder

3.2.1   Tree-Structured Decoder

Following the idea of GTS [16], we introduce a tree-based decoder. GTS uses the pre-order traversal manner to generate the expression tree, and the decoding process can be expressed as:

- **Step 1:** We first use Z_g to initialize the root node vector T_{root}, and then use the attention component of GTS to get the context vector C:

$$C = \text{GTS-Attention}(T_{root}, \overline{Z}) \tag{7}$$

- **Step 2:** Following the top-down manner, the decoder generates left child node T_l based on the parent node T_p and the context vector C, and the token $\hat{y}$ is predicted:

$$\begin{aligned} T_l &= \text{GTS-Left}(T_p, C) \\ \hat{y}_l &= \text{GTS-Predict}(T_l, C) \end{aligned} \tag{8}$$

If the token $\hat{y}$ is an operator, then step 2 will continue. Until the generated $\hat{y}$ is a number, go to step 3.

- **Step 3:** In the process of generating the right child node and the corresponding token $\hat{y}_r$, we utilize the context vector C, the left child node T_l and a sub-tree embedding t_l as the input:

$$\begin{aligned} T_r &= \text{GTS-Right}(C, T_l, t_l) \\ t_l &= \text{GTS-subtree}(\hat{y}_l, T_l) \\ \hat{y}_r &= \text{GTS-Predict}(T_r, C) \end{aligned} \tag{9}$$

where t_l is computed by the sub-tree embedding component of GTS. And if the token $\hat{y}_r$ is an operator, go back to step 2. If $\hat{y}_r$ is a number, we start step 4.

- **Step 4:** In this step, we implement backtracking to confirm whether there is an empty right child node, if not, complete the generation process, otherwise, return to step 2.

3.2.2   Sequence-Structured Decoder

The sequence-structured decoder will be used to generate the suffix expression. More specifically, we decode the vector $\overline{Z}$ by LSTM and an attention layer, which can be expressed as:

$$\begin{aligned} s_i &= \text{LSTM}(s_{i-1}, y_{i-1}, c_i) \\ c_i &= \sum_{j=1}^{n} a_{ij} \overline{z}_j \\ a_{ij} &= \frac{\exp(e_{ij})}{\sum_{k=1}^{n} \exp(e_{ik})} \\ e_{ij} &= v_s^T \cdot \tanh(W_s \left[s_{i-1}, \overline{z}_j \right]) \end{aligned} \tag{10}$$

where s_i is the hidden state vector of the decoder, c_i is the context vector, a_{ij} determines the attention distribution of the i-th output on the j-th input. v_s and W_s are parameter matrices.

3.3   Model Training

For each problem-answer (P, T), the loss function L is defined as a sum of the negative log-likelihoods of probabilities for token y_i, which combine the sequence-structured decoder and tree-structured decoder. The form is as follows:

$$
\begin{aligned}
L &= -\frac{1}{E} \sum_{i=1}^{E} (\log p(y_i | s_i, c_i, T_s) + \log p(y_i | T_i, C_i, T_t)) \\
p(y_i | s_i, c_i, T_s) &= \mathrm{softmax}(W_1 \cdot \tanh(W_2 \cdot [s_i; c_i])) \\
p(y_i | T_i, C_i, T_t) &= \mathrm{softmax}(W_3 \cdot \tanh(W_4 \cdot [T_i; C_i]))
\end{aligned}
\tag{11}
$$

where E is the number of tokens. $\{W_1, W_2, W_3, W_4\}$ are parameter matrices. T_s and T_t respectively denote the output expressions of the sequence-structured decoder and tree-structured decoder.

4   Experiments

4.1   Dataset

We experiment on the Math23K dataset [6], which includes 23162 problems. It is a standard data set for MWPs. We randomly divide the dataset into a training set (90%) and a test set (10%).

4.2   Baselines and Evaluation Metric

We compare with some mainstream baselines and state-of-the-art models: **Math-EN** [7] propose equation normalization to solve the problem of expression diversity. **T-RNN** [8] propose a template-based solvers with recursive neural networks. **GROUP-ATT** [21] introduce multi-head attentions in the model. **GTS** [16] propose a neural model for math word problems by directly predicting an expression tree. **Graph2Tree** [17] propose the model to improve task performance by enriching the quantity representations.

Following previous works, we use solution accuracy as the evaluation metric.

4.3   Implementation Details

We use the pre-trained word embedding with 128 units, a two-layer BiLSTM with 512 hidden units, a one layer graph encoder with 4 GCNs and a group attention with four different functional 2-head attention. We set the number of epochs and batch size as 80 and 64 respectively. The dropout rate for word embedding is set to 0.3 and for other layer is 0.5. Our experimental equipment is NVIDIA GTS 1060 and our experiment is implemented in PyTorch 1.1.0.

Table 2. Comparison of solution accuracy.

Model	Accuracy
Math-EN	66.5%
T-RNN	66.8%
GROUP-ATT	69.5%
GTS	74.9%
Graph2Tree	76.2%
Ours	**77.4%**

Table 3. Accuracy of equation and answer.

Model	Equation-Accuracy	Answer-Accuracy
GTS	62.5%	74.9%
Graph2Tree	66.9%	76.2%
Ours	**67.5%**	**77.4%**

4.4   Overall Results

Table 2 shows the accuracy comparison of different models. We can see that our model performs better than all baselines on Math23K. In addition, the accuracy of Math-EN, T-RNN and GROUP-ATT are all lower than 70%, mainly because they ignore the structural information of text and expressions, which leads to insufficient accuracy of text representation and expression generation. Compared with GTS and state-of-the-art model Graph2Tree, our model also shows better performance, increasing the accuracy by 2.5% and 1.2% respectively. This is mainly because the design of our multi-attention mechanism and multi-decoders structure play a role.

Table 3 shows the equation accuracy and answer accuracy of GTS, Graph2Tree and our model. It can be clearly seen that equation accuracy is significantly lower than answer accuracy. We think this is mainly due to the fact that one answer corresponds to multiple expression forms. In addition, the equation accuracy of our model is also higher than the other two models, which also reflects the effectiveness of the multi-decoder design in the model.

4.5   Experimental Analysis

Figure 2 shows the evolution of the model accuracy with epoch. It can be seen that the accuracy increases as the training epoch increases. And for the initial stage of training, the accuracy increases quickly. After that, the curve gradually becomes flat, and the accuracy remain basically stable. The accuracy of our model is higher than the other two models in most training epochs.

We count the distribution of equations of different lengths, as shown in Table 4. Figure 3 shows the accuracy of the three models on different length

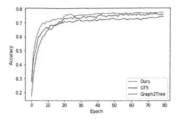

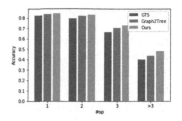

Fig. 2. The evolution of accuracy with epoch.

Fig. 3. Accuracy of different length equations.

Table 4. Distribution of different length equations.

Operator	Test numbers	Proportion
1	486	21.0%
2	1112	48.0%
3	470	20.3%
>3	239	10.3%

equations. It can also be seen that our model performs better than other models in these situations, especially when the number of operators exceeds three. And it is obvious that when the number of operators is greater than three, the accuracy of the three models has a significant drop. This is because longer expressions are usually related to the complexity of the problem. Solving this more complex problem usually requires more auxiliary information, so our model is limited.

4.6 Case Study

Table 5 shows two examples of our model compared with GTS and Graph2Tree. In case 1, the order of GTS operations is wrong, and our model has a correct expression due to the introduction of a quantity comparison graph to constrain the quantity relationship. In case 2, Graph2Tree don't model the multi-features of the text, and make an error when solving for the speed of the truck. Our model, due to the introduction of multi-attention,s model the relationship between the quantity "90" and the quantity "4", and finally get the correct expression.

Table 5. Two examples of our model compared with GTS and Graph2Tree.

Case 1: A train travels 495 km in 4.5 h, and someone rides a bicycle 37.5 km in 3 h. How many times the speed of this train is the bicycle speed?
GTS: 37.5/3/(495/4.5)   **Ours:** 495/4.5/(37.5/3)
Case 2: A passenger train departs from Station A at a speed of 90 km per hour and can reach Station B in 4 h, and a train of trucks departs from Station B and arrives at Station A in 6 h. If two cars depart from Station A and Station B at the same time and go towards each other, a few hours later, the two cars will meet?
Graph2Tree: (90 * 6 + 90) * 4/90   **Ours:** 90 * 4/(90 * 4/6 + 90)

5   Conclusion and Future Work

In this research, we propose a graph-encoder and multi-decoders framework with multi-attention to automatically solve math word problems. We integrate the dependency relationship and quantity comparison information into the model through graph construction, which makes the quantity representation more accurate. Based on the unique language characteristics of math problems, we introduce the multi-attention mechanism according to different feature types, which plays a vital role in bridging the semantic gap. In order to solve the problem of expression diversity, we propose multi-decoders component, which to a certain extent enhance the generation ability of the model. Finally, experiments on the dataset Math23K also proved the effectiveness of our model.

In future work, we will consider the diversity of data sets and strive to solve mathematical problems with longer expressions.

References

1. Zhang, D., Wang, L., Zhang, L., Dai, B.T., Shen, H.T.: The gap of semantic parsing: a survey on automatic math word problem solvers. IEEE Trans. Pattern Anal. Mach. Intell. **42**(9), 2287–2305 (2019)
2. Bakman, Y.: Robust understanding of word problems with extraneous information. arXiv preprint arXiv:math/0701393 (2007)
3. Roy, S., Vieira, T., Roth, D.: Reasoning about quantities in natural language. Trans. Assoc. Comput. Linguist. **3**, 1–13 (2015)
4. Mitra, A., Baral, C.: Learning to use formulas to solve simple arithmetic problems. In: Proceedings of the 54th Annual Meeting of the Association for Computational Linguistics (Volume 1: Long Papers), pp. 2144–2153 (2016)
5. Zou, Y., Lu, W.: Text2Math: end-to-end parsing text into math expressions. arXiv preprint arXiv:1910.06571 (2019)
6. Wang, Y., Liu, X., Shi, S.: Deep neural solver for math word problems. In: Proceedings of the 2017 Conference on Empirical Methods in Natural Language Processing, pp. 845–854 (2017)

7. Wang, L., Wang, Y., Cai, D., Zhang, D., Liu, X.: Translating a math word problem to an expression tree. arXiv preprint arXiv:1811.05632 (2018)
8. Wang, L., et al.: Template-based math word problem solvers with recursive neural networks. In: Proceedings of the AAAI Conference on Artificial Intelligence, vol. 33, pp. 7144–7151 (2019)
9. Mukherjee, A., Garain, U.: A review of methods for automatic understanding of natural language mathematical problems. Artif. Intell. Rev. **29**(2), 93–122 (2008). https://doi.org/10.1007/s10462-009-9110-0
10. Liang, C.C., Hsu, K.Y., Huang, C.T., Li, C.M., Miao, S.Y., Su, K.Y.: A tag-based English math word problem solver with understanding, reasoning and explanation. In: Proceedings of the 2016 Conference of the North American Chapter of the Association for Computational Linguistics: Demonstrations, pp. 67–71 (2016)
11. Guo, C., Xie, L., Liu, G., Wang, X.: A text representation model based on convolutional neural network and variational auto encoder. In: Wang, G., Lin, X., Hendler, J., Song, W., Xu, Z., Liu, G. (eds.) WISA 2020. LNCS, vol. 12432, pp. 225–235. Springer, Cham (2020). https://doi.org/10.1007/978-3-030-60029-7_21
12. Robaidek, B., Koncel-Kedziorski, R., Hajishirzi, H.: Data-driven methods for solving algebra word problems. arXiv preprint arXiv:1804.10718 (2018)
13. Wang, L., Zhang, D., Gao, L., Song, J., Guo, L., Shen, H.T.: MathDQN: solving arithmetic word problems via deep reinforcement learning. In: Proceedings of the AAAI Conference on Artificial Intelligence, vol. 32 (2018)
14. Chiang, T.R., Chen, Y.N.: Semantically-aligned equation generation for solving and reasoning math word problems. arXiv preprint arXiv:1811.00720 (2018)
15. Liu, Q., Guan, W., Li, S., Kawahara, D.: Tree-structured decoding for solving math word problems. In: Proceedings of the 2019 Conference on Empirical Methods in Natural Language Processing and the 9th International Joint Conference on Natural Language Processing (EMNLP-IJCNLP), pp. 2370–2379 (2019)
16. Xie, Z., Sun, S.: A goal-driven tree-structured neural model for math word problems. In: IJCAI, pp. 5299–5305 (2019)
17. Zhang, J., et al.: Graph-to-tree learning for solving math word problems. Association for Computational Linguistics (2020)
18. Hochreiter, S., Schmidhuber, J.: Long short-term memory. Neural Comput. **9**(8), 1735–1780 (1997)
19. Kipf, T.N., Welling, M.: Semi-supervised classification with graph convolutional networks. arXiv preprint arXiv:1609.02907 (2016)
20. Manning, C.D., Surdeanu, M., Bauer, J., Finkel, J.R., Bethard, S., McClosky, D.: The Stanford coreNLP natural language processing toolkit. In: Proceedings of 52nd Annual Meeting of the Association for Computational Linguistics: System Demonstrations, pp. 55–60 (2014)
21. Li, J., Wang, L., Zhang, J., Wang, Y., Dai, B.T., Zhang, D.: Modeling intra-relation in math word problems with different functional multi-head attentions. In: Proceedings of the 57th Annual Meeting of the Association for Computational Linguistics, pp. 6162–6167 (2019)
22. Vaswani, A., et al.: Attention is all you need. arXiv preprint arXiv:1706.03762 (2017)

Heterogeneous Embeddings for Relational Data Integration Tasks

Xuehui Li[1], Guangqi Wang[2], Derong Shen[1(✉)], Tiezheng Nie[1], and Yue Kou[1]

[1] Northeastern University, Shenyang 110004, China
{shenderong,nietiezheng,kouyue}@cse.neu.edu.cn
[2] Liaoning Provincial Higher and Secondary Education Enrollment Examination Committee Office, Shenyang 110031, China

Abstract. Data integration technology can integrate data from different data sources, making it convenient and prompt to use heterogeneous data when processing big data. Therefore, data integration plays an important role in many industries. Recently, more and more work is devoted to data integration for relational data aiming at mining the underlying knowledge from it. Through embedding technology, the features of data can be extracted and expressed in the low-dimensional vectors. Some existing methods took records, attributes and cell values in relational data as various research objects to calculate their embedding representations, but the three types of data objects were trained uniformly in these methods ignoring the differences between multiple types of data. In this paper, we transform the relational data into a heterogeneous graph where different levels of data are treated as different types of nodes. In the training process, different calculation methods are adopted for corresponding node types according to their own characteristics, so that to obtain more accurate embedding representations for data. Then the embeddings are applied to the specific tasks of data integration. The experimental results show that the data embeddings trained by proposed model have good universality and achieve satisfying results in both schema matching and entity resolution tasks.

Keywords: Relational data · Heterogeneous graph · Embedding technology · Data integration

1 Introduction

In the current information age, the amount of data is growing explosively. Various industries obtain useful information through data collection, integration and analysis, and then regard it as reference knowledge for industry development. But data from multiple sources are heterogeneous and inconsistent, there are also more and more useless noise data which makes it difficult to find valuable and usable information. Data integration technology can integrate information from heterogeneous data sources and effectively enhance the value of data. However, data integration is expensive and requires many data scientists to integrate and manage data, including defining rules, extracting features, labeling data, etc. At the same time, extensive problems in datasets, such as data

© Springer Nature Switzerland AG 2021
C. Xing et al. (Eds.): WISA 2021, LNCS 12999, pp. 680–692, 2021.
https://doi.org/10.1007/978-3-030-87571-8_59

spelling errors, missing data and semantic heterogeneity, also restrict the performance improvement of existing data integration technologies.

In recent years, with the development of deep learning, existing work has begun to introduce deep learning technology into data integration, which has effectively improved the performance of data integration by combining the deep perception capabilities and anti-noise characteristics of deep learning. In the existing models based on deep learning, the early methods [2, 3] regarded a row or column in the table as a piece of text data, and directly embedded the word. However, the order of rows or columns does not indicate the semantic relationship, so these methods are not easy to extract valuable information. Later, some methods [15, 16] used neural networks such as RNN to learn the feature representations of the data. Using such models can better capture the hidden semantic and grammatical features, especially in text-type data. In order to more fully consider the possible association between data, some methods [13, 18, 19] of converting tables into graphs appeared. They regarded records, attributes, and cell values as nodes, and then trained the graph to extract the characteristics of the data by graph embedding methods. But the same training method is performed on multiple types of nodes, the heterogeneous graph does not exert its advantages.

Existing models for embedding relational data still have certain limitations. First, most models combined the representations of cell data as the representation of the row or column, instead of directly embedding for rows and columns. Second, in the existing methods of converting relational data into graphs and then embedding, different types of nodes were uniformly trained without considering the respective properties of them. Third, there are more researches on entity resolution, but less on schema matching now. Few universal models for these two tasks still have the above shortcomings.

Therefore, this paper proposes a heterogeneous embedding technology, the calculated vectors can be directly applied to multiple tasks of data integration with good performance. Our contributions can be summarized as follows:

1. A heterogeneous graph structure converted from relational data is proposed, which is constructed according to the correspondence in the table. The subgraphs are then divided based on the aggregation relationship between different types of nodes.
2. A training model is proposed based on the heterogeneous graph, and different methods are used to extract features for different types of nodes. The embedding representation for each type of node is trained through the model.
3. Through experiments, the universal embeddings trained by proposed model are applied to data integration tasks, which have better performance than existing models.

2 Related Work

2.1 Embedding Technology

There has been some work generating embedding representations for relational data. In the relational embedding methods, the relational data was regarded as textual data, and then the word embedding methods were used. Most methods used the Skip-gram model in Word2Vec [1] for training. [2] proposed the idea first, regarding a row of data as a sentence to capture the relationship between and within the attributes. Table2Vec [3] processed different table elements through neural language modeling, embedded table data into vector space and merged the semantic similarity into retrieval models.

Later, some models were proposed to convert relational data into graphs, and then graph embedding technology was used to extract data characteristics. The earliest graph embedding methods were calculated by matrix decomposition. The most classic method is locally linear embedding (LLE) [4]. Subsequent such methods were based on LLE for improvement, such as Laplacian eigenmap (LE) [5], graph decomposition (GF) [6]. There are also some methods that walk randomly in the graph and then introduce word embedding methods. DeepWalk [7] obtained the local context of the graph nodes by the truncated random walks, and reflected the local structure information of each node through a vector. The idea of Node2vec [8] is similar, but it combined DFS and BFS in random walk, considering the local and global features of the node comprehensively to improve embedding quality. In recent years, neural networks have also been introduced into the research of graph embedding. For example, SDNE [9] used a semi-supervised model to capture the graph through the proximity relationship of the nodes in the autoencoder, so as to obtain the vector representation. The GCN [10] has gradually been widely used in graph embedding. It can learn the high-order dependencies between nodes in the training model, which can get better training results.

2.2 Data Integration Tasks

There have been some works that apply the embeddings of relational data to data integration tasks, which mainly include two tasks: schema matching and entity resolution.

Schema matching is the process of capturing correspondence between attributes of different datasets. Traditional methods usually used regular expressions, Google similarity distance [11] and other methods to calculate the data similarity when performing feature extraction and comparative analysis, and then performed schema matching. However, these methods are difficult to successfully match in some complicated situations, so manual processing is still required. Recently, some models have introduced deep learning techniques. [12] combined deep learning and probability graph model to extract the semantic relationship features in the column. REMA [13] used the relational data as nodes to construct an undirected graph, performed random walks in the graph and trained the neural network to obtain the embeddings of nodes. Then, the average of all cell embeddings in a column was used as the column embedding. Finally, the cosine similarity between the column embeddings was calculated for schema matching.

Entity resolution is aimed to find records describing the same entity in the real world from multiple datasets. Magellan [14] applied machine learning methods and achieved good results. More recent work further used deep learning techniques. DeepER [15] used local sensitive hashing and aggregated the representations of words through RNN and LSTM networks. DeepMatcher [16] introduced the attention mechanism [17] additionally and used a bidirectional RNN to extract attribute features. The above methods are centered on attribute level, so manual schema alignment is also required when processing different data sources. As a token-centric method, GraphER [18] first converted the relational data into a graph, and each edge was weighted according to the data similarity. Data embedding was trained through GCN, and then the attention mechanism was introduced to cross-encode the data. However, nodes representing different types of data were trained together, losing the meaning of heterogeneity.

In addition, EMBDI [19] is a relational data embedding model with a certain generality, which can be applied to multiple tasks of data integration. EMBDI converted relational data into a graph including multiple types of nodes and walked randomly. The walk paths were treated as sentences, then word embedding technology was used to get the embedding of each node. But the model applied embeddings to two tasks, schema matching and entity resolution, respectively, without joint training.

Different from the existing work, our model considers the characteristics of different types of nodes in the heterogeneous graph, and divides subgraphs according to the aggregation relationship between nodes. Different calculation methods are used for different node types during training, and embeddings are generated for all types of node. On this basis, the embeddings of attribute nodes and record nodes can be directly applied to schema matching and entity resolution tasks respectively.

3 Method

3.1 Model Overview

In the model proposed in this paper, the relational data is constructed as a heterogeneous graph. Then different methods are used to train the embeddings of different types of nodes. Finally, the embeddings with a certain universality are applied to the specific tasks of data integration. The overall architecture is shown in Fig. 1. There are three layers in the model: data processing layer, feature learning layer and application layer.

In the data processing layer, the relational data in multiple datasets are converted into a heterogeneous graph to represent data relationships. There are three types of nodes in a heterogeneous graph including RID nodes, AID nodes and token nodes. And then the initial vector representation of each node is calculated through pre-training. In the feature learning layer, all nodes are trained in N epochs. In each epoch of training, the embeddings of token nodes are first updated according to the characteristics of their neighbor nodes, and then the embeddings of AID and RID nodes are updated through the aggregation operation. In the application layer, the embeddings of the RID nodes are used for entity resolution and the embeddings of the AID nodes are used for schema matching. In this way, the specific tasks of data integration are realized.

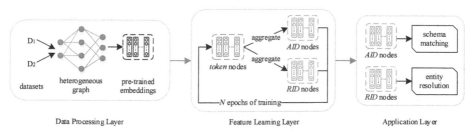

Fig. 1. Model architecture.

The description of heterogeneous embedding algorithm in our model is as follows.

Algorithm 1: *Heterogeneous embedding algorithm*

Input: datasets $D = \{D_1, D_2, \cdots\}$

Output: embeddings $e = \{e_1, e_2, \cdots, e_n\}$

1. define a heterogeneous graph $G = (V, E)$
2. for each D_i in D
3.   add each cell value into V
4.   add each relationship between data into E
5. $e_0 \leftarrow$ pre-train G by Node2Vec
6. repeat
7.   for each epoch
8. $e_{token} \leftarrow$ mean aggregation result
9. $e_{rid}, e_{aid} \leftarrow$ max-pooling aggregation result
10.   loss $\leftarrow$ sum($loss_{sm}, loss_{er}$)
11. until convergence

3.2 Data Processing Layer

In order to show the relationship between the data directly, we use a heterogeneous graph to represent the relational data. The heterogeneous graph is defined as follows.

Definition 1. Heterogeneous Graph. Given a graph $G = (V, E)$, V represents the set of nodes and E represents the set of edges in the graph. The mapping relationship between node types is $\varphi : V \rightarrow \mathcal{V}$, and the mapping relationship between edge types is $\psi : E \rightarrow \mathcal{E}$. When the number of node types $|\mathcal{V}| > 1$ or the number of edge types $|\mathcal{E}| > 1$, graph G is a heterogeneous graph.

In the heterogeneous graph of our model, the nodes are classified according to three node types in the node type set $\mathcal{V}$. The following three types are as: RID (Record ID) node representing a record, AID (Attribute ID) node representing an attribute and token node representing a single word in the cell. The token nodes are obtained by segmenting the attribute value in each cell, where each word corresponds to a token node. After the nodes are created, the edges are constructed according to the relation between the data represented by each node. First, edges are constructed between different types of nodes. According to the row or column correspondence in the table, edges are constructed between the RID nodes and the token nodes existing in the row, also between the AID nodes and the token nodes existing in the column, to retain the structural relationship. Then, edges are constructed between multiple token nodes in the same cell, so that the semantic relationship that may be hidden between adjacent words is preserved. Take a table as an example and transform it into a heterogeneous graph, as shown in Fig. 2.

In order to adopt appropriate training algorithms for nodes, the properties of the three node types are analyzed respectively here. Token nodes represent data content with specific semantics, so they can be trained directly by fusing the features of neighbors into their features. However, RID nodes and AID nodes hardly contain specific semantics, which represent the set of data in a certain row or column in the table. Therefore, our model calculates the features of them by aggregating the features of token nodes. In

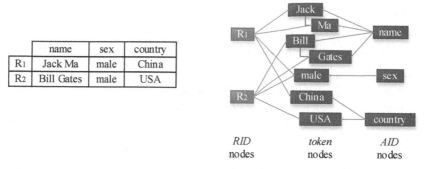

	name	sex	country
R₁	Jack Ma	male	China
R₂	Bill Gates	male	USA

Fig. 2. Example of heterogeneous graph construction.

order to facilitate the operation of aggregation, several subgraphs are divided from the heterogeneous graph. The subgraph involved in our model is defined as follows.

Definition 2. Subgraph. Node n is any node in the graph G, the union $\{\{n\} \cup \Theta\}$ of node n and its neighbor node set Θ is taken as the node set V_n, and the set of edges between node n and all nodes in set Θ is taken as the edge set E_n, then the graph $G_n = (V_n, E_n)$ is the subgraph of the graph G rooted by node n.

In our model, all RID nodes and AID nodes are selected as roots to construct the corresponding subgraphs, that is, the token nodes corresponding to all the words contained in the row or column are included in the subgraph. For example, for the heterogeneous graph on the left side of Fig. 3, two RID nodes and three AID nodes are selected as roots to divide the subgraphs shown on the right side of Fig. 3.

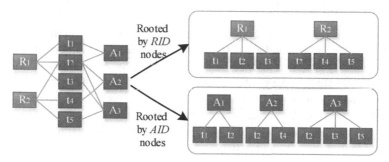

Fig. 3. Dividing heterogeneous graph into subgraphs.

Subgraphs are used in the aggregation operation of RID nodes and AID nodes in the subsequent training. Before the training, the graph embedding method Node2Vec is used to pre-train all nodes in the heterogeneous graph. And the obtained embeddings are input into the training model. In the pre-training, the node types are not distinguished and the training is performed only according to the association between nodes.

3.3 Feature Learning Layer

The main idea is to aggregate neighbor features into the target node feature. As expanding the neighbors iteratively, more and more information is transmitted in the graph.

Neighbor Node Sampling. In the training process, the model learns the node characteristics according to the graph topology. When calculating the target node at level l, the information of all nodes within l hops needs to be aggregated, that is, the number of nodes participating in the calculation increases exponentially with l. Therefore, when the graph scale or the number of training layers is large, the "neighbor explosion" problem will occur [20], which can affect the model effect.

In order to avoid "neighbor explosion", the neighbors of the target node can be sampled before training. In this way, only a part of the neighbors is selected, so that each node has only a fixed number of neighbors to participate in the calculation. In our model, hyperparameters are set in advance to define the sampling number (25 neighbors are sampled in the first layer and 10 neighbors in the second layer). Using the random sampling method, if the number of neighbors is less than the number of samples, some neighbors will be repeatedly sampled.

Embedding Calculation. The feature representation of the target node is obtained by aggregating the feature representation of the sampled neighbor nodes. In order to obtain more accurate embeddings, different aggregation methods are used to calculate different types of node representations, based on their own characteristics.

For token nodes, the embedding of the target node is calculated through the embeddings of this node and its sampled neighbor nodes in the previous layer to aggregate the characteristic information contained therein. The aggregation formula is as follows:

$$e_u^l = \sigma \left(W \cdot \text{MEAN}\left(\left\{ e_u^{l-1} \right\} \cup \left\{ e_v^{l-1}, \forall v \in N \right\} \right) \right) \tag{1}$$

where e_u^l represents the embedding of the target token node u in the l th layer. The embedding e_u^{l-1} of node u at the $l-1$ layer and the embedding e_v^{l-1} of each node v in the set of sampling neighbor nodes N at the $l-1$ layer are summed union. Then, take the average and calculate the product with the weight W. σ is the activation function. Particularly, in the first layer of training, the pre-trained embeddings is input.

For RID nodes and AID nodes, the embeddings of all sampled token nodes in the subgraph rooted by the target node is aggregated. To make the target node retain the significant features of token nodes, the max-pooling aggregation is used as follows:

$$e_u^l = \max \left(\left\{ \sigma \left(W \cdot e_w^l + b \right), \forall w \in \Theta \right\} \right) \tag{2}$$

where e_u^l represents the embedding of the target RID or AID node u in the l th layer. w is a token node in the sampling neighbor node set Θ in the subgraph rooted by node u, and e_w^l represents the embedding of node w at the l th layer. W and b are training parameters, and σ is the activation function.

The model adopts supervised learning method. The loss function $\mathcal{L}_{sm}$ of schema matching and the loss function $\mathcal{L}_{er}$ of entity resolution are calculated as follows:

$$\mathcal{L}_{sm} = CrossEntropyLoss(y_{sm}, z_{sm}) \tag{3}$$

$$\mathcal{L}_{er} = CrossEntropyLoss(y_{er}, z_{er}) \tag{4}$$

where y_{sm} and y_{er} are the results calculated by the proposed model in two tasks, while z_{sm} and z_{er} are the labels in the dataset. The cross entropy loss is calculated for them. The training goal of the model is to minimize the sum of the two loss functions.

3.4 Application Layer

After training the embeddings containing data features, this paper applies the result to specific tasks of data integration, including schema matching and entity resolution. In the training process, RID and AID nodes are calculated by aggregating the characteristics of token nodes, so the overall characteristics of the row or column of data are already included in the embeddings. Therefore, in the specific tasks of data integration, the embeddings of the RID and AID nodes can be used directly, without need to combine the embeddings of the token nodes in the row or column.

In this paper, the embeddings of the RID nodes are used to implement the entity resolution task, and the embeddings of the AID nodes are used to implement the schema matching task.

4 Experiments

In this section, we use the proposed technology to build a heterogeneous embedding model for relational data, and conduct experiments in the specific data integration tasks including schema matching and entity resolution to test the quality of the model results. The experiment uses precision (P), recall (R) and F1 score (F1) as metrics.

4.1 Dataset

This paper tests the effect of the model on Amazon-Google (AG) and BeerAdvo-RateBeer (BR) datasets. The statistical data of the experimental datasets is shown in Table 1, where includes the number of tuples, the number of columns and the number of positive instances in the dataset.

Table 1. Statistics of the datasets for experiments.

Datasets	# tuples	# columns	# Pos
Amazon-Google	4589	3	1167
BeerAdvo-RateBeer	7345	4	68

4.2 Schema Matching Task

In the schema matching task, our model is compared with another model REMA based on deep learning. In order to make a comprehensive comparison, we use two methods of our model to compare with REMA. (1) Ours (tokens): using the same method as REMA, the mean value of the embeddings of all token nodes in each column is used as the embedding of the column. (2) Ours (AID): the embedding of each AID node is used to represent the column directly. The experimental results are shown in Table 2.

Comparing the results of REMA and Ours (tokens), it can be seen that the embeddings of token nodes obtained through the proposed model have better accuracy, so when using the same averaging method as REMA, the column embedding has a slightly better effect. And compared with Ours (AID), the results show that directly using the AID node embeddings obtained by the proposed model can further improve the effect of schema matching. In each epoch of embedding training of the proposed model, the AID node embeddings are aggregated through the max-pooling method. Therefore, after multiple epochs of training, the embedding representation of the AID node aggregates the prominent features in its attribute value, so it can better represent the characteristics of the data in the column, and the embedding is more accurate.

Table 2. Results for schema matching.

Datasets	REMA			Ours (tokens)			Ours (AID)		
	P	R	F1	P	R	F1	P	R	F1
AG	0.83	1.00	0.91	1.00	1.00	1.00	1.00	1.00	1.00
BR	0.84	0.75	0.79	0.84	1.00	0.91	1.00	1.00	1.00

4.3 Entity Resolution Task

First, we compare the proposed model with the RNN and Hybrid models in the classic entity resolution algorithm based on deep learning [16]. The experimental results are shown in Table 3.

Table 3. Results for entity resolution.

Datasets	RNN			Hybrid			Ours		
	P	R	F1	P	R	F1	P	R	F1
AG	0.59	0.48	0.53	0.58	0.64	0.61	0.60	0.64	0.62
BR	0.74	0.70	0.72	0.73	0.70	0.71	0.72	0.69	0.70

According to the above table, the RID node embeddings trained by the proposed model can be used directly in the entity resolution task to obtain results similar to the existing methods. This shows that in the heterogeneous graph, valuable data features can be extracted to a certain extent, but there is still room for improvement.

In addition, the training idea of GraphER is similar to ours. GraphER further processes the embeddings, including steps such as cross-coding, which greatly improved the effect of the model. So we perform the same follow-up processing steps as GraphER on the embeddings trained by the proposed model, and then compare the experimental results with GraphER. The results are shown in Table 4.

Table 4. Results for entity resolution compared with GraphER.

Datasets	GraphER			Ours (Cross-encoding)		
	P	R	F1	P	R	F1
AG	0.69	0.67	0.68	0.72	0.70	0.71
BR	0.79	0.80	0.79	0.78	0.82	0.80

According to the above table, the features can be better extracted during training through our heterogeneous embedding technology. Under the same subsequent processing, our experimental effect is better than GraphER. At the same time, comparing the results in Table 4 with Table 3, it can be shown that the results can be effectively improved by further processing the data embeddings.

4.4 Multiple Tasks

In order to verify that the embedding results obtained by our model have a certain universality, the embeddings are carried out on the two tasks of schema matching and entity resolution at the same time. And the EMBDI model with similar application scenarios is trained on the graph proposed in this paper to compare the experimental results. In addition, two comparison models are set up to test the effectiveness of the heterogeneous embedding technology in our model: (1) Ours (mean): the mean aggregation method is used for all types of nodes in training. (2) Ours (pool): the max-pooling aggregation method is used for all types of nodes in training. The F1 scores calculated by each model in two tasks are shown in Table 5 and Table 6, respectively.

Table 5. F1 scores in schema matching task.

Datasets	EMBDI	Ours (mean)	Ours (pool)	Ours
AG	1.00	0.91	1.00	1.00
BR	1.00	0.86	0.91	1.00

Table 6. F1 scores in entity resolution task.

Datasets	EMBDI	Ours (mean)	Ours (pool)	Ours
AG	0.54	0.55	0.57	0.62
BR	0.63	0.60	0.64	0.70

It can be seen from the table that, compared with EMBDI which does not distinguish the node type, our heterogeneous embedding model achieves better results in both tasks, which proves that the embeddings trained by our model has good universality. In addition, compared with Ours (mean) and Ours (pool) that only use a single aggregator, the heterogeneous embedding model achieves a certain improvement in experimental results, indicating that different aggregation methods for different types of nodes are helpful to better extract data features according to the characteristics of the node type, and more accurate embeddings can be obtained. The heterogeneous embedding technology proposed in this paper is effective.

4.5 Parameter Sensitivity Analysis

In order to set reasonable hyperparameter values so that the model can obtain the highest quality results in the experiment, sensitivity analysis of the parameters is carried out here. In the experiment, the F1 score is calculated in the entity resolution task.

Set different numbers of training layers, the change trend of F1 score is shown in Fig. 4(a). The results show that when the number of training layers increases from 1 to 2, the result is greatly improved, but when it continues to increase, the experimental effect begins to decrease. This implies that only one layer of training cannot learn the characteristics of the data well. But after multiple layers of training, each node aggregates the characteristics of a large number of adjacent nodes, so that the characteristics of each node gradually become consistent and the discrimination of nodes decreases, which affects the experimental effect. Therefore, our model uses two training layers.

Different dimensions are set for the embedding vectors, the change trend of F1 score is shown in Fig. 4(b). It can be seen that when the embedding dimension is low, the embeddings cannot achieve good results in the task. When the dimension increases from 200 to 250, the F1 score has an obvious improvement, but the speed of improvement gradually slows when the dimension continues to increase. As the dimensionality increases, the information contained in the embedding vectors will be richer, so the effect of the model will be better. However, the high-dimensional embedding vectors lead to the training time to be too long, which affects the performance of the model and does not significantly improve the experimental results. Therefore, it is appropriate to set the embedding dimension between 250 and 300. In our model, the embedding vector dimension is set to 256.

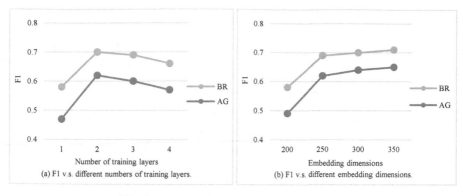

Fig. 4. F1 scores by varying hyperparameters.

5 Conclusion

In this paper, a heterogeneous embedding technology for relational data integration tasks is proposed. In our model, the relational data is first converted into a heterogeneous graph, and then different types of nodes are trained using different aggregation methods, finally the calculated embeddings are applied in two data integration tasks of schema matching and entity resolution. Experiments prove that the relational data embeddings trained by the proposed model have good quality, and the performance in two specific tasks is better than the existing methods.

Our main contribution is to extract data features from relational data and express them in the embedding vectors. The embeddings have high quality and good universality. For the direction that the model can be improved in the future, we can consider extending the model to the task of truth discovery to further complete the data integration.

Acknowledgment. This work is supported by the National Natural Science Foundation of China (62072084, 62072086), the National Defense Basic Scientific Research Program of China (JCKY2018205C012) and the Fundamental Research Funds for the Central Universities (N2116008).

References

1. Mikolov, T., Chen, K., Corrado, G., Dean, J.: Efficient estimation of word representations in vector space (2013). https://arxiv.org/abs/1301.3781
2. Bordawekar, R., Shmueli, O.: Using word embedding to enable semantic queries in relational databases. In: Proceedings of the 1st Workshop on Data Management for End-to-End Machine Learning, pp. 1–4 (2017). https://doi.org/10.1145/3076246.3076251
3. Zhang, L., Zhang, S., Balog, K.: Table2vec: Neural word and entity embeddings for table population and retrieval. In: SIGIR, pp. 1029–1032 (2019). https://doi.org/10.1145/3331184.3331333.
4. Roweis, S.T., Saul, L.K.: Nonlinear dimensionality reduction by locally linear embedding. Science **290**(5500), 2323–2326 (2000). https://doi.org/10.1126/science.290.5500.2323

5. Belkin, M., Niyogi, P.: Laplacian eigenmaps and spectral techniques for embedding and clustering. Nips **14**, 585–591 (2001)
6. Ahmed, A., Shervashidze, N., Narayanamurthy, S., Josifovski, V., Smola, A.J.: Distributed large-scale natural graph factorization. In: Proceedings of the 22nd International Conference on World Wide Web, pp. 37–48 (2013). https://doi.org/10.1145/2488388.2488393
7. Perozzi, B., Al-Rfou, R., Skiena, S.: DeepWalk: online learning of social representations. In: KDD, pp. 701–710 (2014). https://doi.org/10.1145/2623330.2623732
8. Grover, A., Leskovec, J.: node2vec: Scalable feature learning for networks. In: KDD, pp. 855–864, (2016). https://doi.org/10.1145/2939672.2939754
9. Wang, D., Cui, P., Zhu, W.: Structural deep network embedding. In KDD, pp. 1225–1234 (2016). https://doi.org/10.1145/2939672.2939753
10. Kipf, T.N., Welling, M.: Semi-supervised classification with graph convolutional networks (2016). https://arxiv.org/abs/1609.02907
11. Cilibrasi, R.L., Vitanyi, P.M.B.: The Google similarity distance. IEEE Trans. Knowl. Data Eng. **19**(3), 370–383 (2007). https://doi.org/10.1109/TKDE.2007.48
12. Guo, T., Shen, D., Nie, T., Kou, Y.: Web table column type detection using deep learning and probability graph model. In: Wang, G., Lin, X., Hendler, J., Song, W., Xu, Z., Liu, G. (eds.) WISA 2020. LNCS, vol. 12432, pp. 401–414. Springer, Cham (2020). https://doi.org/10.1007/978-3-030-60029-7_37
13. Koutras, C., Fragkoulis, M., Katsifodimos, A., Lofi, C.: REMA: graph embeddings-based relational schema matching. In: EDBT/ICDT Workshops (2020)
14. Konda, P., Das, S., Suganthan, G.C.P., Doan, A., Ardalan, A., Ballard, J.R., et al.: Magellan: toward building entity matching management systems. Proc. VLDB Endow. **9**(12), 1197–1208 (2016). https://doi.org/10.14778/2994509.2994535
15. Ebraheem, M., Thirumuruganathan, S., Joty, S., Ouzzani, M., Tang, N.: Distributed representations of tuples for entity resolution. Proc. VLDB Endow. **11**(11), 1454–1467 (2018). https://doi.org/10.5555/3236187.3269461
16. Mudgal, S., Li, H., Rekatsinas, T., Doan, A., Park, Y., Krishnan, G., et al.: Deep learning for entity matching: a design space exploration. SIGMOD Conf. **2018**, 19–34 (2018). https://doi.org/10.1145/3183713.3196926
17. Vaswani, A., Shazeer, N., Parmar, N., Uszkoreit, J., Jones, L., Gomez, A.N., et al.: Attention is all you need. Nips **30**, 5998–6008 (2017)
18. Li, B., Wang, W., Sun, Y., Zhang, L., Ali, M.A., Wang, Y.: GraphER: token-centric entity resolution with graph convolutional neural networks. AAAI **34**(5), 8172–8179 (2020). https://doi.org/10.1609/AAAI.V34I05.6330
19. Cappuzzo, R., Papotti, P., Thirumuruganathan, S.: Creating embeddings of heterogeneous relational datasets for data integration tasks. SIGMOD Conf. **2020**, 1335–1349 (2020). https://doi.org/10.1145/3318464.3389742
20. Hamilton, W.L., Ying, Z., Leskovec, J.: Inductive representation learning on large graphs. Adv. Neural. Inf. Process. Syst. **30**, 1024–1034 (2017)

Hybrid Checkpointing for Iterative Processing in BSP-Based Systems

Yi Yang[1], Chen Xu[1,2(✉)], Chao Kong[3], and Aoying Zhou[1]

[1] East China Normal University, Shanghai, China
yiyang@stu.ecnu.edu.cn, {cxu,ayzhou}@dase.ecnu.edu.cn
[2] Science and Technology on Parallel and Distributed Processing
Laboratory (PDL), Changsha, China
[3] Anhui Polytechnic University, Wuhu, China
kongchao@ahpu.edu.cn

Abstract. Distributed iterative processing exists in various application scenarios including large-scale graph analytics and machine learning. Many systems employ bulk synchronous parallel (BSP) model to synchronize the iterations. In these BSP-based systems, the long iterative processing time in distributed environments makes the fault-tolerance crucial. Most BSP-based systems write a checkpoint in either blocking strategy or unblocking strategy to achieve fault-tolerance. However, the blocking strategy involves a checkpointing overhead in failure-free cases, whereas the unblocking strategy also incurs a recovery cost if the BSP-based system has not completed checkpointing in failure cases. Motivated by the trade-off between blocking and unblocking checkpointing, we aim to choose different checkpointing strategy when checkpoint is required during iterative processing, in order to reduce the whole execution time. In particular, we propose a *checkpointing choice problem*, i.e., how to choose the strategy to minimize the execution time. The challenge is to make a choice during runtime without future information. To address this problem, we provide a *hybrid checkpointing*, which heuristically chooses either blocking or unblocking checkpointing based on cost evaluation. Our experiments on Giraph, a typical BSP-based system, show that hybrid checkpointing outperforms blocking and unblocking checkpointing.

Keywords: BSP · Iterative processing · Checkpointing

1 Introduction

Distributed iterative processing is widely adopted by various application including large-scale graph analytics and machine learning. Many systems such as Pregel [6], MLBase [5], GraphX [4], adopt bulk synchronous parallel (BSP) [11] to synchronize the iterations. Typically, the BSP-based systems run with a serial of supersteps and take a long time to execute in distributed environments. It is important for these systems to effectively handle failures, since failure usually happens in distributed environments [3]. Alternatively, the design of fault-tolerance mechanism is crucial in the BSP-based systems.

© Springer Nature Switzerland AG 2021
C. Xing et al. (Eds.): WISA 2021, LNCS 12999, pp. 693–705, 2021.
https://doi.org/10.1007/978-3-030-87571-8_60

A common approach is to periodically store the state of the system to stable storage as a checkpoint during normal execution, and then recover from the latest checkpoint upon failure. Usually, the BSP-based systems like Giraph [1] write a checkpoint in a *blocking* strategy which stores the state of system by pausing computation. Clearly, the blocking checkpointing incurs an overhead in failure-free cases. Different from the blocking checkpointing, the *unblocking* checkpointing, employed in [13] as an example, typically stores the system state in parallel with the computation. The unblocking checkpointing might cause a high recovery overhead in failure cases when the system is still writing a checkpoint. Hence, both blocking checkpointing and unblocking checkpointing lead to an additional overhead on the whole execution time. However, the blocking checkpointing involves this overhead during normal execution, whereas the unblocking checkpointing incurs the overhead during failure recovery. Clearly, there is a trade-off between blocking and unblocking checkpointing.

Motivated by the trade-off, the goal of this work is to choose different checkpointing strategy during iterative processing, instead of employing a single strategy over the entire execution, so as to reduce the whole execution time. To achieve this goal, we propose a *checkpointing choice problem* that how to make a decision between blocking and unblocking strategy when a checkpoint is required during iterative processing in BSP-based systems. The challenge of this problem is to make a decision, i.e., during runtime without future information. To deal with this problem, we propose a *hybrid checkpointing* strategy. Our hybrid checkpointing evaluates the cost of blocking and unblocking checkpointing, and heuristically chooses a strategy with a lower cost when checkpoints are required. Moreover, we employ Giraph, a BSP-based iterative processing system, to implement hybrid checkpointing. Our experimental results show that hybrid checkpointing reduces the overhead by 37% compared to blocking checkpointing in a failure-free case, and the overhead of hybrid checkpointing relative to unblocking checkpointing is reduced by 38% in a failure case.

In the rest of paper, we first introduce the background of BSP-based systems in Sect. 2. Then, we make the following contributions.

- We propose a *checkpointing choice problem* in Sect. 3 by analyzing the trade-off between blocking and unblocking checkpointing.
- We propose a *hybrid checkpointing* in Sect. 4 which employs a cost model and a heuristic strategy to address the checkpointing choice problem.
- Our experimental studies in Sect. 5 show that the hybrid checkpointing outperforms the blocking and unblocking checkpointing.

Finally, we list related work in Sect. 6 and conclude our work in Sect. 7.

2   Background and Motivation

In this section, we introduce the preliminary of BSP-based systems, and illustrate two checkpointing strategies adopted by BSP-based systems for fault-tolerance.

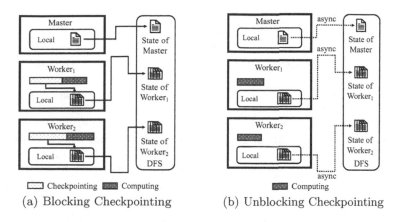

(a) Blocking Checkpointing   (b) Unblocking Checkpointing

Fig. 1. The process of checkpointing

2.1 Preliminary of BSP-Based Systems

Computation Model. In the BSP computation model, a whole computation includes a sequence of supersteps, each of which consists of three phases: local computation, message transmission, and barrier synchronization. This procedure repeats until the convergence.

Basic Architecture. BSP-based systems typically adopt a master/worker architecture. The master partitions the input data and assigns the partitions to workers. After that, the master instructs the workers to complete a series of supersteps. At the beginning of the superstep where a checkpoint is needed, the master coordinates the workers to complete the checkpointing. Then the workers start to complete the partitions computation and continuously report the execution state to the master. Once all workers finish current superstep, the master coordinates the workers to start the next superstep.

2.2 Checkpointing in BSP-Based Systems

Blocking Checkpointing. Most BSP-based systems, e.g., Giraph, achieve fault tolerance in the way of blocking checkpointing. Here, users specify a checkpointing interval and the BSP-based systems write a checkpoint at the beginning of supersteps which meet the user-specified interval. Figure 1(a) shows a BSP-based system with one master and two workers. At the beginning of a superstep with a blocking checkpointing enabled, the two workers directly upload their states from local to distributed file system (DFS) via network. Each worker starts computation for this superstep only if it finishes uploading its state to DFS. Meanwhile, all the workers notify the master and the master uploads its state to DFS. Hence, the state of the workers as well as the master on DFS is a checkpoint written in a blocking manner. Once failure happens, the master and workers read the latest checkpoint from DFS for rollback.

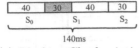   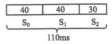

(a) Blocking Checkpointing   (b) Unblocking Checkpointing

Fig. 2. Unblocking checkpointing outperforms blocking checkpointing

Unblocking Checkpointing. Different from the blocking checkpointing, the unblocking checkpointing does not require all the workers to upload their states to DFS before starting computation. As illustrated in Fig. 1(b), at the beginning of a superstep with an unblocking checkpointing enabled, each worker uploads its state from the local to DFS asynchronously via network. In other words, uploading state to DFS is in parallel with the computation if the BSP-based system employs the unblocking checkpointing. Similarly, the master and workers reload the latest checkpoint once failure happens.

3   Problem Statement

In this section, we explain the trade-off between blocking and unblocking checkpointing. Motivated by the trade-off, we propose a checkpointing choice problem for iterative processing in BSP-based systems.

3.1   Blocking Versus Unblocking Checkpointing

Intuitively, under failure-free cases, the system even does not need any checkpoint [8]. In this case, the blocking checkpointing involves a higher overhead than the unblocking checkpointing. That is, it takes a longer execution time if the system explores blocking checkpointing than the one if the system adopts unblocking checkpointing. With failure happens, the system has to recompute from the checkpoints. As described in Sect. 2.2, the blocking checkpointing always ensures a most recent checkpoint is completely written on a reliable external storage, e.g., HDFS, and the system recomputes from a latest checkpoint.

Compared with the blocking checkpointing, the unblocking checkpointing spends more time to completely write the checkpoint from local to DFS. Hence, the unblocking checkpointing might not finish when failure happens, which leads to a longer recovery time since the system has to recompute from an earlier checkpoint or even the beginning.

As described in Example 2(a), the whole execution time of the system with blocking checkpointing is 140 ms, whereas the whole execution time of the system with unblocking checkpointing is only 110 ms. Therefore, under failure-free cases, unblocking checkpointing is better than blocking checkpointing. Different from Example 1, blocking checkpointing might outperform unblocking checkpointing when failure happens. As described in Example 2, the whole execution time of the system with blocking checkpointing is equal to 145 ms, while the whole execution time of the system with unblocking checkpointing reaches 185 ms.

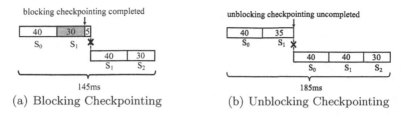

Fig. 3. Blocking checkpointing outperforms unblocking checkpointing

Example 1. In Fig. 2(a), the system with blocking checkpointing consists of three supersteps. In superstep S_0 and S_2, the system takes 40 ms and 30 ms to complete the computation, respectively. In superstep S_1, the system first takes 30 ms to write a checkpoint, and then takes 40 ms to compute. Hence, the whole execution time of the system with blocking checkpointing is 140 ms. For the unblocking checkpointing illustrated in Fig. 2(b), the execution time of the superstep S_0 and S_2 is the same as the blocking checkpointing. However, in superstep S_1, since the system writes a checkpoint in parallel with the computation, the superstep S_1 only takes 40 ms to compute. Hence, the whole execution time of system with unblocking checkpointing is 110 ms.

Example 2. In Fig. 3(a), before the failure happens, the system takes 40 ms and 35 ms to execute the superstep S_0 and S_1 respectively. In particular, the superstep S_1 takes 30 ms to complete the checkpointing in a blocking manner. Hence, after the failure happens, the system recomputes from superstep S_1 until superstep S_2 is completed. As superstep S_1 takes 40 ms and superstep S_2 takes 30 ms, the whole execution time of the system with blocking checkpointing is 145 ms. In Fig. 3(b), before the failure happens, the system with unblocking checkpointing takes the same time as Fig. 3(a) to execute superstep S_0 and S_1. However, the time of superstep S_1 is all used for computation, since the checkpointing is performed in parallel with the computation. When the failure happens, the system has not complete the checkpointing. Therefore, the system recomputes from superstep S_0 until superstep S_2 is completed. Since superstep S_0, S_1 and S_2 takes 40 ms, 40 ms and 30 ms respectively, the whole execution time of the system with unblocking checkpointing is 185 ms.

3.2   Checkpointing Choice Problem

Generally, the computation only starts after the blocking checkpointing, while the unblocking checkpointing executes in parallel with the computation. However, the blocking checkpointing involves an additional overhead on execution time during normal execution, whereas the unblocking checkpointing increases this overhead during failure recovery. Hence, there is a trade-off between blocking checkpointing and unblocking checkpointing. Consequently, employing the same checkpointing manner over the entire iterative processing cannot minimize the whole execution time.

Instead of the same checkpointing manner over the entire execution, we propose the *checkpointing choice problem*. That is, how to make a decision between blocking manner and unblocking manner when a checkpoint is required during iterative processing in BSP-based systems. Given an iterative processing with a serial of supersteps denoted as $S_0, S_1, \ldots$ and a user-specified checkpoint interval τ, the system writes a checkpoint at superstep S_i, where $i \bmod \tau = 0$. In particular, these checkpoints are applied in either blocking or unblocking manner with the object to minimize the whole execution time, including normal execution time, checkpoint time and recovery time. Here, checkpoint time is related to the checkpointing manner, and recovery time depends on when failure happens.

If we know when failure happens in advance, it is possible to enumerate all the combinations of the checkpointing choice and employ the combination with the minimal execution time. However, existing failure prediction models and techniques are still far from providing satisfactory results [7]. Hence, the challenge of checkpointing choice problem is to make a decision in the progress of the iterative processing without future runtime information.

4   Hybrid Checkpointing

To address the checkpointing choice problem, we propose a heuristic *hybrid checkpointing* based on a greedy idea in this section. It evaluates the cost of blocking and unblocking checkpointing at the supersteps with the checkpoints required, and then chooses a checkpointing strategy with a lower cost.

4.1   Cost Evaluation

In order to compare the cost of blocking and unblocking checkpointing, we build a cost model to evaluate the overhead of the two checkpointing strategies. In our model, we assume the system runtime locates at the beginning of superstep S_i and explores either blocking checkpointing or unblocking checkpointing. For a simplicity, this model does not consider the failure-free cases because of the failure in distributed environments is common [3]. Instead, this model assumes the failure happens in a near future and calculates the cost by evaluating the overhead to recover to the state at the beginning of superstep S_i.

As shown in Fig. 4(a), the system currently locates at the beginning of superstep S_i at time t_1. It writes a latest checkpoint at time t_0 which stores the state of superstep S_l. Clearly, the checkpointing which stores the state of superstep S_i would finish till time t_2 if the system adopts blocking checkpointing, whereas a successful checkpoint would not be available until time t_3 if the system utilizes unblocking checkpointing.

According to our cost model, we assume failure happens between time t_2 and t_3. If the system employs blocking checkpointing, it directly recovers to the state at the beginning of superstep S_i after the failure happens, and there is no recovery cost. But in order to achieve such a direct recovery, the system has to take time t_{ck_i} to store the state of superstep S_i before the failure happens. Hence, the cost of blocking checkpointing, $T_{b_i}^1$, is the blocking checkpointing time t_{ck_i},

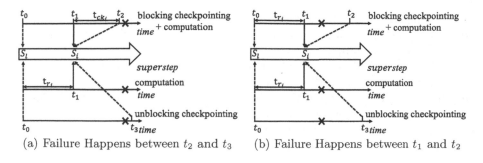

(a) Failure Happens between t_2 and t_3 (b) Failure Happens between t_1 and t_2

Fig. 4. Failure happens at different time

i.e., $T^1_{b_i} = t_{ck_i}$. Different from the blocking checkpointing, the system does not finish checkpointing yet if it adopts the unblocking checkpointing. Once failure happens, the system has to recompute from the state of superstep S_l, so as to recover to the state at the beginning of superstep S_i. Consequently, the cost of unblocking checkpointing, $T^1_{ub_i}$, is the recomputation time t_{r_i}, i.e., $T^1_{ub_i} = t_{r_i}$.

Next, we consider another kind of failure happens in a near future. In Fig. 4(b), the failure happens between time t_1 and t_2. If the blocking checkpointing strategy is adopted, the system does not finish storing the state of superstep S_i yet when the failure happens. Therefore, the system has to recompute from the state of superstep S_l so as to recover to the state at the beginning of superstep S_i. Hence, the cost of blocking checkpointing, $T^2_{b_i}$, is the recomputation time t_{r_i}, i.e., $T^2_{b_i} = t_{r_i}$. Similarly, if the unblocking checkpointing strategy is employed, the system has to recompute from the state of superstep S_l, since the checkpoint which stores the state of superstep S_i is unavailable. Hence, the cost of unblocking checkpointing, $T^2_{ub_i}$, is the recomputation time t_{r_i}, i.e., $T^2_{ub_i} = t_{r_i}$.

Moreover, our model evaluates the cost based on the assumption that failure happens in a near future. However, the unblocking checkpointing usually takes a long time to finish. To simplify our model, we would not consider the case that failure happens after a complete unblocking checkpointing, i.e., time t_3. Hence, we decide to combine $T^1_{b_i}$ and $T^2_{b_i}$ to derive the cost of blocking checkpointing. Here, we assume the probability of failure happens between t_2 and t_3 is p and the probability of failure happens between t_1 and t_2 is q, where the sum of p and q is approximately equal to 1. In the following, we apply the mathematical expectation of $T^1_{b_i}$ and $T^2_{b_i}$ as the cost of blocking checkpointing, i.e., $T_{b_i} = p * t_{ck_i} + q * t_{r_i}$. Similarly, we combine $T^2_{b_i}$ and $T^2_{ub_i}$ to get the cost of unblocking checkpointing, i.e., $T_{ub_i} = p * t_{r_i} + q * t_{r_i}$.

4.2 Hybrid Strategy

Based on the cost evaluation of the blocking and unblocking checkpointing, we propose our *hybrid checkpointing* to heuristically choose a lower cost strategy between blocking and unblocking checkpointing at the beginning of supersteps where checkpoints are required. In particular, the hybrid checkpointing follows the blocking checkpointing if $T_{b_i} \leq T_{ub_i}$. Otherwise, the hybrid checkpointing performs as an unblocking checkpointing. Clearly, according to the cost equation of T_{b_i} and T_{ub_i}, we simplify $T_{b_i} \leq T_{ub_i}$ as $t_{ck_i} \leq t_{r_i}$.

Then, the critical issue of hybrid strategy is to find a way to estimate t_{ck_i} and t_{r_i}. Recall that t_{ck_i} and t_{r_i} represent the time of the blocking checkpointing and the time of the recomputation respectively. Hence, our hybrid checkpointing explores an estimator by the statistics from the runtime of the past supersteps. However, at superstep S_0, i.e., the first superstep requiring checkpointing, the system has no runtime statistic. Hence, we consider the superstep S_0 and the subsequent supersteps, i.e., S_i where $i \neq 0$, respectively.

Initially, the system partitions the input data to corresponding nodes. To avoid a re-partition incurred by job restarting once failure happens, the system usually first stores the partition data and then starts the computation. Following this principle, in superstep S_0, our hybrid strategy employs the blocking checkpointing to ensure that the partition data is stored before starting computation.

In superstep S_i where $i \neq 0$, the estimator collects the time of blocking checkpointing and applies an average on the time of historical blocking checkpointing. Hence, we calculate the average of the time t_{ck_j} spent on blocking checkpointing from the superstep S_0 to the previous superstep S_{i-1} before the current superstep S_i in Eq. (1).

$$t_{ck_i} \approx \frac{1}{N} \sum_{j=0}^{i-1} t_{ck_j} \qquad (1)$$

Here, t_{ck_j} is 0 if the superstep S_j does not apply blocking checkpointing and N represents the total number of blocking checkpointing. In addition, we denote the last superstep with a complete checkpointing as S_l and collect the time of computation and estimate t_{r_i} by Eq. (2).

$$t_{r_i} \approx \sum_{j=l}^{i-1} t_{co_j} \qquad (2)$$

where t_{co_j} is the computation time of superstep S_j.

Putting it together, Algorithm 1 describes the implementation details of the hybrid checkpointing. In particularly, the system employs the blocking strategy to store partition data in the superstep S_0 (line 3). Then, in other supersteps, based on the checkpoint interval τ, the system estimates t_{ck_i} and t_{r_i} according to Eq. (1) and (2) (line 5). With the obtained t_{ck_i} and t_{r_i}, the system employs equation $t_{ck_i} \leq t_{r_i}$ to choose the checkpointing strategy with a lower cost (line 6 to line 10). Finally, the system continuously collects t_{ck_j} and t_{co_j} as historical runtime data at the end of the superstep for estimation (line 13)

5   Experimental Study

We integrate our hybrid checkpointing as well as the unblocking checkpointing [13] in Giraph. The experiments use blocking checkpointing [1] in Giraph

Algorithm 1. Hybrid Checkpointing

1: **while** *convergent condition is not satisfied* **do**
2:    **if** $i = 0$ **then**
3:       *checkpointing strategy of* $S_i \leftarrow$ BLOCKING
4:    **else if** $i \bmod \tau = 0$ **then**
5:       *estimate* t_{ck_i}, t_{r_i}
6:       **if** $t_{ck_i} \leq t_{r_i}$ **then**
7:          *checkpointing strategy of* $S_i \leftarrow$ BLOCKING
8:       **else**
9:          *checkpointing strategy of* $S_i \leftarrow$ UNBLOCKING
10:      **end if**
11:   **end if**
12:   $j \leftarrow i$
13:   *collect* t_{ck_j}, t_{co_j}
14:   $i \leftarrow i + 1$
15: **end while**

Table 1. Dataset description.

Dataset	Data size	# of Vertices	# of Edges
LiveJournal	1.1 GB	3,997,962	34,681,189
Orkut	1.7 GB	3,072,441	117,185,083

and our implemented unblocking checkpointing as baselines. This section shows the efficiency of three checkpointing strategies in the failure-free and failure cases.

5.1 Experiment Setting

Cluster Setup. Our experiments run on a cluster which consists of 7 compute nodes. In this cluster, each node (8 Intel Xeon E5606 CPUs, 30 GB memory, and 2 TB HDD) is installed CentOS 6.5 and Java 1.8. In order to run Giraph jobs, we deploy Hadoop 2.5.1 on this cluster, where one node acts as the master running NameNode and ResourceManager. In addition, we deploy Zookeeper 3.5.5 for Giraph's master election and barrier synchronization. By default, we setup 18 map tasks with 3 GB memory on the cluster, one of which is the master of Giraph and the others are the workers of Giraph.

Workload. We adopt the Single Source *Shortest Path* (*SP*) and *Connected Components* (*CC*) algorithms used in [9,14] for our experiments. These two algorithms run on Giraph with a checkpointing interval as 5. For simplicity, in the rest of this paper, we refer to the Single Source *Shortest Path* and *Connected Components* algorithms as *SP* and *CC*, respectively.

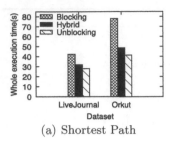
(a) Shortest Path

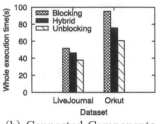
(b) Connected Components

Fig. 5. Checkpointing in failure-free cases

Dataset. We evaluate our workload on two online social network graph datasets called LiveJournal[1] and Orkut.[2] In our implementation, *CC* treats the dataset as an undirected graph and generates the reverse edges before starting computation, while *SP* only applies computation based on the original dataset. Table 1 provides the details of the datasets.

5.2 Efficiency Without Failure

Figure 5 shows the execution time of all supersteps as well as the one with three different kinds of checkpointing strategies. In Fig. 5(a), the unblocking checkpointing of *SP* on LiveJournal decreases an additional overhead of 34% compared to blocking checkpointing on LiveJouranl dataset. The overhead of hybrid checkpointing is decreased by 24% compared to blocking checkpointing. For the Orkut dataset, unblocking checkpointing decreases the overhead by 46% compared to blocking checkpointing. Hybrid checkpointing decreases the overhead of *SP* on the Orkut dataset by 37% compared to blocking checkpointing. We also evaluate *CC* on these two datasets and Fig. 5(b) shows the trend of the execution time of all supersteps follows the one of *SP*.

In general, the hybrid checkpointing as well as the unblocking checkpointing outperforms the blocking checkpointing. Although the unblocking checkpointing achieves the lowest overhead, we cannot always ensure a failure-free environment. Hence, we make further evaluations in failure cases as illustrated in Sect. 5.3.

5.3 Efficiency with Failure

In this group of experiments, we study the impact of failure on three different kinds of checkpointing strategies referring to the failure setting in [10,13]. For the *SP* algorithm in Fig. 6(a) and 6(b), we focus on cascading failure by setting the first failure at superstep S_{13}, varying the second failed superstep from S_{10} to S_{12}, and injecting the third failure at superstep S_{10}. In Fig. 6(a), the blocking checkpointing of *SP* on LiveJournal decreases additional overhead of

---

[1] https://snap.stanford.edu/data/com-LiveJournal.html.
[2] https://snap.stanford.edu/data/com-Orkut.html.

20% compared to unblocking checkpointing on LiveJournal dataset. The overhead of hybrid checkpointing on LiveJoural is reduced by 25% compared to unblocking checkpointing. For the Orkut dataset in Fig. 6(b), blocking checkpointing decreases the overhead by 28% compared to unblocking checkpointing. The overhead of hybrid checkpointing is decreased by 38% compared to unblocking checkpointing. For the CC algorithm in Fig. 6(c) and 6(d), the total number of supersteps is less than 13, since it converges faster than SP. Here, we set the first failure at superstep S_8, vary the second failed superstep from S_5 to S_7, and inject the third failure at superstep S_5. Figure 6(c) shows the overhead of blocking checkpointing of CC on LiveJournal is decreased by 40% compared to unblocking checkpointing. The hybrid checkpointing decreases an additional overhead of 43% compared to unblocking checkpointing. Moreover, the trend in Fig. 6(d) is similar to Fig. 6(c).

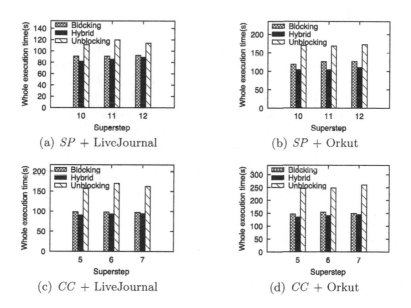

(a) SP + LiveJournal

(b) SP + Orkut

(c) CC + LiveJournal

(d) CC + Orkut

Fig. 6. Checkpointing in failure cases

In general, the hybrid checkpointing outperforms the blocking and unblocking checkpointing in failure cases, since the hybrid checkpointing achieves a balance between the blocking checkpointing and unblocking checkpointing.

6   Related Work

In this section, we discuss some work related to the checkpointing in the BSP-based systems with a focus on checkpoint strategies as well as checkpoint interval.

Checkpoint Strategies. *Blocking checkpointing* employed by the BSP-based systems such as Pregel [6] and Giraph [1] usually takes the method of pausing computation to store the system states. MLBase [5] and GraphX [4] utilize the checkpointing mechanism of Spark [16] in a blocking way. Different from blocking checkpointing, the *unblocking checkpointing* typically stores the system state in parallel with the computation. The work in [13] adopts unblocking checkpointing for iterative processing on Flink [2]. In particular, it materializes checkpoints into external storage within the dataflow pipeline execution during the iteration. However, none of these work combines the two checkpointing strategies together.

Checkpoint Interval. A first-order approximation to the optimal *checkpoint interval* is derived in [15] for general data management systems. The dynamic checkpointing [12] adjusts the interval according to the cost of checkpointing and recovery. Different from these work, our approach focus on the choice of checkpoint strategies under a predefined checkpoint interval.

7 Conclusion

Motivated by the trade-off between blocking and unblocking checkpointing, this paper proposes a checkpointing choice problem to reduce the whole execution time for iterative processing in BSP-based systems. The challenge is to make a decision during iterative processing without future runtime information. In order to address this problem, we propose a hybrid checkpointing which evaluates the cost of blocking and unblocking manner and heuristically chooses the manner with a lower cost. The experimental studies show that the hybrid checkpointing outperforms the blocking and unblocking checkpointing. Presently, we integrate our hybrid checkpointing in Giraph [1]. Nevertheless, it is possible to implement this strategy in other BSP-based systems such as MLBase and GraphX.

Acknowledgments. This work was supported by the National Natural Science Foundation of China (No. 61902128), Shanghai Sailing Program (No. 19YF1414200).

References

1. Apache giraph. https://giraph.apache.org/
2. Carbone, P., et al.: Apache flink_TM: stream and batch processing in a single engine. IEEE Data Eng. Bull. **36**, 28–38 (2015)
3. Dean, J., et al.: MapReduce: simplified data processing on large clusters. In: OSDI, pp. 137–150 (2004)
4. Gonzalez, J.E., et al.: GraphX: graph processing in a distributed dataflow framework. In: OSDI, pp. 599–613 (2014)
5. Kraska, T., et al.: MLbase: a distributed machine-learning system. In: CIDR (2013)
6. Malewicz, G., et al.: Pregel: a system for large-scale graph processing. In: SIGMOD, pp. 135–146 (2010)
7. Natella, R., et al.: Assessing dependability with software fault injection: a survey. ACM Comput. Surv. **48**(3), 1–55 (2016)
8. Schelter, S., et al.: "All roads lead to Rome": optimistic recovery for distributed iterative data processing. In: CIKM, pp. 1919–1928 (2013)

9. Shan, X., Ma, J., Gao, J., Xu, Z., Song, B.: A subgraph query method based on adjacent node features on large-scale label graphs. In: Ni, W., Wang, X., Song, W., Li, Y. (eds.) WISA 2019. LNCS, vol. 11817, pp. 226–238. Springer, Cham (2019). https://doi.org/10.1007/978-3-030-30952-7_24
10. Shen, Y., et al.: Fast failure recovery in distributed graph processing systems. PVLDB **8**(4), 437–448 (2014)
11. Valiant, L.G.: A bridging model for parallel computation. Commun. ACM **33**(8), 103–111 (1990)
12. Wang, Z., Gu, Yu., Bao, Y., Yu, G., Gao, L.: An I/O-efficient and adaptive fault-tolerant framework for distributed graph computations. Distrib. Parallel Databases **35**(2), 177–196 (2017). https://doi.org/10.1007/s10619-017-7192-2
13. Xu, C., et al.: On fault tolerance for distributed iterative dataflow processing. IEEE Trans. Knowl. Data Eng. **28**, 1709–1722 (2017)
14. Xue, J., et al.: Seraph: an efficient, low-cost system for concurrent graph processing. In: HPDC, pp. 227–238 (2014)
15. Young, J.W.: A first order approximation to the optimal checkpoint interval. Commun. ACM **17**(9), 530–531 (1974)
16. Zaharia, M., et al.: Resilient distributed datasets: a fault-tolerant abstraction for in-memory cluster computing. In: NSDI, pp. 15–28 (2012)

Intelligent Visualization System for Big Multi-source Medical Data Based on Data Lake

Peng Ren[1], Ziyun Mao[2(✉)], Shuaibo Li[3], Yang Xiao[4], Yating Ke[2], Lanyu Yao[3],
Hao Lan[3], Xin Li[5], Ming Sheng[1], and Yong Zhang[1]

[1] BNRist, DCST, RIIT, Tsinghua University, Beijing 100084, China
{renpeng,shengming,zhangyong05}@tsinghua.edu.cn
[2] Beijing University of Technology, Beijing 100124, China
[3] Henan University, Kaifeng 475004, China
[4] Northwestern Polytechnical University, Shaanxi 710068, China
[5] Beijing Tsinghua Changgung Hospital, School of Clinical Medicine, Tsinghua University,
Beijing 102218, China
Horsebackdancing@sina.com

Abstract. With the rapid development of information technology, large amounts of multi-source data are constantly being generated in medical field. The automatic visualization system based on them has gained a lot of attention, since the intuitive data presentation can help even non-professional users effectively get the information hidden behind the separate data obtained from different scenarios and make better decisions. In this paper, based on the Data Lake architecture, we improve the performance of an existing novel data visualization recommendation system and resolve three challenges about the processing of multi-source and heterogeneous data. First, we build the framework based on Data Lake to store multi-source and heterogeneous data. Second, we optimize the data manipulation module in the visualization system based on the distributed processing power of Data Lake to get potentially interesting visualization candidates in a short time. Third, we efficiently run exploratory queries on large datasets based on the calculation capability of Data Lake to meet the actual needs of users. According to the experiment results, our system demonstrates a remarkable acceleration effect on the task of automatic visualization of big multi-source medical data.

Keywords: Data Lake · Multi-source and heterogeneous data · Distributed processing · Automatic data visualization

1 Introduction

With the rapid development of "Internet +" and application of technologies such as cloud computing and mobile Internet, all the fields of society have stepped into the era of big data, and the medical field has no exception [1]. Generally speaking, medical big data is the data generated during the entire process from outpatient registration to patient admission, and hospitals' real-time follow-up to patients [1]. Apparently, the hospital has a wealth of data resources, including medical expenses, electronic medical records, medical images, etc. [2]. However, due to the high demand of data processing in the medical field w.r.t. the volume, variety, velocity, and veracity of data, the traditional medical information systems are gradually unable to satisfy the requirements. For instance, the data integrity of the system is not enough, and the data processing speed is slow [3]. Therefore, there is an emerging need to develop a modern system that can efficiently govern and analyze the data with high-performance.

As an important tool for data analysis, data visualization system can present implicit relationships between data variables in an intuitive and easy-to-use manner. According to [4] and [5], it also has powerful advantages of reducing medical waste. However, it is hard to require users to fully understand the underlying heterogeneous data and the visualization skills. In that respect, there is a demand for novel approaches that can automatically visualize the data, including heterogeneous data integration, visualization-aware data discovery, intelligent recommendation and interactive query.

Contributions. In this paper, based on Data Lake, we propose a system called VisLake combining with DeepEye [6] to fulfill automatic visualization of the multi-source medical big data. Data Lake (DL) is defined as a big data analytics solution that can ingest heterogeneous data and allow self-service data processing [7]. The underlying storage layer of our framework is fully compatible with Apache Spark APIs. Leveraging the distributed processing power and powerful calculation capability of Spark, we further optimize the performance of DeepEye by addressing several research challenges. There are the contributions of our work:

1. We build the framework based on DL to store and integrate the multi-source medical data into unified structured dataset. The data from various sources make users more possible to get comprehensive knowledge from the visual representation.
2. Utilizing the parallel processing capability of Apache Spark, we optimize the data manipulation module within DeepEye based on various relational operators (e.g., *groupBy, count, orderBy*) and parallel functions (e.g., *map*), which can significantly accelerate the task for expanding the search space.
3. Taking advantage of the optimized storage and execution plans of Spark, we improve the search performance of DeepEye. It means users can get in-time response through explicitly specifying the range of the values, text searching and so on.

This paper is organized as follows. Section 2 discusses related works. Section 3 presents our architecture and depicts the storage mechanism. Section 4 implements the multi-source data integration. Section 5 describes the optimized techniques used to manipulate data. Section 6 implements in-time interactive query based on Spark SQL. Section 7 concludes this paper and points out future research direction.

2 Related Work

Over the past few years, data visualization is a widely studied field. The high demand for creating good visualizations has nourished a number of state-of-the-art interactive visualization tools (e.g., Vega-Lite [8], Hyper DB [9]). However, automatic visualization of data in medical field has not yet been extensively studied. Besides, most of current visualization systems (e.g., Echarts [10] and D3 [11]) are infeasible for non-technical people to generate both eye-catching and informative visualizations due to complex visualization methods (e.g., various attribute combinations, disparate transformation operations). Furthermore, in order to effectively involve users in the interactive system, researchers have tried both interfacing with powerful data processing engines, and leveraging approximate solutions [12]. Nevertheless, it is prohibitive to materialize the optimization of all results [6], and users may get frustrated once a significant difference between approximate and accurate visualizations happens [13].

DeepEye [6] is an innovative data visualization system which can automate the visualization task that currently requires heavy intervention. Furthermore, it can recommend top ranked visualizations, allow users to clarify needs by text searching and support iterative data exploration. However, it still has room for improvement in terms of integrating and manipulating big multi-source data, which can be solved by the VisLake. We summarize the challenges from the following aspects:

4. **Multiple Data Sources.** Sometimes, attractive visualizations only emerge from multi-source heterogeneous data, such as structured data from relational database, semi-structured data like JSON files, and unstructured data like images. We need to integrate the heterogeneous data into a larger structured dataset with reliability.
5. **Large Search Space.** Usually, the as-is visualizing of a given dataset can only produce boring outputs. Compelling stories may not emerge until the data has been manipulated, such as selecting attributes, grouping by X axis, aggregating on Y axis, which can create a huge search space containing potential attractive visualizations [14]. However, the operations performed to expand the search space may be a time-consuming job, thus it is necessary to optimize the data manipulation module.
6. **Database Engines.** In order to meet the in-time interaction demand, we are supposed to efficiently filter the materialized visualization results according to users' requirements. Typically, it is a database optimization problem, yet still challenging to explore some efficient pruning techniques [15].

3 The VisLake Framework

The framework of VisLake is shown in Figs. 1 and 2. Combining with the DeepEye, VisLake realizes automatic data visualization and recommendation in the Data Consumption module. The DL plays the role of ingesting, processing, and governing multi-source data. In the Raw data zone, all types of medical data are stored in their native formats, allowing users to find the original version of data [7]. In the Process zone, heterogeneous data are fused and stored as intermediate data to facilitate further inquiry. After being automatically manipulated, the visualization candidates are stored as available data in the Access zone to support interactive access. The Govern zone is in charge of insuring data security, data quality, and metadata management [7].

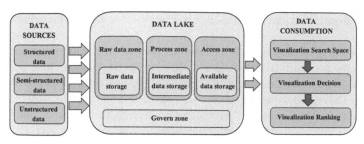

Fig. 1. Functional architecture

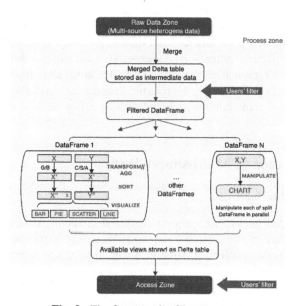

Fig. 2. The framework of Process zone

In medical field, most datasets are continuously updated and users may face the risk to see incorrect data if consistency guarantees are limited. Besides, seeking for low-cost and high-reliability storage approach to insure data quality is still a challenge. By utilizing a specific data storage format called Delta table [16], we can achieve performant and mutable table storage.

The Delta table within Delta Lake (an open source ACID table storage layer [17]) is actually a file system directory including table contents and a log of transaction operations [16]. The table contents are stored as a collection of objects encoded in Apache Parquet format, which can provide higher query efficiency by fetching specific column values and save storage space by column-wise encoding techniques. The transaction log can provide ACID properties, time travel (allowing users to roll back erroneous updates) and faster metadata operations that enables high-performance searching on large tabular datasets [16]. Functionally, this design enables in-time query and update of table in a serializable manner while maintaining high parallel read and write performance similar to the raw Parquet. VisLake leverages this transactional design by invoking Delta Lake APIs.

As is known to all, direct analysis based on the raw data with inconsistent formats may raise some problems like slow data processing speed, low searching accuracy, etc. The main problem is the data integrity is not enough, so we are supposed to fuse all of the raw data and regard them as a whole. We provide a common way to access a variety of data sources by DataFrame API provided by Spark SQL. In essence, the DataFrame is a distributed collection of rows with a homogeneous schema [18], and is equivalent to a table in relational database. Therefore, structured data from external relational DBMS can be directly loaded as the DataFrame. The DataFrame also supports complex data types (e.g., struct, array), thus semi-structured data like JSON files can be transformed to the DataFrame under the support of automatic schema inference. Besides, feature vectors can be extracted from unstructured data (e.g., images) by embedding models (e.g., ResNet), which can eventually be converted to the feature columns of DataFrame. As a result, the DataFrame can be considered as a table holds all of fused data, which will be stored as the target Delta table in the intermediate data storage. We can continuously upsert obtained DataFrame into the archived data by using the *merge* operation within Delta Lake API.

4   The Integration of Multi-source Data

4.1   Integration Mechanism

We consider each data file as a set containing different elements, and the data structure of each element can be represented by a tree graph. Then the integration of heterogeneous data is converted to the problem of union of different sets, which can be illustrated by the fusion of various tree graphs. The root node of the tree graph denotes the primary key that can identify each element in the set, and the sub-nodes denote the features of the element. We outline two tree-based representations. The former has one layer of sub-nodes, corresponding to the structured elements. The latter has multiple layers of sub-nodes, corresponding to the semi-structured or unstructured elements. It is impractical to directly fuse the tree with different forms, since the DataFrame with different schemas cannot be correctly integrated with each other. It is necessary to convert the second type of the tree graph to the first type. We illustrate the process by giving a concrete example about integrating structured dataset with semi-structured JSON document. There are two steps: converting the semi-structured JSON file to the structured dataset and integration of two structured datasets.

Parse Semi-structured Data. As is known to all, the JSON format is basically designed around two types of entities—objects and arrays. Usually, the JSON Object is a set of name-value pairs, where each pair contains identified name along with JSON Value element. The JSON Array is a set of JSON Value elements. We divide the JSON Value element into three types: primitive type (e.g., String, Number or Boolean), object type (a JSON Object element) and array type (a JSON Array element).

Each JSON Object element contained in the JSON file can be represented by a tree graph. There are two types of sub-nodes: (i) The leaf node representing the JSON Pair where the JSON Value is the primitive type, or the array type that consists of the primitive JSON Value element. (ii) The non-leaf node representing the JSON Pair where the JSON Value is the object type, or the array type that consists of the JSON Object element. It is clear that the first JSON Pair has very flat schema—only one level, whereas the latter has nested data structure. In this case, we are supposed to convert the tree with non-leaf nodes to the form with one layer of sub-nodes.

Given a JSON Object element (as shown in Fig. 3) and its tree graph (as shown in Fig. 4), we traverse the whole tree based on the breadth-first approach. The leaf node do not need to be changed. For the non-leaf node, we substitute it by the sub-nodes attached. If the node represents the JSON Pair where the JSON Value is array type, we split the array based on its elements. It means we generate new trees with same schema but identifiable root nodes, and reassign each JSON Value element within array to the corresponding node of new tree. We operate on each node until the tree is converted to the one-level form (as depicted by Fig. 5). Afterwards, the integration of two datasets can be further decomposed as union of different sets.

```
{    "PATIENT":  { "ID": "104891",
                   "ADMISSION_DATE_TIME": "2018-06-08",
                   "DISCHARGE_DATE_TIME": "2018-07-06",
                   "TOTAL_COST": "230772.86" },
     "DOCTOR_ID:  "48",
     "DIAGNOSIS": [
                   { "ORDER_CODE": "102090006",
                     "ORDER_TEXT": "blood routine examination" },
                   { "ORDER_CODE ": "530000055",
                     "ORDER_TEXT": "immunoassay" },
                   ... ]
}
```

Fig. 3. A JSON Object in medical field   **Fig. 4.** The tree graph of a JSON Object

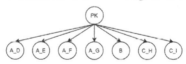

Fig. 5. The tree graph of the parsed JSON Object

Integration of Datasets. Supposing we have set A and B, the A represents the dataset converted from the JSON file, and the B represents the structured dataset (as shown in Fig. 6). The target set S obtained after fusion can be denoted as Eq. (1).

$$S = A \cup B \qquad\qquad (1)$$

There are three situations in the process of fusion: two identical trees, two trees that are partly same, two totally different trees. It is simple to define the first case: if the root nodes and the leaf nodes are same, they are two identical trees, so we only retain one of the elements in the target set. It is also easy to define the third case: the root nodes are different, where we have to retain both of the elements. The second depicts the situation where the root nodes are equivalent yet leaf nodes are different, indicating the element is described by features split in two sets. In this case, we need to get the union of the features from two elements, thus we attach all the distinguished leaf nodes to the root node. The pseudo code below illustrates the fusion of two elements.

Algorithm 1. Fusion of two trees

```
Input:   Tree T1, Tree T2
Output:  Output target tree T
1.    Initialize T
2.    If    T1.root node == T2.root node
3.         If  T1.leaf nodes == T2.leaf nodes
4.              T1 == T2
5.         Else
6.              T1 = T1 + (T2 − T1 ∩ T2)
7.            T = T + {T1}
8.         Else
9.              T = T + {T1} + {T2}
10.   return T
```

After integrating all elements, we fulfill the fusion of datasets. The elements in our set A and B have the same primary key but different features, which is categorized as the second case. The tree graph of the element within target set S is shown in Fig. 7.

Fig. 6. The tree graph of the structured dataset

Fig. 7. The tree graph of the fused data

4.2 Technical Detail

In practice, we utilize DataFrame API to coalesce heterogeneous data into the DataFrame with homogeneous schema. Typically, each element represented by the tree graph denotes a data row in DataFrame, and each node of the tree denotes a feature column in DataFrame. However, considering the JSON Object with feature (the JSON Pair element) consisting of sub-features, the schema of DataFrame may be nested, in which we have to convert it to the form with flat schema first.

Parse the DataFrame with Nested Schema. The structured dataset will be directly loaded as DataFrame with flat schema that contains the information (e.g., name, datatype, nullable type) of each column. Whereas the DataFrame converted from the JSON Object has nested schema, which can be generalized from two angles: column and row. Since the DataFrame can automatically infer the datatype (e.g., StructType, ArrayType, String, etc.) of each column, we can respectively deal with the two situations. In the first case, several sub-features of the JSON Object are compressed in one column of DataFrame, in which we manually disassemble it into the atomic columns. The latter means that several JSON Value elements within the array are compressed in one row of DataFrame, and we split it by the *explode* function within DataFrame API. In short, based on the datatype of each feature column, we parse the compressed row then the column, until the schema of the DataFrame is completely flat.

Integration by *Merge*. We realize the data fusion by utilizing the *merge* function. It is similar to the SQL MERGE command but has additional support for deletes and extra conditions in updates, inserts, and deletes [19].

The *merge* task can be broken down into three key phases. The first phase is finding rows in the target file that satisfy the join condition. It is done via an inner join to find out rows need to be modified. The second phase is reading the touched file again and writing new file with updated and/or inserted rows. In the third phase, it atomically remove the touched files and add the new file. Actually, the new file is obtained via joining the files of target and source. The type of join (e.g., inner join, outer join, left semi join) can vary depending on the conditions of the *merge*. We summarize three typical types as below (The left denotes the source and the right denotes the target):

1. Insert only *merge* (e.g., when not matched): there is no update/delete, but perform left anti join to add the inserted rows from the source file.
2. Matched only *merge* (e.g., when matched): perform right outer join on the target and update the matched rows.
3. Else (e.g., have updates and/or deletes, and inserts): perform full outer join.

Actually, we perform full outer join on the medical data to retain as much information as possible. In essence, *merge* is a combined plan of join including two stages in turn: the inner join and the outer join. During the process, it automatically generates an optimized execution plan that can improve the performance of join. Moreover, some pruning methods like partition prune can also be manually utilized. The advantages are exploited in VisLake by the implementation of applied interface. Based on the integrated medical data, one single program can efficiently query the separate data.

5 Data Manipulation

Usually, the as-is visualization of a given dataset cannot produce impressive outputs. In this section, we introduce the operations performed on the dataset based on Spark SQL, and how to transform the DataFrame obtained from the previous section to the one containing visualization information. We discuss three types of data operations within DeepEye (as shown in Fig. 8) that are optimized by DataFrame APIs.

As the main abstraction in Spark SQL's API, the DataFrame can be manipulated with various relational operators (e.g., *where*, *groupBy*), and also be considered as an RDD [20] of Row objects, enabling users to call procedural Spark APIs (e.g., *map*) as they do for the distributed datasets in Spark. In view of that the DataFrames are evaluated lazily (no computation is launched until users call a certain "output operation" like *count*), and can keep track of data schema, thus enables rich optimizations for relational processing. The data manipulation is implemented by two steps.

Transform: categorize data together based on the following operations.

    (i) Bin: partition the temporal (e.g., hour and date) or numerical (e.g., hospital expenses) values into different buckets;

    (ii) Group-by: group the temporal or categorical values (e.g., specific medicine type) in a column.

Aggregation: compute value for each group or bin by aggregate functions like count, sum, and average.

Sorting: sort the numerical and temporal value in a column.

Fig. 8. Operations in DeepEye

Step 1: Convert the original DataFrame to a set of DataFrames. To be specific, we select one feature column to perform grouping/binning, and the rest of columns to perform AGG (AGG = {SUM,AVG,CNT}), then sort the numerical and temporal values in a column. The operations within DeepEye are executed by a set of relational operators within DataFrame API, including projection (*select*), filter (*where*), transformation (*groupBy*), aggregations (*count, sum, mean*), sorting (*orderBy*), as well as procedural functions like *map* and *reduce*. As a result, each of split DataFrame represents a group of manipulated data categorized by the grouped/binned column.

Step 2: Create one DataFrame that holds the information of all possible visualizations. In practice, for each split DataFrame, we assign chart type and other visualization features for each combination of columns (representing X and Y axes), namely, we create a set of new DataFrames (each row represents one visualization) based on the split DataFrames. In detail, the new schema information can be divided from two aspects. The first is the data features serve as input for subsequent Visualization Recognition model [14], such as the data type (categorical, numerical, temporal) of one column (X/Y axis), the min/max value in one column, etc. The latter is relevant to the final visualization results that mainly serve for interactive query, such as chart name, attribute names of X and Y axes, etc. Next we coalesce all separate DataFrames into one DataFrame and persist it as the Delta table in the available data storage.

It is worthy to note that we retain the optimization methods within DeepEye, such as computing the AGG values on other columns together, utilizing the predefined rules (Transformation, Sorting, and Visualization rules [14]) to directly get relatively valid visualizations, etc. Moreover, we manipulate each split DataFrame in parallel by invoking some procedural Spark APIs like *map*, which can perform distributed computing through task distribution, thus yields significant performance improvement [21]. Ultimately, containing the information of visual charts generated from medical multi-source data, the DataFrame can support high-performance ad-hoc visualization queries with ease.

6   Interactive Query

6.1   Query Implementation

The earliest systems designed for big data applications commonly provide users a powerful, but low-level programming interface [18]. The manual optimization required might be burdensome for large volume of medical data. The Spark SQL module integrates Spark's functional programming API with relational processing, providing a more user-friendly experience by offering relational interface along with automatic optimizations. A novel extensible optimizer called Catalyst has been proposed to improve query performance [18], which can be leveraged by calling DataFrame APIs, and demonstrates a remarkable acceleration effect.

To facilitate the discussion, DeepEye defines a simple visualization language to capture all possible visualizations, including clauses like VISUALIZE, SELECT, and TRANSFORM [22]. Considering a Medical Expense Dataset containing features of expense category and medical expenses, the query Q (see Fig. 9) means grouping by expense category, aggregating medical expenses using SUM and specifying pie as the chart type. The query result shows the expense proportions of each medical category.

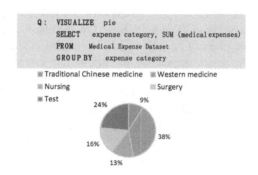

Fig. 9. An example of query and visualization result

In our framework, there are two types of queries summarized from a functional point of view (as shown in Fig. 2): (i) queries on the DataFrame in the intermediate data storage to filter columns and rows intrigued by users, (ii) queries on the DataFrame in the available data storage to obtain a set of visualizations.

In the first case, users can specify the range of data values by the "Filter" panel, then we use the *where* operator to directly filter the DataFrame. In the second case, we register the DataFrame as temporary tables in the system catalog, which can be queried using SQL. When any text query is fetched from the front end, DeepEye first extracts reserved keywords (e.g., table name), then feeds them into the clauses within visualization queries [22]. Actually, the feature columns of the DataFrame (e.g., chart type, grouping/binning operated on X axis) correspond to the fields of a relational table, thus we can easily convert these clauses (e.g., VISUALIZE, TRANSFORM) to the WHERE clause within SQL query. Since both the relational operators and the SQL query (will be passed to SQL parser) can build up an abstract syntax tree (will be passed to Catalyst), optimizations can happen across the relational operations on DataFrame and the SQL. As a result, we obtain a set of visual charts as visualization candidates, which will be discovered and ranked in the Data Consumption module. We recommend the top ranked ones to users based on DeepEye.

6.2 Application

The VisLake has been deployed to XXX hospital to support interactive visualization in knee osteoarthritis domain. Compared with the traditional Hospital Information System, VisLake can query the integrated heterogeneous data ingested from disparate sources with one single program. Moreover, we compare the performance of unified queries on the DataFrame against the queries based on the Oracle RDBMS. The original dataset is 248 GB of osteoarthritis data that contains both semi-structured JSON files and structured data. The experiment results show that the DataFrame-based queries outperform the Oracle-based queries by 1.26 x.

7 Conclusion and Future Work

We present a framework called VisLake based on Data Lake, which can better serve the data visualization needs in medical field. Based on the distributed processing power and optimized execution plans of Apache Spark, we realize medical multi-source data integration, optimize the data manipulation module within DeepEye, and improve the query efficiency to enable real-time interactive data visualization. However, the response time is still subject to the visualization ranking, which is implemented in DeepEye based on a visualization linkage graph [23]. One major future work is to seek for appropriate graph storage mechanisms along with efficient ranking algorithms. Some distributed graph processing frameworks (such as GraphX [24]) are suitable for large-scale graph computing, which can be leveraged to further optimize the performance of visualization recommendation.

Acknowledgements. This work was supported by the National Key R&D Program of China (2019YFC0119600).

References

1. Feng, W., Li, G., Zhao, H.: Research on Visualization and Application of Medical Big Data, pp. 383–386 (2018)
2. Yang, Y., Chen, T.: Analysis and visualization implementation of medical big data resource sharing mechanism based on deep learning. IEEE Access **7**, 156077–156088 (2019)
3. Liu, H., Taniguchi, T., Tanaka, Y., Takenaka, K., Bando, T.: Visualization of driving behavior based on hidden feature extraction by using deep learning. IEEE Trans. Intell. Transp. Syst. **18**, 2477–2489 (2017)
4. Satagopam, V., et al.: Integration and visualization of translational medicine data for better understanding of human diseases. Big Data **4**, 97–108 (2016)
5. Ledesma, A., Al-Musawi, M., Nieminen, H.: Health figures: an open source JavaScript library for health data visualization. BMC Med. Inform. Decis. Mak. **16**, 38 (2016)
6. Qin, X., Luo, Y., Tang, N., Li, G.: Deepeye: An automatic big data visualization framework. Big Data Min. Analyt. **1**, 75–82 (2018)
7. Ravat, F., Zhao, Y.: Data lakes: trends and perspectives. In: Hartmann, S., Küng, J., Chakravarthy, S., Anderst-Kotsis, G., Tjoa, A.M., Khalil, I. (eds.) DEXA 2019. LNCS, vol. 11706, pp. 304–313. Springer, Cham (2019). https://doi.org/10.1007/978-3-030-27615-7_23

8. Satyanarayan, A., Moritz, D., Wongsuphasawat, K., Heer, J.: Vega-lite: a grammar of interactive graphics. IEEE Trans. Vis. Comput. Graph. **23**, 341–350 (2016)
9. Kemper, A., Neumann, T.: HyPer: a hybrid OLTP&OLAP main memory database system based on virtual memory snapshots. In: 2011 IEEE 27th International Conference on Data Engineering, pp. 195–206. IEEE (2011)
10. Li, D., et al.: ECharts: a declarative framework for rapid construction of web-based visualization. Vis. Inf. **2**, 136–146 (2018)
11. Bostock, M., Ogievetsky, V., Heer, J.: D^3 data-driven documents. IEEE Trans. Vis. Comput. Graph. **17**, 2301–2309 (2011)
12. Moritz, D., Fisher, D., Ding, B., Wang, C.: Trust, but verify: optimistic visualizations of approximate queries for exploring big data. In: Proceedings of the 2017 CHI conference on human factors in computing systems, pp. 2904–2915 (2017)
13. Qin, X., Luo, Y., Tang, N., Li, G.: Making data visualization more efficient and effective: a survey. VLDB J. **29**(1), 93–117 (2019). https://doi.org/10.1007/s00778-019-00588-3
14. Luo, Y., Qin, X., Tang, N., Li, G.: DeepEye: towards automatic data visualization, pp. 101–112 (2018)
15. Deng, D., Li, G., Feng, J., Duan, Y., Gong, Z.: A unified framework for approximate dictionary-based entity extraction. VLDB J. **24**, 143–167 (2015)
16. Armbrust, M., et al.: Delta lake. Proc. VLDB Endow. **13**, 3411–3424 (2020)
17. Introduction to Delta Lake — Delta Lake Documentation. https://docs.delta.io/0.4.0/delta-intro.html. Accessed 21 May 2021
18. Guller, M.: Spark SQL. In: Big Data Analytics with Spark, pp. 103–152. Apress, Berkeley, CA (2015)
19. Table Deles, Updates and Merges — Delta Lake Documentation. https://docs.delta.io/0.4.0/delta-update.html. Accessed 21 May 2021
20. Zaharia, M., et al.: Resilient distributed datasets: a fault-tolerant abstraction for in-memory cluster computing. In: 9th {USENIX} Symposium on Networked Systems Design and Implementation ({NSDI} 12), pp. 15–28 (2012)
21. Zhao, X., Lei, Z., Zhang, G., Zhang, Y., Xing, C.: Blockchain and distributed system. In: Wang, G., Lin, X., Hendler, J., Song, W., Xu, Z., Liu, G. (eds.) WISA 2020. LNCS, vol. 12432, pp. 629–641. Springer, Cham (2020). https://doi.org/10.1007/978-3-030-60029-7_56
22. Luo, Y., Qin, X., Tang, N., Li, G., Wang, X.: DeepEye: Creating Good Data Visualizations by Keyword Search, pp. 1733–1736 (2018)
23. Qin, X., Luo, Y., Tang, N., Li, G.: DeepEye: Visualizing Your Data by Keyword Search. In: EDBT, pp. 441–444. (2018)
24. Gonzalez, J.E., Xin, R.S., Dave, A., Crankshaw, D., Franklin, M.J., Stoica, I.: Graphx: graph processing in a distributed dataflow framework. In: 11th {USENIX} Symposium on Operating Systems Design and Implementation ({OSDI} 14), pp. 599–613 (2014)

Improved Raft Consensus Algorithm in High Real-Time and Highly Adversarial Environment

Yuchen Wang, Shuang Li, Lei Xu, and Lizhen Xu(⊠)

School of Computer Science and Engineering, Southeast University, Nanjing 211189, China

Abstract. High real-time and highly adversarial environment put forward higher requirements for the performance of blockchain consensus algorithm. To improve Raft's consensus efficiency and safety, we propose an improved Raft algorithm called "hhRaft" to optimize Raft consensus process by introducing a new role of monitor. In the leader election phase, monitor nodes supervise the candidate nodes by identifying the malicious node's forged Requestvote message. In the log replication phase, monitor nodes supervise the leader node by comparing the computing results of transactions. Through the performance test on the Consortium Blockchain -- Hyperledger Fabric, it is proved that hhRaft is superior to the original Raft algorithm in terms of transaction throughput, consensus latency, and anti-Byzantine Fault capabilities, making it suitable for use in high real-time and highly adversarial environment.

Keywords: Raft · Blockchain · Consensus algorithm · Byzantine fault tolerance · Highly adversarial environment

1 Introduction

The price of Bitcoin has been breaking record highs in recent months, making blockchain a hot topic again. Blockchain is an emerging technology that originated in Bitcoin, connecting data blocks in chronological order to form a tamper-resistant, non-forgeable distributed ledger. With the development of information technology and the exploration of blockchain, the application of blockchain has expanded into new fields such as UAV swarm control [1], missile swarm coordinated guidance [2], electronic medical record (EMR) sharing [3] and so on. These applications require the blockchain to work in a high real-time and highly adversarial environment, and put forward higher performance requirements for blockchain consensus algorithms.

Taking the real-time combat environment [2] of missile swarm as an example, the main features of a high real-time and highly adversarial environment are: (1) Malicious nodes and signal interference may disturb the blockchain network, which requires the consensus algorithm to resist Byzantine fault. (2) Requests from internal nodes should be responded in a timely manner, which requires the consensus algorithm has a low consensus latency. (3) High-speed nodes and changing missile formations make the communication situation extremely unstable, which requires the consensus algorithm to maintain a high consensus efficiency in the dynamic network.

© Springer Nature Switzerland AG 2021
C. Xing et al. (Eds.): WISA 2021, LNCS 12999, pp. 718–726, 2021.
https://doi.org/10.1007/978-3-030-87571-8_62

Raft consensus algorithm has the advantages of high throughput and high scalability, but its flaw is that Raft is a CFT (Crash Fault Tolerance) algorithm. In a highly adversarial environment, Raft has potential security problems. Furthermore, Raft has the possibility of a network split due to network instability issues, which seriously affects the consensus stability and efficiency in a high real-time environment. To solve these flaws, we propose an improved Raft consensus algorithm named "hhRaft" for a high real-time and highly adversarial environment, which introduces a new node role "monitor" to resist Byzantine Fault [4] and improve consensus efficiency. Finally, we evaluated the performance of hhRaft on the Consortium Blockchain -- Hyperledger Fabric.

2   Related Work

2.1   Raft Consensus Algorithm

Raft (Replication and Fault-Tolerant) is a consensus algorithm published by Stanford University professors Diego Ongaro and John Ousterhout in 2014 [5]. It ensures system consistency in a distributed system of N nodes with $(N-1)/2$ nodes or less in the event of non-Byzantine faults (such as node crash, network partitions, message delay, etc.). Compared with the Kafka Consensus algorithm, Raft is more suitable for large networks and better supports decentralization; Compared with the Paxos Consensus algorithm, Raft is as efficient as Paxos but easier to understand and implement. Therefore, Hyperledger Fabric officially made Raft its default consensus algorithm after version 1.4.1.

Raft divides nodes into three roles: leader, follower, and candidate. The leader is responsible for interacting with the outside world, taking out the transaction from the trading pool, packaging the transaction to form a block, and uploading the block to the blockchain. In each consensus process, there is only one leader. Follower synchronizes with the leader and starts an election when it fails to connect with the leader. The candidate is a temporary role when the follower is running for the leader election. To eliminate the adverse effects of newly added nodes on cluster availability and fault tolerance, Raft introduced learner role after version 4.2.1 [6]. A new node will join the cluster as a non-voting learner until it synchronizes with the leader and then converted to the follower. Figure 1 shows the specific node role and transition.

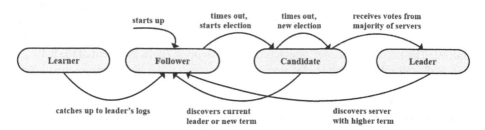

Fig. 1. Role transition in the Raft

Raft divides the consensus process into two stages: leader election and log replication. Time is divided into terms, and each term begins with a leader election. After a successful election, the leader will manage the entire cluster for the whole term. If the election fails, a new term and a new election will begin at once.

1) Leader election. Raft uses a heartbeat to trigger a leader election. When the server starts, the leader will periodically send messages with a heartbeat to all follower nodes. When a follower node can not receive any message from a leader within its specified timeout, it will increase the current term by one and become a candidate. Then it will first vote for itself and then broadcast the RequestVote RPC (Remote Protocol Call) message to other nodes in the cluster. The candidate who gets more than half of the votes will become the new leader and repeat the operation of sending a heartbeat to maintain its dominance.

2) Log replication. Once the leader is selected, it will receive the client's requests. Each request consists of a command to be executed by the replicated state machines in the cluster. The leader appends the requests to its log as a new entry, and then sends AppendEntries RPC message to other nodes in parallel. When this log is copied to most nodes, the leader will apply the log to its state machine and returns execution results to the client.

2.2 Raft's Problem and Improvement

Raft, as a mature and efficient consistency algorithm, has been applied in many scenarios in the field of distributed coordination services and distributed databases. However, Raft has two shortcomings in the high real-time and highly adversarial environment.

On the one hand, Raft is a non-Byzantine Fault Tolerance algorithm, which means Raft can not resist attacks from malicious nodes. Raft was designed not to consider Byzantine fault tolerance in the beginning, but to build the consensus environment based on mutual trust [7]. In a highly adversarial environment, malicious nodes can negatively affect Raft's entire consensus process. In the leader election phase, if the candidate is a malicious node, it can forge the logical timestamp contained in the RequestVote RPC message to get more votes. In the log replication phase, if the leader is a malicious node, it can tamper with log entries sent by the client, causing other normal nodes to receive incorrect logs.

On the other hand, the network split in the leader election phase will affect the efficiency of consensus [8]. As the number of follower nodes increases, the number of candidate nodes and the number of communications will also increase, limiting the speed of elections. In extreme cases, more than half of the nodes in the network are not controlled by the current leader, which is called network split. In the event of the network split, Raft will relaunch the leader election. At this point, the blockchain network will stop accepting new transactions, seriously affecting the consensus efficiency, real-time response, and load balancing in the high real-time environment.

Researchers have done a lot to make up for these shortcomings. For the first problem, Tan D proposed a consensus algorithm VBBFT-Raft combining digital signature and nested hash to deal with Byzantine fault [9]; Rihong Wang combined the threshold signature of BLS signature to design a Byzantine fault-tolerant mechanism of RBFT algorithm [10]; Chenyang Li proposed the BRaft consensus algorithm, which added some restrictions to Raft's consensus process [11]. For the second problem, Wang R optimized Raft by establishing a K-bucket node relationship, increasing Raft's election speed and transaction throughput [12]; Kim D used the method of federated learning to optimize the speed of leader election [13]. However, none of these studies took into account how to balance the security and consensus efficiency of Raft to make it suitable for the high real-time and highly adversarial environment. For this reason, we designed an improved Raft consensus algorithm called hhRaft and applies it to Hyperledger Fabric for experimental verification.

3   hhRaft Consensus Algorithm

3.1   Consensus Process

The hhRaft consensus algorithm proposed in this paper adds a new consensus role of monitor to Raft that enables Raft to resist malicious nodes. The monitor is converted from the candidate who has failed the leader election and the previous leader who has lost the right to interact with the client, and it will convert to a follower when it discovers the next leader. This transition method can maintain the number of monitor nodes in a relatively stable state (usually 2 or 3, slowly increasing as the total number of nodes grows). The node role transition of hhRaft is shown in Fig. 2.

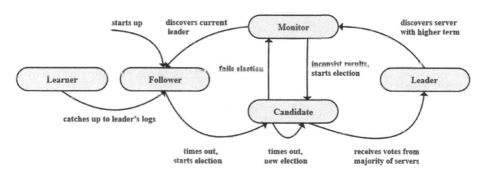

Fig. 2. Role transition in the hhRaft

Like follower nodes, the monitor has the right to vote, but it has the additional ability to monitor other nodes. hhRaft introduces digital signature verification and maintains a blacklist of nodes. The flow chart of hhRaft's consensus process is shown in Fig. 3.

In the leader election phase, the monitor compares the RequestVote RPC messages sent by each candidate. If the monitor discovers a maliciously forged message, it will identify the candidate through the signature of the message and adds it to the node blacklist. Nodes in the blacklist will lose their voting rights and can only be released manually by the network administrator.

In the log replication phase, the monitor calculates requests from the client and uses gossip [14] to receive logs from other nodes. It compares its results with the logs sent by the current leader and other nodes. If the calculation results are consistent, the monitor will send a ResultOK RPC message to the leader. Otherwise, it will add the leader node to blacklist and start a new election to ensure the security of consensus. When the leader receives ResultOK RPC messages of all monitor nodes, it will apply the logs and send the results to the client. In order to reduce the impact of poor communication monitor nodes on efficiency, the leader that has received at least one ResultOK RPC message can submit the log after a pre-set timeout.

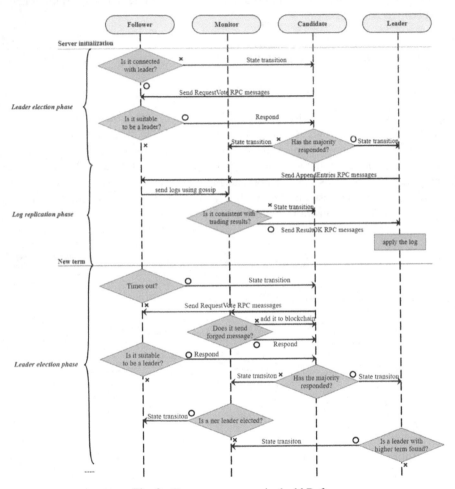

Fig. 3. Consensus process in the hhRaft

3.2 Algorithm Analysis

hhRaft is a consensus algorithm designed for high real-time and highly adversarial environments. Therefore, we will analyze hhRaft algorithm from the aspects of consensus security and efficiency.

Consensus security means all honest nodes can eventually submit consistent results in the presence of malicious or crash nodes. hhRaft should ensure that when there are f

Byzantine nodes if more than $(n-f)/2$ of non-Byzantine nodes have reached a consensus on a log entry, f Byzantine nodes can not affect the decision. Therefore, the total number of hhRaft cluster of nodes needs to be met $n \geq 5f + 1$, which means hhRaft cluster of six nodes should tolerate the Byzantine fault of any one node.

As shown in Fig. 4, if the leader is a Byzantine node, sends a tempered log entry to follower 1. The monitor will receive the tempered log through gossip and discover this inconsistency, then add leader and follower 1 to the blacklist at once and start voting to ensure the normal consensus among honest nodes.

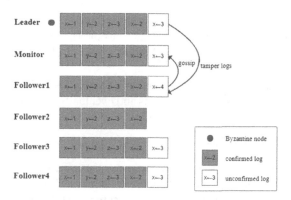

Fig. 4. hhRaft consensus process if leader is Byzantine node

Consensus efficiency depends largely on the speed of the leader election, which is largely negatively affected by the network split. In hhRaft, the monitor can only turn into a candidate when it finds log inconsistency, and candidates who fail in the leader election will be transitioned into monitors and will no longer participate in the next election. Under normal conditions, hhRaft is less affected by changes in the number of nodes and less likely to network split than Raft. Under Byzantine conditions, hhRaft effectively reduces communication overhead and consensus latency by isolating potential Byzantine nodes into the blacklist and preventing them from participating in elections.

4 Experimental Result

4.1 Experimental Environment

hhRaft realizes efficient Byzantine fault tolerance algorithms by introducing the monitor into Raft. We implemented hhRaft using the Golang language based on an open-source library – etcd-Raft. Then, we simulated attacks by malicious nodes to test the security of hhRaft on Docker, recording key data such as transaction throughput, consensus latency. The test platform hardware used an 8-core/16 GB AMD R7-5800H server, and the test platform software development environment used Windows10/Hyperledger Fabric/Golang 1.14.13.

4.2   Consensus Security Testing

We simulated Byzantine attacks on the Hyperledger Fabric which used hhRaft to deploy its order service. The experimental blockchain network contained a total of 6 orderer nodes, and we attacked follower, monitor, and leader separately to see the consensus security of hhRaft.

a.   Attack a follower node, and other honest nodes run normally.
b.   Attack the leader node to send a tempered log to a follower. The monitor finds the inconsistency and starts an election.
c.   Attack a candidate node. The monitor finds the malicious candidate and adds it to the blacklist, then another honest node becomes the leader.
d.   Attack a monitor node. The monitor maliciously starts an election, but it is discovered by another monitor.
e.   Attack two nodes of the cluster. The number of malicious nodes exceeds $(n - 1)/5$. The monitor constantly starts the elections and eventually causes the system to crash.

4.3   Consensus Efficiency Testing

Throughput, as a key indicator to measure the consensus efficiency, represents the number of transactions that blockchain can handle per second (TPS). We chose PTE [15] as the performance test tool for Hyperledger Fabric, setting BatchTimeout to 2 s, MaxMessageCount to 10, AbsoluteMaxBytes to 99 MB, and PreferredMaxBytes to 512 KB. The testing process is as follows: (1) start up the blockchain network locally, configuring the orderer nodes and peer nodes (2) open a new port as the client to send requests (3) test the throughput of hhRaft and Raft at different numbers of orderer nodes. Figure 5 shows both the throughput of hhRaft and Raft decreases as the number of nodes increases, and hhRaft has higher throughput than Raft in the same number of nodes.

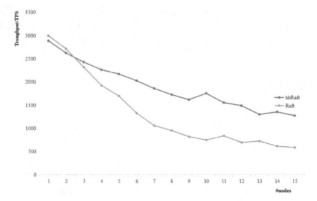

Fig. 5.   Troughput of various nodes

Latency is another important indicator of consensus efficiency, measuring how long it takes the client to receive a response from sending a single request. We used the same configuration to test the latency of the hhRaft and Raft algorithms. Figure 6 shows the latency of two consensus algorithms when the number of nodes varies from 1 to 15 and the client sends 1000 transactions. As the number of nodes grows to 10 nodes, the latency of hhRaft performs better. Because the leader can submit the transaction result to the client after receiving all the ResultOK RPC messages from monitors, reducing the time of commit confirmation.

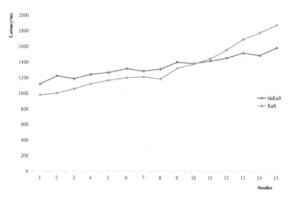

Fig. 6. Latency of various nodes

The speed of leader election can also have an impact on consensus efficiency. We simulated the leader's crash and recorded the latency of selecting a new leader. The results are shown in Fig. 7. From the picture, hhRaft has lower latency and better performance than Raft as the number of nodes increases.

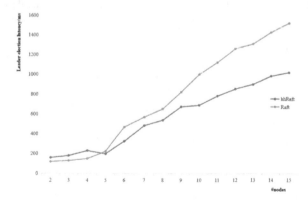

Fig. 7. Latency of leader election

5   Conclusion

The high real-time and highly adversarial environment represented by the real-time combat environment requires the blockchain consensus algorithm to have higher performance and the ability to resist Byzantine Faults. This paper introduces an improved

Raft algorithm called hhRaft to enhance consensus security and efficiency. Experimental results show that compared with the Raft algorithm, hhRaft has a good response to the Byzantine Faults, and has higher throughput and lower consensus latency with the increase of nodes. In the future, we will further use hhRaft in the actual high real-time and highly adversarial environment and explore the new direction for the future of automated military operations.

References

1. Mehta, P., Gupta, R., Tanwar, S.: Blockchain envisioned UAV networks: challenges, solutions, and comparisons. Comput. Commun. **151**, 518–538 (2020)
2. Guohong, Z., Xiong Lingfang, W., Yinghua, H.R.: A blockchain-based missile bee colony coordination guidance mechanism. Tact. Missile Technol. **04**, 100–111 (2020)
3. Sun, Y., Song, W., Shen, Y.: Efficient patient-friendly medical blockchain system based on attribute-based encryption. In: Wang, G., Lin, X., Hendler, J., Song, W., Zhuoming, X., Liu, G. (eds.) Web Information Systems and Applications: 17th International Conference, WISA 2020, Guangzhou, China, September 23–25, 2020, Proceedings, pp. 642–653. Springer International Publishing, Cham (2020). https://doi.org/10.1007/978-3-030-60029-7_57
4. Lamport, L., Shostak, R., Pease, M.: The byzantine generals problem. ACM Trans. Program. Lang. Syst. **4**(3), 382–401 (1982)
5. Ongaro, D., Ousterhout, J.: In search of an understandable consensus algorithm. Annual Technical Conference, pp. 305–319 (2014)
6. Ongaro, D.: Consensus: Bridging Theory and Practice. Stanford University, Stanford (2014)
7. Tian, S., et al.: A byzantine fault-tolerant raft algorithm combined with Schnorr signature. In: 2021 15th International Conference on Ubiquitous Information Management and Communication (IMCOM). IEEE (2021)
8. Huang, D., Ma, X., Zhang, S.: Performance analysis of the raft consensus algorithm for private blockchains. IEEE Trans. Syst. Man Cybernet. Syst. **50**(1), 172–181 (2019)
9. Tan, D., Hu, J., Wang, J.: VBBFT-Raft: an understandable blockchain consensus protocol with high performance. In: IEEE 7th International Conference on Computer Science and Network Technology (ICCSNT), pp. 111–115. IEEE (2019)
10. Zhihong, W., Lifeng, Z., Hang, Z., et al.: A Byzantine fault tolerance Raft algorithm combined with BLS signatures. J. Appl. Sci. (2020)
11. Chenyang, L.: BRaft: A Byzantine Fault-Tolerant Raft Algorithm. South China University of Technology, Guangzhou (2018)
12. Wang, R., Zhang, L., Xu, Q., et al.: K-Bucket based Raft-like consensus algorithm for permissioned blockchain. In: IEEE 25th International Conference on Parallel and Distributed Systems (ICPADS), pp. 996–999. IEEE (2019)
13. Kim, D., Doh, I., Chae, K.: Improved Raft Algorithm exploiting Federated Learning for Private Blockchain performance enhancement. In: International Conference on Information Networking (ICOIN), pp. 828–832. IEEE (2021)
14. Kwiatkowska, M., Norman, G., Parker, D.: Analysis of a gossip protocol in PRISM. ACM Sigmetr. Perform. Eval. Rev. **36**(3), 17–22 (2008)
15. Logan, B.: Performance Traffic Engine - PTE (2020). https://github.com/hyperledger/fabric-test/tree/main/tools/PTE

MHDP: An Efficient Data Lake Platform for Medical Multi-source Heterogeneous Data

Peng Ren[1], Shuaibo Li[2(✉)], Wei Hou[2], Wenkui Zheng[2], Zhen Li[2], Qin Cui[2], Wang Chang[2], Xin Li[3], Chun Zeng[4], Ming Sheng[1], and Yong Zhang[1]

[1] BNRist, DCST, RIIT, Tsinghua University, Beijing 100084, China
`{renpeng,shengming,zhangyong05}@tsinghua.edu.cn`
[2] Henan University, Kaifeng 475004, China
`houwei@henu.edu.cn, lz@vip.henu.edu.cn`
[3] Beijing Tsinghua Changguang Hospital, School of Clinical Medicine, Tsinghua University, Beijing 102218, China
`Horsebackdancing@sina.com`
[4] Viewhigh Technology (Beijing) Co., Ltd., Beijing 100062, China
`zengchun@viewhigh.com`

Abstract. In medical domain, huge amounts of data are generated at all times. These data are usually difficult to access, with poor data quality and many data islands. Besides, with a wide range of sources and complex structure, these data contain essential information and are difficult to manage. However, few existing data management frameworks based on Data Lake excel in solving the persistence and the analysis efficiency for medical multi-source heterogeneous data. In this paper, we propose an efficient Multi-source Heterogeneous Data Lake Platform (MHDP) to realize the efficient medical data management. Firstly, we propose an efficient and unified method based on Data Lake to store data of different types and different sources persistently. Secondly, based on the unified data store, an efficient multi-source heterogeneous data fusion is implemented to effectively manage data. Finally, an efficient data query strategy is carried out to assist doctors in medical decision-making. In-depth analysis on applications shows that MHDP delivers better performance for data management in medical domain.

Keywords: Data Lake · Medical multi-source heterogeneous data · Efficient · Persistence of data storage

1 Introduction

With the advancement of information technology, most hospitals and other medical institutions have realized large-scale informationization. However, the architecture of hospital information system (HIS) is complex. Furthermore, without uniform standard, the information systems used by different medical institutions are distinctive, resulting in different types and structures of medical data. These data include structured data in MySQL, Oracle, SQL Server and other related databases, semi-structured data in the format of CSV, JSON, XML, and unstructured data, such as EMRs, ECGs, CTs, MRIs,

© Springer Nature Switzerland AG 2021
C. Xing et al. (Eds.): WISA 2021, LNCS 12999, pp. 727–738, 2021.
https://doi.org/10.1007/978-3-030-87571-8_63

etc. In front of massive multi-source heterogeneous data in medical domain, how to acquire, store, and provide methods for unified management is the primary key problem [1, 2].

The works of medical data management in the early stage adopted Database [3, 4]. Database technology can cater to the needs of rapid Insert, Delete and Query in the case of relatively few medical data to deal with online transaction affairs in medical domain. However, Database technology cannot well deal with the analysis tasks which are characterized by reading a large amount of data. To address this issue, Data Warehouse technology is applied to medical data management [5, 6]. The medical data management frameworks based on Data Warehouse provide unified data support for medical workers' data management analysis and treatment decision-making. More recently, due to the demand for data flow keeps growing, Data Warehouse technology cannot cope with the remarkable challenge of data management.

Confronted by the challenges brought by Database and Data Warehouse in medical data management, researchers began to pay attention to Data Lake technology. Data Lake can integrate complex data by multiple means, and there are related researches in medical data management [7, 8]. However, the current data management frameworks in medical domain based on Data Lake technology show three challenges as follows:

1. Framework extendibility. In the real scene, the data generated in medical domain is multi-source, multi-structured and massive. Without broader data source range, existing works obtain these original data by a single approach, which demonstrates less extendibility.
2. Data persistence. When data is stored, it is sometimes lost. Though some data storage methods have been well studied, most of them ignore the data persistence over a longer time.
3. Efficiency. Usually, medical workers frequently need to obtain real-time results when analyzing and querying data based on their illness. However, most frameworks leave efficiency optimization to be desired, affecting the real-time analysis of illness for medical workers.

Contributions. The main contributions are summarized as follows:

1. An efficient multi-source heterogeneous data storage persistence method is proposed. With the support of distributed computing, medical multi-source heterogeneous data is converted into one unified data model to store data efficiently and permanently. Meanwhile, it considers recording the data sources and changes to ensure the traceability of the data sources.
2. An efficient multi-source heterogeneous data fusion method is presented. The data fragmentation triggered by the discrete distribution of medical data after storage cannot depict the patient's condition completely, which may lead doctors to make wrong decisions on the condition of patients. The proposed method can effectively tackle this problem and concentrate the data efficiently.

3. An efficient multi-source heterogeneous data query method is put forward. Different from the traditional query method, the proposed method takes into account the distributed computing power as well as space and time required for query, thus achieving high efficiency to underpin doctors' rapid medical decision-making.

The rest of this paper is organized as follows. Section 2 introduces the related work. Section 3 gives the platform architecture. Section 4 presents persistence of data storage. Section 5 describes data fusion. Section 6 implements data query. Section 7 summarizes the paper and points out future research direction.

2   Related Work

The concept of Data Lake was first put forward by Dixon to deal with the challenges brought by Data Warehouse. [9]. In practical applications, Data Lake technology has been employed in academia and industry. The companies that have deployed the Data Lake in the industrial community include AWS, Huawei, Alibaba, Azure, etc. The achievements of the deployment in academia include CLAMS [7], CM4DL [8] etc. Inspired by these Data Lake frameworks, the research work of Data Lake management framework in medical domain began to emerge constantly. Due to the flexibility of Data Lake technology and the diversity of diseases, the existing researches work on Data Lake management framework in medical domain can be divided into two categories, namely, Data Lake technology research and disease analysis research.

With regard to Data Lake technology research, Mesterhazy et al. [10] aimed at the medical image data such as X-Ray, MRI, CT, etc., and adopted cloud-based distributed computing technology to carry out rapid turnover of medical imaging, thus generating a medical image Data Lake processing framework. Hai et al. [11] designed a Web Data Lake framework called Constance, providing users with unified query and data exploration by applying an embedded query rewrite engine. They have been applied in the open source medical engineering project *miMappa*, which demonstrates the practicability of the framework in medical data management. Walker et al. [12] regarded metadata management as the core part of the Data Lake system, and proposed a personal Data Lake framework for personal customization, which could be used to store, analyze and query personal medical data as well as generate personal medical reports conveniently and quickly. Bozena et al. [13] integrated fuzzy technology and declarative U-SQL language into data analysis, and developed a fuzzy search scheme for big Data Lake, which could analyze a large amount of medical data in a distributed way. This scheme is equipped with good scalability, which is a successful step to realize the large-scale medical data declaration.

In terms of disease diversity research, Alhroob et al. [14] designed a data management framework based on Data Lake technology for semi-structured data of cardiovascular and cerebrovascular diseases by using k-means clustering with categorical and numerical data with big data characteristics. Maini et al. [15] proposed a solution to optimize data storage and analysis, applied it to the prediction of cardiovascular diseases, and constructed a prediction framework for cardiovascular diseases based on Data Lake.

Kachaoui et al. [16] came up with a Data Lake framework combining semantic web services (SWS) and big data features in order to predict the case of COVID-19 in real time. This framework extracts crucial information from multiple data sources to generate real-time statistics and reports.

3 The MHDP Architecture

This section introduces the platform architecture from two aspects: software architecture and deployment architecture.

3.1 Software Architecture

As shown in Fig. 1, the software architecture of MHDP is bottom-up, with a total of 5 layers, namely data storage layer, calculation layer, function layer, API layer and application layer.

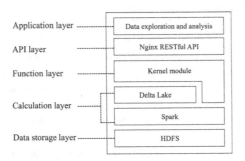

Fig. 1. MHDP software architecture

Data Storage Layer: The bottom layer of MHDP is the data storage layer with storage function provided by HDFS. HDFS is a very popular distributed file system, which provides high fault tolerance and high throughput storage capacity.

Calculation Layer: The calculation layer above the HDFS is supported by Spark which is a fast and general computing engine designed for large-scale data processing, and has formed an ecosystem with rapid development and wide application. Delta Lake is an open source storage layer provided by the Databricks, which brings reliability to the Data Lake. Delta Lake provides ACID transaction, extensible metadata processing, and unifies streaming and batch data processing. Delta Lake is fully compatible with Spark API.

Function Layer: Based on Delta Lake and Spark API, we modify and encapsulate them with code, and form various functional methods, such as multimodal data fusion, data query and other functions. These methods are the core components of MHDP.

API Layer: This layer encapsulates the functions of MHDP into Restful API. Nginx forwards the received requests to the master nodes of different clusters according to their types, and then uses cluster computing power to complete the corresponding tasks.

Application Layer: The top layer is the data exploration and analysis layer, which will package Restful APIs into a visualization module providing various services, interacting with users directly, and assisting users to complete various scientific research tasks of data analysis and data exploration.

3.2 Deployment Architecture

Figure 2 shows the deployment architecture of MHDP. First, a large number of servers are used as DataNode in Hadoop. Meanwhile, the servers are divided into two groups of Worker to provide computing power support for the two groups of Spark clusters. Then, several servers assume the identity of NameNode and Master with Zookeeper, building HA (Highly available) cluster. Once the NameNode and Master are down, the platform will automatically convert the server state from ready to activation.

There are several servers in the Restful Server, which split the services of MHDP. Each server will assume different responsibilities and provide different services. The requests sent by the client will be forwarded by the Nginx cluster to perform reverse proxy according to the category, so as to improve the reliability of the platform by improving the concurrency.

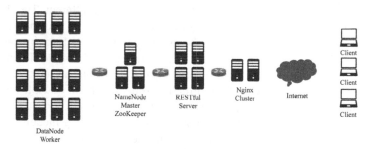

Fig. 2. MHDP deployment architecture

4 Data Persistence

Facing massive multi-source heterogeneous data in medical domain, the primary key problem is how to acquire, store, and provide methods for unified management. In this section, we introduce how to solve the problem mentioned above.

4.1 Data Acquisition Mode

In MHDP, these data sources are integrated through transportation mode and pipeline mode.

Transportation Mode. The interface is used to upload the data files to the reserved space of the Data Lake with SFTP protocol. Meanwhile, during the upload process, the interface will collect the meta information of the data files, such as data source, data structure, data type, data address, etc., and store them in a meta information database. In the future, when the data is traced due to medical safety problems, this information will provide support.

Pipeline Mode. In the pipeline mode, the data will not be store in MHDP. This mode will establish a session for the target data address. When we need to use the data, we will directly put it into the memory of the server cluster by network transmission, not in the disk. When we read the data of a certain pipeline type for the first time, there is also a method to obtain the meta information of the data and save it to the meta information database. The biggest difference between this mode and the Transportation Mode is that we will not save it in the Data Lake. We do not provide storage services for this type of data, and we do not need to be responsible for the security problems such as the leakage of this type of data.

4.2 Efficient Data Persistence

For data from different sources, the persistence methods are different. The data file is uploaded to the HDFS of the Data Lake for persistence with the Transportation Mode, whereas they will be directly loaded into the memory of the Data Lake by the Pipeline Mode when needed. According to different types of data, there are also differences about persistence methods. For structured data and semi-structured data, no additional changes are needed. But for unstructured data, it is difficult to directly obtain valuable information from itself, and we need to further process it. Therefore, in terms of persistence strategy, the original data are first stored. In addition, different embedding models are used to generate different feature vectors according to the requirements, and the information such as the address and size of the original data are stored structurally.

MHDP employs two kinds of models, Dataframe and DeltaTable, as the data models in the data processing stage. These data models can provide unified data models for efficient management of data with different structures and different sources. These data models and methods will transform the computing process of data into the distributed computing task of Spark, and fully mobilize the cluster computing power. DataFrame is a regulation data model in the platform, and DeltaTable is similar to DataFrame. It is the data structure that will be employed when Delta Lake is used. There are mutual transformation methods between DataFrame and DeltaTable. The DataFrame will be discussed as the main body in the following. The Fig. 3 shows an example of using DataFrame as data model for data in the format of CSV, JSON and image data. Whether it is semi-structured data in the format of CSV, JSON, or unstructured data like images, the platform provides an appropriate way to convert them into DataFrame. Structured and semi-structured data are supported by Spark and Delta Lake. For unstructured data, we will convert the feature vectors of these data and the path information for data traceability, as well as data size, data classification and other information into DataFrame for storage.

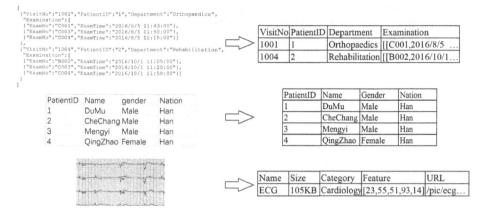

Fig. 3. The conversion from various types of data to DataFrame

5 Data Fusion

The data fragmentation cannot depict the patient's condition completely, which may lead doctors to make wrong decisions on the condition of patients. Aiming at this situation, data fusion is one of the key problems that MHDP solves.

5.1 Integration Method

In general, we can treat data as a set, and every piece of information in the data is an element in the set. This information can be represented by a tree structure, so every piece of data can be regarded as a set of trees. As shown in Fig. 4, it is an ordinary two-dimensional table, and the set of transformed trees is on the right side. The root node of the tree is generally assumed by the primary index.

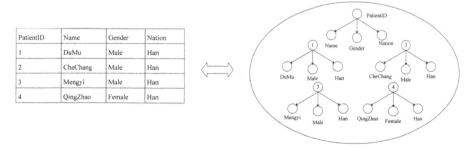

Fig. 4. The transformation between 2D table and tree

We use two cases to illustrate the merging process.

Case I. The first case is tree merging. If the Merged Key is the root node of the normal tree, then the merging process is the same as the special tree. If the Merged Key is the leaf node of the second layer in the normal tree, because the root node is adjacent to the second layer node, simply convert the position between the Merged Key and the root node of normal tree. Taking the fusion of patient information and treatment information as an example as shown in Fig. 5, the two datasets are associated with each other by the PatientId, which is used as the Merged Key. When the PatientID nodes in the treatment information dataset are located in the second layer, they should be converted to root node and be special tree, then the two trees are merged to get the merged result according to the root node.

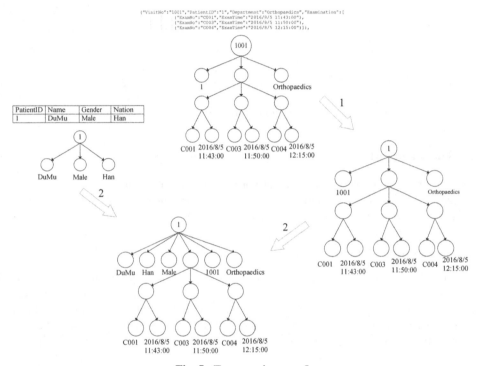

Fig. 5. Tree merging case I

Case II. The case that normal tree with the Merged Key below the second layer, occurring when the Merged Key is nested within a certain data value. The value is often an object or a list, or even a complex structure that both exist. As shown in Fig. 6, taking the merging of treatment information and examination information as an example, the two datasets need to use the examination number ExamNo as the Merged Key. There are three steps:

Step 1: Extracting these nested data. In view of this nesting situation, we design a special method depending on two basic processes. One is to split the Object existing in the nested data to expand the nodes. The other is to split the Array in the nested data to expand the elements. We can always turn the splitting process of a data into a finite combination of these two processes. These two processes can be used to process the data until the Merged Key appears at the root node or the leaf node on the second layer.

Step 2: Like the **Case I** above, changing the position between Merged Key and root node after getting the split tree, a special tree can be constructed.

Step 3: Merging the node to get the final result.

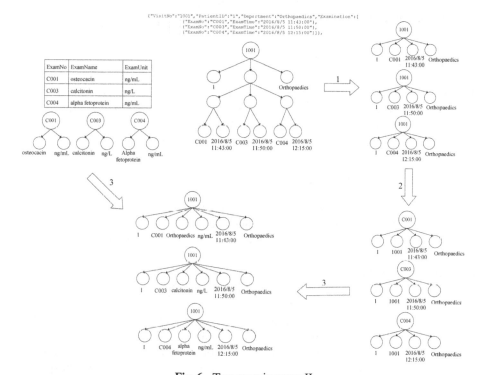

Fig. 6. Tree merging case II

5.2 Implementation Mechanism

In MHDP, we use DataFrame as the collection of trees, that is, the loading container of data. Each row of records in DataFrame corresponds to a tree.

Using DataFrame, we can call computing power of the clusters to speed up the process of data fusion. According to our method, we can solve the problem of merging N sets. No matter whether the merging node is nested or non-nested, we can always merge several data sets into a new data set, and use computing power of the clusters to speed up the process. When the merge node ExamNo in the treatment information is nested in Examination, we use two splitting methods to transform the DataFrame to get the following DataFrame. It can be seen that the ExamNo has been exposed to the outside. Next, only merging with the ExamNo in the examination data can get the final result. In the final result, it can be clearly understood what tests each patient has done and the relationship between these tests in time.

We encapsulate the function of data fusion into a Restful API to support the implementation of various functions of the platform. The API will return different values according to the running situation of the code. When the code runs as expected, it will return the execution situation: true and the corresponding data. Once the code is wrong, it will return the corresponding error information according to the error situation, so as to help medical workers understand the current situation.

6 Data Query

The data management framework in medical domain not only needs to store and manage a large number of multi-source heterogeneous data, but also has the query function to assist doctors to make in-time medical decisions. This section introduces how to efficiently implement data query in MHDP based on above mentioned.

6.1 Query on Medical Data Lake

In the traditional situation, joint query is the most time-consuming whether by Data Warehouse or data container, which will consume a lot of space and time. Most of the inefficient query were due to too many Joins. Our query mechanism is to do single table query based on data fusion. When query conditions exist in both set A and set B, we fuse set A and set B into set C. We only execute single table query for table C. Due to using Spark SQL and RDD mechanism, parallel computing can be used to accelerate the queries, so our query mechanism is very efficient in MHDP.

Generally, the data sets have certain columns. For the data sets with uncertain columns, we design a method. This method obtains the maximum structure of data first, and then uses it as the basis of data query. For the data sets with deep nesting, we also provide the function of data splitting in data fusion, releasing the data to be queried, as shown in Fig. 7.

Department	Examination	PatientId	VisitNo
Orthopaedics	[[C001,2016/8/5 ...	1	1001
Orthopaedics	[[C001,2016/8/23...	2	1003
Rehabilitation	[[B002,2016/10/1...	2	1004
Neurology	[[E001,2018/11/8...	3	1006

VisitNo	PatientID	Department	ExamNo	ExamTime
1001	1	Orthopaedics	C001	2016/8/5 11:43
1001	1	Orthopaedics	C003	2016/8/5 11:50
1001	1	Orthopaedics	C004	2016/8/5 12:15
1003	2	Orthopaedics	C001	2016/8/23 11:38
1003	2	Orthopaedics	C003	2016/8/23 11:45
1003	2	Orthopaedics	C004	2016/8/23 12:10
1004	2	Rehbbilitation	B002	2016/10/1 11:05
1004	2	Rehbbilitation	C003	2016/10/1 11:20
1004	2	Rehbbilitation	C004	2016/10/1 11:50
1006	3	Neurology	E001	2018/11/8 13:56
1006	3	Neurology	E007	2018/11/8 12:30

Fig. 7. The splitting of nested data

6.2 Query API

MHDP encapsulates the query method of data, and packages it into a Restful API to provide services for the exploration and analysis of medical data. This API requires two parameters, one is the address of the data file, the other is the SQL statement. When the code runs as expected, it will return the execution status: true, and the corresponding data. Once there is an error in the code, it will return the corresponding error information according to the error situation, so as to help users understand the current situation.

Taking the query of treatment information and examination information as an example, MHDP will fuse the two first, and then only need to query the results after fusion. When enter SQL standard query codes: 'SELECT ExamUnit, count (ExamUnit) as times from d6 group by ExamUnit order by times', the number of times that the current patient's examination unit appears will be displayed.

6.3 Application

MHDP has been deployed to XXX hospital to support data analysis in knee osteoarthritis domain. These knee osteoarthritis data, spanning from 2008 to present, involving 128,000 people and a total of 230,000 visits, include structured EMRs and unstructured medical images and texts.

Compared with the traditional Data Analysis System (DAS) of XXX hospital, MHDP queries data of different types uniformly and has higher query efficiency. We compare the performance of the Oracle-based DAS and MHDP by comparing the time consumed to execute the same query on the same dataset. The experiment dataset is 200 GB of structured data extracted from EMRs. The query task is to count the effective rate of patients with different genders and ages after PRP treatment. The experiment results show that MHDP-based queries outperform the Oracle-based queries by 3×.

7 Conclusion and Future Work

This paper proposes an efficient and medical-oriented Data Lake platform to manage massive medical data. Different from the traditional medical data management framework, MHDP adopts two unified data models, DataFrame and DeltaTable, to provide scalable and persistent storage capacity for medical multi-source heterogeneous data, coping with the increasing amount of data and the analysis needs of medical related data. During the construction of the platform, Spark and Nginx are used to store, transform and query medical data to provide fast computing capabilities, making the analysis of medical data more efficient. In the future, we will focus on optimizing the traceability of data. Not only do we record when the data is loaded into the Data Lake, but also track the changes of data during data fusion, so as to ensure that each piece of data has a clear provenance.

Acknowledgements. This work was supported by National Key R&D Program of China (2019YFC0119600).

References

1. Lee, C., Yoon, H.: Medical big data: promise and challenges. Kidney Res. Clin. Pract. **36**(1), 3–11 (2017)
2. Kalkman, S., Mostert, M., Beauvisage, N., et al.: Responsible data sharing in a big data-driven translational research platform: lessons learned. BMC Med. Inform. Decis. Mak. **19**(1), 283 (2019)
3. Mitchell, J., Naddaf, R., Davenport, S.: A medical microcomputer database management system. Methods Inf. Med. **24**(2), 73–78 (1985)
4. Mohamad, B., Orazio, L., Gruenwald, L.: Towards a hybrid row-column database for a cloud-based medical data management system. In: 1st International Workshop on Cloud Intelligence, pp. 1–4. ACM, New York (2012)
5. Sebaa, A., Chikh, F., Nouicer, A., et al.: Medical big data warehouse: architecture and system design, a case study: improving healthcare resources distribution. J. Med. Syst. **42**, 59 (2018)
6. Farooqui, N., Mehra, R.: Design of a data warehouse for medical information system using data mining techniques. In: 5th International Conference on Parallel Distributed and Grid Computing, pp. 199–203. IEEE, New York (2018)
7. Farid, M., Roatis, A., LLyas, F., et al.: CLAMS: bringing quality to Data Lakes. In: 2016 International Conference on Management of Data, pp. 2089–2092. ACM, New York (2016)
8. Alserafi, A., Abello, A., Romero, O., et al.: Towards information profiling: data lake content metadata management. In: 16th International Conference on Data Mining Workshops, pp. 178–185. IEEE, New York (2016)
9. Dixon, J.: Pentaho, Hadoop, and data lakes. https://jamesdixon.woedpress.com/2010/10/14/pentaho-hadoop-and-data-lakes/. Accessed 25 May 2021
10. Mesterhazy, J., Olson, G., Datta, S.: High performance on-demand de-identification of a petabyte-scale medical imaging data lake. In: CoRR abs/2008.01827 (2020)
11. Hai, R., Geisler, S., Quix, C.: Constance: an intelligent data lake system. In: 2016 International Conference on Management of Data, pp. 2097–2100. ACM, New York (2016)
12. Walker, C., Alrehamy, H.: Personal Data Lake with data gravity Pull. In: 5th International Conference on Big Data and Cloud Computing, pp. 160–167. IEEE, New York (2015)
13. Bozena, M., Marek, S., Dariusz, M.: Soft and declarative fishing of information in big data lake. IEEE Trans. Fuzzy Syst. **26**(5), 2732–2747 (2018)
14. Alhgaish, A., Alzyadat, W., Alfayoumi, M., et al.: Preserve quality medical drug data toward meaningful data lake by cluster. Int. J. Recent Technol. Eng. **8**(3), 270–277 (2019)
15. Maini, E., Venkateswarlu, B., Gupta, A.: Data lake-an optimum solution for storage and analytics of big data in cardiovascular disease prediction system. Int. J. Comput. Eng. Manag. **21**(6), 33–39 (2018)
16. Kachaoui, J., Larioui, J., Belangour, A.: Towards an ontology proposal model in data lake for real-time COVID-19 cases prevention. Int. J. Online Biomed. Eng. **16**(9), 123–136 (2020)

A Big Data Driven Design Method of Helicopter Health Management System

Xiaoming Xie[1], Tengfei Zhang[2(✉)], Qingyu Zhu[3], and Guigang Zhang[2]

[1] Harbin Aircraft Industry Group Co., LTD, Harbin, China
[2] Institute of Automation, Chinese Academy of Sciences, BeiJing, China
[3] China Aero-polytechnology Establishment, BeiJing, China

Abstract. This article briefly describes the development process of aircraft health management, and discusses the basic principles of helicopter health and use monitoring systems. Then proposed a big data-driven helicopter health management system architecture, analyzed the relationship between the components of the helicopter health management system, and analyzed the sources of the helicopter health management big data, and finally proposed Big data and knowledge-driven helicopter health management system Program. This paper explores the design method of the helicopter health management system driven by big data, and reserves the theoretical basis for the practical research of the future system.

Keywords: Big data drive · Fault diagnosis · Helicopter health and usage monitoring system

1 Introduction

Aircraft health management monitors the working status of each system of the aircraft in real time, analyzes the signals according to the data collected by the sensors, extracts characteristic values, understands the reliability of the aircraft according to the characteristic values, accurately evaluates the performance of the aircraft system, and accurately diagnoses, isolates and predicts on this basis For potential failures or performance degradation parts of the system, contact the ground system and air traffic control system in time to perform necessary maintenance to ensure that the aircraft completes the designated tasks during its life and ensures flight safety. Aircraft health management comes from Built In TestBITand vehicle health monitoring (VHM) technologies. With the development and progress of monitoring technology, the theory of health monitoring has changed from condition monitoring to health management [1], and there have been prognosis and health management (PHM) [2,3], Integrated Vehicle Health ManagementIVHM [4,5], Integrated System Health Management (ISHM) [6,7], Boeing's Aircraft Health Management (AHM) [8], Health and Usage Monitoring System, (HUMS) [6] and other concepts. In addition, in recent years, the US Department of Defense has put forward the concept of CBM plus (CBM+) on

C. Xing et al. (Eds.): WISA 2021, LNCS 12999, pp. 739–746, 2021.
https://doi.org/10.1007/978-3-030-87571-8_64

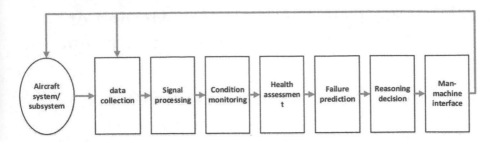

Fig. 1. Aircraft health management OSA-CBM international standard structure diagram.

the basis of Condition Based Maintenance (CBM), which further unifies CBM, reliability management, automatic support and maintenance, The goal is to carry out integrated overall planning and design for the health management capabilities of the equipment system, as well as the condition monitoring, maintenance decision-making, life prediction, logistics support, and cost control of the equipment after it is put into use. Aircraft health management includes intelligent detection, system-level evaluation, control and management functions of the aircraft. According to the OSA-CBM international standard, the aircraft health management system functions are layered into the 7 parts which is shown in Fig. 1:

Helicopter Health and Usage Monitoring Systems (HUMS) is a complex system that integrates avionics, ground support equipment, and onboard computer monitoring and diagnosis products. It contains a series of sensors and monitors the state of the engine, transmission system, rotor system and body structure through various algorithms [9–12].

2   Big Data-Driven Helicopter Health Management System Architecture

The big data-driven helicopter health management system architecture is shown in Fig. 2.

The big data-driven helicopter fault diagnosis and prediction system architecture is divided into three levels, including the basic layer, the functional layer, and the application layer.

At the basic level, first carry out data coordination and the study of the system's full-condition forecast model considering the uncertainty, and then process the overall modeling data on this basis. Then, the refined modeling of components considering the impact of performance degradation is carried out, and on this basis, the degradation mechanism life prediction is carried out.

At the functional level, firstly conduct research on feature extraction and health status evaluation based on multi-source information, and then perform status evaluation fault diagnosis on this basis. Then conduct single unit/multi-unit control scheduling and operation and maintenance optimization method

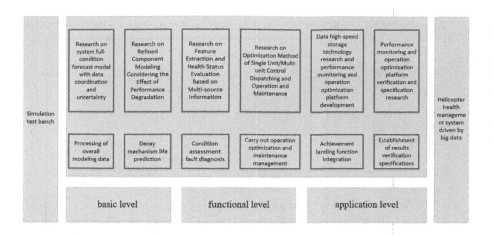

Fig. 2. The architecture of the helicopter health management system driven by big data

research, and on this basis, carry out operation optimization and maintenance management.

At the application layer, first carry out the research of high-speed data storage technology and the development of performance monitoring and operation optimization platform, and integrate the result landing function. Then conduct performance monitoring and operation optimization platform verification and specification research, and conduct the formulation of results verification specifications.

3 The Relationship Between the Components of the Helicopter Health Management System

The cross-linking block diagram of helicopter health management system components is shown in Fig. 3.

The system collects the parameter data transmitted by the system through the integrated data collection and monitoring computer (including the signals of the atmosphere, attitude, sending parameters, etc.), automatic flight control, radio altimeter and control, as well as the three-axis accelerometer and vibration sensor, Data transmitted by the speed sensor, rotor trajectory sensor, etc. These data are sampled and quantized by the computer's internal acquisition and editing device, and the signal is encoded according to a certain frame format. All signals are stored in the recorder and the data recording control box in the form of digital quantities. The data information recorded in the recorder and data recording control box is read, analyzed and processed by portable auxiliary equipment or integrated data processing station, and displayed or printed out in the form of data tables, curves, graphic reports and three-dimensional simulations. Through further analysis of the collected information, functions such

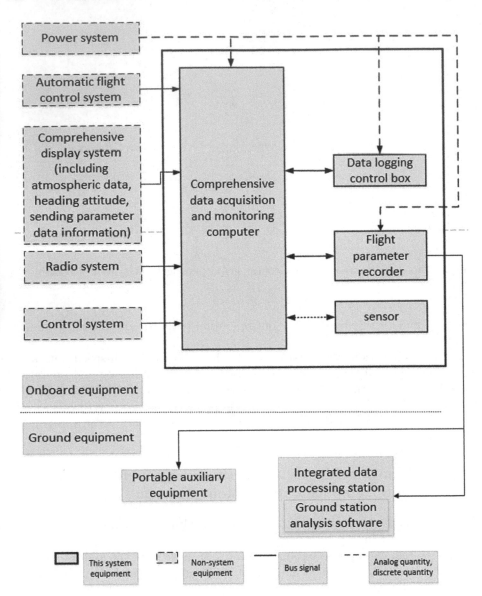

Fig. 3. System cross-linking block diagram

as data playback, rotor cone and dynamic balance adjustment analysis, healthy vibration and trend analysis, rapid data interpretation, usage over-limit statistics, and fleet management can be realized [13–17].

The big data of the helicopter health monitoring system mainly includes those shown in Table 1:

Table 1. The main data sources of the helicopter health management system.

Serial number	Data source
1	Integrated data acquisition and monitoring computer
2	Flight parameter recorder
3	Data logging control box
4	Rotor vibration sensor
5	Rotor speed sensor
6	Rotor trajectory sensor
7	Main reducer vibration sensor
8	Body vibration sensor
9	Engine vibration sensor
10	Drive shaft vibration sensor
11	Tail vibration sensor
12	Tail rotor speed sensor

4 Big Data and Knowledge-Driven Helicopter Health Management System Solution

the overall plan of the helicopter health management system driven by big data and knowledge shows in Fig. 4.

The overall plan of the big data and knowledge-driven helicopter health management system includes three tasks: helicopter health management, helicopter fault monitoring and diagnosis, and helicopter fault knowledge map.

The helicopter health management module contains three tasks: fault diagnosis decision-making, fault maintenance decision-making, and health management evaluation. This module is to build a helicopter intelligent fault diagnosis expert system. First, research on functions, applications, and integrated architecture, and then build an operation and maintenance knowledge graph management platform, which includes operation and maintenance decision-making functions and intelligent scheduling functions, and finally performs actual application verification.

The helicopter fault knowledge map module includes two tasks: acquisition of helicopter fault knowledge and update of helicopter fault knowledge. This module will study the technology of building a helicopter fault knowledge map. First, construct a complex network operation and maintenance map framework and ontology model. On this basis, perform entity identification and relationship extraction of fault knowledge, and finally perform operation and maintenance knowledge map quality verification and automatic Development of update tools.

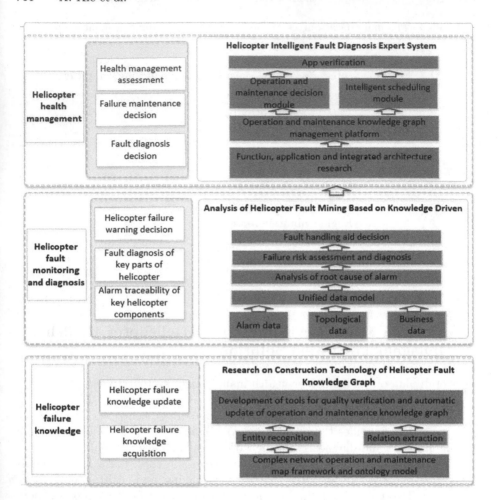

Fig. 4. The overall plan of a helicopter health management system driven by big data and knowledge

The helicopter fault monitoring and diagnosis module includes three tasks: helicopter key component alarm traceability, helicopter key component fault diagnosis and helicopter fault alarm decision-making. This module is to conduct a knowledge-driven analysis of helicopter fault mining. First, a unified data model is constructed from the alarm data, topology data, and business data, and then the root cause of the alarm is analyzed to isolate and locate the fault, then the fault risk assessment and diagnosis are performed, and finally, the fault handling auxiliary decision is made based on the assessment result [18–22].

5 Conclusion

This paper presents a big data-driven helicopter health management system architecture. This framework provides a way of thinking for the health management of helicopters. In the future, we will use this as a basis to develop a helicopter health management system and apply it in practice.

Acknowledgment. This work was supported by Aviation Science Funding (2020 00020M0002).

References

1. Larder, B., Azzam, H., Trammel, C., et al.: Smith industries HUMS: changing the M from monitoring to management. In: Proceeding of Aerospace Conference, Big Sky, MT, USA, pp. 449–455. IEEE (2000)
2. Hess, A.: The Joint Strike Fighter (JSF) prognostics and health management. In: Proceeding of 4th Annual System Engineering Conference. JSF Program Office (2001)
3. Figueroa, F., Holland, R., Schmalzel, J., et al.: ISHM implementation for constellation systems. In: Proceeding of 42nd AIAA/ASME/SAE/ASEE Joint Propulsion Conference & Exhibit, pp. 1–13 (2006)
4. Belcastro, C.M.: Aviation safety program: integrated vehicle health management technical plan summary. NASA Technology Report, pp. 1–53 (2006)
5. Akiyama, M., Makino, T., Morisakih, et al. A solution for vehicle health management. In: AIAA 2003–4889, 39th AIAA/ASME/SAE/ASEE Joint Propulsion Conference and Exhibit Huntsville, Alabama, 20–23 July 2003
6. Zuniga, F.A., Maclise, D.C., Romano, D.J., et al.: Integrated systems health management for exploration systems. In: Proceeding of 1st Space Exploration Conference: Continuing the Voyage of Discovery, Orlando, FL, USA. AIAA, pp. 1–16 (2005)
7. Maggiore, J.B.: Remote management of real-time airplane data [EB/OL]. http://www.boeing.com/commercial/aeromagazine/articles/qtr-3-07/AERO-Q307-article4.pdf.2007-9-1/2008-5-2
8. Fox, J.J., Glass, B.J.: Impact of integrated vehicle health management (IVHM) technologies on ground operations for reusable launch vehicles(RLVs) and spacecraft, pp. 179–186. NASA (2000)
9. Yao, P., Liu, Z., Wang, Z., et al.: Fault signal classification using adaptive boosting algorithm. Electron. Electr. Eng. **18**(8), 97–100 (2012)
10. Li, B., Chow, M.Y., Tipsuwan, Y., et al.: Neural-network-based motor rolling bearing fault diagnosis. IEEE Trans. Ind. Electron. **47**(5), 1060–1069 (2000)
11. Zhao, Q.: The study on rotating machinery early fault diagnosis based on principal component analysis and fuzzy c-means algorithm. J. Softw. (1796217X) **8**(3), 709–715 (2013)
12. Bates, P.R., Cycon, J., Davis, M., Baker, T., Thompson, B., Bordick, N.: Technology development to enable capability-based operations and sustainment. In: Annual Forum Proceedings (2012)
13. Cotter, J.P., Davis, M., Baker, T.: Development and testing of a ground based health and maintenance reasoner. In: AHS 71st Annual Forum (2015)

14. Markos, N., Lefevre, B., Frid, J., et al.: Propulsion system diagnostic and reasoning technology development. In: AHS 20th Annual Forum (2014)
15. Bodes, P., Davis, M., Cycon, J.: Capability-based operations and sustainment-AviationCOST-technology development. In: AHS 71st Annual Forum (2015)
16. Land, J.E.: HUMS-the benefits-past, present and future. In: 2001 IEEE Aerospace Conference Proceedings (2001)
17. Cook, J.R.: Educing military helicopter maintenance through prognostics. In: 2007 IEEE Aerospace Conference Proceedings (2007)
18. Kiteljevich, D., Dupuis, R., Cassidy, K.: Oil debris monitoring for helicopter drive-train condition based maintenance. In: AHS specialist Meeting on condition Based Maintenance (2008)
19. Legge, P.J.: Health and usage monit0ring system (HUMS) in Retrofit and Ab Inito helicopter application. In: AHS 52nd Annual Forum (1996)
20. Land, J.E.: HUMS-the benefits-past, present and future. IEEE (2001)
21. Cook, J.: Reducing military helicopter maintenance through prognostics. IEEE (2007)
22. Chen, H., Dong, Y., Gu, Q., Liu, Y.: An end-to-end deep neural network for truth discovery. In: Wang, G., Lin, X., Hendler, J., Song, W., Xu, Z., Liu, G. (eds.) WISA 2020. LNCS, vol. 12432, pp. 377–387. Springer, Cham (2020). https://doi.org/10.1007/978-3-030-60029-7_35

Business-Oriented IoT-Aware Process Modeling and Execution

Qianwen Li[✉], Jing Wang, Gaojian Chen, Yunjing Yuan, and Sheng Ye

Beijing Key Laboratory on Integration and Analysis of Large-Scale Stream Data,
North China University of Technology, Beijing 100144, China
lqw1597532@126.com

Abstract. Business Process Management (BPM) has a wide range of applications in various fields such as intelligent telemedicine, logistics and supply chain management. The integration of real-time event streams generated by smart sensors from the IoT into business processes (BP) can enable BPM to perceive changes in the physical world more quickly and accurately. However, the Business Process Model and Notation (BPMN) specification does not support the direct modeling of Complex Event Processing (CEP) for IoT event streams. The construction of a BP model that supports CEP has high requirements for business users, and it also brings a high load to the BPM engine. This paper proposes a modeling method by extending the custom attributes of the BPMN activity element, which can support business users to define CEP on IoT real-time event streams. At the same time, a supporting framework that integrates the BPM engine and CEP engine is also proposed, which can effectively execute BP instances that integrate IoT event streams. A case proves that the method proposed in this paper can effectively and friendly construct the IoT aware process model and reduce the load of the BPM engine.

Keywords: Business process · Complex Event Processing · BPMN extension

1 Introduction

There is a large class of Process-aware Information Systems (PAIS) supported by BPM in the industry [1]. Companies use PAIS to design BP models and derive BP instances from the process engine to support their operational activities, manage and deploy its resources and knowledge assets, and achieve their business goals. The I4.0 is defined as an extension of the IoT in the industry sector, which allows industries to improve the effectiveness and trust worthiness in their operations [2]. IoT big data generated by devices and sensors on I4.0 can reflect the state of the physical world in real-time, and has received widespread attention recently [3]. Through the integration of IoT big data into BPM, it can perceive environmental changes in time, improve the flexibility of PAIS, and effectively improve the capabilities and operating efficiency of PAIS through precise resource allocation.

Since the beginning of the 21st century, many related studies [4, 5] have been carried out on the integration of IoT into BP models, and this paper focuses on how business users

C. Xing et al. (Eds.): WISA 2021, LNCS 12999, pp. 747–755, 2021.
https://doi.org/10.1007/978-3-030-87571-8_65

can more intuitively and conveniently define the CEP on the event streams generated by IoT devices in the construction of BP models. The current research work in this area still has the following challenges:

1. In terms of process modeling, business users need to write various complex processing logic codes to define the CEP of IoT event streams according to different types of stream data. This requires business users to be proficient in business process construction and stream data processing technology, which has high requirements for business users.
2. In terms of process execution, traditional BPM engines processing IoT event streams are required to do constant cyclic calculations and storage status comparison, calculations and judgments, which causes a high load on the process engine.

To solve the above challenges, this paper proposes a BP modeling method that can support business users to directly model CEP without having to understand stream data processing logic. In terms of process execution, this paper proposes a support framework to execute BP instances where event processing and BP execution are separated, which can effectively reduce the load of the BPM engine.

2 Related Work

The integration of IoT and BPM can effectively promote a win-win situation and promote their respective development. The BPM provides the business logic for the ever-expanding IoT big data, and the IoT big data enables BP to quickly and accurately perceive changes in the physical world and respond to them. In recent years, preliminary research has been carried out on how to integrate IoT event streams into BPM. From the perspective of BPMN2.0 [6] extension elements, it can be divided into two categories: extensions based on activity elements and extensions based on event elements.

Extensions Based on Activity Elements: Meyer et al. [7] extends a new type of Sensing Task to model the tasks corresponding to IoT devices, which provides an important basis for the representation of IoT devices in BP, but does not mention how to express and execute the event streams generated by the IoT devices in the BP. Event Stream Specification (ESS) and Event Processing Stream Tasks (EPST) are added in [8, 9] to define and execute event streams generated by IoT devices in BP. ESS defines the IoT event stream in BPMN as a display modeling of the event stream, and ESPT encapsulates the logic of streaming event processing. They proposed how to map the BP model to BEPL for execution. In addition, during BP execution, the Eventlet middleware is linked to ESPT execution tasks through ESPT's web service interface. However, this requires developers to develop and define Eventlet, which is not friendly for business users to model CEP in BPMN. Yousfi et al. [10] proposed the extended language uBPMN for BPMN, which introduced event stream as an activity into BPMN, but uBPMN uses predefined event stream task elements, which cannot perform special processing on stream data resources and customize event definitions for different types of event stream data.

Extensions Based on Event Elements: Mandal et al. [11] proposed to extend the event elements combined with the use of annotation elements to support modelers to query and subscribe to complex events registered into the event management platform for modeling. In terms of process execution, a framework for handling events using an external event platform is proposed. However, complex events must be registered into the event management platform by domain experts in advance and cannot be flexibly defined for different types of event streams. Chiu et al. [12] proposed to integrate IoT events into the BPM by extending the conditional event, message event, and error event of BPMN2.0 and creating a type of location event. In addition, the use of OCL expressions optimizes the definition of events, which supports conditional events for averaging, but does not support the definition and execution of CEP for IoT event streams. The execution of the BP models is completely handed by the BPM engine, which causes a large load.

In general, the current related research can't provide business users with a friendly modeling method that supports the integration of different types of IoT event streams into the BPM. Although some work proposes an execution framework that transfers IoT event streams to an external event processing platform, they cannot define and execute CEP for different types of IoT event streams which requires high requirements for business users.

3 Proposed Architecture

This section will present our solutions from two aspects: BP model construction, and BP instance execution. Figure 1 is the overall architecture view of the method.

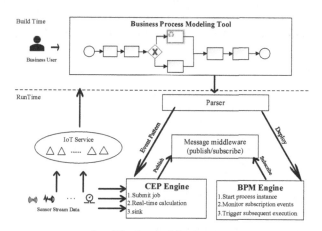

Fig. 1. Architecture

Through the combination of extended BPMN activity element and IoT services, the data source meta-information is made visible to modelers, and the event information of interest can be subscribed to complete the definition of the CEP on the event stream during the construction of the BP models. To improve the engine performance, a combination of BPM engine and CEP engine is used to execute IoT-aware BP instances. When the BP model is deployed, the event pattern executable by the CEP engine is automatically generated through the parser component. During the execution of the BP instance, the CEP engine performs real-time calculation of CEP. After the result event is generated, the BPM engine is notified via the publish/subscribe message middleware component to complete the execution of the subsequent BP model.

3.1 Modeling of IoT-Aware BP

On graphical modeling, a new type of BPMN2.0 activity element *Stream Task* is extended to abstractly represent the CEP of different types of event streams. The IoT Service encapsulates the meta-information of the data sources and provides the custom attributes of the *Stream Task* for its invocation, so that it can be used in BP. The main event stream attributes of the *Stream Task* extension are as follows:

1. IoT Service: provide business users to select specific IoT services, determine the type of event stream, and obtain the corresponding meta information of the IoT event stream.
2. Event Pattern Definition: support two types of CEP definitions: individual pattern and combining pattern.

As shown in Fig. 2, the individual pattern definition can be executed in a single time, or it can be changed to a combining pattern by specifying the times of cycles. The trigger condition of the individual execution patterns is defined by the combination of *Stream Attribute* and *Conditions*. *Conditions* support *simple conditions* ($==, >, \geq, <, \leq$) or *combining conditions* through the *More* operation (*and/or*). Through the *within*, the business user can set the time range of the monitoring CEP. *Greedy* can indicate that the pattern is allowed to be repeated, and *Consecutive* indicates that the pattern can appear continuously.

Fig. 2. Individual pattern

The individual patterns are combined into a complete combing pattern sequence by adding pattern sequence attributes. The selectable pattern sequence attributes are: *next, followedBy, followedByAny, notNext,* and *notFollowedBy.* For example, *A next B* indicates the combining pattern sequence (strictly continuous) where the *B* event occurs after the *A* event.

To ensure the executable of the constructed BP model, the BPMN2.0 extension is shown in Fig. 3. This paper introduces a new subclass *Stream Task* to the activity class.

The *Stream Task* inherits the model associations and attributes of the activity class and is compatible with the graphical model. The *IoT Service* class defines the source and meta-information of the event stream, which can be called by the custom attributes of the *Stream Task*. The *StreamParameterDef* class stores the ID of the *IoT service* and the rule of the CEP event pattern.

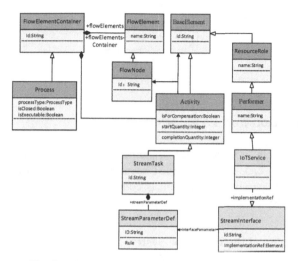

Fig. 3. The extension of BPMN 2.0 specification

3.2 Analysis and Execution of IoT-Aware BP

After constructing the BP model, a BPMN2.0 XML document is generated. When the BP model is deployed to the BPM engine, the parser component determines the data source according to the IoT service selected by the business user in the *Stream Task* and obtains the corresponding meta-information, which is used to register the corresponding external system table connector. Then combine the rule of the pattern sequence in the XML document to generate the executable code of the CPE engine. Figure 4 shows the sequence diagram between the components after the process model is parsed and deployed and the process instance is started.

After the BP instance is started, when the execution flow reaches the *Stream Task*, the BPM engine subscribes to the messaging middleware for the CEP defined in the BP model and starts monitoring the resulting events. At the same time, the CEP engine performs real-time stream calculations and generates the resulting events, and then publishes them to the messaging middleware. BPM engine throws a signal event after the *Stream Task* listens to the calculation result of CEP. When the *Stream Task* ends, the message middleware component cancels the subscription to the result event. After the signal is thrown, the BPM engine automatically captures the event within the scope of the BP instance and executes the subsequent BP model.

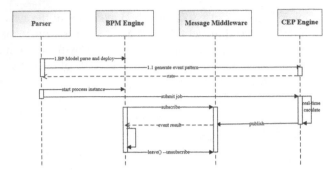

Fig. 4. Sequence diagram between components

4   Implementation and Evaluation

This section first briefly introduces the implementation of the framework proposed in Sect. 3. In addition, a case study in the IoT scenario is proposed. Finally, according to the case study, a comparative experiment is conducted between the proposed method and the traditional BP for IoT event stream processing.

4.1   Implementation of the Prototype System

The prototype system is implemented based on the framework proposed in Sect. 3, in which the Flowable engine is used as the BPM engine and FlinkCEP is used as the CEP engine to calculate CEP. Message middleware occupies an important role in the underlying technical support, and Apache Kafka is used to implementing the communication between the Flowable engine and the FlinkCEP engine.

4.2   Case Study

Scenario: In the field of elderly care and health, the time of the elderly spend in bed reflects some potential dangers. If the elderly spend too much time in bed, they need to notify the caregiver to check and give feedback. The bed pressure sensors in the IoT scenario can reflect whether the elderly are in bed in real-time. By integrating the real-time event streams of the bed pressure sensors into the BP field, it can reflect the status of the elderly in bed in real-time. Once it is monitored that the elderly have been in bed for a certain period, the BP model should give early warning and feedback immediately.

The BP model constructed using traditional BPMN for this scenario is shown in Fig. 5, and the IoT-aware BP model constructed using the method proposed in this paper is shown in Fig. 6.

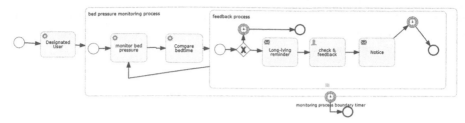

Fig. 5. Traditional BP model of long-lying scenario

Fig. 6. IoT-aware BP model of long-lying scenario

4.3 Experiment and Result

Experiment Setup. The FlinkCEP engine and Apache Kafka are executed on a machine with an Intel Core i5-6500 CPU and 8 GB memory, and the Flowable engine runs on an AMD Ryzen 7 4800H CPU and 16 GB RAM machine.

In the experiment, the basic experiment builds a traditional BP model that simulates a scene of 2 min in bed by writing the handler of the service task, and uses the Flowable engine to execute this traditional BP. The comparison experiment is to use the method and supporting framework proposed in this paper to build and execute the IoT-aware BP, which also stimulates the timeout warning within 2 min.

The evaluation metrics of the experiment are the *CPU utilization* and the *memory usage* of the Flowable engine.

Result and Analysis. The comparison of the CPU utilization and memory usage between using of Flowable engine to execute 200 concurrent traditional BP instances and using the framework proposed in this paper to execute 200 concurrent IoT-aware BP instances are shown in Figs. 7 and 8.

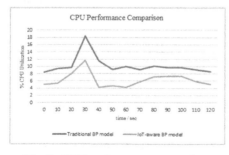

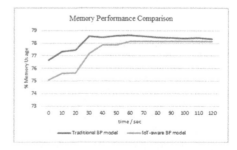

Fig. 7. CPU performance comparison   **Fig. 8.** Memory performance comparison

It can be seen from Figs. 7 and 8 that when 200 BP instances are started at the same time, compared with using traditional BPMN and Flowable engines, using the method proposed in this paper consumes lower CPU utilization and memory usage. Experiments have proved that the modeling method and execution framework proposed in this paper can effectively reduce the load of the Flowable engine.

5   Conclusion

This paper proposes a business user-oriented BP modeling method that supports the integration of IoT event streams into BPM and an execution framework that separates CEP and BP instance execution. This method enables business users to construct BP models in a friendly and convenient manner, and the proposed execution framework can effectively reduce the load of the BPM engine. In addition, the effective management of computing jobs can better improve the performance of the CEP engine. Future work will focus on how to effectively manage the computing jobs in the CEP engine.

Acknowledgments. This work is supported by the International Cooperation and Exchange Program of National Natural Science Foundation of China (No. 62061136006), and the Key Program of National Natural Science Foundation of China (No. 61832004).

References

1. Reichert, M., Weber, B.: Enabling Flexibility in Process-Aware Information Systems. Springer, Heidelberg (2012). https://doi.org/10.1007/978-3-642-30409-5
2. Munirathinam, S.: Chapter Six - Industry 4.0: industrial Internet of Things (IIOT). In: Raj, P., Evangeline, P. (eds.) The Digital Twin Paradigm for Smarter Systems and Environments: The Industry Use Cases. pp. 129–164. Elsevier, Amsterdam (2020)
3. Gao, Y., Yao, L., Yu, J.: Research of query verification algorithm on body sensing data in cloud computing environment. In: Ni, W., Wang, X., Song, W., Li, Y. (eds.) WISA 2019. LNCS, vol. 11817, pp. 176–188. Springer, Cham (2019). https://doi.org/10.1007/978-3-030-30952-7_20
4. Brouns, N., Tata, S., Ludwig, H., Asensio, E.S., Grefen, P.: Modeling IoT-Aware Business Processes, Technical Report IBM Research Report RJ10540 (2018)
5. Fattouch, N., Lahmar, I., Boukadi, K.: IoT-aware business process: comprehensive survey, discussion and challenges. In: 2020 IEEE 29th International Conference on Enabling Technologies: Infrastructure for Collaborative Enterprises (WETICE), pp. 100–105 (2020)
6. Business Process Model and Notation (BPMN), Version 2.0. http://www.omg.org/spec/BPMN2.0. Accessed Oct 2010
7. Meyer, S., Ruppen, A., Magerkurth, C.: Internet of Things-aware process modeling: integrating IoT devices as business process resources. In: Salinesi, C., Norrie, M.C., Pastor, Ó. (eds.) CAiSE 2013. LNCS, vol. 7908, pp. 84–98. Springer, Heidelberg (2013). https://doi.org/10.1007/978-3-642-38709-8_6
8. Appel, S., Frischbier, S., Freudenreich, T., Buchmann, A.: Event stream processing units in business processes. In: Daniel, F., Wang, J., Weber, B. (eds.) BPM 2013. LNCS, vol. 8094, pp. 187–202. Springer, Heidelberg (2013). https://doi.org/10.1007/978-3-642-40176-3_15
9. Appel, S., Kleber, P., Frischbier, S., Freudenreich, T., Buchmann, A.: Modeling and execution of event stream processing in business processes. Inf. Syst. 46, 140–156 (2014). https://doi.org/10.1016/j.is.2014.04.002

10. Yousfi, A., Bauer, C., Saidi, R., Dey, A.K.: uBPMN: a BPMN extension for modeling ubiquitous business processes. Inf. Softw. Technol. **74**, 55–68 (2016)
11. Mandal, S., Hewelt, M., Weske, M.: A framework for integrating real-world events and business processes in an IoT environment. In: Panetto, H., et al. (eds.) OTM 2017. LNCS, vol. 10573, pp. 194–212. Springer, Cham (2017). https://doi.org/10.1007/978-3-319-69462-7_13
12. Chiu, H.-H., Wang, M.-S.: Extending event elements of business process model for Internet of Things. In: 2015 IEEE International Conference on Computer and Information Technology; Ubiquitous Computing and Communications; Dependable, Autonomic and Secure Computing; Pervasive Intelligence and Computing, pp. 783–788 (2015)

Design and Implementation of Annotation System for Character Behavior & Event Based on Spring Boot

Chunying Pi[1,3], Peng Nie[2]([✉]), Yongqi Feng[1], and Licheng Xu[1]

[1] College of Computer Science, Nankai University, Tianjin 300350, China
{pichunying,fengyongqi,xulicheng}@dbis.nankai.edu.cn
[2] College of Cyber Science, Nankai University, Tianjin 300350, China
niepeng@nankai.edu.cn
[3] Tianjin Key Laboratory of Network and Data Security Technology, Tianjin, China

Abstract. To better prevent the spread of Corona Virus Disease 2019 (COVID-19), scientific researchers need the data set to research and predict the trajectories of patients. It is a time-consuming task to extract the data from the web page.

We design and implement an annotation system for character behavior & event for users to extract patients' basic information, trajectory information, and relationship information from the content of the web reports related to COVID-19 patients. Based on basic functions, permission management is added to protect data security and a clear interface is designed to optimize the user experience. To the best of our knowledge, this system is the first annotation system for COVID-19 data. The experiment results prove that our system has played a significant role in promoting the annotation work.

Keywords: Annotation system · Permission management · COVID-19

1 Introduction

1.1 The Need for the Annotation System

In December 2019, a case of COVID-19 was found in Wuhan, Hubei, China. COVID-19 [4] is highly contagious and has quickly developed into a global infectious disease. Many scholars have carried out analysis and research on these data. For example, Liu Yanzhe et al. analyzed the COVID-19 epidemic situation and studied the virus spreading trend in different countries [13]; MSRA (Microsoft Research Asia, Microsoft Research Asia) based on the COVID-19 Open Research Dataset (CORD-19, 2020) data set to visualize the new crown research trend [1], etc., these studies are based on the published data set.

A large number of web reports on COVID-19 have been accumulated in China. The research team of Beihang Big Data Advanced Center has produced

© Springer Nature Switzerland AG 2021
C. Xing et al. (Eds.): WISA 2021, LNCS 12999, pp. 756–763, 2021.
https://doi.org/10.1007/978-3-030-87571-8_66

a data set based on the trajectories of 4634 confirmed COVID-19 patients nation-wide [7]. We hope to achieve a data set that covers a wider area and has more comprehensive attributes than Beihang's data set. It is a time-consuming task to extract the data from the web page. After testing, taking 10 trajectories for a patient as an example (the number of trajectories for each patient is different), it takes about 12 min to label a patient.

Annotation system can facilitate annotation work, while existing annotation systems, such as YEDDA [12] and alpacatag [10], their functions do not completely match our requirement, so we need to design and implement an annotation system for character behavior & event for users to extract the related to COVID-19 patients, and store the data in the database.

1.2   Introduction to Spring Boot Framework

Spring Boot [11] is an open-source lightweight Java development framework with two core functions: IOC (Inversion of Control) and AOP (Aspect-Oriented-Programming, aspect-oriented programming) [15], IOC is used for the management of dependency relationship between objects. AOP separates business logic and general services and uses dynamic agents to implement transactions, caching, logging, and other functions.

Spring Boot also provides a lot of automatic configuration-dependent modules for various development scenarios, such as Logback, log4j, and other log functions and Jdbc Template, JPA, Mybatis to access the database, and inherits security authentication frameworks such as Apache Shiro.

2   Demand Analysis

Annotation System for Character Behavior & Event mainly needs to meet the following requirements:

(1) URL management: add, delete, modify, search URLs imported in batches, assign annotation tasks;

(2) Character basic information management: characters, add, delete, modify and search the character basic information, and bind to the URL;

(3) Character trajectory management: add, delete, modify and search the character trajectories;

(4) Character relationship management: add, delete, modify and search the character relationships;

(5) Authority management and user management: restrict different users from accessing different system resources based on roles and user IDs. The system provides three roles: administrator, auditor, and annotator. Different roles have different resource access rights: administrators can create new users, assign roles, modify permissions, assign annotation tasks to annotators; auditors can view all the data submitted by annotators, audit the correctness of the information and correct the wrong information; annotators can annotate

the URL assigned by the administrator, extract the basic character information, trajectories, and relationships information in the web page, and fill in the corresponding fields, and bind or unbind the URL with the person.

(6) Log record: record the running status of the system.

The whole system is developed around the states of the URL. The URL status is divided into five parts: to be assigned, to be annotated, being annotated, to be audited, and completed. "To be assigned" means that the URL is waiting to be assigned to the annotator by administrators. All the initial state of URLs imported into the database are "to be assigned"; when the administrator assigns the annotation task to a certain annotator, the URL status changes to "to be annotated", waiting for the designated annotator to annotate; the annotator clicks the link to annotate the URL and the URL status changes to "being annotated". The annotation of a URL is essential to extract the basic information, trajectory information, relationship information related to COVID-19 patients on the web page, and fill in the corresponding form. The annotator clicks to submit after finishing the annotation, the URL status changes to "to be audited", waiting for the members of the audit team to audit. After the auditor has passed the audit, the status of the URL changes to "completed", and the annotation of the URL is completed.

3   System Design

3.1   Design of System Architecture

Annotation System for Character Behavior & Event adopts a layered architecture [9], each layer has its specific role and division of labor. We divide the system into the frontend, backend, and database, as shown in Fig. 1.

The frontend uses Vue.js + Element UI + Axios for development, creates a user interface through the Vue.js framework [14], uses the component library provided by Element UI to design a simple and intuitive interface style and interactive effects, and uses Axios for data requests and responses.

The backend uses the Spring Boot framework [3] to create applications, uses maven for version control of dependent packages, and uses the Shiro framework for permission control. The controller layer provides RESTful APIs to the frontend [5], receives requests from the frontend, and calls services provided by the service layer. The service layer is the realization of the specific logic of the backend. Use Spring Data JPA in the persistence layer to correspond the database data table with the entity class in java. The backend also needs to have a log function to record the operation of the entire system, which can be achieved by using the logging tool log4j provided by java.

The database layer is responsible for data storage. We can use MySQL for persistent storage and use Redis to cache system data and temporary data such as sessions.

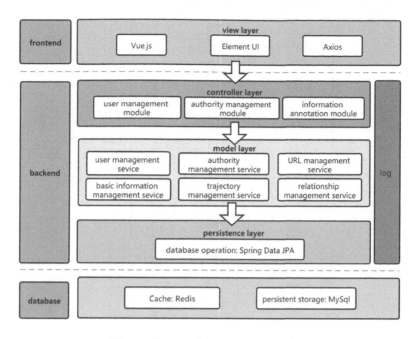

Fig. 1. System layered architecture

3.2   Design of Function Module

The system mainly provides three main functions: user management, authority management, and information annotation, as shown in Fig. 2.

User Management Module. User login, user addition, deletion, modification, and search are implemented in the user management module. To avoid machine login, a picture verification code is used to perform a man-machine operation on the user verification.

The system uses HTTP (HyperText Transfer Protocol) [2] for communication. HTTP is a stateless protocol. After the user is authenticated, the next time a request is sent, it needs to be re-authenticated. To solve this problem, the general solution is to generate a session record on the backend when the user logs in. To mark the currently logged-in user, the session is saved in the memory, disk, or database, and then the sessionID is returned to the frontend, which is stored in the cookie. The cookie needs to be carried in the next request. Then the system can authenticate users through cookies.

After the user successfully logs in to the system, the user's login status is cached in the Redis. Even if the current page is closed, the user can directly enter the system. There is no need to log in again, which improved users' experience.

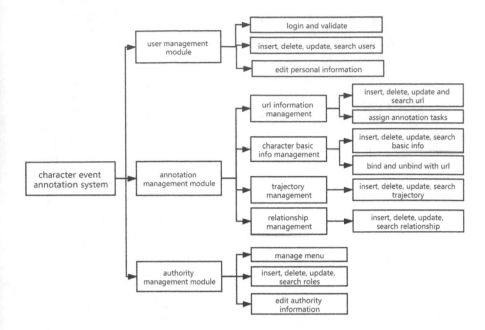

Fig. 2. System function module

Authority Management Module. The authority management module uses RBAC (Role-Based Access Control) technology to associate roles with permissions and uses roles as a bridge between users and resource access permissions. By assigning different roles to users, they can get the permissions corresponding to the roles, which increases the flexibility of the system and is easy to maintain [6].

The system provides three roles: administrator, auditor, and annotator. After the user logs in, the backend server transmits the user's role and authority information to the frontend, and the frontend saves the information in a cookie. For the page-level authority control, a router can be used to assign access permissions to components; for the fine-grained level of authority control, such as button display, can be implemented using Vue custom instructions on the frontend. Only relying on the frontend hidden controls cannot guarantee data security. It also requires authority verification on the backend controller layer interface design. The Apache Shiro framework can be used to handle authentication and access control.

Information Annotation Module. Information Annotation Module can be divided into URL management, basic information management, trajectory management, and relationship management. Each sub-module is responsible for providing the function of adding, deleting, modifying, and search corresponding information [8].

After a successful login, the system dynamically generates a page based on the role of the current user, and the user can only see the content within the scope of his authority, can only operate on the authorized data. The pseudo-code of the annotation module is as follows:

```
1   if (authentication failed)
2       // has no system permission
3       return −1;
4   else {
5       // authentication succeeded
6       if (is admin){
7           enter management page of admins;
8       }
9       else if (is auditor){
10          enter audition page of auditors;
11      }
12      else if(is annotator){
13          enter annotation page of annotators;
14      }
15      return 0;
16  }
```

4   Result

4.1   System Interface

When a user logs in, the system authenticates the user through the correctness of the username and password and performs man-machine verification.

After the annotator logged in successfully, the system jumps to the URL list page, and a button is set behind each URL to jump to the annotation page. The annotation page is divided into two parts, the upper part embeds the web page content so that the annotator can view the content of the page when filling the form; the lower part is used to display the basic information list of the characters, as shown in Fig. 3. Each row of the character list provides buttons for viewing trajectories and relationships. A list of tracks or relationships will pop up from the right when the button is pressed. In this way, the four sub-modules of URL information management, character basic information management, trajectory management, and relationship management are linked together at the front end.

4.2   Effect

By crawling the news released by the Health Commission, we have accumulated 3,000 web reports about COVID-19 patients as a data set that needs to be annotated. Since the system was launched on September 10, 2020, 35 annotators have participated in our annotation work. As of February 20, 2021, the data statistics in the database are shown in Table 1. After review, there is no obvious

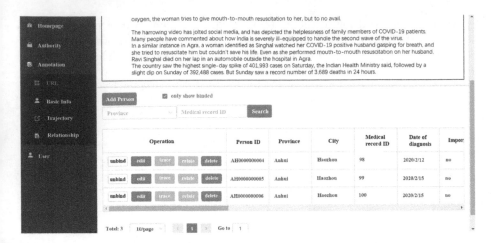

Fig. 3. Annotation interface

Table 1. Statistics of annotation result

	Annotated URL	Basic info	Trajectory	Relationship
Total number	1708	2291	12174	875

difference in the accuracy of the annotation result before and after using our annotation system. When the annotation system is not used, it takes an average of 720.00 s to extract information about patients in the new library on the web; when using the annotation system, it takes an average of 455.09 s, and the efficiency is increased by 38%.

The differences between our system and the related are summarised in Table 2.

Table 2. Annotation tool comparison

Tool	Collaborative	Architecture	Review function	Authority management
YEDDA	×	C/S	×	×
AlpacaTag	√	B/S	×	×
doccano	√	C/S	×	×
our system	√	B/S	√	√

5   Summary

An annotation system for character behavior & event is designed and implemented to assist in extracting COVID-19 patients' related data. The system used the current popular framework Spring Boot and completed authority management, user management, and information annotation functions, and also considered non-functional requirements such as security, maintainability, and performance.

There are many areas that can be improved, such as:

1. The audit function is currently open to the entire audit team, the system doesn't record which auditor passed the URL annotation information. The auditing task can be refined for each auditor in a later period.
2. Add the recommendation function, integrate some natural language processing methods, highlight the keywords and recommend the labeled categories before annotation.
3. Provide the administrator with the function of exporting the annotation progress with one click.

References

1. Microsoft Research Asia — COVID insights. [EB/OL], August 2020. https://covid. msra.cn/ResearchTrend
2. Http. [EB/OL] (2021). https://developer.mozilla.org/en-US/docs/Web/HTTP
3. Spring boot. [EB/OL] (2021). https://spring.io/projects/spring-boot#overview
4. Coronavirus_diseases. [EB/OL], May 2021. https://en.wikipedia.org/wiki/ Coronavirus_diseases
5. Diao, M., Zhao, D., Xue, T., Li, W., Zheng, N.: A highly automatic petrogeochemistry diagram plotting and analysis system based on MVVM. DEStech Transactions on Computer Science and Engineering (2018)
6. He, C.: Design of extended RBAC model based on role-controller under b/s architecture. Mach. Manufact. Autom. **44**(1), 184 186 (2015)
7. Hu, Z., Zheng, C.: BDBC-KG-NLP/COVID-19-tracker. [EB/OL], May 2020. https://github.com/BDBC-KG-NLP/COVID-19-tracker
8. Jiaxiong, H., J.M. : Design of tracing system for patients with new coronary pneumonia based on Wi-Fi probe. Med. Health Equip. **41**(6), 1–4 (2020)
9. Kan, J., Xiaojun, C., Aoming, Q., Jia, L., Jinmei, W.: Design and implementation of teaching quality evaluation system based on spring boot. DEStech Transactions on Computer Science and Engineering (ICCIS) (2019)
10. Lin, B.Y., Lee, D.H., Xu, F.F., Lan, O., Ren, X.: AlpacaTag: an active learning-based crowd annotation framework for sequence tagging. In: Proceedings of the 57th Annual Meeting of the Association for Computational Linguistics: System Demonstrations, pp. 58–63 (2019)
11. Walls, C.: Spring Boot in Action. People Post Press (2015)
12. Yang, J., Zhang, Y., Li, L., Li, X.: YEDDA: a lightweight collaborative text span annotation tool. Computer Products and Circulation (2019)
13. Yanzhe, L., Bingxiang, L.: Evaluation and prediction of COVID-19 based on time-varying SIR model. In: Wang, G., Lin, X., Hendler, J., Song, W., Xu, Z., Liu, G. (eds.) WISA 2020. LNCS, vol. 12432, pp. 176–183. Springer, Cham (2020). https:// doi.org/10.1007/978-3-030-60029-7_16
14. You, E.: Vue.js. [EB/OL], April 2021. https://vuejs.org/v2/guide/
15. Zhang, F.: Design and implementation of physical education video teaching system based on spring MVC architecture. In: Proceedings of the 2019 4th International Conference on Information and Education Innovations, pp. 116–119 (2019)

University Teacher Service Platform Integrated with Academic Social Network

Dan Xiong, Lunjie Qiu, Qing Xu, Rui Liang, Jianguo Li$^{(\boxtimes)}$, and Yong Tang

School of Computer Science of South China Normal University,
510631 Guangdong, China

Abstract. In recent years, with the vigorous development of the construction of colleges and universities, all kinds of innovative colleges continue to emerge, and multi-disciplinary integration categories continue to grow, the traditional school website portal teaching staff construction has been difficult to meet the information needs of modern teachers. Considering that most university teachers have real-time dynamic update of personal profile and academic achievements on the academic social network, and the minimal customized granularity of teacher column display required by different disciplines. This paper proposes a college portal model centered on academic social network, and designs and implements a highly customized college teacher information management service platform, which is associated with SCHOLAT, an academic social network platform, mainly including customized homepage of University Teachers' column, collection of teachers' academic information, teachers' communication and other functions, which is an important window to show the strength of colleges and universities. It has the main advantages of authority, timeliness, security, ease of use, scalability and it truly realizes different construction schemes for different colleges.

Keywords: Academic social networks · Teacher service platform · Web portal · Custom · SCHOLAT

1 Introduction

With the advent of the information age, in the website of colleges and universities, the section of external teacher team display has higher and higher requirements for the timeliness and comprehensiveness of teacher information management. In external publicity, the modern website portal is the most important channel for the outside world to understand the strength of school teachers [1]. However, the traditional college website portals are mostly static display pages, which will increase the difficulty of collecting and modifying teacher information, resulting in uneven data quality [2], different storage methods and incompatible formats. In addition, the teacher information section is not updated in time

Our work is supported by the National Natural Science Foundation of China (No. 61772211 and No. U1811263).

and the notification method is single, resulting in a long data update cycle and lack of timeliness; Once the allocation management of teachers' columns cannot be changed, it lacks personalized customization function and many other shortcomings, which can not meet the modern needs of the college website.

In view of the above problems, this paper aims to design a university teacher service platform integrating academic social network and integrate the basic personal information on the teacher academic design platform, so as to realize the reliable dissemination of data and achieve the purpose of information sharing [3]. Moreover, the platform takes a short time to build and has low requirements on the programming ability of managers. The college's exclusive portal can be established from scratch in about half an hour. In addition, the platform can be customized according to different teacher column needs of each college, which is conducive to the outside world to quickly and accurately obtain relevant teacher information and perfectly solve the above series of problems for teacher information management.

2   Related Work

Teachers' information management in Colleges is mainly faced with the following problems: First, the system is independent of each other, lack of data sharing and interaction effectively, resulting in a large number of redundant data or missing data and other related problems; Second, in the Internet era, updating teachers' information needs to be more fast; Third, different colleges usually have different personalized customization requirements for their own portal websites. Each user group has their own needs and expectations for the website, but they all hope to easily obtain the accurate information they need in a short time. Therefore, usability has become an important issue of College portal [4–6].

Jin et al. [7] designed a teacher's personal homepage system for university teachers with customizable content and fool like operation interface, but it can only meet the needs of their own colleges, and need to spend a lot to update the future needs. Li et al. [8] constructed the teacher's personal homepage system with B/S structure, integrating the unified management of teacher information in the school; Li et al. [9] proposed a school portal based on website group architecture. It reflects the specific advantages of the management level, but each teacher needs to register and submit his own personal information, which greatly increases the tedious process of teacher information collection. As a result, teachers' information is missing and incomplete. Luo et al. [10] put forward an exploration scheme for the construction of personal homepage in colleges and universities based on website group, and build a unified management, unified database, centralized data sharing, safe and controllable personal homepage for teachers. Guo et al. [11] pointed out that the overall design of the Xiamen University portal website is generally lack of connotation and style, which can not meet the needs of humanization, visualization and interaction of users in the modern new media environment.

To sum up, in the above series of college website portal design schemes, it is difficult to meet the needs of teachers' homepage displayed by all colleges through one service platform. This paper proposes a scheme of the exhibition platform of college teachers portal integrating academic design network, and constructs the data stream radiating various universities, taking the information data of the teachers of SCHOLAT as the center. After the audit, it will be shared to different colleges and universities across the country, which greatly reduces the difficulty of collecting teachers' information and the redundancy of information, and realizes the function of real-time updating and feedback of teachers' information. At the same time, a variety of algorithms are used to realize each function module. Finally, the platform[1] has achieved good results in the trial stage.

3   College Portal Model Centered on Academic Social Network

The construction of the traditional college website portal is usually designed with the college as the center, which restricts the interaction with the outside information from the macro level, and is limited to the information interaction within the college. In this regard, the platform is associated with the SCHOLAT,[2] and designs a model of the faculty portal of colleges and universities centered on academic social networks [12].

3.1   Connection and Correspondence of Model Nodes

This model takes the academic social network as the root node, and the maximum of the initial node provides simplicity and rapidity for teachers' information sharing [13]; The administrators of each university register their accounts and apply for the corresponding university domain name and the college domain name, as the second level node and the third level node respectively; And define all sub-columns of the college as the fourth level nodes; All the teachers in this column are the fifth level nodes. Finally, build a website platform that radiates all colleges and universities across the country. The model is shown in Fig. 1. Taking the Computer College of South China Normal University as an example, apply for the corresponding university level domain name and college level domain name, and provide the unique URL[3] as the website address of the teacher portal.

By combining with academic social networks, the model takes the teacher information of scholat as one of the data sources, solves the problems of teacher information collection difficulty and update speed, and gets better feedback in the process of use.

---

[1] http://faculty.scholat.com/.

[2] http://www.scholat.com/.

[3] http://faculty.scholat.com/homepage/scnu/cs.

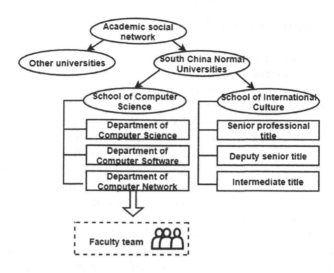

Fig. 1. College portal model centered on academic social network.

3.2   The Dissemination Mode of Data Flow Under the Model

To collect teachers' personal information better, the model integrates teachers' data set in the SCHOLAT and the original teachers' information data set of each college as the online and offline sources. The administrator checks the teacher information according to the uniqueness of the mailbox; And the offline import of teacher data will be compared with the existing teacher information in the college platform, according to the mailbox for comparison and reprocessing; Finally, assign the teachers to the predefined columns to complete the construction of the college portal, and the data flow propagation process is shown in Fig. 2.

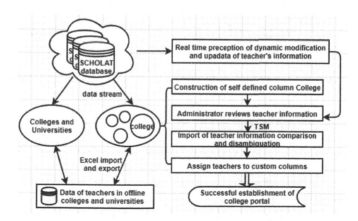

Fig. 2. Data flow propagation flow chart.

4   System Function Design Based on SCHOLAT

The system integrates the academic social network, takes the teacher information data of SCHOLAT as the center, and radiates to various colleges. The system provides highly customized columns for teachers, and truly realizes different construction schemes for different colleges. And it can update the personal profile and academic achievements of teachers in real time, which greatly reduces the difficulty of collecting teachers' information and the redundancy of information.

The system is mainly divided into two modules: external display and background management. After the administrator authenticates the account, the information of the college teachers can be imported through excel template. At the same time, it can also be compared with database of SCHOLAT by one key Association. Teachers only need to manage their own personal space of SCHOLAT, and the system will collect and update regularly, and finally show the unique homepage of teachers. The administrator is easy to operate and more standardized, and each business of the system is no longer independent of each other. The data is effectively shared and interacted, and closely linked. The schematic diagram of university teacher service platform is shown in Fig. 3.

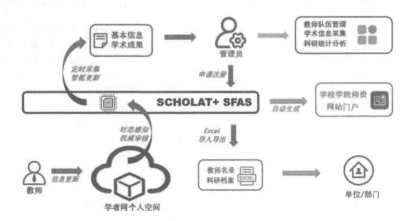

Fig. 3. Schematic diagram of university teacher service platform.

4.1   Dynamic Update of Teacher Information Based on DSPT-ID

With the accumulation of teacher information, there are many disadvantages in the traditional centralized management mode, which is no longer suitable for the development of modern educational administration. The multi-level management mode [14] emerges as the times require, which can improve work efficiency and management quality. At the same time, the system integrates academic social networking sites, SCHOLAT and adopts an efficient shortest path full dynamic update algorithm DSPT-ID to ensure that teachers' information can be updated in time;

As shown in Fig. 4, the system's message reminder adds interactive design, such as generating small marks on the teacher's avatar, associated column text reminder, modifying the button to change the reminder color, to achieve the goal of high efficiency, timeliness and accuracy. At the same time, it is also necessary to do a good job in the unique authentication and verification of teachers' information. Teachers who have registered the SCHOLAT account can not register again; Administrators can invite teachers who do not have registered SCHOLAT account to register.

	黄震华	HuangZhe nHua	2020.06.11 09:59:31 编辑者:asd	jukiehuang 学者网有更新信息
	吴正洋	WuZhengY ang	2020.06.11 10:00:55 编辑者:asd	wukeking
	汤庸	TangYong 2	2020.12.17 21:40:56 编辑者:asd	ytang 学者网有更新信息

Fig. 4. Teacher information dynamic reminder.

This platform uses this algorithm to automatically update teacher information at 1 a.m. every day. And we have tried this function many times in college. The updated information of the platform is very timely and accurate.

4.2   Disambiguation of Teacher Information Based on TSM

Due to the huge amount of information, it is inevitable that there will be "fish in the net". There will be a problem that the same teacher has more than one display, and the subsequent updating of information will not be synchronized. So we introduce information disambiguation mechanism [15]. To maximize the accuracy of the similarity calculations and to reduce the time complexity to an acceptable level, we used the TSM [16] to calculate the similarity of the relevant short texts in SCHOLAT. For teachers without SCHOLAT account, offline Excel can be used to batch import teacher information, and the teacher mailbox will also be used as the only judgment basis. If it is the same, the administrator will be reminded that the teacher already exists and the import fails. After comparing with the existing teachers, the import can continue. The flow chart of teacher information import is shown in Fig. 5.

The platform uses this mechanism to realize the duplication of teacher information, effectively reduce data redundancy and avoid information errors. The effect is very remarkable.

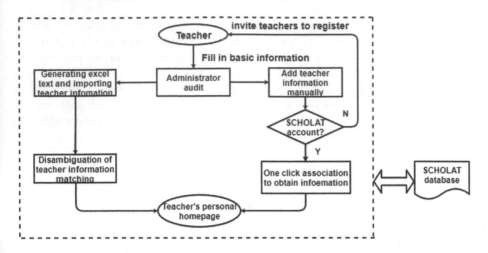

Fig. 5. Flow chart of teacher information import.

4.3 Recommendation of Relevant Teachers with LDA and TF-IDF

In order to make users better access to teacher information, based on the four data visualization research topics proposed by Lee et al. [17]. This system also makes a visual analysis of teacher information. It uses LDA [18] and TF-IDF text similarity hybrid model to calculate the similarity of scholars' interests, and takes the social similarity between scholars as the final similarity. Then the trust degree is calculated according to the multidimensional interaction between scholars. Finally, taking the similarity and trust between scholars as indicators, we recommend 8 relevant scholars with the highest ranking on each teacher's personal homepage. Through background data, users use this function frequently.

5 Conclusions

With the rapid development of Internet technology, network social communication has become the main communication mode of people. This paper makes full use of the basic personal information that scholars update independently, connecting with SCHOLAT., and adopts various algorithms in the function module to realize the big data sharing mode under social network. It provides a good solution and service platform for the construction and renewal of the teachers portal platform of their own colleges and universities. The platform has been put into use in several colleges and universities, and has achieved good feedback effect. In addition, the platform is constantly improving, improving the stability of the platform functions, while increasing the demand for functional diversification. If you are a college related worker, you are welcome to register the platform for testing.

References

1. Zuo, M., Wang, W., Yang, Y.: Promoting high-quality teachers resource sharing and rural small schools development in the support of informational technology. In: Cheung, S.K.S., Lee, L.-K., Simonova, I., Kozel, T., Kwok, L.-F. (eds.) ICBL 2019. LNCS, vol. 11546, pp. 251–264. Springer, Cham (2019). https://doi.org/10.1007/978-3-030-21562-0_21

2. Haegemans, T., Snoeck, M., Lemahieu, W.: Entering data correctly: an empirical evaluation of the theory of planned behaviour in the context of manual data acquisition. Reliab. Eng. Syst. Saf. **178**, 12–30 (2018)

3. Wang, K.: Design of agricultural network information resource sharing system based on Internet of Things. In: Zhang, Y.-D., Wang, S.-H., Liu, S. (eds.) ICMTEL 2020. LNICST, vol. 326, pp. 281–293. Springer, Cham (2020). https://doi.org/10.1007/978-3-030-51100-5_25

4. Akgül, Y.: Accessibility, usability, quality performance, and readability evaluation of university websites of Turkey: a comparative study of state and private universities. Univers. Access Inf. Soc. **20**(1), 157–170 (2020). https://doi.org/10.1007/s10209-020-00715-w

5. Wagwu, V., Obuezie, A.C.: Evaluation on the usability and information content of three university library websites for improved standard of education based on heuristic principles. J. ICT Dev. Appl. Res. **2**(1) (2020)

6. Tian, Yu., Liu, Z.: Usability design study of university website: a case of normal university in China. In: Marcus, A., Rosenzweig, E. (eds.) HCII 2020. LNCS, vol. 12202, pp. 533–551. Springer, Cham (2020). https://doi.org/10.1007/978-3-030-49757-6_39

7. Jin, B., Xu, F., Wu, T.: Design of personal homepage customization system for college teachers. Microcomput. Appl. **32**(21), 78–80 (2013)

8. Li, K., Mao, W., Zhang, Y., Ding, L., et al.: Construction of personal homepage system for college teachers based on B/S structure. J. China's Educ. Inf. (01), 93–96 (2019)

9. Li, S.: Design and implementation of school portal based on website group architecture. J. Sci. Eng. RTVU (03), 32–34 (2013)

10. Luo, J., Gao, Y., Yang, J.: Exploration of personal homepage construction of university teachers based on website group. J. Inf. Commun. (09), 283–284 (2018)

11. Guo, C.: Research on website design of Xiamen University portal-taking Xiamen Institute of technology as an example. J. Comput. Knowl. Technol. **14**(11), 210–211 (2018)

12. Li, L., Sun, L., Yang, J.: Review of social network research. J. Comput. Sci. **42**(11), 8–21+42 (2015)

13. Zhang, Y., Liu, Y., Zhang, H., et al.: Information dissemination model based on online social network. J. Acta Physica Sinica **60**(05), 66–72 (2011)

14. Samia, Z., Khaled, R., Warda, Z.: MultiAgent systems and ontology for supporting management system in smart school. In: PAIS, pp. 1–8 (2018)

15. Sanyal, D.K., Bhowmick, P.K., Das, P.P.: A review of author name disambiguation techniques for the PubMed bibliographic database. J. Inf. Sci. **47**(2), 227–254 (2021)

16. Atoum, I., Otoom, A.: Efficient hybrid semantic text similarity using wordnet and a corpus. Int. J. Adv. Comput. Appl. **7**, 124–130 (2016)

17. Lee, B., Choe, E.K., Isenberg, P., et al.: Reaching broader audiences with data visualization. IEEE Comput. Graph. Appl. **40**(2), 82–90 (2020)
18. Zhao, H., Chen, J., Xu, L.: Semantic web service discovery based on LDA clustering. In: Ni, W., Wang, X., Song, W., Li, Y. (eds.) WISA 2019. LNCS, vol. 11817, pp. 239–250. Springer, Cham (2019). https://doi.org/10.1007/978-3-030-30952-7_25

Author Index

Printed in the United States
by Baker & Taylor Publisher Services